W9-BLL-923

# Geographies of Economies

EDITED BY

ROGER LEE

DEPARTMENT OF GEOGRAPHY, QUEEN MARY, UNIVERSITY OF LONDON, UK

AND

JANE WILLS

DEPARTMENT OF GEOGRAPHY, UNIVERSITY OF SOUTHAMPTON, UK

HF
1025
.G38
1997

A member of the Hodder Headline Group

LONDON • NEW YORK • SYDNEY • AUCKLAND

First published in Great Britain in 1997 by
Arnold, a member of the Hodder Headline Group
338 Euston Road, London NW1 3BH
175 Fifth Avenue, New York, NY 10010

Copublished in North, Central and South America by
John Wiley & Sons, Inc., Third Avenue,
New York, NY 10158–0012

© 1997 Arnold

All rights reserved. No part of this publication may be reproduced or transmitted in any form or by any means, electronically or mechanically, including photocopying, recording or any information storage or retrieval system, without either prior permission in writing from the publisher or a licence permitting restricted copying. In the United Kingdom such licences are issued by the Copyright Licensing Agency: 90 Tottenham Court Road, London W1P 9HE.

*British Library Cataloguing in Publication Data*
A catalogue entry for this book is available from the British Library

*Library of Congress Cataloging-in-Publication Data*
Geographies of economies / edited by Roger Lee and Jane Wills.
p. cm.
Includes bibliographical references and index.
1. Economic geography. 2. Regional economic disparities.
3. Industrial location. I. Lee, Roger, 1945– . II. Wills, Jane, 1965– .
HF1025.G38 1997
330.9-dc21 96-29757
CIP

ISBN 0 340 677163 (Pb) 0 340 67717 1 (Hb)
ISBN 0 470 24424 0 (Wiley) 0 470 23739 2 (Wiley)

Composition by J&L Composition Ltd, Filey, North Yorkshire
Printed and bound in Great Britain by Bath Press, Bath, UK.

# Contents

List of Contributors ix

Prologue **Economic Geographies: Representations and Interpretations** *Roger Lee* xi

Introduction *Jane Wills and Roger Lee* xv

## Section One (RE)CONSTITUTING ECONOMIC GEOGRAPHIES

**Introduction: Cultural Turns and the (Re)constitution of Economic Geography** *Philip Crang* 3

1 **The Dialectic of Culture and Economy** *Andrew Sayer* 16

2 **Economic/Non-economic** *Doreen Massey* 27

3 **The Cultural Production of Economic Forms** *Richard Peet* 37

4 **The Invention of Regional Culture** *Meric S. Gertler* 47

5 **Economies of Power and Space** *John Allen* 59

6 **Nature as Artifice, Nature as Artefact: Development, Environment and Modernity in the Late Twentieth Century** *Michael J. Watts and James McCarthy* 71

7 **Re-placing Class in Economic Geographies: Possibilities for a New Class Politics** *J.K. Gibson-Graham* 87

8 **Local Politics, Anti-essentialism and Economic Geography** *Joe Painter* 98

9 **Rethinking Restructuring: Embodiment, Agency and Identity in Organizational Change** *Susan Halford and Mike Savage* 108

10 **A Tale of Two Cities? Embedded Organizations and Embodied Workers in the City of London** *Linda M. McDowell* 118

## SECTION TWO (RE)THINKING GLOBALIZATION

Introduction: True Stories? Global Nightmares, Global Dreams and Writing Globalization *Andrew Leyshon* 133

11 Globalization, Socio-economics, Territoriality *Ash Amin and Nigel Thrift* 147

12 Unpacking the Global *Peter Dicken, Jamie Peck and Adam Tickell* 158

13 Excluding the Other: The Production of Scale and Scaled Politics *Erik Swyngedouw* 167

14 Globalization and Geographies of Workers' Struggle in the Late Twentieth Century *Kevin R. Cox* 177

15 Notes on a Spatialized Labour Politics: Scale and the Political Geography of Dual Unionism in the US Longshore Industry *Andrew Herod* 186

16 Local Food/Global Food: Globalization and Local Restructuring *Robert Fagan* 197

17 Globalization of R&D in the Electronics Industry: The Recent Experience of Japan *David P. Angel and Lydia Savage* 209

18 Falling Out of the World Economy? Theorizing 'Africa' in World Trade *John Agnew and Richard Grant* 219

## SECTION THREE NEW GEOGRAPHIES OF UNEVEN DEVELOPMENT

Introduction: Theories of Accumulation and Regulation: Bringing Life Back into Economic Geography *Trevor J. Barnes* 231

19 Regional Economies as Relational Assets *Michael Storper* 248

20 Divergence, Instability and Exclusion: Regional Dynamics in Great Britain *Michael Dunford* 259

21 The Post-Keynesian State and the Space Economy *Ron Martin and Peter Sunley* 278

22 Bringing the Qualitative State into Economic Geography *Phillip M. O'Neill* 290

23 The End of Mass Production and of the Mass Collective Worker? Experimenting with Production and Employment *Ray Hudson* 302

24 The Role of Supply Chain Management Strategies in the 'Europeanization' of the Automobile Production System *David Sadler* 311

25 Informal Cities? Women's Work and Informal Activities on the Margins of the European Union *Dina Vaiou* 321

26 Breaking the Old and Constructing the New? Geographies of Uneven Development in Central and Eastern Europe *Adrian Smith* 331

**27 California Rages: Regional Capitalism and the Politics of Renewal** *Richard Walker* 345

**Concluding Reflections on *Geographies of Economies*** *Roger Lee and Jane Wills* 357

Bibliography 359

Index 397

# List of Contributors

John Agnew, Department of Geography, UCLA, Los Angeles, USA

John Allen, Faculty of Social Sciences, The Open University, UK

Ash Amin, Department of Geography, University of Durham, UK

David P. Angel, Graduate School of Geography, Clark University, USA

Trevor J. Barnes, Department of Geography, University of British Columbia, Canada

Kevin R. Cox, Department of Geography, The Ohio State University, USA

Philip Crang, Department of Geography, University College London, UK

Peter Dicken, Department of Geography, University of Manchester, UK

Michael Dunford, School of European Studies, University of Sussex, UK

Robert Fagan, School of Earth Sciences, Macquarie University, Australia

Meric S. Gertler, Department of Geography, University of Toronto, Canada

Katherine Gibson, Department of Geography and Environmental Science, Monash University, Australia

Julie Graham, Department of Geosciences, University of Massachusetts–Amherst, USA

Richard Grant, Department of Geography, Syracuse University, USA

Susan Halford, Department of Sociology and Social Policy, University of Southampton, UK

Andrew Herod, Department of Geography, University of Georgia, USA

Ray Hudson, Department of Geography, University of Durham, UK

Roger Lee, Department of Geography, Queen Mary and Westfield College, University of London, UK

Andrew Leyshon, Department of Geography, University of Bristol, UK

Ron Martin, Department of Geography, University of Cambridge, UK

Doreen Massey, Faculty of Social Sciences, The Open University, UK

James McCarthy, Department of Geography, University of California, Berkeley, USA

Linda McDowell, Department of Geography, University of Cambridge, UK

Phillip M. O'Neill, Department of Geography, University of Newcastle, Australia

Joe Painter, Department of Geography, University of Durham, UK

Jamie Peck, Department of Geography, University of Manchester, UK

Richard Peet, Graduate School of Geography, Clark University, USA

David Sadler, Department of Geography, University of Durham, UK

Lydia Savage, Department of Geography, University of Southern Maine, USA

Mike Savage, Department of Sociology, University of Manchester, UK

Andrew Sayer, Department of Sociology, University of Lancaster, UK

Adrian Smith, School of Social Sciences, University of Sussex, UK

Michael Storper, School of Public Policy and Social Research, UCLA Los Angeles, USA and Faculty of Human and Social Sciences, University of Marne-la-Vallée, France

Peter Sunley, Department of Geography, University of Edinburgh, UK

Erik Swyngedouw, School of Geography, University of Oxford, UK

Nigel Thrift, Department of Geography, University of Bristol, UK

Adam Tickell, Department of Geography, University of Manchester, UK

Richard Walker, Department of Geography, University of California, USA

Michael J. Watts, Institute of International Studies, University of California, Berkeley, USA

Jane Wills, Department of Geography, University of Southampton, UK

Dina Vaiou, Department of Urban and Regional Planning, National Technical University of Athens, Greece

# PROLOGUE

# ECONOMIC GEOGRAPHIES: REPRESENTATIONS AND INTERPRETATIONS

ROGER LEE

## ECONOMIC GEOGRAPHIES OBSERVED

The line of tractors and trailers angle around a corner of the paddock. From neighbouring farms, their drivers, farmers all – there is no space for hired workers in this particular division of labour – are here to buy and to exchange: to sustain their economic geographies. They stand, hunched, capped and agitating against the cold, chatting and watching the fork-lift remove roundels of hay and straw from the barn on to a trailer. Their conversation – intense but founded on a wealth of mutual knowledge and interpretation – makes them oblivious to the work of the labourer tending the heifers in the byre; and he, it appears, is not – or refuses to be – distracted by them.

It is March, a Friday, and this set of transactions is currently a weekly happening. In the distance, silage is being loaded from a clamp into a high-sided trailer waiting to be pulled the 20 kilometres or so to a farm outside Tunbridge Wells in Kent, England; this is a daily – sometimes twice-daily – occurrence.

A woman, scarf around her head, passes by – almost eagerly, certainly independently – pulling a shopping basket down the track to the road, her friends, the bus into Maidstone and the weekly shop; she, too, has no part of these activities. Separated too, an estate car waits at the foot of the stairs to the first floor of a disused barn. Artists' materials are being carried up and work carried down.

Sitting at the word processor, I watch through the bedroom window of the converted milking parlour, hoping that the pretence of my work is not being exposed. James – the fork-lift driver, owner of this farm and seller of the hay, straw and silage – is skilful: short, swift, economical manoeuvres, and the heavy bales are securely in place for their journey. One by one, the tractors pull their load away. The last complete, James jumps to the ground and there is a longer chat before the final tractor too disappears for the week.

A figure cycles, more than a little hesitantly – and not just because of the ruts and the mud – on to the scene, bulky plastic bag over the handlebars: a brace of rabbits. His cap – which may be only metaphorical – is touched, a brief word, a joke, the bag changes hands: James' 'share' from ferreting on his land. It is costless to him and helps to keep a pest under control. But BSE means the price of rabbit meat is soaring; the figure weaves away to sell the rest of his rabbits at the butcher's. He will probably have some distance to go; there is only one in the village.

Much later – it is twilight – and, rather less jaunty now, Vera returns – scarf still in place, shopping trolley heaped and twisting under the weight on the rough track. She makes her way along to the tied cottage to take in the washing left blowing all day, to prepare tea for her husband. As well as his exclusion from the weekly transactions in hay and straw within the farming division of labour, Stan excludes himself – and is actively excluded from – the purchase and preparation of food and most other forms of household production in the domestic division of labour.

I go out to fill the basket with logs; the central heating is designed for chilly summer evenings, not frozen winter nights. The logs, like the rabbits, patronize. The timber is 'free' to Stan, who, in response to our two- to three-weekly order and payment of £10 to Vera, splits it for use on the stove and uses the fork-lift to deliver an apple container-ful outside the cottage. A Thursday or Friday is a good day to order as Saturday afternoons are also 'free' for such work. Sunday is a day off (after the morning feed and daily mucking out of the byre). I too am incorporated in what I observe.

Moments in the construction of economic geographies: production, selling, buying, consumption; a mix of class-based transactions and transactions in kind; inter- and intra-class relations; social and technical divisions of labour informed by property, a particular patriarchy and the need for centrality and control of the production, exchange and consumption of information; the materiality of a market: markets don't just happen, they too involve active agents, knowledge, work before they can take place; the shared meanings which direct the purpose and function of the economic geographies; and the observer: making a living, incredibly enough, from looking and telling.

So what? How are these economic geographies taking place? What, if anything, have they to say to those interested in contemporary economic geography? Why and how am I involved as observer? Why do I find looking at them rather more interesting than what I am failing to write? Singular answers are all that can be possible here. They cry out for a more polyvocal response.

## ECONOMIC GEOGRAPHIES INTERPRETED

Two years before, James had sold dairy herd and milk quota and invested in tourism by converting former farm buildings into high-quality holiday residences. He realized that neither his son nor daughter was interested in taking over the farm and was encouraged in his decision by schemes of rural diversification, the grants that go with it, the opportunity costs of dairying, the realizable value of the herd and, more especially, the EU milk quota (although a number of the herd had already been slaughtered, the BSE affair had not yet blown up into the debilitating scare that it was to become by early 1996), and even local planning regulations.

The particular intersection of space–time that positioned my gaze on this place of production was founded on the collapse of the housing market during the early 1990s. This led to our inability to buy somewhere to live and to the relative cheapness – for this part of south-east England, at least – of a long winter let in a holiday cottage. Much the same forces (borrowing money at low rates of interest) were currently prompting the thought in James of renovating a particularly attractive house and its brick-walled garden on the farm. Certainly the return on the investment looked likely to be greater than that on finishing the conversion of a barn to office use: the office rental market had slumped as a result of the speculative bubble of the mid-1980s. Canary Wharf casts its shadow even this far beyond the M25. The demand for office space is slack; that for cheap, well-lit floor space is readily met by the minimal conversion of a barn to a studio.

A run of hot summers had reduced the productivity of meadows and with it the supply of silage, hay and straw. Not requiring much for use on the farm (other than in an enterprise based on the rearing of calves under contract to the owner of a prize herd of Friesians), James had grown, harvested and stored the foodstuffs and cattle-shed bedding in anticipation of winter shortages. Limited opportunity costs in terms of contract labour, under-used capital equipment and storage space, and the fertility of high-yielding alluvial meadows ('boy's land' James calls it: it is so easy to maintain and cultivate) encourages this speculation in commodities. Nevertheless, deliveries of fertilizer produced in Poland, where environmental regulations are less costly, have already begun. The bright-blue plastic bags occupy the spaces vacated in the barn by the roundels of hay and straw. And, as a late spring exacerbates the shortages, prices rise to the point at which hay, straw and silage are burgled quite regularly from farms with a surplus.

The risks of the trade in which James is engaged are intensified by the lack of a futures market, but their attraction is such that delaying his harvest to maximize volume produces a less than perfect silage. It is this that is finding its laborious daily way to Tunbridge Wells, to the great annoyance, no doubt, of delivery vans, shoppers and travelling workers. The higher-quality silage from the previous year is being fed by Stan to the contract calves in the byre in order to reduce the need for and hence minimize the costs of feed supplements.

But there is a futures market of sorts. How do the farmers whose tractors arrive every Monday get to know of a local supply of silage and hay and how does James advertise the supply? How, more generally, is the framework of expectations – around which so much apparently spontaneous economic activity takes place – sustained? Conversation. A word in the ear of the auctioneer at the Ashford cattle market, some 40 kilometres distant, and, most significantly, in the ear of the feed rep, whose legitimacy is enhanced by a degree in animal husbandry and who is greatly trusted by the farmers on whom he calls in the locality, are the costless ways of supplying and consuming information (now to be enhanced via the Internet for a fee of around £500 p.a.). The quality of the goods can be relied upon: James' own futures (not least with the feed rep and in the market) depend on his spot credibility. The price is negotiated around prevailing demand and supply curves – the price is high – but also around complementary deals in past and future bartering, borrowing, sharing and, above all, a recognition of the need for self-help insurance which such deals provide.

## ECONOMIC GEOGRAPHIES CONSTITUTED

So are these economic geographies constituted, constructed and continuously, sometimes violently, reconstructed. And of what are they so constituted?

Social relations of reproduction within which (local and global) economic meanings and norms are shaped and directed, calculations undertaken, deals struck and, above all, perhaps, relations of power defined and articulated; capitalist enterprises judged by their ability to generate profit; demand and supply curves equilibrating the market but unable to indicate what equilibrium prices should be (their own geometries being shaped not only by countless other transactions but by the norms and evaluations of prevailing social relations of capitalism and interrupted by other more local considerations of individualized and gainful yet collaborative involvement); powers of domination, of mobilization deriving not least from the disciplines, founded on social relations and custom, of everyday life; unpredictable physical geographies; levels of effective and grossly ineffective state regulation and intervention; dynamic social geographies of communication and of the production and transmission of information; divisions of labour, highly sensitized to the need of capital to remain close to the production and consumption of information, to gender relations and to uneven (international) geographies of cost; geographies of trust created and sustained by places of legitimation and discipline (the local cattle market; the food rep) and by the discipline of frequent face-to-face contact – a mixture of hard business, collective engagement and sociability; work, not only in production and consumption but in making markets happen in material as well as in virtual terms; the pleasure and the work in consumption; influential intersections of time and space, of moral geographies based on immediate relations of need, trust and knowledge and political-economic geographies based on calculation and material gain, of global and local geographies, and of cultural and agricultural production; the conditions of inclusion in and exclusion from formal economic relations; the consequences of such conditions.

## ECONOMIC GEOGRAPHIES AND SOCIAL RELATIONS

Is it possible for an observer – in this case, me – to make some kind of sense of all of this, to mount some kind of critique? Aside from the epistemological and methodological issues raised by such questions, at the heart of any answer to them lies the question of value and the prevailing norms of evaluation. And at the heart of the question of value and evaluation are social relations: the bases of sociability, communication and shared or disputed understanding. Economic geographies are shaped and informed by social relations which offer direction, the means of communication and the evaluative criteria of good and bad, better and worse. They shape the criteria of evaluation which etch out the contemporary maps of the unequal struggle to make a living. In this sense, then, economic geographies reflect the playing out of, and the competition between, systems of value and evaluation.

Such are the criteria which influence not only the economic geographies observed but both the possibility and interpretation of such observation. Given the capacity to think about consequences, of going beyond purely instinctive behaviour, social relations begin to establish templates of meaning. Without such templates, the trajectories of social life would oscillate between a crude determinism – beyond individual or social influence – and a formlessness and randomness which, at least over the spaces and times of human existence, are difficult to accept as a complete story.

But particular forms of social relations may emerge by chance as much as by design or struggle. Nevertheless, their acceptability across time and space has to be granted through experience or imposed in some way: they are social constructs. And, as such, they contain a powerful formative influence: the norms that we accept or reject and the language that we speak rest for social comprehension – if not necessarily acceptance – on the existence and practice of social relations. In this sense, then, it is true that individuals have no prior existence other than that engendered through discourse. But discursive strategies and practices are themselves not simply free-floating, capable of being continuously invented anew by discursive individuals. They are shaped by social relations which are in turn both validated through, and serve to validate, social reproduction in the construction of economic geographies. The material matters. We are not constituted simply through discourse.

However, social relations do not emerge simply out of some determinative material reality. Nor are they simply handed down successively – and rarely, if ever, unproblematically – through historical geographies. They are continuously reproduced and transformed through the intersection of discourse, struggle and materiality across economic geographies. In this context, then, the economic is not and cannot be autonomous. It is always socially and culturally constituted. It is possible, therefore, to challenge notions of determinism or randomness: we *can* buck the market; there *is* always an alternative; human beings *cannot but* live in society; history – and geography – are *far from* ended. It is the constitution of economic geographies by social relations of meaning – including relations of power, of belief and myth and of trust and convention as well as of gender, race and class – which drives and shapes the construction of those geographies. And it is possible to observe them from a particular if nevertheless inadequate vantage point of some kind of comprehension.

## ECONOMIC GEOGRAPHIES ASSERTED

Not mutually autonomous, the economic, the social and the cultural are interdependent; so too are the mental and the material, the virtual and the 'real', quantity and quality, local and non-local, value and prevailing social and cultural relations. The what, the where and, above all, perhaps, the how – collectively defining context – shape cause (why?) and consequence (so what?). And geographies are central to this: the context or 'arenas' within which economic practice is conducted, the social construction and representation of nature, the multiple – often cross-cutting – paths of surveillance and the performances involved, the proximity and distance which curtail and enhance the transmission of trust as well as of ideas and material objects, and the imagination of the possible as well as of the real, are powerful formative influences upon the economic. They are the spaces through which it takes place. There are, in other words, no such things as economies, only economic geographies.

# INTRODUCTION

JANE WILLS AND ROGER LEE

This book aims to explore the intersections of economy and geography. Geographies of economies do not merely undermine the assumptions of economic theory but demand critical reconstructions of what the 'economic' may mean. Such a reconstruction is central to the emergence of 'new' economic geography (*see* Barnes, 1995a; Thrift and Olds, 1996), and the chapters in this book illustrate and respond to this 'widening' of the agenda. Drawing upon contemporary research and debate, the three sections of the book focus upon the cultural and social constitution of economic geographies; upon debates about the emergence of processes of globalization; and upon new forms of uneven development and regulatory practice. A detailed introduction to each part of the volume has been written by a leading practitioner in the field, assisting the reader to place these chapters in the context of wider academic debate.

For our part, the Introduction sets out to explain the rationale for this work and to illuminate the ways in which it has come about. Recent developments in economic geography are outlined before a number of concerns are raised about the way we read our own disciplinary histories. In the final pages of the Introduction, the origins and structure of the book are explained before you are urged to read on, embracing the debates which are preoccupying economic geographers towards the millenium.

## 'OLD' AND 'NEW' ECONOMIC GEOGRAPHIES?

In recent years economic geography has been shaken by the winds of economic upheaval, by the challenges made to political-economic analysis and by the 'post-modern' flavour of much contemporary theorization, prompting many to chart a new course for future research and enquiry. Economic geographers are currently researching themes as diverse as finance capital, globalization, new patterns of consumption, identity and representations of the economy. The chapters of this volume are illustrative of this 'new' research in economic geography. In presenting some of this 'new' scholarship, however, we are keen to emphasize its links with the past, as the 'cultural turn' has reshaped, rather than replaced, the traditional issues of enquiry (production, circulation and exchange). In this volume, the chapters in Section One explore the ways in which research in economic geography has been enhanced by a renewed interest in culture and social theory. The authors in the second section of the book then illustrate these 'new' concerns by attending to the processes of globalization. By exploring the connections between discourse and material practices these scholars are taking on new ideas, but often couch their debate within the brief of traditional political economy. In the third, and final, section of the book this continuity with the rubric of the past is clearer still as the tools of political economy are used to dissect the changing nature of uneven development, the role of the state and political practice. As will be illustrated in many of the chapters that follow, the geographical engagement with social and cultural theory is proving very productive. And beyond this refiguring of existing subject matter, economic geographers are also opening up new questions for study, putting questions of meaning, discourse and identity on to the research agenda. Our purpose here, in exploring some of this 'new' economic geography, is to illustrate the extent and depth of these new bodies of work.

But this exploration of the productive encounter between economic, cultural and social theory also highlights the need for further critical debate in this field. For in the shift away from classical political economy, questions of political agency have seemingly been overlooked. Although 'new' economic geographers have rightly alerted us to the dangers of metamorphosing a transcendental political subject for whom research outcomes are intended (be this a tokenistic, pragmatic or impassioned gesture of support for the working class or other social groups and movements), questions of social and political action are now much more complex and difficult to deal with. Attention to the contingency of the subject is invaluable as a counter to the sometimes aspatial, atemporal and even deterministic assumptions of some political economic analysis, but there are implications of this contingency which are yet to be clarified in economic geography. Indeed, it is perhaps this question, more than any other, which differentiates those writing in the tradition of political economy from those schooled in social and cultural theory. While Karl Polanyi's (1944) work speaks to many of the themes in this book, for example, the solutions offered ring hollow in the light of much contemporary theoretical predilection. In the Foreword to the *The great transformation,* Robert MacIver summarizes the contribution Polanyi makes to understanding the social implications of particular economies and declares: 'What our age needs [is] the reaffirmation for its own conditions and for its own needs, of the *essential values of human life*.' (op. cit., x, emphasis added). And here is the rub, for contemporary analysis has a space where that 'essential subject', or 'human agent' and 'species-being', used to sit. Many twentieth-century philosophers – and the contributions from structural and post-structural thinkers in particular – have rightly torn any notion of the disembodied, ahistorical and group-based subject asunder. But while this draws attention to the particularities of events (and in our case, economic geographies) in time and place, the political implications of particularity are often left unexplored. More especially, when the older certainties and stabilities of the political order have been questioned, there is a need to explore the possibilities for political intervention in the processes researched.

The political implications of much contemporary analysis are thus difficult to chart, and it is not clear how the 'new' economic geographies will shape the political weaponry of the future. Indeed, it is by no means certain that political economy cannot be reconfigured to deal with the challenges of the 'cultural turn'. Many of the contributions in Sections Two and Three of this book, for example, would suggest that such processes are certainly possible and are already under way. And as Barnes (1995a: 428) concludes in his analysis of the post-structural challenges to political economy: 'political economy's greatest assets as an intellectual framework are its ability to adapt, and to accommodate and respond to criticisms'.

## MATERIAL AND IDEOLOGICAL EXCEPTIONALISM IN ECONOMIC GEOGRAPHY

There are, however, other reasons for being cautious of making too grandiose a claim for what is 'new'. One of the underlying arguments which weaves its way through the chapters in this volume is that economic geographies are the product of complex, unevenly developed, historical processes. Taking a historicized view of economic geographies in practice, should prompt us to use the same approach in understanding the subdiscipline of economic geography itself. Described by Hobsbawm (1994) as 'the age of extremes', the twentieth century has witnessed profound and devastating political-economic change. Arguments between proponents of free-market capitalism on the one hand, and interventionists of various hues on the other, have been revisited more than once in the past hundred years and the century looks set to close with the former still in the ascendancy. As the balance of power shifts between those who benefit from the pursuit of profit and those who suffer the consequences of competition, global accumulation, deregulation and the erosion of welfare protection, the political-economic landscape is being constantly redrawn. And it is in this geohistorical context in which the discipline of geography has itself taken shape. In order to understand the trajectory of our own discipline then, we need to consider the lessons of economic and political change in the past. Indeed, in the Introduction to his overview of the twentieth century, Hobsbawm (1994: 3) argues that part of the rationale for his scholarship is to keep the lessons of the past alive for new generations. As he explains:

> The destruction of the past, or rather of the social mechanisms that link one's contemporary experience to that of earlier generations, is one of the most characteristic and eerie phenomena of the late twentieth century. Most young men and women at the century's end grow up in a sort of permanent present lacking any organic relation to the public past of the times they live in.

Echoing this sentiment, part of the purpose of this book is to ensure that the attraction of all things 'new' does not detract from the legacy of the past, the insights that were gained and the importance of contemporary connections between past and present.

If we take this challenge seriously, it becomes important to place the disciplinary debates of the present in their historical setting. So, for example, while the theoretical and epistemological uncertainties triggered by the collapse of the post-war boom in the early 1970s might be regretted, it is important to recall the very exceptionalism of that particular phase of political and economic history and the economic geographies through which it took place. Moreover, the view taken from the heart of Western economies (built on the profits of empire, protectionism and military-economic supremacy) is necessarily exceptional. Although imperialism, trade and production have drawn links between peoples across the globe, integrating the histories of one place with those of another, the uneven nature of power and experience makes economic geographies very particular to place. The years of post-war boom between the 1950s and early 1970s were the time, and the Western economies were the territory, in which the subdiscipline of economic geography as we have known it came to the fore. Scholars of today are grappling with ideas and methods forged in exceptional times and exceptional places. Acknowledging this, we argue, makes it easier to celebrate the diversity in contemporary analyses of the changing political-economic environment.

It has been common, however, for authors to begin any description of recent developments in economic geography with the word 'crisis'. Ron Martin (1994a), for example, argues that a crisis has taken root in economic geography since the mid-1980s, owing to rapid change in the nature of the economic world (particularly with the development of information technologies, tertiarization, hyper-consumptionism, globalization and re-regulation), the collapse of the economic certainties detailed in neo-classical, Keynesian and Marxist theorizations, and the challenge of 'post-modern' ontology and epistemology. While such changes have undoubtedly shaken the foundations of thought, however, we are less convinced that there is a crisis. There is a real danger of seeing the new as a total break with the past. Just as the transition models promoted by regulationist thought in geography tend to lever the 'post-Fordist', 'flexible accumulative' and 'post-industrial' off a caricature of the 'Fordist', so today there is much connecting the threads of economic enquiry past and present.

Indeed, 'crisis' implies that there used to be a unity and coherence of thought before the mid-1980s, something which has never really existed: the regular journal reports of 'progress' in this field of geography are littered with alternative approaches to understanding economic change (for examples *see* Keeble, 1979; Taylor, 1986; OhUallachain, 1992; Barnes, 1995a; Thrift and Olds, 1996). The traces of neo-classical, Marxist and Keynesian approaches have long coexisted within geographical analysis, and much recent scholarship in this field has been located firmly within the traditions of political economy (*see* Scott and Storper, 1986; Peet, 1987; Allen and Massey, 1988; Sayer and Walker, 1992; Amin, 1994). But more than this, many contemporary economic geographers are connecting economic change to questions of culture, discourse, and subjectivity.

Rather than 'post-modernity' or the 'cultural turn' being emblematic of crisis, then, we hope this volume bears testament to new interdisciplinary research exploring the interconnections between processes of economy, culture, society and politics at a variety of spatial scales. In this context, as Martin (1994a: 45) so cogently argues, economic geography needs to be reconstructed as a 'more multidimensional, multiperspectival and multivocal' set of approaches (for a similar agenda, *see* Thrift and Olds, 1996). This process, we believe, is already under way and the chapters of this book illustrate both the enormity of the field of enquiry and the breadth of research undertaken. The point is to contextualize rather than to undermine the economic, by locating it within the cultural, social and political relations through which it takes on meaning and direction.

## THE ORGANIZATION OF THE BOOK

This book is organized in three sections, each with its own introductory chapter written by a leading practitioner in the field. Delimiting between these sections was not easy, however, and there are many other lines of communication between chapters in the book. They will, we hope, enable readers to reach across our boundaries and read as they please, following themes across different chapters and between different authors. In our order of things though, Section One seeks to rethink the meaning of 'economy', broadening our understanding of the term to include questions of culture, gender and power. In this vein the chapters illustrate the imaginative ways in which scholars have responded to the challenges of new social theory. Section Two focuses upon the meaning of globalization and its implications for the state, labour and geopolitics at a variety of spatial scales. These sections cover two areas of debate which clearly illustrate the new ways in which economic geographers are embracing change, developing new theoretical approaches, new subjects of study and scales of enquiry. And finally, the chapters in Section Three focus upon the uneven nature of economic processes and regulation, drawing upon new empirical work conducted in the UK, Europe (West and East) and the USA. In particular, the chapters in Section Three demonstrate the importance of geography and political regulation to economic activity and change. Clearly, some aspects of recent research in economic geography have been omitted from consideration here, but it is hoped that the epistemological and empirical insights offered will prompt others to develop new lines of thought and enquiry in the future. Given its objectives, this is a book which cannot be anything other than partial, suggestive rather than definitive, exploratory rather than complete.

We hope then, that *Geographies of economies* contributes to continuing debate and exchange about the changing world economic order. As suggested above, the book echoes themes and ideas which have been widely discussed for many years. Reflecting a collective endeavour such as this, the plurality of voices on the pages that follow bears testament to the health and dynamism of this branch of academic enquiry.

## ACKNOWLEDGEMENTS

The organizational task of compiling and editing material from 39 authors living in three continents has been surprisingly easy and extremely enjoyable, providing us with a remarkable learning experience. We are very grateful to everybody who has contributed and co-operated with our many requirements.

The broad agenda of *Geographies of economies* reflects the origins of the book. Like economic geographies themselves, it is a collaborative product – not merely because it is the work of many hands and minds – but because it emanates from discussions within the Economic Geography Research Group (EGRG) of the former Institute of British Geographers (now the Royal Geographical Society with the Institute of British Geographers (RGS-IBG)). In recent years the individuals involved in these debates have been preoccupied with the need to recast economic geography, to recognize the socially and culturally constituted complexity of the economic, and to revisit the intersection of the political and economic. This book stems from these discussions and was manifest in six conference sessions held at the 1995 IBG Conference at Northumbria University in Newcastle. Members of the EGRG have given enormous service in conceiving and realizing both the successful conference sessions and this publication, which includes many who were unable to participate in the conference sessions. Our aim has always been to stimulate debate and interest in economic geography and, in the collective spirit of this endeavour, all royalties from *Geographies of economies* will go back to the EGRG, to promote further research and enquiry in the field. We are thus very grateful to all the contributors for writing in the knowledge that there was no material reward for their labours.

Laura McKelvie at Arnold has been nothing short of angelic in her indulgence of us during the editing process, combining boundless enthusiasm with the practical tasks of getting the book into print. Our sincere thanks are due to her, not least for believing that it was going to happen.

# Section One

## (Re)constituting economic geographies

# CULTURAL TURNS AND THE (RE)CONSTITUTION OF ECONOMIC GEOGRAPHY
# INTRODUCTION TO SECTION ONE

**PHILIP CRANG**

In their own varying ways, all the chapters in this opening section reflect on the identity of economic geography as an academic subdiscipline. They do so at a moment when that identity is as fluid as it has been at any time since the early 1970s, a period that saw the beginning of the rise to intellectual hegemony of Marxian political-economic approaches. In the mid-1990s, this political-economic tradition is on the defensive. The content, concepts and approaches of (political) economic geography are all being actively reassessed. Content is being rethought in terms of what social and spatial portions of life count as economic, what portions (if any) are therefore non-economic, and how these designated spheres of the economic and non-economic interrelate. Concepts are being reflected on, both in terms of the definitions and utility of key political-economic theoretical constructs such as class, production, labour, capitalism or the economy itself, and in terms of the relevance of theoretical concerns that are conspicuous only by their absence from such a list (identity, representation or meaning for instance). And approaches are being reconsidered through debates over the kinds of theorization and methodology needed to make sense of whatever empirical and conceptual concerns economic geographers decide they are interested in.

The reasons for this uncertainty over the nature of economic geography are multiple. Practically, over the past decade the difficulties of elucidating the complicated geographies of transition in contemporary economies – especially the 'new' times and spaces of capitalism proclaimed by a host of mostly 'post-Marxist' commentators – have led economic geographers into new areas of substantive interest and, on occasion, into searches for fresh intellectual resources. Conceptually, such resources have been made available as human geography more generally has enagaged with a number of post-structuralist, feminist and ecological literatures. And politically, radical political culture – which in varying degrees of commitment and tokenism has long inspired and legitimated left-leaning political economic geographers – has witnessed a mutation of the emancipatory politics of class struggle into the representational politics of political, cultural and environmental recognition. Add to this mix the institutional dynamics of a field of enquiry in which academic capital was, and is, produced primarily through intellectual innovation rather than loyalty to existing luminaries, and one has a powerful pressure for subdisciplinary reinvention.

The forms of that reinvention have been varied but, as a number of the chapters in this section note, increasingly a particular shorthand has come both to represent and further stimulate economic geography's self-reconstitution. That shorthand is the 'cultural turn' (*see* Crang, 1994b). Noted more generally within the human sciences (Chaney, 1994) and human geography (Philo, 1991), this turn to culture is apparent in the framing of intellectual gazes both beyond the academy – suddenly culture and the cultural are absolutely everywhere – and within it – particularly through the emergence of cultural studies as a central interdisciplinary field. Indeed, for David Chaney, the turn has been such that 'culture, and a number of related concepts, have become simultaneously both the dominant topic and most productive intellectual resource in ways that lead us to rewrite our understanding of life in the modern world' (1994: 1).

Of course, the precise forms of such turns have been multiple and contested. There is no single cultural turn; no single school of 'new' cultural geography (Duncan, 1994) upon which a singular 'new' economic geography draws; no single understanding of just what culture and the cultural might be. None the less, culture – in a variety of substantive or conceptual guises – is increasingly key to economic geography's research agendas. As Andrew Sayer suggested in a wonderfully pithy editorial for the journal *Society and Space*, and as he reiterates later in this section, 'one of the most striking features of the last decade of radical academic literature has been the shift from economy to culture' (1994: 635). And the fact that this is so poses particular dilemmas for the subdisciplinary self-identity of economic geography. For the economic and the cultural have long been cast as 'self' and 'other', each defined as what the other is not. Consider, for example, one of the most interesting commentaries on the cultural turn within British social geography, a manifesto for 'delimiting and de-limiting' social and cultural geography set out by the Committee of the Social and Cultural Geography Study Group of the Institute of British Geographers (1991). In this discussion document, a range of new research areas is outlined under the broad remit of a concern with 'moral geographies'. And these questions of moral geography are quite explicitly defined in opposition to economic geographies:

> Taken as a whole, the above discussion signposts the initiating of a more moral lens on the human geography of the world than has been present before . . . . Initiating this moral lens is not necessarily to deny the validity of all previous human-geographical work. . . . Indeed, we would suggest that the lens which has prevailed throughout recent decades – a lens that has been predominantly economic, whether the neo-classical economics of spatial science or the (loosely) Marxist political economy of much radical geography – constitutes 'the other' of our moral lens; an other that is indispensable, but which is maybe modified by looking at the world in terms of its moral geographies. (Committee of the Social and Cultural Geography Study Group of the Institute of British Geographers, 1991: 17)

Economic geography is clearly differentiated from social and cultural geography. Whereas it contributes to, say, an understanding of residential segregation through the 'economic dimensions . . . of housing, land and labour markets, as keyed into the dynamics of capital accumulation', it necessarily neglects 'the moral frameworks underlying group formation and its spatial expression' (ibid.) that social and cultural geography might analyse. This is a particular construction of the subdisciplinary landscape, but it is far from unusual. Its distinction of the moral and the economic has parallels in much more widely used dichotomies of the mental and the material, the ideal and the real, life-worlds and conditions of life, or symbolic and instrumental interaction. So, if economic geography has established its identity – and had its identity established – through an opposition to the cultural, what happens to that identity through a cultural turn? What becomes of the 'economic' in an enculturated economic geography? To what effect does this economic self confront its other?

These are questions that recur throughout this section's chapters. Sometimes they are addressed head-on (most explicitly in the chapters by Sayer, Peet and Massey). Sometimes they are explored in terms of specific enculturations of the economic (for example, by Gertler, Watts, McDowell, Halford and Savage). In other essays they are approached slightly more tangentially: through an examination of how to conceptualize power in a reconstituted economic geography (Allen); through an analysis of particular keystones of political-economic approaches (for instance, Gibson-Graham on class); or through the parallel question of how the economic and the 'political' interrelate (Painter). This Introduction tries to offer a way into these various themes by developing a schematic overview of the responses by economic geographers to their cultural turn. This overview is structured around five options in thinking about the relations between the economic and the cultural: first, that *the economic and the cultural continue to be opposed* – indeed, if anything, to be reinforced as distinctive entities; second, that *the economic is exported to the cultural*, as existing economic forms of analysis are applied to cultural life; third, that *the economic is understood as contextualized or embedded in the cultural*; fourth, that one views *the economic as represented through cultural media* of symbols, signs and discourses; and fifth, that *the cultural is seen as materialized in the economic* (i.e. economies are seen as involving the production, circulation and consumption of 'materials' that are cultural). As I progress through these alternatives I will be referring to the other chapters in this section of the book, though, to be fair, I should stress that all offer the kinds of textured analysis that show just how schematic my schema is (so I am afraid that reading my annotations of these chapters is no excuse for not tackling the full versions). By way of conclusion, I then try to pull the preceding discussion together in order to reflect on just what should be 'economic' about an economic geography engaging with the cultural turn.

## THE CULTURAL

How one responds to the so-called cultural turn depends in part, of course, on what exactly is being turned to. I want to begin, therefore, with a brief consideration of how we might define the cultural. In such discussions it is pretty much *de rigueur* to cite Raymond Williams' observation that 'culture is one of the two or three most complicated words in the English language' (1983: 87), and certainly suggestions for a cultural turn in economic geography have drawn upon a variety of understandings of just what culture is. Broadly these understandings have been of two main types. First, there are those that cast culture as a 'generic' facet of human life, bound up with human competencies to make the world meaningful and significant. Here, then, the cultural turn involves economic geographers taking seriously questions of meaning and value. Second, there are those who stress culture as a 'differential' quality, marking out and helping to constitute distinctive social groups each with their own meaning and value systems. Here, the cultural turn involves economic geographers analysing the interrelations between these multiple cultures and economic conduct and regulation. Personally, and drawing in particular on Jonathan Friedman's critical reassessment of the culture concept in anthropology (1994), I am happier with the first of these definitions, but only if culture is cast less as a thing which all human beings possess than as a process that we are all involved in. The cultural, then, concerns the meaningful mapping of the world and one's positionings within it. It concerns practices of identity, meaning and signification – practices which are not inevitably closed around the assigning of an aesthetic sign value, but which also always, at the same time, have the potential for involving a moral-ethical attribution of significance. In turn, this cultural activity is very much a question of practice, best viewed as it occurs rather than through the lens of some metaphorical end product. So, 'culture is [and our maps of meaning are] not something out there we seek to grasp, a text or hidden code. It is [they are] a relatively instable product of the practice of meaning, of multiple and socially situated acts of attribution' (Friedman, 1994: 74; with my additions).

Of course, one potential result of such cultural practice can be the production of seemingly 'distinct' cultures and culture areas, each providing a particular piece of a broader human mosaic. In consequence, and as Pam Shurmer-Smith and Kevin Hannam neatly express it, 'not only is culture a process and not a thing but . . . it is a process which is often treated as if it were a thing' (1994: 79). Hence it is precisely this construction of cultural things which needs explaining. Cultures as things are the starting point of analytical endeavour, not the end point. In and of themselves they explain nothing.

Any cultural turn in economic geography therefore requires a careful and critical deployment of culture concepts (*see also* Gertler later in this section on the dangers of using ideas of national, regional or organizational culture to explain economic performance). But, taking that for granted, we still have to decide whether this careful and critical deployment is a good thing for economic geography, and if it is (at least in some respects), how that deployment should be fashioned.

## OPPOSING THE ECONOMIC TO THE CULTURAL

Of course, one response – both potential and actual – is to resist the cultural turn, defending the purity of existing economic self-identity by continuing to oppose the economic and the cultural. This opposition involves defending and reiterating the existing strengths of political-economic analysis. Such a defence may be developed on the grounds of substantive concerns. Concerns with material realities rather than imaginative geographies. With costs rather than culture. With social action rather than texts. With worlds rather than words (Thrift, 1991). Or it may be mounted in terms of intellectual approach. Following Bryan Palmer's onslaught in social history, the 'descent into discourse' of cultural studies can be deplored (1990). After all, 'language is not life', and hence '[c]ritical theory is no substitute for historical materialism' (p. xiv). What is more, too overwhelming a concern with meaning and language can soon lead to a style of analysis and writing that is marked by an 'obfuscating aestheticism' (p. 188), a degeneration into 'an endless and meaningless . . . playing with words' (p. 30). Indeed, Palmer argues, at its worst a concern with questions of the discursive constitution of social and economic phenomena has produced '[m]uch writing that . . . is, quite bluntly, crap, a kind of academic wordplaying with no possible link to anything but the pseudo-intellectualized ghettoes of the most self-promotionally avant-garde enclaves of that bastion of protectionism, the University' (Palmer, 1990: 199). Now, if we agree – even a little – with Palmer's blunt verdict, then economic geography's tradition of historical-geographical materialism should be seen as an important resource. It resists the idea that 'there is nothing outside of discourse to which it can refer' (Sayer and Walker, 1992: 12). It keeps awkward issues like determination (*see* Painter in this section) and, for some at least, reality on the table (a table which, of course, is not only produced through language but is material enough that it hurts if I bang my head on it). It also provides a connection into a political project that has relevance beyond the academy, and into questions of everyday life like job security or poverty that matter to people other than those 'all in black', polo-neck wearing intellectuals who like to justify being pointlessly clever by calling it deconstruction.

I am sure you have heard all this before. And in many ways quite rightly so. Some of the social scientific work that covers itself under the rubric of a cultural turn is pretty awful (I should know – the little I've 'contributed' probably falls into that category). And any major reconstitution of an intellectual field always runs the risk of just abandoning hard-won theoretical insights and important empirical concerns in a quest for novelty and innovation. But although they provide important reminders, I am not sure that blanket defences of the economic against the cultural are a particularly productive response to the cultural turn. They can mistake, as all good realists ought to know, the contingently poor quality of particular bits of intellectual output for a necessary problem with cultural analysis. They can distract us from continuing dilemmas faced by political-economic analysis, whether that be the relations between capitalist and non-capitalist economic practices (for example, between wage economies and non-wage domestic economies) (*see* Gibson-Graham in this section), the theoretical centrality given to class at the expense of other social divisions and differentiations (Gibson-Graham again and also Massey), the conceptual peripheralization of communicative action, or the productionist emphasis of Marxian thought. And perhaps most importantly, they can end up simply reproducing – and indeed amplifying – some unhelpful dualisms apparent in less interesting cultural studies: such that culture and economy become easy ciphers for further oppositions of the mental and material, the ideal and the real, or rhetoric and reality. This is, I think, a great pity. Just to take the first two of these dichotomies, we have to be open to the possibility, for example, of what the economic anthropologist Maurice Godelier terms 'ideal realities'. We need to recognize, and conceptually account for, the fact that

> no material action of human beings upon nature . . . can be executed without setting to work mental realities, representations, judgements, principles of thought which can under no circumstances be simply reduced to reflections in thought of material relations originating outside it, prior to and independently of it. (Godelier, 1986: 10–11)

What matters, then, is not the trumpeting of the mental over the material, the text over social action, the signifier over the signified, or vice versa. What is required is a critical reflection on these very dualisms.

Of course, that reflection need not necessarily find them to be illegitimate. It may – and this is another, more subtle way in which we might oppose the economic to the cultural – result in the identification of an analytical distinctiveness between economy and culture. This possibility is intriguingly pursued by Andrew Sayer in his contribution to this section. Sayer argues that the cultural and the economic operate to rather different logics. Cultural practice, he suggests, has a primarily 'intrinsic' orientation, operating not as a means to an end but as an end itself. The meanings and normative values that pattern and orientate our social behaviour engage us so that we value them for themselves rather than for what they provide us with. In contrast, 'economic activities and processes involve a primarily instrumental orientation; they are ultimately a means to an end' (Sayer, Chapter 1 of this section, p. 17), that end being the reproduction of social life. So, while we need to examine the substantive articulations of the economic and the cultural, we also need to keep an analytical distinction between them, a distinction made 'in terms of a difference between intrinsically meaningful activities, artefacts and relationships whose value is primarily internal, and instrumental activities directed towards the external goal of the reproduction of social life' (this volume, p. 17) Sayer is not arguing for separate economic and cultural realms. Economic life is always culturally inflected, such that '[t]he economy [has always been] as much a cultural site as any other part of society, such as the family, community or school' (this volume, p. 17). Nor is he suggesting that questions of economic geography can therefore only be understood through uniquely economic forms of analysis, though those forms of analysis are a necessary part of any economic geography. What he is suggesting is that it is not enough simply to proclaim some fusion of the economic and the cultural, and then proceed to analyse the economy as just another arena for cultural interpretation.

Practically, this seems to me to be eminently sensible. It is clearly rather dangerous, especially at a time when the virtual worlds of economic modellers carry such power through structural adjustment regimes and the like (Miller, 1995), for economic geography to be full of people – like myself – who are woefully ignorant of economic reasoning. More generally, though, I am not sure that this practical need is best linked to a fundamental analytic distinction between economic and cultural logics. For a start, at a high level of abstraction we need to recognize that our instrumental logics of economic practice (assuming that Sayer is right to attribute these to us, of which more later) are themselves culturally constructed, part of the development of some sense of economic life as outside the forms of regulation governing more cultural reflections of self and cosmos. As Doreen Massey puts it, we need to 'understand the economy, and the geography of the economy, as [themselves] fully culturally constituted' (Chapter 2 of this section p. 35). Moreover, and more specifically, Sayer's equation of cultural practice with only non-instrumental forms of valuation carries with it the danger of sidelining all questions of the relations between culture and power. Perhaps a better place to start is precisely with those relations. Handily, John Allen's contribution to this section does precisely this, providing a clear and much-needed overview of different conceptualizations and modes of power and its spatialities. One of the things such an overview helps us with is an alternative to searching for distinct logics to economic and cultural practice, an alternative that instead involves examining the varying modes of power found in both cultural practices – the production of significant meanings – and economic practices – the reproduction of social life (*see also* Clarke, 1991). What I am suggesting, then, is that the logics of cultural and economic practices are less an analytical given than a matter for empirical investigation. And in turn it seems to me that one very promising route for such investigations to follow is the analysis of how both those sets of practices involve, to use Allen's rubric, processes of 'power over', 'power to' and 'constitutive power'.

## EXPORTING THE ECONOMIC INTO THE CULTURAL

So while sympathetic to the concerns of those who see the cultural turn as a profound loss for economic geography, I do not think a defensive putting up of the shutters through recourse to oppositions of the economic and the cultural is any sort of proper response. But if not defence, what about attack? Or, to get away from military metaphors and to dig myself into even more clichéd commercial ones, rather than consuming flashy imported Continental cultural durables, how about a full-blown export push for the economic, aiming to sell it well beyond the boundaries of what Dick Peet calls 'Econoland' (Chapter 3 of this section, p. 45)? The two main prongs of the promotional campaign are clear enough. First, political-economic theorization is actually a handy helpmate for all those cultural jobs about the academy ('You need your cultural change explaining, madam? Why, our new flexible accumulation theory is just what you've been waiting for'). And second, the economy is already well installed in cultural territory anyway – to quote Andrew Sayer (this volume), the realm of culture has been well and truly 'economized' – so proper protective political-economic clothing is essential for the conscientious cultural analyst. These two assertions are frequently interrelated, so let me consider them through three rather different arguments: (a) that there is an economic *determination* of culture; (b) that there is an economic *operation* to culture; and (c) that increasingly there is an economic *colonization* of culture.

When dealing with the relations between culture and economy, questions of determination have most commonly been associated with the role of an economic infrastructure in the determination of a cultural superstructure, such that political-economic processes are the explanation for cultural experiences and forms. The already classic reassertion of this position is David Harvey's account of the post-modern condition (1989a), in which postmodern cultural forms are traced back through the experiences of time and space that shape them (and which they try to represent) to the political-economic systems, and the spatial and temporal dynamics of capitalism, which in turn produce those experiences of time and space. As befits such a powerful argument and influential book, critical reviews of *The condition of postmodernity* have themselves already become citational hit-listers (*see*, for example, Deutsche, 1991; Massey, 1991; Morris, 1992), and their arguments are fairly well known. Three criticisms, though, are especially relevant here. First, we can question whether the three entities of Harvey's chain of determination (capitalist political economy – experience of time and space – cultural forms) are really as separate as his analysis suggests. Does experience sit outside of cultural forms? Is the capitalist space economy insulated off from ideas, beliefs and their cultural expression? If not, then

> since thought is not an instance separate from social relations, since a society has neither top nor bottom, since it does not consist of superimposed layers, we are forced to conclude that . . . the distinction between infrastructure and superstructure . . . cannot be taken as a distinction between levels or instances, any more than between institutions. (Godelier, 1986: 18–19)

Second, we can see how one of the consequences of Harvey's economic determinism is the weakening of his cultural analysis. This is not through any intrinsic lack of respect. As Godelier stresses, especially in their original German of *Grundlage* and *Überau*, notions of infrastructure (foundations) and superstructure (the house lived in) do not ascribe any greater importance to the former (Godelier, 1986). Rather, the problem is a tendency towards a reductive reading of cultural forms into only those aspects of themselves which might be economically determined, a reading that leaves so much about them – in Harvey's case especially their sexualizing and gendering – out of the analytical frame. Third, and more abstractly, Harvey's analysis epitomizes and reaffirms Marx's own philosophical 'reduction of the self-generative act of the human species to labor'. For while

> [i]n his empirical analyses Marx comprehends the history of the species under categories of material activity *and* the critical abolition of ideologies, of instrumental action *and* revolutionary practice, of labor *and* reflection at once . . . [he] interprets what he does in the more restricted conception of the species' self-reflection through work alone. (Habermas, 1978: 42; emphases in original)

Are there, then, alternatives to this singular determination of culture by economy? This issue is addressed most directly in this section by Joe Painter's parallel consideration of the relations between economic geographies and local politics. Rather than opposing any notion of determination, he emphasizes the reciprocal determinations established within the variable 'structural couplings' of these two fields. Moreover, drawing on the work of Bob Jessop, he argues that both local politics and the space economy are what he terms 'autopoietic systems' (Chapter 8 of this section, p. 102). That is, they exist as independent (but interrelating) systems that set their own boundaries for their activities. So, local politics have no predetermined remit; what counts as local politics is established through local political action and discourse, as in turn multiply determined by and determinative of other systems such as the space economy. Of course, in the abstract this kind of 'anti-essentialist' determination is horribly vague. Indeed, how could it be anything else when it is only within the actual flow of political and economic life that particular forms of reciprocal determination and boundary construction are drawn? But in principle, it suggests that in thinking about the relations between the economic and its various constructed others we do not have to choose simply between determinism and an absence of determination.

In many ways this analysis parallels the work of Pierre Bourdieu on the 'field of cultural production' – defined in terms of the sphere of recognized cultural productions such as art and literature – and its relations to the broader 'field of power' established within capitalist societies (Bourdieu, 1993). Bourdieu's intention is to provide an alternative to both internalist analysis of cultural products and externalist reductions of them to mere reflections of broader contours of power. To achieve this, he argues that cultural production operates within a distinctive 'field'; that is, 'a structured space with its own laws of functioning' (Johnson, 1993: 6) and its own distinctive forms and valuations of capital. But the cultural field is situated within a broader field of power; that is, within 'the set of dominant power relations in society' (ibid.: 14), the dynamics of which it refracts rather than reflects. So, in the case of the cultural field, one key structuring dimension is marked out by the nature of this refraction. Autonomous principles of cultural legitimacy emphasize the reversing of external logics of valuation (for example, by devaluing economic or market success); while heteronomous principles of legitimacy appeal to those externally sanctioned valuations. The balance between these principles, and the precise forms they take, is dependent on agency both within and beyond the cultural field, and is reflected in trends towards either the 'autonomization' or the 'deformation' of a distinctive cultural sphere (Featherstone, 1995a). Those trends in turn help to shape the relations between the economic and cultural fields. Relations that may, for example in situations of strong cultural autonomization, be characterized by contradictions between practices of signification and practices of social reproduction (*see* Bell, 1979, on the contradictions between counter-cultural hedonism and the work ethics of late capitalism); and/or by the influence of cultural specialists on everyday life (*see* Featherstone, 1995b, on cultural specialist ideologies of the 'heroic life' and their impact on accounts and practices of everyday life); and/or by the restructuring of cultural production through the dynamics of economic organization (for example, through the commodification of specialist cultural production).

Bourdieu, like Painter, therefore outlines how questions of determination can be approached without crude attributions of economic causation through an analysis of the two-way, mutually constitutive interrelations between constructed economic and cultural spheres. In this he also parallels the research agendas set out in the 'cultural sociology' of Raymond Williams (1981). Of course he also exemplifies another, rather different, way of thinking about the theoretical importance of economic analysis to cultural studies, one that is less about determination than about 'operation', and less about causal links than about metaphorical productivities. Throughout his writings on the cultural field, but especially in his critical analysis of aesthetics and taste (Bourdieu, 1984), Bourdieu is keen to explore cultural practice through the metaphorical application of economic concepts – of capital, value, accumulation, inflation and so on – to the cultural field. The details of Bourdieu's application of these concepts are well known and need not detain us here. Suffice to say, it is governed by three central arguments: first, that cultural practices of aesthetic and ethical appreciation are not disinterested, but combine questions of self-constitution and expression with questions of social reproduction; second, that there are (complex) processes of exchange and conversion between the economic and cultural realms of life, processes that Bourdieu attempts to grasp through examining the interrelations between various forms of economic, cultural and social capital; and third (to quote from the opening page of *Distinction*), that while 'there is an economy of cultural goods . . . it has a specific logic' (1984: 1). Bourdieu's argument is not, then, that cultural practice can be reduced to a narrowly defined economic rationality. Rather, it is that the economic rationality of cultural practice needs to be analysed and understood. Indeed, as I have already noted, in his more recent work he makes this even clearer by elaborating on the possibility of multiple rationalities, of more than one logic for the economies of cultural goods (1993).

Bourdieu's conceptual economization of the cultural realm is therefore based on the claim that cultural goods and meanings are produced, circulated and consumed/used in fields which inevitably have logics that govern this production, circulation and consumption. These logics can be termed economic in so far as (a) they concern processes of production, circulation and consumption; and (b) they have an important part to play in processes of social reproduction. This is a rather different claim than that which often accompanies accounts of the 'economization' of the cultural sphere, namely that cultural life is increasingly being 'colonized' by an invasive capitalist economy. In the latter vision, economy is equated with capitalism and a particular economic logic is seen as increasingly predominant in the cultural field. The understanding of this capitalist logic is in turn usually grounded in concepts of profit, instrumentality and control. So, for Theodor Adorno, the conversion of cultural life into the 'culture industry' is accompanied by the triumph of extrinsic exchange value over intrinsic use value (*see also* Sayer in this section), a connected de-differentiation of unlike things (which may be masked by illusory pseudo-redifferentiations), and the organized closure of so-called 'free time' around incarcerating forms of instrumental rationality (Bernstein, 1991). More recently, much the same arguments are simplified in George Ritzer's critique of the McDonaldization of mass culture, when, using McDonald's as an exemplar, he breaks down this cultural industrialization and rationalization into ideologies of efficiency (defined as the speedy fulfilment of needs), calculability (the ability to quantify everything), predictability (a standardization we can rely on) and control (by the organizations of mass culture) (Ritzer, 1993).

There are, then, a number of ways to argue that economic geographers should 'export' their expertise into questions of culture, and that what we need within the human sciences and human geography is less a cultural turn than a political-economic return. I have labelled these as concerns with economic determination, operation and colonization. Taking these in reverse order, let me summarize and also draw out some limits. First, economic colonization. A concern with capitalist culture industries rightly highlights trends in some times and places for cultural life to be increasingly incorporated within the webs of capitalist organization (*see* Willis, 1991, for a fascinating story about leisure, working out and incorporation). However, all too often the metaphor of colonization is deployed in such a way that this incorporation is seen as a one-way process in which the economic transforms the cultural. And yet, as analyses of less metaphorical colonizations increasingly demonstrate, colonization is never a one-way process. It always transforms the colonizer as well as the colonized. To use Nicholas Thomas' terminology from his account of colonialism and material culture in the Pacific, colonization is a process that 'entangles' identities, goods and modes of operation (Thomas, 1991). So, for example, and responding to Adorno's analysis of popular music, Bernard Gendron (1986) examines, through the case of doo-wop, how a cultural industry's producers, consumers and artefacts work with and through different dynamics of standardization, individualization and innovation than those of, say, the car industry, precisely because of their positioning within expressive cultural life. Such entanglements of the cultural and the economic are also demonstrated in those innumerable accounts of the reinsertion of particular use-values into exchanged commodities by consumers. So, for instance, at the scale of the individual, Susan Buck-Morss argues that industrial production and exchange:

> does not prevent them [commodities] from being appropriated by consumers as wish images within

> the emblem books of their private dream world. For this to occur, estrangement of commodities from their initial meaning as use-values produced by human labour is in fact the pre-requisite. (Buck-Morss, 1989: 182)

And at the scale of national incorporation, Danny Miller (1992) examines the specific Trinidadian appropriations of the American soap-opera *The Young and the Restless*. Or, alternatively and more vaguely, we could think of how these entanglements of the economic and the cultural are made apparent in analyses of workers' inhabitations of employments and workplaces, inhabitations that respond to organizational colonizations of the self (Casey, 1995) in part through bringing more of the self to work (at times to the frustration of organizational discipline – when that chatting, gossiping and telephoning-home self arrives; at times to its satisfaction – when that rounded worker self is precisely what the organization wants to construct and display) (Crang, 1994a). But in all these cases, there is no one-way economization of cultural practice; economic practice and product is also reworked as it gets entangled in cultural life.

An emphasis on the capitalist colonization of culture also runs the risk of equating the economic and the capitalistic. To use Gibson-Graham's terminology from her contribution to this section, there is a problem here of 'capitalocentrism', of viewing cultural economies only in terms of a capitalist culture industry. In contrast, to stress the economic 'operation' of culture, as Bourdieu does, helps to resolve this by allowing, in principle, space for multiple operational economic logics. For instance, and to use Bourdieu's examples, it caters for both the use of cultural creativity to maximize economic capital and the determination to signal cultural creativity and capital precisely through the absence of economic rewards (the logic of not selling out, of staying with the indie label, and so on). However, while Bourdieu emphasizes how there are different economic games being played in different fields, he still tends to fall back on an assumption that all these games are, albeit with different rules and prizes, played by people wanting to win. There is still a singular economic logic here of maximizing capitals, even though those capitals vary (Honneth, 1986). In consequence, the co-ordinates of Bourdieu's cultural field have difficulty locating the more nihilistic impulses that can characterize cultural production (*see*, for instance, Marcus, 1989). There is little place in his schema for those who do not want to win in anyone's terms; those who just want to irritate, upset and generally provoke by refusing to play whatever game governs a field. And there is little place, to screech along with Cyndi Lauper for a moment, for those girls (and boys) who just want to have fun. In short, the expressive capacities of cultural practice are hard to conceptualize through the economic lexicon of Bourdieu's analysis. This does not mean he is wrong; nor that cultural life, at least if it is to be properly cultural, cannot be about capital accumulations. Rather, it is to point to the possibility that there are some aspects of cultural practice that economic metaphors cannot grasp (even when that cultural practice is thoroughly commodified).

With regard to questions of determination, I have argued that there are clear possibilities for going beyond accounts of a uni-directional causality in which economy determines culture, in particular through an analysis of the two-way, mutually constitutive constructions and reconstructions of economic and cultural spheres. However, this takes us only so far. This is partly because such a claim says very little until empirically worked through. But it is primarily because a focus on distinct cultural and economic fields deploys a rather partial systemic imagination. Rearranging and recoupling the Parsonian cybernetic hierarchy or the Marxist base and superstructure has its value. This is especially so when those and other systemic imaginations are partly what is under analysis. But its concern with the nouns of economy and culture, even when understood as cultural constructions themselves, means we learn little about the adjectival questions of how the economic is cultural and the cultural economic. It is therefore to these adjectival questions that I now want to turn, by thinking about how economies are culturally embedded, represented and materialized.

## EMBEDDING THE ECONOMIC IN THE CULTURAL

One way of moving beyond analysing just the structural couplings of distinct economic and cultural systems is to think about how economic activity always takes place and is embedded in a culturally constructed context. Karl Polanyi's outline of a substantivist economic anthropology in the 1950s began this process, based as it was on the argument that the economy is not a universal entity, governed by an always-the-same formal logic, but a set of activities differently instituted in different times and places. In particular, Polanyi argued that the place of the economy in 'modern' and 'non-modern' societies was rather different, as too, in consequence, were their economic rationalities, with only the former having purified spaces of economic maximization. More recently, substantivist economic anthropologies have stressed how this dichotomy of modern and non-modern needs to be refined. 'Modern' economies may not be as 'pure' as Polanyi supposed and 'non-modern' economies may deploy logics of maximization more than he thought (Orans, 1968); and notions of the modern and the non-modern may be becoming increasingly unhelpful in a world of many, different 'global modernities' (Featherstone *et al.*, 1995). Instead, more complex maps of 'local, situational [economic] logics' (Prattis, 1987: 20) may be required, maps that represent the multiply scaled embeddings of economic practice and organization.

Such maps are sketched out in a number of the contributions to this section, especially those of Meric Gertler, Dick Peet, Linda McDowell, and Susan Halford and Mike Savage. Let me pull out two aspects of these accounts. The first concerns the appropriate scale of analysis. This might start at the level of the individual, in so far as economic 'rationalities derive from the broader experiences of generally socialized and encultured persons' (Peet, Chapter 3 of this section, pp. 37–38). We might then progress to the level of the organization or firm, not by appealing rather blandly to some organic organizational culture of shared values and norms, but more particularly through a focus on how firms are able to establish a particular way of doing things – a particular economic rationality – through the way they make up organizational identities (du Gay, 1996; Rose, 1990), whether that be via the purifying filter of recruitment policies (McDowell) or the disciplining, but resisted, discourses of organizational change (Halford and Savage). We might also consider the local or regional scale, and the possibility of distinctive cultural and economic formations marked by their own social and environmental imaginaries as reproduced through local or regional institutional forms (Peet). Finally, we might look to the national and supranational scales of political regulation, to trace out their role in shaping what counts as economically rational within different legal and institutional environments (Gertler). Just how these differently scaled contextualizations interact is of course the really fascinating agenda. To rehearse some questions that are already being answered: how do individuals and companies negotiate constructions of organizational identity with more generally apparent norms, and abnorms, of individual identity practice; to what extent do companies regionalize their corporate self-fashionings; and what happens to regional or national institutions, identities and imaginaries when constituted within globalized spaces of flows?

Answers to these questions are further complicated by a need to clarify just how economic practice is embedded. In setting out some agendas for economic sociology, Sharon Zukin and Paul DiMaggio – in some ways helpfully, in other ways not – distinguish cultural, structural and political forms of embeddedness (1990b; *see also* Granovetter, 1985). They define these as follows:

> When we say that economic behavior is 'culturally' embedded, we refer to the role of shared collective understandings in shaping economic strategies and goals. Culture sets limits to economic rationality: it proscribes or limits market exchange. . . . [C]ulture may shape terms of trade. . . . Culture . . . prescribes strategies of self-interested action . . . and defines the actors who may legitimately engage in them. . . . Finally, norms and constitutive understandings regulate market exchange, causing persons to behave with culturally specific definitions of integrity. . . . '[S]tructural embeddedness' [is] the contextualization of economic exchange in ongoing patterns of interpersonal relations. . . . By 'political embeddedness' we refer to the manner in which economic institutions and decisions are shaped by a struggle for power that involves economic actors and nonmarket institutions, particularly the state and social classes. (Zukin and DiMaggio, 1990b: 17–20)

I say these distinctions are helpful, in so far as they (a) demonstrate that economic embeddedness is not only culturally constituted, but has social and political dimensions too; and (b) articulate an understanding of the cultural – as the shared understandings and norms of some social group – that is widespread in conceptions of the cultural turn. They are unhelpful, though, to the extent that this definition of culture is a rather limiting one. At the least, if we adopt such a culture concept, then it becomes vital – as Meric Gertler argues later in this section – to analyse how these understandings, norms and social collectivities are produced through social relations and political action. Culture must not become some *deus ex machina* to be dragged on stage as the final word of explanation ('it's all because of their culture' . . . ). To position culture as cause, or at least as explanatory context, simply replicates uncritically the crudities of the worst managerial culture theory, and taps into a long and undistinguished history of substituting notions of culture for notions of race or *Volk* (Young, 1995). But we can, as the contributions to this section show, go beyond this definition of culture. We can do more than reverse Bourdieu's situating of cultural production in a broader economic field of power by locating economic activity in a broader cultural order. Instead, we can think of how political and social/structural embeddedness are bound up with cultural practices of identity formation and meaningful signification. But to do that is to shift our focus away from concerns of contextualization and embedding and towards the representational and discursive constitution of economic life.

## REPRESENTING THE ECONOMIC THROUGH THE CULTURAL

Let me elaborate. To speak of embedding is to speak of situating the economy in a cultural context, in a placed culture, whether its cultural place be the individual (as representative of some wider spatial scale), the firm, the region or the nation-state. To speak of representing the economic in the cultural is to highlight how all these 'places' are themselves cultural constructions, and how as constructions they sit alongside and help to constitute a host of other constructed economic entities that make up what an economy is (labour, work, home, the harvest, profit, and so on).

Perhaps the easiest way to get at such constructions is through the analysis of the rhetorical devices of economic analysis itself. So, Donald McCloskey has set out to show how economics is part of the 'conversation of mankind' (McCloskey, 1988: 282), and as such how its tools of analysis and forms of expression are not just a stylistic gloss jazzing up a substantive object, but constitutive of that object themselves. Arguing directly against the kind of materialism advocated by Bryan Palmer, he deploys concepts of rhetoric to defend a concern with styles of speech and thought:

> The distinction between style and substance has burrowed like a worm deep into our culture . . . yet it has few merits. It is all style and no substance. Consider. What is the distinction of style and substance in ice-skating or still-life painting or economic analysis? . . . The 'substance' of a cake is not the list of basic ingredients. It is the style in which they are combined. (McCloskey, 1988: 286)

In economic geography, Trevor Barnes has similarly argued for an analysis which recognizes that 'theories . . . in part . . . create the reality they seek to interpret' (1992: 118), exploring this in part through a reading of the physical and biological metaphors underlying neo-classical and Marxist economic geographies respectively.

Now, in so far as economics, and economic geography, are not activities confined to ivory towers but practices and institutional sites that help to regulate and fashion economic policy and activity beyond the academy, understanding this rhetorical (McCloskey) or metaphorical (Barnes) construction of the economic is not just the stimulus for a (depending on your point of view) helpful or disabling academic self-reflexivity. Lurking here, there are clearly much wider questions about the forms of understanding used to govern economic life, and the role played by academic economic understandings in these. Trevor Barnes signals as much when he emphasizes how singular academic metaphors leave 'no room for the diversity of individual places' (1992: 134). Again, though, it is within economic anthropology that this agenda is perhaps best progressed, for me most strikingly in Stephen Gudeman's outlining of a 'cultural economics [based] upon the direct comparison and contrast of [different local] metaphors and models of livelihood' (1986: ix). For Gudeman, scientific economics operates as one or more of these local models, to be relativized alongside swidden agricultural understandings of crops as hair and harvesting as barbering or any other lay understandings of economic action. Such a cultural (geographical) economics is precisely Dick Peet's concern in his contribution to this section. He too argues that 'economic rationalities are culturally created, take diverse forms, and have distinct geographies' (p. 38). Moreover, his account of a New England 'discursive formation' offers not only a fascinating exemplification of Gudeman's concern with how the cultural representation of the economic is both shaped by and shapes economic life, but also develops the notions of 'discursive regulation' and the 'cultural order of production' as ways into these mutual constitutions. Both, it seems to me, are particularly productive in so far as they suggest a move beyond just the linguistic or rhetorical fashioning of economic realities, and towards the both told and embodied ordering of economic practice. Indeed, following John Law's thin but wild and wonderful ethnography of Daresbury research laboratory, its networks of relations, and the representational acts involved in sustaining such networks (1994), what is opened up here is a concern not just for different cultural orders of production, coexisting and interrelating in some form of synchronic cultural-economic mosaic and/or diachronic cultural-economic transformation, but for the multiple 'modes of ordering' that are performed and slipped between by organizational actors. To put it another way, what begins to emerge in these kinds of analyses is something more than a contextualization of economy in a place and time specific form of cultural representation (which is, I think, what Gudeman gives us), or discursive formation (which is where Peet begins). Instead, the whole notion of context is dynamized and respatialized, to allow for the possibility – and maybe it is only that – of multiple discursive orderings of economic practice, orderings that are performed in particular times and spaces in more or less routine ways.

## THE CULTURAL MATERIALIZATION OF THE ECONOMIC

Which brings me on to one final way of thinking through the relations between the economic and the cultural, one in which economies are seen as involving the production, circulation and consumption of 'materials' that are cultural in character. In this portrait, then, the discursive orderings of economic life are not just matters of regulation; they are (the point of) that economic life in the first place.

One of the more sustained attempts to make this argument is that of Scott Lash and John Urry (1994). Generally, they argue for a recognition of the de-differentiation of economy and culture:

> economic and symbolic processes are more than ever interlaced and interarticulated; that is, . . . the economy is increasingly culturally inflected and . . . culture is more and more economically inflected. Thus the boundaries between the two become more and more blurred and the economy and culture no longer function in regard to one another as system and environment. (Lash and Urry, 1994: 64)

More specifically, they illustrate this de-differentiation with regard to: (a) the 'discursive reflexivity' of informational economies (such that '[i]n terms of the discursive nature of the knowledge that flows through their "arteries", production systems have become, not so much dependent or interarticulated with expert systems, but expert systems themselves') (ibid: 108); (b) the 'aesthetic reflexivity' of the culture industries (where, contra Adorno, they argue that '[t]he growth of . . . "cultural industries", . . . organized as they are partly around notions of the aesthetic, reflects the increasing culturization of economic life') (ibid.: 109); (c) the combined informational and aesthetic interactions of various kinds of service economy, such that '[t]o talk of services is to talk of information and symbol and of the increasing importance of both within many diverse kinds of post-industrial space' (ibid.: 222); and (d) the growth of touristic forms of consumption as part of a more general aestheticization of everyday life, forms of consumption that mean that codes of cultural geographic difference no longer operate as an external barrier to capitalist homogenization, but are actively produced through the cultural-geographic differentiations of capitalist economies (Grossberg, 1995; *see also* Cook and Crang, 1996; Crang, 1996).

As Andrew Sayer rightly points out in his chapter in this section, the empirical significance of these touristic forms of consumption, and these informational, aesthetic and interactional economic sectors, needs careful consideration. Clearly, Lash and Urry offer a partial portrait of contemporary late capitalist economies. But at the same time, it would be a shame to let some well-worn concerns over the conceptual chaos and statistical measurement of, say, service employment smother their provocative analysis. Rather, it might be better to pick up on three more general and implicit suggestions. The first of these is that the materialization of economic activities matters for the character of those activities themselves. That this might be so was flagged some time ago by Daniel Bell in his (still deeply flawed) proclamation of post-industrial society, when he asserted, for example, that 'the fact that individuals now talk to other individuals, rather than interact with a machine, is the fundamental fact about work in a post-industrial society' (Bell, 1973: 163). One does not have to endorse Bell's historical analysis to accept his contention that the means and materials of production matter to the character of that production process and its social relations. Indeed, to pay attention to the quality of these materials is to recognize the role of such materials as important components in the organization of economic activity in their own right (Law, 1994). So, whether one is producing/circulating/consuming crude oil, or a piece of music, or a car, or some polite conversation, matters to those moments of production, circulation and consumption, and, indeed, to the character of the cultural circuits through which those moments are interconnected.

Second, Lash and Urry's analysis highlights just how limiting the reduction of economic activity to one particular kind of material – say, to physical consumer goods – is. It is not that the alternative examples they offer exist in isolation from such physical goods. For example, interpersonal service is often staged around the delivery of a physical product (the meal in the restaurant, for example) and performed on a physical set (the restaurant interior). Cognitive labour needs physical materials for information storage (disks, for example). And aesthetic labour – say in the fashion industry –

obviously deploys physical materials to achieve stylistic effects, and has to pay a great deal of attention to that physicality (for example, to the properties of cloth, dye, magazine paper, and so on) if its creative aesthetic ambitions are to be realized. But to recognize this provides no justification for closing analysis around those goods. Rather, what is required is a recognition of the materially heterogeneous character of economic geographies and their processes of production, circulation and consumption. To also think, for example, of the economies of talk worked through in interpersonal conversational product elements or in interpersonal production processes (*see*, for example, Drew and Heritage, 1992; Leidner, 1993, on the routinization of talk in standardized service outlets; or Ventola, 1987, on 'service encounters' as speech acts). Or to consider the economies of love worked through in caring labour, whether in the domestic economy or the public sphere (*see*, most famously, Hochschild, 1983, on the emotional labour of flight attendants). Or somehow to evoke the varying economies of bodily performance found in consumption and production practices like dancing, hill walking, door to door sales negotiations, attending a meeting, changing a tyre, or getting in the cows for milking (*see*, for example, Crang, 1994a; Crang, in press; McDowell, 1995).

Third, it is not sufficient – though at times it may be necessary – simply to document the many different 'materials' used in economic practice. We also need to grasp something of their differing economic geographies, and the active construction of these in economic practice. Something, then, of their potentialities for mobility and fixity. Of their fit with differing forms of surveillance. And of their geographies of display, or their 'displayability': their economies of the seen and the unseen; of trust and distrust; and their modes of display (for example, whether screened, performed, or touched).

## CONCLUDING COMMENTS

By way of conclusion, let me take some steps back. I have been arguing that economic geography is reconstituting itself in the context of a number of cultural turns within the human sciences. Initially, the impact of these cultural turns was simply to bring what were still rather separate economic and cultural approaches to bear on broadly the same areas of interest. Questions of consumption, for example, could be approached through political-economic analyses of provision systems and their regulation and also through cultural analyses of consumer practices, consumption arenas, and the imaginative geographies bound up with consumed products and places. These two analyses might be seen as complementary, or as in competition. But they kept their distance from each other. It was easy to differentiate who the 'cultural' and 'economic' geographers of consumption were.

Increasingly, though, the waters are becoming a little more muddy. Accounts of systems of provision and regulation examine the representational politics and poetics of interest constructions within those systems (*see*, for example, Marsden and Wrigley, 1995). And slightly less often, accounts of consumers and their usages of commodity systems move away from questions of identity to think of issues of value for money, thrift, household ethics and domestic love (Miller, 1995b). In this review I have concentrated on how economic geographers might embrace – and are embracing – forms of cultural analysis. I have argued that simply rejecting questions of culture for the sake of defending a pure political-economic project, while not without its appeal, betrays a self-satisfaction that sits uneasily with a number of real difficulties faced by political-economic approaches. Instead, I have tried to outline some of the possibilities produced through responding more openly and with a greater willingness to destabilize the identity of economy geography. While reviewing a number of options, I have been particularly keen to flag some issues around the discursive ordering and the cultural materialization of economic activity, but the chapters that follow provide both critical analyses of such an emphasis and other alternative ways for reconstituting economic geography that I might have spent longer on (in particular, the discussions of power, difference and the imaginary in the chapters of Allen, Gibson-Graham and Watts have shadowed rather too implicitly my argument here).

Of course, one outstanding question is just what the kinds of cultural-economic geographies I have been advocating might look like and just what is economic about them. The following chapters combine both exemplary cases and more general reflections on this. For my part, I am unsure about defining the economic in terms of distinctive subject matter, approach or internal logics. On all three counts, economic questions are inevitably and quite properly entangled with a host of other geographies. Instead, perhaps economic geography's value lies in its motivation as much as its methods or concepts; that is, in its emphasis on understanding differentiated spaces, places and practices of production, circulation and consumption, and the forms of surplus extracted within and between them. These spaces, places and practices are never purely economic, and nor are the surpluses they produce. But the commitment to study these vital economic moments – production (in its broadest sense), circulation and consumption – and their regulation, by whatever means necessary, wherever they take place and whatever materials are produced, circulated and consumed in them, is a constitution of which economic geography can be proud.

# CHAPTER ONE

# THE DIALECTIC OF CULTURE AND ECONOMY

ANDREW SAYER

## INTRODUCTION

Many social sciences, particularly sociology, human geography and urban studies, have undergone a 'cultural turn' in recent years, evident both in method and content, with the emergence of a long-overdue dialogue with literary studies, and an increased concern with cultural phenomena. The flip-side of this has been a decline in interest in political economy following its peak during the late 1970s and early 1980s. Urban studies exemplifies the shift; even though economic work continues, it is more about institutional forms than about basic aspects of capitalism. This raises some questions: Where does all that old-time political economy of urbanism stand now? Is the city now any less dependent on the production and circulation of capital than hitherto? Has culture come to dominate economy?[1] What discussions there have been so far concerning culture and economy tend only to have been implicit in arguments about more concrete topics, such as the debates over the interpretation of gentrification in urban studies and over Japanese work practices in industrial sociology. Though interesting, these fail to bring out more general issues of culture and economy.

The reasons for the cultural turn and the decline in interest in economy are difficult to explain. We might be tempted to invoke the sociology of knowledge, which points to the need of a new generation of scholars to distinguish themselves from the old, political economy-dominated academic hierarchies and to the attractions of increased cultural capital associated with studies of culture compared to studies of economy. More convincing is the argument that the cultural turn reflects a change in political culture, involving a shift from the 'politics of distribution', dominated by economic matters and associated with labourism and traditional conservatism, to a new 'politics of recognition' which is more cultural in character, being about identity and respect. Political movements such as feminism, anti-racism and the greens combine *both* types of politics (Fraser, 1995). In addition there are new political currents, particularly the politics of sexual identity, which are entirely about recognition. But if both types of politics are involved in many of the new political movements, one would expect this to be reflected in academia by more attempts to combine economic and cultural interests rather than largely abandoning or diluting the former.[2] I would suggest that the decline of the politics of distribution and the cultural turn in social science could be a consequence of the defeat of the Left's alternative economic agenda and the search for areas of social change which offer more hopeful prospects. However important the economy remains, the scope for progressive change looks greater in more cultural directions where the politics of recognition are prominent.[3]

The focus of this chapter is prompted by another possible explanation of the shift – namely that culture has become more important than hitherto, and that the economy is becoming 'culturalized'. While evidence of this can be found, I want to argue that much of what might appear to indicate a culturalization of the economy can often be more convincingly read as an 'economization of culture'. Economic forces continue to dominate contemporary life, and thus, however unfashionable, economic analysis cannot be sidelined. However, this is not a simple either/or matter. The relationship between culture and economy is highly complex, and some authors believe they are increasingly becoming indistinguishable. In order to assess these matters, this chapter discusses the relationship between culture and economy, first in general, and then in relation to employment, consumption and politics.

## ARE CULTURE AND ECONOMY (STILL) DISTINGUISHABLE?

There is little doubt that arguments about an increasingly close relationship between economy and culture can point to processes which, if not entirely new, are growing. However, it is one thing to note the new ways in which economy and culture interact and combine, but quite another to say that the distinction is no longer tenable (e.g. Hall, 1988; Jameson, 1990; Lash, 1990). I shall argue that there are still politically and theoretically crucial differences between culture and economy, and that it is important to understand how one affects the other. In order to develop this position, it is necessary to define the basic terms.

Though 'culture' is formidably polysemic, I want to focus on a feature central to all uses of the term: that is, practices and relationships to which meanings, symbols or representations are central.[4] These pattern and orientate social behaviour within particular groups. They include – but are certainly not exhausted by – normative aesthetic and moral-political principles or judgements, and such cultural values are fundamental to the constitution of identities. Although they are hardly ever constructed under egalitarian conditions, cultural phenomena must in some sense be shared by the members of the culture, even if they are contested; they cannot simply be imposed. This is because the realm of meaning is at least immanently dialogical. As critical theorists have argued, even where there are attempts to impose meanings, the processes of communication cannot reduce wholly to monologic transmission, and the same must go for culture (Lash and Urry, 1994; Williams, 1958).

A crucial feature of many of the goals or goods associated with culture is that they are primarily internal. For example, the elderly or a certain kind of music might be valued, but this respect or value is not accorded merely in order to achieve some external goal, but because the elderly or that kind of music are valued in themselves. (However, it is possible that they may *also* be cultivated instrumentally, for other ends, such as distinction (Bourdieu, 1986).) In saying that these values are intrinsic I do not mean that the objects are beautiful or good in themselves, independently of a subject, for value is always relational. By intrinsic I merely mean non-instrumental. Although some things – such as a BMW – may sometimes be valued as a means to an end, often of distinguishing ourselves from others, they may also be valued for their own particular qualities. Sometimes their instrumental functions may be conditional on their intrinsic qualities; the BMW would not be a source of prestige for its owner if it was unreliable and awful to drive.

Cultural norms and values regarding actions have, at least in part, a deontological character, according to which actions are seen as good or bad in themselves rather than in terms of their consequences.[5] However, in stressing the normative and particularly the moral element within cultures I do not want to idealize them; some of the norms involved may actually be repressive. The intrinsic values of sexism and racism are cases in point. The relationship between culture and economy should therefore not be coded: culture (good), economy (bad).

By contrast, economic activities and processes involve a primarily instrumental orientation; they are ultimately a means to an end, satisfying external goals. Economic work may of course be satisfying in itself as well as a means to an end, and while this is obviously desirable, the work itself is rarely more important than its product, be it material production or interpersonal work such as childcare.[6] The needs which the economy provides for include not merely transhistorical physical needs but ones which are geohistorically specific, such as beer-drinking or wearing jeans, and even the transhistorical or species-wide needs are always met in culturally specific ways.

Economic activities are always culturally inflected. There is no way in which they could be conducted independently of systems of meanings and norms. The economy is as much a cultural site as any other part of society, such as the family, community or school, but where mainstream economists abstract from the cultural side, political economists and institutional economists have been increasingly willing to consider it. The form of the union of the economic and the cultural is almost certainly changing but there has always been some such union; it is a transhistorical rather than a post-modern phenomenon. However, this does not imply that the distinction between the economic and cultural is untenable. Certainly if we define it as above, in terms of a difference between intrinsically meaningful activities, artefacts and relationships whose value is primarily internal, and instrumental activities directed towards the external goal of reproduction of social life, then the fact that they are universally combined in various ways does not challenge it. Moreover, while all economic activities have a cultural dimension, the converse does not apply, for not all cultural activities are directed to reproduction; for example, a birthday party is not an economic activity. That such cultural activities do not escape economic implications (e.g. buying presents) does not make them in any meaningful sense economic.

This association of the cultural and the economic implies that it is wrong to think of the relation between them as simply external. In answering questions like 'How has the economy been influenced by culture?', we need to remember that there was not first a pristine economy which somehow later got influenced by culture, for the economy was always culturally influenced, both from within and without, from the start. On the other hand, this does not make the question meaningless, for at least some cultural influences are indeed external to the economy. Nor does the fact that economies are always culturally inflected mean that there cannot be tensions rather than harmony between them. At times, the logic of one may dominate the other, as when cultural practices are subordinated to economic demands, though this does not mean that the domain of culture shrinks, but rather that the culture changes character, possibly becoming less norm-driven.

If we move to a more specific level to consider advanced economies, especially capitalism, then the relationship between economy and culture takes on another form, in which economy extends beyond the life-world to become a major social system standing to a certain extent opposed to it and dominating it. The combination of unprecedented degrees of division of labour and knowledge with dependence of economic survival on competition for the spending of often distant and unknown others, makes the influence of individuals over their own life-chances more indirect and uncertain than ever before. This leads to the situation in which we speak more readily of the human problems of economic activities than the economic problems of humans. What matters is the product and price, and the abstract or formal rationality of exchange value.

Many have followed Polanyi in arguing that this preoccupation with 'economizing' in terms of economical use of labour and resources is restricted to capitalism, and is absent in pre-capitalist societies where economy is indistinguishable from culture (Polanyi, 1957a, 1957b). However, if we make the distinction, as I have suggested, it is clear that in many such societies, the need to economize is much more vividly appreciated. The relationship is summed up nicely by a telling moment in an excellent film on the Hamar people in Ethiopia. The Hamar are mostly self-sufficient in food, and women are primarily responsible for growing crops. In the film, a young woman who is about to be married is being counselled by an elder sister, who tells her that when she is married she must be sure to keep the field well weeded, or else the good name of her family and the honour of her father will be damaged. The younger sister acknowledges this. This, of course, is just what culturalists like to hear: the economic being subordinate to the symbolic order. But wait – the bride then says, 'yes and I realize of course that we'll starve if I don't'. In other words, she well understood the relationship between economy and culture.

Within capitalism, some authors point to recent developments as evidence of a collapse of the culture–economy distinction. The rise of the service sector has been cited as indicating the emergence of a post-industrial society in which traditional material production is becoming secondary to a more strongly culturally inflected service economy. But it is not clear that much has changed here. First, the expansion of services is widely exaggerated since many of the activities classified as service production, such as catering, have a major element of manufacture (Sayer and Walker, 1992). True services involving interpersonal communication, provision of information or provision of ambience, such as teaching or counselling, do indeed involve stronger cultural elements than material production, and have a dialogical and performative character in which the quality of the service is affected by the 'consumer' as well as the producer. Moreover, in the case of the professions, the work has a normative character in so far as it involves evaluating the situations and behaviour of clients/patients/students and deciding what they need. Yet all these remain activities pursued for economic reward and subject to economic constraints. Alongside the expansion of non-material commodities in the form of service work proper, the wealth of material commodities continues to expand relentlessly, even though fewer people are involved in making them. Thus the air hostess's smile presupposes planes, in-flight meals, baggage-handling systems, radar and airports, and the hundreds of thousands of material components that go into them.

Another line of argument popular among cultural analysts takes up Baudrillard's emphasis of the growing importance of the 'sign value' of commodities; that is, their symbolic significance as means by which lifestyles and identities can be constructed. While this aestheticization is probably increasing in consumer products, two things have to be remembered. First, sign value has certainly not replaced exchange value as the regulator of economic activity in capitalism – company accounts or bank balances are not assessed in sign value! Second, the majority of commodities in a modern economy are not consumer commodities but intermediate products like oil, computer chips or bearings, which do not need to be aestheticized, even if some of them do end up in consumer products.

A stronger argument for a fusion of culture and economy could be drawn from Bourdieu, who analyses culture as having an economic logic (1977, 1984). Bourdieu sees almost every act in instrumental, indeed explicitly economic, terms; the pursuit of honour or status, expressions of goodwill and especially gift-giving are seen as disguised strategies of exchange through which symbolic, social or cultural capital is accumulated. This is indeed illuminating for a wide range of actions, both for pre-capitalist and capitalist societies, but not all actions are instrumental; some are done for their own sake, and moral acts are a particularly important case.

So, in summary, we can agree with Stuart Hall that culture is not 'a decorative addendum to the "hard world" of production and things, the icing on the cake of the material world' and that 'through design, technology and styling, "aesthetics" has already penetrated the world of modern production'. But it does not follow from this that the distinction between the economic and cultural 'is now quite useless', as he claims (Hall, 1988): indeed, it is striking that those who would fuse the two concepts into one find themselves still needing to use them separately a few pages later! Similarly, I accept that 'the economy is increasingly culturally inflected and . . . culture is more and more economically inflected' (Lash and Urry, 1994: 64), but despite the inflections, economic and cultural logics remain different and often pull in opposite directions. Thus, as the range of commodities grows, the 'imperialism of instrumental reason' threatens the immanently dialogical qualities of cultural values (Lash and Urry, 1994: 83).

## EMBEDDING AND DISEMBEDDING

One of the main ways in which the association of culture and economy has been theorized is through the concept of 'embeddedness', interest in which has become a defining feature of recent economic sociology (e.g. Dore, 1987; Granovetter and Swedberg, 1993; Crouch and Marquand, 1993). Economic processes have always been embedded in social relations, which have a cultural content in terms of meanings and representations.[7]

It is through this embedding that the trust required by economic activities in advanced economies is established. An economy which was not socially embedded and instead conformed to the Hobbesian model of rational choice theory, with mutually suspicious, asocial, atomistic individuals pursuing their self-interest, would never get off the ground. The lack of mutual trust and shared norms would embroil everyone in cripplingly inefficient attempts to contractualize, insure and police economic relations to cover every possible form of deceit or eventuality. In addition, without any patterning and institutionalization of actions to serve as templates for future action, there would be chronic uncertainty and anarchy. Trust is an essential public good in economic interaction and concerns both competence and honesty; cultural norms reduce doubts about both through their moral content and through providing familiar and reassuring systems of signs – particularly the forms of acceptable dress and address – which reduce uncertainty. Thus the embedding of economic processes always has a cultural dimension, as studies such as Nigel Thrift's on the old and new cultures of the City of London show (Thrift, 1994b).

While all economies are socially and culturally embedded, in advanced economies, division of labour, markets, money and capital are also *dis*embedding forces; economic processes work largely behind the backs of individual agents, so that outside the household economy, the economy becomes a separate realm on which we depend rather than one which we can straightforwardly produce. We therefore need 'system trust' as well as interpersonal trust.

The increased autonomy of the economic dimension of social life under industrialism and the loss of control that individuals have over their survival means a reduction in the influence of moral considerations in economic behaviour – a decline in the 'moral economy'. It is one thing to produce for those one knows according to moral obligations, but quite another to do so for unknown others according to market signals or a central plan. Yet such normative considerations are not left behind altogether because the amount and distribution of paid and unpaid work, and the pattern of taxation and transfer payments, are still influenced by social norms – albeit strongly gendered ones increasingly under challenge in their present form – and not merely by market forces. Moreover, as we have seen, economic systems require a significant degree of honesty and trust. Beyond this the logic of market processes is fundamentally amoral (Sayer, 1995).

However, markets are just one sphere of modern life, and individuals are concrete beings who have to act within many overlapping spheres. Consequently, as Etzioni argues in his 'I and We paradigm', individuals experience perpetual inner tension among their various moral commitments, and between these and their desires (many of which might be met through markets) (Etzioni, 1988). For instance, according to our moral commitments we may feel we ought to work hard and not let down our colleagues because doing so is right in itself, while according to instrumental, calculative rationality we may decide to work hard and not let down our colleagues if it pays to do so. People do not merely seek to maximize pleasure but balance this with acting morally.

While this account of embeddedness and morality in industrialism in general may be accepted, it has recently been implied – if not stated categorically – that so-called post-Fordist developments include an increase in the embeddedness of economies and perhaps a limited culturalization and remoralization of economy, associated with increased co-operation and trust among related firms. Even if this is happening through the expansion of relational contracting and networking, it does not amount to a remoralization of the economy, just a more sophisticated and effective way of meeting the external goals of business. Embedding may enable and indeed facilitate economic activities, but it is costly. (That it can also facilitate the creation of value does not contradict this.) In all but the most abundant economies, embedding is limited by the need to get on with the work and to produce enough. In capitalism, competition forces firms to try to reduce avoidable or unnecessary embedding of transactions. Both in its extent and form, embedding is shaped by the costs and benefits it produces.

It is normal for embedding and disembedding forces to interact and indeed to be interdependent. The operation of financial markets presupposes strongly embedded relationships between key actors. At the same time, their operations produce unintended and only partly registered effects on other market actors and on the livelihoods of vast numbers of unknown others. Despite all the (often unacknowledged) labour that is required to make markets work, their operation still, crucially, includes powerful disembedding mechanisms – glossed as the invisible hand – which are not the conscious product of any agent. The disembedding forces are particularly strong in financial markets because through them exchange value has split off from particular use values and their producers and users so that it circulates without regard for the economic security of individuals, households, firms, regions or whole countries.

The levelling effect of money, markets and bureaucratization and the disembedding effect of capital in many ways make capitalist mechanisms *in*sensitive to particular cultures, as Marx and Engels so famously saw in the *Communist manifesto*. The culture of the person with money does not matter to the seller, and in the case of the client of the bureaucracy it should not matter. Capitalism selects out aspects of cultures for commodification according to their economic rather than their cultural value.

## THE INSTRUMENTALIZATION OF WORK CULTURE

Culture is everywhere, including within economic institutions, but it has often been instrumentalized for economic gain. This is nothing new; norms regarding gender roles have long been manipulated for economic purposes, and nineteenth-century social reformers encouraged church attendance among the working classes for economic reasons. Indeed, the making of the working class was as much a cultural change as an economic one. However, the instrumentalization of culture appears to be increasing, as evidenced by the importance attached to corporate culture, the 'value-driven company' (Peters and Waterman, 1982) and 'Japanization'. Thus many employers increasingly value and try to develop their employees' communication skills, for facilitating both their internal operations and their relations with suppliers and customers (Morgan and Sayer, 1988; V. Smith, 1996).

In one sense, the subset of cultural values constituted by moral values are difficult to instrumentalize precisely because of their intrinsic character. Capitalism certainly does not fit easily with this. The rhetoric of the market is based not upon following intrinsic values associated with particular kinds of activity, but upon consequentialist reasoning by self-interested individuals regarding the pay-offs for undertaking such activities; according to Adam Smith, in a 'commercial society' you are likely to do more good as regards others by pursuing your own economic self-interest in the market than by doing what you believe to be in the public interest (A. Smith, 1976). Moreover, one of the general features of moral reasoning is that it involves 'reciprocal recognition', in which we judge how to treat others in terms of how we would like them to treat us (Benhabib, 1992: 52). Yet reciprocal recognition and mutual commitment are inconsistent with the radical inequality of the capital–labour relationship.

Motivating people in purely instrumental ways – 'do this and there will be a pay-off' – may sometimes be successful on its own, but often, external goals such as profit can be attained more effectively by harnessing cultural norms. The more liberal economies may be suffering not from having overestimated the importance of market ends in economic success, but from having underestimated the possibilities of more dialogical forms of organization as means for meeting those ends. Sometimes the instrumentalization of values is completely transparent, as in exhortations to salespersons to believe in their products, but often the exhortations appeal to the intrinsic value of particular ways of working with others rather than to the economic consequences. Some instrumentalized cultures are constructed openly, as in 'designer work cultures' (Casey, 1995); others disguise the element of construction and appeal to existing values outside, as in the case of large Japanese companies adopting the rhetoric of family values (Eccleston, 1989).

When practices influenced by moral and aesthetic values become means to ends which have nothing to do with the moral, the good or the beautiful, those qualities are arguably degraded or tainted to some degree. The conflict between integrity and personal or corporate gain and the respective rewards they are likely to receive invites cynicism and reaction. Nevertheless, it must be remembered that although new 'human resource management' and the like are introduced for instrumental reasons, they may be better than what preceded them. Studies by Victoria Smith of low-status workers who had undergone training in social and communication skills so that they could handle clients more effectively showed that the skills could be valuable to them both inside and outside work, and that some felt empowered by this aspect of the instrumentalization of work culture (V. Smith, 1996). People do not work only for money; the intrinsic quality of the work itself and the social relationships of the workplace matter too. In this sense, popular cultural values may not be difficult to instrumentalize. The extent to which instrumentalization or economization of culture degrades the latter should therefore not be overstated. In any case, whatever the appeal of the rhetoric of culture managment, its effects are likely to be limited, for 'the so-called sub cultures found within some organisations are stronger and more enduring than transitory managerial cultural espousals that would overcome them' (Anthony, quoted in Thompson and Ackroyd, 1995; *see also* Alvesson, 1993).

It is clear that the economization of culture, rather than the reverse, is the dominant process here. This is also the case with the attempts of many governments to use quasi-markets to force a logic of economizing upon professional work in the public services, such as teaching, which frequently leads to professional ethics being compromised (Sayer, 1996). However, I do not want to argue that economization is always bad. Where intrinsic cultural values may be reactionary, economic interests may sometimes prompt a reduction in discriminatory behaviour. Companies that, on racist grounds, refused to sell to certain people or employ them would be at a disadvantage relative to non-discriminatory competitors, who would be able to operate in larger product and labour markets.[8] The neutrality of markets and money may therefore have 'civilizing effects' (Hirschman, 1982).

## CASH, COSTS AND THE CULTURAL TURN WITHIN POLITICAL ECONOMY

Firms are almost certainly paying more attention to corporate culture as an aid to profitability, but, as Karel Williams and associates have argued, the fate of firms depends ultimately on *cash* – on costs and revenue (Williams *et al.*, 1994). No matter how sophisticated the management systems, the work culture, the design of the product, including (for consumer products) its targeting at particular lifestyles and identities, firms do not survive without cash, and the latter is not necessarily influenced much by the former. To a certain extent management can keep down costs, and manipulating work cultures may help, yet labour costs – ultimately the main component of all costs – depend primarily not on managers but on country-specific social settlements and levels of development. Thus, within the UK or the European Union generally, it may be possible for firms to lower their labour costs somewhat, for example by hiring illegal immigrants in place of nationals, but they cannot expect to lower them to the levels dominant in, say, Indonesia unless they move there. Similarly, while managements can raise the sales and value-added of their product and hence get higher revenues, a majority of firms cannot do this simultaneously unless macro-economic circumstances are such that aggregate demand is increasing. To forget this is to invite a fallacy of composition. As Williams *et al.* argue, new management philosophies, designer work cultures, networking and the like all operate within constraints which are *beyond* managerial control. The former make small differences within the larger differences of economic costs which are mainly external to individual firms. After all, globalization is not primarily to do with responses to differences in work culture and the institutional form of firms around the world (although it affects and is affected by these), but is predominantly a response of capital to the uneven geography of costs and revenues.

One of the most influential cases that has generated support for the idea of culturalization of the economy, particularly through new management systems and work cultures, is that of Japan, particularly its electronics and car industries. This is an example of economic forces putting culture on the agenda: there would be little interest in the West in Japanese culture if Japan posed no economic threat. But it turns out that culture is less important in this case than is often supposed. Williams *et al.* show that the much-vaunted productivity advantage of the Japanese car firms over their North American and European rivals has been greatly exaggerated. Their main advantages lie in low labour costs and rising demand. Japanese industry is far more vertically disintegrated and reliant on small firms with low wages than are its rivals. This lowers the cost of inputs considerably. On the demand side, Japanese firms in the leading sectors have until recently enjoyed a steadily expanding domestic market and often rapidly increasing overseas demand, while European and North American competitors have at best faced intermittently rising demand. This makes a huge difference to the economics of major investments in product and process innovation as regards cost recovery. While an optimum work culture might facilitate innovation, external market and macro-economic conditions are crucial for determining whether it pays.[9] This is not to deny that Japanese work culture is distinctive but merely to recognize the limits to the difference it makes and the continuing importance of economic variables. A further implication of this is that the exportability of Japanese work relations to the West is limited not only by the different cultural contexts of Western countries but by their different, more cyclical, *economic* character.

Understanding of the institutional characteristics of firms has improved considerably in the past two decades, but unfortunately sometimes at the expense of economic analysis. Thus instead of assessing the success or failure of firms through an economic analysis of what ultimately determines their viability – their costs and revenues – the matter is often addressed obliquely by studies of their work cultures or management systems, coupled with attempts to classify them as 'Fordist' or 'post-Fordist'. Indeed, there is a danger of economic analysis being displaced in some fields, such as urban studies, by the crude metanarrative of Fordism/post-Fordism – whose 'binary histories' obscure more than they illuminate about the determinants of economic success and failure (Sayer and Walker, 1992; Elger and Smith, 1994). The victory of this metanarrative itself bears witness to the success of management science in promoting its concerns as *the* crucial determinant of economic performance and to the extent of the displacement of economic studies by cultural studies; indeed, Stuart Hall claims that post-Fordism is as much a cultural phenomenon as an economic one (Hall, 1988). Even 'political economy' has become more cultural with recent emphases on networks, communication, learning, management 'philosophy' and 'corporate culture' (Barnes, 1995a). When researchers explain the changing fortunes of industries without bothering to assess costs and cash, something is seriously wrong. Having once challenged the monopoly of economics in the study of economic phenomena, sociology, urban studies and human geography are now in danger of giving it back by reducing their studies of economics and industries to matters of work cultures and management systems.[10]

## CULTURE AND CONSUMPTION

Consumption has been a major focus of cultural studies of late, countering the dismissive and often negative treatment of consumption in political economy and earlier cultural materialism. Consumption or reception of cultural 'products' can be active and creative rather than passive, social rather than individual. On closer inspection, some of the innovations associated with 'mass culture', especially those involving commercial electronic transmission and consumer recording, which construct their audience as consumer rather than public, can allow people more rather than less room for creative use (Hebdige, 1988; Miller, 1989). At the same time, if there has been a shift in the culture industries from a treatment of the audience as public, in which the authority of the artist was paramount, to one in which the audience is conceived as market or consumer, with sales rather than internally generated standards of aesthetic quality determining success or failure, then this represents a clear case of the economization of culture (Keat *et al.*, 1994; Lury, 1993).

There is a danger of uncritically reflecting or even celebrating the commodification of culture; this is evident in the way social relations move to the background in some of this literature while relations between people and things (usually commodities) come to the fore (e.g. Featherstone, 1994). To a certain extent this reflects what has been happening: national, class or local cultures have weakened and a more individualistic 'stylization of life' has developed; formerly dominant, supposedly universal judgements on intrinsic values have been weakened by more subjectivist and relativist conceptions.[11] All this is as one would expect in an increasingly commodified and pluralistic society. Aesthetic values can also be instrumentalized by the users of goods and services, certain kinds of object (clothes, decor, holidays, art) being preferred not only or not at all because of any intrinsic qualities but because, as Adam Smith feared, they aid strategies of distinction (Smith, 1759).

The commodity may be valued by the user for its intrinsic use value, but to the seller it is unequivocally a means to an end, to the achievement of the external goal of making a profit, and if it is unlikely to make a profit it will not be offered for sale. This remains the case no matter how aestheticized, how saturated with cultural values the product is. In one sense this means that economy trumps culture and involves an economization of culture. In another sense, since the profitability of such products depends on their being culturally valued, albeit only if those who want them can afford to buy them, then culture leads economy.

Advertising and product design targeted at specific identities and providing means by which identities can be constructed are undoubtedly increasing and gaining in sophistication, as consumers become more reflexive: the hot hatch for young men and the chic hatch for young women (each of a certain class). Nevertheless, it is not in the seller's interest to refuse to sell such a targeted product to someone whose identity does not fit. Thus if a nun wants to buy a hot hatch the seller would be losing an advantage to competitors if he refused. Therefore running counter to the culturalization of economy with respect to consumer goods and services, and associated product differentiation according to differences among consumer lifestyles, is the familiar effect of money levelling difference and transgressing cultural boundaries.

Commodification allows products with strong cultural identities and content to break out of their originating cultural contexts. As a middle-aged white man I am unlikely to find acceptance in an Afro-Caribbean youth subculture centred on a certain kind of music, but if the music is commodified in the form of tapes, members of that subculture are unlikely to stop me buying a tape if I so wish, (a) because it need not result in my intruding into their lives, and (b) because my money is as good as anyone's.[12] Marx and Engels' famous comments in the *Communist manifesto* on the corrosive effects of commodities on local cultures still apply, but they were always double-edged. While the 'heavy artillery of cheap commodities breaks down all Chinese walls', capitalism not only floods markets with products of its own cultures of origin but also selects 'exotic' products which it can commoditize profitably, like Chinese meals, and diffuses them to other countries. The other side of the coin to the levelling, standardizing tendency of imperialism is therefore its destruction of parochialism ('rural idiocy') and a consequent economically driven pluralization and *enrichment* of cultures.

What appears to be happening is not merely a growth of individual materialism (in the sense of acquisitiveness). In the case of the culture industries the influence goes beyond the direct effects of purchasing commodities. Pervasive media images also provide a prism through which a great number of objects and events can be classified. As regards pop culture, Lash and Urry argue:

> The negative consequences of this are that the ubiquity and centrality of such popular culture objects to youth lifestyle can swamp the moral-practical categories available to young people. And entities and events which would otherwise be classified and judged by moral-political universals are judged instead through these aesthetic, taste categories. (1994: 133)

Thus, here, as in the instrumentalization of work cultures, contemporary trends in consumption point not only to a domination of the dialogical and moral aspects of culture by economic imperatives, but to their attenuation to an individualized, subjective 'feelgood factor' (plus a streak of 'style fascism'!). Again, this 'de-valuation' is entirely in line with what one would expect of a deeply commodified culture.

## POLITICAL CULTURE AND LIBERALISM

In the Introduction to this chapter I commented on the rise of a (cultural) politics of recognition alongside the (economic) politics of distribution. Here I want to make some different comments on the economy–culture relationship with respect to politics, particularly regarding liberalism. Liberalism is highly relevant to the preceding arguments about the effects of economic logic, especially commoditization, on culture, for it endorses markets and individualism. Its central belief is that individuals are the best judge of their own interests, and that they should be allowed maximum freedom to pursue their private projects (subject to the constraint that they do not harm others). Markets are ideal for liberals because, arguably, they allow maximum liberty to individuals to choose their way of life, and the price system co-ordinates their activities without political coercion. Hence the neo-liberal project of constructing people as consumers rather than political citizens. As Etzioni argues, liberalism is based upon relations between self and others, and has difficulty recognizing a 'we' (Etzioni, 1988). It therefore tends to erode the shared cultural values of the public domain.

Social democratic critics of liberalism, such as David Marquand, bemoan the decline of a political culture animated by 'active citizens' concerned with the public good (Marquand, 1988). Politics always involves debate over the nature of the public good, but liberalism refuses to prioritize these conceptions over individual conceptions of the good. As consumers, individuals are treated as bundles of preferences, and their values regarding the public good are reduced to thinly disguised self-interested desires. Thus parents are increasingly encouraged to think of themselves as consumers of education for their own children rather than consider what would be the best education system for everyone. These represent fundamental social changes of the late twentieth century, and ones which uncritical variants of post-modernism evade.

The relentless march of commodification is especially clear in cities. Increasingly there are attempts to convert the cultural capital of particular cities into economic gain through 'place-marketing' (Myerscough, 1988; Lury, 1994). Behind the playful façade of the post-modern city is a society dominated by private, and especially private capitalist, interests, in which a public culture is being eroded (Davis, 1990; Harvey, 1989a; Christopherson, 1994). As Susan Christopherson writes,

> Issues such as environmental quality, safety and freedom from urban congestion are not conceived of as urban qualities that citizens create but as commodities. . . . Normative claims to social justice such as those for universal health care or for the homeless are equated with particularistic claims. . . . The street is no longer perceived as public in the sense that it is owned by all people who use it . . . [but has become] a gauntlet to run between safe places. (Christopherson, 1994: 415–16, 421)

The 'active citizen' is as much a prescriptive as a descriptive concept. He or she must be capable of rising above his or her particular interests in order to make a disinterested judgement of the general interest and be willing to revise opinions in light of arguments of others, assuming a burden of social choice and decision instead of leaving them to leaders or others. This concept of politics as mutual education is utterly foreign to neo-liberalism (Marquand, 1988). Yet there is a danger of idealization in these sorts of critical commentary; the active citizenry is or was overwhelmingly white, male and often middle-class,[13] both in its concerns and in its membership.

The longing for active citizens and a sense of the public fund evident in critiques of neo-liberalism and the post-modern city are associated with the desire for a radically democratic society. Yet as Dunn puts it, the 'central paradox' is that 'we have all become democrats in theory at just that stage of history at which it has become virtually impossible for us to organise our social life in a democratic fashion any longer' because of the enormous division of labour and knowledge which any extension of democracy must face (Dunn, 1979: 28). Unless we grasp this paradox, critiques of the neo-liberal or post-modern city and both its economy and culture are doomed to romanticism. The idealized 'we' of democratic politics was undermined from the start not only by commodification and liberalism but by deep divisions of class, race and gender.[14] On top of this, the deepening of the division of labour also erodes a sense of the public by allowing what were formerly responsibilities towards others to be off-loaded on to private firms or the welfare state (Ignatieff, 1984). This creates an opening in which the Right can argue that we should be moving from 'passive dependency' on the state to 'self-reliance', from the citizen as client and passive funder of the state, nominally approving the state's role, to 'active consumer', choosing between services, and forcing laggard providers to improve or go out of business. While a sense of the 'we' remains important as an ideal, it is vital to recognize that for all these reasons – division of labour, class, gender and race – especially in multicultural societies, there is no simple 'we', but large numbers of people who are neither 'us' nor 'them' but in between, neither wholly inside nor outside the dominant culture (Young, 1990a; Bhabha, 1994).

In 1958 in *Culture and society*, Raymond Williams drew attention to the competing pulls of a conception of society as an indivisible whole in which people go forwards or backwards together, which was still evident in working-class culture, and a more bourgeois conception of society as a backcloth against which individuals make their fortunes. Williams's valuation of the communal conception is flawed by his silence regarding the presence of gender divisions and inequalities within the community. Yet while the attraction of the communal conception (with the inequalities removed) as an ideal remains strong, the terrain of the argument has now been radically changed by continued commodification and globalization, multiculturalism and increasing hybridity. How can we go forward together when the basis for a 'we' has so diminished? At the same time, despite the proliferation of subcultures and worries about fragmentation, there are cultures in common, and economically differentiation is combined with increased, not decreased, interdependence between different groups.

Ironically, answers to these kinds of problem can lead back to liberalism, for when it comes to suggestions of what kind of political order is best for multicultural societies, both Taylor and Gutman (1992) and Young favour a kind of liberal order or something very like it as the most benign available political shell (*see also* Kymlicka, 1989), with the state merely providing a framework in which different groups can pursue their own goals without mutual harm. But once we have appealed for a liberal political order, we can hardly continue to lament the decline of a sense of a common good. Liberalism and a post-modern celebration of difference can easily converge.

## CONCLUSIONS

Economy and culture do not impose upon each other as wholly external forces but are always intimately associated. Nevertheless they have different logics, the one dialogical and including certain intrinsic or non-instrumental values, the other having instrumental values related to external goals of reproduction. Their interactions are complex and their effects a mixture of good and bad. Culture can be instrumentalized to serve economic ends, but this can be accepted precisely because of the value of those cultural elements. Commodification can degrade cultural values, by subjecting them to an economic calculus, but it can also enrich cultures.

In addressing the different spheres of paid work, consumption and politics, I hope to have indicated how economy and culture interact without licensing a kind of analysis which allows one to swamp the other. Rather than simply following the cultural turn, we need to consider whether it adequately grasps the changing relationship between economy and culture, and indeed politics. To give the impression that economic logic has become subordinated to culture is to produce an idealized picture of an often brutal economically dominated world.

Finally, these views of the instrumentalization of culture and the degradation of political culture have an important critical implication for certain tendencies in British cultural studies. It implies that celebrating – or even just uncritically reporting – the *sign value* of cultural practices, relations and products, forms of consumption and lifestyles, or 'the stylization of life' as Bourdieu (1986) and Featherstone (1994) call it, is complicit in the very erosion of cultural values I have been criticizing. If culture, as an object of study, is treated as *no more than* the stylization of life, then it is worse, since it ignores the possibility of another side to culture. And if that is what is actually happening in the modern world, then we are in trouble. While dominant conceptions of culture differ by country and discipline, it is no accident that this impoverished concept of culture is to be found in the most economically and politically liberal societies. Equally, it is no accident that in more social democratic societies, where there is a stronger sense of the public or the common, such as the Scandinavian countries, a more anthropological and moral-political way of understanding culture (e.g. as 'life-form') which goes far beyond the stylization of life is still strong.

## ACKNOWLEDGEMENTS

Earlier versions of this chapter were presented to the 10th Urban Change and Conflict Conference, 5–7 September 1995, Royal Holloway College, University of London, and to seminars at Lancaster, Bristol, Sussex and Liverpool Universities. The current version owes much to comments from participants in these places and to the editors.

## NOTES

1. For example, *see* Andrew Leyshon's comments on some recent literature from the IBG Social and Cultural Study Group Committee (Leyshon,1995; also Sayer, 1995).
2. There may also be a gender subtext to the turn; political economy was – and is – dominated by men, and this has been reflected in its concerns, particularly in the relative neglect of the domestic economy. Perhaps, therefore, the turn is a reflection of the feminization of social science. But this is a strange reaction, for

women fare worse in economic terms than men, so one might have expected far more of a feminization of political economy than has been seen.

3. Thatcherism also had a cultural side, albeit one shaped mainly by neo-liberal economics, particularly 'enterprise culture' (Keat and Abercrombie, 1991; Heelas and Morris, 1992). More recently, 'New Labour' has taken a cultural turn, hoping to effect change by cultural as well as economic means (DEMOS, 1995), though of course this may merely be an acknowledgement of economic impotence.
4. On the whole I intend to exploit rather than limit the polysemic qualities of 'culture'; any strictures I have about particular, restricted definitions are not so much about their content as about attempts to displace other uses of the term.
5. While there is much to admire in Bourdieu's *Distinction*, I wish to resist his refusal of any notion of intrinsic value or of any way of combining the treatment of cultural values as intrinsic and as means of distinction. I think it should be possible *both* to accept that cultural practices and artefacts are often (more unintentionally than intentionally) a means by which particular groups distinguish themselves from others and make the most of what they have (or 'refuse what they are refused'), *and* to argue that certain practices and objects are valued as intrinsically just/unjust, good/bad or beautiful/ugly, etc. There are differences in the cultural capital associated with the Beatles and Beethoven, respectively, though not independently of the social position of the subject, but they are preferred to anything I could compose not merely because they afford more cultural capital but because they are better.
6. Marxism claims that work is and should be an end in itself. While I would not wish to deny that work should be satisfying, I agree with John O'Neill in disputing the Marxist assumption that it takes priority over the nature and quality of what is produced (J. O'Neill, 1994). The relationship of producer to product in Marxism is highly narcissistic, for the product is valued only as an objectification of one's labour. This is ecologically dubious, for no value is ascribed to nature independently of being an object of one's labour; moreover, applied to the work of childcare, the interests of the child would be secondary to those of the carer. One of the sources of Marxism's problems is a consistent thread running through it which elevates labour over the use value of what it produces, ignoring consumer interests, needs and wants, the problems of co-ordinating production and demand, and allocative efficiency (Sayer, 1995).
7. However, if we are to use the metaphor 'embedding', it should not be construed as implying that the economy is 'founded on' or 'based on' culture; an inversion of the base–superstructure metaphor is no better than the original.
8. This classic liberal argument was used by John Stuart Mill as a central element of his critique of discrimination against women in the labour market (Mill, 1869). At the same time, markets also have inegalitarian effects in so far as competition encourages actors to take advantage of and exploit to the full inequalities in society, for example by paying subordinated groups below-average wages (Sayer, 1995).
9. As Berggren observes in the case of Womack *et al.*'s advocacy of lean production, they have wrongly attributed the 'stable production expansion and lack of cyclicality to the production system [with its associated instrumentalized economic culture] and completely overlooked the importance of macro-economic factors' (Berggren, 1994: 89; Womack *et al.*, 1990).
10. Having noted this culturalization of economic analysis, it must be said in its favour that it treats culture as being centrally concerned with social relations, including norms and codes of behaviour, rather than about discourses and the relationship between people and things, as appears to have happened in Anglo-American cultural studies proper.
11. Naturally, the former period reflected the power of cultural authorities rather than consensus.
12. However, if the musicians change from amateurs to professionals they become subject to external influences on the kind of music they play; what was formerly a practice governed by its internal values and standards becomes subordinate to the demands of consumers who may be ignorant of that practice.
13. It is interesting to note that in recent years a new, feminized kind of active citizen has emerged. Instead of men organizing through formal representation and public speaking, we are seeing women organizing self-help activities, often informally in neighbourhoods (*see* Campbell, 1993).
14. Although Marquand's critique of Britain's neo-liberal political culture is powerful, its emphasis on the role of the professional and near silence on class, gender and race gives it a middle-class, patriarchal and patrician character.

# Chapter Two

# ECONOMIC/NON-ECONOMIC

**DOREEN MASSEY**

## INTRODUCTION: HISTORIES OF THE PRESENT

Imagine a high-technology workplace devoted to research and development. It is a classic space of economic geography.

We know, of course, that it is not *only* an economic space/place. It is also 'social' (and the need for scare-quotes at once indicates the difficulty of distinguishing in any simple way between the economic and the social). It is a 'social' space which is devoted to a single activity (R&D), which is agreed to be of high status, and which is largely populated by men. These characteristics, we also know, are intimately interwoven through sets of relations which go way beyond what is conventionally termed 'the economic'. The high status and the masculinity are bound up together, and both are interwoven with the nature of the activity which is carried on here, and with its compartmentalization into a specialized time-space. And there are aspects of social class, as well as the calculations of economics, in the separate, and particular, location of this element in the social division of labour (*see* Massey, 1995b).

Such interconnections run deep. Within Western economies these are places of 'science'. They are places of Reason, not emotion. And – reflecting this – they are separated off, specialized, defended (Massey, 1995c). They are also 'masculine' spaces, not in the sense that it is mainly men who work here, but in the sense that their construction *as spaces* embodies the elite, separated, masculine, concept of reason dominant in the West.

This, already, is to go beyond the 'normal', economic, characterization of these workplaces. But the aim of this chapter is to push such thoughts further: to ask *what histories lie behind the constitution of this object of study of 'economic' geography*?

How did this object of study come to be constituted in the first place? Such a history could be investigated from many angles, but this chapter concentrates on one: the history of the constitution of elite, exclusive, masculinized spaces of the production (and protection) of knowledge.

Thus, one can begin by questioning the very process by which R&D became spatially separated from physical production. Why did it happen? For this is a pushing forward of the division of labour in a particular direction: there are, presumably, many ways in which tasks may be broken down. This particular way, however, represents a moment in a wider, and again widely recognized, process: that of the separation of conception from execution. We could then, in the spirit of pursuing the analysis further, see in this space-time/place of Western commercialized science elements of a manifestation in physical form, and in the particular character of the division of labour within late-twentieth-century capitalism, of a separation between Mind and Body, between the purity of mental reason on the one hand and the fleshly materialities of the world on the other, which is a reflection of one of the great Western philosophical dualisms.

But to write that things are 'reflections' is to skip very lightly over processes. Ways of thinking, of apprehending the world, have to be produced and maintained. They are products of real histories. Moreover, such a history happens through processes which involve conflict and struggle. The fact of 'reflection' is thus certainly not pre-given. This high-tech palace, it will be argued here, is the product of a history which has been struggled over. The first object of what follows, then, is to unearth some of the contested history of the present of the high-tech workplace.

Moreover, it will be argued that history is far wider in its scope than anything that can meaningfully be included under the rubric of the 'economic'. What will be explored is the breadth and complexity of the contested cultural history which lies behind this apparently simple object of study of 'economic geography'.

It has struck me often, as I read about such R&D workplaces or find myself, seeking an interview, in a palace of high technology in Cambridge, how much they resemble monasteries. It is not, of course, that one comes across people singing matins. But there *are* similarities: the dominant maleness of the places, the dedicated nature of the spaces, the awe – even reverence – in which they are widely held. High-tech workplaces and monasteries, both in the character of their spaces and in the apartness of their geographies, but most importantly in terms of what they represent, are expressions of a similar social phenomenon. Both are (or, perhaps more accurately, profess to be) at the conceptual end of the conception–execution divide. They are spaces dedicated to intellectual endeavour. Even in a very immediate sense the high-tech location often reflects a desire to be associated with the university world, and the self-characterization of those working there a desire to be connected to that world. And in their turn, of course, the old European universities have themselves grown through and out of monasticism. The time-spaces of monasteries, the old universities and high-tech R&D share three outstanding characteristics: they are devoted to (the socially legitimated form of) knowledge production; they are inhabited by a caste which specializes in, and monopolizes, the definition and production of that knowledge; and they are masculinized.

David Noble (1992) argues that Western science, of the sort which the physicality of this celebration of high-status high technology applauds, grew historically out of the older European clerical culture. In a marvellous account, he traces the social bases of the organization of much of modern Western science to the clerical culture of the Latin church. In particular he points to the establishment of priestly castes, to the separations made between Reason and non-Reason, to the struggles to establish rules of gender, and to the intimate relations between all these things. He writes: 'the scientific enterprise [now, our high-tech workplace] . . . bore within it the enduring and deforming scars of a more ancient rupture, in the relations between the sexes. The scars are visible still' (Noble, 1992: xiii).[1]

The relation of 'big dualisms' to sex and gender has of course been widely analysed. What Noble does is to provide a grounded history of its practical social establishment in the Western world. There are, then, lines of historical continuity or connection, even if – as we shall see – it is the continuity of struggle.

As his title (*A world without women*) indicates, Noble concentrates on the expulsion of women from this masculinized culture. However, there are other subthemes implicit in his story which will also be explored here. First, I want to argue that the story is not only one of the expulsion of already-constituted women (and the retreat of already-constituted men to their safe places) but also that it contains elements of a struggle to produce aspects of 'intelligible genders'. In other words, this process of the historical construction of what is to be meant by religious orthodoxy/science was bound up with elements of the construction of particular forms of gender dichotomization.[2] Second, discussion of the construction of 'intelligible genders' refers to the work of Judith Butler (1990, 1993). However, in her analyses, Butler pays no attention to space. Another aim of this chapter, therefore, is to emphasize the critical role of spatiality in this tale of the construction of identities, genders and social orders. It is an evolving spatiality which today can be read in the form (among others) of a workplace devoted to high-technology R&D.

# FROM CLERICALISM TO HIGH TECHNOLOGY

## Propositions

The propositions are therefore the following: that an examination of the history of a Western high-tech research workplace reveals a long story behind some of its seemingly natural characteristics. Its high status and its masculinity are the (hopefully precarious) outcomes of centuries of struggle over the relationship between the nature of knowledge, its appropriation by an elite caste, and the dichotomous definition of genders. Moreover, the space/place of the high-tech workplace is also a product of a history of geographies: spatiality has been a significant force in all these contests. We have here geographies of gender construction, geographies of social hierarchy, geographies of knowledge production and definition – far more, in other words, than 'economic geography' alone.

## Sexual and social dissent

Throughout their histories, both the Western Christian church and Western science have been involved in struggles to establish and maintain their social and ideological hegemonies. And these struggles have in turn affected the nature of each. The established church which handed on its inheritance to science was hierarchical and masculine (of a particular sort), with a well-established orthodoxy and a caste of priests who were guardians – and monopolizers – of that orthodoxy. These things gave the church both its identity and its power.

But things were not always so. Early Christianity took root in the Jewish and Roman household. It was a beginning which was both spatially dispersed and relatively unhierarchical and which offered a role, sometimes prominent, to women (Noble, 1992: 5). 'From this familiar foundation arose two lasting legacies which would endure for a millennium: a married priesthood and the presence (and at times prominence) of pious women as clerical and lay wives and widows' (note that the priesthood – even if a dispersed stratum – was none the less already all-male). However, alongside this basis of Christianity grew up another: what Noble designates as 'the androgynous ideal of Christian piety' (ibid.). Thus, as he argues:

> If the household institution guaranteed the presence of both sexes within the elite circles of the emerging church, the androgynous ideal promised an unprecedented equality between the sexes, and indeed, in the view of many Christian ascetics, a transcendence of the sexes altogether. Thus, from the outset of Christianity, both the married household and its opposite, celibate asceticism, fostered a noticeable female presence. (ibid.)

Women played an important role in the church throughout the first millennium of the Christian era. They founded abbeys and led them; they were prominent in the dispersed religious institutions which existed during the whole period of what we commonly refer to (coincidentally?) as 'the Dark Ages';[3] and even the more centralized 'peak institutions' often took the form of 'double monasteries' where men and women worked together in the pursuit of knowledge.

Throughout its history, however, 'the church'-which-was-to-become-hegemonic was faced with threats to its existence and power – threats from without and threats of dissent from within, from 'heretics', dissidents, 'sects' and non-conformists. Over and over again, the radicalism of these dissident groups – the nature of the threat which they posed to the existing or emerging order – revolved around three terms: sexuality and the constitution of gender difference; democracy as opposed to centralized hierarchy; and – as a means, product and reflection of these things – spatialities.

Even in the very first centuries the battle-lines emerged which would continue to be fought over (and which would be modified and redefined in the process) for centuries. That early strand of asceticism, referred to above, came to find itself in radical opposition to the institutionalized base of the church in the (heterosexual) household. Two particular, and continuing, characteristics stand out. First, there was a challenge to the gender assumptions which underlay the institutional basis in the heterosexual household. It was a challenge at two levels, which it is important to distinguish. On the one hand, all these groups, from – for instance – the early ascetics to the twelfth-century Cathars, were far more welcoming than the formal church to the equal participation of women. On the other hand, and potentially of far greater significance, they challenged notions of gender differentiation itself. The early ascetics renounced the flesh, and especially sexuality and hence reproduction.

> In renouncing their sexual lives, the Christian celibates renounced as well their sexual identities. They thus became an altogether new type of being – a 'third race', according to pagan critics. The celibates themselves believed that, in Christ . . . 'man is no longer divided – not even by the most fundamental division of all, male and female'; through baptism 'the believers were considered to have recaptured a primal, undifferentiated unity'. (Noble, 1992: 12, citing Meeks)

Endless examples from various groups may be cited, and Clark (1986) argues that the practical expression of this 'androgyny' was 'a close spiritual companionship between men and women in which sexual identity had all but lost its significance' (Noble, 1992: 16). Origen, one of the most prominent early Christian philosophers, believed that the body was a transitory and fluid phenomenon, and sexual differentiation an equally passing phase (Brown, 1988, cited in Noble, 1992: 18). Priscillian asserted that all division – including sexual division – denied the unity of God (he was executed!). The Cathars denied the reality of the sexes as they denied the reality of all life in the flesh. And so on. As Meeks (1974) argues, the differentiation of male and female became 'an important symbol for the fundamental order of the world, while any modification of the role differences could become a potent symbol of social criticism' (cited in Noble, 1992: 12).

Second, these dissenting groups presented a wider social critique. Brown argues, for instance, that sexual renunciation could lead to a challenging of 'the ancient city', with its hierarchy of social relations and expectations (Noble, 1992: 11). The challenges to social hierarchies and sexual dichotomies went together.

* *

That same combination of social and sexual/gender radicalism characterized the multiplicity of dissensions which arose from the time of Paracelsus and Agrippa in the sixteenth and seventeenth centuries: the Lollards, the Anabaptists, the Muggletonians, the (early) Swedenborgians, the Brownists, the Baptists, the Quakers, the Ranters. . . . This proliferation of passionate and committed groups shared a number of salient beliefs and attitudes. First, again, there was the issue of gender. All these groups were (relatively) welcoming to, and in consequence attractive to, women. Hill, for instance, writes that 'women had played a prominent role in the heretical sects of the Middle Ages and this tradition came to the surface again in revolutionary England. Sects allowed women to participate in church government, sometimes even to preach' (1972: 250, cited in Noble, 1992: 194). Moreover, to some degree, though in a way which was less radical than in previous centuries, they challenged the division into two sexes. Again this was sometimes associated with sexual renunciation and voluntary celibacy, and Noble writes of 'the hermaphroditic paradigm' of the Paracelsian tradition (p. 177). And elements similar to the earlier ideal of androgynous asceticism reappeared: Quaker founder George Fox maintained that 'Christ was one in male and female alike' (cited in Noble, 1992: 196).

Second, this challenge to existing gender identities and relations once more went hand in hand with social radicalism. Within religious circles, all this ferment of activity constituted a challenge to the – by now – established hierarchy of the church, and this not only because of the flourishing of the independent sects themselves. What was also at issue was the nature of religious experience and knowledge. For many of these groups the heart of religious understanding was direct experience unmediated by a priestly caste. They

> placed stress upon the primacy of revelation over reason, of Scripture . . . over commentary, of the book of nature over the books of schoolmen, of the lessons of practice and experience over the formal education of the universities. In all of these ways, their teaching defied the established clerical and scholastic order, offering the unordained – including women – direct, unmediated means of spiritual enlightenment. (Noble, 1992: 179)

Paracelsus spent his life among the people rather than in the company of churchmen and academics, and when he did once hold a (brief) post at the University of Basle he 'dramatically broke with tradition by lecturing in the vernacular rather than Latin [we shall see later why this was so important] and inviting the attendance of nonacademic barbers, surgeons, and apothecaries' (Noble, 1992: 183)

That notion of the nature of knowledge (more related to material practice, attainable by each individual rather than monopolized by a caste) arose in threatening opposition to the established elite who believed they alone could interpret the Scriptures. It spread widely, and was linked to alchemy and to popular magic (in the latter of which, particularly, women played a highly prominent role). As such, it was linked to a challenge, not just to religious knowledge but to established 'science' more generally. The dissenting sects, as Hill demonstrates, 'attacked the monopolization of knowledge within the privileged professions, divinity, law, medicine' (Noble, 1992: 186), and the association between these views and a more general social radicalism was very strong (as, for instance, is shown by Hill for the period of the English Civil War). What had been created, writes Noble, was 'a new, enlarged social space for religious and philosophical inquiry. Moreover, to the extent that it had become connected with popular magic and the radical religious revival, it had also become identified, by association, with women' (p. 187).

Challenges to the current gender order, and radical and more democratic alternatives to the nature of knowledge and who could legitimately possess it, thus frequently went together. Such challenges have risen and fallen through the past two millennia of Christian religion and Western science. Nor have they always been peripheral. Indeed, Noble argues that it was only at the beginning of the second millennium that what we now know as the hegemonic model (hierarchical and masculine) achieved its dominance; and the challenges have continued since then. Had such challenges succeeded, today's high-tech workplaces might not have the characteristics they do: as the elite spaces of a masculine caste of the producers of authorized knowledge. Each time, however, the dissenters were virulently opposed. And over and over again these struggles interwove issues of gender, of identity and authority, and of spatiality.

## Backlashes

The asceticism of the early centuries was resisted in just such terms. The early years of Christianity were times of persecution, and

> leaders of the church struggled to formulate a coherent and unifying Christian identity and to consolidate their own position within pagan as well as Christian society. In both regards, by the mid-second century the role of women in the church had become a focal point of their effort. . . . Out of this struggle, there emerged not only a powerful new ecclesiastical institution and a canon of orthodoxy which prescribed the Christian woman's proper place, but also a new and different social caste – an orthodox ascetic male clergy – intent upon remaking the world in its own peculiar image. (Noble, 1992: 43–44)

Issues of identity, social hierarchy, masculinization and orthodoxy were inextricably interrelated. The early church had expanded and diversified. To protect such a dispersed and differentiated community from persecution (the threat from outside), existing leaders attempted to unify (through orthodoxy) and to centralize (under their own authority) the organization. Charismatic and communal authority began to be supplanted by the authority of an increasingly centralized caste (Fiorenza, cited in Noble, 1992: 44). Against orthodoxy was defined a thing called 'heresy', 'an entirely new kind of internal enemy, invented by Christianity' (Ste. Croix, 1981: 452, cited in Noble, 1992: 45). Canons of orthodoxy and hierarchies in the definition and possession of that knowledge emerged together. Those groups which believed that they did not need a clergy to mediate with God suffered persecution from the emerging church itself.

In this process, the household church was superseded by the 'Church as the household of God' (Fiorenza, 1983: 301–2, cited in Noble, 1992: 46) and this in turn resulted in a greater restriction of church leadership to men. The Christian patriarchal household, and the marriage of priests, were fiercely defended against the horror of the celibacy, androgyny and independence of women found among the ascetics. The question of gender became central to the issues of identity and authority.

> In the second and third centuries, the Christian church was engaged in the quest for its own self-definition. Striving to define itself over against non-Christians without and dissenters within, the church drew firm lines, precise boundaries, between itself and these heretical . . . movements . . . the mainstream church's limitation of women's roles can be understood in part as an aspect of its quest for self-definition – that is, for an identity that clearly distinguished it from rival movements. (Clark, 1986: 33, cited in Noble, 1992: 46–7)

Ironically, the attack on independent women which resulted from all this, and the desire to increase the distinction between clergy and laity (and thereby increase the authority of the former), together with a host of other factors, led the Christian clergy to adopt an element of their enemies' own character: they adopted celibacy.

> But in their exaltation of virginity the clergy had come to resemble the very heretics they continued to condemn. Thus, whereas before they had pitted the sanctity of marriage and the patriarchal dignity of the household against ascetic rigorism to distance themselves from heresy, they now pitted their own ideal of asceticism against another, an orthodox clerical asceticism against the androgynous asceticism of the heretical sects.
>
> If the latter was an ideal shared by both men and women alike in their common pursuit of Christian salvation, the former was to be an ideal, like orthodox clerical status itself, for men only. And if sexual temptation posed a threat to androgynous asceticism, the greatest danger to clerical asceticism was more narrowly defined: the presence of women. Whereas the androgynous ideal had fostered a chaste mingling of men and women, the clerical ideal instead drove men into frightened flight from women. As never before, then, the struggle against the heretical groups identified the threat of heresy with the proximity of women. At this early stage in the formation of the church, the contrast of orthodoxy from heresy already implied a world without women. The inherited patriarchal assumptions of the household-based clergy had subordinated women; the new ideals of the ascetic clergy eliminated them. (Noble, 1992: 49)

Whereas 'heretical' celibacy had drawn the sexes together in androgyny the clerical form divided them. Women and men were now defined as essentially different, and in that very process hierarchical structures became affirmed and their upper echelons confined to men. And in all this, the spatiality of the church was reinvented. The uneven, differentiated democratic organization of the early years was drawn into a hierarchical space over which was spread the uniformity of orthodox 'truth'.

Spatial definition of an even sharper kind was to follow, and once again it was bound up with the distinction of genders and the expulsion of women, the question of authority and the legitimacy of (particular forms of) knowledge. Monasticism grew slowly, and in its long history took a variety of forms, but the early ideals, and the ones which were to win out in the second millennium, 'reflected and reinforced the chief characteristics of the ascetic orthodox clergy: sexual renunciation, a disciplined bond of brotherhood, and . . . distrust of women' (Noble, 1992: 53).

Distance from, and dread of, women. In the early years the space of seclusion was the desert. Brown tells of 'a fog of distrust gathered along the edge of the desert. . . . Fear of women fell like a bar shadow across the paths that led back from the desert into the towns and villages' (198: 244, cited in Noble, 1992: 55). Spatial separation was crucial: what Noble terms 'the male monastic flight from women' (p. 77).

Yet the fortunes of this version of religious study fluctuated through history, from insignificance to dominance. And each resurgence involved space, gender and hierarchy. In the Carolingian reorganization, women were excluded from areas of the church and from particular functions as part of the 'organization of the clergy as an exclusive male hierarchy' (cited in Noble, 1992: 101). By law, women were excluded from pursuing scholarship; the new cathedral and monastic schools were restricted to celibate men. Effectively, women were excluded from mainstream education, which was confined within spatially exclusive homosocial communities.

When this reinvigorated organization was itself faced with decline it was revived through the activity of Benedictines, and yet again in the eleventh and twelfth centuries with the Gregorian reforms. Each time, the position of women suffered a decline. Discipline, orthodoxy, celibacy and the spatial exclusion of women – as a package of things which constantly recur together – were each time reinforced. Thus

> In its attack upon clerical marriage, the Gregorian Reform was a continuation, and culmination, of the clerical ascetic movement of the early church. As such, it reflected the deep-seated male monastic anxieties, not merely about the alienation of church property or the loss of ecclesiastical prerogatives, but about the corruption of the world, concupiscence, heterosexuality, and women (Noble, 1992: 127)

– what Gregory himself called 'the foul plague of carnal contagion' (Emerton, 1960: 52, cited in Noble, 1992: 129).

## 'Science'

It was within this kind of socio-spatial and gender order that Western science emerged, with *its exclusive spaces marking out the monopolizers of knowledge, its binary gender rigidities and masculinity, and its priestly caste*. Moreover, the dominance of science in this form, just as in the case of what became the hegemonic church, has only been maintained through active and militant diligence. Once again the battles revolved around gender, space, knowledge and hierarchy.

What Noble terms this culture's 'own distinctive puberty ritual, the cult of Latin' (1992: 151) marked out clear boundaries of both gender and space. Latin was taught only in all-male schools and universities:

> Closed to girls and to women . . . [they] . . . were male rendezvous strongly reminiscent of male clubhouses in primitive [*sic*] societies. . . . This *specially closed environment* of the university was maintained by a long apprenticeship or bachelorship. . . . But in helping to maintain the *closed male environment* the psychological role of Latin should not be underestimated. It was the language of those on the 'inside' . . . the first step toward initiation into *the closed world*. (Ong, 1959: 109, cited in Noble, 1992: 151; emphases added)

This was not just a geography of social inequality, nor even solely of intellectual elitism and legitimization. With its oft-expressed dread of women and fear of the flesh it was also a geography of fear, of anxiety and of the expulsion of unwanted desires; a geography of aversion (*see*, for instance, Sibley, 1995). It was also, very precisely, an active production and use of particular spatialities in a long historical process of production of what Butler (1990) has termed 'intelligible genders'. The troublesome problem of woman/the body/heresy within was actively expelled in the construction of masculine knowledge. It was another step towards the constructions of space which today typify the high-tech workplace: masculinization, elitism and exclusivity consolidated around the social monopolization of the production (and ratification) of 'truth'.

But, once again, it was not an untroubled history. Such spaces of knowledge production did not emerge without a fight.

One of the greatest challenges to this hegemonic culture of science came from that mixture of alchemy, reformation and dissension which bubbled up in intellectual iconoclasm and cultural revolts from the sixteenth century to the eighteenth. As we have seen, it regularly entangled together religious unorthodoxy, social radicalism and challenges to the gender order. There was for a time a real democratization of thought and knowledge in which women participated at all levels. '[T]he walls of established ecclesiastical and academic institutions' were no longer able to ensure the monopolization within them of religious and intellectual legitimacy. The backlash was – as usual – intense. As Noble puts it, 'the times had begun to demand, yet again, a reliably masculine identity. The challenge that had confronted the clerical ascetic authors of Christian orthodoxy now arose once more for the men of modern science' (1992: 205).

Most famously, most widespread and most horrifically, 'witchcraft' was created, defined (another enemy within) and persecuted. More parochially, in England, Parliament restricted the reading of the Bible (newly vernacularized under Henry VIII) along class and gender lines, and Elizabeth I insisted on academic celibacy and the total exclusion of women from universities. (Not until 1882 were Fellows of Oxford and Cambridge allowed to marry; Noble, 1992: 154.) Noble well describes the response of the men of science:

> In their flight from heresy, they took flight from women. They identified female witchcraft in order better to distance themselves from it, and condemned the magi and alchemists for the same reason. They discarded the animism of popular magic and ridiculed the hermetic philosophers' alchemical resolution of body and spirit. They denied the possibility that nature might be, like Scripture, a medium of revelation through which regenerated men and women could become again as one before God. In the place of such revivalist notions, they substituted mechanism and dualism, purging the natural world once and for all of its unseen sympathies and irrevocably divorcing spirit from earth, mind from body, subject from object, male from female. They withdrew from the disorders of enthusiasm and experience to the abstract certainties of mathematics. In league with papal and secular rulers, and a reinvigorated state, these new Augustines, Benedicts, and Dunstans sought a restoration of spiritual and intellectual order to match the restoration of political order. And like their glorious forebears, they found such order, and a lasting refuge, in a world without women. (Noble, 1992: 211)

A new 'academic asceticism' gained ground, of which the personal lives of some of the 'great men' of science and the antics of institutions such as the Royal Society are well-documented and almost parodic exemplars.

> With the advance of science, then, the clerical culture gained new territory, extending even further its world without women.
>
> . . . The universities, the chief locus of this extended church, would remain exclusively male well into the nineteenth century, and the scientific academies would hold the line against female enthusiasm until the middle of the twentieth. (Noble, 1992: 242)

Spatialities, identities, genders and social orders – such has been the long history of just one aspect of their co-constitution. It is a history even now embedded in the laboratories, the excluding masculinities, the elitist and specialized time-spaces of today's new industry of high-technology R&D.

## FURTHER REFLECTIONS

A number of points can immediately be drawn out of the preceding discussion. First, that – as Noble insists – 'a world without women did not simply emerge, it was constructed'. Second, and to take Noble's argument even further, that process of construction of 'a world without women' involved the social moulding of the categories 'woman' and 'man' themselves. It involved the creation of intelligible genders. Third, that this brief history indicates how important spatialization was in the construction of those worlds, genders and identities. The reverse is also true: that the analyses by Noble and Butler can provide insights into what lies within the long historical production of twentieth-century spaces – such as the spaces of high tech. Fourth, that long historical production has not been a simple linear inheritance 'from the Greeks'. On the contrary, it has involved constant struggles: over the meaning of genders, the articulation of spaces, and the construction of knowledge and the rights to its possession.[4] Indeed, Greek learning was, of course, lost to the West for a long period, and was only reintroduced, in the thirteenth century, via Arabic schools and in particular by Ibn-Rushd (Averroës). Moreover, Noble demonstrates that Greek philosophy, rather than being the root of Western monastic misogyny, was seized upon as a *post hoc* justification for the already established fear of women and that, in order for that to be possible, it had to be cleansed of the relative sexual egalitarianism of Ibn-Rushd himself. He himself was 'a strong believer in the natural similarity, and hence social equality, of men and women' (Noble, 1992: 157) and, as such, his work was condemned as heretical and repudiated. With this, 'the misogynous classics were shorn of their heterodox Arab commentary, and henceforth unambiguously provided belated scientific fortification for the established European world without women' (Noble, 1992: 159). It is, perhaps, in the light of this long struggle that we can – fifth – read today's feminist challenge to the culture and nature of 'science'. It is in a grand tradition.[5]

Furthermore, this historically grounded story of the development of Western science indicates how closely are and have been related the culture of scientific production/possession and ideas about the nature of science itself. The masculinization of the scientific culture, and the contemporaneous definition of (a specific form of) masculinity, went hand in hand (if that is not too bodily an expression!) with the development of the notion of science/rationality as necessitating separation from nature/the body/ sexuality. Moreover, the material spatialization of these separations both depended upon and reinforced the discursive counterpositions.

This double level of culture and content, and the simultaneously material, social and discursive spatialities to which both related, point indeed towards a third: to the body-maps and the implicit geographies of subjectivity and identity bound up in the Western-masculine notion of Reason. Thus Seidler (1994) has written not only of how the construction of this masculine rationality sets it apart from nature/sexuality (which therefore act as a threat to the rational identity), but also of how this relates to what he terms a Kantian mapping of the self. In this geography of rational-masculine subjectivity, Reason – the source of the identity of the rational self – is located in the mind, while emotions and feelings are more dissipated, and located simply 'elsewhere in the body'. These different locations, and their meanings, have effects. There is seen to be no positive connection between the two spheres. Indeed, their separation and the prioritization of one over the other leads to their being in conflict. Reason, the supposed source of freedom and autonomy, is distracted by emotions and feelings which are seen as threats from 'outside', as sources of unfreedom and determination. The fortress of Reason (just like the monastery, the early university or the masculine high-tech workplace) must be defended against these potential threats. (At this point the lives of a number of great male scientists are often cited in evidence. But so can be evidence from recent empirical work on what today's high-technology R&D researchers see as the essential features of the scientific identity. Such features can include not only positive attributes of scientific ability, but a distancing of other, apparently conflicting, potential aspects of identity; *see* Massey (1995b).)

At the age of 30, around the year 500 CE, Benedict of Nursia went out into the desert to become a hermit. It was not easy:

> The allurements of voluptuousness acted so strongly on his excited senses, that he was on the point of leaving his retreat to seek after a woman whose beauty had formerly impressed him, and whose memory haunted him incessantly. But there was near his grotto a clump of thorns and briars. He took off the vestment of skins which was his only dress, and rolled himself among them naked, till his body was all one wound, but also till he had extinguished forever the infernal fire which inflamed him even in the desert. (cited in Noble, 1992: 83)

The mind, the desert, the monastery, the ancient university, the workplace devoted to the labour of Reason can all be read as part of imaginary geographies in which they are to be purged of the distractions of nature, carnality, the body. These are geographies which are instrumental in the construction of the identities and attributes themselves. Moreover, they are geographies which are constructed both materially and discursively. Thus, just as Benedict had trouble with invasions of unwanted thoughts, so too the high-tech workplace is 'in fact' not a palace of pure reason. It has to be (materially and discursively) constructed as such, and that attempt at construction does not fully work. The scientific workers have considerable emotional investment, paradoxically, in the rationality of their work, for instance (*see* Massey, 1995c). In spite of these intrusions, however, the hegemonic discursive construction of such places is as spaces of pure reason. And the legitimation of the status of such high-tech places comes in no small part from this powerful discursive construction.

## CONCLUSIONS: CULTURES OF THE ECONOMY

The most basic proposition of this chapter is that the high-technology workplace of today embodies much of this history. These workplaces are not simply a recent product of economic calculation on the part of the owners of capital. Spatialities and temporalities are both more complex than this.

The *times* of these places – as of any places – are many. Not only are they products of present calculations, but their character also reflects long histories of social struggle, over the nature and ownership of knowledge, over the meaning and delineations of gender, over the material establishment in lived relations of the 'philosophical' divide between Mind and Body. These workplaces are physical and social precipitates of particular intersections of a multiplicity of varied histories. This point in itself indicates another: that it is not only capitalism which has a history; so also do the establishment of gender relations (as we know) and the maintenance as hegemonic of particular ways of thinking about the world. The 'Mind–Body' dualism is not a fixed idea, given for all time or somehow inherited in pure form 'from the Greeks'; it has a history – material, discursive and embattled. And in those histories, as I have tried to show, spaces and places and their construction have been of signal importance.

This raises the question of what might be the meaning of *current* changes. Throughout the history of these distinctions there have been, as we have seen, oppositions and rebellions. Their force and dominance have ebbed and flowed. Are we presently witness to another rebellion? Two movements seem to be on the horizon.

First, there is a growing critique within some parts of the scientific community itself (primarily within academe rather than the commercial sector) of some of the separations which have become central to Western scientific practice, and in particular the separation of abstract thought from other modes of understanding. Thus Ho (1993) writes of a manner of knowing – with one's entire being, rather than just the isolated intellect (p. 168) – which is, she says, currently foreign to the scientific tradition of the West but which is on the agenda for the future simply by following the emerging logic of that science itself.

Second, and far more prosaically, there may be moves within high-tech industry to reduce the spatial separation of R&D and physical production. It is a shift which, to the extent that it is happening, results from the recognition of some of the practical disadvantages (for industry) of their mutual isolation.

The first of these moves has been propelled by a recognition of the ultimate dead-end nature of isolated abstraction, and has been crucially informed by both non-Western philosophies and Western feminism. The second has been stimulated by problems thrown up for capitalist production which relate, precisely, to a need to reintegrate production and research; but it has no philosophical or gender content at all that can at present be identified. However, the consequences for the nature of the spaces and for the hegemony of the power relations on which they depend may, once again, be far wider than the 'economic' alone.

Moreover, understanding the complex temporalities of these places also tells us something about the nature of their *spatialities*. Most obviously, these are more than economic spaces and they are spaces of more than economic power. They are spaces expressive of class distinction, spaces exultant with the supremacy of Reason, spaces which express in their very nature deeply embedded assumptions about masculinity and gender. (The 'gendering' of these spaces, in other words, does not rest on the fact that it is mainly men who work there. This in itself is an outcome of a deeper process, of the construction in the first instance of specialized, defensive 'places of knowledge'. Allowing women into these places would not suddenly demasculinize them. It is the nature of the spaces themselves which is at issue.) And, of course, not only do these places result from the operation of a range of forms of power, they also in their high-status existence reinforce them. The R&D workplace is thus certainly a space of social relations which transmit the relations of power inherent within capitalism. But it is also a space formed by and exemplifying categorizations and differentiations which are part of what holds dichotomous genders in place and defines them as such. These are spaces which reflect and reinforce a range of kinds of power. It is a form of socio-spatial demarcation which expresses the dualisms fundamental both to the long historical process of the dichotomization and counterpositional characterization of genders, and to the delineation of what is legitimate 'science'.

These are spaces thoroughly imbued with a variety of logics and histories. It is possible to begin to see here the potential depth of the connections between what we are accustomed to think of as economic geography and other aspects of geographical and wider analysis. 'Putting culture into economic geography' is (or it could be) to understand the economy, and the geography of the economy, as fully culturally constructed. What we think of as 'the economic' is itself expressive of other aspects of our culture/society.

A high-tech R&D workplace is indeed an economic space. It is also a single-activity space, a gendered space and a space of well-defined status. But all those characteristics are themselves the complex outcomes of long historical processes. Today's high-tech palace bears within it the fruits of victories and defeats in long struggles over the meaning of science and masculinity. So, next time you come across such a place, think of it as one of the many precipitates of all of that. To call it an economic space alone is seriously to diminish both the complexity of the processes of which it is a product and the range of social identities and structures for which it stands.

## NOTES

1. This connection between science and religion is different from and additional to other ways in which a faith in rationality may be likened to religion; *see*, for instance, Mouffe (1995). I should like to thank David Noble, not only for the stimulating ideas in his book, but for his permission to quote from it so fully.
2. The emphasis here on 'particular' forms of gender dichotomization is significant and twofold. First, it refers to aspects of gender dichotomization which were slowly constructed around such dualisms as Reason : Emotion and Mind : Body. Second, through much of this long history these aspects were very much focused on constructions of gender

among the elite sections of society – from priests, to learned men and aristocratic women, to today's intellectuals and academics. The class dimension of all this is taken up again later. (The research from which these ideas were developed was funded by the ESRC, project no. R000233004.)

3. These 'Dark Ages' were, in Noble's terms, a period 'in which women came to play a far more prominent part in the culture of learning than they did in either ancient Greece or the High Middle Ages' (Noble, 1992: xv).
4. Grimshaw (1986) also emphasizes the inadequacy of assumptions of continuity; of the Greeks as the direct forebears of all the characteristics of what has become the dominant form of Western philosophy. In her argument the emphasis is on the changes in approach which have taken place over the centuries.
5. This 'list of points' is shortened because of space constraints. But it is important to add that not *all* of the mechanisms through which women came to be excluded from the church/scientific establishment resulted directly from concerns over gender/knowledge. In the Carolingian and Benedictine reforms, for instance, the exclusion of women occurred in part (though only in part) through the side-effects of other manoeuvres to consolidate power. And with the return of the monarchy after the Interregnum in Britain, the anxiety of scientists to conceal any revolutionary elements in their past may have reinforced their impulses towards the repression of alternative knowledges (*see* Noble, 1992: ch. 9, and Hill, 1961).

# CHAPTER THREE

# THE CULTURAL PRODUCTION OF ECONOMIC FORMS

RICHARD PEET

Suppose economy were to be approached from a different vantage point? The intellectually taken-for-granted world of analytical categories might shift, change, appear differently, lose certainty, expand to include new aspects, new things. Economic categories emerge from an intense scrutiny of the production of existence which pretends to the status of science. But this claim to scientific certainty is possible only by taking economy as the pure sign of history, by separating the economic from the social and cultural (to use equally confining categorizations), closely defining 'economy' around the immediate activities of production, distribution and exchange, thereby creating from the real mass of barely distinguishable, interrelated things, processes, ideas and emotions a realm of objects theorizable in objective terms. Moreover, economy is seen as a self-determining discrete entity, creating its own institutions, rationalities and conditions of existence. Economic motives such as profit-making, and overall economic logics such as capital accumulation, appear to be created by systemic functional efficiency, rather than historical, social and cultural agency. In bourgeois materialism, economic rationality is the direct impress on the mind of a seemingly natural efficiency.

Yet functions are not agents. Immersed in social relations, bearing and making culture, people with socio-cultural identities are agents. This was recognized theoretically in the debates over structure and agency. And the concept of agency has subsequently been deepened and transformed by post-structural and feminist notions of multiple identities. But there is a broader and more radical project inherent in this reconceptualization. In creating economic forms agents do not act freely, as though motivated entirely by personal will, but act as socially constructed or culturally produced identities. Yet this broader theoretical integration of economy with culture has barely begun, even through the single, most obvious route of agency/identity. Culture is brought *into* economic analysis, as though culture is a separate and subordinate realm, mainly at moments of theoretical crisis, when purely economic analysis obviously cannot suffice: a motive here, a type of behaviour there, a consumer preference every so often, after which the analysis returns as quickly as possible to 'objective certainty'. This continuing lack of integration, for example the new cultural with the new industrial geography, is partly because a synthetic project is complicated. But it is also because the movement for theoretical reform got stuck on 'the complexity of agency' or 'the multiplicity of identity' rather than engaging in a complete rethinking of the social and cultural origins of economic institutions, behaviours and actions (cf. Granovetter, 1985, 1990; Swedberg and Granovetter, 1992).

That, however, is not all. There is a yawning gap between the objectivity of economically derived logics and the extreme, even excessive, emotional subjectivity of agency/identity. This needs bridging by intermediate concepts which, at the present time, in the prehistory of a new cultural-economic geography, can only be guessed at. So, let us therefore speculate. A link between culture and economy through the mediation of subjectivity might be specified by the concept of economic rationality. By this is meant the 'logic of economic action', understood as the reasons people have for behaving in certain ways as economic agents. This logic is formed, and displayed discursively, in the rationalized structures of explanations agents offer for their economic behaviour and the reasons they give for their overall success or failure. More generally, rationality implies the motives which drive economic actors, even the ethical systems to which they must constantly appeal; ethical justification is an integral part of rational explanation in the post-Enlightenment world. In both senses, as logic and motive, rationalities derive from the broader experiences of socialized and encultured persons; they are signs of identities, rather than being divined from the inherent structural logics of economic systems. Economic rationalities are culturally created, take diverse forms and have distinct geographies. A geography of economic rationalities awaits our exploration!

Yet rationalized actions create, through repetition, the systemic logics of economic forms. The operation of such systems 'disciplines' rationality, 'proves it to be rational' through efficacy, in a dialectically functional way; that is, a definite system of functional connections open still to dramatic transformations. Furthermore, economic rationality has its own, relatively autonomous dynamic within, and between, economic systems. That is, once formed, a type of rationality has a consistency and a momentum which prove hard to break, if only because agents derive power from their positions within existing economic forms, or because elites culturally represent economic logics as naturally optimal. Thus economic forms are culturally created, economically maintained, interact with subjective processes of agents' identity formation, and to a degree are self-generating or at least self-maintaining. No wonder theorists have avoided the topic!

This chapter tries to bridge the gap between culture and economy by exploring the middle ground of economic rationality. It assumes that social forms of the production of existence are 'imaginary economies'; that is, social constructions. It assumes further that economic imagination derives from the cultural history of a people. 'Imagination' in this formulation should not be interpreted idealistically as Spirit-derived consciousness, but in Castoriadis' (1991) terms of 'social imaginary' or Peet and Watts' (1996a) 'environmental imaginaries'. The idea is to connect social forms of imagination with specific reasons for economic action. The chapter uses the discourses of the nineteenth-century New England economic elite as empirical case material. It asks how economic elite identities were formed and how a distinct economic rationality was 'discursively' constructed in the early days of a significant new social form

## THE NEW ENGLAND DISCURSIVE FORMATION

Between the late eighteenth and early twentieth centuries, a regional discursive formation written by New Englanders played a powerful, even hegemonic, role in the cultural and political regulative order of the northern USA (Peet, 1996). For Anglo-Americans, especially those of New England descent, the ethical/religious/rational belief system of the Puritan settlers, the culture created by the New England literary elite, the aesthetic representations of the region's landscape and history, framed the ideals of the social imaginary. Distilled down for farmers, mechanics and native-born labourers, disseminated through the popular media of the time (sermons, lectures, books, monthly magazines and newspaper articles), the New England discursive formation served morally and ethically to regulate a class system structured by relatively undeveloped but rising social contradictions (Brooks, 1936; Bercovitch, 1975; Buell, 1986; Commager, 1967; Miller, 1953).

Pre-Fordist capitalism was predominantly regional in terms of modes of regulation and other sociocultural aspects of reproduction. The USA had several regional discursive formations, or modes of regulation, each spreading from cultural bases on the eastern seaboard as settlement proceeded inland: hence a New England system with Boston as its cultural and economic capital – this was the hegemonic cultural order of the northern tier of the USA; a Virginia system extending across the southern Middle West and South; a Philadelphia-Quaker system in between; and so on. These regulatory/discursive systems existed in hierarchical relation, one with the other. Hegemonic relations were expressed culturally and artistically, but also had a definite economic power basis.

By comparison, Fordism originated in the Middle West, within the New England regulatory formation, but quickly became national in scope, with New York and then Los Angeles as its cultural capitals, its makers of ideals. Late Fordism and post-Fordist systems are international in scope, although Los Angeles remains their cultural centre, the place from which post-modern ideas and images are mass-produced like computer chips and mass-distributed like detergents. Discursive formations and regulatory regimes are changed through class, gender, ethnic and inter-regional struggles which form aspects of economic strife but also have autonomy: there is a cultural struggle over the power to form social imaginaries. Regulatory modes change reluctantly, with old regimes never disappearing entirely. Indeed, elements from previous modes are continually reprocessed in the collective memory.

# DISCURSIVE REGULATION

New England has long been a centre of economic initiative in the capitalist revolution. It was the North American centre of mercantile trade in the eighteenth century until Boston was overtaken by New York, itself dominated by New England merchants (Albion, 1932), between 1800 and 1820. It was the first region in North America to industrialize on a mass scale, with the growth of the textiles and other consumer goods industries in the late eighteenth and early nineteenth centuries. In the mid- to late nineteenth century it developed a broad base of tool, machine and metal industries, sources of the defence industries of the twentieth century. After the decline of the older components of this industrial base, parts of the region were revitalized by the computer industry in the 1960s, 1970s and 1980s. Substantial parts of the wealth generated by workers in this sequence of economic activities are recycled by elites and local states through social and cultural institutions, including one of the largest concentrations of major universities in any world region, to produce a cultural infrastructure supportive of further economic growth.

## Cultural Regulation Theory

The cultural hegemony of the New England discursive formation is related to the region's economic development: the power of one underlies the force of the other. This interrelationship can be seen in terms of the economic determination of culture: the New England mercantile and industrial bourgeoisie financially supported cultural production, culture is the persuasive arm of economy. From this perspective discourses are agencies of social regulation. That is, discursive formations relate not just to other discourses, nor to power in general, but to the regulative requirements of definite geohistorical, political and economic formations organized by specific social relations (Foucault, 1980; Aglietta, 1979; Lipietz, 1987): at certain scales (national, local) and in phases of hegemonic calm, crisis and change. Here the attraction of regulation theory, as Jenson (1989) points out, is its negotiation of the straits between 'agentless structures' and 'structureless agents', in this case self-reproducing economies and apparently free actors. Jenson adds to the concept of 'mode of regulation' (i.e. the institutional forms, procedures and habits which cohere to persuade agents) the further notion of a 'societal paradigm' which sets out meaning systems and orders collective identities discursively constructed under conditions of unequal power relations. Rose and Miller (1992: 175) add 'problematics' (in their case, problematics of government) analysed in terms of the changing discursive fields in which the exercise of power is conceptualized, the moral justifications for using power, notions of the forms, objects and limits of politics, and conceptions of the proper distribution of tasks among secular, spiritual, military and familial sectors – change 'politics' to 'economics' and this would specify a 'cultural problematics of economic power'.

We can add to this cultural specification the notion of *discursive regulation*, indicating the limited opportunities discourses offer for expression and representation within modes of regulation; the conditioning effected by regulatory imperatives on discursive contents and styles; and the multiple interplays between the two as discourses are used to achieve social harmony (i.e. regulation). Yet discourse is a medium in which ideas, imaginaries and rationalities are not only displayed, but are worked out, practised provisionally through the fantasy of literature, and later examined critically with the wisdom of hindsight. Discourse is an active medium of regulatory power. The aim, then, is to open regulation theory to advances in discourse analysis, to the processes of identity formation, and other areas of post-structural theory, in a project combining cultural with economic modes of theorization.

Yet despite the added complexities, this position still basically sees economy creating culture, a regime of accumulation 'requiring' a mode of regulation, a New England discursive formation 'suited to', 'appropriate for', 'fulfilling the requirements of' capitalist industrial development. In this view, regimes of accumulation are initiated by 'purely' economic and technical processes, and agents, coerced or persuaded by regulatory institutions, remain relatively passive. The straits between Jenson's 'agentless structures' and 'structureless agents' are not completely bridged. Modes of regulation and their constituent discursive formations determine the components, paths and types of economic development by creating the personalities of the economic actors, fabricating the rationalities which guide economic action, setting the cultural context in which social relations form, and putting content into these relations in terms, for example, of the mutual obligations deemed ethically acceptable between groups of actors (capitalists and workers, patriarchs and wives . . . ). The dominant discourses of a time are prime agencies in the cultural formation of economic rationalities. In such ways may discourse theory and regulation theory be conjoined.

The early years of something new, in this case industrial capitalism in New England, present an opportunity for catching institutions, economic personalities, rationalities and logics as they are created. A glimpse into this process of the cultural creation of economic forms is provided by the discursive record left as relics and icons in the landscape and as written materials in libraries and museums. The New England elite typically left an extensive record of their own discursive construction in a vast and multifarious cultural infrastructure (not a village without a library – like pubs in Australia!). Two fragments of the New England regional discursive formation are presented here: some early ideas on industrialization by New England entrepreneurs; and the ideologies propagated by the New England Societies.

## THE CULTURAL CONSTRUCTION OF NEW ENGLAND INDUSTRIALIZATION

Much of the eastern seaboard of the USA was conquered and seized from its indigenous inhabitants as part of the expansion of international Calvinism: this brought the Dutch to New Amsterdam (later New York), French Huguenots to New York, South Carolina and Massachusetts, Scottish Presbyterians to the Middle Colonies, and the Puritans from England through The Netherlands to New England in 1620 (Prestwich, 1985). As disciples of Calvin, the Puritans enforced religious principles in everyday life through a Congregational system of church government. While the Puritans were soon joined by followers of other religious beliefs, New England for the first two hundred years of Anglo-American rule was overwhelmingly a Protestant, and very much a Calvinist, domain.

### From Calvinism to Unitarianism

Yet the New England discursive formation rests on a dual base of religion and reason. For New England and Pennsylvania were the heartlands also of the American Enlightenment. From the Puritans came (supposedly) the Calvinist notion that a rational intellect is necessary for knowing the word of God. This reached its ultimate form in Unitarianism, particularly prevalent in New England after a split in the Congregational Church in 1816, which saved religion for an age of science by appealing to Enlightenment rationality (hence *Uni*tarian rather than the more mystical *Trini*tarian Christianity; Cooke, 1910; Szasz, 1993). A Unitarian belief in the innate goodness of humankind and its progress towards a higher moral life, together with a desire to make religion practical, made support for justice, liberty and purity a matter of moral obligation, while Unitarians were active too in literary culture and movements for social and economic progress (Robinson, 1985). Yet the New England Unitarians were 'religious liberals and social conservatives' and an 'instructive example of a patrician class giving leadership to its community . . . reformers who feared change' (Howe, 1970: 11–12). Especially when combined in enlightened ways, reason and religion were powerful ideologies used in the social production of behaviour within the 'Anglo-Saxon' community of New Englanders and, when mass immigration threatened social and cultural order, served well enough until Fordist cultural technologies were added to the regulatory system in the early twentieth century.

### The Protestant Ethic

This culture of religion and reason formed the ideological basis of rationalities used in all aspects of the New Englanders' lives. In many ways New England's Congregational–Unitarian–Enlightenment culture is a regional case study of Weber's thesis. In *The Protestant ethic and the spirit of capitalism*, Weber (1958 [1904–5]) explores the combination of circumstances leading to the singular appearance in Western civilization of science, rationality and capitalism (the pursuit of profit by means of continuous, rational enterprise employing free labour). The Western path of development, Weber (ibid.: 40) says, is the product of a specific form of economic rationalism which Protestants showed a special proclivity for developing. The Calvinist notion of predestination eliminated magical means of attaining the grace of God (ibid.: 105). Combined with harsh doctrines of the corruption of the flesh, the inner isolation of the individual explains the negative attitude of Puritanism to all sensuous and emotional elements in religion. Intense worldly activity was the way of developing the necessary self-confidence for Calvinists to believe in their eternal salvation; economic performance was a way not of *attaining* salvation, but of eliminating doubts about it. From this follows the further belief that labour in the material world has the highest ethical evaluation, while profit is morally recommended, so long as it does not lead to luxury. Weber uses a statement by Benjamin Franklin (1806), born and raised in Boston but living in Philadelphia, to exemplify this capitalist spirit: Franklin advocated making money as the ultimate purpose of life, combined with the strict avoidance of all spontaneous enjoyment wherein this money might be squandered. Hence the origins of the capitalist spirit (i.e. rationality) are to be sought in Calvinist religious ethics (Weber, 1958 [1904–5]; Giddens, 1971). By extension, the economic development of New England, and the recycling of profit through education, museums, concert halls and other 'non-frivolous' institutions, is based in the pre-existing dominance of Calvinist beliefs, updated in this regional case by Unitarianism.

What can we learn from this? Rather than seeing economic rationalism, productive behaviour, the profit motive, perhaps even the accumulative logic of production, as functionally self-generating, Weber attributes them to a prior system of religious beliefs. This contextualization projects 'economy' back towards culture (in the form of religion). But Weber does not explain religion with the same depth, as an attempt to understand nature, social relations, existence, in ways determined, as Marx (1967: 79) would have it, by the forces and relations of production: 'The religious world is but the reflex of the real world. And for a society based upon the production of commodities . . . Christianity with its *cultus* of abstract man, more especially in its bourgeois developments, Protestantism, Deism, etc. is the most fitting form of religion.' This prevents Weber from seeing (or at least saying) that the doctrine of predestination is just as magical as the notion of earning salvation through good deeds recorded by a heavenly calculus. From this comes the uni-causal structure of Weber's explanation, the way religion solely is connected directly with economic rationality, which makes for a compelling thesis, but is overly simplistic, as Weber realized in his later work (Weber, 1968; Schlucter, 1989). While highly suggestive, therefore, Weber's thesis shows how *not* to construct a cultural order of production, which is inevitably multi-causal, full of indirect influences, with many feedbacks between economic and cultural aspects.

## The Protestant Ethic and the Morality of Capitalism

In New England, imported Calvinist beliefs interacted with the circumstances of settlement and cultural characteristics derived from regional experience, including an 'environmental imaginary' (Peet and Watts, 1996a), to form a multi-layered, multi-faceted cultural order of production. New England culture comes from a *radical* bourgeois project, in which the government *was* the people via the town meeting, in an agricultural society where every individual had the right to own (be proprietor of) means of production (land), yet the towns were collectivities organized around the church and the school: there was an intense individualism and yet a strong sense of collective order – a moral economy (J.C. Scott, 1985), yet one with developing class divisions. New England remained, until the Civil War, predominantly a society of Calvinists, Unitarians and other Nonconformists. But also the American Enlightenment, bringing rationality and reasoned (rather than divined) ethics, quickly became centred in the region. Hence New England merchants and entrepreneurs recoiled in horror from the social consequences of the British industrial revolution and consciously tried a different industrial system which, they convinced themselves, had more in common with Robert Owen's social experiment than Engels' Manchester. They thought that the harsh circumstances of the New England environment and the early difficulties of settlement produced a special people, a 'peculiar race' of hard-working, innovative, practical folk, an imagined 'Yankee ingenuity' repeated so often it became a reality. Such regional cultural characteristics, projected into dominance by an active, successful people, produced the American entrepreneurial system, morally honest and hard-working, intensely profit-seeking, innovative and mechanically inclined, conservative yet socially conscious, rationally enlightened yet religious, a cultured entrepreneurship radically believing in equality through opportunity. New England's culture and discourses produced a particular version of bourgeois rationality, a moral reasoning put into economic terms, which was essentially, archetypically modern yet drew on the prior moral economy of the region (Goodman, 1966; Jaher, 1982; Horne, 1983; Dalzell, 1987).

## Moralizing Industry

In the USA, industrialization was not automatically accepted as economic necessity, the wave of the future, but was subject to intense debate (Folsom and Lubar, 1982). On the one side were people like Thomas Jefferson (1788), who thought that:

> Those who labour in the earth are the chosen people of God, if ever he had a chosen people, whose breasts he has made his peculiar deposit for substantial and genuine virtue. . . . While we have land to labour then let us never wish to see our citizens occupied at a work bench.

On the other side there were those like David Humphreys (1794), who thought that public morality would only be improved by industry and were willing to wax almost eloquent on its behalf:

> From industry, the sinews strength acquire,
> The frame dilates, the bosom feels new fire.
> Unwearied INDUSTRY pervades the whole,
> Nor lends more force to body than to soul.
> Beyond all other aid, this Pow'r alone,
> Gives to man's character a manlier tone,
> Exalts the purpose, dignifies the mind,
> And adds unconquer'd firmness to mankind.

In New England this moral schism over industrialization produced specific institutional forms. Zakariah Allen (1795–1882), mechanic, mill proprietor, entrepreneur and horrified visitor to Manchester (England), advocated a particular 'Rhode Island system' of textile manufacture which would avoid similar social and environmental mistakes (Allen, 1835). Hence the 'mill village' (Kulik *et al.*, 1982) was an intermediary geo-economic form between the pre-industrial and industrial systems – an attempt to guide industrial urbanization which left hundreds of small mills at virtually every fall in the New England river system, producing a distinct industrial landscape under the guidance of an environmental imaginary (villages clustered at falls and separated by fields and forests) and a class-conscious rationality (preserving the moral order of the Calvinist pre-industrial village literally watched over by pastor and professional, including an enlightened but profit-seeking bourgeoisie). About this, Zakariah Allen (1835: 153–5) said:

> In most of the manufacturies in the United States, sprinkled among the glens and meadows of solitary watercourses, the sons and daughters of respectable farmers, who live in the neighborhood of the works, find for a time profitable employment. The character of each individual of these rural manufacturing villages, is commonly well scanned, and becomes known to the proprietor, personally; who finds it for his interest to discharge the dissolute and vicious. . . . God forbid . . . that there ever may arise a counterpart of Manchester in the New World. . . . Whilst a cold climate and an ungrateful soil render the inhabitants industrious, thus distributed in small communities around waterfalls, their industry is not likely to be the means of rendering them licentious; and of hampering the purity of those moral principles, without which neither nations nor individuals can become truly great and happy.

The mill village was thus the socio-spatial form by which pre-industrial values were to be retained in the industrial age by a morally concerned, conservative group of enlightened entrepreneurs. New Englanders did not accept the economic rationality, the economic (institutional and urban) forms, or the social relations of an already powerful, 'functionally proven' industrial system from the UK, but attempted, with partial success, to form their own industrial system. Eventually Manchester, New Hampshire, did come to resemble Manchester, Lancashire (Hareven and Langenbach, 1978), indicating that modern industry has a systemic logic which produces similarity out of heterogeneity through the disciplining of competition. But the American entrepreneurial system on which this logic operated already had distinct characteristics. The hybrid resulting from interaction eventually became hegemonic. In this way culture interacts with economy to produce space.

# THE NEW ENGLAND SOCIETIES

In the eighteenth and nineteenth centuries the pool of immigrants backed up in New England for a hundred and fifty years after colonization was released to form a main source of migration into the North American interior (Mathews, 1909). New England migrants showed a distinct tendency to become the economic and cultural elites of what some called 'Greater New England', the northern tier of states where New England migrants, ideas and models predominated. As part of this, they founded a loose network of New England Societies where they talked about their region of origin, their evident economic and political success, and the connections between the two. These orations, as they were called, employed the finest artistic and political talents of their time. The nostalgic but often critical speeches and sermons (Brainerd and Brainerd, 1901) given before the New England Societies were important expressions, but also active creators, of the New England discursive formation. Through this formation they entered the national ideological structure, becoming particularly important for the connections drawn between effort and reward, and between individual striving and spiritual community.

## Reading Ideology Sympathetically

The discourses of participants in a historical process have to be read critically, with an elaborated hermeneutic forming part of any theorization; in this case we might read the New England Societies discourse through the lens of ideology. However, there are problems with the very strength of an ideological reading. When accounts given by an elite are shown to legitimate power, all traces of other bases for behaviour disappear. So the elite's appeals to god, environment, place, ethics are mere ideological pretence which legitimate economic and political power; they employ destiny to excuse power. But when explanatory themes are repeated for a century and more by hundreds of speakers, as they were in the discourse of the New England Societies, they were also deeply held beliefs, forms of understanding which *in*formed the economic and political practices of powerful agents, rather than just being ideologically formed *by* economic practices, as their justification. Hence we need a 'sympathetic' reading which employs the concept of ideology but is not confined by it; here Marxian theories of ideology and Foucauldian discourse theory might be merged.

Throughout the nineteenth century, the New England Society discourse returned hundreds, if not thousands, of times to the same linked obsessions: the character of the New Englanders as a distinct people; and the origins of their economic and political power. The oratorical tradition began with Daniel Webster's speech to the bicentennial celebration of the landing of the Pilgrims at Plymouth in 1820 in which the dynamic and affirmative value of the original New England colony was restated as prelude to a new era, 'a new impulse to the human mind' (Vartanian, 1971: 25). A generation of New Englanders was taught to recite this speech from memory so that, as Vartanian (ibid.: 27) says, the Puritan that Americans came to know was the symbol created by Webster's address. The New England Societies were subsequently prominent in transforming the Puritan from regional ideal into national symbol: 'The Puritan has been the main factor in producing the mightiest republic upon which the sun has ever shone' (T. Roosevelt in NESNY, 1898: 42–3). Yet while apparently speaking historically of their Puritan origins, the emigrant New Englanders were also speaking of themselves as leaders of the USA economy and polity: 'The cheerful industry, the hardy enterprise, the ingenuity, the calculation, the self-reliance, the thrift, which distinguish the occidental form of Saxon civilization, have, beyond dispute, their seat and their source, chiefly in the land of the Pilgrims' (Hadduck, 1841: 273). Speakers continually stressed the ethical basis, founded in Protestant beliefs, of their economic behaviour; one speaker listed '[t]hrift, industry, self denial, a sound appreciation of the relative values of immediate indulgence and future happiness' (NESCB, 1920: 53).

Looking back on their region of origin they embellished this story (much of which is a regional version of Weber's Protestant ethic) with accounts of environmental determination and the effects of place, time and circumstance: 'Upon every homogenous nation, Providence impresses distinctive moral and intellectual traits, through the agency of natural causes, and of these the influence of climate, soil, and the configuration of the earth's surface, is the most active and conspicuous' (Marsh, 1844: 382); and 'If the character of the people has a larger range and a greater versatility, perhaps they may thank their climate of extremes' (Emerson, 1870: 379). From such conditions, they thought, 'springs a life with its own distinctive characters for good and evil – sturdy love of liberty, thrifty ways of life, habits of methodical industry, reverence for religion and education, respect for law. . . . Each man acts always with reference to a social organization which exists ready formed in his head' (Holmes, 1855: 274). Puritan ideals and the New England environment were considered to be mirror copies, as one speaker said: 'the very bleakness and severity of the coast and climate of New England . . . reflected the sombre mood of the stern idealists who had come to live among its granite rocks' (Hill, 1897: 28). Also, in founding a frontier colony, they believed people had to be enlightened, their morals stricter, their reasoning powers more acute and discriminating, so they could be self-governing, fix their own articles of belief, and assume the awful responsibility of reasoning for themselves. Such characteristics, typical of the English mind (so they thought), had developed more intensely in the clear mountain air of New England, under social conditions unencumbered by the dead weight of pre-existing institutions. As a result, 'the moral, intellectual, religious and political seed sown on these northern shores was as pure and as full of life as any ever sown on any soil in any age of the world' (Knapp, 1829: 150). New Englanders, they believed, made the USA 'a vast empire of intelligence, of science, of invention, of political wisdom' (ibid.: 153; Hatch, 1957).

## Imagined Economies

The idea, as one speaker (Winthrop, 1839: 243) put it, was that 'the rock-bound region of New England . . . produce[d] a wealth richer than gold, and whose price is above rubies – the intelligent and virtuous industry of a free people'. A combination of physical hardship, an original 'unparalleled equality in the distribution of property', access to free education for everyone, and the Calvinist religious system produced a system of social relations 'where every man is . . . the arbiter of his own destiny' and yet also 'every man's lot is determined by others' (Bacon, 1838: 205). The belief system emphasized that the discipline of the Puritan ancestors for 'self-denial is . . . the ground of all virtue' (Cheevers, 1842: 293). The private individual, without aid from government, with a 'free arm working the will of a free spirit' developed more strongly in the New England colony than in any other community (Upham, 1846: 435). From the moment the Pilgrims stepped on to Plymouth rock, the energies, rights and dignity of man as an individual were secured for ever: 'The highest possible development of each individual . . . is essential to the progress of the race. The advancement of the individual and of civilization are wrapped up together' (NESNY, 1920: 55). The colonist found a land stretching for ever – 'there was none to dispute his possession, or interfere with his movements, or in any way restrain or affect the exercise of his will' (Upham, 1846: 438). Of fundamental significance was the private ownership of land: 'The moment man emerged from savagery the instinct of property developed in his heart . . . all progress and all civilization were based upon that notion' (NESNY, 1913: 79). The power of character growing from this freedom of mind, the feeling that each person could determine their destiny, was the 'secret magic' by which the sons of New England commanded success and wealth wherever they went. This force of individual character, this consciousness of inherent power, became habitual when exercised, entered the frame of the mind and clothed people with their true strength:

> The impulsive projection of each individual, according to his peculiar nature, into the engagements and struggles of business and of life in all its forms; this self-originating, and self-stimulating earnestness of pursuit, taking effect upon a whole people . . . this spirit of enterprise rises into greatness, and becomes truly imposing. It secures perpetual and boundless progress. It diffuses prosperity. It evokes all latent power. It silently and by a most benignant process, wins for a nation nobler victories, and a greater dominion than the mightiest armies could have achieved. (Upham, 1846: 441–2)

The idea was to combine intelligence and ingenuity with hard work, to be open to science and the arts, to be moral and ethical in one's concerns, but most of all to trust the individual's energy of character and enlightened mind: such would turn barrenness into fertility, straighten winding and crooked paths, accelerate speed, reduce cost, multiply business and create wealth (Upham, 1846: 443–4).

## Hermeneutics of Economic Imaginary

What was this group of elite men doing by telling such stories? In part, ideologically justifying their success through implying choice by Providence and spinning tales of hardship, denial, effort, achievement doubly deserved because of the enormous effort necessary just to survive, let alone prosper, in a glaciated land where winter lasts fiercely for five months, and intermittently for seven months of the year. But also, from their belief system, grounded in the Puritan ethic, modified by a particular regional colonizing experience, they were actively creating from the raw materials of history a rationality suited to an emerging form of economic life. They were imagining an economy and inventing their personalities as leading economic and political agents of a system still owned by individuals or small groups of entrepreneurs. The living myths of freedom and American entrepreneurship were being culturally created via continual retelling of the story of the Puritans and the history of the New Englanders. They were both making a kind of generalized rationality, a mixture of belief and economic motive, and outlining a set of specific instructions to guide economic behaviour, a kind of declaration of honest dealing among people with hardship common to their pasts, one rooted in the ethics of New England's religious rationalism but suited to a modern world of agreements between equals and contracts between strangers.

How did this process of the cultural invention of rationality unfold? Was it an economic-discursive process? The New England elite clearly was not satisfied with immediate, functional explanations, like the oft-quoted thrift and hard work. Instead they reached into mythical history, rehearsed symbols, told stories about themselves. Rather than dismissing these orations as after-dinner exaggerations, we might sometimes (but not always!)

remember that members of elites are human beings too. So their discourse may be read as an ideological justification for power, but also as an existential quest for meaning, the two as interacting moments. Power in this case takes the particular form of *control*, over fierce nature, over other people, over the course and causes of life, while the search for meaning eventually is about *significance*, here an intense desire to place one's life within some broader scheme, which they called Divine Providence or Destiny depending on their specific mysticism; a synthesis of power and meaning is made possible by the similarities (in existential terms) between control and significance. So the discourse of the New England emigrant elite might be read as an appeal for significance riddled with the desire for control, a search for meaning as well as an ideology of class, gender, regional and ethnic power.

In this 'sympathetic' reading, economic action (entrepreneurship, innovation, management) is a striving for control over the course of one's own life, over other people's lives and over nature. Economic success is both a sign of significance in the present life and, in the Calvinist system and its religious derivatives, a way of alleviating those awful, nagging doubts about the utter long-term insignificance, the total absence of meaning (which the shallow fiction of 'heaven' is supposed to alleviate), of the existence of the individual person. 'Success' prevents the full force of the realization that, a century later, few signs remain that an individual life once happened – so even prominent members of the Societies are commemorated now in dusty books, many not taken out of libraries since 1938 or earlier (as this writer can personally attest), their lives marked physically only by quick summaries hidden on long-forgotten, grass-covered gravestones. (Strange quirk of ideological fate that I, a Marxist, am the only person who knows of their bourgeois lives.)

These discourses reveal not only justifications made up after the fact by an already successful elite but initial, continuing, updated motivations for economic action. These migrants tried hard to be good New Englanders, they believed their myths, they lived them, exemplified them. The important thing about the stories New Englanders told about themselves is not whether they were 'true', for clearly they were not – witness the total absence of the American Indian – but whether they were believed and how they were acted on: people do not live by truth, but by myth and symbol. People far prefer lies over truth. Lies make existence possible. In such ways do subjective myths become objective economic forms.

We can look through economy into culture. We can read economy as a cultural sign.

# THE CULTURAL ORDER OF PRODUCTION

What if we were now to put our economic spectacles back on? The idea of these brief historical exercises is to see economy as part of a larger cultural order. Economy can be thought of as a system organized by relations of production and driven by rationalities/logics. Institutions, economic agents, groups of agents, such as classes or genders, behave in certain ways, in terms of motives for economic action, preferences and attitudes, and think in terms of economic rationalities, which are reinforced by the overall logics of economic systems. In this perspective, the central dynamic of economy is the interrelation between systemic logic (e.g. capital accumulation) and the rationalities of economic agents (e.g. efficiency to make profit), the two forming each other (economic logics formed by, and reproducing, appropriate forms of economic reasoning).

## Reading Economy

Yet this is not an isolated, purely 'economic' system of relationships played out in a separate realm called Econoland, a kind of austere Disney World, with cartoon characters dressed as Mr Gross Profit and Ms Marginal Cost solemnly walking around, shaking hands and generally pretending to have a thoroughly bad time. Economy interacts not only internally but also with a broad array of sociocultural forces theorized by a rich array of concepts developed by recent social theory. 'Reading' the economy in this way involves a kind of uncovering, seeing through, expanding from economic instance to cultural totality. So we might gaze at the surface patterns of economy intending to see through them, like one of those seemingly one-dimensional pictures (stereograms) which, with the right fixation, suddenly evaporate to reveal their three-dimensional interior. The capitalist economy is one product of a wider phenomenon, the cultural order of modern religious rationalism, expressed archetypically by New England's Calvinist bourgeoisie. Economic forms, such as rationalities for action and the collective logics reproduced in economic systems, originate as aspects of this broader cultural order under definite regional circumstances. In turn, the cultural order consists of the belief systems, ethics, symbols, philosophies, myths and lies with which people explain their existence and experiences. It consists also of a geomaterial dimension of substances and types of product which form the physical structures and implements within which existence unfolds. And it consists of the interaction between beliefs and materials in the aesthetic, artistic and literary realms where representations of existence and economic imaginaries are made and appreciated. 'Economic' personalities and rationalities are component parts of this cultural order.

Yet economy is hardly a passive moment of culture. With its power of conferring life or death, economy re-creates the cultural order. Economy is the arena in which alternative rationalities and identities engage in practical, material struggle. Efficiency or inefficiency in production are main criteria (under capitalism) reinforcing or weakening, on pragmatic grounds, identities and rationalities chosen to be centrally significant in the creation of continued livelihood. Perhaps even more importantly, economy, as a central agency of power in society, has the partly autonomous capacity to create aesthetics, styles, consumer preferences and other aspects of the cultural order, or choose, from an erupting imaginative display, which survive in what forms. This complex of interactions between culture and economy, this matrix of relations and dynamics, can be understood only by a formal dialectics of the cultural order of production.

## Producing Economic Identities

Looked at in this way, much of contemporary social theory can be employed in conceptualizing the mediations between nature and economy on the one hand, and between power and meaning on the other, the general term for the complex of such mediations being culture. That is, economic motives and logics, consumer preferences and worker attitudes, are not created by economy alone, but instead derive from the identities of agents and the discursive formations and social and environmental imaginaries of their culture and time. This emphasis on the socialized and encultured economic agent involves most centrally notions of subjectivity, rationality, imagination and identity. Here there are at least two fundamental aspects of the formation of identity, alluded to earlier as significance and control. First, identity (being an agent, a person, a self) means moral consciousness and beliefs, by which is implied concern for others and questions of human dignity, what makes lives meaningful and fulfilling (Taylor, 1989). Such considerations may be seen as deriving from the quest for meaning which existential phenomenology tries too earnestly to understand. Second, identity is constituted relationally through involvement with significant others, through integration into communities, so that group categorizations and norms are major constituents of individual identity, the connections being to social relations, institutions and power (Epstein, 1987; Young 1990b). Marxism and Nietzschean post-structural theory try too competitively to comprehend this. Such sources of identity are penetrated by the cultural production of the imagination in systems of power.

Here we could follow Foucault (1979), deconstructing the modern economic subject in terms of the disciplining and normalizing technologies of power. But this would be only one of several equally intriguing, intertwining paths. We could read Castoriadis (1991), who says that humans organize their worlds, including their economic worlds, through social imaginary significations which invest everything with meaning; this underlies Touraine's (1988) notion of social struggles in the production of society. Or Habermas' (1984) communicative action and rationality, tracing the conflicts between system (the sphere of material reproduction) and lifeworld (the symbolic space of collectively shared background connections). Or Bourdieu's habitus: 'the mechanism by which meanings of the cosmos are internalized and incorporated', or, more simply, 'cultivated dispositions to act' (Robbins, 1991: 84, 109). This last notion is particularly interesting because it puts together the following: space experienced as what Giddens (1984: 188), in a similar argument, calls 'locale' – 'the settings of interaction . . . essential to specifying its *contextuality*'; beliefs, cognitive understanding and primary, phenomenological knowledge of the social world; and the notion of behaviour as an expression of a class ethos which reinforces a social trajectory. Yet it also sees people not as dupes of power systems but as agents creating the conditions within which they are conditioned. And there are many other possibilities, merely in the area of the cultural making of the social and economic subject. A project linking economic with cultural analysis, industrial geography with landscape studies, for example, is supported by an embarrassingly rich array of intellectual resources, which only the blinkers of conventional economic thinking prevent us from fully using.

# CHAPTER FOUR

# THE INVENTION OF REGIONAL CULTURE

MERIC S. GERTLER

## INTRODUCTION

'Culture' has re-entered the lexicon of the economic disciplines with a prominence not seen for some time. With the growing interest in the social nature of production systems, signified by the use of terms such as 'industrial networks', 'industrial districts', and especially the 'new social economy' and 'socio-economics', a new significance has been ascribed to socio-cultural context. Hence, in emerging production systems in which the social division of labour is recognized as being of increasing importance, social and cultural characteristics have begun to figure prominently in the work of economic geographers (Storper, 1992; Sayer and Walker, 1992), industrial economists (Lundvall, 1988; 1994), political economists (Putnam, 1993) and management theorists (Kanter, 1995). The inter-firm relations which have come to dominate the 'new competition' (Best, 1990) are said to be based increasingly upon non-market forms of interaction bound by trust, in which cultural commonality between co-operating and transacting partners is seen as an advantage.[1]

Extending this line of thinking, a position which is gaining currency in the growing volume of work on the adoption and propagation of network relations between firms holds that Anglo-American 'business culture' is not favourably predisposed to the idea of co-operation. The claim is made that the ethic of rugged individualism and the rhetoric of 'dog-eat-dog' competition are so strong as to discourage Canadian, American or British firms from participating in (or reaping the full benefits from) inter-firm co-operation and collaboration. The converse of this argument is that particular national or regional cultures are inherently more predisposed to co-operation, or that cultural ties are coming to dominate all others in shaping the emerging alliances and partnerships between businesses.

A related argument is the idea that certain cultural traits – for example, the 'traditional' Japanese values of dedication to education and hard work, devotion to higher authorities such as one's employer, and a sense of social cohesion, or the 'typical' German predisposition towards all things technical and complex – explain the success with which particular national economies have adopted 'post-Fordist' production methods, including new forms of complex production technologies and modes of workplace organization *within* the individual firm. Conversely, the absence of these traits in other cultures explains the failures of their own indigenous firms to adopt these new practices with the same degree of success.

Hence, a new variable has entered the debate on regional and national competitiveness, and the prescriptions for policy flow directly from the diagnosis: in the absence of a naturally inherited 'manufacturing culture', the state must attempt to create or 'manufacture' one, by exhorting firms to 'co-operate to compete' – that is, to change firms' behaviour by convincing them that it is in their own best economic interest to co-operate with other firms. Furthermore, to help them along in this process, the state should train individuals to act as 'brokers', to bring reluctant firms together by helping them recognize complementarities they may share with other firms (normally within the same region) (Rosenfeld *et al.*, 1992; Bosworth and Rosenfeld, 1993).

I wish to argue in this chapter that the role of culture in this debate has not been adequately specified, on either a theoretical or an empirical level, and that it needs to be thoroughly rethought. I hope to demonstrate the need to examine the process by which industrial cultures – whether at the level of the workplace, the region or the nation – are themselves constructed by social practices. For the purposes of this analysis, I shall focus on one small but significant part of this process, by examining the role of economic institutions and regulatory frameworks – primarily public ones, and operating at both the regional *and* national scale – in shaping practices, customs, norms of economic behaviour and even what appear to be individual traits. In so doing, I hope to take issue with the notion that certain regional or national cultures are somehow more naturally predisposed to engaging successfully in post-Fordist manufacturing activities than others – that their success is due to their naturally endowed 'manufacturing culture'. I also wish to assess critically the recent arguments that such manufacturing cultures can themselves be readily manufactured in particular places where they were previously underdeveloped. Here, I shall argue that the process by which industrial practices are produced is more complex, involving forces of regulation operating not only at the level of the individual workplace, corporation, community or region, but also at the spatial scale of the nation-state.

At the same time, and somewhat ironically, I shall argue that this process is considerably more transparent than the current literature would have us believe. When economic analysts resort to 'cultural' influences to explain the behaviour of managers, firms and workers, this is normally tantamount to an admission of ignorance.[2] It is as if the processes at work arise from some timeless, primordial traits whose formation mystifies and confounds understanding. Instead, I shall attempt to demonstrate that the motivations underlying many of these practices within and between individual firms can be seen to arise quite directly from the structure of the macroregulatory environment in which these entities function. In doing so, then, I hope to begin the process of demystification in our study of contemporary economic relations. I also wish to argue that those prescriptions for regional economic renewal which focus on the need to correct the dysfunctional behavioural tendencies and culturally shaped attitudes of individual firms, managers and workers are usually based on a misdiagnosis of the problem. Consequently, these prescriptions are misdirected at changing only the attributes and traits of managers and workers, when they should also focus on the broader, systemic characteristics of the regulatory environment.

The principal empirical basis for this critique and reconstruction is my continuing research on the production and use of advanced manufacturing technologies and processes – the technologies and work practices said to be at the very heart of the transition to 'whatever it is that comes after Fordism' (Gertler, 1992). The work arising from this project, which is studying the production and implementation of such practices in North America and Europe, reveals the role that both national and regional institutions and regulatory features continue to play, in contrast to the almost exclusive focus on regional institutions evident in the industrial networks literature (Storper and Scott, 1989; Pyke and Sengenberger, 1992; Scott, 1996). Hence, by demonstrating the role that this regulatory framework plays in shaping the development and use of new manufacturing technologies, I am attempting to understand and outline the spatial foundations for the construction of capital used in production, but also of 'industrial culture'.[3]

In the following section I shall review briefly some of the leading arguments arising from the recent literature in economics, economic geography, economic sociology and related disciplines, concerning the growing role of culture in regional economic relations. I shall then present some of the more salient findings arising from my own work, in which particular forms of economic behaviour related to machinery use appear to arise from some organically conceived notion of cultural differences. Upon further investigation, these are in fact shown to arise from other sources. I shall then conclude the chapter by offering up some implications for current debates within economic geography, extending the argument to other aspects of economic behaviour, including employment relations within the workplace and inter-firm collaboration.

## CULTURE AND REGIONAL SYSTEMS OF INNOVATION AND PRODUCTION

Integral to the claim that the nature of capitalist competition has shifted in the late twentieth century is the key idea that systems of innovation and production have become more social in nature.[4] This assertion has two distinct but related components. First, production systems are coming to be characterized by a more finely articulated social division of labour, achieved through the process of vertical disintegration of large firms and the growing use of various forms of outsourcing, including subcontracting to smaller supplier firms. This externalization of the production process is said to offer the chief advantage of agility in meeting the needs of ever more rapidly changing and fragmented markets. As market demands shift qualitatively, producers are able to respond more effectively in such 'open' systems because (a) they can more readily absorb the innovative ideas of supplier firms to help them devise new products and improvements, and (b) they can rework their sources of supply to match the particular attributes of the 'product of the moment', in both cases drawing upon the rich resources of a large collection of suppliers.

The second component is that, as individual firms come to rely more heavily on their relations and exchanges with other firms, *non-market* forms of interaction become more important. Viewed in terms of the Williamsonian continuum between public markets and private hierarchies, much of the interesting action is seen to be taking place in the middle ground: relations are social, but are increasingly buttressed by trust. In particular, as Harrison (1992) has pointed out, for these innovative production systems to function properly, firms must develop a considerable degree of interdependence on one another (including surrendering proprietary information), but will do so only when a relationship of trust has been established. Such relations are more likely to arise when firms interact with one another directly and repeatedly over time, as they are more likely to do when they are located in the same region (*see* Crewe, 1996). However, as sociologists such as Granovetter (1985) have pointed out, this interaction takes place through informal as well as formal mechanisms, and is reinforced by shared histories and cultures.

Focusing particularly on innovation, Lundvall (1988) has drawn attention to the importance of social (including cultural) context in contributing to the success of interaction between the producers of new advanced process technologies and their users. This interaction is said to be especially important at times when technological development crosses the threshold to a new paradigm. When the technology in question is particularly complex, expensive, and subject to rapid change, then 'closeness' – in both a physical and cultural sense – is crucial to successful interaction leading to effective innovation. Spatial proximity facilitates the easy, frequent face-to-face contact necessary for the exchange of detailed technical information. Cultural commonality further reinforces this link, since it is easier for producer and user to understand one another at deeper levels of meaning:

> When the technology is complex and ever changing, a short [geographical and cultural] distance might be important for the competitiveness of both users and producers. Here, the information codes must be flexible and complex, and a common cultural background might be important in order to establish tacit codes of conduct and to facilitate the decoding of the complex messages exchanged. . . . When the technology changes rapidly and radically . . . the need for proximity in terms of geography and culture becomes even more important. A new technological paradigm will imply that established norms and standards become obsolete and that old codes of information cannot transmit the characteristics of innovative activities. In the absence of generally accepted standards and codes able to transmit information, face-to-face contact and a common cultural background might become of decisive importance for the information exchange. (Lundvall. 1988: 355)

Lundvall goes on to note that, because cultural attributes are still most frequently mapped on to nation-states, these political entities retain enduring significance as sets of institutions shaping economic processes (a theme to which I shall return later). Hence, proximity at two distinct spatial scales – the region and the nation – are argued to be important.

Storper (1992) proposes a conceptualization of the geographical foundations of innovation which is strongly consistent with Lundvall's emphasis on user–producer interaction. Storper describes a phenomenon he dubs 'product based technological learning' or PBTL, which he observes to be occurring most commonly in dynamic, subnational agglomerations he calls 'technology districts'. This phenomenon, according to Storper, is underpinned crucially by what he refers to as 'conventions', which 'structure the participation of agents' in such districts. Furthermore, these conventions are 'territorially bounded' and serve to 'define the qualitative basis of the external economies of PBTL systems' (p. 62). Storper's concept of conventions is rich and multifaceted. In essence, they amount to a set of acknowledged and shared rules 'that mobilize resources and regulate interactions so as to make PBTL possible' (p. 90), and that create 'localized expectations' and 'preference structures' concerning concepts such as 'time horizons, payoff points, etc.' (p. 85). He goes on further to note:

> Conventions lie beneath the regularized social interactions that sometimes appear as formal rules or institutions, and at other times appear simply as routines or unwritten 'rules of the game'. Conventions describe the underlying forms of collective order of the production system. (p. 86)

Storper is more specific about the *effects* that conventions have on economic interaction in PBTL-based technology districts than he is about their *sources or origins*. Nevertheless, in explicitly citing Lundvall's work, he places considerable emphasis on (a) the importance of shared conventions for facilitating technology-based communication and learning through interaction, and (b) the positive role that shared cultural attributes ('territorially based and noneconomic forces'; p. 62) will exert on this process. Hence, by implication, one chief source of convention is a common culture between transacting agents. Note, however, that on the basis of the same arguments, Storper sees the subnational district as the most important scale at which conventions are defined, while Lundvall sees culture's influence as operating primarily at the national level.

More recently, Saxenian (1994) has introduced culture as a key variable in her analysis of the reasons for the widely diverging performances of two regions producing innovative products such as semiconductors and personal computers: California's Silicon Valley and Massachusetts's Route 128. Indeed, the very subtitle of her book (*Culture and competition in Silicon Valley and Route 128*) makes clear the central role that culture plays in her story. In attempting to explain Silicon Valley's continued technological success and the failure of Route 128, despite the fact that they competed in the same product markets and as recently as the 1970s boasted comparable levels of economic activity, Saxenian attributes causality to the divergent 'industrial systems' that characterized these two different regions. This regional industrial system is said to have three closely interconnected dimensions: 'local institutions and culture, industrial structure, and corporate organization' (p. 8). The first of these elements is described as follows:

> Regional institutions include public and private organizations such as universities, business associations, and local governments, as well as the many less formal hobbyist clubs, professional societies, and other forums that create and sustain regular patterns of social interaction in a region. These institutions *shape and are shaped by* the local culture, the shared understandings and practices that unify a community and define everything from labor market behavior to attitudes toward risk-taking (p. 8; emphasis added).

Note here that the 'culture' at work is explicitly local or regional in character. Furthermore, while Saxenian draws attention to the reflexive interaction between regional institutions and regional culture, national institutions and culture do not figure in this discussion.[5]

More recently still, Kanter (1995) has extended the application of cultural ideas to the process of regional economic growth by arguing that communities that wish to serve as successful destinations for foreign direct investment need to create, among other things, a local culture of collaboration. As she puts it (p. 160), 'In addition to the physical infrastructure that supports daily life and work – roads, subways, sewers, electricity, and communications systems – communities need an infrastructure for collaboration to solve problems and create the future.' According to Kanter, this infrastructure is largely informal in nature, with the chambers of commerce in rapidly growing communities such as Spartanburg and Greenville in South Carolina acting to provide the 'social glue' that fosters co-operative action and joint learning in the region.

## THE SOCIAL FOUNDATIONS OF ECONOMIC PROCESSES

The growing interest in the role of cultural phenomena in regional economic systems of production and innovation, as documented above, itself reflects a parallel development within other social sciences, where the influence of social processes in the economic sphere has come to be much better appreciated. This is particularly evident in the field of economic sociology or 'socio-economics', which has drawn much inspiration from the pioneering work of Polanyi (1944).[6] The central argument in this work is that social and economic systems are inextricably bound up with one another – that one cannot understand the workings of economic processes in a given place and time without simultaneously interrogating the social systems which underpin them (Block, 1990b).

Hence, while economists are wont to regard the market as the single, transcendent form of economic rationality (as a way of organizing production and distributing its benefits), economic sociologists reject such universalist claims, noting instead how social and political processes have created the institutions (e.g. property rights, contract law) whose existence underpins the very workings of markets. Furthermore, they show that social (including cultural) influences are *not* separate from, or exogenous to, the operation of the economic sphere, or mere trivial epiphenomena, but constitute real and significant elements of our understanding of the economic performance and other characteristics of particular economic systems (whether national or regional). As Block notes, 'the "economistic fallacy" imagines that capitalist societies do not have cultures in the way that primitive or premodern societies do' (1990: 27).

However, social influences act in tandem with other forces to shape economic systems and outcomes. Block contends that the forces at work arise from three distinct sources: individual choices (which may or may not be structured by markets and the single-minded pursuit of economic rationality); state actions that structure an economy (by, for example, creating the institutions that enable the working of markets or other systems of production and allocation); and social regulation ('the social arrangements that condition and shape microeconomic choices', and through which 'individual economic behavior is embedded in a broader social framework'; ibid.: 42). While Block contends that it is necessary to understand how all three of these elements work together to produce economic outcomes, he also argues that the last of these – social regulation – has been least appreciated until recently.[7]

Hence, we can conclude from this work that social (including cultural) processes are centrally implicated in economic life. However, what remains unspecified, either by Block or by those expressly interested in regional industrial systems, is the precise nature of 'culture' and its relationship to economic processes. This is somewhat ironic, for (despite repeated urgings that economic analysts should regard social and cultural phenomena as central to the operation of economic systems), proponents of this argument have adopted a surprisingly unsophisticated understanding of how culture is formed and changes over time. Indeed, as a result of the neglect of this question, the work reviewed above has often employed a rather wooden, static and timeless view of 'culture' – much akin to what McDowell (1994) describes as culture 'as an abstraction' (p. 149). McDowell notes that practitioners of the 'new cultural geography' have criticized similar tendencies within the earlier work of cultural geographers to see culture as something of 'an all-powerful entity, subject to its own logic, which people inherit and diffuse'. Hence, the traditional conception of culture as 'superorganic' 'sees culture as having causal powers . . . detached from and determining the actions of people in a locality. . . . Culture is seen as a totality, almost as a "black box" rather than as a pluralistic set of social practices' (p. 149).

So, while it is important to assert that cultural characteristics are much more than 'mere epiphenomena', it is also important to examine the process by which cultures are actively produced and reproduced by social practices and institutions over time. Sayer and Walker (1992) have reached a similar conclusion, noting that 'culture is often misrepresented as something ethereal and eternal, divorced from historical material practice, or misconstrued as a self-perpetuating tradition that determines contemporary actions' (p. 178).[8] Instead, they argue, 'productivity depends not just on individual attitudes or even on types of authority relations, but on specific forms of material social organization which make these qualities yield economic results' (p. 178). Furthermore, they suggest provocatively that 'every culture surrounding economic behavior and social order rests on material origins, whether long-standing traditions such as Confucian morality or recent innovations such as corporate paternalism' (p. 118).[9] Saxenian (1994) comes close to the same position by asserting, 'A region's culture is not static, but rather is continually reconstructed through social interaction' (p. 8). Nevertheless, though her rich case studies hint at how this process might unfold, nowhere does she again address this question explicitly.

## MACHINERY CULTURE: THE SOCIAL CONSTRUCTION OF 'OVERENGINEERED'

It seems from the above review that, while we have accumulated some helpful general ideas about how culture ought to be incorporated into a 'new' economic geography or industrial economics, we still do not have a very clear understanding of precisely how cultural forces both produce, and are produced by, social practices. In the following sections, I employ an analytical device that yields some useful insights into this question by studying what happens when obviously distinctive 'industrial cultures' *collide* – that is, when economic actors arising from different 'cultures' interact with one another.

Since 1991, I have been studying the process by which manufacturers in Ontario have acquired and implemented their new process technologies. The study has focused on the relationship between these Canadian 'user' firms and the companies which produce these technologies for them ('producers'). Given the somewhat underdeveloped state of the Canadian advanced machinery industry (itself the result of decades of dependence on machinery imported first from the UK and then from the USA – two sources which have themselves been in decline for the past 15 to 20 years), many of the leading producers of advanced manufacturing technologies in use in the Canadian plants are now found in Japan, Germany, Italy, Sweden and other European and Asian countries. After surveying and interviewing a selection of these users in Ontario,[10] I was able to identify the sources of advanced machinery and equipment used by these firms. In a subsequent phase of the study, I then visited and interviewed a sample of producers of these technologies in Germany, one of the leading offshore sources for such production systems.[11] In addition, I conducted interviews with a number of representatives and suppliers of these foreign machinery producers, residing in Canada and serving as intermediaries between overseas producers and local users.

Interviews with the technology users in Ontario sought to determine the nature of the user's relationship with the producer, including the history of this relationship, the frequency and types of contact, the implementation history of the machinery or system in question, the nature of the firm's own workforce, training efforts, and the outcome of the implementation process (performance relative to expectations, unexpected developments, and so on). The German producers were asked to reflect on many of the same issues from their own perspective. However, they were also asked to discuss how, if at all, their relationships with North American customers (Canadian and American) differed from the sorts of relations they had developed with their customers in Germany and other parts of Europe.

Generally speaking, the findings indicate that many users in Canada continue to experience significant problems of implementation and operation long after the installation is completed (for a more complete discussion, *see* Gertler, 1995a, 1997b). Hence, even after being given time to 'work out the kinks' and 'move along the learning curve', users (including some large, relatively sophisticated operations with deep financial resources and in-house technical staff) have had a difficult time achieving effective implementation. The machinery and systems, once installed, failed to live up to the user's expectations (or the salesperson's claims) for product flexibility, speed of production and change-over, quality, ease of use and reliability. Breakdowns and malfunctions were frequent and downtimes were lengthy and disruptive. In general, the returns from such costly and difficult investments were often disappointing. Furthermore, and crucially (given the theme of this chapter), these problems seem to have been particularly likely to arise (and to be especially acute) when the technology in question originated in 'far-off' places such as Germany, Japan, and many other overseas sources.

When users (or producers) were asked to explain the reasons for and sources of these difficulties they pointed first to the minor but significant complications introduced when trying to carry out communications and transactions involving complex technical subjects (including both the initial specification of technology requirements and the subsequent problem-solving and 'trouble-shooting' procedures) over long distances. These relate to the delays introduced by intervening time zones, the difficulties of technical problem-solving without face-to-face contact (despite the widespread use of information and telecommunication technologies to connect users to producers), and problems of comprehension which may arise owing to differences of language. These concerns address many of the issues raised by Lundvall (1988) in his analysis of the need for 'close' interaction between users and producers to facilitate contact and communication.

However, subsequent discussions revealed that a deeper source of these problems lies in the fundamental differences in expectations, characteristic workplace practices and norms, managerial routines, transactional behaviours, and understandings of key concepts such as 'technology' itself – in short, what appear to be substantially different industrial or business cultures in Canada and Germany respectively. Indeed, interviewees on both sides of the Atlantic readily identified differences in 'culture' or 'mentality' (a term used far more frequently by German respondents than by Canadian ones) as the root of their problems in dealing with one another. Nevertheless, notwithstanding this diagnosis, what I have been able to show, by examining some specific instances in which these differences have become salient, is that underlying these apparently cultural gaps are *fundamentally different regulatory regimes and institutional structures* which are themselves *instrumental in reproducing these 'cultural' differences*. Presented below is a sample of two specific symptoms – expressed as differences in expectations, attitudes, accepted business customs and practices – which have led to misunderstandings, disappointments, conflict and, in extreme cases, termination of the relationship between machinery producer and user.[12]

## Maintenance

One of the clearest differences to emerge from this study was in the contrasting practices of German and North American users regarding machinery and equipment maintenance. German producers remarked (usually with disbelief and more than a little disdain) that North American industrial culture did not seem to assign much value to the importance of regular, preventive maintenance. As a consequence, production systems in Canadian and American plants would, in the view of the producers, fail with predictably greater frequency. This stood in sharp distinction to the dominant practice in German plants, where not only managers but also the operators themselves would maintain and service the machinery on a regular basis. More than one German producer commented on how, in their German customers' plants, the operators were 'married to' or 'owned' their machines, and would lavish attention upon them. In the words of one German manager, 'German workers . . . have the feeling, "that is my machine, and I am responsible for it"'.

## Machine Complexity and Ease of Operation

Canadian users complained that the production systems supplied by German producers were considerably more difficult to operate effectively than they had been led to believe at the time of sale. They may have held this impression despite having travelled to another user's plant (often in Germany) to observe the operation of a similar system in real time before deciding to make the purchase. A frequently heard comment (both from those users that did buy German machinery and those that did not) was that German technology was 'overengineered' and 'too complex'. This was usually accompanied by remarks to indicate that this was due to cultural traits that predisposed German producers and users to overly complicated technical solutions. The German producers had difficulty knowing how to regard such complaints, since they were aware that similar systems worked perfectly well and with little difficulty in the plants of their German customers. Instead, in the face of criticism from users that producers were 'rigid', 'unbending', or trying to 'dictate' inappropriate technical solutions to their precise production problems, the German firms would tend to place the blame with the user, accusing it of not doing enough training of its workers and managers, or of investing insufficient attention and resources in maintenance ('the problem must be yours').

As I have indicated above, the distinctive differences between German and Canadian practices were most frequently comprehended and described by those interviewed as arising from cultural dissimilarities. Indeed, the two sets of characteristic practices, expectations, attitudes and norms documented above *might themselves be viewed as constituent parts of distinct industrial and business cultures*. However, this diagnosis begs the obvious question, namely: how are such differences produced? More to the point, if one accepts that 'culture' (industrial or otherwise) is not some natural, prior, unchanging and inherited whole, then how does it interact with contemporary social practices in its own production and reproduction? One way of answering this is to set these cultural characteristics within their broader social and political context, by examining their relationship to readily identifiable institutional and regulatory features. Given that many aspects of this context *also* differ markedly between Germany and the Anglo-American economies (*see* Christopherson, 1993; Keck, 1993; Goodhart, 1994; Herrigel, 1994; Wever, 1995), it should come as no surprise that these larger, background differences might play a role.

In fact, I would argue that the differences observed above can be linked quite directly to the nature of social institutions which regulate capital markets and business finance, labour markets, labour relations and the employment relations of user firms. Beginning with the issue of sharply divergent maintenance practices, much of this can be explained by examining the enduring differences in capital market structures in the two countries, which create marked differences in time horizons between the German and North American machinery users. Canada and the USA have created business environments, based on the classic Anglo-American system of public capital markets for equity investments, in which there is a strong division between financial and industrial capital. Shareholders usually exert significant power, creating strong pressures to produce short-term returns on investment (Canada Consulting Group, 1992). In contrast, German businesses raise the bulk of their equity capital through private investments. In a system in which financial institutions and industrial firms are closely linked, and in which (as a result of the labour relations institutions described below) a broader array of stakeholders (including workers and unions) are routinely represented on boards of directors, investment objectives are longer-term. The pursuit of short-run returns is tempered by sources of capital which are patient or 'quiet', and by a stronger voice in favour of social returns, resulting from the direct representation of workers on the managing boards of many larger German firms. As a consequence, German industrial firms have considerably more latitude to wait longer periods of time for investments to bear fruit, explaining their considerably longer managerial time horizons, relative to their North American counterparts.

Hence, the stark differences in maintenance practices can now be understood as arising, at least in part, from the structure of industrial investment finance and the institutions shaping capital markets. When investment capital is acquired on terms that are so strongly skewed in favour of quarterly returns, it should come as no surprise that Canadian (or American) users treat their capital equipment in a manner consistent with the prevailing truncated time horizons.[13] When their decision-making horizon stops at two to three years, and their expectation is that a machine will be in active service only this long, it is understandable that managers will undervalue regular expenditures for the purpose of longer-term machine and system maintenance.

This tendency is further reinforced by sharp distinctions between the German and North American institutions and systems of regulation shaping labour markets and the employment relation. One of the most distinctive features of the German economy is its system of labour relations based on the principle of 'co-determination'. Under this system, workers – both directly through firm-based 'works councils' and indirectly through national unions – have a significant and institutionalized role in many aspects of the firm's decision-making, including training, technology acquisition and implementation, and day-to-day operations. Furthermore, and as a result of labour's institutionalized power, there are serious curbs on employers' ability to fire or lay off workers. Instead, the system works to encourage a stable employment relation characterized by long length of employment tenure and the active use of internal labour market practices to manage firms' personnel needs. Furthermore, with a much greater degree of centralization of wage determination, and strong concordance between wages in union and non-union workplaces, inter-firm competition based on wages is held in check.

All of this stands in sharp contrast to the Anglo-American norm, where employment relations are far less stable over the long term, where employers make far more extensive use of external labour market practices (hiring and firing), leading to the high turnover rates discussed earlier. Furthermore, apart from some key sectors such as automotive assembly, unionization rates are low and (at least in the USA) declining, as is labour's power in the workplace in general. As a result, the degree of inter-firm variation in wages and working conditions is significantly greater than in Germany, and employers are encouraged to view labour cost as one of the chief dimensions of inter-firm competition (O'Grady, 1994).

These fundamental differences in the institutional and regulatory framework surrounding employment play a large role in producing the practices and attitudes documented earlier and described so frequently as being cultural in origin. Hence, it should not be surprising that North American workers do not develop the same sense of 'ownership' of their machinery as was seen to be the case in Germany, and do not engage in the same kind of lavish maintenance behaviour that the German producers so admired in the practices of their domestic customers. Furthermore, when you have a system in which machine operators are much more likely to participate in the decision to purchase the machinery in the first place (including the process of deciding on technical specifications), this is a powerful force in the development of the sense of 'ownership' of a machine that was referred to earlier.

The differing perceptions of machinery complexity and ease of use are also understandable in these terms. It is clear that the entire constellation of regulatory features and rules in German industry foster a set of incentives and imperatives for a very different type of competitive outlook by its users of advanced technologies. When wages are removed from competition, when unions are strong and more centralized bargaining systems ensure a high prevailing wage rate, and when lay-offs are discouraged by labour market institutions, firms naturally turn to other means of competition – especially the technological capabilities or qualitative aspects of their products.[14] Thus, the allegedly cultural foundations of the German machinery builders' penchant for designing technically superior or 'overengineered' production systems should instead be understood as a rational response to such a competitive regime.[15] So too does it become clear why their German customers not only demand such process technologies, but also invest heavily in the kinds of practices required to make these systems function at the peak of their potential capabilities. And in a setting where labour market institutions promote stability and minimize turnover, there are equally powerful incentives for businesses large and small to support the much-admired, highly developed national system of technical education, training and apprenticeship. Further, when competition is based less on prices than on technical and qualitative aspects of products, worker involvement becomes a key factor in the firm's success, including its success in implementing advanced process technologies.[16]

Small wonder, then, that advanced machinery designed and built in Germany has proven to be so difficult to implement with the same degree of effectiveness in North American plants. In the absence of a supportive social and institutional matrix – indeed, in the midst of what is clearly an antithetical regulatory regime – it should come as no surprise that Ontario users encountered such difficulties and frustrations.[17]

## CULTURE, INSTITUTIONS AND INDUSTRIAL PRACTICES: IMPLICATIONS FOR THEORY AND POLICY

I have endeavoured to show how the traits and attitudes we commonly understand as being part and parcel of inherited cultures are themselves produced and reproduced over time by day-to-day practices that are strongly conditioned by surrounding social institutions and regulatory regimes. Hence, we can see that workplace practices, attitudes and norms in the use of advanced machinery in Germany or North America do themselves constitute distinctive industrial cultures – but ones which are actively shaped by the prevailing macro-regulatory context. By demonstrating the impact of the institutional setting on the formation of industrial culture, I have hoped to convey something of the perils arising from the more prevalent approach to the question of culture's influence in national and regional economic systems. The argument advanced here implies strongly that the very practices we take as signifiers of distinct cultures are themselves influenced by a set of institutions constituted outside the individual firm. Moreover, in the story told here, the institutions that seem to matter most are largely *national* in origin.

This implies that we as analysts need to be much more careful in our use of cultural concepts to 'explain' differences in the performance of local or regional production systems. Culture is not a static, analytically prior concept, which 'produces' these differences. To a very significant extent, it is the outcome of regulatory forces emanating from a set of socially constructed institutions for the governance of investment and the use of labour. A further implication is that what we have sometimes taken to be organic, *sui generis* behaviour – among, say, the artisanal firms of the Third Italy or the mechanical engineering firms of Baden-Württemberg – is to an important extent strongly consistent with the overarching national system of regulation (akin to what Nelson and others have referred to as national systems of innovation; *see* Nelson, 1993). Therefore, it is much easier to understand how the technical excellence of German engineering firms is *produced rather than simply 'inherited'* when one examines the broader social context within which these firms operate.

Pushing this line of argument somewhat further, one could extend the analysis beyond the specific study of machinery implementation to consider how other allegedly 'culturally shaped' forms of economic behaviour – such as the 'negotiation culture' so prominent in the German or Japanese employment relation, or the much celebrated 'collaboration culture' of firms in Baden-Württemberg and other industrial districts – might themselves be strongly shaped by broader regulatory and institutional frameworks. Space does not permit me to develop such arguments fully here. Nevertheless, it is important to acknowledge the recent work of Wever (1995), who has explored the experiences of American firms attempting to institute 'US-style' employment relations and human resource practices in the operations of their German branches, as well as German firms attempting the analogous feat in their US branch operations. Her findings are strongly consistent with those reported in this chapter, and contrast sharply with the oft-heard pronouncements in the business press about the extent and inevitability of 'convergence' of corporate governance systems across nation-states (note, for example, the triumphant tone of *The Economist*, 1996a, 1996b). Despite their best efforts to supplant the domestic business culture and practices at their overseas operations, substituting instead their own practices and procedures from abroad, the firms involved were ultimately frustrated in their attempts. Wever's conclusion is that such experiments are doomed to perpetual uncertainty and only limited effectiveness in the absence of a set of nationally regulated reforms to support a 'negotiated' model of firm adjustment. So long as the national context remains unfriendly, such reforms will fail to take root.

Similarly, going beyond the realm of practices and relations within the individual workplace or firm, one can also show that (at least in particular cases) *inter-firm* relations might also be enabled (or, as the case may be, constrained) by the character of national systems of regulation. Hence, in some of the most celebrated instances of spatially clustered inter-firm collaboration, supraregional institutions may create a supportive regulatory context to facilitate the development of trust-based, open relations between firms – to create at least the necessary, if not sufficient, conditions for collaboration and learning-by-interacting to flourish. Thus, to provide one brief but suggestive example, the well-documented inter-firm co-operation in Germany's Baden-Württemberg can be shown to have been enabled by a sympathetic set of national institutions regulating labour markets and industrial relations. These institutions, by minimizing wage competition and employee turnover, create strong incentives for employer-provided training. This, in turn, reduces the prevalence of practices such as poaching of skilled labour which, owing to its common occurrence within Anglo-American labour markets, serves to undermine the desire to collaborate with other firms and the accumulation of 'social capital'. Furthermore, because technical excellence (rather than low wages or poor working conditions) constitutes the basis of competition between firms in Germany (for reasons outlined earlier), there is a powerful incentive to co-operate vertically with specialized suppliers of key inputs, such as machinery and production systems.[18]

These insights provide both an optimistic and a pessimistic prospect for regional development policy. On the up side, they demystify the hitherto murky origins of successful economic systems, showing how they can in fact be produced by deliberate state action.[19] On the down side, those policy-makers who would wish to intervene *solely* at the regional scale (or, for that matter, at the level of the individual firm) in order to alter industrial practices will be discouraged to know that their initiatives will be somewhat futile in the absence of generally supportive (or at least, not actively antithetical) national regulatory features. Thus, it will be difficult to get Canadian firms to co-operate more with one another while they compete with one another over wages and poach each others' skilled workers. So too will it pose a challenge to get British or American firms to invest in the training necessary to implement advanced technologies effectively, when labour market institutions foster instability and systems of industrial finance reward the pursuit of short-term returns.

This insight has considerable relevance to contemporary political debates in the Anglo-American countries concerning the virtues of 'stakeholder capitalism' and the mechanisms one might adopt in order to promote a shift to such a system (*The Economist*, 1996c). To this point, the most common approach – employed for example by the Clinton administration in the USA, and by Tony Blair, the Prime Minister – has been simply to exhort firms (especially large ones) doing business within a particular national jurisdiction to change their own corporate *cultures* in order to place more value on long-term strategic considerations, longer-term, more stable employment relations, and the like. Clearly, as the preceding analysis has shown, exhortation to change cultures is inadequate as a strategy for effecting such a transformation.

This raises another issue of significance for theory and policy: namely, the relative importance of regional versus national institutions in the production of 'favourable' industrial practices. It is clear from the preceding analysis that the most telling and significant differences between the German machinery producers and their Ontario customers originate from national-level distinctions: in systems of labour market regulation, in training systems, in industrial relations, and even in the systems of industrial finance and capital markets. As such, the arguments in this chapter stand in marked contrast to much recent work in economic geography and related disciplines (such as that of Storper and Saxenian reviewed earlier) which has accorded causal significance to *regional* institutions. Indeed, so little of the difficulty arising in this bilateral relationship appears to be regional in origin that it is worth reflecting on this issue at greater length.

I would not wish to argue – even on the basis of the evidence provided in this chapter – that regional institutions are unimportant in the creation of conditions conducive to innovative and otherwise progressive industrial practices. They are crucial, for example, in helping us understand the very distinctive trajectories of different regions within the same nation-state (for example, Germany's Ruhrgebiet compared to Baden-Württemberg, or Silicon Valley versus Route 128 in the USA). It should also be recognized that collective bargaining and/or broader political action within particular regions are often responsible for essentially setting national standards (although, in cases like Baden-Württemberg, which has often acted in this fashion, the ability of regional bargains to influence national standards relies to a very large extent on the nationally defined industrial relations system). Thus, regional institutions *do* indeed matter. The point is simply that those analysts who emphasize the role of local or regional institutions to the exclusion of all others – or who fail to consider how these two levels of regulation interact to produce outcomes in particular places – do us a disservice that is as great as the denying of any role for regional institutions.

Nor do I wish to deny a role to be played by corporate practices and strategies (including the explicit and subconscious shaping of corporate cultures) in further differentiating the performance of one firm from another. To deny such a role would be to eliminate a crucial source of economic agency from our analysis. We already know much about the capacity of different firms in the same industry and country (even region) to respond in very different ways to a common set of competitive challenges and regulatory incentives. For example, in the case of the German machinery producers, the more progressive firms have successfully fashioned strategies to overcome the difficulties encountered by their customers in North America (Gertler, 1996b). But many others have failed to make the crucial adjustments to corporate organization, product design and market relations needed to meet this challenge.

Hence, it remains important for economic geographers, other social scientists and policy-makers to appreciate the importance of nation-state institutions in creating the enabling, accommodative space within which particular regional growth phenomena may arise. In this sense, then, we can understand the spatial construction of industrial practices as occurring through the interaction of local, national and subnational regulatory forces as well as corporate strategy. However, it is equally important to consider the provenance of the very institutions which we have implicated as having so much power to shape corporate and regional practices. Just as it is crucial to espouse a dynamic conception of culture, so too is it important not to treat institutions as if they were 'carved in stone' or inherited from on high. Indeed, it is likely that the relationship between institutions and practices is fundamentally dialectical in nature, with the latter possessing the potential to reshape the former over time.[20]

In concluding, it would seem that a major task ahead for economic geographers is to interrogate the dynamics of these various dialectics – between economic institutions and culture, between the institutions and forces at different spatial scales – if we are to succeed in understanding how economic relations in particular places are embedded within (and not separate from) a broader social and political matrix.

## ACKNOWLEDGEMENTS

The support of the Social Sciences and Humanities Research Council of Canada is very gratefully acknowledged. The author would also like to acknowledge the very helpful discussions with and comments from David Craig, Phil Crang, Roger Lee, Anders Malmberg, Peter Maskell, Kevin Morgan, Richard Munton, Jamie Peck, Sue Ruddick, Tod Rutherford, Andrew Sayer, Erica Schoenberger, Gavin Smith, Jane Wills, Peter Wood and Neil Wrigley.

## NOTES

1. Martin (1994a) offers an equally plausible alternative interpretation for the growing interest in local social context in regional economic geography, based on the rise of a post-modern perspective within human geography. In his view, the post-modernist critique has contributed to the growing awareness that 'economic events are necessarily contextual, that is embedded in spatial structures of social relations' (p. 42). Hence, it seems that one can arrive at this point by pursuing either the post-Fordist or the post-modernist line of argument.
2. Sayer and Walker (1992) concur: 'When insight into the specific business methods of foreign competitors is lacking, the fallback explanation is the "national culture" of the mysteriously successful strangers' (p. 118). Later, they add: '"culture" can easily become a "dustbin category" for anything we can't explain' (pp. 177–8).
3. For earlier findings of this work, including detailed descriptions of the structure and findings of the study, *see* Gertler (1993, 1995a, 1996, 1997b).
4. These arguments have become so commonplace within the literature that no detailed attribution of sources is given here. However, for the uninitiated, the chief elements of the arguments outlined below may be found in Piore and Sabel (1984), Scott (1988b), Sabel (1989), Storper and Scott (1988), Best (1990), and Sayer and Walker (1992).
5. This is no doubt the result of the structure of Saxenian's problem: to explain the divergent economic performance of two region systems located in the same nation-state.
6. Of course, Granovetter (1985), already cited, represents one of the most influential sources of inspiration for much of this work. His paper is cited by Harrison, Sayer and Walker, Storper, and Saxenian, as well as by many other social scientists working on local economic systems.
7. It is on the basis of this omission that Block launches into a devastating 'critique of economic discourse'. Indeed, the bulk of Block's book is dedicated to the task of restoring to their rightful place of prominence the influence of social context and process within economic systems.
8. They also echo the sentiments found in Block (1990b) when they observe the following in their penetrating analysis of Japanese capitalism: 'Characteristically, westerners bracket it as a special case, distorted by the peculiarities of Japanese culture. But this is thoroughly ethnocentric, for if Japanese capital is shaped by its cultural contexts, then so too is American, British, etc. Once we recognize this, we can ask whether certain characteristics we had previously assumed to be normal, or intrinsic to capital as such, are really effects of parochial, national, or regional contexts' (p. 164).
9. Morgan (1996) has recently made a similar point in assessing the well-known practices of close inter-firm cooperation and relational contracting within Japanese manufacturing. Rejecting 'culturalist' interpretations of these practices, he observes (drawing on the work of Nishiguchi, 1994) that 'trust and reciprocity (i.e. social capital), far from being pre-existing cultural assets, hardly existed prior to the war economy' (p. 4). Instead, they were 'actively constructed' by corporate strategic decisions and state intervention to promote small

firms, associations representing small suppliers to large firms, and the adoption of non-exploitive subcontracting practices.

10. Specific details of the survey and interview samples can be found in Gertler (1995a). Briefly, 170 plants, representing over 400 different cases of advanced technology implementation, were surveyed by mail in the summers of 1991 and 1992. These included Canadian- as well as foreign-owned plants, distributed equally over three different size categories, and drawn from four industries: transportation equipment, electrical/electronic products, fabricated metal products, and plastic and rubber products. Detailed follow-up interviews and site visits were then conducted with 30 user plants in the summer of 1992.
11. I conducted some 17 interviews in Germany during the summers of 1993, 1994 and 1995. Machinery producers interviewed were, without any prior selection on my part, found to be clustered in three geographical areas: in the Stuttgart region of Baden-Württemberg (particularly for producers of machinery and systems for the automotive industry) in the southern fringe of North Rhine-Westphalia, on the border with Hesse (especially in the area surrounding the city of Siegen, home to many producers of machinery for handling, finishing, cutting and shaping metals); and in Lower Saxony (near the cities of Bremen and Hanover, the site of firms producing specialized machinery for the rubber and plastic products industries).
12. A more complete discussion of these findings is provided in Gertler (1995b). In addition to the two specific dimensions examined in the current chapter along which German and Canadian industrial practices ('cultures') diverge, Gertler (1995b) also documents similar differences in contracting procedures, payback periods, operating manuals, training and other labour market practices, service expectations and conceptions of 'technology'.
13. In fact, the comments of the German machinery producers indicated a strong degree of similarity between Canadian, American and British users in this regard. As one very tangible manifestation of these differing time horizons and their impact on managerial decision-making, German machinery producers reported that they are routinely asked by their Anglo-American customers to demonstrate that the purchase cost of their machinery can be paid back within periods as short as two years. Meanwhile, German customers of the same machinery producers rarely make such requests, recognizing that they have considerably more time to recoup their initial investment.
14. As others have noted before, institutional arrangements which promote high wages and labour market rigidities also foster technological upgrading and skill development by producers, and make worker involvement logical, in order to achieve the productivity gains necessary to offset high wages. *See*, for example, Streeck (1985).
15. As one vivid example of the often direct relationship between high labour costs and machinery design, one German producer of large machine tools for the shaping of heavy metal tank parts noted that, for his domestic customers, it is extremely important that his machines have advanced capabilities for the handling of parts. As he notes, 'One reason for that is that handling time is very expensive. If you can shorten the handling times here in Germany, it might be possible for you to produce one [unit] more per hour. The difference between the German salary and the American salary is enormous, so in their [US customers'] point of view, the payback from reduction of handling times is not important.'
16. Thus, as Gordon (1989) has observed, it should follow that 'German machine tool design concepts . . . assume that the most sophisticated performance levels can only be achieved through a combination of advanced technology and human labor, not through automation alone' (p. 21).
17. By extension, the same kinds of difficulties ought to be expected in other Anglo-American industrial regimes, such as the UK. *See* Gertler (1993) for a review of studies documenting similar technology implementation problems among British users. For two rather different appraisals of the importance of social (and geographical) context in shaping the trajectories of implementation of capital goods and other complex commodities, *see* Appadurai (1986) and Thomas (1994).
18. For another example of how national regulatory frameworks can serve to shape the character of inter-firm relations, *see* recent work in the 'new retail geography', especially Wrigley (1992), Wrigley and Lowe (1996) and Hughes (1996).
19. It should be remembered, after all (Keck, 1993), that Germany was considered an industrial laggard through the last decade of the nineteenth century. Ironically, nowhere was this more true than in industrial machinery, where Germans were heavily dependent on technology imported from the UK and the USA (*see* Gertler, 1996).
20. Gertler (1996b) documents such a relationship within the German machine tool industry. Certain firms have pursued overseas strategies based on foreign direct investment and the full-scale production of new models for the North American market. In pursuing these practices, they have created a set of conditions which have the potential to undermine the institutions that shape industrial relations in the German economy. As simplified, US-built versions of their machinery come to be adopted in the plants of some of their customers back home in Germany, the power of skilled trades and unions is undermined. Furthermore, as new investments by the machine tool producers are made not in Germany but abroad, this exerts a 'disciplining' force on IG Metall, the dominant union in the industry, giving employers' associations more leverage in seeking to 'reform' industrial relations and labour market regulation.

# Chapter Five

# ECONOMIES OF POWER AND SPACE

JOHN ALLEN

## INTRODUCTION

This chapter is concerned with the different theorizations of power in play within economic geography and, more explicitly, with the difference that space makes to our understanding of power relations. Power has long been central to economic concerns within geography, although with the growing interest in the work of Foucault, Lefebvre and others it has increasingly come to the fore. More often than not, however, it is under-theorized, with little attention paid either to the different *modes* of power – domination is different from authority, which is different from coercion, which is different from seduction, and so forth – or the different *conceptions* of power – as a capacity or as a practice, for example. It is not uncommon in accounts, for instance, to find domination acting as a kind of shorthand for all kinds of power relations, or for those very same power relations to appear as resources one moment only to dissolve into practices or even discourses at the next. Equally important, for our purposes, geographers have rarely thought through sufficiently what it is about space, or more precisely about spatiality, which adds something to our understanding of how power 'works' – for example, in the ways that distance affects its reach or capability, or how the 'stretching' of power relationships over space often involves a combination of different modes of power rather than an endless play of domination.

In this chapter, an attempt is made partially to redress this state of affairs by exploring three broad theorizations of power currently in play within economic geography. In order to avoid an 'A to Z' of power, the theoretical readings of power examined possess either implicitly or explicitly a spatial vocabulary of power. The first is a loosely based realist account of power which has influenced, both directly and indirectly, attempts to explain changes in the spatial structuring of economic production and circulation. The second account takes its cue from collective or network notions of power which have recently been drawn upon to explain the growth of strategic alliances between firms, joint ventures, supply agreements and the like. In contrast, the third is an explicitly Foucauldian account of power which has been used as a means of regulation and control at the workplace and beyond. Following that, attention is drawn to the ways in which spatiality and power relations are considered within each approach and, significantly, how distance, movement and mobility actually alter our understanding of the ways in which different modes of power operate and the different sets of relations through which they are effective.

It should perhaps be stated at the outset that no one approach is deemed preferable, nor is an eclectic synthesis of resources, practices and capacities of power proposed at the expense of the structures of meaning inscribed in each approach. Indeed, the aims of the chapter are more modest and relate to two concerns in the way that power is conceived within much of economic geography. The first is a concern, in general, to go beyond 'domination/resistance' models which can only conceive of power as constraint or as always negative in its effects. Both associational and governmental forms of power take us in a more positive direction, although they have as yet had only limited influence in economic geography.[1] A second concern is to show that the extension of power in space–time not only calls forth different modes of power, but that such modes actually make a difference to the nature of economic relationships, whether between firms or between firms and their workforces or, more generally, in the economic culture of daily life. As Hannah Arendt (1970) has pointed out, the failure to distinguish between modes of power, to use them as synonyms for power, 'not only indicates a certain deafness to linguistic meanings, which would be serious enough, but it has also resulted in a kind of blindness to the realities they correspond to' (1970: 43). She might also have added that the 'stretching' of power relationships over space actually produces the need for the mediating relationships of authority, manipulation, inducement, seduction and the like, but that would be to rush ahead of ourselves.

## CONCEPTIONS OF POWER

As indicated, the three theoretical readings of power chosen are not intended to convey an exhaustive impression of the different conceptions of power drawn upon within economic geography. Rather, their selection, and indeed their merit, is based upon the fact that – with varying degrees of abstraction – they each include spatial characteristics within their frameworks of meaning. Space in one form or another is integral to their conception of power. At a glance this may seem an odd statement, for although space is almost always implicated in relationships of power, it does not follow that all social theorists acknowledge this, or indeed recognize space as significant to the maintenance, establishment or even the outcome of power struggles. Realism, in Roy Bhaskar's critical or transcendental form, is one such approach that has the ability to do so.[2]

### Power as a Capacity – a 'Centred' Conception

Strictly speaking, Bhaskar's particular variant of realism has had little direct impact on economic geography, and what influence there has been owes much to the 'translations' provided by Andrew Sayer (1982, 1992). In terms of power, however, a loose sense of realism has been evident across a range of enquiries for some time. It is perhaps appropriate at this point, therefore, to set out what this conception entails – in both its loose and its more restrictive, transcendental sense. Jeffrey Isaac's book *Power and Marxist theory: a realist view* (1987) offers a useful way into the issues and, in particular, to the twists and turns that Bhaskar's views add to a realist notion of power.

At the core of a realist notion of power is the assumption that power is an *inscribed capacity* of either individuals or institutions – inscribed in the sense that power is something that is possessed by virtue of the social relationships which constitute you or an institution. It is located: it is centred in institutions or other agencies because of the very structure of relations out of which they are produced. To speak of multinational firms in this context is thus to speak of the powers they possess by virtue of their capitalized multi-country operations and the workforces which comprise them, as well as the web of nation-state and market relationships which envelops them. The ability to relocate an operation, to switch investments from one place to another or to exercise corporate powers of acquisition and merger across borders is thus part of what it is to be a multinational firm. Or rather, they are specific capacities that they possess which, in turn, enable them to secure such outcomes.

That said, the view that power may be considered as the capacity of an individual or an institution to realize their will or to secure certain outcomes is a feature of a great many accounts of power (including that of Max Weber), and is not especially realist. In general, it refers to the dispositional nature of power; that is, to the likelihood or tendency of individuals or institutions to act in certain ways because of, say, their interests or objectives. In Bhaskar's realist terms, however, such dispositions are not varying in their account, but rather spring directly from the intrinsic nature of such bodies. Their capacity is an inscribed capacity which may or may not be realized depending upon the context in hand. Power, in this scenario, is thus always *potential*; it is not only evident or effective when it is exercised or displayed. Hence, for realists such as Isaac, the possession of power is not the same thing as the exercise of power. The two are conceptually and empirically distinct.[3] We do not have to calculate the number of winners and losers in the struggles over economic space in various countries to know that multinationals are potentially powerful: they are so by virtue of their capabilities and the resources available to them.

This is not to suggest, however, that 'potentiality' is all. On the contrary, as intimated above, it is only under particular circumstances and conditions that an inscribed capacity of an institution to *dominate* – that is, to control, to command, to direct the actions of others – will be realized. Relations of domination are thus considered to be open-ended, dependent upon the particular conditions in which control is exercised or directives issued. Contingency is 'built into' the process, and so too is space in so far as outcomes relate to the unevenness within and between economies. The ability to realize certain outcomes or to maintain power relations is regarded by those such as Isaac as inherently problematic, precisely because contexts and places are multiple and varied.

Isaac is also careful to stress that it is domination rather than simply power relations in general which frame his analysis. Following Max Weber's usage of the term 'domination', Isaac wishes to treat domination as a 'special case of power', in so far as the former draws attention to the *asymmetrical* nature of the relationship between two parties.[4] On this view, one side of the relationship has the 'power over' another by virtue of their structural capacity to command. This unequal distribution of power is itself mirrored in the inequality of outcomes, whereby one side invariably gains at the other's expense, especially if the relationship is one of interdependence. As such, power relationships are generally conceived as zero-sum games, with only a fixed amount of power or resources in play.[5]

In a strict sense, as noted earlier, this account of realism and power has had little direct influence on the work of economic geographers. Indirectly, however, notions of power as a capacity inscribed within social relationships which have the potential to alter the economic landscape in uneven and unequal ways are not that difficult to find. The work of David Harvey is a case in point.[6] His account of the production of the built environment in *The limits to capital* (1982), for example, or the commodification of urban space in nineteenth-century Paris in *Consciousness and the urban experience* (1985a) are replete with instances of the (cap)ability of capital to dominate space. In *Limits*, it is the mobilities inscribed within different kinds of capital which determine the production of unevenness, whereas in nineteenth-century Paris it is the properties of money capital which appear to shape much of the city's urban and social development. Similarly, in *The condition of postmodernity* (1989a), it is the capacity of capital to speed up and to accelerate economic processes which has produced the latest 'round' of the annihilation of space by time and the consequent domination of regional or place-based movements by the manoeuvrings of capital. Capital, as he puts it, 'continues to dominate, and it does so in part through superior command over space and time' (1989a: 238).

On this view, therefore, although it is significant whether or not such domination is realized, it appears to be more important to recognize the potential power that capital holds over people and places. As Derek Gregory (1994) has noted of Harvey's account of nineteenth-century Paris, it is his insistence on treating power as centred, marshalled and possessed which actually carries the narrative. Indeed, much the same could be said of the 'world cities' literature and its stress upon such locations as centres of control and command in the global economy, or about Manuel Castells' globalization of power flows (1989).[7]

What is virtually absent from such accounts, however, is a clear recognition of the conditionality of outcomes, of the way in which power works with and across differences between places. Contingency, if acknowledged in such accounts, rarely fulfils the explanatory expectations placed upon it by realists. In Sayer's work on the distanciated nature of the division of labour (Sayer and Walker, 1992), for example, or in Doreen Massey's work on the spatial structuring of production (1988, 1995a), it is possible to trace the more direct influences of realism. Massey's work, in particular, is an interesting illustration of how relationships of domination and subordination inscribed within firms can be seen to reproduce unequal geographies in a contingent, open-ended fashion.

In *Spatial divisions of labour* (Massey, 1995a), for instance, the contours of the UK space economy are drawn from the complex bundle of managerial and technical hierarchies which comprise firms within different industries. At the top end of such hierarchies are to be found the powers of strategic investment – the (cap)ability to open or to close down operations – as well as the ultimate powers of control over production and the labour process. Further down such hierarchies, different degrees of administrative and technical control are in evidence, with those with limited autonomy in the production process subordinate to the rest. In short, power relations are mapped across the hierarchies, with specific organizational and structural capacities inscribed at different levels.

Central to this argument, however, is that such relations also construct economic space in an asymmetrical fashion. This, in turn, can be said to produce something akin to a zero-sum geography, whereby certain areas of the UK benefit at the expense of others depending upon their prior economic, political and cultural histories. To put it another way, those regions which attract HQs and R&D establishments at the expense of other regions effectively construct the latter as less developed. As a result, not all places can be 'winners'; some must 'lose' in the struggle for the types of investment which raise both the economic profile of a region and the number of higher-level functions (powers) possessed. What influences the outcome, however, is conditional upon the particular industry in question, its locational requirements and the specific histories of places.

In this type of realist account, therefore, space is conceived as integral to the way that firms operate; the hierarchies themselves work through the differences between places. It is, none the less, a rather limited spatial vocabulary, especially in relation to power. In such accounts, for instance, it is rarely considered whether distance may actually undermine the effectiveness of power relations or whether different modes of power come into play as social relationships are 'stretched' across space. In the case of Harvey, how far does unmediated domination reach? And what role do relations of authority or coercion play in Massey's spatial hierarchies? Before we can pursue these and other questions, however, there are two further conceptions of power to consider.

## Power as a Medium – a 'Networked' Conception

The idea that power may be conceived as some kind of medium for securing certain ends represents a rather different way of thinking about power from that of realism. On this view, power is not so much held over others, as it is a resource for achieving diverse ends. This is a conception which stresses the 'power to' rather than the 'power over' dimension of things; that is to say, the emphasis is placed upon how power is generated to achieve certain outcomes – be they concerned with entering hitherto closed markets, exploiting the latest technologies or whatever – rather than upon how power constrains social or individual action.

Within economic geography, variations on this conception of power are broadly to be found in debates over the nature of markets, institutions and, in particular, organizations. As before, however, there is no pristine version of this conception of power which has been bequeathed, as it were, to economic geography. Its traces are often indirect rather than direct – the result more often than not of successive 'translations' from sources as far apart as Talcott Parsons' functionalist accounts and Peter Blau's exchange theory.[8] A more recent translation of this approach, albeit eclectic in its conceptual debts, is to be found in Michael Mann's two-volume *The sources of social power* (1986 and 1993 respectively).[9] It is useful here primarily because it adopts an explicit spatial vocabulary.

The claim that power is *produced* or *generated* by groups or institutions distinguishes this conception of power from conceptions of it as an inscribed capacity. Power is thus not something which inheres in certain social relationships; on the contrary, it is produced by a process of mobilization whereby firms, say, reflect upon their own resources to achieve certain goals and, realizing their limitations, attempt to pool their resources with like-minded organizations as a means of securing what is now a common goal. Joint ventures, for example, where a multinational teams up with a local firm to gain access to a new market, may be successful at exploiting opportunities at the expense of other, less collective-minded, rivals. In the process, a new collective power base is realized which, although of uncertain status, amounts to an enhancement of power to those in alliance.

From this example too can be seen the possibility of power as more than a fixed sum; that is, as more than a game in which there are only winners and losers. Power, in this context, is a fluid *medium*; it can expand in line with collective ventures or it can diminish once collective short-term goals have been achieved or, as often happens, alliances fall apart. The scenario in this sense is less deterministic than zero-sum conceptions of power. This is not to suggest that power relationships never involve mutually exclusive objectives or never lead to an inequality of outcomes; merely that the amounts of power in circulation are variable and context-bound.

The plausibility of such a view largely turns on the assumption that power is actually a medium, or rather, following Giddens, 'resources are the media through which power is exercised' (1979: 91). The ability to mobilize *resources* and to use them to secure specific outcomes lies at the heart of this conception. Resources, which for realists would be generally considered as extrinsic, such as access to finance and technology, or control over information and knowledge, in this account provide the means, quite simply, for institutions of whatever kind to get things done. To adapt a line of thought from Hannah Arendt (1970), resources, when pooled collectively, can actively empower groups and organizations, and do so only as long as such resources are used in concert.

In Mann's (1986, 1993) writings, this focus upon the associational aspects of power takes an interesting spatial turn. While Mann's concern is an immodest attempt to track the history of organizational power from the time of the first settled societies onwards, what is of value to us here is how his attempt to think through the scope and intensity of different forms of power leads him to adopt a conception of power based upon *networks* which cross-cut and overlap one another. Arguing that it is through the networks themselves that power is generated by the mobilization and deployment of resources, he wishes, as one of his pressing concerns, to demonstrate how such resources are organized and controlled over space. I will not go into detail here, but Mann describes how power can be centred or decentred in such networks and that willed commands (domination) have only a limited spatial reach. In such complex circumstances, effective organizations may generate other sources of power, more collective, diffuse and less coercive in their impact, to achieve far-flung goals. The more effective the organization, the greater the number of ways in which resources are controlled through the networks.

This is an excessively truncated reading of Mann's work, but it does convey something of the complex networks of power relationships which can bind together and influence the fortunes of economic organizations and institutions. As noted above, neither Mann's networked version of power as a medium to secure outcomes, nor a less elaborate version of it, can be said to have directly influenced thinking in economic geography. The traces of this conception of power, however, are certainly evident in the discipline.

Perhaps one of the best examples is the ongoing work of Peter Dicken and Nigel Thrift (separately, as co-authors, and with other authors). In an early series of publications, Thrift (Taylor and Thrift, 1982b, 1982c) was concerned to show how power in relation to industrial organizations may be gauged by their use and control of resources, rather than by any intrinsic capacity. Access to resources was seen to hold the key in what was, for them, a segmented economy in which organizations were positioned in local and global networks of power. Although characterized by asymmetric power relations between firms, the meaning of domination as simply 'power over' was called into question in the knowledge that firms engage in a wider variety of interrelationships than simply zero-sum affairs. Similarly, Dicken in *Global shift: the internationalization of economic activity* (1992) and elsewhere has pointed to the collective aspects of power that can arise from networks of externalized relationships when firms engage in strategic alliances, research consortia, joint ventures or subcontract relationships. In such cases the amounts of power generated through collaboration in, say, the exploitation of complementary technologies remain only so long as the organizations act in concert. The power of all those involved is thus momentarily enhanced in relation to their competitors.

Dicken and Thrift (1992) have elaborated and extended their analysis of organizational networks to incorporate a more decentred view of corporate power, which allows for the possibility of networks themselves generating sites of power. On this account, relationships of power may be characterized by varying degrees of symmetry rather than by the mutual exclusion of interests. Above all, there appears to be a dynamism built into the networks which reflects the role of power as a fluid medium.[10]

In Amin and Thrift (1992), this dynamism is also apparent in their account of 'networked' forms of corporate organization, although in this instance the stress is placed upon the ability of the larger multinational firms to achieve their collective goals at the expense of smaller or local players in the network. Agglomerations of power and authority, referred to as neo-Marshallian nodes in global networks, are identified in which interaction between firms, institutions and social groups mobilizes resources on a collective basis to maintain competitive advantage within a particular industry. Such neo-Marshallian nodes are seen to act as a pool of resources which firms of all shapes and sizes can dip into as and when it benefits them. Questions of access and the control and co-ordination of resources over space are thus prominent concerns, as is the maintenance of flows through the networks.

Recognizing that spatiality enters into the way that power is produced, however, tells us little about the modes required to sustain relationships across the length and diversity of the networks. The intermediary arrangements of power, the processes of mobilization entailed, are likewise assumed in Mann's diffuse model of power. If the proxemics of power add a new twist to the dynamism of organizational networks, how does this affect more distanciated contacts? These and other aspects of this conception of power will be returned to after an outline of the third, Foucauldian-inspired account.

## Power as a Technology – a 'Diagrammatic' Conception

Power for Michel Foucault is neither a means to an end, nor an inscribed capacity, but essentially a series of strategies, techniques and practices. Perhaps the key point which flows from this conception of power is that it is not something which is held or possessed, but rather something which is exercised. Thus the question of who has power, either as a property or as a resource to be mobilized, is misleading in so far as power is something which passes through the hands of the powerful no less than through the hands of the powerless. Power is exercised through groups and institutions, but it is not centred in them. This, then, is a conception which holds that power works on subjects, not over them or simply as a resource to meet their various needs.

In that sense, power is a more elusive phenomenon than, say, realists would have us believe.

Within economic geography, it is Foucault's initial work on power, the techniques of domination and surveillance in *Discipline and punish* (1977), which has tended to find a voice, rather than the more positive, self-regulating techniques explored in volume 3 of *The history of sexuality* (1986) and in his later work on governmentality (*see* Foucault 1988, 1991). Both formulations are in evidence, however, although rarely in a direct sense.

As mentioned above, it is the exercise of power, its *operation* or *practice*, which holds Foucault's attention. The intentionality of actors or the interests that they possess are of little concern to him. Rather, it is the techniques by which the conduct of specific groups is moulded and their range of actions limited which lie at the heart of his analysis of power. Such techniques, although mobile and diffuse, are seen to limit the scope of action through strategies of incitement, seduction or simply constraint (Foucault, 1982). So, for example, power and control at the workplace does not so much emanate from a central point of administration, as circulate through the organizational practices of a firm. The imposition of a form of conduct is secured through an almost endless play of micropractices which involve the detailed organization of work space, the laying out and serializing of tasks in time, and the composition of the labour process in ways which promote self-regulation and control. Not surprisingly therefore, given the emphasis on the fragmentation of space/time, the Tayloristic overtones of the disciplinary process at the workplace have been duly noted (*see* Dreyfus and Rabinow, 1982).

The clearest illustration of the kind of disciplinary technologies of control which Foucault has in mind is that of Bentham's Panopticon. The characteristics of panoptic technology are well known – the rings of backlit cells encircling a central observation tower which affords the maximum of visibility and surveillance for the minimum of effort and direct control – and require little elaboration. More importantly, for Foucault, the architectural and organizational merits of the Panopticon rest with their generalizability as a schema of power for the control of social life in all its shapes and sizes. Panoptic technology is represented as an ideal form for the pervasive control of everyday life, a *diagram* of power which sets out the particular points through which it passes and which enables everyday life to be regulated down to its smallest detail.

Thinking of power in this way, as diagrammatic in form, also enables Foucault to develop a particular spatial vocabulary of power: the continuous circulation of power through the relations of everyday life, its embodiment in particular sites and local institutions, and the particular domains or fields of its operation.[11] On this view, power is anchored in *institutional space*, with different types of disciplinary institutions, the workplace among them, governed by an assembly of practices which map on to one another but also display their own specific diagrammatic quality. Panoptic technologies are thus only one way of exercising power through space; one way of locating and placing people and, more broadly, attempting to constrain their movement and mobility.

Concomitant with this conception of power as mobile and diffuse is the claim that disciplinary power, if it is to be effective, is a constitutive rather than a repressive force. In contrast to asymmetrical views of power which tend to regard the act of power as wholly repressive and resistance as something which is rallied against it from the outside, Foucault wishes to stress the *productive* side to power which, in Ian Hacking's (1986) telling phrase, 'makes up' people. Put another way, people are subjectively constituted in such thorough ways by the routinized and ritualized practices etched in disciplinary modes of power that they know no other. In Lois McNay's (1994) terms, Foucault posits a set of normalizing forces at work within society which operates at the level of mundane, everyday experience – the micro-physics of power – which enable individuals to fashion themselves through techniques of self-regulation. In short, people constitute themselves, but not necessarily in all possible ways of choosing.

McNay is also careful to point out, however, that there is a tension in Foucault's conception of power between its positive and negative dimensions; that is, between the indirect techniques of self-regulation and the more repressive, disciplinary side. On this interpretation, despite the emphasis placed by Foucault on power as a productive, heterogeneous force, there is a tendency in his work 'to fall back into a negative view of power as a unidirectionally imposed monolithic force' (McNay, 1994: 3). Others such as Burchell *et al.*, (1991) and Hindess (1996) would probably take issue with this view, preferring instead to emphasize the room for manoeuvre built into Foucault's notion of governmentality, which enables individuals to exercise a degree of choice over how they act and make use of their skills and attributes.

Of late, Foucault's ideas on power have been the subject of much discussion and application within geography (Driver, 1985, 1993, 1994; Murray, 1995; Matless, 1992; Philo, 1989, 1992), although this trend has been slow to develop in economic geography. What little application there has been has focused on the disciplinary side of power and is perhaps best exemplified through the work of Philip

Crang (1994a), who has explored issues of surveillance and self-regulation in the context of workplace geographies.

Drawing upon personal experience as a waiter in a corporate chain restaurant, Crang's account of managerial surveillance and control demonstrates the significance of visibility, or 'the gaze', as a means of organizing the workforce on a discontinuous and intermittent basis. Two forms of visibility were shown to be in operation. First, the synoptic visibility of the waiting staff was achieved through the organization of working space: the spatial arrangement of tables, the subdivision of the restaurant floor and the location of vantage points from which managers could 'gaze' on the setting. As a diagram of power, its various points obviously owed much to the specific nature of the service institution. Second, individualizing visibility – that is, the detailed observation of employees – was primarily achieved through the 'order of service', a series of operations laid down by management to standardize the movement of staff and programme the timing of service from the first introduction to the customer to the moment of their departure. In both kinds of gaze, therefore, an array of disciplinary micro-practices generated information which the management could then use to induce or incite particular forms of behaviour (in this instance, a waitress's or waiter's performance).

More importantly, the one-sided nature of the gaze, its asymmetrical character whereby management view the waiting staff but not vice versa, produces its own power effect. As with panoptic technology, the one-way visibility relation offers no clue to the waiting staff as to when or whether they are under observation, with the result that they may internalize the gaze and, to all intents and purposes, discipline themselves. How far this is the case, as Crang recognizes, is dependent upon how overbearing we take the force of domination to be. Certainly in Crang's account, the gaps in power, the possibilities to circumvent the gaze, are more evident than in Foucault's *Discipline and punish.*

Not all technologies of power, it should be noted, were considered by Foucault to rest on the principle of visibility.[12] What is evident in his writings, however, is the significance that he attaches to the organization and use of space as a means of exercising power. It is in this sense that economic institutions, from the monolithic multinational to the local firm, are sites of power circumscribed by various technologies and practices of regulation. What is less obvious in this account is why spatial strategies of surveillance and regulation should effectively deliver the kind of disciplinary control that Foucault has in mind. As Michel de Certeau has commented, beneath 'the "monotheistic" privilege that panoptic apparatuses have won for themselves, a "polytheism" of scattered practices survives, dominated but not erased by the triumphal success of one of their number' (1984: 48). Equally, the strategies used to achieve domination are not necessarily the modes of power required to secure it. Surveillance is rarely wholly effective and domination never total, especially when issues of distance or co-ordination and control across sites enter the frame.

It is to these and other aspects of the spatiality of power that we now turn.

## THE SPATIALITY OF POWER

Each of the three conceptions of power possesses its own spatial vocabulary of power, and in that sense we have already begun to unpick the relationship between spatiality and power. In a realist account, broadly speaking, power works with and across the differences between places; in a networked conception, the control and co-ordination of resources over space is of paramount importance; and in a Foucauldian approach, power centrally involves the use and organization of space. Thus all three incorporate a spatial dimension in their understanding of how power operates. In this section, the intention is to push the analysis of spatiality and power a step further, first, by considering how *distance* problematizes the establishment and maintenance of power relations, and second, by briefly addressing the common view that *movement and mobility* may overcome whatever problems distance poses to the effective flow of power.[13] Central to the line of thought developed here is that the extension of power in space–time actively produces the need for mediated relationships which involve different modes of power in overlapping and coexisting spatial arrangements. Domination, to make the point, is a one-dimensional view of power that only exists on the head of a pin. It is always mediated in space–time.

### Distance Matters and Matters of Distance

As a general observation, the power of an organization or perhaps even a bloc of capital may be enlarged by its extension over space. The more extensive its reach, the greater its geographical spread, the more comprehensive may its power appear before us. As with the almost seamless circulation of money, information or knowledge, the flow of power is likewise frequently portrayed as a straightforward, unremarkable process. However, while distance may enhance power, it also problematizes its effectiveness.

Unmediated domination of the type that Harvey appears to subscribe to, for instance, is unlikely to give capital of whatever inscription the command over space that he would have us believe, no matter how sophisticated its manoeuvrings. In the first place, as Mann (1986) fully recognized, power as a centralized, marshalled force has a limited spatial reach. Domination, in the sense of command over space by virtue of the capitalist relations of production and circulation, is therefore not a particularly effective mode of power when stretched across space. More significantly, there is nothing in the social relations of capital which, in itself, automatically secures certain sets of power relations. Such relations presuppose the capacity of power over space, but little else.

In Massey's (1995a) account of the spatial structuring of capitalist relations, a greater sensitivity is displayed to the degree of control appropriate to the HQs of firms and their dispersed branch plants. From the full control over production, investment and the labour process at the HQ level down to the partial control over labour and production at the branch plant, the spatial division of powers appears to mirror that of the spatial division of labour. Less apparent in this spatial hierarchy of control and administration, however, is how the delegation of control is transmitted down the levels without loss of meaning, ambiguity or reinterpretation. The decentring of power, or its delocalization as Barry Barnes (1988) refers to it, is a double-sided affair whereby the power of the centre is enhanced by delegation and so too is the power of those in receipt of power at the dispersed levels of the hierarchy. Discretion and autonomy are thus 'built into' the supervisory process in ways which may frustrate or even distort the aims of a central HQ.

From this consideration, it is possible to identify what is missing in Massey's account of economic space and power; namely, a more nuanced conception of power which recognizes that the extension of power over space–time involves relationships other than domination–subordination. Effective power at branch plant level, for instance, involves the co-ordination of clusters of workers across and within sites, as well as the motivation and control of the workforce. Domination, in the Weberian sense that there is every probability that a given command will be obeyed, is unlikely to secure such co-ordination and control. Other modes of power, such as *authority*, which is conceded by others on the basis of a legitimate claim to manage and co-ordinate, are likely to be more effective. In such contexts, authority relations may secure a willingness to comply with, say, a particular set of production changes, whereas direct domination from the centre would not.

Such a view is consistent with Burawoy's (1979) stress on social relations *in* production as an effective means of producing active co-operation and consent at the workplace. In his study of factory life, the autonomy ceded to workers by plant management, the relaxed enforcement of certain managerial goals and the seduction of all concerned into the game of 'making out' generated an acquiescence to the broader relations of economic control. Systems of motivation, bound in some cases by co-operation and respect, help, in the words of Sayer and Walker (1992), to knit together divisions of labour.

This is not to suggest, however, that below HQ level, relations of authority necessarily come into play. On the contrary, the co-ordination and integration of the division of labour over space is more likely to involve *overlapping modes of power* rather than simply various states of domination. It is conceivable, for example, that *coercion* rather than authority may be exercised by management to maintain control at branch plant level. The ability to secure acquiescence by the threat of redundancy is, after all, a basic feature of the employment relationship. Its effectiveness as a practice, however, is constrained by its acute visibility when used and the intense hostility it tends to generate in direct response. As such, coercion is often effective only in the short term, as one mode of power exercised in combination with other modes by management at different levels.

Interestingly, a similar line of critique can be directed towards a Foucauldian approach to power. As indicated earlier, there is a debatable tendency in Foucault's writings – highlighted by McNay – to fall back on a negative view of power. Fraser (1981) has taken this insight further by arguing that Foucault, in his earlier works at least, calls all manner of different kinds of social practices 'power' without discriminating between domination and other modes of power such as authority, force and violence. 'Phenomena which are capable of being distinguished via such concepts are simply lumped together under his catch-all concept of power' (1981: 286).

While this overstates the point, especially when Foucault's later work on governmentality is taken into consideration, there is none the less little offered in his analysis of how, for instance, relations of manipulation or those of seduction are involved in techniques of self-regulation. The *manipulation* of workers, for example, which may entail management concealing their actual intentions from the workforce or selectively restricting the information that they receive about a new set of work practices, is altogether different from the use of seduction as a strategy. As a mode of power, *seduction* is a renunciation of total domination in that it always leaves open the possibility that a subject will choose not to be seduced. In contrast to the obligation to comply engendered by authority relationships, seduction is broadly indeterminate as a mode of power – although it is no less effective for that in certain contexts, especially in relation to 'techniques of the self'.

The significance of these points is perhaps best understood when the question is addressed of how the different 'diagrams' of power (of which panopticism is only one) are generalized at the wider society level. The co-ordination of particular disciplinary techniques within and across sites – for example, across a chain of service establishments – and the ability to secure the kinds of conduct required at a distance is a more complex task than that delivered by surveillance and self-regulation. Technologies of power have their limits, in terms of both their social and their spatial reach. Again, a spatial overlap in modes of power is more than likely to be in operation in such institutional contexts.

The complexity of co-ordinating and controlling economic activities and resources across space is perhaps most clearly revealed in the case of networked firms. Ranging from joint ventures and strategic alliances, on the one hand, to controlled franchising and subcontract relationships on the other, the issue here is not so much about how power is maintained over others, as it is about how alliances based on collective power are sustained across space. The distanciated nature of many alliances, themselves often the result of firms' recognizing their geographical limitations, raises the potential for surplus or ambiguous meaning to be disruptive in such networks. The instability of such ventures, in particular the frequency with which alliances between global firms have broken down, points perhaps to the difficulty that distance poses for the constitution of these kinds of networks.

Rather than dwell upon the global character of such alliances, however, it may be more fruitful, following Bruno Latour (1987, 1991, 1993), to look at the *intermediate* arrangements responsible for the construction of networked associations. From this vantage point, it is possible to unravel how agents 'translate' phenomena into resources and resources into networks of power by the active mobilization and enrolment of others around 'mutual' objectives (Callon, 1986, 1991; Callon and Latour, 1981). On this view, power works through a process of translation which, broadly speaking, 'fixes' a collective orientation through a complex process of constructed meanings which, to all concerned, appear to be indispensable and irreversible. A particular configuration of power is stabilized by the enrolment of 'other wills by translating what they want and by reifying this translation in such a way that none of them can desire anything else any longer' (Callon and Latour, 1981: 296).

When fluid resources are positioned in networks in this way, however, it is not entirely evident which modes of power are responsible for stabilizing the lateral powers of association. While not explicitly concerned to distinguish the different modes of power in operation, Amin and Thrift (1995) draw upon the work of Callon, Latour and others to show how networks can be built, maintained and extended through the strength of associations between actors. Negotiation, collaboration, reciprocity and other forms of lateral association are regarded as central to the ability of actors within networks to act 'at a distance'. Notwithstanding this recognition, Amin and Thrift wish to argue that such forms of networked association should none the less be seen within the broader context of strong and weak networks, where networked forms of corporate governance continue to be largely accountable in terms of domination by the already powerful, global firms over the smaller or local players in the network. From their standpoint, it would appear there has been no genuine spread of authority downwards, and co-operation within the networks continues to be maintained on an unequal basis (*see also* Amin and Dietrich, 1991; Amin and Malmberg, 1994; Amin and Thrift, 1992). Such a view, however, is arguably reliant upon a vertical language of power which acknowledges power as something held over others, which excludes rather than includes and comes close to a zero-sum model of power as a fixed rather than a fluid resource. Granted, there will be instances of domination between strong and weak industry networks, as well as manipulation through the reification of wants by the more powerful actors within networks, but there will also be relationships of co-operation between partners which effectively stabilize resources to the benefit of all involved. Moreover, these relationships may not be characterized by equality in terms of the resources at their disposal, yet small firms, say in subcontract relationships, may still resist the demands of the larger players simply because they hold something of value which is to the latter's advantage. Fluid power relationships are in play in such a scenario, but they are not reducible to those of domination or even those of authority.

In fact, small or local companies do not necessarily seek delegated powers of authority from the larger, global players; on the contrary, they seek to enhance their power through association with them. Such relationships may be unequal, but neither party is powerless. In thinking through the nature of these networks it is perhaps useful to acknowledge the *lateral* as well as the *vertical* modes of power in operation which can fix a collective orientation. Lateral modes, such as *negotiation*, *persuasion* and *inducement*, although less familiar as a vocabulary of power, are none the less a significant part of what holds networks together over space, as indeed Amin and Thrift have done much to show. Moreover, the distanciated character of economic networks, the fact that what happens in one part of them is affected by actors absent as well as those present, is likely to involve coexisting modes of power at different points in a network. Distance matters in this instance more for its potential to introduce instability into networks than for its ability to problematize the extension of power.

## Mobile Powers and Powers of Mobility

If it is valid to argue that distance may problematize and destabilize power relations, it is equally valid to point out that distances are collapsing around us as the result of rapid transformations in transport and communication technologies. At the risk of caricature, the speed-up of communications, information, people and whatever else is required to mediate power relations across space is now a resource available to the command centres of all kinds of transnational institutions. What were once conceived as the distant powers of a multinational HQ, for example, are now, it would seem, capable of almost simultaneous effect at even greater distances. It is not necessary to subscribe to this view of telematics and buzzing electronic spaces (*see* Graham and Marvin, 1996), however, to broadly accept the proposition that power may be enhanced by the immediacy brought about by speed and mobility. In a way similar to Giddens' (1984) direct linkage of power to the process of time–space distanciation, it is conceivable that as the flow and circulation of power accelerates, so the likelihood of ambiguity or surplus meaning in distanciated contacts is reduced. As with distance, however, the impact of speed and mobility is not so transparent.

Harvey (1982, 1985a, 1989a), for instance, has frequently drawn attention to the social power that emanates from the interlocking relationships of money, space and time. In *The condition of postmodernity* (1989a), as mentioned earlier, the rapidity and scale of capital flows lies behind the latest 'round' of the annihilation of space by time. It is also said to lie behind a speed-up in the commodification of culture and the associated turnover and disposability of images, information, ideas and other cultural products. Consumers in the spaces of the advanced economies are supposedly swept away by the power of this economic logic, as permanence gives way to ephemera and the self is manipulated to respond to pre-orchestrated needs.

Leaving to one side the likelihood of such a response, the assumption of equivalence of speed and rapidity with power over a distant audience is itself flawed. The pace at which information is shared is dependent not upon the speed of transmission, but upon how it is *decoded* and *interpreted*. Rapid communication, say in the form of the power of advertising images, is a hit-or-miss affair which allows for the possibility of refusal or indifference and also of misunderstanding or distortion. As a practice of *seduction*, advertising is not a particularly uniform mode of power at a distance, no matter how rapid the circulation of signs and symbols (*see* Lipovetsky, 1994).

A similar set of points could be directed to those who anticipate that the new technologies of speed and surveillance will deliver an effective form of domination over space. At one extreme, there is the work of James Der Derian (1989, 1990) on the (s)paces of international relations.[14] In an attempt to extend Foucault's vision of panoptic technology as a generalized schema of power, he argues that modern panopticism is already in place – with political and military domination established through a range of intelligence technologies such as radar, telemetry and photo-intelligence. 'That power is here and now, in the shadows and in the "deep black". It has no trouble seeing us, but we have had great difficulties seeing it. It is the normalizing, disciplinary, technostrategic power of surveillance' (Der Derian, 1990: 304). There are echoes here of Mike Davis' vision of downtown Los Angeles as a virtual 'scanscape' where electronic surveillance techniques secure the core districts and its smart buildings for its middle-class workforce against the wanderings of a dispossessed urban strata (Davis, 1992a).

Coming down to earth, we could say that, like Foucault's early analysis of power, Der Derian's account overestimates the effectiveness of disciplinary technologies and, in particular, conflates their installation with their impact. The pervasive potential of such technologies may be high, but their intensity is likely to be low in relation to the control achieved. Indeed, much the same could be said of technologies of surveillance at the workplace which, at best, appear to represent a rather superficial form of visibility (*see* Sakolosky, 1992). If de Certeau (1984) is correct, the 'strategic gaze' of management at the workplace is more likely to reveal its formal, quantitative side (the production rate for individual workers, for example, or the number of workers on site), rather than the fragmented and mobile practices which comprise daily working life (*see* du Gay, 1996; Jermier *et al.*, 1994). Power, in this context, is bound by its very visibility, regardless of the speed at which images and information may be transmitted to various points of regional control within an organization.

Clearly surveillance, along with other techniques of control which rely upon the transmission of information over space, may enhance power in general (Lyon, 1994). The crux of the matter, however, is not that of speed, which, after all, may involve slow as well as fast movements, but the *production and circulation of meaning* through any number of symbolic mediums. One of the best illustrations of this point in economic geography is the Latour-inspired work of Nigel Thrift (1994b, 1996a) on the City of London, which shows how the City acts as an interpretative and communicative 'centre of calculation', condensing the flows of information in such a way as to enable those at the 'centre' to control a realm of processes distant from it. Included here, of course, is the meaning attributed to mobility itself and how different groups and institutions endow speed with certain powers (*see* Thrift, 1996c). Control over mobility, as Doreen Massey (1993a) has argued, is an issue of power geometry, with some groups in control of the process of movement, others on the receiving end of it, and others simply trapped by it. Part of this control, therefore, is the ability to persuade an audience that speed today has a particular meaning; namely, that the far-off workplace or market-place is no longer inaccessible, nor what happens there difficult to read. This is the power to *define* distance and proximity which organizes both space and language in hegemonic ways.[15]

## CONCLUSION

In a chapter of this length and a subject matter as broad as power, it is possible only to signal some of the ways in which the stretching of power relationships across space produces the need for mediated relationships which involve different modes of power in overlapping and coexisting spatial arrangements. An understanding of the realities that such modes correspond to, as Arendt prompts, alerts us, for instance, to the limited spatial reach of domination; the forms of concealment entailed by manipulation; the acute visibility and episodic nature of coercion; the spaces of broad indeterminacy opened up by seduction; and the simple fact that authority is conceded, not exercised like power. Knowing the meaning and use of these modalities not only provides a more critically informed sense of power, it also suggests the ways in which they are inextricably wrapped up with their extension in space–time.

This, however, is only the starting-point for an examination of modes of power which, either separately or in combination, go beyond simple constraint and domination models in economic geography. In so far as we have begun to develop a more nuanced spatial vocabulary of power, so too do we need to develop a more nuanced language of power within geography that, among other things, entertains a productive view of power whereby resources are mobilized to enhance rather than diminish the potential of groups and institutions.

## ACKNOWLEDGEMENTS

Jo Foord, Doreen Massey and Andrew Sayer, as well as the editors, Roger Lee and Jane Wills, commented on an earlier draft of his chapter. Needless to say, I have not been able to satisfy all their concerns, although the chapter has benefited greatly from their provocative questions and insights.

## NOTES

1. Associational forms of power stress the significance of collective agency and are evident in the work of Amin and Thrift (1995) and also in their chapter in this collection. Their treatment of assocationalism is discussed later in this chapter. An insightful account of the productive side of power as a positive, collective force which only exists in its actualization is to be found in the work of Arendt (1958, 1970). Foucault's notion of governmentality has to my knowledge had no direct influence on economic geographers to date. For an assessment of the latter, *see* Burchell *et al.* (1991), Hindess (1996), Miller and Rose (1990), Rose and Miller (1992).

2. I refer here to his earlier statements on realism outlined in Bhaskar (1975, 1979). For those wishing to follow up the trajectory of his thought, *see* Bhaskar (1993, 1994).
3. For a dissenting view, *see* Hindess (1996), who regards any conception of 'power as capacity' as tantamount to a pre-ordained script in which 'those with more power than others will invariably prevail over those who have less' (1996: 138).
4. According to Weber (1968), 'The concept of power is sociologically amorphous. All conceivable qualities of a person and all conceivable combinations of circumstances may put him [*sic*] in a position to impose his will in a given situation. The sociological concept of domination must hence be more precise and can only mean the probability that a *command* will be obeyed' (1968: 53; emphasis in original).
5. It should be noted, however, that within realism there is no necessary connection between asymmetrical views of power and zero-sum games which presume a closed system and a single scale of measurement.
6. Harvey would resist the description of his work as realist in approach and indeed has been openly critical of realism (*see* Harvey, 1987). Even those sympathetic to Harvey, however, have recognized the resemblances between Bhaskar's critical realism and Harvey's interpretation and use of Marx (*see* Castree, 1996).
7. Such a centred view of power is evident in the work of Friedmann (1986) and Friedmann and Wolff (1982). Sassen's (1991) account of global cities demonstrates an awareness of the limitations of this view and stresses instead the production of post-industrial inputs as the basis of a city's global capability for control. There is, however, little sense in which this capability is mediated across space in her work.
8. *See* respectively Parsons (1960, 1963) and Blau (1964).
9. A third and a fourth volume are planned by Mann which, in turn, bring the analysis into the twentieth century and provide a broader set of theoretical conclusions. In volumes 1 and 2, both a dispositional and a collective conception of power are interwoven, although he follows Giddens (1979) in recognizing resources as a medium of power.
10. This example is not intended to give the impression that networks are the only organizational form through which resources act as a medium of power. Much of Ohmae's work (1990, 1995a), for example, is concerned to show that, in an interdependent global economy, there are no absolute winners and losers, and that co-operation between companies and governments in the establishment of 'growth triangles' across national borders may enhance the pay-offs to all concerned.
11. The diagrammatic aspect of Foucault's account of power and the significance of space is brought out clearly by Deleuze's (1988) account of Foucault's ideas on power. It also has to be said, however, that there is as much, if not more, of Deleuze's ideas in this assessment than there is of Foucault's actual thinking.
12. In an interview entitled 'The eye of power', Foucault (1980) acknowledges that the procedures of power 'at work in modern societies are much more numerous, diverse and rich. It would be wrong to say that the principle of visibility governs all technologies of power used since the nineteenth century' (p. 194).
13. This is not to imply that distance, movement and mobility exhaust the language of spatiality. They have been chosen here because of their relevance to the extension of power in space–time. Equally, if length had permitted, borders and boundaries would have been considered too, along with other aspects of spatiality.
14. Der Derian has been chosen here because of his influence in the field of critical geopolitics in both the USA and the UK.
15. The obvious reference point here is Said's (1985b) work on *orientalism*.

# CHAPTER SIX

# NATURE AS ARTIFICE, NATURE AS ARTEFACT: DEVELOPMENT, ENVIRONMENT AND MODERNITY IN THE LATE TWENTIETH CENTURY

MICHAEL J. WATTS AND JAMES McCARTHY

> A 'global ecological crisis' is a crisis the causes of which are diffuse and the effects of which are universal. From the economic point of view, a global crisis is much different from local crises. In local crises . . . local agents are usually directly accountable for damages to local victims . . . we are dealing with a . . . 'stabilized universe', where people agree upon basic goals, duties and rights. By contrast in the global ecological crisis, the 'culprit' may be nothing less than the model of development encompassing whole continents and 'victims' may be on other continents with other styles of living. We are in a controversial universe involving debates about national models and international justice. (Lipietz, 1995b: 1)

Twenty-five years after the first stirrings of Earth Day and notions of Spaceship Earth, environmentalism and green capitalism – now couched in the language of sustainable development – are back on the political agenda, albeit in a far different ideological climate. Some 19 green parties are active in a dozen West European states, and environmental movements dot the landscape of the former socialist bloc, while the consolidation of civil society associated with widespread democratization in the South has created new spaces for robust environmental movements. Whatever its ambiguities, the current lexicon of sustainability links three hitherto relatively disconnected discourses. It is now taken for granted that the *global environmental crisis* (that is to say, distinctively new global environmental problems such as ozone depletion), a renewed concern with *global demographic growth* (the return of neo-Malthusian thinking, often linked to the purported security threats posed by high-fertility regimes), and the terrifying map of *global economic inequalities* (blandly documented for us each year in the World Bank and UNDP development reports) are necessarily all of a piece. This new 'environment–poverty–population' consensus (*see* DasGupta, 1994; World Bank, 1992) confirms that the eradication of poverty through enhancing and protecting livelihood strategies is as much a green and a fertility (i.e. women's) issue as a narrow resource endowment or growth question.

It is tempting to see the new visibility of green thinking as history repeating itself, but the current conjuncture is quite different from that of the 1960s and early 1970s. First, the restructuring of capitalism in the North Atlantic economies radically transformed the regulatory environment, while new institutional forms of globalization and market integration (WTO, NAFTA) coupled with new and more destructive technologies and substances in a climate of aggressive deregulation, suggest a quite different world from Earth Day 1970. Second, so-called 'industrial compression' (Wade, 1990) associated with the extraordinarily high rates of industrial growth in some of the 'late developers' (Brazil, China, Korea, Taiwan) has exacted a heavy environmental toll. At the same time, the terrifying environmental record in the former 'socialist' bloc is now slowly becoming public knowledge. Indeed, there is a profound sense in which the very crisis of socialism was precipitated by serious environmental and resource problems generated by the economics of shortage. Third, the recognition of new long-term catastrophic *global* tendencies (global warming, ozone depletion, biogenetic hazards) has spawned new efforts at multilateral and transnational institutional regulation and governance: witness UNCED in Rio, the Montreal protocols on climate change, and the efforts to green GATT (Esty, 1994; Sand, 1995). Fourth, the collapse of many actually existing socialisms and the rise of a neo-liberal hegemony in policy circles has signalled the exhaustion, if not the extinction, of socialist and, in many cases, national inward-orientated import substitution models of development. In this sense the neo-liberal counter-revolution presents a quite different ideological atmosphere of what one might call high or ultra-modernism quite unlike that of the 1960s: as a World Bank official put it, 'The world knows much better now what [development] policies work and what policies do not. . . . [Now] we almost [never] hear calls for alternative strategies based on harebrained schemes' (cited in Broad, 1993: 154). And finally, there is a distinctive line of apocalyptic (and quite crude) thinking flying under the banner of 'environmental security' (Homer-Dixon *et al.*, 1993). Here the conflation of Malthusian-style demographic crises coupled with environmental degradation is organically linked to group-identity conflicts and civil strife (Rwanda is, of course, the oft-cited paradigmatic case). Kaplan's (1994) highly influential *Atlantic Monthly* article, in which Africa, wracked by environmental insecurity, decay, anarchy and high fertility, is seen to be sliding back into its Victorian-era 'darkness', is simply the most egregious instance of this type of thinking.

The theoretical firmament of the past 15 years also differs markedly from that of the first environmental wave of the 1960s, which was dominated by human and cultural ecology, both drawing heavily on metaphors of adaptation and the logic of systems ecology and cybernetics. Perhaps the most fertile line of new social scientific thinking about environment and development falls under the banner of 'political ecology'. This term can be traced with some certainty to the 1970s, when it emerged as a response to the theoretical need to integrate studies of land-use practice with local–global political economy (Wolf, 1972) and as a reaction to the growing politicization of the environment (Cockburn and Ridgeway, 1979).[1] Subsequently taken up by geographers, anthropologists and historians, it is perhaps most closely associated with Blaikie (1985) and Blaikie and Brookfield (1987). In their view, political ecology combines the concerns of ecology with 'a broadly defined political economy' (1987: 17). Accordingly, environmental problems in the Third World, for example, are less problems of poor management, overpopulation or ignorance than of social action and political economic constraints. Standing at the centre of Blaikie and Brookfield's political ecology is the 'land manager', whose relationship to Nature must be considered in 'an historical, political and economic context' (ibid.: 239). Blaikie and Brookfield's political ecology investigated the dialectical relationship between poverty and degradation through three major axes of analysis. The first is the concept of *marginality* and ways in which political, economic and ecological marginality can be self-reinforcing; in other words: 'land degradation is both a cause and a result of social marginalization' (Brookfield and Blaikie, 1987: 23). Second, the *pressure of production on resources* is transmitted to the environment through social relations that compel the land manager to make excessive demands ('the pressure of deprivation' as Brookfield and Blaikie call it). And finally, in an unacknowledged nod to post-structuralism, they recognize that the *facts of degradation are contested*, and that there will always be multiple perceptions (and explanations): one person's degradation is another's soil fertility.

Collectively, this body of work punched huge holes in the 'pressure of population on resources' view and in the market distortion or peasant irrationality/mismanagement theory of degradation. In their place it affirmed the centrality of *poverty* – or more properly, the social relations of production – as a major cause of ecological deterioration (de Janvry and Garcia, 1988; Martinez-Alier, 1990; DasGupta, 1994), and pointed to the need for a fundamental rethinking of both conservation and development. At the same time, political ecology opened a number of avenues which it did not adequately explore. While recognizing that environmental facts are typically contested (e.g. the dimensions of the Himalayan environmental crisis), political ecology was relatively mute about the actual production of different environmental truths or about the social constructions of nature itself. Similarly, its very theoretical identification – *political* ecology – belied a singularly impoverished sense of politics and a surprising neglect of the profusion of environmental movements worldwide. And while political ecology may have recognized local environmental knowledge and local systems of regulation, this awareness was rarely employed as a critique of the development project itself or of those forces (transnational capital, structural adjustment) which effectively produced or reproduced poverty in the first place. It was precisely these lacunae, coupled with the growing visibility of green movements at a time of ever more global and corporate practices of environmental management (i.e. Brundtland, the Rio Declaration), which lent a major impulse to the emergence of what Escobar (1996) calls 'poststructural political ecology'. As a theoretical project, it engages with wider debates within social theory on the one hand and the rise of an activist, post-modern approach to 'development' (the alternatives to development school of Wolfgang Sachs (1992)) on the other.

In this chapter, we will explore three fundamental aspects of this post-structural turn in environment–development thinking. These three broad themes – sustainable development as discourse, environmental movements and visions of alternative development ('imagining post-development', in Escobar's language), and the capitalization of nature associated with a new greening of capitalism – represent in our view the centrepieces of current debates over environment, development and social theory (*see* Crush, 1995; Benton and Redclift, 1994; O'Connor, 1994; Leff, 1995). Running through our chapter is a direct engagement with the work of Arturo Escobar, whose book *Encountering development* (1995) is the central text in the burgeoning new field of alternatives to development and discursive approaches to development as knowledge–power. We have chosen his work because it articulates clearly the distinctive theoretical position of (a) critical thinking about development (as an alternative to the high modernism of neo-liberal theory) (b) the role of green movements and politics in sustainable alternatives (the Achilles heel of political ecology to date) and (c) the links between political ecology and theories of capitalist transition (notably debates regarding 'post-Fordism' and a new international division of labour). Escobar's work, however, can and should be situated on a larger landscape which includes the work of Wolfgang Sachs, Gustavo Esteva, Stephen Marglin and Vandana Shiva, among others (Sachs, 1992, 1993; Schurmann, 1993; and Booth, 1994, are key texts here).

## DEVELOPMENT AND ENVIRONMENT AS DISCOURSE: DECONSTRUCTING SUSTAINABILITY

> [A]mbiguity runs through all of the most important discourses on economy and the environment today. . . . Precisely this obscurity leads so many people so much of the time to talk and write about 'sustainability': the word can be used to mean almost anything . . . which is part of its appeal. (O'Connor, 1994: 152)

According to the United Nations Development Program (UNDP, 1992, 1996), the polarization of global wealth doubled between 1960 and 1989. In the *fin de siècle* world economy of the 1990s, 82.7 per cent of global income is accounted for by the wealthiest 20 per cent, while the poorest 20 per cent account for 1.4 per cent. In 1960, the top fifth of the world's population made 30 times more than the bottom fifth; by 1989 the disparity had grown to 60 times. This massive shift in global income distribution is unquestionably rooted in the period of adjustment and stabilization – the so-called Lost Decade – since the debt crisis of the early 1980s. Attending these global inequities is the bleak and horrifying prospect of the deepening of poverty around the world. In its new annual report, the World Health Organization – hardly the voice of a global red brigade – paints a devastating picture of increasing global poverty and the devastating impact of diseases of the poor (WHO, 1996).

Against this backdrop of global inequality and intractable Southern poverty, it is perhaps inevitable that development theory and practice is mired in debate and controversy. For good reason, many intellectuals and activists from the South have come to see development as a cruel hoax – a 'blunder of planetary proportions' (Sachs, 1992: 3) – which is now coming to an unceremonious end. 'You must be either very dumb or very rich if you fail to notice' notes Mexican activist Esteva (1992: 7), 'that "development" stinks.' It is precisely the groundswell of *anti-development thinking* – oppositional discourses that have as their starting-point the rejection of development, certain forms of rationality and the Western modernist project – at the moment of a purported Washington consensus and free-market triumphalism (*ultra-modernist thinking*) that represents one of the striking conjunctures of the 1990s.

Another line of critical thinking starts from the purported *'impasse' of Marxian political economy* (Booth, 1985, 1994; Schurmann, 1993). Geographers have been quite central to this impasse debate[2] which represents an effort to extend the post-modern project to development (*see* Brass 1995 for a review). In Booth's view, the heart of the development theory impasse is the reductionist, economistic and epistemologically flawed nature of Marxism itself, which ignores complexity, diversity and non-class movements from below.[3] Encompassed in this failure is another, namely the failure of radical development theory to engage with development practice since 'we still do not know how to solve the problems of poverty by means of "applied" development' (Booth, 1994: 7). Marxism has failed, in other words, to provide practical assistance to those on the front lines of development.

The confluence of these two interpretations of development in crisis – anti-development thinking and impasse theory – turns out to be the common ground of post-structuralism, which attacks both neo-liberal orthodoxy *and* Marxism, both as theories and as complex sets of practices, tactics and strategies. For Booth (1994) this project takes the form of the celebration of a 'post-Marxism' which seeks out diversity, resistance/empowerment from below, and choice. Anti-development efforts to imagine a 'post-development era' start from the capacity of the development imaginary to shape identity and produce particular sorts of 'normalized subjects' in the South. Development here threatens diversity, homogenizing local traditions through the apparatuses of the state (investment, measurement and planning). The local is eclipsed by 'the use of general conceptual categories and Western assumptions' (Manzo, 1991: 6). A post-development alternative, then, depends fundamentally on local spaces of self-determination and autonomy (Sachs, 1992). Both articulations are draped in the presumptions of anti-foundationalism and of a resistance to what are seen as universalizing discourses of the West.

Escobar's work (1995) is central to imaging post-development because it links both of these lines of thinking and provides an explicitly theoretical framework for the study of development and environmental sustainability. Specifically, he finds modern development discourse to be an invention – more properly, a 'historically produced discourse' (p. 6)[4] – of the post-1945 era, the latest insidious chapter of the larger history of the expansion of Western reason. This discourse is governed by the 'same principles' as colonial discourse but has its own 'regimes of truth' and 'forms of representation' (pp. 9–10). Development is about forms of knowledge, the power that regulates its practices, and the forms of subjectivity fostered by its impulses. Hegemonic development discourse appropriates societal practices and meanings into the modern realm of explicit calculation, thereby subjecting them to Western forms of power–knowledge. It ensures the conformity of peoples to First World economic and cultural practices. Development has, in short, penetrated, integrated, managed and controlled countries and populations in increasingly pernicious and intractable ways. It has produced underdevelopment, a condition politically and economically manageable through 'normalization', the regulation of knowledges and the moralization and technification of poverty and exploitation. The new space of the 'Third World', carved out of the vast surface of global societies, is a new field of power dominated by positive and normative development sciences. These political technologies which sought to erase underdevelopment from the face of the earth have succeeded only in converting a 'dream into a nightmare' (p. 4).

Following many of the contributors in Sachs' volume (1992), Escobar sees poverty as invented and globalized with the creation of a battery of transnational 'welfare' institutions at Bretton Woods and in San Francisco following the signing of the United Nations charter. The discourse of national and international planning and development agencies was able to constitute a reality 'by the way it was able to form systematically the objects of which it spoke, to group and arrange them in certain ways, and to give them a unity' (Escobar, 1995: 40). Patriarchy, ethnocentrism, gender, race and nationality were embraced in this discourse (pp. 43–4), at the same time that economists were privileged within its ranks. This rule-governed system (p. 154) has remained unchanged at the level of practice, although the discursive formations have been unstable. In all of this, modernity's 'objectifying regime of visuality' (p. 155) turned people of the South into 'spectacles', and the 'panoptic gaze' of development became an apparatus of social control (pp. 155–6). Institutions such as the World Bank thus embody what Donna Haraway (1991) calls the 'God trick' of seeing everything from nowhere. Development is constructed in large part through keywords – 'toxic words' – which really mean something else: 'planning' normalizes people; 'resources' desacralize nature; 'poverty' is an invention; 'science' is violence; 'basic needs' are cyborgs, and so forth (Sachs, 1992). The Third World came to believe what the First World promulgated: development as a technical project, as rational decision-making, as specialized knowledge, and as normalization (pp. 52–3). What was and is missing from development, according to Escobar, is *people*.

The new greening of development and the genesis of an ecologically informed global managerialism within the development apparatuses lends itself, of course, to Escobar's post-structural analysis. Sustainable development can be deconstructed as yet another instance of a discourse – rooted in the same old transnational development institutions as the World Bank, now armed with new instruments such as the Global Environmental Facility (GEF) – that serves the West's global ambitions and interests and tends to systematically erode and/or neglect local environmental knowledges and capacities. Shiva's (1989) account of the Green Revolution in the Punjab is precisely about eco-cultural loss: the loss of local economic autonomy, the loss of knowledges/practices at the hands of a Western capitalist science, and the loss of cultural and political resources at the hands of the Indian (neo-colonial) state. Ecological and cultural integrity is replaced by the region's chaotic integration into a global division of labour. Likewise, she argues that global environmental discourses (Shiva, 1993: 150) conceal how local (i.e. Western) interests are framed as universal and in so doing 'destroy the environment which supports subjugated local peoples' (p. 151). Escobar sees sustainable development in the same way. *Our common future* (the so-called Brundtland Report, World Commission on Environment and Development, 1987) marks, in his view, the entry of the globe into rational discourse: global resource management to ensure the survival of a white, masculinist and neo-colonialist 'we' (p. 193). Sustainable development emerges then to save the global ecosystem according to a perception of those who rule it. Hence sustainability is framed by 'liberal ecosystems professionals' who see the problem as global (not local), by a discourse in which the poor are deemed irrational and without ecological awareness, and by experts who manipulate the discourse in a way which renders it compatible with economism and developmentalism (pp. 195–6). Nature is converted into the 'environment' in sustainability discourse to ensure that sound ecology becomes good for capitalist profit. *Our common future* is, in Escobar's vision, a tale of disenchantment told by the West about the West (*see also* Vishvanathan, 1995), in which the prospects for sustainability are seen to rest once more on the contract between the nation-state and modern science (p. 198).

What, then can be said about the turn to discourse and the deployment of post-structural tools in the understanding of development and environment? On the one hand it is important to examine the history of development and to take seriously the ways in which what passes as development knowledge and practice are institutionalized and with what effects. The language of development is not neutral and neither are its institutions. To deconstruct sustainable development – to interrogate the assumptions of particular institutionalized visions of green development – is useful on many counts. To see how and with what consequences 'ecological economics' has been employed by the Global Environmental Facility might shed much light on the sorts of projects funded by the World Bank and the consequences of specific types of large multilateral programmes. Conducting ethnographies of development institutions, as Ferguson effectively does in his account of a project in Botswana (1990), is uniquely helpful in understanding how particular places and problems are constructed and legitimated by experts and managers. But a singular focus on the discursive aspects of knowledge–power, on populist senses of empowerment and resistance, and on cultural diversity and difference carries its own freight. Indeed, one can see a number of grave weaknesses in Escobar's book, suggesting perhaps an old conservatism in 'new' clothes (Brass, 1995).

What is striking about Escobar's book is its similarity to 1960s dependency theory, a perspective largely discredited on the grounds of its simplistic theory of power and crude sense of political economy (*see* Kay, 1991). Like the Latin American *dependentistas*, *Encountering development* (Escobar, 1995) confers enormous power on an external world-system, privileges local autonomy and cultural identity, and sees sovereignty as a central plank of its own vision of development (Cardoso and Faletto, 1979; Furtado, 1970; Sunkel, 1969). Like dependency theory, the nature of external power is often crudely articulated in bold outline – which in Escobar's case is ironic in so far as he employs Foucault to express an alternative theory of capillary power. Escobar's account of the World Bank as an instrument of modernity's power, for example, reads like Third World nationalism of old. No attempt is made to lay out the complex internal divisions within the Bank or the important reversals made by the Bank around some environmental and dam projects (Fox, 1995), or (on the basis of data he himself provides) the relative insignificance of World Bank resources in relation to other capital flows to the South. Indeed, the anti-development communities focus on keywords and the representation of the Third World could be culled from a cursory reading of Andre Gunder Frank in the late 1960s or indeed Che Guevara's meditation on the condition of underdevelopment (the Third World as a 'stunted child') penned at least 30 years ago. Unlike dependency theory, Escobar privileges (as we shall see) a different sort of community (local not national), and seems to reject development as modernity *tout court*. But the theory of underdevelopment and power contained within *Encountering development* is as simple and blunt as early dependency thinking (albeit with the addition of Foucault's notion of biopower). Curiously, it also represents precisely the sort of metanarrative and totalizing discourse decried by post-structuralism itself.

A second weakness is the distorted and truncated history of development discourse itself (Cowen and Shenton, 1996; Cooper, 1996). Throughout much of the anti-development literature, the Third World and underdevelopment are assumed to have been invented or discovered by President Truman in his famous 1949 speech on 'fair dealing' and undeveloped areas. The key conjuncture is the decade after the end of the Second World War (Escobar, 1995: 4), in spite of the fact that the South was seemingly of marginal concern in relation to the reconstruction of Europe, the revivification of the world financial system or the threat of communism. Escobar (ibid.: 27) does acknowledge that development discourse has a pre-1939 history but he has almost nothing to say about it. Conversely, in his book *Keywords*, Raymond Williams (1976: 104–6) notes that the historically complex genealogy of development in Western thinking can 'limit and confuse virtually any generalizing description of the current world order'. Rather it is in the analysis of the 'real practices subsumed by development that more specific recognitions are necessary and possible'. The history of these 'real practices' is, however, long and complex, longer than Escobar seems prepared to admit. Development came into the English language in the eighteenth century with its root sense of unfolding and was granted a new lease of life by the evolutionary ideas of the nineteenth century (Rist, 1991; Watts, 1995). As a consequence, development has rarely broken from organicist notions of growth or from a close affinity with teleological views of history, science and progress in the West. By the end of the nineteenth century, for example, it was possible to talk of societies in a state of 'frozen development'. There is another aspect to the genealogy, however, traced by Cowen and Shenton (1995) to eighteenth- and nineteenth-century notions of Progress, and specifically to development as a sort of theological discourse set against the disorder and disjunctures of capitalist growth. Classical political economy – including Smith, Ricardo, Mathus and the like – is suffused with the tensions between the desire for unfettered accumulation on the one hand and unregulated desire as the origin of misery and vice on the other (Herbert, 1991). Development in Victorian England emerged in part, then, as a cultural and theological response to Progress. Christopher Lasch (1991), for example, has described a late nineteenth century obsessed by cultural instability and cataclysm. Saint-Simon himself devoted himself in his last years to a new creed of Christianity to accompany his industrial and scientific vision of capitalist progress. Trusteeship, mission and faith were, according to Cowen and Shenton, the nineteenth-century touchstones of 'development'.

Of course, there is a sense in which Third World development – in its specific forms of state and multilateral policy harnessed to the tasks of championing economic growth, 'catching up', improving welfare and producing governable subjects – is of more recent provenance. These origins of development theory and practice as an academic and governmental enterprise – and of development economics as its hegemonic expression – are inseparable from the process by which the 'colonial world' was reconfigured into a 'developing world' beginning in the 1930s but especially in the aftermath of the Second World War. Africa, for example, became a serious object of planned development after the Great Depression of the 1930s. The British Colonial Development and Welfare Act (1940) and the French Investment Fund for Economic and Social Development (1946) both represented responses to the crises and challenges which imperial powers confronted in Africa, providing a means by which they could negotiate the perils of independence movements on the one hand and a perpetuation of the colonial mission on the other. But the process by which – to use Escobar's language – this produces 'development as a historically singular process' requires a sensitivity to regions, political economy and politics which *Encountering development* does not have. The significance of this misreading resides not only in the production of poor (and impoverished) history but also in its failure to realize a much more complex and nuanced way in which global discourse and local power, and local discourse and global power, actually intersect and reproduce particular regional formations. Development was necessarily Eurocentric because its origins lay in the European efforts to deal with the 'essential fact' of capitalism, as Schumpeter put it, the tragedy of underdevelopment and the paradoxical unity of modernity (Berman, 1982). In the same way that Polanyi (1944) saw the later eighteenth-century welfare debates in England as the discovery of the 'social', so the invention of development in nineteenth-century Britain was about the 'failure' of free-market capitalism. Within the belly of modernization has always resided its utopian alternatives.

It is surely incontestable that knowledge can be a source of power, but the danger of a turn to discourse is that development ideas simply become (and remain) narratives or stories. Development narratives remain, in other words, *only* narratives. In Roe's (1995) work, for example, development problems are converted into stories or Proppian folktales. Stories underwrite or stabilize the assumptions of policy-makers. What is required in narrative policy analysis is, then, good, better or different stories. In their account of environmental crises in Ukambani in Kenya, Rocheleau *et al.* (1995) identify differing accounts by peasants, experts and so on of the environmental history of the region, from which they conclude that 'multiple stories' have to be incorporated before problems can be solved. This seems to leave open the question of better or worse stories, of a 'politics of naming'

(p. 1038) which does not permit such incorporation, or indeed the conditions under which such a multiplicity of stories could be heard. Development as narrative or story-telling (*see* Watts, 1995 for a review of this literature) runs the risk of excluding politics, interest, institutionalized authority and legitimacy and putting in their place a naive sense of sitting around the camp-fire telling each other stories. Escobar does, of course, root development discourse in the post-1945 nexus of development institutions, but he makes extraordinary claims about the power and efficacy of these discourses in shaping subjectivity ('how people come to recognize themselves as underdeveloped' (p. 10)). To this extent the linkages between regimes of truth or development representations and actual subjectivities seem weak to say the least. As a consequence, Escobar's claims about subjectivity and the power of discourse – 'discursive homogenization of the Third World' (p. 53) or 'all encompassing systems of control' (p. 52) – carry all of the flaws of the orientalist-type historicizing that post-colonial and post-structural work typically rejects (Said, 1985b).

Finally, in turning to an analysis of environmental and development discourses, the massive machineries of modernity seem capable of spinning off infinitely more subtle and complex narratives which leave everything of substance intact (i.e. structured inequality). But the anti-development community has little to say about exploitation or class (Escobar (1995) makes a passing reference to this 'materiality' on p. 53); it seeks rather to 'shift the ground' so that we can look at that materiality with different eyes (Escobar, 1995: 53). But when all is said and done, and after the dust of regimes of truth, dispositions, visibilities and logocentrism has settled, are we provided with a radically different or more powerful account of the Green Revolution (pp. 156–63) or of the World Bank (pp. 163–71) than that provided by, say, Marxian political economy? In our view the answer must be no, even if the questions posed by Escobar and others potentially open up another way of thinking about power. What remains in the alternative development community is a series of ambitious claims about the non-colonized spaces, and the resistances to the hegemonic discourses of development. Here, as we see in the next section, the 'castaways shipwrecked by modernity' (Latouche, 1993) have repositories of tradition to pull upon, and it is these capacities which reside within the populist discourse ('the discursive insurrection of Third World peoples' (Escobar, 1995: 17)) of new social movements of the Third World 'other'.

## ENVIRONMENT, NEW SOCIAL MOVEMENTS AND ALTERNATIVES TO DEVELOPMENT

> Like other ideologies environmentalism is socially constructed but it is an especially indeterminate, malleable ideological form. (Buttel, 1992: 15)

> Ecological movements are not creating a new economics for a new civilization, they are not presenting a solution for the crisis of the modern world, and they do not have the capacity . . . for ending development. But they can show the difficulties, shortcomings and limited scopes of the dominant as well as the alternative models for development at the level of action. (Linkenbach, 1994: 81–2)

A striking consequence of radical economic restructuring in the South in the wake of the 1980s debt crisis (and in the post-socialist bloc after 1989) has been 'the resurrection, reemergence and rebirth of . . . civil society' (Cohen and Arato, 1992: 29). Partly in response to the collapse of state resources, and partly as an outcome of an uneven democratization process, various forms of local and community movements have emerged in the interstices of the state–society nexus. Buttel (1992) sees these new social movements (NSM) as new in so far as they represent a sort of post-modern politics outside of and in many respects antithetical to class or social democratic party politics. Many are also new in that they are an integral part of a widespread 'environmentalization' of institutional practices. Enormously heterogeneous in character and scale (anti-dam movements, squatter initiatives, minority cultural movements), these grassroots movements often focus on efforts to take resources out of the market-place, to construct a sort of moral economy of the environment (*see* Martinez-Alier, 1990). Broad (1993), for example, documents what she calls a new citizens' movement in the Philippines (five to six million strong) consisting of 'mass-based organizations' which arise from the intersection of political-economic plunder and local demands for participation and justice. In much of this literature the label 'environmental' is misleadingly narrow, since the proliferation of grassroots and NGO movements often focus more broadly on livelihoods and justice. Indeed, it is striking how indigenous rights movements, conservation politics, food security, the emphasis on local knowledges, and calls for access to and control over local resources (democratization broadly put) cross-cut the environment–poverty axis. This multi-dimensionality is, according to Escobar (1992a), indicative of a new mode of doing politics, so-called 'autopoietic' (that is to say, self-producing and self-organizing) movements which exercise power outside the state arena and which seek to create 'decentred autonomous spaces'.

The *local community* and *grassroots initiatives* loom very large in post-structural approaches to development (Parajuli, 1991). What they represent is certainly a form of collective action, but more specifically and profoundly a 'resistance to development' (Escobar, 1995: 216; also Routledge, 1994) which attempts to build new identities. The implication is, of course, that these identities fall outside of the panoptic gaze of the hegemonic development discourse as new forms of subjectivity which stand opposed to modernity itself. As Escobar puts it (1995: 216), these movements are not cases of 'essentialized identity construction' but are 'flexible, modest, mobile, relying on tactical articulations arising out of the conditions and practices of daily life'. To the extent that these movements are 'environmental' or 'green', vast claims have been made on their behalf: they are a 'revolt against development' (Alvarez, 1992: 110), 'a new economics for a new civilization' (Shiva, 1989: 24), 'learning to be human in a posthuman landscape' (Escobar, 1995: 226). What, then, is the new content of such movements and what are their relations to anti-development?

There are, in our reading, at least five fundamental and overlapping aspects attributed to grassroots movements as vehicles of counter-modernity. First, they purportedly contain new sorts of politics and new sorts of political subjectivity. They are typically local, outside of the organized state sphere and 'without one particular ideology or political party' (Escobar, 1992: 422). They are 'self-organizing and self-producing', exercising non-state forms of power. Second, 'cultural difference is at the root of postdevelopment' (Escobar, 1995: 225) and hence the movements are, above all, examples of popular cultural discourse. Minority cultural communities[5] figure centrally in both green and anti-development movements; indeed, the 'indigenous' becomes the lodestar for the 'unmaking of the Third World'. Indian confederations in Latin America or 'ethnic' green movements in Africa often turn on the ways in which cultural identity is mobilized as 'a transformative engagement with modernity' (Escobar, 1995: 219). Third, the movements employ, in creative ways, local or subaltern reservoirs of knowledge. Paul Richards' (1985) invocation of 'inventive self-reliance' rooted in local African peasant knowledge is an influential case in point of this line of thinking. The proliferation of the field of 'indigenous technical knowledge' and the so-called actor-orientated interface analysis is another (*see* Arce *et al.*, 1994). In singing the praises of this subaltern science position, women's knowledge and nature are often central. In Shiva's words, 'women as victims of violence of patriarchal forms of development have risen against it to protect nature' (1989: xvii)', and by virtue of their organic relationships to things natural have a 'special relationship with nature' (ibid.: 43). Indeed, for Shiva, feminine/ecological ways of knowing are 'necessarily participatory' (ibid.: 41). Fourth, local community and 'tradition' are neither erased nor preserved as the basis for alternative development but are refashioned as a hybrid: hybridity entails 'a cultural (re)creation that may or may not be (re)inscribed into hegemonic constellations' (Escobar, 1995: 220). This is the heart of the new political subjectivity which speaks to a 'transcultural in-between world reality' (ibid.). And finally, these movements produce a defence of the *local*: such a defence is a 'prerequisite to engaging with the global . . . [and represents] the principal elements for the collective construction of alternatives' (ibid.: 226).

These are, to put it mildly, bold and ambitious claims, and they demand careful scrutiny. Clearly, there is a long history of grassroots initiatives, and these initiatives have been emboldened by the new spaces opened up in the 1980s by the contraction of the state and the democratic opening worldwide (Seabrook, 1995). Equally, some (but by no means all) of these movements in the South are 'new' in the sense that they do not conform to a simple Eurocentric model of single-issue struggles. As an Indian worker quoted in Seabrook (1995: 103) says, 'you cannot separate work from all other aspects of life: all must be integrated into a single struggle'. As Escobar properly points out, the new 'green' movements are more than 'environmental' in so far as they seem to link in complex ways a number of social justice and cultural issues. Many of these NGO activities are contributing to a new sort of internationalism through global electronic networking and solidarity activity. But as a human rights activist in Brazil comments (cited in Seabrook, 1995: 59), this internationalism is not yet a movement and is in no sense systematized by key issues. A central weakness of the social-movements-as-alternative approach is precisely that greater claims are made for the movements than the movements themselves seem to offer. To put it bluntly, there is a strong sense in which the very fact of being from the South, being female and being rural is often seen to confer a sort of all-knowing, omniscient political virtuousness. Indeed, the sorts of claims made by Shiva – that ecological knowing [and local knowledge in general] is 'necessarily participatory' or that women have a special relationship with nature which organically produces conservation and protection of it – express exactly the sorts of essentialisms that she (and others) attribute to retrograde Eurocentric discourses. In much of this work, culture and popular discourse from below are privileged in quite uncritical ways; identity politics is championed by Escobar because it represents part of an alternative reservoir of knowledge and because such ideas stand against the 'axiomatics of capitalism'. But there is surely nothing necessarily anti-capitalist or particularly progressive about cultural identity: calls to localism can produce Hindu fascism as easily as Andean Indian co-operatives. Running through much of the social movements as alternative to development literature is an uncritical appeal to the 'people' – that is to say, populist rhetoric – without a sensitivity to the potentially deeply conservative aspects of such local particularisms.

Insights into the weaknesses and ambiguities contained within the new social movements canon can be gleaned from an analysis of three such movements – Chipko in north-west India, the Brazilian rubber-tappers and the Ogoni in Nigeria – which are often held up as prototypical cases of the new visions of anti-development. The Chipko movement, the tree hugging, anti-logging movement in Garawhal India, is seen by many as an ideal-type case of a local counter-hegemonic movement from below (Shiva, 1989: Guha, 1990). Amazonian rubber-tappers (Keck, 1995) are another instance of a case picked up by the international community in which local subaltern knowledge, acting as a counterpoint to the World Bank's vision in Acre, is a pivotal source of political resistance in reshaping the modern. And the case of the Ogoni people (Naanen, 1995), thrown into international relief by the hanging of their leader Ken Saro-Wiwa in 1995, is a compelling instance of how cultural politics (ethnic identity) is fused with green concerns, in this case a direct assault on the ecological devastation wrought over three decades by the oil companies (Shell in particular) in the delta region of Nigeria.

At first glance all three cases seem to exemplify the sorts of new politics and new visions articulated by Escobar: local visions, articulated with the global in hybrid ways, rejecting the modern, and infused with self-organizing capacities rooted in local practice. While these elements are vividly present in each of the movements, such an account is incomplete and obfuscatory. To take Chipko, Linkenbach (1994) notes that the movement has at least three differing models of development within it – eco-development, survival economy and selective economic and social improvement – and that the movement is as much pro-modern as it is anti-development. Rangan (1995) has shown how the Garawhal region was a site of earlier Gandhian and Communist Party organizing which shaped Chipko (i.e. this is hardly a purely 'indigenous' movement), which in turn has gradually developed into a regional autonomy initiative. Likewise, the crisis in Ogoniland in Rivers State (Nigeria), which generated MOSOP (the political entity pursuing Ogoni self-determination and environmental compensation) and culminated in November 1995 in the hanging by military authorities of Ken Saro-Wiwa and the Ogoni Nine, was part of a longer (pre-colonial and colonial) history of inter-ethnic conflict, particularly in relation to the dominant Ibo community against which they fought in the Civil War (Naanen, 1995; Watts, 1997). The movement itself emerged in 1990 as a response to the devastating environmental consequences of oil drilling but it expressed a sense of separatist identity and a need for self-determination that long pre-dated oil exploration, which began in the 1950s. Moreover, MOSOP was riven by generational, clan and political conflicts – no simple, harmonious community solidarity here – led by an intellectual of a quite modern sort, and it sought a share of petroleum's modernity and capital accumulation, not its rejection (Welch, 1995). And not least, the rubber-tappers' political activity emerged in part from the efficacy of the Workers' Party activity in Brazil and received a large fillip from an international green community with a quite ambiguous stance regarding alternative development rhetoric. All of these movements, then, while in some senses new, also undercut Escobar's account in so far as they also throw into question the rejection of modernity, the idea of an unproblematic community vision, or of untainted local knowledge, and untainted authentic forms of agency.

A striking feature of Escobar's work, and the work of much of the alternatives school, is the constant uncritical appeal to the local, to the popular, and to the cultural (where 'cultural' is synonymous with a local sense of community). Yet as Pierre Bourdieu has noted, in discussion of 'the people' and 'popular' discourse, what is at stake is the struggle between intellectuals (1990: 150). These debates among intellectuals celebrate, in a quasi-mystical way, the efficacy of all action or knowledge from below; they contain a rejection of a *fin de siècle* modernity rooted in the losses and reaffirmations of local particularisms which are and always have been the accompaniment of capitalism. Moreover, they often forget that the 'local' is never purely local, but is created in part by extra-local influences and practices over time. And to this extent there is the danger that the alternative to development school has, to quote Marshall Berman, 'lost touch with the roots of its own modernity' (1982: 17). In portraying development as a discourse of control only, they forget that it can also be a tremendously powerful discourse of entitlement, and that the coexistence of these two aspects must be kept in sight in a dialectical understanding of it.

## GREENING CAPITALISM? DEVELOPMENT AND THE CAPITALIZATION OF NATURE

> [T]he primary dynamic of capitalism changes form, from accumulation and growth feeding on an external domain, to ostensible self-management and conservation of the system of capitalized nature closed back upon itself. (O'Connor, 1993: 8)

The past three decades have witnessed remarkable developments in the perception and management of the environment at a global scale – not least the very ascendance of the term 'environment' over 'nature'. Two trends – increasing scientific consideration of the environment at a global scale, and growing attention to whether capitalist growth and environmental quality were mutually sustainable – were pivotal in the discovery (or creation) of 'global environmental problems' such as global warming and in the corresponding growth of global (or at least multilateral) institutions to manage the global environment (Taylor and Buttel, 1992). Technological advances, especially in the biological and ecological sciences, have led to both a more complete understanding of environmental problems and an increased ability to regulate and manipulate nature; in short, to both an increased need and an increased capacity to consciously manage the 'global environment'. These developments contributed to the rise of 'sustainable development' as a discourse that purports to reconcile the historically opposed goals of environmental protection and economic growth (the recent wave of interest in linking rainforest and biodiversity conservation and commodification being perhaps the clearest example). They also contributed to new institutions and regulatory agendas: for example, various attempts to make World Bank priorities and projects more environmentally sensitive, the Montreal climate change protocols, UNCED in Rio, the creation of the Global Environmental Facility, and recent efforts to incorporate environmental concerns into GATT (Esty, 1994; Sand, 1995).

Escobar and others have extended the post-structural critique of development and environment to these multilateral institutions and discourses of governance and regulation, and in doing so make epochal claims regarding the purportedly new dynamics of capitalism at a global scale (Sachs, 1992; Escobar, 1996). In this examination of the transnational and global levels, a complement to the focus on the local movements, Escobar posits that environmental and developmental agendas have converged – indeed merged – in the workings of capitalism at a global scale. In other words, the preservation of environmental quality and resources (conservation) has become an *essential precondition* of sustained capital accumulation. His suggestion overlaps with recent work on Marxian approaches to environmental issues (*see* Altvater, 1993, 1994; O'Connor, 1993), much of which makes comparable claims. They share an underlying vision of capitalism as an inherently expansionary system that has historically continually incorporated and conquered new frontiers as inputs (raw materials) and as outlets (for commodities and surplus capital). While this incorporative process is incontestable, the spatial definition of capitalism is taken too far and reduced to a simple binary logic: places are either inside or outside of an international capitalist economy in this view, and so sooner or later capitalism encircles the globe and meets itself coming the other way: there are no more 'exterior' areas to incorporate. Proponents argue that this limit point is now being reached with regard to external nature; capitalism must begin intensively reworking what is already 'inside'. Nature thus becomes something intentionally *produced* and treated as an element *internal* to capital – somewhat akin to fixed capital – and less something simply *extracted* as raw material for exploitation. In short, *nature itself becomes capital.* Sustainability and careful regulation therefore emerge as essential to contemporary capitalism, explaining their recent prominence in environment/development discourse and institutions.

This viewpoint has received its fullest articulation in recent work by Escobar (1995, 1996), who refers to capitalism's allegedly new form, based on the capitalization of new aspects of nature and society, as its 'ecological phase'. Breaking his periodization into 'modern' and 'postmodern' forms (1996: 47, 54–7), Escobar says the former relates to nature as a set of conditions of production. Theoretical efforts here focus on the capitalization and politicization of these conditions, and on the existence of a 'second contradiction' between their reproduction and sustained capital accumulation (O'Connor 1988). It is the latter, post-modern form that is supposedly new and different in the contemporary period. Escobar advances a number of propositions regarding what exactly is new and different about post-modern capitalism. First, he claims (somewhat implicitly) that capitalist control over nature is both distinctively new *and* virtually complete in this period. Second, he makes a series of claims regarding the origins of value in contemporary nature/society dialectics. Third, he asserts that it is the joining of the knowledge, conservation and capitalist use of nature that is unique to the present conjuncture. And fourth, he makes an explicit link to post-structuralism, arguing that the discursive construction of nature as capital is what is both new and essential in the 'post-modern form'. In our view, however, the fundamental 'newness' of many of these capitalist relations to nature, and therefore of the need for a new analytical category such as the 'postmodern form of capitalism's ecological phase', is unconvincing. Further, many of these arguments rest on a significantly flawed understanding of capitalism as an economic system, and in particular on questionable periodizations of economic history. Escobar's interpretations of capitalism and its relations to nature, like his understanding of development, are far too monolithic and insufficiently dialectical.

First, at the heart of Escobar's argument is a series of strong claims regarding increased control over nature. His most basic claim is that human control over nature has now extended to be virtually complete, with a number of major consequences. Nature can, for example, be far more intensively regulated, with the primary dynamic of the dialectic between capital and nature becoming 'the sustainable management of the system of capitalized nature' rather than 'exploitative accumulation' (1996: 47). Technology promises a time when 'nature will be produced to order' (1996: 65).[6] But Escobar and other proponents of this view have confused a difference in degree for a difference in kind: efforts to control nature of the sort they examine are neither *new* nor *complete*. What seems to be at issue in the more measured presentations of this argument (e.g. O'Connor, 1993; Altvater, 1994) is the degree to which 'nature' is at least in part a result of previous productive labour that requires some ongoing application of human labour to maintain it in its desired state. Technological advances – those in biotechnology being perhaps the most profound in recent years – continue to extend the range of 'nature' that is subject to intentional human control, and struggles and outcomes surrounding these processes are unquestionably a vital area for social science research. Yet it is too easy to exaggerate such advances and to argue that they constitute an epochal shift without specifying what is analytically distinct in the 'new' period. Nature has been 'produced to order' within existing cultural parameters for a very long time, as is demonstrated by the briefest glance at agriculture (the very essence of which is controlled transformation of nature for sustained production) or at rural landscapes (which have been consciously designed to meet cultural criteria since classical times). In modern capitalist societies these processes have of course accelerated enormously and are inseparable from the dynamics of capital accumulation and class struggle, but that should not be confused with any essential *newness* to the intentional human transformation of or control over nature.

Nor is such control over nature *complete*, as Escobar seems at times to imply. Uncontrolled diseases abound, meteorological phenomena can enhance or destroy a capitalist enterprise's profitability overnight, and gravity and sunlight remain exogenous to capitalist management and regulation, to name just a few examples (*see* Benton, 1989, 1992).[7] Moreover, there is reason to suspect that it never *can* be, because any transformation of nature inevitably will produce unforeseen, unintended environmental consequences: with intensive breeding and management comes mad cow disease; with fossil fuels comes the possibility of global warming (*see* Benton, 1989; Lewontin, 1992; Levins and Lewontin, 1985). The relationship of capitalism to nature is, then, not one of ever-increasing control of a finite sphere but one of an ever-unfolding dialectic with unpredictable outcomes. Sachs recognizes this at times, suggesting that 'the purpose of global environmental management is . . . to control the consequences of the control over nature' (1996: 20). But the extent to which control over nature would include control over its consequences was directly addressed by Marx and Engels one and a half centuries ago, and has been explored more recently in debates between Benton (1989, 1992) and Grundmann (1992) (*see also* Smith, 1984). Even for those capitalist uses of nature apparently mostly under conscious control, it is hardly plausible that sustainable management has become the norm. As regards inputs, the economy is still fundamentally dependent upon the use of prodigious quantities of non-renewable fossil fuels, the mining of ores, the harvesting of timber at non-renewable rates, and forms of agriculture that amount to a mining of the soil. Modern industrial societies still produce staggering quantities of waste, much of it toxic, and neither its physical nor its social disposition – in the oceans, or in underdeveloped areas desperate for any source of income – appears particularly 'sustainable'. In short, the exploitative use of prodigious 'external' natural resources in a non-sustainable fashion still appears to be quite characteristic of capitalism at a global scale.

Second, Escobar engages a fundamental theoretical debate concerning the source of value in a capitalist economy and nature's contribution to it. Value is apparently produced and conceived of differently in the 'postmodern form' of the 'ecological phase' of capitalism: 'Nature and local people themselves are seen as the source and creators of value – not merely as labor or raw material. Species . . . are valuable . . . as reservoirs of value . . . that scientific research . . . can release for capital' (1996: 57). This complicated set of statements demands careful reading and analysis. It runs entirely against classic Marxian value theory, according to which the inputs described are, in fact, precisely raw materials and labour, with specific sorts of rents.[8] The 'inherent values' referred to here would seem to be entirely *use* values; Escobar completely ignores fundamental distinctions between *use values* (some inherent in nature) and value *per se* as a *social* relationship that cannot exist a priori in natural resources but is rather created by transformative human labour, measured and in part created by the circulation and *exchange* of commodities in a larger capitalist economy (Smith, 1984). In fact, what is striking about Escobar's conception of value in this 'new' 'postmodern form' of the 'ecological phase' of capitalism is its *archaic* character: without saying so, Escobar here reverts to an essentially *physiocratic* notion of value, in which value 'contained' in nature is 'released' by capital and labour.[9] This is perhaps not surprising given his advocacy of indigenous rights to resources: physiocratic theories of value logically and emotionally lend support to populist visions tied to land-based production, explaining their persistent historical association.

Third, one of the most important claims regarding the distinctiveness of contemporary capitalist use of nature advanced by Escobar (1996: 58) is that the *knowledge*, *conservation* and capitalist *use* of nature are now all linked projects (witness attempts to catalogue and preserve rainforest biodiversity in order to find new products) (1996, p. 58). There are undeniably important overlaps in contemporary efforts in these three spheres, and their convergence is arguably accelerating. However, this specific constellation of relations to nature is not new either. Parallels can be seen in imperial science and conservation programmes dating back to the seventeenth and eighteenth centuries (cf. Grove, 1992), and in some of the first formal efforts of the 'conservation movement' in the nineteenth century. Timber, to provide just one example, was a vital natural resource in the USA in the late nineteenth century, absolutely essential to sustained national economic growth. Massive quantities were required in the construction of a rail network, mining technology at the time demanded extensive supplies, and it was still the major building material and a significant fuel source. Up until the 1880s, timber production in the USA tended to clearcut – if not devastate – areas and move on. By the late 1800s, the nation's last remaining untouched supplies of timber were being approached. Fears quickly grew of a 'timber famine' that would hobble the nation's economy. In response, and also as part of the era's anti-monopolistic efforts, the federal government began to withdraw extensive lands from the public domain into 'forest reserves', which would later become the national forests. This undertaking, which represented a monumental shift away from the standing policy of privatizing the public domain, required massive research and mapping expeditions throughout the US West. From the beginning, it was clear that timber corporations would have privileged access to these forests for private commodity production, and forest regulation and management evolved over the decades primarily through close working relationships between private corporations and the new government agencies created to administer the public forests (Steen, 1992). Regulatory and 'conservation' policy often explicitly favoured and served private capitalist interests. Increased knowledge, conservation and commodification of nature were, therefore, tightly interlocking goals.

Fourth, post-structuralism's emphasis on discourse figures prominently in Escobar's conception both of what is new in contemporary shifts in capital–nature relations and of how these shifts are effected. The 'postmodern' form is, in his view, apparently distinguished by 'the progressive semiotic conquest . . . [of] the very heart of nature and life'; it is 'a new process of capitalization, *effected primarily by a shift in representation* (1996: 56, 47; emphasis added). But what is the precise meaning of this shift? The point seems to be that new material relations with nature – new technologies, new forms of appropriation and substitution – rely fundamentally on the discursive construction of nature *as* capital. Contemporaneous with this process has been a shift from 'nature' to 'environment' as the dominant term used to describe the non-human physical world in the post-war period. Escobar equates this with the 'symbolic death' of nature: a shift from seeing nature as 'an entity with its own agency' to seeing it as an inert supply of resources (1996: 52). To claim such an ideological shift is surely untenable. Merchant (1980) traces precisely such a 'death of nature', but over centuries, beginning at least as early as the empiricism of Francis Bacon in the early seventeenth century (*see also* Smith, 1984).

In moving from specific discourses to broader ideological claims, Escobar asserts that 'the notion that nature and the earth can be managed is an historically novel one' (1996: 49). While the hubris exhibited in the discourse surrounding the institutions of global environmental management and regulation is admittedly enormous, it is far from novel. Rather, it is best understood as a contemporary articulation of an ancient theme: that of nature as a garden to be improved by human agency. Glacken (1967) traces this theme from classical times up to the modern period, showing both its continuity and – contra Escobar – centuries-old instances of the belief that this process of controlling and improving nature had been completed, or had proceeded so far as to leave no 'untouched' nature left.[10] Likewise, contrasting recent environmentalist discourse of 'global change' with the earlier 'limits to growth', Sachs asserts that the former 'puts mankind in the driver's seat and urges it to master nature's complexities with greater self-control' (1992: 14). While certainly an accurate reading of the global change discourse, this mandate is not new in the modern period, nor was it ever in serious jeopardy even at the height of the 'limits to growth' discourse.

As we saw in his discussions of development, Escobar often overestimates the importance of discourse (e.g. 'It is no longer capital and labor that are at stake *per se*, but the reproduction of the code' (1996: 56)). While discursive shifts are doubtless vital to extending, reproducing and legitimating particular power relations, the political economy of changing material relations between capital and nature remain central to the process of capital accumulation. Somewhat ironically, Escobar's view of the global management and capitalization of nature makes sense only if the discourse of the institutions he examines is fairly closely related to their actual programmes. As Escobar himself has frequently argued, there is good reason to be profoundly sceptical regarding such a confluence. If, conversely, the new 'environmental' initiatives of the World Bank and other multilateral institutions do not represent a real break from their dominant historical agendas, as often seems to be the case (Peet and Watts: 1996b, 18), then the case for a distinctive new epoch in the relations between capital and nature, characterized by the sustainable management of nature, is correspondingly weakened.

The flaws in Escobar's argument reveal an odd and dangerously misleading understanding of capitalism as an economic system rooted in recent economic theories positing an epochal, post-modern refiguring of late twentieth-century capitalist dynamics. Escobar accepts uncritically concepts such as a 'post-Fordist regime of accumulation' and 'flexible labor' as clearly existing new phenomena (1996: 59) which underlie and shape his analysis. He implies that his project is theorizing the consequences for nature of this epochal shift in the workings of capitalism. Yet categories such as 'post-Fordism' exaggerate the 'newness' of many long-standing features of capitalism and overlook major continuities between economic eras, thereby often obscuring as much as they reveal about actual capitalist historical geographies (Goodman and Watts, 1994; Walker, 1995). In addition, Escobar's central trope – that of capitalism having encircled the globe and begun to rework itself internally – is a strikingly *organic* one: the metaphor that leaps to mind is, of course, that of the mythological snake that circles the world and begins to consume its own tail. In fact, images of global 'tentacles of productivism' appear frequently in the literature regarding the 'internalization of nature to capitalism' (e.g. Sachs, 1993: 12). Capitalism becomes, then, a *single* organism regulating its own *internal* metabolism.

There are profound problems in thinking of the global capitalist economy in this way. First and foremost, while it is an extremely useful and well-founded abstraction, 'capitalism' as such, as a single entity or entirely consistent set of rules and practices, does not exist. Two of the most salient features of capitalist economies are ongoing, ruthless *competition* between individual capitals (Marx's 'warring brother' theme) and *struggles between* classes. While there are certainly cases in which the capitalist class shares some overall interest that may be pursued through the state or other forms of collective action, such projects are always partial and temporary. Therefore, the thesis that capitalist use of nature is or *can be* rationally planned, managed or regulated at a global scale runs counter to some of capitalism's most basic features. Second, the image of a single organism regulating its internal workings suggests *homoeostasis* rather than *growth*. Organisms do of course grow, but they do so in a somewhat predictable fashion, according to some pre-existing plan, and above all to a *limited* degree. Capitalism as an economic system conversely grows in ways precisely opposed to organic growth and maturation. Escobar's vision of capitalism as a global system of sustainable environmental management and regulation thus in an odd way reproduces, on a vast scale, political ecology's earliest homoeostatic interpretations of local human–environment relations. Third, the vision of a single entity constrained by space lends itself too readily to simplistic spatial metaphors of capitalism running up against external natural limits: 'Economic expansion has already come up against its biophysical limits; recognizing the earth's finiteness is a fatal blow to the idea of development' (Sachs, 1993: 6). While the geographic incorporation of new areas has been of course a central feature of the historical geography of capitalism, as Sachs suggests (Harvey, 1982, 1985b), the unceasing revolution of forces and relations of production *within* capitalist economies remains much more fundamental. The need to rework what is already 'inside' is neither new nor, if history is any guide, an insurmountable 'limit' to the continued expansion of capital. Likewise, natural 'limits' are rarely absolutes: limits change as technology and social relations change (*see* Sachs, 1993: 12; cf. Benton, 1989; Harvey, 1993). Capitalism is not usefully conceived as an organism filling a defined space.

Finally, Escobar's theory – like his view of development – tends to reify capitalism into a single thing rather than positing a complex and often contradictory maelstrom of social relationships. It is the very complexity of these relations, and the absence of a single managerial hand, that makes the analytical task of untangling the relations between the environment and development, and the political task of creating more viable and desirable ones, so extraordinarily intractable. These latter tasks are, in our view, the appropriate agenda for a 'political ecology' worthy of the name; a powerful theory of capitalism and a measured appreciation of the dialectical relationships between discourses and material practices and social relations are essential prerequisites to it. Exaggerated claims about the causal powers of discourse alone or the newness of capitalism's transfigurations run the risk not simply of blinding us to historical continuities but of undermining our ability to understand and change the world.

## CONCLUSION: POLITICAL ECOLOGY REDUX

> To champion postmodernist fragmentation and dispersion does sometimes deflect attention from realities that *should* be brutally (rather than strategically) essentialized. (Treichler, 1992: 62; emphasis in original)

The post-modern critiques of development, and the move toward a post-structural political ecology, potentially open the way for a fuller understanding of the multiplicity of ways to comprehend the extraordinarily complex nexus of development–environment relations. There is no question that social theory's engagement with post-structuralism and linguistic analysis over the past two decades has produced important insights which have only begun to be explored with regard to human–environment relations. Recent developments in political ecology that encourage us to consider both *a broader conception of the forms of contention* (from class struggle to social movements to everyday resistance) and *a deeper conception of what is contested* (from ownership of productive resources to control over the human imagination) can be seen as a direct fruit of this engagement. Escobar's work, in this regard, is provocative and exciting. To recognize that development institutions, theories and experts hold the symbolic power to in part construct reality – a major dimension of political power – is an important insight. Furthermore, one of the great merits of the discursive turn within political ecology is the demand it makes for nuanced, richly textured empirical work to match the nuanced beliefs and practices of the world. Not least, it encourages attention to discourse and ideology as significant mechanisms and arenas in struggles over development and the environment. Such work at its best locates specific sorts of movements emerging from the tensions and contradictions of capitalist economies and seeks to understand the subjective basis of their oppositions and visions of a better life and the discursive characters of their politics, and to see the possibilities for broadening environmental issues into a movement for livelihood, entitlements and social justice. In short, serious attention to the imaginaries embedded in alternative discourses, in social movements and in critiques of environmentally 'sustainable' capitalism can have profound emancipatory potential.

But as we have shown, this post-structural turn – and its political project of anti-development – also carries its own burdens, its own mythos and its own politics. We see three grave dangers in some of this work. First, post-structuralism construes language as essentially a system of *exchange* and then extends an analysis based on this to society and the economy. But Perry Anderson has properly argued that these analytical analogies are flawed, that society and the economy are based on *production* and thus are fundamentally not much like systems of exchange at all (1984: 40–46): they diverge sharply from language in the presence of material scarcity, in their most relevant units of social analysis, in their susceptibility to conscious change, and in their maximum possible rates of change. Escobar (1996: 46–7) unreservedly advocates the application of post-structuralist analysis to political ecology, following *precisely* the chain of flawed analogies that Anderson so incisively critiques. Social theory's insufficiently critical adoption of post-structuralist linguistic analysis has been in many respects disabling. We urge far greater caution and critical reflection before taking political ecology along the same road. Society and the economy are unlike language in critically important ways that are too frequently overlooked in the enthusiasm for post-modern analysis; how much *less* like language are material interactions with myriad physical environments, in which scarcity remains highly relevant – contra Escobar (1996: 53) – and social relationships must contend with structures existing well outside their bounds?

A second grave danger in much post-structuralist work resides in its tendency to vastly exaggerate the causal primary, power and influence of what Bourdieu (1990) calls symbolic power (or of discourse generally). To decode World Bank narratives or discourses of neo-liberal structural adjustment – its regimes of truth and market imaginaries – is an important exercise and exposes a certain (if limited) sort of politics and power. But how these discourses have power in Ministries of Planning or why the World Bank can impose such orthodoxies can surely only be grasped in relation to the ways in which that august institution is organized, funded and is inserted into circuits of capitalist circulation and accumulation (in short, political economy of the old sort – about which, it needs to be said, much more needs to be known). And third, in some (but not all) of the post-development imaginary literature, there is a sort of rush to judgement in which the field of what it deems to be deterministic, Eurocentric or antiquarian is hastily jettisoned. Yet we would argue that post-structural theory, which owes much of its appeal to the deconstruction of the Western myths of science, truth and rationality, itself has fabricated its own mythology about political economy and the baggage of modernity. To this extent there must be a path back out of the cul-de-sac of post-modern politics, and from the often romantic, incomplete and populist visions of anti-development:

> there is surely something wrong with summarily dismissing the language and politics of rights, equality and collective solidarity with the poison-skull labels of Eurocentrism and masculinism and leaving it (and them) with that. For insofar as they are also, and indeed, pre-eminently concepts formed out of the long experience of capitalism . . . itself, precisely as the always contested, uncertain terrain of counter-cultural resistance to the exploitation . . . intrinsic to that culture's workings, then . . . some transliterated version of them will become necessary in any place or situation in which capitalism is found at its deterritorializing/reterritorializing work – which is to say very nearly everywhere on the planet. (Pfeil, cited in Waterman, 1996: 175)

In short, we believe that it is essential for political ecology – and for any investigation or evaluation of development, social movements, or capitalist uses of nature – to retain as central the analytical insights of political economy. A compelling and liberatory political ecology must begin with an accurate understanding of capitalist dynamics for the simple and profound reason that they lie at the roots of most problems with which political ecology concerns itself. Accounts of environment and development therefore should begin still with the overall contradictory character of relations between societies and natural environments (Harvey, 1993; Leff, 1995; O'Connor, 1988).

If one is to pay heed to Escobar's and others' call for a post-structural political ecology or a liberation ecology, there are perhaps three avenues along which such a project might proceed. First, a critical approach to nature has been one of political ecology's central weaknesses, and there is surely need for a more social relational understanding of natural science itself (of the institutions of science and scientific regulation (Latour, 1993)), including a sensitivity to what one might call nature's agency or causal powers. Included in this re-examination of nature is (and here some aspects of post-structuralism are exceedingly useful) a grasp of the complex imaginaries and social constructions of particular environmental ideas – conservation, risk, erosion – as they emerge in (typically) local formations with circuits of capital (Neumann, 1995; Lash *et al.*, 1996; Beck, 1992; Brechin, 1996). Second, we are only now beginning to have ethnographically complex accounts of the new green or alternative development movements, accounts which reveal the genesis, history, dynamics and political strategies of liberation ecologies (Rangan, 1995). But here again, the multiple types of transnational green movements – and the social capital by which local and transnational actors build networks and alliances – are barely understood. And finally, a political ecology worth its salt must be rooted in the long *durée* of capitalist dynamics in such a way that it produces some insight into the ways in which successive rounds of accumulation and crisis are in some way shaped by nature's agency (Lipietz, 1995b; Benton, 1989). Whatever the inadequacies of Cronon's (1990) account of Chicago as 'nature's metropolis', it has moved some way towards that goal. Critical science studies, environmental history and social movement theory are critical ingredients of political ecology that, to return to Blaikie and Brookfield (1987), truly engage the dialectics of nature and society.

## ACKNOWLEDGEMENTS

The authors wish to thank Dick Peet, Dick Walker, Allan Pred, Michael Johns and Louise Fortmann for encouragement and critique, and Liz Oglesby for a careful critique of the manuscript.

## NOTES

1. Political ecology had its origins in the critique of ecological anthropology and 'cultural ecology' of the 1970s (Watts, 1983). This earlier theory drew attention to the adaptive capacities of indigenous societies both in the efficacy of their 'cognized models' of the local environment (for example, farmers in Ivory Coast possessed a sophisticated understanding of local soil conditions and botanical relations), and in their *structural similarities* to all biological populations and living systems. Rappaport's (1967) classic account of the role of ritual pig killing among the Tsembaga Maring in local environmental regulation of fragile tropical ecosystems was a model of this ethnographically rich 'systems thinking' about human adaptation to the environment. Culture – for example, ritual practices or social structure – was seen to function as a homoeostat or regulator with respect to environmental stability. These studies took concepts derived from ecological theory or cybernetics and applied them directly to the sphere of social life; peasant societies were adaptive systems just like any other biological population, and culture was posited as an ecologically functional attribute of the evolutionary demands of the environment. Societies were closed homoeostatic systems populated, as Jonathan Friedmann caustically observed, by 'cybernetic savages'. Typically working in rural and agrarian Third World societies, cultural ecologists none the less unearthed important data on local ethnoscientific knowledges and the relations between cultural practices and resource management (*see* Peet and Watts, 1996b).
2. In an astonishingly arrogant and self-serving review of this body of work, Tom Brass (1995) has argued that the debate is impoverished in large measure *because* of Geography; as he puts it, a 'safe' discipline characterized by 'nice maps, pity about the theory'. Sociologists, anthropologists and economists who contribute to the debate are castigated as 'good people fallen amongst geographers'. Geography is also held accountable for the epistemological origins of the moral economy notion, which is, for Brass, all that is bad in development theory.
3. There is insufficient space to respond to this critique but in our view it replaces the essential concepts of class and accumulation with more dubious (and, as Booth notes, untheorized) populist and conservative appeals to popular culture and civil society (*see* Watts, 1995).
4. There seems to be some confusion over the nature of the development discourse in *Encountering development*. Escobar claims that the discourse 'forms systematically the objects of which it spoke . . . as a unity' (p. 40) but later says it does not have a unity but is characterized by 'the formation of a vast number of objects and strategies' (p. 44).
5. 'The greatest political promise for minority cultures is their potential for resisting and subverting the axiomatics of capitalism and modernity in their hegemonic forms' (Escobar, 1995: 224). Escobar, however, has little to say about what constitutes minority (Hindu nationalism? the Islamic Salvation Front in Algeria?) or what indeed is non-hegemonic modernity or capitalism.
6. Escobar also contends that this increased control over nature has blurred the boundaries between nature and society, destabilizing allegedly sound ontological categories: 'Nature is now modeled as culture; sooner or later, "nature" will be produced to order' (1996: 65) (*see* Haraway 1991 on 'cyborgs' for a somewhat different approach to these arguments). Yet, as discussed above, conceptual categories being far less clear in empirical instances is hardly new either – in fact, Latour (1993) argues persuasively that it is merely the *belief* in this ontological divide, as opposed to its actual existence, that characterizes the modern period. Escobar even goes so far as to claim that such material and conceptual blurrings of the boundaries of the nature/culture dichotomy may even lead to 'the end of the ideologies of naturalism' (1996: 61). Given the extraordinary durability of these ideologies in the face of long-standing complex forms of mutual influence between nature and culture (*see*, for example, Glacken, 1967; Smith, 1984; Williams, 1980; Worster, 1977), this prediction seems premature at best.
7. Interestingly, Smith (1984) is rarely mentioned in the recent literature regarding the 'internalization of nature to capital', although its main arguments closely parallel his thesis regarding the 'production of nature' (not least in being overstated).
8. For example for local knowledges, perhaps for Southern states, for those who control access to a geographically specific resource, and so forth.
9. Although it should be noted that Escobar apparently contradicts himself, saying later in the same chapter that 'new biotechnologies . . . further capitalize nature by planting value into it' (1996: 65).
10. For instance, describing the work of Sebastian Münster in the first half of the sixteenth century, Glacken writes: 'In his outline of cultural history, Münster says that as civilization advances, clearing and draining go on, towns are born, castles rise on the hills. Earthworks and dams control the water. Man finishes the creation. Gradually by cultivation, with settlements, castles, villages, fields, meadows, vineyards, and the like, *the earth has been so changed from its original state that it can now be called another earth*' (p. 365: emphasis added).

# CHAPTER SEVEN

# RE-PLACING CLASS IN ECONOMIC GEOGRAPHIES: POSSIBILITIES FOR A NEW CLASS POLITICS

J.K. GIBSON-GRAHAM

And finally, there is the matter of difference and politics: how identity which rests on difference and division, can produce a political common ground. How . . . can we think about forms of political action that enable us to understand these processes of identification rooted in difference, rather than try to transcend them? And it seems to us that this project requires not a retreat from class but a desperate need to retheorize where class has gone to, and to rethink, and reassert it in nonessentialist terms. (Pred and Watts, 1992: 198)

The 'new' economic geographies produced over the past two decades have increased our understanding of global political and economic forces, of spatial divisions of labour, of the heterogeneous nature of corporate practices and of the interaction of economic with social, political and cultural processes, particularly at the local and sectoral scales. A complex knowledge of differentiated economic landscapes has been generated within the overarching project of describing and explaining the restructuring of space economies. In service to this discursive mapping, we have proliferated categories by which the features of variegated economic landscapes may be distinguished. But in so constituting geographies of economic difference we have also and simultaneously produced a geography of sameness. By constructing a vision of an expansive and exploitative global capitalist economy we have created, in the terms by which we construct our class analysis, a geography of class homogeneity. For just as the economic landscape is dominated by capitalism in geographic representation, so capitalism hegemonizes the representation of class. Whatever non-capitalist class processes are brought to light by economic geographers are either relegated to quaintness and unimportance, or become conceptually folded into the hegemonic narrative of capitalist success/sameness.

We would like to suggest that the concept of class could be used as a much more effective tool for constituting landscapes of economic differen-ce,landscapes in which capitalist *and* non-capitalist class relations are visible and vibrant. In what follows we make a rudimentary and experimental attempt to 're-place' class in recent economic geographies, stressing the dimension of difference that certain geographers have recognized but not developed or pursued. Our hope is to open up a discursive and material space for thinking and enacting class differences, providing yet another way of configuring the relationship between economic geography and political intervention.

This project of class differentiation is not easy to undertake within the discursive frame of economic geography. Given the epistemological realism of this tradition and the forceful narrative of capitalist restructuring it has established as a 'reality', we may expect to encounter in our readers both antagonism and incredulity. We therefore take our contrary steps quite warily and respectfully. We are aware that what we see as theoretical choices, others may see as theoretical givens. For us a 'given' that stands in the way of a new politics of class is the vision of the economy as a capitalist totality.

Rather than accept this vision of capitalist hegemony, we would like to call it into question and to put forward a class-differentiated and anti-hegemonic economy. We do so in the spirit of other radical theorists who theorize diversity in the social realm and locate hegemony in the field of discourse or the symbolic order (e.g. Butler, 1995). Some feminists, for example, have theorized a hegemonic discourse of binary gender that overdetermines social existence, while depicting the social world as home to a diverse array of gender practices and embodiments. They pursue this discursive strategy as an alternative to theorizing a social structure of gender domination such as patriarchy (Pringle, 1995). While the reductive and hierarchical binary is both prevalent and powerful, and while it deeply affects social existence, it does not (unlike patriarchy) fully capture the social field. Gender as a social process exceeds the binary formulation that is one of its constituents. Gender may be seen as continually and performatively reinvented, in radical and even revolutionary ways, and power and political efficacy are not confined within a hegemonic social structure or hierarchy. This theoretical strategy offers greater visibility for social difference and a broader range of political possibility.[1]

In theorizing the economy, as in other kinds of theorizing, there are always theoretical options, and for us in this case it is a question of where to locate the hegemony. Like the feminists who theorize a hegemonic binary gender discourse rather than a society structured by gender domination, we wish to represent 'capitalist hegemony' as a dominant discourse rather than as a social structure or systemic unity. This move allows us to depict the economy as always already differentiated, but also to understand why there should be so few representations of a class-differentiated economy. Our project in this chapter is to prise open the potentiality of certain minoritarian strains in economic geography and read them perhaps against their grain, to catch some glimpses of a diverse class economy. We undertake this project in the hope that a vision of class diversity, even one that is faint and hard to discern (lurking in the shadow of a radiant capitalism), even one that speaks softly (muffled under the weighty discourse of capitalist hegemony), might inspire others to depict class difference with more boldness and clarity. If non-capitalist class practices could be represented as prevalent and persistent, if we could see them as proliferating in the face of capitalism, perhaps we could lend both inspiration and credibility to an anti-capitalist politics of class transformation.

As feminists have struggled not only against male domination but to liberate gender difference from the phallocentric discourse of the heterosexual binary, perhaps we can struggle not only against capitalist exploitation but to liberate class from the capitalocentric discourse of capitalist hegemony.

## GEOGRAPHIES OF ECONOMIC DIFFERENTIATION: THE CREATION OF A HOMOGENEOUS CLASS LANDSCAPE

In at least one sense, the project of economic geography has always been to constitute a differentiated economic landscape. The term 'constitute' is a loaded one, however. Many economic geographers would see their research activities as projects not of 'discursive constitution', but of 'uncovering' or 'exposing' what has really happened in the changing world economy. Thus the categorizations of economic differentiation that have been generated by economic geography are usually represented as increasingly accurate reflections of transformations occurring in the economic field. In contrast, to talk of 'constituting' an economic geography evokes the performativity of social representations, encouraging us to consider the ways in which our choices and desires as theorists are implicated in the worlds we ostensibly represent.[2] It is precisely this performativity that we are concerned in this chapter to highlight and explore.

Traditionally, the practice of economic geography was concerned with specifying the categories by which we might map changes over space and time in the form and function of economic activity. We are all familiar with the old distinctions between primary, secondary, tertiary and quaternary economic activities and the ways in which such a classification focused upon the physicality of the product produced *in situ* (its closeness to nature, degree of transformation by technology, tangibility, life span and so on). Whole landscapes of sameness defined by specialization within these rules of differentiation emerged: resource and agricultural regions, industrial heartlands and urban complexes. The behavioural turn in economic geography focused upon the micro-differentiation of these functionally specialized landscapes brought about by individual decisions of managers acting as optimizers or satisficers of least-cost location.

Under the influence of Marxian political economy, new categories of economic differentiation based upon different relations were introduced into the lexicon of economic geography. The landscape could be divided into that devoted to production (collapsing the distinction between primary and secondary activities and selectively incorporating particular tertiary, or service, industries) and non-production, including so-called 'unproductive' commercial and financial activities, as well as the activities of consumption and 'reproduction' conducted in domestic and neighbourhood spheres. The centrality of production to the Marxian problematic prompted new classifications based upon sectoral and skill-linked divisions of labour, which allowed for the emergence of an economic landscape structured by 'spatial divisions of labour' (Massey, 1995a). These new, theoretically informed mappings were often seen to be the spatial manifestations (in part at least) of economic logics – the capitalist imperative to accumulate and the tendency towards capitalist crisis.[3] Industrial restructuring theory, which took these logics of accumulation and crisis as central organizing narratives, focused attention upon the changing spatial linkages and scope of firms (and the regions in which they were based) and the generation of landscapes of globalization and localization. It also defined different organizational regimes of production and consumption, thereby constituting geographies of Fordism, neo-Fordism and post-Fordism. New 'spaces of production' built upon post-Fordist organizational principles were identified, for example, as the emerging 'growth centres' of global capitalist development (e.g. Scott, 1988b).

Economic geography in its 'new' guise constituted a landscape in which relations of power, exploitation and oppression became visible as part of geographical (not just political or historical or sociological) reality. The nature of ownership of industrial property and control over technology, and the type of labour process employed, became as important as, if not more important than, the extractive (primary), fabricative (secondary) or interpersonal (tertiary) nature of economic activity. And the peculiar 'behaviours' of corporate decision-makers were read in terms of their degree of adherence to the discipline of the market and the dictates of capitalist profitability. Thus landscapes of different types of capitalist production and capitalist employment preoccupied the new spatial vision. And concern over heightening conditions of exploitation and oppression motivated much of the first-wave work on the geographies of deindustrialization, internationalization of production, labour market segmentation, outwork, subcontracting and export processing zones.[4]

As the relations of production were so central to this Marxian vision, the social geographical world was seen to be increasingly defined in class terms. Empirical research gravitated to those regions where a 'new' working class was being recruited, or where an 'old' working class was being abandoned owing to processes of capital mobility.[5] The 'localities research' agenda explored the details of working-class capacity *in situ*, focusing either explicitly or implicitly upon distinguishing and explaining the degree of working-class mobilization against or co-operation with capitalist processes (Murgatroyd *et al.*, 1985; Bagguley *et al.*, 1990; McDowell and Massey, 1984; Parr, 1990).

On the other side of the class divide, industrial geographers studied differentiation within the capitalist class primarily in terms of corporate structure – defining the very different intra- and inter-firm organizational forms that coexist in the contemporary economy. Recently there has been an attempt to moderate the accentuated economism of most corporate analyses by exploring the cultural and social 'embeddedness' of enterprises (e.g. Mitchell, 1995). Work by economic geographers on producer services or on non-production fractions of capital (including finance, trade and real estate) has differentiated the economic landscape in terms of national and international affiliation, size, market power and orientation, management style and masculinist practice (Thrift and Leyshon, 1994; McDowell, 1995), among other things. In these enquiries into the capitalist landscape, the language of class is somewhat attenuated. While it is clear that capitalist enterprises and practices are under examination, class has little purchase as a tool of analysis.[6]

Where economic geography overlapped with development geography and the study of the 'Third World', early Marxist attempts to describe the economic landscape employed the concepts of class and the articulation of different modes of production – capitalist and peasant or traditional modes, for example. Interestingly, as analysis of the internationalization of capital has developed and been succeeded by the language of globalization, the landscape of class diversity in the 'Third World' has been cast into shadow. More recently the prior, if less precise, categorizations of informal, traditional and capitalist 'economies' have assumed dominance over the now out of favour structuralist imagery of articulation. On the terrain of globalization, the relationship between capitalist and non-capitalist practices in the 'Third World' has largely been represented in terms of a one-way flow of influence and power, and if class is mentioned it is linked to the growth of a new proletariat and 'middle' class.[7] In the face of an increasingly networked and integrated world economy, the difference posed by Third World economies is more likely to be apprehended in terms of the specificities of culture, race, ethnicity and post-coloniality than of class (Gibson-Graham, 1996: ch. 6).

Though the term 'economic difference' has not been widely employed as a descriptor of what economic geography actively constitutes, from this brief and inevitably idiosyncratic overview it seems clear that we have been prolific in representing a differentiated economy.[8] Interestingly, though, it would seem that what has emerged from economic geographic discourses of restructuring has been a landscape of class that foregrounds one class relationship: that between capital and labour. We have depicted a class panorama of intensities and passivities, of old and new, of skilled and unskilled, of organized and disorganized, of First and Third World labour. It is a world of difference undoubtedly, but difference within the same – the 'working' class.[9] At the same time, we have produced a class landscape of large and small capitalist enterprises, Fordist and flexibly specialized industrial producers, mass banking and niche-orientated financiers, new business services and outmoded family enterprises. It is also a world of difference, but again difference within the same: the capitalist class. The multiplicity of possible class relations has been reduced to one key relationship, largely because class has become associated with a solidified and singular categorization of social and economic identity: a capitalist social formation, a capitalist world economy, or some other version of capitalist hegemony.

Geographic representations of capitalist industrial restructuring have created an increasingly unified economic world within which difference is produced and reproduced. As Pred and Watts put it,

> [W]e want to stress the importance of understanding the historical production of difference within a *unified system*. Difference and identity are produced and reproduced within a field of power relations rooted in interconnected spaces linked by political and economic relations. (1992: 198; emphasis added)

For us, the field of power relations that 'contains' difference is not only capitalism but capitalocentric discourse; that is, discourse that represents capitalism as (naturally) fuller, more real, more central, or more powerful than its non-capitalist others.[10] Capitalocentric discourse defines capitalism as the economic standard, and defines non-capitalist forms with respect to capitalism: as the same, the opposites, the complements, the subordinates, or the deficient others of capitalism.[11] Thus, capitalism may be seen as an integrated and functional economic system that possesses its own internal laws or logics and is capable of reproduction, whereas coexistent forms of non-capitalist activity may be seen as incapable of self-reproduction and as dependent for their continued viability upon capitalism. When capitalism is represented as the economic container or the dominant element of the economic totality, everything within that is not part of capitalism loses its economic autonomy. The classes defined by non-capitalist production are positioned as marginal, residual, insufficient and incomplete, becoming in effect parts of capitalism or necessary for capitalism (e.g. relations of exploitation in the household that are seen as capitalist reproduction). Their contribution to a landscape of class diversity is thereby rendered relatively insignificant and ineffective. The capitalocentrism of restructuring discourse has created a landscape of class sameness, eclipsing the terrain of non-capitalism. And it is this effacement of class diversity that we wish to begin to redress, via the strategic intervention of a different conception of class.

## THE DIFFERENCE DEFINITION MAKES

In economic geographic discourse, class is usually defined as a composite structural category denoting a specific relationship to (a) property ownership; (b) power and control over the labour process; (c) exploitation (the appropriation of surplus labour); and (d) organizational capacity. Taken together, these things are seen to confer membership in a social grouping, or class. By the term 'structural' we mean that classes and class positions are situated within an overarching social structure or system (usually capitalism). This structure is usually dominant with respect to other class structures, such as slavery, feudalism and independent commodity production, and associated with economic closure and social hegemony.

Underlying the composite structural class definition is the vision of a unified class subject for whom the essential four relationships line up in a coherent and self-reinforcing way. For example, when an individual owns no property, has no control over the labour process that they are engaged in as a wage labourer, and is a member of a mass-based organization of similarly placed workers, it is easy to classify this individual as a member of the working class. But where these relationships do not all line up neatly, analytical and political problems emerge. For instance, when an individual owns a house and some land and runs a plant nursery open for sales at weekends, works as a wage employee as a computer programmer in a capitalist firm where the labour process is completely in her own control, manages the labour of other computer staff, and belongs to a white-collar union, their class-ification is difficult to accomplish. The individual owns the means of production of plant propagation, but this does not make her a capitalist; in this activity she is engaged in independent commodity production, appropriating and distributing her own surplus labour. As a computer programmer she sells her labour power and produces surplus value and in those acts could be seen as a member of the capitalist working class, but she chooses when and how to work and supervises the labour of others, which places her in a specific segment of the working class – perhaps the labour aristocracy. Her membership in a white-collar union that operates as a professional association and does not engage in oppositional politics means that her working-class capacity is diminished and she is a fragmented and 'suspect' class actor/subject.[12]

By using a composite categorical definition of class we are forced, in situations where classification is not straightforward, to weight one or two of the essential characteristics of class more heavily than others. The contradictory interaction of an individual's multiple dimensions of social and economic identity risks being understated or conceptually papered over. In this example we might represent the worker as a member of the new flexibly specialized post-Fordist workforce whose allegiance to class politics has been disappointingly diluted despite her structurally defined membership in the capitalist working class. Her economic role as an independent commodity producer might be ignored, or might be seen to confirm rather than contradict this classification. So we could be led to assume that her ownership of a small business contributes to the dilution of her political energy as a worker, due to her petty bourgeois consciousness and tendency to identify with the capitalist class, whereas in fact it may lead to her resistance to increased working hours and compulsory weekend shifts in the computer business.

Our argument is that the capitalocentrism of most economic discourse has significant class effects because it leads to the privileging of capitalist class relations. And since the composite categorical definition of class rests upon the notion of a unified class subject, subjectivity becomes a conduit by which class homogeneity rather than diversity is discursively produced. Capitalocentrism not only positions the capitalist economy as the principal social and economic identity (as in the term 'capitalist society') but designates the classes of capitalism as the principal form of class subjectivity. So when confronted by a fragmented class subject, it is the participation in capitalist class relations that is valorized while the participation in non-capitalist class relations is devalued or discounted. And when confronted by a diverse class landscape it is the capitalist and working classes who are represented as capable of innovation, self-reproduction, progressive opposition, collective organization and class transformation while these capacities are denied to other, non-capitalist classes.[13]

One of the important (though unintended) effects of the 'new' economic geography and its constitution of a homogeneous class landscape in which certain relations of power, property, exploitation and capacity are highlighted at the expense of others has been the damping down of an anti-capitalist politics of economic intervention. In the face of a capitalist hegemony that is almost unquestioned, that veils the practice of non-capitalist class processes and the contradictory class outcomes of restructuring, and that locates progressive political will in a collective subjectivity that is putatively undermined, it is no surprise that projects of non-capitalist economic development are thin on the ground.[14] Nor is it surprising that much of the left (more or less resignedly) has embraced post-Fordist strategies of economic rejuvenation as the way to patch up or improve capitalism (because we cannot get rid of it) (Gibson-Graham, 1996: ch. 7). The question we are motivated to ask is, can there be other visions that allow for economic difference in class terms both for the here and now and for the future? And what kind of class politics could be imagined and enacted in the presence of such alternative visions?

## CLASS REDEFINED

In order to reinvigorate class as a dimension of economic difference, we advocate unburdening it of some of its freight and associations. We prefer to define class not in terms of an amalgam of characteristics (property ownership, control over the labour process, appropriation of surplus labour) or as a social group that obtains its identity and political role within a structural or systemic totality, but more simply as a social process involving the production, appropriation and distribution of surplus labour in whatever form (Gibson-Graham, 1996: ch. 3; Resnick and Wolff, 1987). When individuals labour beyond what is necessary for their own reproduction (however defined), surplus labour is produced and appropriated by those individuals or by others in a process of exploitation. The appropriators then make the decisions about where and when and how the produced wealth is distributed.

One reason we are interested in this definition of class as a process with two moments (one of exploitation and the other of distribution of appropriated wealth) is that it enables us to highlight not only the violation and 'unfairness' that is exploitation but also the social and political potential of distribution. Distributions of surplus labour feed and invigorate certain social arenas and activities, and bypass and foreclose others. It makes a great difference whether a particular share of appropriated value is distributed to shareholders in the form of dividends or to the union in the form of severance benefits or to the community in the form of a fund for alternative development (Gibson-Graham, 1996: ch. 8). A politics of class centred on displacing one appropriator – a capitalist, perhaps, or a board of directors of a firm – and supplanting that appropriator with the producers or their representatives and/or members of a given community not only changes relations of exploitation but may enable a new set of surplus distributions. These distributions will contribute to a different set of activities and perhaps even to the creation of a new social order. It is in part to enable a vision of surplus labour distributions as a potent social constituent that we define class in terms of the production, appropriation and distribution of surplus labour rather them in terms of power or relations of domination (though power is an important overdeterminant of relations of class).

In our vision, multiple class processes – capitalist, independent commodity, slave, communal, feudal, for example – can coexist in any society, since none is associated with a systemic totality. The economy and social formation can be seen as fragmented and dispersed or as provisionally and hegemonically unified in the dimension of class. The question then presents itself in any theoretical project, of whether and when to theorize the dominance of one class process (Gibson-Graham, 1996: preface). Class subjectivity is similarly not closed or unified around a particular class process but is structured by a variety of class experiences and positions as well as other relations; for example, of gender, ethnicity, sexuality, age or locale. Again, what this means is that the understanding of any class subject is a theoretical project, in which specific relations of power and property and particular experiences of consciousness and capacity must be theorized rather than presumed.

To empty class of much of its structural baggage and prune it down to one rather abstract process concerning labour flows might seem rather reductive. Our purpose is, following the plea by Pred and Watts quoted at the outset of this chapter, to adopt an anti-essentialist[15] definition of class; that is, to allow the conditions of existence of any class process (such as a particular geography or constellation of property and power relations) to assume specific importance in the formation of class societies and subjectivities without presuming their presence or role. For us, one of the liberating effects of employing the notion of class as a process is that we can abandon the concept of a class as a group of unified economic subjects and the associated expectation that such subjects should or could act in their own class interest (defined as something singular and obvious).[16] Most importantly, though, by divorcing the concept of class from a structural or systemic totality, we are able to challenge the vision of a single axis of class antagonism and to examine the economic landscape with an eye to perceiving the heterogeneity and multiplicity of class relations.

## CLASS AND NON-CAPITALIST PRACTICES

We have sketched out the broad contours and lineage of an economic geographic discourse that naturally contains its own critics and even counter-discourses. In this section of the chapter we would like to build upon the work of two economic geographers whose critical voices have done much to push the boundaries of thought within the field. The work of Doreen Massey and Michael Watts is interesting in the context of the present argument for the ways in which it has problematized and documented non-capitalist production and classes and explored the relationship between class and non-class processes, specifically those of gender and kin.[17] Their work provides a rich source of raw materials for a rewriting of the landscape of class in terms of diversity rather than sameness. What follows is an attempt to reread selected aspects of their empirical research projects somewhat against the grain.

### Class-ifying the Locality

In *Spatial divisions of labour*, Massey presents a close examination of the evolution of new social and spatial structures of production resulting from capitalist restructuring in the UK. The research is politically motivated by an implicit interest in the implications of these changes for the organizational capacities of trade unions and other expressions of working-class solidarity. Thus she is keenly observant of the processes by which 'both the decline of the old and the form of the new composition and geography of the working class present difficulties to the construction of a coherent workforce organisation' (Massey, 1995a: 284).

Massey's analysis pays particular attention to the relationship between these new structures and the 'pre-existing' socio-economic spatial structure in selected localities. In the farming and tourist region of Cornwall she highlights the presence of classes other than the working class. Women, for example, have been 'directly involved [in] a kind of petit-bourgeois self-sufficiency and independence' (1995a: 217) in the self-employed hospitality sector, and a new brand of 'small entrepreneurs' has grown up among newcomers to the area attracted by the environment and lifestyle (ibid.: 220). Cornwall is represented as a region of class heterogeneity in which a new band of migrant independent entrepreneurs has begun to infiltrate the established landscape of independent commodity producers.[18] It is also portrayed as a place that is resistant to a working-class politics of opposition and where the capitalist class practice of competition for survival is absent:

> While many married women in Cornwall may not have been confined to the home and to domestic labour, and may even have appeared in official statistics as 'economically active', they have nonetheless rarely been working in capitalist wage relations and even more rarely within any labour process likely to provide the basis for solidarity and workplace organisation.
>
> . . . although there were certainly contrasts in attitudes to growth, *none* of the 'independent small-firm manufacturing sector' conformed to the classic model of capitalist accumulation. Certainly, the traditional sector tended to be of the 'simple reproduction' variety, but the new entrants do not live up to the fashionable image of the striving entrepreneur either. (Massey, 1995a: 217, 220; emphasis in the original)

Massey's empirical analysis employs a composite categorical definition of class (1995a: 30–44). The members of a petit-bourgeois independent producing class in Cornwall own their own means of production (such as a bed-and-breakfast establishment or a home-based office with computers), control their own labour process, are self-appropriating and are seen as incapable of either the working-class capacity of collective organization and solidarity or the capitalist class instinct of competitive self-reproducing behaviour:[19]

> [T]he implications of this kind of [new] entrant for employment *growth* in the future, and therefore a further expansion of the working class, are not major . . . and for all the same reasons, it does not provide much basis for self-reliant regional growth.
>
> . . . Certainly, a strategy based on these kinds of firms is unlikely to be a viable basis for regional economic resuscitation. (Massey, 1995a: 221; emphasis in original)

Cornwall is interestingly constituted by both class absence and class presence, and Massey's analysis provides a heterogeneous terrain to be productively reread in the terms of an alternative class analysis.[20] If we were to step outside of a capitalocentric discourse dictating that capitalist enterprises have the unique or pre-eminent capacity to grow and become self-reliant, that only members of the working class can organize effectively at the workplace, and that non-capitalist economic practices are incomplete and dependent in some inherent way, we might be able to conceive of a conceptual or political space in which a different kind of (non-capitalist) regional resuscitation could possibly emerge. The possibility of promoting collectivities among self-appropriating businesswomen or environmentally aware small businesspeople presents itself for consideration by economic activists interested in developing a self-reliant non-capitalist economic sector. A decentred and inclusive conception of economic subjects and economic activism could unleash a new politics of 'development' which is not dependent upon the presence of capitalist class relations or doomed or truncated in their absence.[21]

In chapter 3 of *Reworking modernity*, Watts (1992) constitute, in very different circumstances and with different emphasis, a knowledge of non-capitalist labour practice and struggle. His fascinating study of contract agricultural labour in the Gambia focuses upon changes in the pre-existing production relations prompted by the introduction of a state-sponsored rice irrigation project. Watts employs a composite categorical definition of class in which power mediated by exchange relations plays a determining role. He describes the contract agricultural labourers in his study as an emerging global proletariat (1992: 104), a distinctive class fraction 'captured into the capitalist working class' (p. 207). While these independent commodity producers, or peasant farmers, own their own land and means of production, control their own labour process and are able to collectivize their farming labour if they wish, the agricultural products of their labour are contracted to a centralized commodity buyer (the state) which is also the provider of production credit. The state authority sets a production schedule to which the growers must conform: 'Under contract, the technical division of labor and the overall production process is regulated, directly and indirectly, by the contractor, who controls a dispersed and typically unorganized class of growers' (p. 104).

In Watts' reading of the class landscape in the Gambia, certain power relations exercised by the state (over the financing of irrigation, production and exchange of agricultural produce and thereby the distribution of the surplus produced by independent commodity producers) come to dominate the classification of these workers as 'unfree' and therefore members of the working class:

> Authoritarian and despotic forms of contracting establish and regulate work conditions in a manner that renders household labor in effect unfree. In other words, grower labor is unfree in the sense that it is directly distributed, exploited and retained by politico-legal mechanisms. . . . Nominally independent growers retain the illusion of autonomy but have become in practice what Lenin labeled 'propertied proletarians', de facto workers cultivating company crops on private allotments. (Watts, 1992: 82)

Whereas we might be interested in classifying these workers as self-appropriating (though regulated and constrained by the state), Watts' definition of class in terms of power relations makes him see them as *de facto* members of the working class. In the case of our classification, we are motivated by the desire to keep open the possibility of a politics of self-appropriation, through which growers might improve the conditions of non-capitalist production or increase the amount of surplus they appropriate (Hotch, 1994). Watts' interest is very different. He wishes to illustrate empirically the ways in which 'capitalism may contribute to the reproduction of nonwage labour' (1992: 105). This desire results in the description of an intensely varied terrain of production relations, property relations, oppositional struggles and symbolic conflicts – a heterogeneous landscape that, incidentally and fortunately for us, provides wonderful raw materials for thinking through the implications of a different conception of class.

Prior to the introduction of the irrigation project, rice production took place in lowland areas reclaimed from mangrove swamps. Rice production was women's work and swampland was owned by women by individual right. Household production in the local Mandinka society is based upon the cultivation of both individual fields and collectively owned familial property and, under customary law, the rights of ownership and distribution of the crop produced on each type of property are different. The product of collective land is controlled and distributed by the senior male in the household and the product of individual land is appropriated individually. In the terms of our class analysis we have here two different class processes with different conditions of existence – a collective class process in which the distributional moment is controlled by the patriarch (we could call it a patriarchal collective class process), and an individual self-appropriating class process.

With the introduction of the rice irrigation project and the movement of men into rice production, this property complex and mix of household class processes appears to have been altered.[22] The sequestering of land to the project and associated rearrangement of property rights meant that women's access to their traditional land and to an independent self-appropriating class process was largely destroyed:

> Women . . . lost their individual rice fields, and were in effect proletarianized by project development. The lack of compensation, in other words, the failure to establish a conjugal resolution, compelled women to both withdraw from domestic farming responsibilities on irrigated plots and simultaneously sell their labor power or engage in other income earning enterprises. (Watts, 1992: 96)[23]

One response made by women who had been rendered landless was to join together with other similarly dispossessed women to sell their labour power. Drawing upon traditional organizational practices and 'customary social relations as a basis of recruitment' (ibid.), women formed groups of similar age to work in gangs in the rice paddies. The labour team (kafo) utilized reciprocal labour practices and negotiated a collective wage which was distributed equally among the members:

> Women are fully cognizant of both the need for skilled paddy field labor, which they can supply, and the growers' dependency on prompt in-field operation to meet the rigorous schedules imposed by project management. The kafo groups bid up their collective wage accordingly. . . . Working as a group, the women are able to transplant two plots a day, the proceeds from which total roughly 60 per cent more than the individual daily wage. (Watts, 1992: 209 n52)

In our understanding, these women were forced off the lands on which they had previously engaged in a non-capitalist class process of self-appropriation and have entered into a capitalist class process[24] in which they collectively receive a distribution of appropriated surplus value in the form of a wage premium. This latter relation indicates their considerable power with respect to the growers (male peasants in the role of capitalist appropriators of the surplus generated from the women's 'proletarian gang labor' (ibid.: 96)). In this instance, then, the women are exploited but powerful, in the sense that the growers are dependent on them. Their power is not infinite but partial and particular, as is the growers' power with respect to them. In this context of shifting and differentiated power relations, many possible transformations of relations of exploitation could be explored. Just as these rice fields have been fertile ground for shifts in class relations in the distant and recent past, they could be the ground in the future of an intentional politics of class transformation.

The question of how to treat this complex of capitalist and non-capitalist class relations in the Gambia rests upon our theoretical purpose here – whether to highlight difference and diversity or to establish sameness by anchoring differences within a given unity. Watts is interested in the latter project (he sees the contract farming and women's kafo labour as 'unfree' and as therefore basically subsumed to capitalism (1992: 103–5)). We respect this theoretical decision though we cannot entirely divine the reasons for it. For our part, we are interested in a different project, that of creating the discursive conditions for a politics of class transformation. If we can see class as always already complex and differentiated, and class difference as not (or not necessarily) produced and contained within a hegemonic totality, we may be able to imagine and, indirectly, to enable a new class future, one in which class differences of a positive and progressive sort are produced politically.

The heterogeneous economic landscapes of Massey's Cornwall and Watts' Gambia provide a rich set of examples of non-capitalist economic activity. Though the representation of each area is shadowed by capitalist class relations (their local absence in the case of Cornwall and their highly mediated but ubiquitous presence in the case of the Gambia), these places are shown in such sensitive detail that we have been enticed and provoked to illustrate (rudimentarily) our conception of class on their terrain. Our definition of class is nothing more than a strategy by which we can attempt to think economic difference given the dominance of capitalocentric discourse. There is no inherent ability of this concept of class as opposed to another to liberate difference. But there is the likelihood that by highlighting the very contextual and overdetermined, in fact non-essential, relationship between power, property, exploitation and organizational capacity, class identity can be decentred and multiplied and the unified identity of the economy can be discursively fractured. Under such theoretical conditions, possibilities of economic and class transformation become more visible or imaginable.

## CONCLUSION: ON POLITICS

Our class process perspective is not tied to elaborating class relations in terms of the dominant cleavage between capitalists and proletariat. It can therefore make visible a landscape of class difference and bring to light the intense class struggles being fought in many sites (including, as in Watts' work, between men and women within the household).[25] Slave, capitalist, collective feudal, and independent relations of surplus appropriation coexist on a terrain of class differentiation in which these and other production relations are enacted in both market and non-market spheres, and in which the identity of the economic 'totality' is not only unfixed but up for grabs, conceptually and politically (Gibson-Graham, 1995).

It is the contribution of an anti-essentialist class analysis to the project of alternative economic activism that attracts us most. While it might appear an empty task to rename class homogeneity as class heterogeneity, or a fraction of a proletariat as many different classes, such a naming exercise has important implications for both understanding political subjectivity and for envisioning or constituting spaces of economic possibility. Such a class approach can be used to challenge the capitalocentrism of economic discourse, which forecloses the many opportunities that might exist for seeing class capacity in action or for constituting different paths of economic development – alternatives, that is, to the royal road of capitalism.[26] Our political and theoretical interest is in creating (constituting) alternative economic futures in which class diversity can flourish. Thus we are attracted to explicating class as a process and to highlighting its many different contemporary and potential forms.

In our current research, we have begun to explore regions as class (and otherwise) differentiated economies. This project we see not only as theoretical and empirical but also as political. Research, the activity most economic geographers are engaged in, involves the production of knowledge, and knowledge is performative and powerful. Alternative knowledges interpellate (in the Althusserian sense of calling into being) alternative political subjectivities and thereby enable new forms of agency. By engaging in an action research project with local interviewers who are economic actors trained to investigate the variety of regional class relations (including but not limited to self-employment, domestic labour arrangements among household members, collectives, capitalist businesses, slave or feudal relations (often in illegal or non-market sectors of the economy)), we produce an alternative knowledge of regional development, outside the capitalocentric discourse that perceives these regions as declining resource extraction sites suffering a prolonged withdrawal of capital. This alternative discourse provides new subject positions for economic agents (for example, as creators of local instances of socialism, defined as surplus appropriation by a collectivity) and new visions of the 'economy' incorporating various systems of social reproduction and value formation. In this sense, the research provides a basis for a new kind of economic activism and contributes to the emergence of a different class politics.[27]

Producing representations of class relations and of differentiated economies is something economic geographers have done over the past 25 years with skill and adventurousness. We hope that our thoughts on alternative representations of class will be seen as part of this lineage.

## NOTES

1. A recent article in *Social Text* provides a biological example. Hubbard (1996) notes that into the world are born all sorts of babies whose genitals do not conform to binary gender. In our society, these 'failures' are quickly rectified, while in others they may become part of a three- or multi-sex spectrum. In this example, binary gender is a discursive construct that leads to the heterosexing of 'ambiguous' babies, obliterating the field of difference that biology offers.
2. As Foucault states, 'It does not matter that discourse appears to be of little account, because the prohibitions that surround it very soon reveal its link with desire and power. There is nothing surprising about that, since, as psychoanalysis has shown, discourse is not simply that which manifests (and hides) desire – it is also the object of desire' (1981: 52–3).
3. One of the points of difference among economic geographers interested in restructuring was the degree of determination granted to this logic versus other more contingent, local or particularist features (*see* Massey 1995a: 301–26 for a clear discussion of this axis of theoretical 'difference').
4. A seminal work of economic geography that brought together a description and theorization of these new features of spatial–economic differention with concerns over class was Doreen Massey's *Spatial divisions of labour*, subtitled *Social structures and the geography of production* (published in 1984 and republished in a second edition (Massey, 1995a)).
5. Somewhat more problematic was work on the new 'middle class' being formed by the same processes (Thrift, 1987). Indeed, it is this work on, for example, the banking and financial sectors, or the new managers, which has looked to theories of gender and consumption identities as a way out of the confusion that surrounds class theories of the non-classic working class (*see* McDowell, 1995).
6. Perhaps because of the definitional difficulties involved, but also owing to the political orientation of the researchers, the landscapes of capitalists as individuals are rarely distinguished or differentiated whereas the differential geographies of the working class have been the focus of much attention by the localities researchers.
7. This has prompted Pred and Watts to make the plea for retheorizing 'where class has gone to' in the passage we have chosen as our epigraph.
8. Interestingly, when 'difference' is referred to within economic geography (for example, in the outline of this book) it more often refers to so-called non-economic axes of identity such as race, gender, ethnicity or sexuality. Difference appears to be the domain of culture or the social or the subject.
9. Perhaps we would not mind so much about this sameness if it were not for the fact that this landscape was not also one loaded with political expectation and disappointment. The intense pressure and judgement associated with the contours of class and 'its' capacity is enough to make many retreat from this concept as a liberating and useful analytical tool.
10. Of course, some might say that if we live in a capitalocentric world, we need to recognize it as such. We would argue that this view exemplifies what Althusser called 'empiricism' in which the truth of the object of discourse is assumed to lie in the object itself rather than to be produced through the contentiousness of the discursive process. Similarly, some have expressed the concern that by downplaying the power of capitalism, we expose ourself to the political danger of hiding the gross nature of capitalist exploitation. We would respond that the gross nature of capitalist exploitation can be recognized (if indeed that is what the theorist wishes to do) without conflating it with capitalist power or hegemony, just as domestic violence can be perceived as a terrible and widespread problem without associating it with a stable hegemonic structure called patriarchy.
11. The definition of capitalocentrism is based on an analogy with phallocentric discourse in which woman is represented as the same, the opposite, or the complement of man (Grosz, 1990: 150; Gibson-Graham, 1996: 35).
12. Wright (1985) would see here an example of a contradictory class location.
13. It appears that the capitalist class division alone has the power to affect the course of economic history. Thus the working class, remnant and dispersed though it may have become because of restructuring, remains the expected vessel of progressive class capacity. Such a view ordains certain social sites with the capability to reshape the class

contours of society, ignoring the possibility that classes positioned as residual, archaic or dependent may also have the capacity for progressive social change (*see*, for example, the positioning of informal workers by Pahl and Wallace (1985)).

14. This is not to say that an oppositional politics that effects change in the economy does not exist, but that such movements are often not identified by either themselves or political commentators as economically targeted or anti-capitalist (Kauffman, 1995).
15. We prefer the term 'anti-essentialist' to 'non-essentialist' as there is no way that any definition or representation of identity can completely avoid employing the strategy of essentialism, no matter how mediated or momentary its use. Our commitment to anti-essentialism signifies an epistemological pledge to fight many of the effects of essentialist thought without making the essentialist claim that non-essentialism is possible (Fuss, 1989).
16. Biewener (1995) argues, following Joan Scott, that even E.P. Thompson, who set out to free class from the categorical prison of Marxist structuralism, ultimately sees class as an identity rooted in structural relations that pre-exist politics: 'in his use of [experience], because it is ultimately shaped by relations of production, [class] is a unifying phenomenon, overriding other kinds of diversity' (Scott, 1992: 29).
17. Both researchers are interested in the interaction of power relations constituted outside the workplace class relation, for example in the household, upon class dynamics.
18. Of course, it is difficult to know whether Massey's petit-bourgeois producers are capitalist or independent commodity producers, since her class analysis was conducted in slightly different terms. We would classify firms that appropriate surplus labour from wage labourers as capitalist and classify as independent commodity producers (and therefore non-capitalist) those firms in which an individual appropriates her or his own surplus labour.
19. These attributes of successful resistance to individuation and exploitation or successful adoption of accumulationist growth strategies are projected on to the absent working and capitalist classes.
20. Clearly we are directing our class analysis to a different end than that of Massey. While we greatly value the work that is seeking to explain the uneven development of working-class organizational capacity, we are interested in exploring the variety and diversity of possible non-capitalist class relations in the interests of promoting an anti-capitalist politics of economic difference.
21. Hotch (1994), for example, treats self-employment as a viable non-capitalist economic strategy, rather than as a precarious form or failure of capitalism.
22. Land for the project was sequestered from collective household property as well as from individual women, and in addition was newly cleared by men who argued, drawing upon customary law, that this labour conferred ownership on the clearer of land, and that they were now the traditional owners.
23. Note here that the 'other income earning enterprises' are not separated out from selling their labour power in terms of the class positioning of the women. In Watts' reading, both of these are associated with proletarianization, whereas for us only the latter is clearly so associated.
24. Women not only withdrew from working on their individual land (this had been taken) but also withdrew their labour from collective household production in order to work in the kafo. This resulted in domestic violence and divorce (Watts, 1992: 96).
25. In Watts' words, 'the struggles over work obligations and the disposition of rice surpluses – in short, production politics – were typically conducted in the idiom of conjugality, that is to say, the customary reciprocalities and exchanges between husbands and wives' (1992: 94). It is the capitalocentrism of most economic discourse that leads to the common representation of production politics in the domestic sphere as gender and kin conflicts while production politics in the extra-household realm are represented as class conflicts. *See* Gibson-Graham (1996: ch. 2) for an extended discussion of the 'sphere of (capitalist) reproduction' and the discursive obliteration of domestic labour as an axis of class difference or revolutionary organization.
26. This means breaking with the representations of unity, singularity and totality that are associated with capitalism (Gibson-Graham, 1996: ch. 11). In contrast to Watts, we believe that it is only outside of capitalocentric discourse that a landscape of difference in class terms can be imagined and realized.
27. And not just class politics. We would hope that it would contribute to a social and economic politics in which gender, race, and other dimensions of difference were as important as, or even more important than, the dimension of class.

# CHAPTER EIGHT

# LOCAL POLITICS, ANTI-ESSENTIALISM AND ECONOMIC GEOGRAPHY

**JOE PAINTER**

## INTRODUCTION

Local politics affects the geography of economic activity in many ways. It is perhaps rarely the most important influence, but its implications can scarcely be ignored. In most countries, city and local governments are major economic actors, employing large workforces, spending significant budgets and providing the local population with goods and services. Investments in particular places by both private and public sectors are influenced by planning, environmental, infrastructural, educational and cultural circumstances, which are in turn controlled or affected by local political institutions and processes. Local political campaigns around issues as diverse as environmental protection, labour relations, housing, service provision, education and many others have consequences for the direction of economic change. At the same time, the influence does not run only one way. Firms and business organizations use a variety of techniques to influence local political decision-makers, while local political institutions often depend directly or indirectly on the prosperity of the local economy for their resources.

However, recognizing that there are important relationships between local political institutions, actors and processes on the one hand and economic geography on the other, reveals nothing of the nature of the links, and nothing of how they should be interpreted and conceptualized. Both the economy and local politics are highly complex and the relationships between them correspondingly so. Among other things, this means that local politics is not a simple reflection of local economic activity, and local economic activity is not controlled directly by local political institutions. In this chapter I shall argue that local politics and economic geography have to be understood both on their own terms and in relationship to one another. I define 'local politics' as politics whose object is to affect 'local society' (the local economy, the local community or local institutions) but where defining local economy, community and institutions and so on *is itself a political process* involving the development of 'discourses of the local' and of political practices which constitute 'the local' as an object of political intervention. This kind of approach is very much in tune with the emergence of new ways of thinking about economic geography that no longer take objects of analysis (such as 'local economies') for granted, but increasingly question the nature of 'the economic' and the nature (and even the existence) of a separation between the economy and other aspects of society.

## MICHEL FOUCAULT'S 'GENERAL HISTORY'

In investigating the interconnections between local politics and economic geography, I wish to avoid according ultimate explanatory primacy either to economic processes or to politics (of which local politics is a part). Instead, the approach I have in mind is akin to Michel Foucault's description in *The archaeology of knowledge* of the 'new history' which he says will be a 'general' rather than a 'total' history. For Foucault (1972: 9), traditional 'total' history 'seeks to reconstitute the overall form of a civilization, the principle . . . of a society, the significance common to all the phenomena of a period, the law that accounts for their cohesion'. In constructing a 'total' history,

> it is supposed that between all the events of a well defined spatio-temporal area . . . it must be possible

> to establish a system of homogeneous relations: a network of causality that makes it possible to derive each of them, relations of analogy that show how they symbolize one another, or how they express one and the same central core. (Foucault, 1972: 9–10)

This is the approach adopted implicitly or explicitly by much writing on the relations between politics and the economy. For some, the 'central core' is the economy (sometimes more specifically the process of capital accumulation), around which all other aspects of society are thought to revolve and in terms of which they are explained. Conversely, others organize their accounts around a 'political' core, such as the 'nation-state' or a cultural-political construction such as the nation. In these accounts, the relative prosperities of different countries are understood to derive from different institutional arrangements, or differences in some imagined 'national character'. While in one sense these contrasting accounts are completely opposed to one another, from the perspective of Foucault's critique of 'total history' they are united in adopting a particular form of explanation, namely that the diverse and complex mixture that makes up society is thought to be explicable in terms of forces, causes and relationships which can be traced back to a central core. This explanatory core may be different in different accounts, and this has been the source of much academic dispute, but the approach to explanation is the same.

By contrast, the new history, Foucault writes, will be 'general' history. The postulates of 'total' history are challenged by 'general' history

> when it speaks of series, division, limits, differences of level, shifts, chronological specificities, particular forms of rehandling, possible types of relation. This is not because it is trying to obtain a plurality of histories juxtaposed and independent of one another: that of the economy beside that of institutions, and beside these two those of science, religion, or literature; nor is it because it is merely trying to discover between these different histories coincidences of dates, or analogies of form and meaning. The problem that now presents itself – and which defines the task of a general history – is to determine what form of relation may be legitimately described between these different series; . . . what interplay of correlation and dominance exists between them; what may be the effect of shifts, different temporalities, and various rehandlings; . . . A total description draws all phenomena around a single centre; . . . a general history, on the contrary, would deploy the space of a dispersion. (1972: 10)

This passage makes it clear that while Foucault rejects the idea of a single central core which animates all other features of the social world, he insists that there are linkages and connections between different spheres. While the history (and, I would add, geography) of political institutions cannot be explained in terms of the history and geography of the economy, this does not mean that they do not affect one another. Politics cannot be reduced to economics (or vice versa) but nor are they, in Foucault's view, independent. The task is thus to explore the nature of their connections, without making prior assumptions about them.

Having established that economic geography and local politics (among many other facets of social life) are related but not reducible to one another, I now want to consider how these relations might be understood. As I shall suggest below, one set of relations which is particularly important (and often, I think, underestimated in the past) is the connection between ideas, discourse and forms of knowledge. These concerns are also central to Foucault's work, but while I have found his ideas in this area useful, there are other areas I want to consider where Foucault's writings are of less help. This is partly because in *The archaeology of knowledge* he moves quickly on from the discussion of general history to discuss his approach to the history of ideas, and his hints at the character of general history are not elaborated in further detail. It is also partly because, in thinking about the connections between local politics and economic geography, I want to defend the concept of determination. To be sure, in my view, determination is multiple, complex and reciprocal, and I want to insist that accounts of determination *can* be partial, provisional and anti-essentialist (because they need not be written as if 'ultimate' or 'fundamental' causes have been discovered). Nevertheless, although this approach to causality does not appear markedly at odds with Foucault's sketch of 'general history', it is clear that elsewhere Foucault has considerable reservations in principle about the notion of causation. Given the absence of space to discuss these reservations in detail, it seems appropriate to turn to other sources to develop my ideas further. To begin with, I will draw on David Harvey's account of the conditions within which a relatively autonomous urban politics can arise. I then want to push the argument further using Bob Jessop's work on the concept of 'autopoiesis' (self-reference) in social systems.

## MAKING SPACE FOR LOCAL POLITICS

In his essay 'The place of urban politics in the geography of uneven capitalist development',

David Harvey (1985c) takes issue with a number of writers who, he says, claim that the complexity, contingency and fluidity of contemporary urban politics cannot be interpreted within the framework of the Marxist theory of capitalist accumulation, and that Marx's ideas should therefore be abandoned. Harvey goes on to argue that the Marxist theory of accumulation not only is *compatible* with the existence of a dynamic urban politics which has (in the short run, at least) autonomy from economic processes, but actually provides an *explanation* of how such autonomy arises.

The cogency of Harvey's proposals depends on conceptualizing the process of accumulation as inherently spatial. The process of capital accumulation, according to Marx, depends on the appropriation of labour power and its combination with other means of production, raw materials and so on, in the labour process. However, as Harvey points out,

> labor power, Marx emphasized, is a peculiar commodity, unlike others in several important respects. To being with, it is not produced under the control of capitalists but within a family or household unit. There also enters into the determination of its exchange value a whole host of moral, environmental, and political considerations. And finally, the use value of that labor power to the capitalist is hard to quantify exactly because of the fluidity of inherently creative labor processes. (1985c: 128)

The process of accumulation must be understood as inherently spatial partly because it depends on labour power which, in the short run at least, is place-bound. On any given day, workers can only travel a limited distance to their places of work. This produces a local labour market whose qualities (in terms of the supply of labour) are fixed in the short term, though liable to change over the longer term. In the case of capitalist production, local labour markets tend to be *urban* labour markets, because concentrating labour power in limited geographical areas produced more job opportunities for workers and a greater and more varied supply of labour for firms.

According to Harvey, such place-bound urban labour markets provide the conditions which give rise to forms of urban politics that are relatively autonomous from the process of accumulation. That is, they open up a space for political activity, institutional formation and policy-making which are not simply direct products of economic processes, but have their own dynamics and patterns of development which may run counter (at least in the short run) to dominant economic interests. This is the consequence of processes of both production and consumption in the urban area involved. On the production side, capital investment often involves the development of fixed infrastructures such as buildings, communications networks and service facilities which tie particular firms to particular labour markets for a period of time. There may also be 'soft' investments, like links with suppliers and product markets, or skills training. Moving away would mean these investments being written off – a potentially costly exercise. On the consumption side, a particular labour market, with its specific mix of skills, wage levels and workplace cultures, generates a particular mix of consumption activities (purchases of goods and services, collective provision, self-help networks and so on). These consumption activities involve not only necessities (the minimum requirements for the reproduction of labour power) but also discretionary and luxury goods and services. (Among other things, this means that it is not necessarily sensible for local firms to try to pay the lowest wage possible, since this will depress the demand for such discretionary consumption.)

Over time, therefore, urban labour markets become differentiated from one another. They develop along different paths, each of which reflects a different mix of production and consumption activities and associated social, cultural and political norms and behaviours. The result, according to Harvey, is that each urban region (based around a particular urban labour market) exhibits a tendency towards 'structured coherence', at the heart of which

> lies a particular technological mix – understood not simply as hardware but also as organizational forms – and a dominant set of social relations. Together these define models of consumption as well as of the labor process. The coherence embraces the standard of living, the qualities and style of life, work satisfactions (or lack thereof), social hierarchies (authority structures in the workplace, status systems of consumption), and a whole set of sociological and psychological attitudes toward working, living, enjoying, entertaining, and the like. (1985c: 140)

Harvey emphasizes that urban regions exhibit a *tendency* towards coherence, rather than established coherence, because both capital and labour are more place-bound over shorter time-scales and more mobile over longer ones. This means that the development of structured coherence is liable to disruption in due course, which in turn explains why Harvey believes that urban politics is *relatively* rather than wholly autonomous. Autonomy is evident in the short term; in the longer term the character of urban politics is likely to be affected by the economic consequences of the mobility of labour and capital.

The tendency towards structured coherence 'spawns a distinctive urban politics' (Harvey,

1985b: 140). This happens because (again in the short to medium term) capital and labour share a dependence on the particular mix of skills, wages, investments, infrastructures and production and consumption activities in the urban region. As Harvey puts it (1985c: 141), 'cooperation, cooptation and consent are also a part of class struggle . . . accommodation [between capital and labour] plays a vital role in giving a relatively stable structured coherence to production and consumption within the urban region'. The result of this, Harvey argues, is to promote an urban politics based around shifting (and often unstable) urban or regional 'class alliances'. This is a politics which is relatively autonomous from the process of capital accumulation because it can involve alliances and coalitions between groups whose interests appear to be fundamentally opposed when the abstract features of the accumulation process are considered in isolation and without reference to their necessarily spatial manifestation.

## BREAKING WITH ESSENTIALISM: LOCAL POLITICS AND AUTOPOIESIS

Harvey's account speaks directly to the problem I outlined at the beginning of the chapter; namely, how can we understand the links between local politics and economic geography (Harvey's 'uneven capitalist development') in ways which see them as related, but which do not reduce local politics to a direct expression of economic processes? It therefore overcomes many of the problems of more reductionist accounts of the relations between local politics and economic processes, such as Cockburn's (1977) assessment of the local state as mainly an instrument of capital. However, in one respect, Harvey's account is not wholly compatible with the criteria I specified in the introduction above. This is because although his account of urban politics is a rich and nuanced one, it does retain a vestige of essentialism.

An essentialist argument is one which grounds its explanation in a fixed foundation which is permanent (and thus provides a firm basis for knowledge) and which lies outside whatever is being explained (and can thus be claimed as the ultimate cause or end-point of the explanation). Once the essence has been uncovered, the explanation is held to be complete. In classical Marxism, labour power provides the essence underlying explanation. As Trevor Barnes (1996: 57; *see also* Graham, 1990) points out, in the Marxist schema the labour theory of value meets both the above criteria; the nature of labour power (as a 'physiological fact') is held to be invariant over time and place, and labour values (defined in relation to amounts of human energy) are independent of economic processes and therefore lie outside of the thing they explain (the process of capital accumulation).

According to Trevor Barnes, essentialist explanations should be rejected for two reasons. First, 'as soon as we accept essences, we have a closed system, a system impervious to the dynamics, diversity, and difference of the changing contexts in which social practices are embedded' (1996: 56). Second, as a number of thinkers (including philosophers Richard Rorty and Jacques Derrida) have shown, essentialism is logically contradictory. Essentialism seeks to provide a foundation for explanation which lies outside of the thing being explained. But this means that logically these foundations cannot themselves be explained without contradicting the essentialist method. If they are explained (using the approach of essentialism) by reference to something other than themselves, then by definition they were not ultimate essences at all, but the product of something else. Conversely, if they are explained by reference to something within themselves, then the essentialist argument that the source of explanation must lie outside the thing being explained is contradicted.

Harvey argues that a relatively autonomous urban politics is a necessary product of the combination of two things: the need for labour power in the process of capital accumulation and the constraints on the ways in which labour power is used that arise from its organization in geographically bounded labour markets. This approach is not inevitably essentialist, and in some places there is a distinctly non-essentialist cast to Harvey's account. Thus he points out that both labour power and labour markets are themselves socially conditioned and produced. In the case of labour power, for example, 'there also enters into the determination of its exchange value a whole host of moral, environmental, and political considerations' (1985c: 128), while 'work habits, respect for authority, attitudes towards others, initiative and individualism, pride, and loyalty are some of the qualities that affect the productivity of labor power as well as its capacity to engage in struggle against capitalist domination' (ibid.: 133). In the case of labour markets, 'segmentations may exist in which certain kinds of jobs are reserved for certain kinds of workers (white males, women, racial minorities, recent immigrants, ethnic groups, etc.)', while the 'nature of skill distributions in relation to the mix of labor processes within the urban region' (ibid.: 128) affects the smooth operation of the urban labour market.

Thus Harvey certainly does not regard labour power as an unproblematic category. On the other hand, he quotes approvingly Storper and Walker's claim that 'labor differs fundamentally from real commodities because it is embodied in living, conscious human beings and because human activity (work) *is the irreducible essence of social production and social life*' (Storper and Walker, 1984: 22–3; emphasis added). This suggests that the social conditioning mentioned above has the status of a contextual influence on an essential core, an interpretation which is supported by the words (such as 'imperfections' and 'modifications') which Harvey chooses to describe the factors that affect the character of labour power in particular labour markets.

It would, though, be misleading to suggest that Harvey's account of urban politics is irredeemably flawed by its essentialist elements. On the contrary, it reveals that it is not necessary to adopt an explicitly anti-essentialist position to generate considerable space for diversity, difference and socially embedded political practices. However, the fact that the notion of structured coherence is derived ultimately from the essential characteristics of human labour power does leave its mark on Harvey's argument. For example, the autonomy of urban politics in Harvey's account is only ever 'relative'. In the end, the temporary fixity, on which the possibility of such a politics depends, will be swept away by the turbulent upheavals of the accumulation process. In the medium and long term the possibility of labour migration, and the search by capital for new sources of accumulation and new sites for investment start, constrain the range of political outcomes that are possible at the urban level. Furthermore, within the space afforded by the tendency to structured coherence and the differentiation between urban regions, Harvey privileges political alliances based on class above those based on other social divisions.

Equally, it is important to be clear that one can be anti-essentialist and still see an important role for capital–labour relations. The mobility of capital may well have important effects on the options available to political actors in urban regions, migration may undermine the tendency to structured coherence, and class alliances may often be of crucial significance. However, from an anti-essentialist perspective, such claims can only be made on the basis of empirical investigation. They cannot be derived by deductive reasoning from the first principles of Marxist theory. Anti-essentialism holds that there are no first principles; behind every apparent first principle lies another set of questions. Substantive outcomes observed in any given urban region may well be those proposed by Harvey, but this is not inevitable; rather, as the overdetermined products of multiple causal processes their character needs to be investigated, not assumed.

A non-essentialist approach, which provides a way of conceptualizing the relationship between economic geography and local politics without according ultimate explanatory power to one or the other, can be developed using the insights of a version of systems theory which examines what are known as 'autopoietic' systems. 'Autopoiesis' may be translated as 'self-reference', and refers to the way in which certain systems exhibit paths of development which are independent of, and without reference to, their surrounding environments. What this means in the social world is that 'society' is made up of numerous systems, each of which operates and develops according to its own logic and independent dynamic. The internal norms, structures and practices of the system are seen as determining its future behaviour; it therefore behaves with reference to itself, rather than to outside 'determining' forces (hence 'self-reference' or 'autopoiesis'). Autopoiesis has its own problems, and does not provide the basis for a full theory of local politics. However, it serves to highlight the differences between essentialist and anti-essentialist approaches, and it is therefore interesting to outline the autopoietic approach, even if the insights it provides have to be incorporated rather selectively into any understanding of local politics.

Although autopoietic theory was developed initially in evolutionary biology, it can be applied to social systems.[1] This produces a formulation which has striking parallels with the notion of 'general history' developed by Foucault:

> [Autopoiesis] emerges when the system in question defines its own boundaries relative to its environment, develops its own unifying operational code, implements its own programmes, reproduces its own elements in a closed circuit, obeys its own laws of motion. When a system achieves what we might call 'autopoietic take-off', its operations can no longer be controlled from outside. Autopoieticist social theorists argue that modern societies have seen many such systems develop along functional lines and have therefore become so highly differentiated and polycentric that no centre could coordinate all their diverse interactions, organizations and institutions. Nor is there any functionally dominant system which could, *pace* Marxists, determine societal development 'in the last instance'. (Jessop, 1990b: 320)

From an autopoietic perspective, therefore, local politics and the urban or regional economy may be seen as two separate systems (coexisting with many other systems) with neither controlling the other, and neither explicable in terms of the other. Local politics has its own sets of institutions, rules, codes of conduct, logics of development and patterns of behaviour which have to be understood mainly in terms of the path taken by the 'local politics system'. The urban and regional economy, organized around the urban labour market, has a different set of rules, norms, institutions and practices which develop according to its trajectory. In practice, in most countries, it is doubtful whether local politics is fully autopoietic. This is not because its development is governed by the logics of the local labour market, but because many of the institutions and activities which make up local politics form part of another system: the national political system of government and the state. Local government organizations are usually part of the national state. However, the work of Niklas Luhmann suggests that this national political system *can* be thought of as autopoietic (*see* Jessop, 1990b: 320–7).

Among the defining features of autopoietic systems, according to Jessop, are that they set their own boundaries and define their field of activity in their own terms. Within the national political system, for example, many areas of social life are defined as lying beyond the boundaries of the system and therefore no concern of politicians. The operation of the (supposedly free market) economy is one such, as are religion, sport and many others. This shows actors within the political system defining its limits – autopoieticism in practice, as it were. However, this is a contested terrain. The setting of the limits to politics is itself a matter of political conflict. I will return to the implications of this in relation to local politics later on. As well as the substantive boundaries, the system also *operates* according to its own *sui generis* rules and procedures. In the case of the central state this involves the legislature, the executive and the judiciary, elections, party competition and forms of bureaucratic procedure. Local government similarly operates its particular systems for policy formulation, decision-making and implementation.

The notion of autopoiesis provides a break with essentialism, by allowing us to interpret urban and local politics on their own terms: to trace their own paths of development, institutional practices and norms, and the histories and geographies these produce, without assuming that these are ultimately limited by the constraints imposed by necessity from other spheres. However, while it is important to break the necessary, logical, essentialist connection with the labour theory of value, it is also important to understand that there *are* connections between economic processes and urban politics, albeit ones which are ultimately contingent. Here Luhmann's autopoietic theory is of less help. As Jessop points out (1990b: 327), by understanding social systems as radically autonomous, autopoiesis cannot explain the relations between them except by resorting to the idea that there may be co-evolution of two or more systems purely by coincidence.

To solve this problem, Jessop uses the concept of 'structural coupling'. Although autopoietic systems define their own boundaries with their environment and operate according to their own rules, the nature of their environment is not irrelevant to their behaviour. They may deal with it in ways which are self-defined, but deal with it they must, especially if it is a source of disruption to, or constraint upon, the reproduction of the system. In the modern polycentric societies it is likely that any two systems will form part of each other's environment. Thus the urban and regional economy is part of the environment in which local politics operates, and vice versa:

> Since any system necessarily exists (and is reproduced) in an environment, its own development is always related to that of its environment. In the case of autopoietic systems, however, it is their internal operations which determine how they will react to exogenous events. . . . Thus the current operations and organizations of a system are always a joint result of its own dynamic and that of its environment. Through this sequential, path-dependent interaction the system comes to be structurally coupled to its environment. (Jessop, 1990b: 328)

## CONSTITUTING THE LOCAL POLITICAL ARENA

I have suggested that the relationships between local politics and its economic environment need to be conceptualized in a non-essentialist manner which respects the dynamics at work in each and which does not allocate ultimate explanatory priority to either. I have also suggested that Jessop's use of the idea of structurally coupled autopoietic systems is a helpful starting-point. I now want to explore the implications of these ideas for the study of local politics in more detail.

Just as politics consists of much more than elections and the activities of governments, so local politics is not limited to local elections and the activities of elected local government. Going beyond this narrow interpretation, though, will cause problems for those who like their definitions discrete and unambiguous. We might define local politics as political processes which have impacts on local areas, but it would be difficult to find any which do not. Alternatively, we might identify local politics as politics which 'happens locally'. However, in one sense all political activities take place in specific local settings; all political agents and institutions occupy particular places, and, as geographers frequently argue, are influenced by context. More problematic still are approaches which define local politics around a particular substantive political issue or conflict, such as Peter Saunders' (1981) proposal of what became known as the 'dual politics thesis' in which urban politics and the local state are identified as the realm of consumption conflicts while the nation-state is regarded as the realm of the politics of production issues.[2] Similarly, Peter Taylor's (1985) suggestion that the scale of 'locality' represented the sphere of experience in contrast to the spheres of ideology and reality represented by the scales of the nation-state and the world economy seems to draw too direct an association between analytical categories and particular spatial scales.

In place of these problematic a priori conceptions, I want to work with a non-essentialist definition of local politics which begins with the proposition that the category of 'the local' is itself constituted politically. I define local politics as politics whose object is to affect 'local society' (the local economy, the local community or local institutions) but where defining local economy, community, institutions and so on *is itself a political process* involving the development of 'discourses of the local' and of political practices which constitute 'the local' as an object of political intervention. This approximates to the autopoietic process outlined above of a system setting its own boundaries, but, in contrast to some versions of systems theory, such boundary-setting needs to be understood as a social and political process.

# KNOWLEDGE AND LOCAL POLITICS

Interpreting local political practice in relation to the economy thus needs to begin with the conflicts over what the local economy is and where the legitimate role for political activity lies. Such conflicts arise from differences between the institutions and actors involved in local politics. These differences, of course, include differences in material interests and political strategies adopted, but of most relevance here are differences in the understandings, knowledges and ways of knowing which different groups of actors bring to the political arena. To clarify the significance of 'ways of knowing', it is helpful to consider examples. A brief comparison of two recent, but very different, studies will illustrate the point well: Nick Fyfe's account of the planning of Glasgow and Irene Bruegel and Hilda Kean's reflection on municipal feminism (Fyfe, 1996; Bruegel and Kean, 1995).

## Post-war planning in Glasgow, Scotland

Nick Fyfe's outline of the contested visions of Glasgow embodied in the ideas of competing planning authorities and in the writings of local poets resonates with the argument I have been developing about the significance of the struggle to define and represent localities. As Fyfe puts it himself (1996: 394), 'there can be few better examples of a struggle for the definition and the making of the built environment than the events in postwar Glasgow'. Drawing on Henri Lefebvre's (1974) work on the production of space, Fyfe proposes that the rationalist conceptions of the city which dominate the designs of the planners constitute what Lefebvre calls representations of space, while the work of the poets produced mainly in response to these technocratic visions forms Lefebvre's spaces of representation – the field of lived experience in which human understandings of spaces and places are developed, contested and transformed.

In his study, Fyfe discusses two plans for Glasgow which were prepared soon after the Second World War, and which contain contrasting (and as it turned out, politically competing) visions of the future of the city. While both plans were agreed that a radical transformation of the built environment of the city was needed, their strategies for bringing this about were markedly different. The local state in Glasgow, in the shape of the then Glasgow Corporation, produced a plan for the reconstruction of the city in the high modernist tradition of Le Corbusier (the Bruce Plan, after its author and City Engineer, Robert Bruce). For Bruce, 'planning is an instrument, the skilful use of which can, in time, secure order out of disorder, efficiency in many spheres where efficiency cannot under present conditions be attained, and beauty which in uplanned areas exists only by accident' (cited in Fyfe, 1996: 390). Bruce's vision of the city involved massive redevelopment and rationalization of the core, a fast network of urban motorways to take traffic and people quickly and easily around the city, and the clearance of large areas of working-class housing and the relocation of their residents away from the redeveloped centre and traffic corridors to suburban locations on the edge of the built-up area.

The Clyde Valley Regional Plan (CVRP), produced under the auspices of the Scottish Office (that part of the UK's central state responsible for the government of Scotland), shared Bruce's diagnosis of Glasgow's problems (especially its overcrowding and chaotic spatial arrangement). However, by contrast with Bruce, the CVRP proposed relocating those displaced in the slum clearance programme not to the city's suburbs, but to a series of new towns lying beyond a green belt, which would be protected from development. In this, Fyfe implies, the authors of the CVRP were drawing on an anti-modernist discourse which sees large cities as pathological and in need of control and spatial restraint, and which harks back to the much smaller settlements of pre-industrial Britain as the ideal form of spatial organization.

In the event, according to Fyfe, the result lay some way between the two plans. The city centre and transport infrastructure were redeveloped in much the way proposed by Bruce, while the CVRP's new-town strategy was also partially implemented. What the two plans shared, of course, was a distinctively modern confidence in the ability of rational planning to create a more ordered, and by implication a better, world. Where they differed was in the extent to which Glasgow's social and economic infrastructure should develop into a high-modernist mega-city. The key points for the argument of this chapter are, first, that the two plans expressed contrasting understandings of what Glasgow as a social and economic space should be, and second, that the conflict between these different understandings was the basis for political conflict between different institutions: Glasgow Corporation (in favour of the Bruce Plan), and the Scottish Office (in favour of the CVRP) (Fyfe, 1996: 394–6).

## Municipal Feminism in British Local Government

The second study I want to consider is also concerned with changing the definition of the locality and local economy, albeit in a very different way. In this case the interest is less in the shape of the built environment of the local area (the economic infrastructure, as it were) and more with the operation of the labour market and the links between the social relations of the labour market and those within the local state.

In their interpretation of the role of feminist thought and practice in British local government, Irene Bruegel and Hilda Kean argue that the period 1983–87 marked the 'moment of municipal feminism'. Municipal feminism in the 1980s, they suggest, drew on, but differed from, earlier and ongoing traditions of socialist feminism in its more critical conception of the state and local state. It was not enough for women to use the local state as a tool in their pursuit of feminist objectives. Rather the local state would also itself have to be transformed along feminist lines. This position arose from the recognition that while the local state might formally adopt feminist policies, these would often be ineffective if the informal social relations within the local state were left unchanged. The local state is not a neutral instrument capable of being used equally by all social groups and political movements. On the contrary, the local state, like the central state, is itself constituted around conflictual and unequal social relations. As a gendered institution in its own right, the local state is an arena, as well as a tool, of feminist struggles.

For many feminists, including Bruegel and Kean, much of the labour movement (the traditional source of political strength for the British left) was dominated by sexist attitudes and practices. These tensions grew on the back of processes of economic restructuring in which traditional sources of male employment were in dramatic decline. For many in the traditional labour movement, manual employment tended to be seen as providing 'real jobs' only in so far as they were full-time jobs in traditionally male occupations. By contrast, the growing numbers of part-time women manual workers were often felt to be a threat to full-time, male employment. White- and 'pink'-collar jobs in offices, another major source of growth in female employment, were also seen as problematic. Bruegel and Kean (1990: 155) recall 'a G[reater] L[ondon] C[ouncil] billboard slogan calling for "jobs not offices", which took quite a lot of hard bargaining to get altered'.

Thus for Bruegel and Kean, as for many other feminists, municipal feminism involved much more than the promotion of women's interests within a local labour market which was unproblematically taken for granted. It rather involved questioning the constitution of that labour market, and particularly its restructuring, in gender-sensitive rather than gender-neutral terms. This in turn involved questioning the way the local state worked. In many localities, local government was one of the largest local employers and certainly among the largest employers of women. This opened up the possibility of using the apparatus of the local state as a means of influencing employment practices in the wider local economy. For example, if the wages of low-paid women manual workers in the public sector could be raised, the numerical domination by local government of the market for this type of labour might force local employers in the private sector to pay higher wages too (Bruegel and Kean, 1995: 152). In addition to pay, the character and status of traditionally female occupations were also the targets of municipal feminist questioning of labour market relations. In the 1980s the Conservative-controlled central government forced local authorities to invite private-sector tenders to undertake services then provided directly by in-house workforces. The services involved included major employers of women workers such as the school meals services and the cleaning of council buildings. Several campaigns to retain such services in-house were influenced by the municipal feminist approach. These involved service users (such as parents and children in the case of school meals) and women workers working together to improve the quality of service provision and training programmes for staff to improve the productivity and career prospects of the workforce (Bruegel and Kean, 1995: 153; *see also* Painter, 1991).

The ideas and political practices of 1980s municipal feminism represented the mobilization of a particular set of understandings about, and definitions of, the local economy in the process of political struggle. These included the definition of the public sector as a key part of (rather than parasitic upon) the local economy; the recognition of the gendered nature of the impact of economic change on local labour markets; the attempt to overcome the differences of interest between the producers and consumers of services through co-operative working; and a focus on networks as a form of organization between the hierarchical form of the traditional public sector and the market form of the traditional private sector.[3] In each case the taken-for-granted nature of the local economy is being problematized on the basis of distinctively feminist ways of knowing.

## LOCAL POLITICS AND THE ECONOMY: STRATEGIES FOR INTERPRETATION

I have suggested that analysing the struggle to constitute the local economy as an object of political practice is one of the keys to an anti-essentialist interpretation of the relationship between local politics and economic geography. Further, I have argued that the struggle to define the locality and its economy is a struggle between different sets of ideas, which reflect different forms of knowledge and ways of understanding the relationships between economic relations, the state and the local state, civil society, and the built environment. On the other hand, neither economic nor any other outcomes are determined by the struggle of ideas. It is important not to substitute an essentialism grounded in the nature of labour power with an equally essentialist view that identifies conscious human understandings as the foundation of social life. Social and economic outcomes are determined, but that determination is multiple and complex. First, no one set of understandings wins any battle of ideas outright. Second, the production of strategies informed by particular ways of knowing often has consequences which were not intended or foreseen by the actors involved. Finally, in line with the notion of autopoiesis outlined earlier, discursive, material and institutional processes and practices in other fields of social life have profound effects on the impact of political strategies.

If the starting-point is the discursive constitution of the local economy, this kind of approach means that the interpretation of local politics can be taken further in two directions.[4] Moving in one direction, the analysis can trace the political processes to the struggles between ways of knowing and acting and the different social and political groups and actors involved. As the interpretation proceeds, an increasing array of other influences and processes (each with its own material and discursive conditions of existence, which could in turn be unpacked) are brought in from other social spheres. This direction provides some assessment of the effects of local politics on the geography of the economy and leads to an analysis of outcomes. In the other direction, the interpretation traces the conditions of existence of the competing knowledges themselves. To take the examples used above, what processes of production gave rise to the contrasting discourses of rationalist planning and municipal feminism? What political identities and social groups provided the settings for their construction?

It will be apparent that these two (apparently opposite) directions intersect to form a perpetual circle, or, better, a set of intersecting (and no doubt sometimes incomplete) circuits or networks.[5] The preconditions for one set of phenomena are themselves phenomena with their own pre-conditions. This ontology is consistent with anti-essentialism, in that it is never possible to claim to have found an ultimate cause. The process of explanation can never rest secure on a set of first principles or an eternal essence because each step in the explanatory process simply raises more questions to be answered.

## NOTES

1 There are, of course, major difficulties with importing ideas from the natural sciences into social theory. However, I suggest that the link here is one of analogy which provides an interesting tool to think with, rather than a substantive claim that interpretations of social processes can be modelled directly on natural processes.
2. Saunders subsequently revised his arguments in the wake of criticism. Note, though, that criticism of the dual politics idea should not be taken to mean that consumption issues are unimportant in local and urban politics, merely that they cannot be used to define its limits.
3. While noting the connections between feminist arguments and the process of networking, Bruegel and Kean also point out that many networks are patriarchal in their operation, while rule-bound hierarchies (based on formal political authority) and market relations (based on exchange) might be seen as gender neutral. However, in practice, they argue, all social relations are gendered (1995: 160–5).
4. I have suggested elsewhere how Pierre Bourdieu's concepts of *habitus*, field and strategy may be of help in this process (Painter, 1997).
5. At first glance it might appear that this approach runs the risk of generating a 'totalizing account' contrary to the argument made at the start of the chapter. Such fears are, I think, misplaced, for four reasons. First, although the interpretation of local politics can move in both directions, it seems impossible that a total account could, in practice, be produced. The multiple determinations involved produce such an extraordinarily complex web that a complete map of all its interconnections must lie beyond the scope of understanding. Our accounts of the social world are always partial and provisional, and recognizing the existence of complex relationships is not at all the same thing as being able to define what they all are simultaneously. Second, there can be no privileged location from which the analyst or observer of the social world can undertake the task of tracing all the interconnections. Rather, the analytical and self-consciously theoretical 'way of knowing' typical of academic accounts of local politics is just that: another way of knowing. Third, the idea that interpretation involves tracing the intersecting circuits or networks of social and political processes should not be taken to mean that such circuits or networks are whole and perfect. There is still scope for broken circuits and fractures in the networks to allow for the discontinuities and breaks of social life. Finally, by adopting an anti-essentialist approach to explanation, the interpretative strategy I have outlined eschews the search for the unitary organizing logic which Foucault identifies as the hallmark of total history.

# Chapter Nine

# RETHINKING RESTRUCTURING: EMBODIMENT, AGENCY AND IDENTITY IN ORGANIZATIONAL CHANGE

**SUSAN HALFORD AND MIKE SAVAGE**

## INTRODUCTION

This chapter explores the ways in which gender is embedded in organizational restructuring. Our specific focus is the British banking industry and recent attempts to restructure high-street banking. Here, as in many other sectors, important challenges have been made to long-established organizational paradigms. New forms of management have been privileged, emphasizing innovation, flexibility and the need for cultural change. Alongside this, much has been made of the need to elevate the status of the customer, necessitating new organizational practices of 'customer care' (*see* Halford and Savage, 1995, and Halford *et al.*, 1997, for further discussion). Our investigation of the gendering of these changes raises a number of issues which resonate with wider debates about the nature of 'economic restructuring' and organizational change. In the first part of this chapter we examine these issues through a sympathetic critique of approaches used by geographers and economic sociologists within 'the restructuring paradigm' (Bagguley *et al.*, 1990). This approach directed attention to the specificity of restructuring processes and delineated some of the varied strategies which produced changes in employment relations. However, we will argue that it took an over-economistic view of the dynamics of restructuring, treating organizations as merely instrumental in the process of change and treating people as simply passive recipients of structural organizational change. We will suggest that these interconnected problems diminish the ability of the restructuring paradigm to conceptualize the place of gender in the restructuring process.

In developing an alternative perspective we situate our arguments in the context of other recent research generated by the restructuring paradigm. Few of the writers we draw on are concerned directly with gender, but we attempt to pull these strands together in a way which offers new insights into the gendered nature of restructuring and which, in turn, may contribute to the development of a new perspective on restructuring more generally. We emphasize two points. First, we argue that restructuring must be (at least partly) understood in terms of the social and cultural processes internal to organizations which construct particular personal qualities as desirable or undesirable and, therefore, that restructuring is tied up with redefinition of the sorts of personal qualities which individuals are expected to possess. Second, we argue that people themselves are not simply passive recipients of strategies for organizational change and are able to deny, contest or reinterpret such strategies.

## THE RESTRUCTURING PARADIGM

The 'restructuring' paradigm marked a major advance in the analysis of economic and social change, highlighting the significance of employers' changing demands for particular sorts of labour as a cause of local social change. Massey and Meegan (1979) and Massey (1984) emphasized the variety of restructuring strategies affecting the fates of different sectors within manufacturing. They showed that firms located specific processes and forms of production in areas where they could draw upon 'appropriate' labour, whether they be software houses searching for (male) professional workers in England's Home Counties or multinational manufacturing firms searching for semi-skilled or unskilled female part-time labour in the old industrial regions. Thus, this framework emphasized the way that firms' *demand* for labour changed as a result of restructuring and the social impact of this, especially on local labour markets. It also indicated how local labour markets themselves might influence employers' choice of restructuring strategy and spatial location (Cooke, 1983; Massey, 1984; Murgatroyd *et al.*, 1985; Cooke, 1990).

In this formulation, gender is considered only as an exogenous factor, which might, for instance, affect the supply of labour, but is not intrinsically involved in the restructuring process itself. Gendered considerations may affect decisions made by firms (for example, where cheaper women workers are used to replace more expensive male workers) and economic restructuring may in turn have an impact on gendered jobs and gendered social relations. None the less, the core of the restructuring process remains fully economic: capital's demand for labour is entirely based around cost considerations. However, whilst restructuring is usually presented by employers as the only economically rational action to take – this does not, in itself, constitute a satisfactory explanation of restructuring strategies. While the *claim* of economic rationality may place a given restructuring strategy within a privileged discursive position, 'the claim and the ability to enforce the claim should not be mistaken for the reality of rationality' (Morgan, 1990: 80). Rather, 'the economic' and 'economic rationality' need to be placed within a cultural and social context. Particular restructuring strategies may be adopted because they become fashionable in management cultures, because they protect the interests of powerful groups within an organization or because managers find it hard to see beyond established social norms and practices. These points are illustrated by Schoenberger's study (1994) of the failure of certain US industrial giants to take on restructuring strategies which senior managers had evidence would improve their economic performance. Schoenberger explains this resistance through reference to the ways in which managers' own interests and identities, and their sense of the firm's interests and identity, were embedded in established forms of organization. She cites Pascale's (1990) analysis of General Motors:

> How does one explain this persistent tendency of GM managers to ignore compelling evidence . . . ? The answer lies in its collective identity and deeply etched social rules. It is almost beyond comprehension for GM management to contemplate the full-blown changes in status, power and worker relations that adaptation to the Toyota formula would entail. It is easier to install robots. (in Schoenberger, 1994: 438)

Clearly, it is inaccurate to impute pure economic rationality to management restructuring strategies, however they are presented. This is not to deny that chosen restructuring strategies may benefit firms economically – robotization may improve long-term profits – but to explain their introduction in these terms is to commit the functionalist fallacy of deducing causes from effects. Rather, economic action should be seen as 'culturally embedded' whereby collective understandings shape economic strategies and goals (Zukin and DiMaggio, 1990a), and restructuring can only be understood in this context. These reformulations offer promising opportunities for thinking about the gendering of restructuring.

At least we might argue that the boundaries between economic, social and cultural processes are blurred. More radically, it can be claimed that no purely economic, social or cultural relations are distinguishable but, rather, that each is already embedded within the other (Connell, 1987). This is not to deny the significance of the economic but to indicate that the economic is already cultural (Allen and du Gay, 1994). This enables us to transcend earlier accounts which held gender apart from (economic) restructuring, offering instead the possibility that gender is central to the restructuring process itself. Thus, we can overcome some of the limitations experienced by those researchers who were interested in gender but retained the economic-rationality model of restructuring. For example, while Bagguley *et al.* (1990) drew attention to the limitations of existing research on gender and restructuring, their main suggestion was to refine analyses of the spatial specificity of gender relations in local labour markets (Walby and Bagguley, 1991). Although such a development is to be welcomed, it demonstrates the persistent tendency to treat organizational restructuring as separate from gendering processes, with the result that gender can only 'impact' on gender relations (here, in the local labour market). This obscures the ways in which gendered processes may themselves contribute to restructuring.

One way in which gendered dynamics may enter the restructuring process is via organizational cultures and management decision-making about particular strategies for change. Many accounts of restructuring have tended to treat organizations as simply instrumental in the working out of wider capitalist economic imperatives. For example, Walker's (1989) 'requiem for corporate geography' dismisses the possibility that organizations themselves may have any causal effects on the geography of production. In his formulation, it is capitalism and place which matter. Organizations become merely incidental conduits for processes which originate elsewhere. However, as Dicken and Thrift (1992) argue, Walker can make these claims only because he conceptualizes organizations in an instrumental Weberian way. That is, Walker understands organizations to be rational economic machines in which defined rules and procedures ensure the strictly bureaucratic functioning of the enterprise. Within organizational theory, such instrumental and rationalist notions have sustained severe attacks in recent years. Writers such as Clegg (1990) and Morgan (1986) have emphasized that organizations construct their *own* activities, meanings and cultures rather than simply responding to an external environment. Similarly, Dicken and Thrift's (1992) 'defence of corporate geography' argues that rather than being instruments of external capital logic, organizations have an internal ability to amass and wield social power: 'seen this way, the large corporation is not just an instrument it is also a cause' (p. 183). A crucial rendering of this point is that the inequalities evident inside organizations are, in part, the result of organizational processes and not simply derived from the wider economic and social environment. This point has been elaborated by writers on class who have discussed the significance of 'organizational assets' (Wright, 1985; Savage *et al.*, 1992) and by feminists pointing to the way in which organizational processes are routinely implicated in the construction and maintenance of male domination (*see* articles in Savage and Witz, 1992; Halford *et al.*, 1997). That is to say, gendered relations inside organizations are not merely the reflection of wider social relations of gender but are caused, in part, by specifically organizational dynamics. This point offers us the possibility of further insight into the ways in which gender may be bound up with processes of organizational restructuring. If gender is embedded within organizational forms and practices, then gendered processes are inherently implicated in the construction and implementation of strategies for organizational change.

This discussion of the causal powers of organizations raises a further point. When we say that organizations wield power, or construct meanings, or that organizations contribute to gender inequalities, what precisely do we mean? What are 'organizations' and how are these properties which we have described manifested? This brings us to the place of *human agency* in organizations. We should be wary of reifying organizations in a way which suggests that they have powers independent of the people who constitute organizational activity. Few writers within the restructuring paradigm have paid attention to this form of social action. One example of this is to be found in Bagguley *et al.* (1990). While wishing to avoid an economically determinist view of restructuring (e.g. Bagguley *et al.*, 1990: 211), their endorsement of the structures/agents dualism gave them no real alternative to it. Thus their chapters on restructuring in manufacturing and services did not refer to the activities of the people who were devising, implementing or responding to the restructuring. It was only in a later chapter that they studied the changing position of different types of employees, which they present as a study of 'the experience of restructuring'. In this way restructuring is seen as something which happens to people, rather than as something which happens because some people chose (albeit not necessarily under circumstances of their own choosing) to make it happen. But as Schoenberger says, corporations are run by real people! Organizational change is animated, resisted or modified by the actions of organizational members. This is not to say that all organizational members have the *same* power, but to recognize that restructuring is a human process, requiring human action for change to take place. As such, restructuring strategies and their outcomes are never simply predictable on the basis of an abstracted capitalist or organizational logic.

The restructuring approach tended to assume implementation of particular strategies for change through the type of instrumental organization we have discussed above. Processes which revised, modified or undermined various restructuring strategies were generally ignored. Admittedly, there was some interest in cases of explicit conflict between employers and trade unions, for instance over redundancies, wage cuts, increases to the pace of work and so on (e.g. Beynon *et al.*, 1990; Bagguley *et al.*, 1990). However, these were all overt conflicts which took place within formal organizational hierarchies. There was little consideration of how informal interests and practices within organizations might affect implementation. Halford (1992) has shown the importance of this level of analysis in her studies of the implementation of equal-opportunity policies in local authorities. Such policies have been undermined by internal organizational structures, bureaucratic inertia and the politics of empire-building as well as gender-specific forms of resistance embedded in the cultures and routine practices of local authority workers. Consequently, few policies were implemented straightforwardly, if at all. Similarly, Pendleton (1991) found that the introduction of flexible rostering on the railways actually reduced rather than increased flexibility because of the way managers introduced it. Thus, we are claiming, far from organizational structures and organizational members being separate, organizations are animated by the actions of organizations' members, actions which are informed not solely by purely bureaucratic procedures and logic but by complex individual and collective interests and articulated through various social resources.

Another limitation of the separation of 'structures' from 'people' prevalent within the restructuring paradigm becomes apparent when we consider what, precisely, is being restructured. It has been commonly assumed that it is only organizational structures which undergo change during restructuring, but studies of service-sector restructuring have increasingly questioned this assumption. In the service sector, the product (service) is often intimately connected to the identity and performance of the person providing the service (Urry, 1987). Illustrating this, McDowell and Court's study of merchant bankers concludes that gendered identities are an essential element of the service to clients, and that '[t]he disembodied ideal of the male bureaucrat in which rational advice was constructed as a cerebral product, purportedly unconnected to the purveyor of that advice has been displaced' (1994a: 247).

Recently, Urry (1990; also Lash and Urry 1994) has emphasized that within service provision it is difficult to disengage the restructuring of work from the restructuring of the people who do that work. Since the quality of service provision is bound up with the characteristics of the service providers, restructuring does not simply have an *impact* on people, but involves redefinition of the workforce. One example of this is du Gay and Salaman's (1992) study of recent trends in organizational restructuring which promote the cult(ure) of the consumer, both in internal organizational relations and relations with external customers. Here the authors show that restructuring involves the reconstruction of employee identities and, further, they argue that the new 'enterprising firm' operates through the soul, giving people a stake in the firm and a responsibility to achieve. Workers internalize new identities in accordance with organizational restructuring, 'becoming' what the enterprising firm requires: 'The discourse of enterprise brooks no opposition between the modes of self-presentation required of managers and employees and the ethics of the personal self' (du Gay and Salaman, 1992: 626) Thus, the enterprise culture merges with individuals' sense of self-fulfilment, offering the organization a near-perfect mode of worker control and surveillance (*see also* Crang, 1994a; Lash and Urry, 1994). Following this argument it becomes impossible to define jobs or the restructuring of jobs without recourse to the attributes of the people who carry them out and the ways in which these are being changed. This is the very point which feminist writers on gender segregation within the workplace have been making for some time. Pringle (1989) and Crompton (1993) have both shown that the gender of specific types of workers affects how jobs are defined. While jobs may appear to be defined in gender-neutral abstract terms, independently from the gender of the workers who fill them, this conceals the inherent gendering of work and, in the context of this chapter, the restructuring of work.

So far we have argued that restructuring should not be conceived of as a disembodied abstract process, as a process which is somehow 'above' the individuals and collectivities which comprise organizations. Rather, restructuring is simultaneously an economic, social and cultural process, mediated through complex organizational dynamics and intimately tied up with redefining the personal qualities which are deployed inside organizations. However, unlike du Gay and Salaman (1992), our approach does not lead us to assume that individuals will necessarily adopt newly privileged qualities and identities. Rather, organizations must be seen as a terrain on which various organizational members can mobilize, with the result that restructuring is not passively 'experienced' but is interpreted, sometimes fought over and resisted both by individuals and by groups of people who may have very different assumptions and agendas about what changes should occur and how.

In the remainder of this chapter we will explore these suggestions using in-depth interview material from our study of banking. Through these accounts we will show how restructuring in banking has attempted to transform the attributes which a 'good banker' (especially a manager) should embody, shifting from paternalistic, masculine qualities towards a new and apparently gender-neutral range of attributes. The accounts we use show the unevenness with which these efforts to restructure the culture of banking have been greeted and the ways in which individuals resist, challenge or opt out of these changes. These accounts vividly demonstrate some new paradoxes and tensions which restructuring has generated for the bank.

## LIFE HISTORIES

We have argued above that restructuring involves changes to the embodied characteristics and attributes of workers and that restructuring requires actions by various organizational employees in order for any change to take place. Restructuring is not carried out 'on high' and 'implemented' down through organizations but rather, at every level, different sorts of contestation and accommodation with organizational activities take place. These two points, that restructuring is 'embodied' and that restructuring is 'agentic', are intertwined with one another. Embodied aspects of restructuring may prompt individuals into action; that is, to (try to) alter themselves in line with newly privileged qualities and attributes. Equally, individuals may resist the new culture, promote alternative visions of change or construct their own individualized escape attempts which do not involve direct challenge or confrontation but, none the less, constitute a rejection of the new identities and values central to bank restructuring. In all cases – endorsement, rejection or avoidance – we can see the centrality of individual agency to the progress of cultural change.

In pursuing these arguments further, individual accounts of restructuring, placed in the context of workers' life or work histories, are especially illuminating. Thus, although we begin in the next section, 'Old times', with some brief contextual information about recent changes in banking, in the remainder of the chapter we draw intensively on the life histories of seven banking workers. From these accounts we can distil an experiential sense of bank restructuring, explore how demands for change pose challenges to the identities and values of these workers, and examine their responses to these demands. Our aim is to show the close articulation between individual identities and organizational change, in order to demonstrate how restructuring is an embodied process.

Our research was carried out between 1990 and 1992 and involved an extensive questionnaire survey of over 300 'Sellbank' employees and interviews with 35 of these workers. In what follows we use the accounts of three women and four men in a range of clerical and managerial positions. Brief pen-portraits of these individuals are included as an appendix to this chapter.

## OLD TIMES

Throughout the 1980s British banking was in a state of turmoil as it attempted to respond to an environment dramatically changed by the globalization of financial markets and legal deregulation of financial services (Cressey and Scott, 1992; Leyshon and Thrift, 1993; Halford *et al.*, 1997). Working within the restructuring paradigm, we would expect that banks would respond by changing their *demand* for labour, and there is certainly evidence that this has happened. Workforces have been segmented, especially by using tiered recruitment, to allow specialization of services and products, and there has been increased use of casual and part-time female clerical staff in both regional service centres and high-street branches (*see* O'Reilly, 1992a, 1992b). At the same time as banks have changed their demand for labour, the *supply* of labour has changed; in particular, there has been a rapid increase in the numbers of graduates – especially female graduates – entering the labour market.

However, when describing recent changes, bank workers made most reference neither to changing supply nor to changing demand, but to efforts to change the 'culture' of the bank. More specifically, workers referred to changes in the style and atmosphere of the bank and to the qualities and attributes which bank workers, especially managers, are now expected to possess. These points are best made through comparison with the older banking culture.

The traditional banking ethos was built on an ascriptive culture in which organizational positions and activities were based upon explicitly classed and gendered criteria. Women and men were explicitly recruited into different jobs, tied into clearly defined promotion opportunities and barriers, and the class background of aspiring workers was openly investigated. Grammar school and professional parents were a passport to promotion and a lifelong career. Stuart has worked for Sellbank since the early 1970s, when he joined the bank following A-levels. Now a branch manager himself, he describes his first manager in the 1970s:

> 'this was in Corden, and Corden being a market town the manager was still someone who walked along the high street, a Captain Mainwaring type, raising his hat to everybody and they would all say, "Good morning Mr Jerome" . . . everybody knew him, he was a character in a small market town . . . and I remember him saying to me, "any young man joining this bank – two thirds of them make manager". Nothing about women joining the bank in those days.'

Sylvia also began working for Sellbank during this era. Her position as a secretary, not integrated into the mainstream clerical grades until the late 1980s, added a double segregation to her position as a woman in the bank. By the time we met, Sylvia had been 'promoted' by being made secretary to ever more senior male managers. Sylvia described a traditional set of assumptions about managerial–secretarial relations in the older culture in which the gendered, aged and familial characteristics of the boss–secretary relationship were fundamental:

> 'The managers used to be real gentlemen, most of them. Some were jumped up, you know, but the older ones in their forties, they had got families, because they weren't appointed managers until they were married and probably had got children, because they thought they couldn't manage a branch if they had not got children.'
> *Q: And how did they treat their workforce?*
> 'Like fathers. The old secretary . . . she used to say to me, "well, when you are a secretary the men were like fathers, and then they become brothers, and then like sons".'

In the old culture, succession within the banking hierarchy was heavily managed by senior management. Jobs were not advertised and promotions were not applied for. Individual men could influence this process, for instance by their choice of a 'suitable' wife, but they could not plan careers. Stuart described his promotions and geographical moves:

> *Q: Did you ever have a choice about these moves?*
> 'Well, no. The choice was "do you want promotion or not?" And in actual fact when I was at West Horton I didn't enjoy it . . . at that point I applied for a job with the Halifax . . . and while I was waiting for the result . . . Sellbank offered me a transfer nearer home . . . which suited me, so it never came to anything.'

Similarly, Doug, who has worked for the bank since the early 1960s, and is now a branch Operations Manager, insisted that he had had little control over his career:

> 'I didn't know what was in store for me. I had been to a grammar school . . . I had a nice stable family background. I didn't do anything silly . . . I didn't really think too far ahead. . . . It all virtually just happened. I can't say I ever made any decision that has affected my job or career because those sorts of things are made, they are quite honestly made for you.'

Clearly, Doug recognized the value of his personal background but also saw no point in being proactive about his career, which would, he believed, be taken care of elsewhere. Insomuch as he is now in a managerial post, this faith proved to be justified. Doug had no complaints about these practices and remained highly committed both to his work and to Sellbank as an organization.

## CHANGING THE CULTURE

Over the past decade Sellbank has made sustained efforts to challenge the traditional banking culture. The old authoritarian and paternalistic culture is seen as inappropriate in a climate where the bank must woo its customers and establish new services in order to compete effectively. Just as individuals described the traditional culture through reference to the personal qualities and identities of banking staff, the new culture is described in embodied terms. The principal changes concern re-evaluation of the traditionally masculine ethos of branch banking and of the place of individual agency within bank work and careers. Describing recent restructuring, Sylvia's principal focus was the way in which the ascriptive culture of banking was being undermined by these changes. In particular, she was saddened by the way that age and familial responsibilities were no longer a criterion for promoting men into managerial positions (*see above*). For her, this weakened their managerial competence, which she saw as intrinsically connected to age and experience. Doug, emphasizing another aspect of recent change, pointed to the shift from the traditionally paternalistic but very formal relationships between managers and staff to more 'friendly' and informal relations:

> 'it's not just branch managers either, it's actually senior managers . . . and it's been done on purpose in Midcity . . . they have made it that way. . . . Let me tell you something about what happened to me when I was in a branch not very far from here in the early 1960s. I was called into the manager's office and he said, "Mr Williams, when you refer to me in the office I am Sir." I don't know what had happened, I must have called him "Mr Whatever" and he just called me in and said, "I am Sir. You refer to me as Sir." That just about sums it up.'

Stuart echoes these observations with a specific example from his own career. Describing his second managerial appointment, he explains:

> 'Queen's Street branch had been a very, very old-fashioned branch and they shut it down for three months while they redid it. They gutted the branch and they rebuilt it to the modern image, a bit like McDonald's, you know, with chrome and everything and the flowerpots in the banking hall . . . and one of the junior managers then had been there since demob in 1954 and they basically swapped us . . . he was a bit of a miserable old beggar, not what they wanted in a modern branch that they had spent a lot of money on . . . because of my personality I was moved sideways.'

Sellbank wanted open, friendly staff, accessible to the public, placed in open, accessible buildings with bright logos and identifiable uniforms. This is closely tied in with a new emphasis in Sellbank on 'sales' and, in turn, on *individual* performance. Where once individual proactivity was unimportant, the new culture encourages this by setting sales targets and using performance-related pay. Mark, who has worked for the bank for 20 years and is now a senior regional manager, explained:

> 'I do believe that one of the big changes, there is a big recognition that we must achieve added value from our customers and that is the sales ethos, that we must be much more proactive . . . the culture has moved on now to a sales-orientated, performance-related culture'

Sylvia experiences this on a day-to-day level in her branch:

> *Q: How have managers changed more recently? Have they changed their style?*
> 'Well, the job has changed. They now have to meet targets and they get paid if they meet their targets and the targets are set for them by the higher authorities.'

Sylvia is critical of these changes (*see below*), but others are far more positive. Justin, a graduate entrant to the bank and now the manager of a large branch, describes the new culture:

> 'It used to be very much that . . . the bank chose for you. "What is the next job coming up? Oh, it's account manager at Aberdeen. Oh, you are the next one on the list, you go whatever your skills may be." Now it is very much for you to choose where you want to go.'

The 'skills' which Justin refers to in describing his own career include quality management, human resource management, motivation and empowerment, and establishing 'synergy' in the branch – quite different from the past, and far less tied into banking as a specific occupation than a reflection of generic developments in management. Justin himself makes this distinction and explains how being a graduate is an asset:

> 'the one that has gone through the ranks, he probably, may well be a little bit blinkered in approach. They might be very banky, banky, banky . . . someone, say, who has got other vocations or been through college or done a few other [things] is probably going to have a fresher approach, be less inhibited . . . have more lateral thinking about them.'

While he recognizes that these may be 'gross generalizations' they none the less stereotype the contrasts between the new and the old culture and the way in which these contrasts reside in the actual qualities and assets of staff themselves.

Restructuring in banking has undermined the older, paternalistic culture in which careers were based on classed, aged and gendered characteristics and managed from above. Attempts have been made to replace this with a new culture where success is determined by effort and performance. The new culture is presented as both meritocratic and individualized. In what follows we will explore two points further: first, whether the new culture has really disassociated 'performance' from ascribed characteristics linked to class, gender and race, and second, the degree to which the new culture has been adopted across bank branches.

## PERFORMATIVE MERITOCRACY?

The management style now favoured within the new banking culture is presented as a 'performance', in which anyone with merit can do well. However, in their accounts of the new culture many interviewees expressed considerable doubt about this. Rather, it was suggested, an implicit ascribing persisted, albeit with major differences from the past.

The first dimension to this concerns gender and physical appearance. As we have described above, one of the goals of restructuring was to increase the openness of bank branches. This cultural shift has clearly gendered concomitants, highlighting the 'feminine' qualities of accessibility in place of the distant, authoritarian image associated with the male branch manager. Sellbank has tried to put women in visible positions in the bank (*see also* Morgan and Knights, 1991; Kerfoot, 1993) in order to encourage people into the banking hall and increase sales of new services. But gender alone is not sufficient to place women in these newly important 'front-stage' positions. Mark explains:

> 'if you have got an irate male customer coming in – and this is where we get really sexist – if you have got one of your staff who is a rather good-looking girl, you send them over to the counter to start to deal with it, and it is amazing how that male's complaints start to collapse. . . . So what I am saying is that we are actually using our staff and our clerical people to the benefit of the organization.'

Atthia, still in junior clerical work after several years' service, feels that this is one reason why her career has faltered. She knows that a key post in her career progression is branch receptionist – a new front-stage position created by restructuring – but, as she explains:

> 'I have never been put on reception and sometimes I think that is because of my colour. Maybe that is just me but none of the Asian girls, well none of us have, and I think that is strange . . . because I love meeting people and helping them and I like to sell. . . . I asked if I could be on there [reception] because that's the next step out but he has put me back upstairs.'

Sellbank employs very few Asian workers and even fewer Afro-Caribbean staff. 'Race' is a second ascribed characteristic which persists in the new banking culture. Atthia is reluctant to articulate this, but clearly she does have these doubts about her career.

Another issue for the new meritocratic culture was motherhood. Although none of the six workers we draw on here mentioned this specifically, it was commonly recognized that motherhood was totally incompatible with a managerial career in the bank. This was linked to expected hours of work, especially overtime, but it was also clear that the category 'mother' attracted a range of *assumed* characteristics (unreliability, reduced career commitment and so on) which excluded mothers from the new meritocracy.

Like the traditional culture before it, full participation in the new culture is open only to some, and not others. This continues to be prescribed by ascribed characteristics although in a far more implicit way than was the case in the past. Thus, the new meritocracy remains a mystery to Atthia, however much she places this within tacit faith in the new culture: 'I suppose if you stand out at a job, doing better than anybody else, you get on better but . . . you can be just as good as everyone else and you just don't get [on].' This scepticism about the new meritocracy was widely echoed across the bank.

## IMPLEMENTING CHANGE?

Because ascribed characteristics continue to shape careers opportunities, some workers feel excluded from full participation in the new banking culture. Try as they might to perform according to new banking criteria, they remain tied into embodied attributes and were excluded from them. Not all banking staff were even keen to take part. Some were actively hostile. For them, the new culture represented such a challenge to their established beliefs and identities that they were not willing, or in some cases able, simply to change their attitudes to work and organizational activities.

For Sylvia, the displacement of the 'father-figure' manager has undermined management–staff relations (*see above*), while for Doug this had impacted negatively on customer relations. Doug felt that the old-style, fatherly manager commanded respect and he was unconvinced by the trend towards younger managers:

> *Q: Do you think it is a good thing?*
> 'The general view is, I mean it was always my view, no, it can't be. You do need certain experience. I mean, if a couple come in and they are discussing their future, their finances and they find somebody on the other side of the desk who is half their age, I just wonder what they will think.'

Many agree with him, pointing particularly to graduate entrants who usually enter management at a relatively young age. Commonly, non-graduates would argue that graduate trainees were 'butterflying' – flitting from one area of banking to another – failing to learn enough 'banking basics'. Many staff were also actively hostile to the new emphasis on sales, targets and performance-related pay. This was reflected in the surprisingly common choice by clerical staff not to compete for managerial appointments. Sally explains:

> 'I felt that there was no way I was going to get on, because they were all into sales and although I can talk the hind leg off a donkey, selling is just not my thing. If somebody wants something I believe they will come and they will ask but you can't force it. . . . I am not so career minded any more. I am not so keen to go and do my exams, or whatever; as long as I come and they pay me I am fairly happy.'

Sylvia describes the same phenomenon among male staff, who in the past would have almost automatically entered management:

> 'they see the way that managers are treated and they don't want to be a manager because . . . they don't get much promotion and . . . [t]hey have to meet all these targets these days, the targets are set for them, they have to sell so many life policies.'

There is a strong feeling that the targets are unachievable. Stuart, who entered management before targets were introduced, described his targets as 'a nonsense' and refused to agree with the objectives set for him by senior management. Others raise similar concerns in terms of the bank's wider fortunes. For Doug, whose loyalty to banking and Sellbank are unquestionable (*see below*), restructuring is undermining traditional branch banking. He explains this not simply as a personal preference but as his belief that the best interests of the bank are at stake.

There are, then, several different forms of resistance to recent changes to banking culture and practice. As we have seen above, Mark has some major reservations about the new culture, especially about generic management and graduate trainee schemes. He is wearied by the constant changes at Sellbank, but also confident in his right and his ability to challenge restructuring:

> 'we have to do what we are told, *given that we don't believe that what we are being told is wrong*, but if we think it is wrong there are certain ways of getting it changed and I am aware of changes to certain directions that we have gone as a result of people coming back and saying, "look, this ain't working this way".' (Emphasis added)

While Mark and Stuart have taken direct action, challenging the validity of change itself, others exercise more covert forms of agency in resistance to new expectations of them. Some staff 'opt out' altogether, relinquishing their career ambitions where promotion is no longer seen as desirable and/or clashes with individuals' sense of self-identity.

Of course, not everyone resists change. Some positively embrace the new culture but – and this is the key point – only some staff are actually in a position to fully embrace the culture because only certain types of people are allowed access to it and because only some staff feel 'personally' comfortable with the restructured organization. Justin is one such person. He endorses the new culture and is clear about his place within it. However, Justin's rejection of traditional 'banky' qualities (*see above*) also means that he has disassociated his own goals from those of the bank:

> 'I don't owe anything to Sellbank. If I saw something else [outside the bank] I would have no qualms about doing it. I wouldn't apologize to anyone about it. I work to live, I don't live to work.'

Compare this with Doug, a non-graduate, long-serving and more traditional manager:

> 'I have worked for Sellbank now, I am in my 33rd year. And I am proud, I don't hide the fact that I am proud to work for Sellbank. . . . I think it is pretty evident that I would kill for Sellbank. Well, that's a bit over the top. I mean I am very loyal to Sellbank.'

Herein lies an unexpected paradox of recent bank restructuring: the danger that the more staff endorse the qualities newly privileged, the less loyal they may become to the bank *per se*.

Clearly, not all managers take the same line, and restructuring has *divided* management. Rather than there being a coherent new managerialism, managers find themselves in conflict over the sorts of business they are supposed to be carrying out. Furthermore, while at the time of our research the 'performative' managers were in the ascendancy, there is more recent evidence suggesting that things have changed. As Sellbank was forced to adopt better credit and risk controls in order to reduce bad debts, it placed more weight on the people 'on the ground' who were thought best able to judge clients. Over the past few years, active pressure by managers such as Doug has resulted in a partial reinstatement of the powers of the branch manager (stripped away by restructuring in the late 1980s) as some management functions have been devolved to the larger urban branches in 'networking' systems. This points to possibilities for resistance to change as well as perpetuating divisions within the management cadre.

## CONCLUDING COMMENTS

Clearly, restructuring 'Sellbank' has not been conceived of simply in terms of a new economic rationality. The organization and practice of banking have always been tied into embodied characteristics of gender, class, age and race, and to notions of appropriate behaviour and style. Just as in Schoenberger's (1994) study, senior managers at Sellbank have conceptualized the possibilities for change in these terms. 'Economic' restructuring is inevitably embedded within 'social' and 'cultural' dynamics. Change has not been implemented across Sellbank in an unmediated way. Organizations are not merely instrumental conduits for change, and many of the people who constitute Sellbank have acted to opt out of, ignore or contest change. The values endorsed by the most senior managers at Sellbank have been resisted by many different types of workers. Older men and women are hostile to the attack on 'experience', feeling this as an attack on both their personal identities and their understanding of the bank's identity; many younger women feel excluded by the implicit assumptions made about desirable attributes; and some (though certainly not all) young men find the competitive culture unattractive.

This reveals as a fallacy the notion that 'culture' can simply be restructured by management. Such notions can be found even in sophisticated analyses of contemporary change. Du Gay and Salaman (1992), for instance, claim that restructuring has captured the soul of organizational members (and *see also* du Gay, 1996), while Grey (1994) claims that in accountancy the 'career' has been constructed as such a compelling game that young recruits become totally absorbed by its competitive culture. As Cohen (1994) notes, such assumptions rest on an impoverished notion of 'culture' as though it resides simply in management wishes and can be imposed on to an infinitely malleable staff. To the contrary, '[c]ulture has none of this rigidity; rather it is infinitely readable and interpretable. It is constantly being made and remade through people's behaviour' (Cohen, 1994: 98). Top-down efforts to restructure cultures will fail since organizations

> 'are composed of individuals as active agents who process their relationships and determine their movements through different social milieu. They are not simply acted on'. (Cohen, 1994: 99)

Notions of embodiment, identity and agency are crucial to understanding recent organizational restructuring at Sellbank. Many of the same issues are also clearly apparent in other sectors (*see*, for example, Halford *et al.*, 1997 on local government and nursing; du Gay, 1996 on retailing). Thus, we suggest that our argument has far broader implications for research into and understandings of restructuring more generally.

## APPENDIX: PEN-PORTRAITS OF INTERVIEWEES

**Atthia** – a junior clerk who has worked for the bank since leaving school about five years ago. Although she has received several clerical promotions, she is unhappy with her progress at the bank. She is British Asian and single with no children.

**Stuart** – manager of a medium-sized urban branch who has worked for Sellbank for around 15 years since leaving full-time education. He is a non-graduate. His career has followed a traditional pattern of promotion up through the clerical grades to management in his early thirties. He is married with children.

**Mark** – a senior regional manager, in his early forties. An A-level entrant, Mark has enjoyed rapid promotion to a managerial position. He is married with children.

**Sally** – a junior clerical worker in her late twenties. Having joined the bank from school, she has had few recent promotions. Sally is single.

**Doug** – a junior branch manager in his early fifties, Doug has worked for Sellbank since leaving school. For many years his career appeared to have 'stalled' but he has recently earned promotion to management.

**Sylvia** – a secretary for over twenty years, Sylvia has worked for Sellbank since leaving school. She is single with no children.

**Justin** – manager of a large urban branch, Justin is a graduate trainee. He is in his early thirties, and earning rapid promotion to senior management. He is married with children.

## ACKNOWLEDGEMENTS

The research presented in this chapter was part of a larger project funded by the ESRC (Research Grant R00023277301) which included the study of local government and nursing and involved Anne Witz. For further details of this research, *see* Halford *et al.* (1997).

# Chapter Ten

# A TALE OF TWO CITIES? EMBEDDED ORGANIZATIONS AND EMBODIED WORKERS IN THE CITY OF LONDON

LINDA McDOWELL

The City of London is a tale of two cities. The one was traditional and stuffy, with codes of conduct reflecting a close-bound world of long-established firms and personal relationships built on trust. . . . The financial revolution of the eighties marked the emergence of the new city. (Budd and Whimster, 1992)

## INTRODUCTION

In this chapter, I want to explore the class and gender implications of the changes that apparently occurred around and after the Big Bang in labour markets in the City of London. After the Financial Services Act of 1986, which extended deregulation in the City of London, employment in the financial services sector expanded even more rapidly than it had in the early 1980s. A rhetoric of fast money and fast living developed around this part of the labour market, which seemed to be the apotheosis of Thatcherite success. Indeed, the 1980s as a whole have been dubbed the 'sexy/greedy' decade by the geographers Daniels, Thrift and Leyshon – years immortalized in celluloid by Gordon Gekko and Michael Douglas, who in the film *Wall Street* (1988) pronounced that 'greed is good'. These attitudes seemed to be shared by Thatcher and her Cabinets and by the high-flying young men and women who entered the City and Wall Street in their thousands. Champagne flowed, oysters were consumed and white powder was snorted in lavatories. These dealers, traders, corporate raiders, analysts and bankers became the heroes, and to a less noticeable extent heroines, of films, plays, TV serials and novels, as well as the key movers in 'factions' about the world of banking such as Michael Lewis' *Liar's Poker* and *The Money Culture* in which the cultural attitudes of the 'new' world of money were compared to those of the stuffy bourgeois world of earlier decades. These new entrants to the City and Wall Street were the movers and shakers in the bull markets of the latter years of the 1980s.

The generally accepted overall impact of the expansion in the fiancial sector in the City is summed up in Budd and Whimster's assertion that the world of money is a 'tale of two cities': one replacing the other in the fast and furious 1980s – perhaps a decade too soon at the end of the twentieth century. One hundred years ago it was the nineties, not the eighties, that were naughty. The 1990s, however, in the financial world at least, are marked by more of a return to the values of the 'old' city than authorities mentioned above, be they academic or cultural commentators, perhaps expected. The early 1990s were marked by a service-sector recession and retrenchment in the banking world, as well as by a number of scandals, collapses and bankruptcies. Although merchant banks in general were far less hard hit by employment retrenchment than the retail banks that Halford and Savage examine in their chapter, the days of almost reckless expansion, when it seemed as if almost anyone could get a job in the City, had passed. And bankruptcy, takeover and scandal in some of the most august institutions of the City may have fatally weakened its reputation.

I want to focus here, however, on employment practices and argue that a more detailed and nuanced investigation of the practices and cultures of merchant banks is necessary to uncover what actually happened to the class and gender composition of bank employees in the glory years at the end of the 1980s and since. The gendered nature of the 1980s expansion has not so far been examined in any detail and it is my contention here that the changes in class composition have been greatly exaggerated. I also want to suggest that aggregate changes disguise significant differences between institutions when the finer spatial scale of the firm, in this case individual banks, is examined. It is also here, at the firm level, that recent theoretical innovations in economics, economic geography and economic sociology that have insisted on the importance of understanding the embedded nature of institutional action and the significance of the culture of organizations might be assessed.

## THE 'NEW' ECONOMIC SOCIOLOGY, THE CULTURE OF ORGANIZATIONS AND EMBODIED WORK

The shift towards a new set of questions has recently directed geographers to new literatures – in the main, the 'new' economic sociology, the sociology of organizations, feminist studies of work and gender stereotyping, and cultural anthropology – in analyses of the changing structure of labour markets. Attention has moved away from the restructuring/division of labour paradigm towards approaches that conceptualize both organizations and employees as actors with sets of cultural attributes which are constituted in, affected by and affect the huge range of interactions that take place at a variety of spatial scales in any form of economic interaction.

For geographers and sociologists interested in the impact of current labour market changes in a variety of industrial sectors and individual organizations, there has been a fruitful coincidence of two sets of literatures. The first set is that group of various theorists jointly labelled the 'new economic sociologists' (Granovetter, 1984; Granovetter and Swedberg, 1993; Smelser and Swedberg, 1995; Zukin and DiMaggio, 1990a), from whom the concept of embedded organizations is drawn. In analyses of the ways in which national and local factors have an influence on the development of those areas of the economy that were expanding in the 1980s, a number of geographers, Schoenberger (1996), Storper (1994) and Thrift (1994b) among them, have turned to the revitalized area of economic sociology, which not only seemed able to shed insight on the location of these industries and their position within national and local systems of political and financial regulation but also seemed to offer a purchase on new questions about the cultural meaning of the new products and the workers who were providing them. It is in this latter area – the meaning of services and service work – that economic sociology coincides with both organizational sociology and its investigation of the gendered nature of institutions and with feminist analyses about work as performance.

The 'new' economic sociology (*see* Ingham, 1996, for a useful review) became influential again from the 1980s as it seemed to offer a way of understanding the radical shifts in the financial and business environment of that decade in the USA and UK. This body of work focuses on power, markets, business–government links, social networks and the culture of organizations. The concept of economic rationality is rejected, and instead it is argued that economic action is embedded within the social context and the specific institutions within which it takes place. Like all social interactions, economic decisions are as much affected by tradition, historical precedent, class and gender interests and other social factors as by considerations of efficiency or profit. Firms are complex organizations that operate in an environment of uncertainty and, within this, companies rarely operate a maximizing strategy in taking decisions. Instead they rely on organizational routines and experience-based formulas. In addition, the behaviour of employees affects how decisions are made. There is, for example, often conflict between different levels of managers, between staff and workers, and between managers and workers. Top managers are said to put sales and growth above profitability, middle managers strive to boost their department's budget, professional staff may care more about their reputations outside the firm than their contributions to it, whereas workers often combine to resist manangement attempts to boost productivity (Zukin and DiMaggio, 1990a). Chance effects and constant change are also important. Schoenberger (1996) adopted this framework in her investigation of the seemingly inconsistent decision-making by senior executives in major US corporations.

At the more macro level, similar factors lead to a complex and variable environment. Markets, as well as organizations and bureaucracies, are social institutions with histories, traditions and taken-for-granted rituals. Typically, several functionally differentiated levels of embeddedness are distinguished which it is possible to map on to a descending hierarchy of places, spatial scales or levels of analysis. Zukin and DiMaggio, for example, distinguish between individual, cultural, structural and political forms of embeddedness. Political embeddedness operates at the most aggregate scale down to individual embeddedness. The categories are relatively self-explanatory. Included as 'political' actors or influences are a range of institutions from international and national legal frameworks to systems of collective bargaining, policies of national and local state, the social balance between regional employers, and the willingness of the local labour force to tolerate change. Structural embeddedness is clearly related to political embeddedness but is used by Zukin and DiMaggio, and within the wider literature of economic sociology, to refer not to the wider structural phenomena of capitalist societies but in a more restricted sense of social networks. In the specific case of merchant banking, for example, analyses of familial patterns of power, the significance both of Jewish immigration and Jewish social isolation (Michie, 1992) and the networks of social elites that link directors of British firms and the Conservative Party (Stanworth and Giddens, 1974a, 1974b; Scott, 1991) might be included.

DiMaggio and Zukin's third category is cultural embeddedness. Although the 'cultural turn' in economics and economic geography may seem a radical shift and is hotly disputed (Sayer, 1994; Barnes, 1995a), cultural analysis has had a long history in economic sociology from Marx's analysis of commodity fetishism and reification and Durkheim's comments on the non-contractual elements of contracts. More recently it has become a central issue in the rapid growth of organizational theory focusing on the level of the firm. Cultural embeddednesss is defined by Zukin and DiMaggio (1990b: 17) as

> the role of shared collective understandings in shaping economic strategies and goals . . . Culture provides scripts for applying different strategies to different classes of exchange. Norms and constitutive understandings regulate market exchange, causing persons to behave with institutionalised and culturally specific definitions of integrity even when they could get away with cheating. On the one hand, it constitutes the structures in which economic self interest is played out; on the other, it constrains the free play of market forces.

Thatcher's achievement in the 1980s was to reverse these constraints and to create the circumstances for the spread of a new set of cultural assumptions in the City and in wider society as the ideology of individualism – that people are solely motivated by pecuniary gain – and the associated claims for the efficacy of deregulation in freeing the market from the stranglehold of the state and assuring economic efficiency and success. How these notions progressively infiltrated discourse and practice in a range of economic institutions is a major research challenge for economic sociologists and anthropologists.

As merchant banking in particular, and the City as an institution, was one of the prime locations for the successful promulgation and diffusion of these new cutural assumptions, it seemed an exemplary site for research. Of all the middle-class occupations in that new service or new cultural class that expanded in the 1980s, it was bankers who were characterized as the personification of the era: the apotheosis of individualistic, profit-orientated 'Yuppies'.

Moving to a finer spatial scale, that of individual firms, Zukin and DiMaggio's notion of scripts for different types of exchanges is close to social interactionist and ethnomethodological approaches in sociology, and has a particular purchase on the ways in which everyday social interactions in the City are constructed to include and exclude different social actors. Finally, the concept of cognitive embeddedness is, according to Zukin and DiMaggio, the 'ways in which the structured regularities of mental processes limit the exercise of economic reasoning' (1990b: 15–6). Cognitive psychology and decision theory are influential here, and the main focus has been on the level of an individual organization, looking at the transaction costs of bureaucracy and the ways in which uncertainty, complexity and costs of information combine to restrict attempts to achieve rationality.

These levels of embeddedness are clearly interrelated, and to some extent their differentiation is more of an explanatory device than a reflection of their separation in everyday and less frequent economic interactions and exchanges. Further, Ingham has argued that the continuing separation of 'the economic' from the 'social' in this 'school' as a whole is problematic. The implication is that the economic is where the action is and the social is 'context', leaving in place economists' imperialist claims that the social is merely that which prevents the rational operation of 'the market'. Zukin and DiMaggio, for example, clearly argue that culture sets limits to economic rationality rather than, as might be preferable, theorizing the constitution of the economic as part of the social/cultural realm. Nevertheless, this approach is extremely valuable in demonstrating that economic decisions, like other forms of interaction and behaviour, are socially constructed, made and carried out in a set of social circumstances and institutions which have a history and a present, and are based on sets of vested interests and alternative power bases. In this view instead of being a rational mechanism for the efficient allocation of resources, the economy, like other social institutions, is constructed through interactions within social networks that are the product of particular cultural formations. This conception of the City, the history of particular investment banks within it and the behaviour of specific social actors, lays behind the empirical work on recruitment and employment procedures in merchant banks that I shall discuss in the next section.

The second group of theorists who are having an increasingly visible impact on the analyses of economic geographers are a more disparate set of organizational sociologists and anthropologists, many of whom have an interest in gender relations. These theorists more commonly focus on the level of the firm itself and its employees than do the multi-level analyses of the new economic sociologists. In particular, new studies of the embodied characteristics of work have become influential, in which the ways in which gender (and class) attributes are part and parcel of the formation of new ways of working in the post-industrial service economy are examined (Casey, 1995; du Gay, 1996; Holliday, 1995; Leidner, 1993; Wright, 1995). Halford and Savage's chapter, is an excellent illustration of this type of approach. As they argue here and elsewhere, the changing gender composition of the workforce in particular organizations is not an outcome or consequence of economic restructuring but part of the process of restructuring itself. As feminist economists have long argued, the gender characteristics of both workers and occupations are mutually constituted. This recognition about gender *per se* has been extended to the very structure of employees' bodies. In a growing range of occupations in service-based economies, from fast food to fast money, the service or the product has became inseparable from the person providing it. Workers with specific social attributes, from class and gender to weight and demeanour, are disciplined to produce an embodied performance that conforms to idealized notions of the appropriate 'servicer'. In this normalization, the culture of organizations, in the sense of explicit and implicit rules of conduct, has become increasingly important in inculcating desirable embodied attributes of workers, as well as in establishing the values and norms of organizational practices as recognized by the new economic sociologists. Here, in the coincidence of embeddedness and embodiment, the separation of structures from agents is overcome.

It has become increasingly clear from a growing number of fascinating studies that organizational structures, institutional practices and employees' attitudes, social characteristics and bodily forms are restructured during periods of rapid economic change. It is in such periods of change or crisis that the social and cultural practices that define such matters as recruitment and employment procedures become clearest. If workers are either being recruited in growing numbers or alternatively made redundant, then organizations are forced to make more transparent than normal their criteria of employment. In order to establish the attributes of desirable employees of merchant banks, and to assess the assertions that the class and gender composition altered from the mid-1980s onwards, an empirical survey was carried out in merchant banks in the City of London.

## THE SURVEY

The research which I draw on here is part of a larger piece of work which was carried out between 1992 and 1994. It involved a questionnaire survey of 360 merchant banks in the City of London to establish the class and gender composition of their workforce, and detailed interviews with a sample of 50 professional and 25 non-professional workers in three then British-owned banks, as well as a small number of interviews elsewhere and interviews with a number of human resources professional workers in the three case study banks. The survey of professional workers included a number of pairs of men and women in the same position in the occupational hierarchy to compare their career histories. In this chapter, which focuses on recruitment issues, I use the data from the personal interviews as well as historical material about the class and gender composition of City employees.

The three banks within which detailed interviews were carried out varied in the date of their foundation, their status, size and culture, ranging from the investment arm of a high-street 'name' to an old family, blue-blooded bank, with a smaller, more recent second-tier bank somewhere between the two extremes. I have called these banks, respectively, by the pseudonyms Merbank, Bluebros and Northbank to indicate their status and origins and to protect the confidentiality of my sources. All individual informants have also been given pseudonyms.

# EMBEDDED ORGANIZATIONS: OLD AND NEW TIMES IN THE CITY

Of all the institutions and all the areas that most clearly exemplify the trends towards economic globalization, merchant banking in the City of London perhaps qualifies for the prize. The City is the powerhouse of the national economy and a site of agglomeration in the centre of the new world money market that circles the globe, as well as an important command centre for transnational corporations – an exemplary site in the space of flows, as Castells (1989) has dubbed the new global economy. But the City is also a distinctive place, a spatially bounded location and a prestige address for the global banking corporations that are situated there that differs from the other key nodes in the space of flows. The City of London is an integral part of a capital city within which many of the growing number of City employees live as well as work, even though a proportion of these people may well be members of a new mobile middle class: the new service economy workers whose interactions may be as much with their peers in New York, Tokyo or Singapore as with the employees of other firms and banks located in spatial proximity within the City itself. For these employees, however wealthy, Greater London provides a particular spatial mix of activities, separated in space and time and whose accessibility and use still depend in part on overcoming the friction of distance. Indeed, these short, everyday or local distances, as it were, are often a great deal more difficult to compress than the expanses of the global financial economy.

But the City itself, in the sense of the Square Mile and its recent extensions, is also a place or a location, a socio-spatial institution as well as an economic organization, characterized by a distinctive set of social and cultural practices that have developed during its particular history. These past practices, which are embedded in the peculiarities of the British economy, have resulted in the development of a set of class-based masculinist institutions characterized by peculiar ways of doing business that depended on the development of face-to-face contact and talk, and a bourgeois ideal of trust between equals. After the 1974 crash in financial and property markets, for example, Lord Poole, then the chairman of Lazards, was able to state, quite unselfconsciously, that 'I never lost any money, because I never lent any money to anyone I didn't go to school with.' This notion that financial dealing is a business conducted by a bourgeois elite, brought up and socialized to accept particular ways of behaving, who, in the business world, all play according to unwritten rules – 'a gentleman's word is his bond' – meant that regulatory mechanisms were late in developing in British financial markets. The corollary was that the system was open to large-scale abuses, which, ironically, only became visible as more formal regulatory mechanisms were both introduced and evaded in the period following the Financial Services Act 1986.

Although financial scandal cannot be directly attributed to the changing class (and gender?) composition of the City, in popular discourse the opening up of City institutions to a different fraction of the middle class is clearly associated with the 'new' practices. After the Barings scandal in 1995, when this most blue-blooded of banks was bankrupted by Nick Leeson's gambling in derivatives on the Singapore Exchange, the press made great play with his class origins: a grammar school boy, Essex man, brought up in a council house, implying that if only he had attended Eton and Oxbridge, the cultural attitudes imbued there might have saved him from his errors. While Leeson is a single example, it has been widely suggested that one of the consequences of more open markets, and the movement of the City to an international (and more latterly global) position, has been a pronounced shift in the accepted culture of the City of London (Budd and Whimster, 1992; Pryke, 1991; Thrift, 1994b). In explaining the overturn of a leisured, bourgeois and 'gentlemanly' form of capitalist social practices and the rise of a so-called 'new City', a number of factors have been suggested. These range from the changed economic circumstances including expansion and growth, the penetration of the London financial markets by foreign banks, brokers, securities houses and legal firms, and the increasing size of firms, as well as the rise of a neo-liberal state and new regulatory systems combined with the recruitment of growing numbers of men and women from non-elite backgrounds. New circumstances, it is argued, may have led to the demand for new types of employees who in turn might challenge the accepted conventions of City business practices. As Budd and Whimster (1992) argued in their initial and stimulating assessment of the extent and depth of the changes over the 1980s, a great deal more empirical research is needed to establish whether 'the newly wealthy of the international City salariat [are the] old middle class or upwardly mobile from lower down the class hierarchy . . . and whether a process of differentiation in the formation and transmission of cultural and material capital may be occurring' (p. 25).

## THE OLD CITY

The basis features of the social composition of the 'old' City are well established. It was an elitist and masculinist environment into which professionals were recruited through personal networks, school and family ties, and an extreme gender division of labour separated women from men. According to a respondent to the City Lives Project (1991), it was not until the 1960s that there were any women at all in professional occupations, and the exchanges and trading floors remained off-limits for women for a further decade.

In their investigation of the development of a class-based business and financial elite in Britain, Stanworth and Giddens (1974a, 1974b) argued that the elite that came to dominance towards the turn of the nineteenth century remained in power throughout most of the twentieth century. It was, as Scott (1979) argued in a later study, an elite with 'a common background and pattern of socialisation, reinforced through inter-marriage, club memberships etc. [which] generated a community feeling among members of the propertied class and this feeling could be articulated into a class awareness by the most active members of the class' (p. 125–6). According to Unseem (1990), 'the solidarity [of this class in Britain] is underpinned by a unique lattice work of old school ties, exclusive urban haunts, and aristocratic traditions that are without real counterpart in American life' (p. 265). Indeed, the specificity or peculiarity of the British financial elite with its tradition of 'gentlemanly capitalism' has been seen as one of the reasons for Britain's relatively poor industrial performance in the twentieth century (Hutton, 1995a). At the core of this power bloc was the

> group which came to be termed the 'establishment' . . . an exclusive upper circle of the status hierarchy, a group rooted in their common education in the public schools and their commitment to the maintenance of the established traditions of old England. This establishment dominated political power within the national and local state apparatus, and its power and influence stretched out to all the salient institutions of British society. The establishment was a tightly interknit group of intermarried families who monopolised the exercise of political power in Britain. (Scott, 1991: 61–2)

Family dynasties retained a key significance in the banking world. At the end of the 1980s, for example, among the largest companies in entrepreneurial rather than institutional control, Kleinwort, Benson Lonsdale, Robert Fleming Holdings, Schroders, N.M. Rothschild and Barings were listed (Beresford, 1990). The significance of an elite education also appears to have retained a tenacious grip. In an investigation of the backgrounds of 97 senior partners or chief executives of top City firms (banks, stockbrokers, life insurance companies, investment trust managers, solicitors and accountants) in 1992, Bowen (1992) found that 10 went to Eton, 9 to Winchester and 37 to other public schools. While 30 of the 97 had no university degree at all, confirming the continuing significance of family background and experience in the banking world, only 12 of the 52 British graduates had attended a redbrick university. The rest were Oxbridge products. As Bowen points out, given that 'an 18 year old Etonian who had joined a stock broker in 1962, would after all only be in his late forties now' (p. 14), the influence of these men (only 2 per cent were women) will persist well into the twenty-first century.

But to what extent have these old patterns of class and gender reproduction retained their significance in the new City? The British university education system expanded significantly in the 1960s and again in the 1980s and early 1990s, and the elite institutions, Oxford and Cambridge supreme among them, while they may not have expanded as rapidly as the system as a whole, have experienced a marked shift in their gender composition. Almost a half of these two universities' undergraduates are now women, compared with less than 12 per cent at the end of the 1960s. How successful have the women graduates of these universities and the men and women from the more meritocratic redbrick and new universities been in obtaining employment in the 'new' City? And have their cultural attributes and attitudes altered the old elitist practices of City institutions? At present we know relatively little about whether and how older attitudes have been challenged and almost nothing about the detailed lives and social networks of individual employees within merchant banks in the last decade of the twentieth century.

## THE 'NEW' CITY

Impressionistic evidence of the changing class composition of the City was found in Lewis' account of working for an American organization, Salomon Brothers, in London (Lewis, 1989, 1991). Lewis (1991) suggested that the 'Americanization' of the banking world increased class diversity in the City of London, not only in the

American-owned finance houses. He argued that the drift of European financiers towards the American value system undermined the established class pretensions.

> Even though the American investment banks tended to hire young men [*sic*] with country homes, they at least paid lip service to the idea of equal opportunity. At Salomon Brothers, anyway, it was not uncommon to find two Englishmen peddling bonds side by side who wouldn't even have *seen* [original emphasis] each other had they passed on the street. (Lewis, 1991: 155)

I asked all my interviewees about the culture of their organization, phrased in terms such as 'what is it like working here?' and 'how has the City changed recently?' to establish whether there was evidence for 'new' practices or for the continuation of 'old' ways of doing things. The majority of my respondents agreed with Lewis' suggestion. Stuart, a 31-year-old working as a manager in financial control at Northbank, explains his understanding of the changes as follows:

> 'The City has changed slowly although it is still very cliquey. I mean, you just have to go out for lunch around here, especially the traditional areas, and look around. But I think that that's decreased since Big Bang and so on. I think that the blue tie [did he mean blue tie or was this a confused but neat amalgam of old school tie and blueblood, I wonder?] my-father-was-a-stockbroker image, and the progression of the family through a particular area of financial services, I think that has changed a little bit now. I think Big Bang opened things up, and indeed just the fact that you have a lot of American and Japanese banks, who maybe don't have, dare I say it, the same class structure – my father was a banker so this is what you do – I think they, the Americans and Japanese, are such that they just recruit the best people for the job, no matter what, and I think some of that has rubbed off on to the UK banks, so they've opened up a little bit more and therefore those stuffy lunches, hand over the port and so on, I think they've also decreased a little bit. Having said that, I do think it definitely goes on.'

And 'it' did seem to. Here is another respondent, Judith, a 25-year-old woman working in the capital markets division of Bluebros, reflecting on the class composition of her colleagues:

> 'It's a job which appeals to and gets done by a certain sort of person. Probably the criteria that go with it almost inevitably mean public school, or if not actually public school, then that sort of person, if you see what I mean. I suppose it's all to do with the hard work, the team playing, the families, the ethics of what we're doing. There is an implicit code of honour there which all fits the sort of public school training. Hence the sort of people who are recruited fit a certain sort of mould and they have to be people who are highly presentable. . . . I've been told that the criteria for success here are availability, affability and ability – in that order.'

And another respondent in this bank, Isobel, an older woman (aged 37), who also worked in the capital markets division at Bluebros, made a similar point, albeit initially intimating that the bank had become more meritocratic:

> 'There's a fair spread of educational backgrounds among the people I work with. There's a hard core of Etonian, Oxbridge types and, in fact, the bank still I think . . . it's almost carried on as almost a personal crusade among the very dedicated alumni of Eton; the bank still sponsors I guess the visits of these young men who are in their late twenties to the school to tell their sixth-formers about the bank. . . . I think it's a narrow and self-selected band that turns up here and it's self-perpetuating, a system of patronage develops when they turn up here.'

The net result, according to Isobel, who was not English, was that her immediate colleagues 'are traditional English bourgeois with the habits of a particular, very self-selected, reinforcing group of people. They are quite theatrical to watch.' Crystal was employed, however, by Bluebros, the most traditional of the three banks in my sample, which still has the narrowest class representation among its employees, or at least among the sample whom I interviewed. As a manager in the Human Resources division of this bank remarked, 'there are quite a lot of old Etonians here, I have to say'.

My respondents from each of the banks were at pains to point out to me the significance of both the differences between British- and US-owned institutions and the differences among British-owned merchant banks themselves. The history and ownership structure of a bank makes a difference to its culture, its recruitment procedures and its everyday practices. A simple tale of two cities is an inadequate description of the changing social composition of banks in the City. Isobel explained, 'The point is that ownership structures may mean different things for the type of people they attract in the first place and also perhaps for the expectations these people have of how their career development might advance.'

US-owned banks, in the view of my respondents, corresponded most closely to the stereotypical views of the new City as a cut-throat competitive environment where merit rather than connections is rewarded. I therefore expanded my investigation by interviewing a small number of bankers employed by US banks in London, as well as questioning in detail my respondents who had worked in other banks, especially US-owned ones, before joining their current employers. The results were unambiguous. The environment in British banks is still distinctively different: more leisurely, better-mannered and slower-paced than in US-owned institutions. According to a salesman at Bluebros who had worked in a US financial institution earlier in his career, 'It's very different in American firms. There there's much more pressure, it's not so friendly, people were more brash and the working environment's not so pleasant.' A current employee of a US institution, an Englishwoman of 27, confirmed his view: 'It's terribly Darwinian around here, chuck people in and see who comes out on top.' And a woman who was now a director of corporate finance at Merbank explained why she rejected offers from US banks a few years earlier in her career:

> 'I talked to all the American banks but I decided not to apply for them basically on the grounds that I didn't like the culture. The very macho culture of, you know, "we're all big swinging dicks here, if you come and be a bigger swinging dick, then that's fine, otherwise forget it". Anyone who works less than 24 hours a day is you know a failure, etc. and I thought, I just don't want to do that with my life.'

I shall return to the gendered connotations of this statement later.

In English banks, on the other hand, respondents were as likely to argue that their place of employment

> 'is definitely on the caring end – that's not to say it's totally caring, because there's obviously uncertainties at the moment, but the atmosphere is a lot better than in some US banks. . . . I don't think people feel the same need to assert themselves here.'

The voice is that of a 30-year-old woman analyst at Northbank. The history of individual British banks also affects their policies and the scope for changes in their social composition. Although employees clearly have a degree of agency, banks are generally conservative institutions, and long-established ways of 'doing things' led to marked continuities in, for example, their recruitment strategies and social interactions in the workplace. Peter, a senior director in the corporate finance division at Northbank, said:

> I think each merchant bank has an ethos, particularly in the corporate area. . . . Kleinwort has a fairly arrogant self-confidence about itself, partly though success. Warburg's is very professional; if you've got a cup of coffee at Warburg's you're doing very well; it's that sort of very spartan and rather a grim place to work. Probably not in fact – this is the sort of reputation. The more blue-blooded sort of backgrounds are clearly the most professional, disciplined organizations, I suppose it's the more successful, it's sort of . . . we, I think, have probably got a relatively friendly ethos – not aggressive – but straight down the middle. But innovative.'

This ethos is reflected in the ways in which my respondents spoke about their own organizations, as I illustrate in turn. Interestingly, the interviewees at Northbank and Merbank explained the culture of their own organizations by an indirect or direct comparison with Bluebros. As a woman assistant director explained:

> 'It is a lot more egalitarian than a lot of other English merchant banks. Most are still very stuffy and public school dominated. . . . We mould ourselves into a type of work but by no means all Oxbridge, by no means all public school, or all landed gentry. There is a difference between Merbank and a blue-blooded bank. They give an air, a greater air, of pomposity and arrogance.'

Another woman in corporate finance, a director with almost a decade's more experience, outlined the same distinction:

> 'The cultures in Merbank and the blue-blooded banks are perceived to be different. In the blue-blooded merchant banks there are a lot more independently wealthy people. Here at Merbank most of the men are from minor public schools compared with Eton and so on in blue-blooded banks. Generally the culture would be more old-fashioned, more traditional and more male-dominated. Somewhere like Barings probably has a very male-dominated culture and that's very blue-blooded. Or Lazards, that's very old-fashioned, more hierarchical. Here we mix freely on a social level; even a junior person would go out for drinks after work with the directors. Colleagues at Schroders and Morgan Grenfell are very surprised by this. This place was and is much less hierarchical, much more friendly and much more a meritocracy I think also.'

Northbank, by contrast with Merbank, is a small, originally family-owned bank, relatively blue-blooded but less 'traditional' or 'prestigious' than Bluebros. According to a young male executive, 'it's, umm, relatively blueblooded . . . it's not one of the extreme examples, which I quite like, as I have this belief that some of the blue-blooded characters that fill some of the merchant banks are twits'. The bank expanded quite rapidly in the 1980s, after a merger and another change of ownership, finally moving into European hands in the early 1990s. By repute Northbank also has an elitist atmosphere, dominated by employees from prestigious universities:

> 'the recruitment policy seemed to be, when I joined, very Oxbridge dominated. I think that's true about most merchant banks but I was surprised actually here how many people had degrees from universities that didn't seem to be of the supposed top rank and by that I mean the Oxbridge, Bristol, Durham

type. I like to think that London is among the stronger universities [his degree was from London] in terms of a name, but I mean there are people here who've come from Cardiff University . . . but the thing is, every time you make a presentation the client wants to see the credentials of the team so it will say where you went to university.'

The speaker is a young man employed as an executive in the Mergers and Acquisitions department. A senior and relatively long-standing employee confirmed, however, that there had been a shift away from Oxbridge over the 1980s:

'I've noticed a very radical change in hiring since I've been here – I've been here since 1982. I can actually remember somebody saying in 1984 "Well, he wasn't, I mean he wasn't at Oxbridge; we are not going to take them, are we?" We shouted him down at the time. But two or three years later you'd never hear anybody saying that, never, not in this bank.'

In fact, when I explored the recruitment procedures of these banks in 1992 and 1993, I discovered that the move away from Oxbridge candidates had been halted by the effects of the retrenchment in employment. The three banks had concentrated their efforts on the elite universities in these years. (For further details of their policies, *see* McDowell (1997)). Among the 50 professional employees whom I interviewed, 11 had degrees from Oxbridge and a further 13 from that elite group of 'top' universities including Durham, Bristol, London, Exeter and Trinity College Dublin (48 per cent of the sample). That almost half this sample had degrees from only seven universities (from a total of over 80 in the UK alone) reveals the remarkable persistence of selection from among an elite group of institutions. As this group of people were relatively young – most of them between 24 and 37 – the continued dominance of the City by this group will stretch well into the early decades of the next century. Eight of those among the professional sample had no degrees at all (3 men and 5 women, although one of these women had attended a Swiss finishing school), and only 2 had attended one of the former polytechnics. The remainder had been to a variety of redbrick universities or to universities outside the UK.

## GENDER AND CLASS CONNECTIONS

If the class composition of banks remained solidly biased towards the top end of the middle class, in the opinions of employees as much as in reality, what had happened to the gender division of labour in these banks and in the masculinist culture of these institutions? Space precludes a thorough consideration of the changing gender composition of City banks and I have anyway discussed the details elsewhere (McDowell and Court, 1994a, 1994b; McDowell, 1995). In brief, women are still under-represented among the applicants and the recruits to merchant banks in comparison with their growing numbers in the undergraduate body. Within the subjects from which recruitment was most concentrated, women made up almost 50 per cent of those leaving Cambridge in 1993 with a law degree, although still only 25 per cent of economists, but figures from Cambridge University Careers Service, for example, revealed that approximately twice as many men as women among the year's leavers were working in professional positions in banking six months after their graduation.

As my Bluebros respondent remarked, 'there's still a heavy dominance of males applying in comparison to women' and, as all three representatives from the human resources divisions of the three banks were at pains to emphasize, the environment of an investment bank is a tough one which may intimidate women. At the stage of recruitment, there is a predominance of men among the interviewers, and traits such as 'arrogance' and 'ruthlessness', which might be characterized as masculinist, as well as the undefinable 'fit', are emphasized in selection procedures. In terms of everyday interactions, many of the women respondents referred to their exclusion through a sort of 'mateyness' among their male peers which excluded women. Others mentioned the intersections between class and gender discrimination. Thus, for example, 'It's a sort of preference for the old sort of environment; I think, too, clients are very keen on seeing traditional and male corporate finance teams working with them.' This connection between class and gender divisions was brought home to me by the honest, if shocking, comment of a woman director at Northbank, who suggested that women from the right class background had been recruited during periods of expansion because 'otherwise we would have had to have taken men from the polytechnics'.

I had assumed that the dominance of these 'old' attitudes and practices might mean that women would thrive and find the atmosphere more favourable in the more meritocratic US-owned banks. In fact my assumption was corrected by a number of respondents. These views are summed up here in the words of Bethany, a 31-year-old Englishwoman who works in the Human Resources department of a US bank, which she referred to as a 'minefield':

'Two friends of mine have gone to a very English institution, Barings. It's as English as it comes and we were talking about this and they said, "It's OK. It's hierarchical but at least you know where you are. They say – men say – "oh my dear, may I open the door for you" which may look incredibly heavy-handed but they say that there's this sort of English respect for women. At least you actually know there's an old-school, old-style network so you know where the barriers are. But in a firm like this it's all about who you know, stab you in the back when you are not looking, do it in a locker room. In the old-fashioned hierarchical organization you know from day one what you are getting into and you could probably say to someone, "look, you know, this old school network thing isn't on", whereas here you have got to side-step hidden mine-fields all the time. People say one thing but the messages are very confused . . . if you want to know the rules you have to find out for yourself. So it's actually quite difficult to negotiate, I think, for women.'

It seems, therefore, that both in the old-style class-based masculinism that continues to dominate many British institutions and in the cutthroat meritocracy of other institutions, women are at a disadvantage. I want to suggest, however, that when the intersections of embedded cultures and new forms of embodied work are considered together, the prospects for women are less bleak.

## EMBODIED WORKERS: SELLING ONESELF OR WORK AS A PERFORMANCE

As Halford and Savage argue in their chapter, 'restructuring is tied up with the redefinition of the sorts of personal qualities which individuals are expected to possess' (p. 108). Thinking this through in terms of the concept of embodiment and performance in the workplace enables the analysis of the embedded character of organizations to be deepened. The sorts of employees who are recruited by merchant banks are distinguished not only by the combination of class and gender characteristics but also by a further set of attributes that partly shape and partly are shaped by their everyday performance in the workplace. It was continually emphasized to me by recruitment specialists that 'they [applicants] basically have to sell themselves to us' (Bluebros Human Resources department member). This metaphor was extended by my interviewees, currently employed in banking, who also recognized that 'all we have to sell is ourselves'.

Merchant banking in its many aspects – selling, dealing, purveying advice to corporate clients – is an occupation that demands an interactive performance between client and banker. Physical appearance, weight and bodily hygiene, dress and style were all mentioned by interviewees as a crucial part of an acceptable workplace persona. As I have also demonstrated elsewhere, there is a complex interaction between class, gender and the nature of employment in banks in which different aspects of hegemonic masculinity are valued. The stereotypical world of fast money which was fictionalized so often in the 1980s is the sphere of trading and dealing, ideally floor-based rather than electronic trading in which classic 'macho' masculinity – an unbounded and unbridled, noisy and uninhibited masculine-embodied performance – is commonplace, absolutely opposite to the idealized view of the masculine public arena of middle-class occupations as disembodied rationality. Here women are excluded by the crudest representation (mentioned above) of the successful male trader as a literal 'dick' or virtual phallus.

Elsewhere in banks, however, a more complex set of negotiations between class, gender and embodied masculinity and femininity are in process of restructuring. Interestingly, it was the 1990s recession rather than the mid- to late 1980s expansion, when 'work walked in off the streets', that has made the renegotiation clear. As each team in each corporate division competed for clients, a new focus on the personal-service nature of the interaction became evident. Although in the 'old' City there had always been close relations between clients and bankers (and the significance of the old school tie needs no more labouring), in the 'new' City, successful bankers had to, in the words of one of my female respondents, 'get inside the skin of their clients'. Here the common masculine strategy was to develop personal relationships in the gym or on the sports field, in clubs and bars in the capital. Women, both the youngest and the more senior among my sample, reported numerous difficulties in adopting a similar set of behaviours. But as the carnival atmosphere of the 1980s gave way to a new austerity in the 1990s, women began to report that what might be termed a more feminized way of interacting, and certainly a style that women found more congenial, became more highly valued. Careful attention to the needs of the individual client, accurate and personalized research services, detailed record-keeping and careful monitoring of time on clients' behalf brought the women in my sample growing recognition and financial rewards. Male recognition of women's success brought forth overly simple and essentialized explanations in terms of women's 'natural' empathy with men, their greater decorativeness and other stereotypical views. For both sexes, however, the growing awareness of the importance of workplace performance was bringing a changing view of acceptable workplace behaviours and a re-evaluation of the supposedly naturally superior attributes of conventional masculinity.

## CONCLUDING REMARKS

I began this chapter with the desire to deconstruct the simple narrative tale of two cities, arguing that within each bank a combination of political, structural, cultural and individual factors produced a variety of embedded organizational behaviours. The illustration is, inevitably, limited, and the relationships between the different cultures of the case study banks in particular and that wider set of banks currently operating from the City of London, their own individual levels of profitability and the success of the City of London at large cannot be assessed here. Recent events, however, might lead to certain assumptions. Clearly the old-style patriarchal, class-based management of Barings was unable to cope with the seemingly unaccountable runaway performance of its employee Nick Leeson. Here lack of internal regulation and the general absence of controls by the Bank of England are a partial explanation. The disclosures in the summer of 1996 of speculative investments by Peter Young, a pension fund manager at Morgan Grenfell, which is owned by Deutsche Bank, also raise interesting questions about the relationships between the high profit–quick returns culture of the City, compared with the longer horizons of German-owned institutions, as well as the reported arrogance of the Oxford-educated Young, who apparently believed he knew better than his superiors and failed to conform to the rules regulating investments.

In my British-owned case study banks, what is interesting is not only their emphasis on personality and the possession of cultural capital – what the human resources and personnel staff referred to as 'fit' – but also the ways in which recruits from elitist backgrounds, schools and universities seemed able to retain their dominance throughout the period following deregulation. This finding counters the argument of those who suggested that the social base of recruitment in the banking world expanded in the post-regulation period. The gender of applicants, and perhaps their ethnic background (although my entire sample bar one was white), may be changing, but their class composition remains solidly bourgeois. Although all three of the banks suggested that they operated a form of equal-opportunities policy, it appeared to be in theory rather than in practice. The continuing dominance of graduates from a small number of elite universities in three different types of institutions seems remarkable, given the enormous changes in the nature of provision of higher education in Britain since the 1960s. It is difficult, of course, to draw direct links between these findings and the contention of Hutton that the reproduction of the values of a financial elite, entrenched in a specific class-based social network, and its associated social practices and attitudes, have paralysed the British economy. However, my work has revealed few signs of a cultural revolution in City institutions and little evidence of either a 'new' or an 'American' City. But, as a growing number of institutions in the City pass into the ownership and control of foreign, especially European, capital, perhaps the next century may witness a more substantial challenge to the hegemony and cultural capital of a particular fraction of the British bourgeoisie, established in the nineteenth century and barely breached in the twentieth. Certainly, recent attempts to strengthen the systems of regulation in the City seem to indicate the end of the 1980s' unthinking adherence to the superiority of deregulation.

A final comment remains about the changing nature of economic geography. There are a number of the interesting aspects of recent analyses of global economic restructuring. First, the recognition by Harvey and others that globalization paradoxically increases the importance of local differences as mobile capital is more able to differentiate between and move between places and local labour forces is borne out in a growing number of studies of different sectors. Here, it seems to me that what is needed is a re-emphasis on comparative analysis, not only between banks at a local scale, as I have done here, but also between places, between financial centres and global cities. Here, the productive coincidence of economic sociologists, social economists (if such they may be designated) and social, cultural and economic geographers in rethinking the links between, indeed the perhaps unhelpful divisions between, their subjects and objects of analysis is remarkably optimistic. It is increasingly clear that the 'social' is more than the context in which the economy operates (what Ingham has referred to as the 'tosh'!) but that economic processes themselves are socially constituted. Second, there has been a new and welcome move to examine the links between restructuring, new forms of work, especially in the service sector, and the constitution of subjectivity or identity. Here, the coincidence of new forms of work, of casualization, the growing feminization of the labour force, of deregulation and lack of commitment to old forms of industrial discipline are raising new issues about the inculcation of workplace cultures and about workers' conformity and resistance to them. Here a fascinating coincidence of interests among geographers, sociologists, psychologists, organizational theorists and others is raising new questions about the place of work in formation of individual, class and community identities. In complex and challenging ways, the economic and the cultural are also becoming indivisible.

## ACKNOWLEDGEMENT

The research described in this chapter, which was carried out with the assistance of Gill Court, was funded by the ESRC (grant no. R 00023 3006). The chapter draws on material from my book *Capital culture: gender at work in the city*, Blackwell, 1997.

# Section Two

# (Re)thinking globalization

# TRUE STORIES? GLOBAL DREAMS, GLOBAL NIGHTMARES, AND WRITING GLOBALIZATION
# INTRODUCTION TO SECTION TWO

ANDREW LEYSHON

Globalization has been described as 'a fallible attempt to capture something fundamental that is happening across the globe, much of which we can only understand in a partial and incomplete manner' (Allen, 1995: 58). The 'something fundamental' referred to here is usually taken to be what Dicken (1992) describes as a 'global shift' of economic activity, whereby an international economy has increasingly given way to one which is global in scope. Globalization is a product of the second half of the twentieth century, so that whereas economic activity has been international in scope for at least three hundred years, only relatively recently has economic activity interrelated and interconnected on a 'global' scale. Economic globalization, then, is seen to involve 'an interpenetration of economic activities: that is, an ever-tightening mesh of networks which strengthens the interdependencies between different parts of the globe and, in so doing, helps to undermine the ability of nation-states to manage their own economic affairs' (Allen, 1995: 60–1).

Although the concept is now an established part of the disciplinary lexicon, it is often forgotten how recently economic geography has started to take the issue of globalization seriously. Thus, while Peter Dicken's landmark text *Global shift* was first published as long ago as 1986 (Dicken, 1986; *see also* 1992, 1998), work on economic globalization remained very much on the margins of economic geography for quite some time. Globalization only really seemed to be of interest to those who had an existing attachment to the issue through an interest in the geography of multinational corporations (e.g. Taylor and Thrift, 1982a, 1986), or those who were investigating the geography of producer and financial services, which during the 1980s began to 'go global' at a rapid pace (e.g. Daniels, 1987; Daniels *et al.*, 1988a, 1988b, 1989). The gaze of economic geography during the mid- to late 1980s was for the most part focused elsewhere: upon theories of economic transition, such as regulation theory, flexible specialization and flexible accumulation, and upon empirical investigations of industrial districts and 'localities'.

However, in the 1990s globalization did at last become the focus of wider critical attention within economic geography, and there is clear evidence that a 'globalization' debate is taking shape (*see*, for example, Allen and Hamnett, 1995; Amin, 1993; Amin and Thrift, 1994; K.R. Cox, 1992, 1993; Dicken, 1994a; Harvey, 1996a; Hirst and Thompson, 1992; Olds, 1995a, 1995b; Thrift, 1994a; cf. Roberts, 1995). That such a debate is now beginning to emerge signals that more economic geographers *are* now taking globalization seriously, and many of the key themes in this debate are developed further by the contributors to this part of the book.

However, not everyone is happy about the way in which the subject of globalization has been placed on the economic geography research agenda. For example, in the view of Roberts (1995), geographers have generally failed to get to grips with globalization. In an imaginative and stimulating paper she argues that while economic geographers have described the ways in which capitalist firms and institutions have stretched across, speeded up and transformed the economic world, much of this work has been undertaken in a relatively uncritical manner. Economic geographers have, she argues, 'tended to take globalization as a rather unproblematic term and have tended to produce works that are focused empirically on some aspect of the economic (rather than the social or cultural) geographies of globalization'. In other words, while economic geographers have been happy to attend to the material outcomes of globalization, they have failed to appreciate that the process of globalization is both material and discursive at one and the same time; globalization is both a material reshaping of scalar relations . . . through various socio-spatial practices, and a set of discourses. The practices and the discourses are not really separable since practices are discursive and *vice versa*' (Roberts, 1995: 9). This call for a geographical reconsideration of economic globalization is an important one; it resounds through much of the rest of this chapter. Her critique of extant geographical treatments of globalization is also fairly emphatic. She gives the distinct impression that economic geography is the domain of the 'epistomologically challenged', populated by individuals who are, for the most part, happier with facts and figures or with familiar, but outdated, theoretical frameworks. It is as if economic geographers have been caught in the headlights of the advancing juggernaut of globalization (cf. Giddens, 1990), which has had the effect of freezing their critical faculties.

While there might be some truth in Roberts' indictment, it could perhaps be criticized for being a little overdrawn, and at least two points need to be made in mitigation on behalf of economic geography. The first point relates to the relative novelty of the globalization debate within economic geography. While Roberts' observation that the global 'has only very recently been subject to *rigorous examination* by geographers' (1995: 7) is again broadly accurate, it is worth noting that one could dispense with the 'rigorous' in her sentence and the statement would still hold true, for all the reasons given above. The second point revolves around Roberts' criticism of extant accounts of globalization for their overwhelmingly materialistic focus. In making this criticism, she seems to suggest that materialism is all of a piece with a kind of 'naked empiricism'. This is clearly an unsustainable accusation. Many accounts of globalization have indeed been undertaken from a theoretical position which could be described as broadly 'historico-geographical materialist' in scope, and in doing so have furnished us with extremely valuable insights into the contemporary reorganization of global economy *and* culture (Harvey's (1989a) concept of time–space compression is just one example that comes to mind). Indeed, even from such approaches it is possible to see attention to the role of language and advocacy in the ordering of the contemporary global economy (*see*, for example, Harvey, 1996a).[1]

These points notwithstanding, it should also be pointed out that in recent years, the subdiscipline has gradually begun to explore new theoretical approaches that are less tied to a broad historical materialism, mainly through an engagement with the ideas of the likes of Foucault, Derrida, Rorty, Latour and others, the result of which has been that economic geographers have begun to pay greater attention to the importance of discourse, texts and the ideological realm more generally. This engagement means that economic geographers are acquiring the theoretical and epistemological knowledges and practices which will enable them to add a discursive dimension to materialistic accounts.

Therefore, there are good reasons why economic geography has dealt with globalization in the way that it has, but it should also be recognized that the ways in which economic geographers are approaching the subject are increasingly moving in the direction recommended by Roberts and others. Indeed, a good deal of interesting work has being going on at a fertile interdisciplinary meeting-point at which there has been an exchange of ideas between political economy, geopolitics, cultural studies, economic sociology, and even science studies (for example, *see* Agnew and Corbridge, 1995; Gill, 1993, 1995; Herod *et al.*, 1997; Knights, 1992; Leyshon and Thrift, 1997; Leyshon and Tickell, 1994; Miller and Rose, 1990; Popke, 1994; Sinclair, 1994; Thrift, 1994b, 1996b, 1996c). While this work has yet to acquire a label, it is clear that it is animated by what Barnes (1996: 8–9) describes as '"post"-prefixed theory', which is an ensemble of anti-Enlightenment views that have in common a rejection of the idea of 'progress' and associated notions of rationality, reason, truth and the inviolate human subject.

This emerging body of work is therefore informed by a broadly 'constructivist' agenda which denies any possibility of an underlying foundational order to social and economic life and resists any notion of an *ex-ante* rationality. Therefore, geographical epistemologies which promise to reveal pre-ordained orders and rationalities are rejected as misguided and as ultimately futile, because

> there is neither a single origin point for inquiry nor a single logic, spatial or otherwise. The best we can hope for are shards and fragments; there is not one economic geography but many economic geographies, not one complete story but a set of fragmented stories. (Barnes, 1996: 250)

Therefore, rather than attempt to uncover some underlying order upon which economic geographies are founded, 'post'-prefixed economic geographical enquiry seeks to pay attention to the ways in which economic and social order is constructed. Although Barnes describes such 'post'-prefixed economic geographies as 'conservative', because they are sceptical of the progressive ambitions of traditional social theories, these geographies nevertheless have considerable radical potential through the 'epistemological violence' they can do to prescriptive accounts of economic order and rationality, which in effect tend to serve the interests of particularly powerful and influential economic and social groups. In other words, economic geographers are beginning to heed John Law's advice that there is perhaps more to be gained from seeing 'order' as a verb than as a noun (Law, 1994).

In the remainder of this introductory chapter I want to consider ways in which a discursive approach to economic transformation might add to the established material strengths of economic geography approaches to globalization and studies of global economic transition. The chapter proceeds in four stages. The next section turns towards post-structuralist international relations and to critical geopolitics, because an approach is emerging from this body of work which is attentive to the ways in which economic and political change is 'scripted', and which is worthy of far greater attention by economic geographers. The chapter then examines different ways in which economic change is being scripted. 'Geo-economic intellectuals and post-Cold War economic discourse' looks at 'geo-economic' discourses, which are concerned with the relative competitiveness of national economic spaces in a post-Cold War economy. 'Global dreams, global nightmares' goes on to examine discourses of economic globalization. Although these two sets of discourses are closely related and overlap in many ways, they can be differentiated both in terms of the communities that help to script them and in terms of the audiences which consume and relay them.

## TOWARDS A CRITICAL GEOPOLITICAL ECONOMY

The intellectual roots of a critical geopolitical economy may be found in the academic discipline of international relations (IR). IR has been defined as the study of all social phenomena not confined within a single state, such as the relations of states with one another, the operations of non-state actors such as international organizations, multinational corporations and religious movements, and so on (Ogley, 1994). One of its overriding concerns has been to explain the nature of international order and stability, although much attention in the discipline has been directed towards understanding disorder and instability.

The discipline began to develop during and after the First World War, and was part of a broader movement to inculcate a 'cosmopolitan internationalism', the purpose of which was to bring about a 'civilizing' of inter-state relations which might avert future wars. This perspective came to be known as 'idealism', and placed great store on the formulation of international rules and procedures of inter-state relations, to be overseen by an international governing body such as the United Nations.

However, the advent of the Cold War led to the decline of idealism and its replacement by a 'realist' perspective. Realists argued that multinational quasi-state organizations such as the United Nations were doomed to failure because it was impossible to reconcile the competing interests of a constellation of nation-states. Informed by a Hobbesian world-view of a 'war of all against all, realists argued that states would in the end be unable to act altruistically and forced to take decisions which reflected "the national interest"' (Ogley, 1994). For realists and neo-realists, then,

> international life is inherently conflictual, with anarchy the rule, stability and justice the exceptions. In this world of conflict, power is the ultimate arbiter of politics, and the distribution of national power resources determines the pattern of relations between rival states in the inter-state system. By anarchy is meant the absence of a global force, such as that which would be provided by a world state, which can impose order on nation-states. (Gill and Law, 1988: 25)

Realist, and more latterly neo-realist, approaches have been dominant in IR since the early 1950s. They are also widely adhered to within practical political communities, and in particular within the security complexes of most nation-states.

It was as an attempt to undermine the dominance enjoyed by neo-realist approaches within IR that a post-structural approach emerged within the discipline during the 1980s (Ashley, 1987, 1988; Ashley and Walker, 1990; R.B.J. Walker, 1990, 1991, 1993). Post-structuralist IR is constituted as a disruptive force, a movement of resistance which was established in opposition to normative political science approaches. According to Dalby, post-structuralist IR

> issues a challenge to all students of world politics to look at the historically constituted practices that impose and restrict power, that discipline both the understandings of politics and practices. [In doing so, such analyses attempt to] subvert the discursive practices of conventional politics, calling into question all the silences and taken-for-granted constructions on which they are based . . . [by] refusing to accept reality as presented by the dominant discourse, numerous ways of looking at politics are opened up. (Dalby, 1991: 268, 269)

The objectives of post-structuralist theorists such as Ashley and Walker are to place political ideas in their historical context. In doing so, their purpose is to undermine the power of neo-realism by revealing its essentially ahistorical nature, and showing that the assumption of neo-realists that the state is a 'natural' container for society is actually nothing more than an act of reification. They seek to expose the state as a quintessential modernist entity, which is seen to have evolved as a historically specific 'solution' to an Enlightenment quest to propagate 'justice' and 'progress'. Such ideals were to be pursued within the recognized and nominally impenetrable borders of states. These borders replaced the overlapping and shifting jurisdictions of the pre-modern feudal era, and their creation was in part symbolic of the abandonment of myth and religious authority in 'modern' Western society.

While the precise timetable of events which led up to the formation of the international system of sovereign states remains a subject of heated historical debate, the point is that the formation of such a system was the result of contingent and contested historical processes. Put another way, the idea of state sovereignty and the autonomous state 'did not appear out of thin air' (Walker, 1993: 168), but was the outcome of a historical 'demarcation of political space/time' (ibid.: 169).

Thus, post-structuralist political theory strives to surround the self-confident edifice of neo-realism, which takes the existence of states for granted, and ascribes to them an unwarranted degree of transcendence and permanence, with an air of uncertainty and doubt. It attempts to expose the ultimately contingent character of the taken-for-granted structures which govern the operation of the international political system. In so doing, theorists such as Ashley and Walker rely in large part upon a Foucauldian strategy. For them, 'order' within the international political system is brought about by dominant discourses, or 'plays of power which mobilise rules, codes and procedures to assert a particular understanding, through the construction of knowledge within these rules, codes and procedures' (Dalby, 1988: 416).

According to Ashley, Foucault's genealogical approach, whereby history is both considered as, and understood by, a layering of multiple interpretations and understandings one upon another, has particular relevance for the analysis of the international political system:

> From a genealogical standpoint, international community can only be seen as a never completed product of multiple historical practices, a still-contested product of struggle to impose interpretation upon interpretation. In its form, it can only be understood as a network of fabricated understandings, precedents, skills, and procedures that decline competent international subjectivity and that occupy a precariously held social space pried open amidst contending historical forces, multiple interpretations, and plural practices. (Ashley, 1987: 411)

Thus, a Foucauldian genealogical approach is deployed to 'do interpretive violence' to the justificatory claims of neo-realist political theory, to reveal that international relations is not a 'natural' field of study or practice, but is a historical fabrication, the survival of which has much to do with its ability to serve the interests of particularly powerful social groups.

Within geography such ideas have been used to open up a terrain of critical geopolitics and have resulted in the proliferation of 'critical political geographies'. These seek to do the same sort of 'interpretive violence' to 'pregiven, taken-for-granted, commonsense spaces', via investigations of the 'politics of the geographical specification of politics' (Dalby, 1991: 274). Much of the work within this field has been concerned to deconstruct the traditional concerns of political geography through critical considerations of the discourses associated with geopolitical militarism and violence (Dalby, 1988, 1990; Dodds, 1993a, 1993b, 1994; Ó Tuathail, 1992a, 1993b, 1996; Ó Tuathail and Agnew, 1992).

More recently, however, a more overtly political economy dimension has begun to emerge from within critical geopolitics. The possibilities of moving in this direction have been there from the very beginning. For example, consider this passage from Simon Dalby, which deals with a debate which revolved around the technicalities of fighting a nuclear war during the Cold War:

> [It] is a debate that excludes politics by reducing the possibilities of discourse to a number of intellectual specializations . . . which monopolize that which may be discussed. The expert, equipped with theoretical knowledge derived from theoretical work, versed in its techniques and competent in the rituals of the discourses, is the only competent participant in the process. Wider political participation is denied or coopted within the strategic discourses. Learning the languages is not unduly difficult, but having learnt them they in turn delimit what it is possible to discuss. (Dalby, 1988: 437)

What is interesting about this commentary is that it could just as well be applied to discussions of contemporary economic debates, such as those which revolve around issues like monetary policy, competitiveness, economic regulation, and so on. There are clearly strong parallels here with the ways in which economic debates are framed around and through specialisms and expertise, which can have the effect of closing off wider political participation within the economy (Gill, 1992). Therefore, if critical geopolitics is able to destabilize normative political theory, then it surely also has the potential to do the same for normative economic theory, to 'historicize' and contextualize forms of economic thinking. In so doing, such an approach may help to introduce doubt and uncertainty into otherwise self-confident economic epistemologies.

Developments in this direction are indeed already under way. In particular, Ó Tuathail's attempt to construct what may be described as a *critical geopolitical economy* is deserving of further attention in this regard (Ó Tuathail, 1992b, 1993b, 1996; *see also* Popke, 1994). Ó Tuathail seeks to construct a critical geopolitical economy by making a bridge between more materialist political economy approaches on the one hand, and the textual strategies employed by critical geopolitics on the other. For a materialist approach, Ó Tuathail turns towards the historico-geographical materialism of John Agnew and Stuart Corbridge, and in particular their conception of geopolitical economy (Agnew and Corbridge, 1989, 1995; Corbridge and Agnew, 1991), which he then aligns with a critical geopolitical approach. The two approaches are seen to complement one another in various ways. Thus

> critical geopolitics needs to be situated within a materialistic context whereas geopolitical economy needs a means to deepen its analysis of 'the active process of constituting the world order'. Critical geopolitics offers the means to do this by the analysis and deconstruction of competing forms of reasoning in global politics and the means by which certain forms became persuasive and others are subordinated or rendered ineffective. A comprehensive geopolitical economy approach should have a place for the ideological deconstruction of critical geopolitics within its larger project of developing a coherent account of the material and institutional dimensions to global change. What a critical geopolitics can provide is a necessary distance from the analysis and self-serving interpretations of elite institutions, states, and intellectuals in the contemporary world order. Rather than following the narratives and concepts generated by participants (bureaucracies within states, intellectuals within civil society, politicians within political society) in the complex processes of world ordering these ideologies themselves become the object of one's analysis. (Ó Tuathail, 1992b: 978–9)

In the remainder of this introductory chapter, I wish to explore the possibilities of a critical geopolitical economy approach by examining two sets of discourses which provide related but ultimately contrasting interpretations of economic and political change at the end of the twentieth century. The first of these are 'geo-economic discourses', which can be described as 'an elite set of statist discourses on world politics which establishes a series of priorities and agendas for state management' (Ó Tuathail, 1992b: 991). Propagated by a cohort of 'geo-economic intellectuals', such discourses are important because they translate a geopolitical sensibility into the economic sphere, and produce economic prescriptions which lean towards defensiveness, closure and containment (Ó Tuathail, 1992b, 1993b). These discourses are considered in more detail in the next section of the chapter.

They are opposed by a second set of discourses which, although they interpret the nature of contemporary economic and political change in much the same way, contain a very different set of economic and political prescriptions. 'Discourses of globalization' argue that global flows of capital and money should not only be accepted but embraced (Luke and Ó Tuathail, 1996; Ó Tuathail and Luke, 1994), for to do otherwise would be to invite economic and political disaster. These discourses are considered in the section after next.

## GEO-ECONOMIC INTELLECTUALS AND POST-COLD WAR ECONOMIC DISCOURSE

The term 'geo-economics' has arisen from an ongoing attempt to understand the logic of the post-Cold War global economic and political order. Whereas during the Cold War states were strongly animated by geopolitical concerns, some commentators have argued that we have increasingly moved into a geo-economic world, where the relative success of national economies within this global economic space has now become a central arbiter of political success for governments and states, leading to the development of what some political scientists have described as 'competition states' (Cerny, 1990, 1991).

The emerging discourse of geo-economics has a number of different elements, of which three are particularly noteworthy. First, it includes a splash of millenarian speculation, with deliberations about the relative standing of different nation-states after the end of the 'short twentieth century'

(Hobsbawm, 1995; Kennedy, 1988).[2] Second, it also contains a measure of neo-liberal celebration over the consequences for world order of the collapse of the USSR and Eastern European communism, which, in the words of one such celebrant, represents 'an unabashed victory of economic liberalism' (Fukuyama, 1989: 3). In such a world, geopolitical concerns will lose their ideological edge, bringing about a '"common marketization" of international relations' (ibid.: 18). This will call into existence 'a world dominated by economic concerns on which there are no ideological grounds for major conflict between nations, and in which, consequently, the use of military force becomes less legitimate' (op. cit.: 17). Third, and finally, also thrown into the mix we find a rather belated recognition that there exist a number of different 'economic worlds', many of which conform to the boundaries of national economies, the recent history of which seems to suggest that some 'worlds' are more successful than others (Albert, 1993; Hutton, 1995a; Thurow, 1992; *see also* Blim, 1996).

The figure who has given the name to this growing body of work is Edward Luttwak, who has been described by Ó Tuathail (1996: 231) as 'a quintessential "defence intellectual"', influential in Washington circles over many years but who, with the end of the Cold War, 'discovered "geo-economics" and retooled and remarketed himself as a new geo-economic strategist who could expertly read the armory, battle formations, and strategic stakes of an emergent worldwide struggle for industrial supremacy among states' (ibid.: 232). The essence of Luttwak's argument is that the end of the Cold War means that the world will no longer be ordered according to the logics of geopolitics but instead by geo-economics, which he sees as being produced through the transmogrification of geopolitical conflict into an economic medium (Luttwak, 1990, 1993):

> Everyone, it appears, now agrees that the methods of commerce are displacing military methods – with disposable capital in lieu of firepower, civilian innovation in lieu of military-technical advancement, and market penetration in lieu of garrisons and bases . . . as the relevance of military threats and military alliances wanes, geo-economic priorities and modalities are becoming dominant in state action. (Luttwak, 1990: 17, 20)

In many ways, Luttwak's is an anti-globalization tract. He takes issue with those who would argue that we are moving from a world dominated by politics to one dominated by business and global corporations (for example, *see* Barnet and Cavanagh, 1994). He points to the continuing importance of the state, which retains a key regulatory role:

> World Politics is still not about to give way to World Business, i.e. the free interaction of commerce governed only by its own non-territorial logic. . . . Instead, what is going to happen – and what we are already witnessing – is a much less complete transformation of state action represented by the emergence of 'Geo-economics'. This neologism is the best term I can think of to describe the admixture of the logic of conflict with the methods of commerce – . . . the logic of war in the grammar of commerce. (Luttwak, 1990: 19)

Luttwak argues that such a propensity to geo-economic competition has remained latent in nation-states over a long period of time, but during the Cold War was subdued by its 'adhesive properties', which helped to keep US–European–Japanese commercial tensions in check because of the overwhelming imperatives of strategic geopolitical concerns. However, as geopolitical fears have subsided in the wake of the collapse of the USSR, Luttwak argues that commercial tensions will once more return to the fore, bringing with them a form of global neo-mercantilism:

> fundamentally, states will tend to act 'geo-economically' simply because of what they are: spatially defined entities structured to outdo each other on the world scene . . . their raison d'être and the ethos that sustains them still derive from their chronologically first function: to provide security from foes without. (Luttwak, 1990: 19).

This is clearly a controversial reading of contemporary processes of economic and political transformation. And indeed, there are at least two serious problems with Luttwak's account (Ó Tuathail, 1992b, 1993b, 1996). First, as the above quotation reveals, he tends towards a rather deterministic and transcendent reading of political and social action. Second, his account is firmly rooted in the realist and neo-realist tradition of international relations. Although his interpretative framework argues that the 'logic of order' has changed from one of politics to one of economics, his notion of economic space is one which is taken directly from the realist canon; thus, for Luttwak it is the space between states, or economies, which demarcates (Walker, 1993). But the trouble with this particular geographical imaginary is that it denies the growing transnational and global character of much economic activity. As Ó Tuathail (1996: 239) argues,

> What Luttwak describes but does not grasp is that space is no longer primarily mastered by nation-states. The territoriality of global affairs is no longer one of competing, segmented, and discretely sover-

eign nation-states but a territoriality shaped by global flows. . . . Thus, while American corporations may be moving overseas and forming strategic alliances with other foreign corporations, [Luttwak] nevertheless speaks about 'American' corporations, 'American' industry and 'American' technology. Despite the evident deterritoriality of 'American capitalism,' he still assumes the fictional unities of an 'American economy' and 'American national interests'.

In other words, Luttwak glosses over material evidence pointing to the globalization of economic activity, for to do otherwise would destabilize his established and comfortable ontological understanding of the shape of the world. The scripts and discourses propagated by geo-economic intellectuals such as Luttwak are revealed to be influenced by the material structures and institutions which are themselves legacies of earlier discourses. Although the dilemmas of a geo-economic struggle would seem to be very different from those of a geopolitical one, geo-economic intellectuals seem unable to escape the 'discursive horizons' of the geopolitical imaginary, which is testament to the survival of Cold War sensibilities into a post-Cold War era. Or perhaps it is that such intellectuals do not want to. Realist-mercantilist perspectives are strongly embedded within the influential bureaucratic communities that allow modern states to operate (Gill and Law, 1988). Therefore, it may be, as Ó Tuathail (1996: 239) suggests, that the recasting of geopolitics as geo-economics provides a route of continuity for such communities and 'offers state bureaucrats a means of relegitimating themselves and their authority over their citizens'.

Despite such criticisms, it is important to note that similar geo-economic discourses are gradually securing a much wider constituency, well beyond those bureaucratic communities that have the greatest self-interest in their propagation and survival. Tim Luke and Geróid Ó Tuathail have illustrated the ways in which 'defensive', state-centric economic discourses surfaced in both the 1992 and the 1996 US presidential election campaigns through the campaigns of candidates like Ross Perot and Pat Buchanan (Ó Tuathail and Luke, 1994; Luke and Ó Tuathail, 1996). Both candidates propagated a determinedly national vision, and in doing so translated geopolitical concepts such as containment into the realm of the economy. This defensive vision garnered considerable popular support among certain groups of Americans who considered themselves to be increasingly economically disadvantaged by a perceived erosion of US economic and political strength. According to Luke and Ó Tuathail (1996: 30), what Buchanan and his supporters embody is 'the politics of nationalistic possibility – of secure nation states holding, containing, keeping, or retaining the power to choose identities that remain fixed, stable, and certain'. Thus, Buchanan gives voice to a fear held by many within the USA that 'global flows are erasing/effacing/eroding American prosperity and power, causing Americans to lose their life, liberty and prosperity to the mysterious multilaterisms of the UN, IMF, NAFTA, or the WTO' (ibid.).

A geo-economic sensibility has also begun to resonate elsewhere, as a growing body of popular economics seeks to draw attention to a new age of intensified economic competition. The geopolitical undertones of such books are clear from their titles. For example, in *Head to head: the coming economic battle among Japan, Europe, and America*, Lester Thurow (1992) sets out the problems facing 'the US economy' in the context of a three-way struggle for global economic supremacy, while a similar story is told from a European perspective by Michel Albert, who in *Capitalism against capitalism* (1993) argues that the coming years will see an economic struggle between a 'neo-American' model of capitalism and a more communal form which he describes as the 'Rhinish model'.

Albert's ideas were particularly influential in the formation of what has probably been the most successful popular geo-economic publication to date. In 1995, for the first time that anyone could remember, a book which its author described as a work of political economy topped the best-seller lists in the UK. This book, *The state we're in*, was written by Will Hutton, at the time economics editor of the centre-left daily British newspaper the *Guardian*, and latterly editor of its sister Sunday publication, the *Observer*. The book has two main aims: first, to identify the causes of long-term British economic and political decline, with a particular focus on the effects of the 'Thatcherite revolution' of the 1980s; and second, to present an ambitious blueprint for the economic rehabilitation of the British economy (Hutton, 1995a).

His diagnosis of economic decline contains within it strong echoes of debates carried out over many years within the pages of *New Left Review* (e.g. Anderson, 1964, 1987; *see also* Ingham, 1984). Hutton points to the problems caused by the feudal remnants of the British *state*, with its lack of accountability, unwritten constitution and highly centralized political system. He also points the finger at the UK's *education system*, which he likens to one of apartheid, with its strong and growing divide between a large but increasingly impoverished state sector, and a small, but far better resourced, private sector, which has the effect of locking individuals into particular paths of personal and career development, and which ensures that

British society is not as meritocratic as it could be. The final part of his diagnosis is to argue that the path of British *economic development* has been blighted by the dominant role, over a long period of time, played by the City of London. Borrowing from the economic historians Cain and Hopkins (1986, 1987, 1993a, 1993b), Hutton argues that it is from their vantage points in the City that Britain's 'gentlemanly capitalists' have long held sway over the British economy. The influence of the City has tended to overshadow more 'productivist' forms of economic developments and British industry has been unusually influenced by financial considerations. Hutton portrays the City as the most powerful collective actor in the British economy. This may be seen in the way in which the Treasury is forced to navigate a perilous course between inflation, growth and interest rates, all of which is played out largely on behalf of a critical City audience (*see* Lee, 1995). The consequence of this, Hutton argues, is that 'dividend-driven' corporate behaviour encourages corporate short-termism within the British economy, which in turn militates against long-term investments which might not reap benefits for many years.

Hutton's prescription for these economic and political ills is actually fairly straightforward: it is that Britain should in effect undergo a 'Germanification' of its economy and polity, and he develops an extensive list of policy options which he recommends that an incoming Labour government should adopt.[3] In one sense, it is here that Hutton has been most successful, for his ideas have already entered the formal realm of political discussion, to be roundly denounced as dangerous and radical by those on the right, and to be celebrated on the centre-left as a possible route to a working raft of ideas which might provide some alternative to New Right policies. Indeed, Hutton's idea of a 'stakeholder society', the inspiration for which is rooted in the more associative form of capitalism found in Germany, has already been appropriated by the British Labour Party, albeit in a rather toned down and less radical format than Hutton intended it.[4]

However, while Hutton may have confirmed his status as an emerging guru for the centre-left in British politics, his ideas have been rather less warmly received within the academic community. They have been criticized on at least four counts.

First, his reading of economic history is flawed, particularly in relation to the City of London. According to Barratt Brown (1995b), Hutton relies too heavily upon a 'revisionist' version of British economic and political history (Barratt Brown, 1988), which downplays the role of the industrial revolution in Britain to argue instead that British capitalism remains founded upon merchant and commercial capital, so that real power in the British state resides in the City–Treasury–Bank of England nexus and that Britain is little more than a *rentier* economy. However, while there are elements of truth in the revisionist account, it is too simply drawn (Chapman, 1992). Moreover, there are serious difficulties in extending this historical analysis forward to the present day, so that as Barratt Brown (1995b: 34) points out,

> it has to be recognised that it is absurd to speak, as Hutton does, of a continuing *rentier* economy, and of the banks setting the conditions for investment, when the overwhelming part of the savings available for investment are not in the hands of persons (*rentiers*) and their bankers, but of large transnational trading and production companies.

Second, and in related vein, Hutton is similarly too dependent upon Cain and Hopkins's thesis of gentlemanly capitalism, which, although useful in conveying the prevailing mores and values of the City during the eighteenth and nineteenth centuries, is less appropriate to the City of London in the twentieth century, and particularly the late twentieth century, during which its social and cultural make-up has been transformed through internationalization. The City is a far more culturally diverse and meritocratic place than it ever was in the past (Jones, 1996; Leyshon and Thrift, 1997).

Third, Hutton has been criticized for overestimating the success of the German model, and for underestimating the difficulties of transplanting institutions from one context to another. Thus, Hutton chooses to gloss over the very real problems that the German economic and political system is facing in the 1990s, which have been caused in part by the pressures of the single European market programme, in part by German unification. According to Radice (1995: 18), 'meeting the costs of German unification has awoken the German middle classes to the raw deal they now have as savers, in terms of individual "freedom of choice" and real rates of return'. The massive process of restructuring brought about by the incorporation of the five eastern *länder* has brought about a rupture of the closely regulated German financial system that will prove difficult to repair (Leyshon and Thrift, 1995; Pratt, 1995). Moreover, the belief that institutions which work in one context can be cloned and then introduced successfully in another shows a rather impoverished understanding of economic history and the importance of embeddedness and social capital (Amin and Thrift, 1992, 1994, 1995; Amin and Thrift, this volume; Putnam, 1993).

Fourth, and finally, Hutton can be criticized for his traditional 'geographical imagination'. For all his avowed radicalism, he resembles Luttwak in being firmly wedded to a neo-realist conception of global order, which is seen to be unproblematically made up of nation-states, the borders of which conform to 'economies'. Therefore, that the focus of his critical attention is upon the City of London and what he describes as 'global financial markets' is not surprising, for these are institutions that escape the spatial bounds of the neo-realist world: they occupy that space which exists between states; they move across them, and in so doing avoid being fully captured and controlled by the political authorities of such states.

But to argue in this manner against Hutton would in many ways miss the point of what he is trying to do. His is not meant to be an 'objective' academic treatise: it is better seen as a purposive intervention within British economic and political debate, and in particular as an effort to influence the policies of a political party – the Labour Party – before it came to power (Barratt Brown, 1995b). Hutton's intervention should be read as an attempt to script economic change so that it conforms to his own economic and political visions and ambitions. Moreover, his account also conforms to his own belief that a broader understanding of political change is constructed both by and for those with power in order to serve their interests. Thus, Hutton's main purpose in writing the book was to counteract the Thatcherite myth that 'there is no alternative' to neo-liberal doctrine and policies in Britain. Part of this myth includes a representation of economic globalization as a 'natural' consequence of the free-market progress. Thus, Hutton's overtly geo-economic intervention can be seen as an attempt to develop a counter-script to a neo-liberal discourse of globalization. This discourse is seen as being particularly powerful and influential within Britain (Hutton, 1995b), and is so taken for granted that 'those who advance a more progressive agenda . . . [are] dismissed as outdated nationalists' (Radice, 1995: 14). Hutton argues that this view is a myth, constructed in order to serve the interests of those aligned to transnational capital.

In this, Hutton is broadly in step with a growing band of critics and commentators who have sought to shatter what they see as the myth of globalization (K.R. Cox, 1992, 1993; Cox, this volume; Hirst and Thompson, 1992, 1996). These critics see the 'global economy' as being as much constructed as real, but as being conjured into existence by a highly successful and increasingly influential discourse of globalization (Roberts, 1996; Gibson-Graham, 1996). It is to the idea of 'globalization as discourse' that we turn in the next section.

## GLOBAL DREAMS, GLOBAL NIGHTMARES: SCRIPTING GLOBALIZATION

Discourses of globalization may be seen as existing in an obverse and dialectical relationship to geo-economic discourses. The two sets of discourses may be seen as the flip-side of one another. According to Luke and Ó Tuathail (1996: 3), accounts of globalization can be seen as comprising a '"discourse of pace" [which is] linked to accelerating transitions, speeding flows, overcoming resistances, eliminating frictions, and engineering the kinematics of globalization'. Meanwhile, geo-economic discourses may be seen as part of a broader 'discourse of place', which seeks to resist the erosion and erasure of borders, identities and differences which is predicted, promised and celebrated by prophets of globalization.

Therefore, just as there are 'geo-economic intellectuals', there are also 'intellectuals of globalization'. However, there are clearly many more of the latter than there are of the former. The greater numbers of globalization intellectuals can probably be explained by two factors. First, they are drawn from a wider number of *interpretative communities*, and second, they have a larger and more diversified set of audiences that are receptive to their views.

Globalization discourses are produced in a number of different arenas. First, as with geo-economic discourses, they are produced within political communities. For example, Luke and Ó Tuathail (1996) contrast the place-based geo-economic discourses of US presidential candidates Pat Buchanan and Ross Perot with an opposing interpretative framework developed by the Clinton administration Labor Secretary, Robert Reich (Reich, 1991). Reich's views are of particular interest as an example of a politically grounded 'globalization discourse' and so are worth considering in more detail.

Reich's celebration of globalism is informed both by a fear of the implications of the kind of geo-economic populism whipped up by the likes of Perot, Buchanan and other geo-economic intellectuals, and by a belief that the USA contains a number of inherent comparative advantages that favour it as a economic space in a more integrated global economy (Reich, 1991). Thus, Reich admits that the end of the Cold War means that the temptation to turn towards geo-economics is an understandable one, and at some levels even appealing, given the co-ordinating role that the Soviet Union played as a 'nemesis', or 'other', against which the USA could project itself and receive a sense of 'moral authority' and purpose (Reich, 1991: 322; *see also* Agnew, 1993; Dalby, 1988; Ó Tuathail and Luke, 1994). By moving on to a geo-economic footing, the USA could seek to mobilize around an economic adversary (Reich suggests that Japan is the most likely candidate in this regard; *see also* Morley and Robins, 1991; Ó Tuathail, 1992b, 1993b), a move which might well have some short-term economic benefits, because

> an economic war with Japan could function as a pretext for directing America's newly freed resources toward the health, nutrition, and schooling of all of America's children (we must invest in the future generation of Americans, lest the Japanese overtake us!), and our infrastructure (Americans must be linked together by fiber-optic cable, so we can meet the Japanese challenge!). (Reich, 1991: 322)

But Reich advises against such a course of action, for such an 'economic Cold War' would only damage all parties involved in the long run. He argues that geo-economic perspectives, through emphasizing dangers inherent global fluidity and porosity, effectively overplay the loss and lack brought by the movement of money, technology and production across borders. Reich admits that such developments have called into doubt 'the very idea of an American economy . . . and notions of an American Corporation, American capital, American products, and American technology', and raise the spectre of an even bigger question, 'who is "us"?' (Reich, 1991: 8). But, in answering his own question, Reich argues that an essential core of the economy, the US labour market, remains in place. Thus, in an economic world which is increasingly defined in terms of flow and movement, it must be remembered that for the most part workers and employees are locked into places through all manner of social and cultural ties. Moreover, Reich argues that the US workforce possesses considerable competitive advantages in terms of skills, which means that through such embodied skills and competencies, the economic space of the USA is well placed to win out in a round of global economic competition. It is for this reason that Reich encourages the USA to fully engage with the globalization of economic activity.

For Reich, then, the issue of 'who owns what' is less important than 'who does what, where'. It is for this reason that he is far more relaxed about the loss of production jobs to, say, the Mexican *maquiladora* zones and other developing countries than are the likes of Buchanan and Perot (Ó Tuathail and Luke, 1994), because these are not the sort of jobs that really 'add value' to the place where the work is done. According to Reich, the majority of jobs in the USA fall into one of three functional categories. The first category of job is described as *routine production services*, which involves routine, low-level work, 'the kinds of repetitive tasks performed by the old foot soldiers of American capitalism in the high-volume enterprise' (Reich, 1991: 174). The second type of job is described as *in-person services*, so called because they must be provided person-to-person, and are for the most part low-status and low-paid jobs. The third and final category of job is what Reich describes as *symbolic analytic services*, which are essentially concerned with the identification and solving of problems the manipulation of concepts and ideas. Symbolic analysts may be found in the guise of designers, engineers, consultants, bankers, lawyers, accountants, advertisers, and other similar workers. For Reich, what distinguishes such workers is that their stock-in-trade is the use of symbols and expertise in analysis, and this may be seen in the form of mathematic algorithms, legal arguments, financial techniques, scientific insights, research methodologies, and so on. In addition, whereas the pay of workers in routine production services and in-person services is a function of the number of hours or the amount of work and effort expended, the income of 'symbolic analysts' is not really related to how much time they put in or the quantity of the work they put out. Rather, income depends upon the quality, cleverness and speed with which they solve or identify new problems.[5]

It is these sort of jobs, then, that best 'add value in place', and so Reich's prescription for the USA is that it must 'get smarter', through a general upgrading of education, and encourage its workers to enter the realm of symbolic analysis. This is where the greatest volume of value-added can be garnered within the global economy and where, importantly for Reich's argument, the USA already has a competitive advantage, as illustrated by a balance of trade surplus with the rest of the world in such services.

Reich's analysis might be seen as relatively unusual, on at least two counts. First, it is an attack on protectionism which puts a positive gloss on the process of globalization from a broadly centre-left political standpoint. As we saw in the previous section of the chapter, such standpoints can encourage a fairly defensive, geo-economic sensibility to contemporary global economic change. However, Ó Tuathail and Luke (1994) argue that Reich's prescription is nothing more than a slight modulation of the policy of transnational liberalism which has formed the basis of US international economic policy since the end of the Second World War (*see*

Cox, 1987; Gill, 1990, 1992; Leyshon, 1992; Leyshon and Tickell, 1994). Thus, in the past the US commitment to maintaining an 'open international economy' was in part sustained by the belief that such a policy was beneficial to the US economy, because of the perceived comparative advantages of US firms and US-based production in an export-driven trade system. Reich's analysis, and his insistence that the US workforce holds a place-based comparative economic advantage through its possession of high-level 'value-adding' skills, provides a justification for maintaining a commitment to an open economy despite the internationalization of ownership and production systems both within and beyond the US economy. Thus, although the landscape of corporate America has been internationalized to an unprecedented extent, it remains a good place to do business in the global economy owing to the particular skills of its workforce.

Second, as an example of a globalization discourse emerging from within a political interpretative community, Reich's analysis is also unusually careful and considered, and backed up by a good deal of documentary and even theoretical evidence. Most of the globalization discourses which emerge from within political communities tend to be far more simplistic in their general tenor, if not evangelical in tone, and display a burning faith in the 'natural' benevolence and 'obvious' utility of markets (for example, *see* Redwood, 1988).

Although tracts like Reich's are important, the political community is not the most important arena for the development of globalization discourses. There is a very different interpretive community which is mostly responsible for the production and dissemination of such discourses. This is a business-orientated interpretive community that produces scripts and understandings of global economic and political transformation for a largely corporate audience. It is at this stage of the argument that we return to the work of Roberts (1995), for she has argued that it is this community which has helped produce what she describes as the 'strategic globalization discourse', which she insists 'is central to the totalizing strategic gaze through which transnational corporations represent, frame (and claim) their world' (Roberts, 1995: x). The strategic globalization discourse is seen to be made up of

> scholarly journals and general magazines devoted to management, trade publications, and books or reports published as management texts or business institute/think tank reports. The literature is part of influential debates over corporate management and policy. . . . The institutional sites of this discourse may be identified and mapped . . . [and] include particular business schools (Harvard, MIT, and U. Penn's Wharton School and salient examples) [as well as] particular authors, journals and presses. (Roberts, 1995: 17)

Key figures in the strategic globalization discourse, therefore, include academic specialists in business administration and management theory (e.g. Levitt, 1983), management consultants (e.g. Ohmae, 1990, 1995a), as well as large numbers of 'hero-managers', usually successful heads or former heads of successful companies, who write books and articles containing their recipes for business success (*see* Huczynski, 1996; Luke and Ó Tuathail, 1996, for examples). Also important in the propagation of this discourse are companies themselves, which depict themselves as straddling the global economic stage through media advertisements and corporate literature (Roberts, 1997).

Therefore, this is largely a private-sector discourse produced by and for a community concerned to develop scripts surrounding corporate strategy. Roberts is interested in this discourse as part of her wider project to destabilize such interpretations of economic and political change, for she sees the strategic globalization discourse as less of an explanation than a justification for the remaking of the world along paths which favour some over others. In other words, it is a partial, exclusionary vision, in which 'the world of the majority is hardly mentioned', and which is less an objective depiction of the world and more a 'manifesto' put forward by those who would like to change the world along the lines of the strategic globalization discourse (Roberts, 1995: 48).

Thus, Roberts has much in common with geo-economic intellectuals such as Will Hutton who wish to challenge the hegemony of globalization discourses, and adds a post-structuralist dimension to the growing academic assault upon the idea of globalization which is already being undertaken from a broadly political economy perspective (K.R. Cox, 1992, 1993, 1997; Hirst and Thompson, 1992, 1996).

However, while it is indeed important to recognize that globalization discourses provide only a partial view of the world, and can be seen as mutually reinforcing representations of a 'small world-within-a-world' at that (Roberts, 1995: 48), this recognition does not necessarily blunt their power, and may not destabilize these discourses to the extent that Roberts would like. Although her critique may well have the effect of forcing economic geographers to rethink the way in which they think about 'globalization', and to pay more critical attention to the production and dissemination of this discourse, it is likely that the strategic globalization discourse will continue to gain momentum and spread. Although the globalization thesis may well be flawed and based upon a mixture of poor social science, hyperbole, exaggeration and corporate desire, it works as a discourse because it has a highly receptive audience within the offices and boardrooms of the international business community. It is in this sense a self-affirming and self-propagating discourse.

As Huczynski (1996) has observed, once established, key management ideas and principles appear to be immune to academic criticism, even if this is sustained over several years. The main reason for this would seem to be that 'the field of management treats words like "theoretical" and "academic" with disfavour, if not contempt' (Huczynski, 1996: 102). The preference is instead for practical-based ideas which bear some relation to their own experience, and which successfully express what managers felt 'they half knew . . . but never knew how to put into words' (Huczynski, 1996: 114). Therefore, if one accepts that the success of management ideas is based on 'not offering their audience anything too outlandish to accept' but requires 'only a limited rearrangement of existing views' (Huczynski, 1996: 102), then one can see why the notion of globalization has become such a popular idea within management circles:

> Popular management ideas wrap simple propositions up in a package form and attach impressive sounding labels . . . the small amount of effort that managers [have] to expend on 'unwrapping' the idea-parcel [gives] them a feeling of achievement and satisfaction. (Huczynski, 1996: 85–6).

Thus, it could be that the strategic globalization discourse has become so successful because it successfully articulates a feeling within management circles that 'something has changed' in the global economy, and which needs to be responded to by rethinking the way in which their businesses are organized. Thus, it is both an explanation and a programme of action.

Thrift (1996b, 1996d), borrowing from Jowitt, sees this shift in management perspective as the result of the decline of a prevailing 'Joshua discourse' within management circles, which was 'founded on the idea of transcendental rationality, on the notion of a single, correct, God's-eye view of reason' (Thrift, 1996b: 13). To replace it there has emerged a 'Genesis discourse'. The difference between the two discourses is that while the Joshua discourse proclaims that 'order is the rule and disorder is the exception' (Thrift, 1996b: 14), in the Genesis discourse the reverse is true. Thrift (1996b, 1996d) argues that the rise of this new discourse is rooted in material experience, and in particular, by the economic and political changes which surrounded the breakdown of the Bretton Woods system in the 1960s and 1970s. This had the effect of introducing new levels of uncertainty and unpredictability into economic life.[6] The outcome of these material changes has been a growing receptiveness to a broad managerial discourse which stresses the need for adaptability, manoeuvrability and the ability of organizations to 'go with the flow' (Thrift, 1996b: 19).

A recognition that the material conditions within which managerial practice takes place have changed in part helps explain the rapid growth in managerial ideas during the 1980s (Huczynski, 1996). 'Globalization', of course, in all its variants and subforms, is just one of many such ideas.[7] Although one of many management ideas, it is an important idea and one which, significantly, has clearly paid dividends for many aspiring 'global corporations', because for many such companies the concept of globalization fulfils the crucial determinant of success of any management idea: it successfully articulates a felt understanding of change while at the same time delivering a practical guide to action which for many managers is seen to 'work', in that it brings increases in turnover and profitability.

Therefore, to return to Roberts' original thesis, she is correct to argue that globalization is a process that works both discursively and materially at the same time. But in recognizing this it must also be acknowledged that the reason that the strategic globalization discourse has been so effective is that it 'plays' well among its most important audience: a class of extant and aspiring corporate managers and their neo-liberal supporters within political communities. It is significant, then, to note also that the audiences that are most receptive to geo-economic discourses are very different, tending to be based on the centre-left or within highly conservative nationalist wings of political communities, and within domestically focused businesses which have most to lose from increasing levels of 'global competition' (Luke and Ó Tuathail, 1996).

However, that these discourses have been successful in generating appreciative audiences is due in large part to their relative simplicity and the ease by which their messages may be grasped. They manage to convey *an* understanding of what are highly complex and fast-moving economic and political processes as well as providing a strategy for action. But in doing so, such discourses give the impression that the processes they depict are self-evident conditions; that is, we are living now in either a geo-economic or a globalized world. However, as the contributors in this part of the book suggest, it is safer and probably more accurate to see geo-economics and globalization as *processes* (Allen, 1995; Harvey, 1996a), which exist alongside and in a complex relationship with one another.

## CONCLUDING REMARKS

This introductory chapter has sought to provide a summary of some new approaches to globalization being explored in economic geography, by looking at the way in which the ideas of geo-economics and globalization are scripted and transformed into generalizable discourses of economic and political change. The chapters in this part of the book follow their own particular paths in their consideration of economic globalization.

Kevin Cox in his chapter speaks directly to the concerns developed above, in that he too focuses on the idea of 'globalization as discourse', the purpose of which he sees as 'naturalizing' capitalism as a necessary and inevitable social relation. He sees the propagation of globalization discourses as part of an international class struggle, and feels that they overstate the power of 'mobile' capital and the relative weakness of 'place-bound' communities and workers. Cox is concerned to point out that globalization is a response to earlier territorialized resistance to capital, and so, like Roberts, he stresses the importance of 'distinguishing between a reality and a discourse aimed at becoming a reality'.

Erik Swyngedouw too, in his chapter, focuses upon the discursive dimensions to economic order, arguing that concepts such as the 'local' or the 'global', notoriously present in much of contemporary geographical debate, are often merely speculative, discursive – but eminently powerful – vehicles which are used to order political, social and economic processes in particular spatialized kinds of way. Swyngedouw focuses on discursive and other struggles surrounding issues of scale, and this theme is also picked up in the chapters by Amin and Thrift and by Herod. Amin and Thrift focus on the complex interrelationship between the 'local' and the 'global'. As they point out, it is not only that the local has gone global, as most conventional accounts of globalization insist, but that in many places the global has gone local. Through an analysis of Putnam's concept of social economy they point to the possibilities of localities and communities 'making up' their own scripts of economic and political organization through networks of governance and learning. Meanwhile, Herod's chapter focuses upon the ways in which representations of scale are constructed by way of a historical analysis of a conflict between two rival dockers' unions in the USA during the 1950s. Even for capital, the dialectic of global and local is hardly straightforward and is shaped by culturally complex readings of globalization. David Angel and Lydia Savage point to the constrained globalism of Japanese firms in the USA, while Robert Fagan argues that simple economistic notions of globalization are undermined by geographies of corporate restructuring.

In their chapter, Peter Dicken, Jamie Peck and Adam Tickell attempt to summarize the state of play in the 'globality debate' as it exists within economic geography and its related disciplines. They are hardly any more charitable than Roberts, and they describe this debate as 'polarized, unconstructive, and somewhat ill-disciplined', although they make it clear that they do not believe that globalization is merely a myth put about by transnational capital and its neo-liberal 'spin doctors'.

Similar sentiments are expressed by John Agnew and Richard Grant in their review of the changing position of Africa in the global economy. The economic and political problems of this continent have become manifestly worse in recent decades. Therefore, somewhat perversely, the rise of a global economy has led to many spaces being economically written out of the global economy, and so written off. The economies of Africa are perhaps the most notable and tragic examples of this.

It is through examples such as this that one comes to realize the limitations of both geo-economic and globalization discourses. In the former, we encounter a defensive, mercantilist mindset which effectively condemns economies such as those in sub-Saharan Africa to their fate, as Western economies seek to defend what they do have at the expense of weaker economies. In the latter, we see a more expansive attitude to economic development, but one which accepts a deepening of existing levels of uneven development in the global economy. In the light of the tragedy of Africa, neither vision is particularly appealing.

## ACKNOWLEDGEMENTS

I would like to thank Roger Lee and Jane Wills for their patience and forbearance. I would also like to thank them and Adam Tickell for the extremely helpful and insightful comments they made on an earlier draft of this chapter. They have made this chapter a better one than it otherwise would have been, although they will all still disagree with parts of it. The responsibility for the remaining errors and misinterpretations in the chapter remains mine alone.

## NOTES

1. For example, consider the reflections of David Harvey – whose materialistic credentials can hardly be called into question – on the collapse of the Bretton Woods system and the rise of a more decentralized system of global financial regulation co-ordinated through markets: 'what really happened here was a shift from one global system (hierarchically organized and largely controlled politically by the United States) to another global system that was more decentralized and coordinated through the market, making the financial conditions of capitalism far more volatile and far more unstable. *The rhetoric that accompanied this shift was deeply implicated in the promotion of the term "globalization" as a virtue*. In my more cynical moments I find myself thinking that it was the financial press that conned us all (myself included) into believing "globalization" as something new when it was nothing more than a promotional gimmick to make the best of a necessary adjustment in the system of finance' (Harvey, 1996a: 10; emphasis added).
2. This work is paralleled by a strong political economy of writing upon processes of 'hegemonic transformation' (*see*, for example, Arrighi, 1993, 1994; P. Taylor, 1996).
3. The political changes that Hutton recommends are a written constitution, proportional representation, republicanism and devolution of most of the social functions associated with government to regional and local authorities with responsibility for their own finances (as is the case with the German *länder*). In education he recommends education for all up to the age of 18, with a particular focus on vocational training in the workplace. The economic policy recommendations include an independent but regionally integrated central bank (i.e. like the Bundesbank); a banking system integrated with trade associations that directly supports long-term investment in small and medium enterprises as well as large companies; and a partnership between companies and trade unions (Hutton, 1995a; *see also* Barratt Brown and Radice, 1995).
4. The 1996 Labour Party Conference seemed to confirm further infiltration of Hutton's ideas into the heart of policy. In his conference speech, Labour Party leader Tony Blair argued that his three main priorities for policy were to be 'education, education and education'. However, all the evidence suggests that 'New Labour' sees Hutton's prescription as too radical for it to adopt as it seeks to position itself in the middle ground of British politics.
5. Reich makes an important observation here, for not only are these sorts of jobs well paid, they are also becoming increasingly important to the fabric of contemporary capitalism, which is a product of, first, the 'value added' which such workers can add to production chains (through design, engineering, advertising expertise, etc.); second, the role such workers play in facilitating the circulation of capital between different firms, different parts of the economy, and between different countries (for example, as in financial services, and accountancy); and third and finally, a growing recognition of the importance of culture and knowledge within contemporary capitalism, which can be seen in the rise of cultural industries, but also more generally (*see* Miller, 1995a; Thrift, 1996b, 1996d; Thrift and Olds, 1996).
6. A key objective of the Bretton Woods system was to control the more speculative and volatile aspects of international capitalism. *See* Helleiner (1994) and Leyshon and Tickell (1994).
7. Other key management ideas popularized in the 1980s include quality circles and total quality management, lean production and just-in-time, strategic alliances, and business process re-engineering (*see* Thrift, 1996b).

# Chapter Eleven

# GLOBALIZATION, SOCIO-ECONOMICS, TERRITORIALITY

**ASH AMIN AND NIGEL THRIFT**

## INTRODUCTION

'Globalization' has become a catchphrase in the 1990s. Most commentators accept that the world has 'gone global' in one way or another – that money, markets, firms, politics, people and cultures now transcend territorial boundaries, that the influences and problems of the world are becoming one, and that access to once remote parts of the world has become easier. Geographies seem to be shrinking, perhaps even disappearing.

One common interpretation of globalization is that it is an 'exogenous' force which threatens local and national identities, integrities and autonomies – a force to be resisted, through the pursuit of autarchic or separatist strategies to protect the domestic heartland. Another interpretation is that globalization offers the opportunity for local or national renewal – the promise of new ideas, new people, new investment, new opportunities for embarking upon a different development path. The rhetoric of free trade, multiculturalism, continental federalism and internationalization often accompanies this perspective.

Thinking about links between the global and the local in these terms, however, is to force discussion into a dualist opposition between globalism and autarchy, between global 'flows' and local 'fixities'. We wish to propose that the nexus between the global and the local can be conceptualized differently: as a dialectical relationship, composed of multiple and asymmetric *interdependencies* between local and wider fields of influence and action. Such a proposition means seeing global forces (such as transnational corporations) as fields of action producing a differentiated geography of potentialities, rewards and penalties. It also means seeing territories as integrally a part of a global political economy, and not separate from it.

In this view, to think of globalization only as a threat or opportunity for local or national communities is to miss the point. Nations and regions are now an intrinsic part of, and dependent upon, the increasingly globalized economic, political and cultural forces which define them, in combination with local attributes. A key practical question for nations and regions seeking a place in global capitalism might be how best to develop within a global political economy, how best to combine global and local potentialities, and how to avoid being bypassed or damaged by the mainstream.

The purpose of this chapter, which focuses on issues of economic development, is to identify the feasible areas of national action serving to enable a pattern of development based on *cumulative* and *endogenous* strengths. It therefore draws on insights from the emerging subdiscipline of socio-economics to identify two territorially bounded 'externalities', each with distinctly positive effects for national or local communities. The first concerns learning capability and learning-based competitiveness. The second refers to the encouragement of diffuse institutional governance of the economy, involving – in open and associationalist interdependence – a decentralized and democratized state, intermediate organizations between market and state, and active citizenship. The argument is that it is the ability to internalize knowledge to competitive ends and to upgrade institutional capacity that will help nations and regions to become, or remain, *self-regenerating growth poles* in the global economy. In contrast, the use of 'territorially disembedded' policy weapons such as cost or price adjustments, or incentives to firms, will fail to provide unique, sustainable or cumulative advantage to territories, and they will fail to achieve such goals. What we are therefore proposing is a reconsideration of debates on the global and the local which is simultaneously a reconsideration of what counts as 'economic' and as 'geography' (Thrift and Olds, 1996).

To set the context, the chapter therefore begins with a summary of some of the key aspects of contemporary economic globalization and a discussion of their significance as a set of place-transcending forces. Then, focusing on the national arena, but not implying any primacy over subnational regions, the chapter disentangles the implications of a 'relational' or 'interdependency' perspective on global–local relations. It emphasizes the changed but enduring significance of the national (or local) arena and the nation-state as meaningful spheres of economic organization and collective political action. The final part of the chapter then develops some of the arguments on territorial pathways to growth rooted in encouraging learning capability and institutional associationalism.

## THE GLOBALIZING ECONOMY

Although various forms of globalization can be traced back at least four centuries from the rise and subsequent expansion of capitalism across the world, there is some agreement that a decisive transition towards intensive global economic integration took place in the early 1970s with the break-up of the Bretton Woods system of regulatory financial control of national economies. We can now see some of the tendencies arising from the subsequent reorganization of the world economy. There are, of course, many of these, but five can perhaps be counted as important in terms of the rise of *new trans-territorial* influences.

The first is the *increasing centrality of the financial structure* through which money is created, allocated and put to use, and the resulting increase in the power of finance over production. Thus Harvey (1989a) and S. Amin (1996) write of the degree to which financial capital has become an independent force in the modern world, notably at times of generalized crisis, when the link with industrial accumulation is weakened, while Strange (1991) writes of the increased 'structural' power exercised by whoever or whatever determines the financial structure. It is particularly the global reach of finance which is striking today, as global money, in a variety of guises, straddles and regulates the world's national economies and business transactions.

The second tendency is the increasing importance of the *'knowledge structure'* (Strange, 1988) or 'expert systems' (Giddens, 1990). In economics, sociology, economic sociology and so on, more and more attention is being paid to the importance of knowledge as a factor of production. Lash and Urry (1994) write of a new phase of 'reflexive accumulation'. Storper (1995a) refers to the rise of the 'learning-based economy' out of the ashes of Fordism. And so on. The debate can be found in the form of arguments over whether an increase in the educational base of workforces produces higher GNP, in controversies concerning the importance of learning as a factor in efficient production, and in arguments over the importance of cognitive and aesthetic reflexivity as corner-stones of modern business conduct (Lash and Urry, 1994; Boden, 1993). Whatever the form of the debate, it seems clear that the production, distribution and exchange of knowledge is a crucial element of the global economic system on a scale that was never the case before.

A third and related tendency is the *internationalization of technology*, coupled with an enormous increase in the rapidity of redundancy of given technologies. The development of micro-electronics applications has quickened the diffusion of standards and know-how. At the same time, however, the increasing reliance on technological innovation has forced the speed with which technological trajectories are required to change (Dosi *et al.*, 1990; Metcalfe, 1988). Thus at one level, the globalization of technology represents a levelling of access, while at another level, the greater uncertainty produced by this more complex environment militates against any except those institutions with considerable resources and continuous learning capacity. In other words, the new institutionalized risk environments (Giddens, 1991) reward only the adaptable.

The fourth major tendency is the *rise of transnational oligopolies*. There is a sense in much recent writing that we have now reached a point at which corporations have no choice but to 'go global' very early on in their life cycles (Strange, 1991). The result is that national measures of concentration and market share have become less relevant as corporations manoeuvre in global markets and perceive national markets as one component of a wider frame of action and opportunities. Today, more than half of all economic activity in Belgium, Canada, The Netherlands, Switzerland and the UK, and more than 30 per cent of all economic activity in Australia, France, Italy and Germany, is composed of the combined value added of foreign transnational corporations and the foreign output of home-based transnationals (United Nations, 1992).

Finally, and running parallel to the globalization of production, knowledge and finance, there is the *rise of transnational economic diplomacy* and the *global orientation of state strategies*. There is clearly a sense in which we have entered a new era in which governments and firms bargain with themselves and one another on the world stage (R. Cox, 1993; Cerny, 1995). In addition, transnational, 'plural authority' structures like the UN, G7, the European Union (EU) and so on have become increasingly powerful (Held, 1991). The result appears to be the reorganization of the hitherto prevalent system of international stabilization and rule formation based upon the unchallenged might of nations with the greatest 'structural' power (e.g. the USA), but in new directions which as yet remain unclear.

These tendencies, taken together, represent the consolidation and extension of an increasingly global sphere of organization and influence, which, as MacLean (1994: 2) puts it, has 'something to do with matters located at the level of the global system, or at least with things international'. Thus we agree with McGrew (quoted in Dunning, 1994) that globalization captures the rise of worldwide processes as well as inter-territorial dependence:

> Globalisation refers to the multiplicity of linkages and interconnections between the states and societies which make up the present world system. It describes the processes by which events, decisions and activities in one part of the world come to have significant consequences for individuals and communities in quite distant parts of the globe. Globalisation has two distinct phenomena: scope (or stretching) and intensity (or deepening). On the one hand, it defines a set of processes which embrace most of the globe or which operate world-wide; the concept therefore has a spatial connotation. . . . On the other hand, it also implies an intensification of the levels of interaction, interconnectedness or interdependence between the states and societies which constitute the world economy. Accordingly, alongside the stretching goes a deepening of global processes.

In contemporary international political economy (i.e. heterodox international relations), two major fears have become associated with the phenomenon of globalization, and particularly the 'stretching' connotation. One is that the rise of a global field of action represents the *end of the nation-state* or, more precisely, the significant erosion of national economic sovereignty. The second fear is that the rise of a global field of action represents the *end of geography* – a delinking of development processes from their local or national settings.

# GLOBALIZATION AND NATION-STATES

In this section, against the first of the above positions, we wish to argue that globalization is neither a *substitute* for territorial forms of governance, nor a 'space-of-flows . . . which exists *alongside* the space-of-places that we call national economies', as Ruggie puts it (1993: 172). Instead, we wish to propose that the new global forces and global interconnections might, paradoxically, imply a heightened influence for national systems of development and governance. Anderson and Goodman (1995: 25) have argued, in recognition of the evolutionary and historical nature of societal change, that

> States, once born, are continuing to exist rather than being tidily removed to clear the ground for new polities. Their powers and roles may be changing, but they continue to co-exist and interact with a plethora of other, different kinds of political communities, institutions, organizations, associations, networks. Globalization is overlaying the mosaic of states with other forms of political community, non-state polities and non-political market relations which are shaped by different forms of authority, 'territorial', 'non-territorial' and 'functional'.

Against such an idea of increasing layers of governance, within international political economy in particular, there is an emergent view that the role of the nation-state as the principal instrument for national cohesion is now under threat. Many argue that global transformations from above, such as those outlined earlier, together with the growth of regionalist movements from below and horizontal inter-regional coalitions from the side, are said to be challenging the traditional authority and reach of the nation-state (Cerny, 1995; Jessop, 1994a; Harvie, 1994; Hirst, 1994). This process of the 'hollowing-out' of the nation-state is said to represent, in essence, two transformations: the reduced significance of the nation-state as the principal or sole governance authority, and the reorientation of national state priorities in the direction of securing global competitiveness.

As regards the first transformation, the argument is that globalization and regionalism are forcing the nation-state to simultaneously concede power upwards to supranational institutions (the EU is an obvious example) and downwards to local institutions with detailed knowledge of the strengths and weaknesses of individual regions. Thus principles of subsidiarity create pressures for certain functions concerned with issues articulated on a transnational scale (e.g. security, migration, competition, financial markets, industrial organization) to be transferred to the appropriate international organizations, or executed on an intergovernmental basis. Similarly there is pressure for other functions, notably the provision of supply-side support for entrepreneurship (e.g. infrastructural improvements, labour market adjustments, support for inward investors, technical and financial assistance to firms), to be devolved to local organizations with better on-the-ground knowledge (Begg and Mayes, 1993; Cooke, 1994; Ohmae, 1993).

The second transformation concerns the reorientation of national economic policy priorities. Globalization is said to have weakened the ability of national governments to rely on traditional measures such as trade and exchange controls, monetary or credit policies and national industrial policy as a means of protecting the national economy. At the same time, globalization appears to have made it an imperative for nation-states to develop economic policies orientated towards securing global competitiveness, at the expense of growth models orientated towards the domestic economy (Reich, 1992).

Thus, the 'competition' or 'entrepreneurial' state is constrained to introduce a number of changes in economic policy. First, domestic economic policy is shifted towards supporting internationally competitive entrepreneurship and innovation. Second, the state is compelled to divert resources to boost competitiveness through such schemes as training programmes, innovation and technology transfer support, help for small firms, and infrastructural improvements. Third, globalization turns state policy towards initiatives designed explicitly to make the supply side of the domestic economy more attractive for inward investment, and to force macro-economic policy (especially monetary policy) to be attentive to externally influenced interest rates, monetary fluctuations and national indebtedness (R. Cox, 1993). In other words, state policy becomes more and more driven by external forces.

## AN ENDURING ROLE FOR NATIONAL ECONOMIC GOVERNANCE

Much of the above chimes with contemporary experience in free-market economies, and the critical question to be asked is whether the hollowing-out thesis is perhaps more accurately a description of neo-liberal politics and policies than of a global structural tendency to which all nations will have to conform (Jessop, 1995a). The experience of economies such as Germany, Korea and Japan, in which economic success has relied on active and domestically orientated state intervention, suggests that, given the political will and appropriate national policy action, the forces of globalization can be harnessed to viable national development programmes. This is not to argue that all nations are capable of turning globalization to their advantage, simply to claim that globalization need not necessarily crush all national governance autonomy. Nor are we trying to claim that national politics and policies will suffice for economic success; in the absence of an appropriate set of competitive assets, nothing much will happen. This said, there remain at least three autonomous areas of national policy as far as economic development is concerned.

The first of these is in the area of corporate governance. As Lazonick (1993) notes in his critique of Reich (1992), the governance of corporate investment strategies can remain a *national* phenomenon. He points out that while a liberal market for corporate ownership and control has arisen in the USA and UK, it has not done so in Germany and Japan, where mergers and take-overs remain restrained by appropriate anti-trust rules as well as the restricted public flotation of companies. Relatedly, it might be added that the national financial system, and the extent to which it supports the lending of long-term and industry-sensitive capital, also remains an important aspect of successful economic development, with obvious contrasts again to be drawn, say, between the UK and Germany or Japan (*see* Hutton, 1995a; Albert, 1993). Such an argument also endorses Porter's (1990) strictures about the importance of the 'home base', composed of an adequate supply base, domestic demand and institutional support, for international business competitiveness. More widely, it supports Dunning's argument (1993) that in the contemporary context of intensified global competition, governments can make a huge difference by helping firms to overcome or counteract market failure resulting from uncertainty, information irregularities, product deficiencies and so on, via schemes to reduce transaction costs (e.g. tax concessions, access to finance and information, subsidies).

Second, there is now emerging a forceful body of literature, arguing against the globalization thesis, that emphasizes the enduring importance of *national* systems of innovation (Lundvall, 1992; Humbert, 1994; Zysman, 1994). The idea of disembodied technology and knowledge networks that transcend national boundaries, as articulated earlier, is not judged to be convincing. For instance,

Archibugi and Michie (1993) and Patel and Pavitt (1994), while recognizing the increased global exploitation of technology and the increased international technological collaboration associated with transnational corporations and the rise of global communications networks, find less evidence showing the global *generation* of technology and know-how. Both sets of authors conclude that, in spite of 'globalization', national innovation systems continue to pay a crucial role in the production, organization and first-chance appropriation of research and know-how. Indeed, firms appear to be heavily influenced by national capabilities when taking strategic decisions concerning international joint ventures or the internationalization of their R&D facilities, thus making *national* innovation systems *more*, rather than less, important.

National innovation systems are, of course, composed of technical and non-technical aspects of knowledge creation and learning – formally and informally constituted (*see later*). But recognizing this point does not downplay the decisive influence that continues to be exerted by the state. The most obvious field of action, of course, is the national (or local, for that matter) educational and training system. In innovation-intensive economies such as Germany and Japan, a unique feature is the public provision of high-quality general and specialized education as well as ample opportunities for retraining within and between jobs. Another field of state action is science and technology policy, where national differences in state support for R&D in the public and private sector are considered to be of decisive influence on the technological competitiveness of industry (Fransman, 1995; Nelson, 1993). The depth and durability of public-sector support for industry, especially at the interface between the development and application of science and technology, is said to be assuming heightened significance as the pressure for, and cost of, technological renewal among firms continue to rise, precisely as a consequence of increasing global competition.

The third area where the national arena remains all-important is in relation to the labour market and the industrial relations system. Labour market issues such as the social security system, wage negotiation, and training and education continue to remain a national responsibility, with states reluctant to transfer regulation either downwards to local institutions, or upwards to a supranational body such as the EU. Teague (1995) has argued recently that problems of scale and heterogeneity – radically different scales and systems of wage determination, for example – make supranational regulation highly unlikely, while the evidence for the reconstitution of wage bargaining or industrial relations systems at the regional level remains scant.

As with the question of innovation, issues concerning the cost, quality and availability of labour, together with those concerning attitudes to work and industrial relations in general, are not reducible to national state action. The industrial relations climate of a country, for example, though influenced by state regulations and government policies, is the product of the historical evolution of the capital–labour relation. But this point acknowledged, the role of state action should not be underestimated. There is, for instance, a very clear division between neo-liberal pathways to economic prosperity rooted in state efforts to depress wages, deregulate labour markets or curtail social expenditure, and social democratic pathways rooted in investment in skills, training and education, active industrial policies, and state arbitrage between labour and capital.

In summary, the scope for national state economic action remains considerable, notably on the supply side (*see* Amin and Tomaney, 1995b). More importantly, we wish to suggest that differences between nations concerning crucial supply-side regulatory arenas such as corporate governance, innovation and labour markets might make the difference in terms of the development and sustainment of the competitive advantage of nations (more on this later). How else are we to distinguish economies such as Germany, Japan and the East Asian tigers, which have embarked on development trajectories that are quite distinct from those of less interventionist nations in the present phase of globalization?

## SOCIO-ECONOMICS AND TERRITORIALITY

We now wish to turn to the second of the fears associated with globalization mentioned earlier, namely the idea that globalization represents a supersession of the twentieth-century economic 'space of places' by a new, deterritorialized 'space of flows'. We would contend that such an idea is sustainable only on the assumption – not uncommon in neo-classical or other rigidly formal and decontextualized methodologies in economics – that economic behaviour is not rooted in territorially articulated social, cultural, and institutional settings, and that such 'embeddedness' has no role to play in shaping economic outcomes. However, any cursory observation of the workings of firms and markets confirms that such an assumption is hard to sustain, and indeed, a serious attempt is now being made across the social sciences to redefine what can be regarded as the 'economic', precisely in reaction to current economic orthodoxies concerning what is and what is not 'economic'.

## Perspectives from Socio-economics

This new economics might be described as 'socio-economics'. It stresses the relational, social and contextual aspect of economic behaviour. Implicit in this is, as we shall see later, a different understanding of territoriality (in the face of globalization). Socio-economics is composed of a variety of strands, which together make up a brand of economics that is much broader than the moral communitarian socio-economics associated with the ideas of Amitai Etzioni (1988).

One strand of work, found especially in economic sociology (*see* Smelser and Swedberg, 1994), has its roots in the rediscovery of Polanyi's vision of the economy as an instituted process that is 'clearly embedded in networks of interpersonal relations' (Granovetter, 1985: 506). Then, with the actor-network theory of Latour (1986), Callon (1986), Law (1994) and others working on the sociology of science comes the idea that 'actors define one another in interaction – in the intermediaries that they put into circulation' (Callon, 1991: 135), where intermediaries are usually considered to be embodied subjects, texts, machines and money. Thus, interaction and the technical intermediaries mobilized influence economic capabilities in networks.

A third strand of work comes from economics. The rise of institutional economics has been rapid and still continues (Hodgson *et al.*, 1993). This new work takes the view that economics must be about the study of institutions, broadly defined. Hamilton's (1963: 84) definition is perhaps the best known:

> It connotes a way of thought or action of some prevalence and permanence, which is embedded in the habits of a group or the customs of a people. . . . Institutions fix the confines of and impose form upon the activities of human beings.

Most particularly, institutional economists tend to distinguish between habits, which involve individuals; routines, which involve groups; and institutions, which are composed of habits and routines. Again, as in actor-network theory, institutions are viewed as able to stabilize over a certain period, against a background of uncertainty, characteristics which, most importantly, will include skills, tacit knowledge and formalized information. Thus institutional economists stress the evolutionary nature of economic change, against the rational abstractions of neo-classical economics as well as the structural teleologies of ahistorical strands of Marxist economics.

A fourth strand of work can be found within organization theory. A new organizational theory is currently being built up which relies, to a much greater extent, on the notion of organizational cultures, and the idea of organizations as 'many different things at once' (Morgan, 1986: 339), as well as the notion that organizations are modes of representation (Cooper, 1992; Hassard and Parker, 1993), in order to stress the role of (institutionalized) cultural and cognitive forces in explaining success or failure in economic competition (Hodgson, 1994).

These are strands of work with diverse origins, but they have certain elements in common, which have particularly important implications for how we should understand economic behaviour. First, there is an emphasis on the locking in of economic agents into networks of association, into institutionalized relations of one sort or another which are necessary for economic survival. Networks are modes of transaction which presume some form of *mutual orientation* and usually *obligation*.

Second, the binding agent of networks, and therefore also individual success, is usually held to be *information*. This information can be tacit and informal (like certain kinds of skill), or explicit and formal (like many kinds of management procedures). It can be in the form of habits, or routines, or conventions, or narratives.

Third, and as a direct corollary of the emphasis on information, the socio-economic approach also tends to emphasize the *learning* abilities of networks (Avadikyan *et al.*, 1993, Storper, 1994). Considerable effort has gone into investigating the degree to which certain networks, for a variety of reasons, are able to store information and generate innovation, as well as reflect on them rather better than others (Dosi, 1988).

Fourth, the socio-economic approach considers in some depth the importance of the *thickness* and degree of *openness or closure* of networks. The approach is concerned with the extent to which networks of governance can interweave and the ways in which this interweaving can be beneficial or detrimental to economic 'performance' (howsoever defined) by allowing or preventing transfers of appropriate knowledge and information (Sabel, 1994; Bianchi and Miller, 1994; dei Ottati, 1994). Thus socio-economics also recognizes the *asymmetry of power* between networks of interdependencies.

Finally, central to socio-economics is the recognition of *path-dependent evolutionary change*. That is, it emphasizes the degree to which networks constructed in particular contexts for particular purposes are consolidated over time. Networks are thereby relatively invariant but they are still subject to change because networks remain *context-dependent*. Thus the socio-economic approach tends to assume that the category of the economic is a diachronic rather than synchronic one, with different sites locked into specific trajectories of economic evolution and development.

Taking these commonalities together, socio-economics explains economics and differences in economic outcome in terms of the dynamics of networks of association, path dependency, evolutionary impulses, manipulation and management of networks, and access to and control over information, innovation and learning capability.

## Socio-economics and Territoriality

From the perspective of socio-economics, it is difficult to see why *globalization should represent the end of geography*; that is, the deterritorialization of the economy. The core concepts of socio-economics summarized above – from the centrality of innovation, learning and information to network building and path dependency – are all profoundly territorial concepts. These 'soft' but crucial foundations of economic behaviour are the characteristics of living and evolving communities – including regions and nations. Certain economic or governance networks of the global political economy (for example, international corporate systems or international governmental coalitions) draw upon networks and foundations which are not automatically place-bound, and this is an important trans-territorial development. But these as well as other networks still continue to draw on the attributes of national and local societies. Habits, customs, informal learning environments, industrial cultures and economic expectations are not the characteristics of distanciated global networks, but the essence of *entrenched* institutional and social arrangements within different local contexts. Perhaps this is why, despite globalization, we continue to distinguish between Japanese, German, Scandinavian, Korean, American or British capitalism, and why we feel unconvinced by the idea of a *homogenizing* global logic advanced by transnational corporations and banks, international organizations, globally mobile professionals and executives, and 'Coca-Colarized' consumption patterns.

With socio-institutional influences in mind, John Zysman clarifies why, even in the context of the vexed issue of technological globalisation, national 'roots' and national models of innovation remain salient. It is worth citing a recent piece in some detail:

> National systems of innovation and development should endure because they are rooted in entrenched national institutional and social arrangements. The argument runs something like this: Technology, like all market processes, is not disembodied. It develops in communities; it has local roots. Technological trajectories can only be defined in reference to particular societies. Consider, first, that technological knowledge and know-how is transmitted through at least three mechanisms: individuals, organisations and communities . . . the character of these organisations and communities gives particular focus to the process of technological development and innovation. . . . Over time, the bets accumulate . . . trajectories that emerge in one country cannot be easily copied . . . the result of cumulating know-how and investment are lock-ins to development, to trajectories. The lock-ins . . . are not simply matters of learning and intellect. Rather, the cumulation of technological bets underpins these trajectories. Entrenched, enduring and significant variations in national lines of technological development . . . (Zysman, 1994: 9–10).

To emphasize that individuals, organizations and communities provide 'particular focus', and to recognize 'lock-ins to development', is a first step towards acknowledging territoriality. Zysman's acceptance of the power of intermediaries in the arena of innovation, and in such broad terms as those above, forces recognition of the contextual circumstances in which knowledge and know-how are produced, transmitted, assimilated and renewed. These circumstances might include formal settings such as research centres, training institutions and schools, and how their products are assimilated within the industrial environment. They might also include, importantly, the quality of informal settings such as meetings, conversations, workplace cultures and national attitudes towards learning, all of which have a profound influence on the propriety to innovate. Similarly, Zysman's reference to 'lock-in' stresses the historical legacy of the latter institutions, and, more pertinently, their combined effects in shaping a distinctive national or local 'system of innovation' that fixes the economy to a particular trajectory. The innovation–development link defined in these terms is an invitation to consider the raft of societal influences on innovation deeply rooted in the institutions and cultures of individual nations and regions.

Thus far, we have used the term 'territoriality' to signal the institutional and cultural specificity of nations and regions, against the idea of a decontextualized global political economy. But geographers and regional analysts working on individual cities and regions also suggest an alternative meaning, associated with the relevance of links of *proximity* and spatial agglomeration, even in our globalizing age. The survival of highly localized industrial clusters such as industrial districts in Italy, high-tech agglomerations in California or centres of manufacturing excellence in southern Germany is forever quoted as evidence for the continuing powers of local linkages within otherwise global interconnections. This paradox, which elsewhere we have described as the possibility of localities being able to remain as centres of knowledge and excellence in otherwise global industrial networks (Amin and Thrift, 1992), needs to be explained.

As the most obvious level, regionalization can be put down to the well-known advantages of industrial clustering identified within the economic orthodoxy (at least, those versions that acknowledge life beyond individual firms and the market). Thus, from a lineage of classical economists that runs from Adam Smith to Myrdal, Hirschman, Perroux and Kaldor, we may adduce advantages of specialization accruing from spatial clustering of firms and industries, together with reductions in search costs for firms, cost advantages of proximity, and so on. More recently, researchers working on industrial districts have extended the ideas of Alfred Marshall at the turn of the twentieth century to emphasize the efficiency gains and flexibilities associated with task specialization between co-operating firms, the depth of know-how and expertise resulting from area-based product specialization, and, in general, the 'industrial atmosphere' of such areas that secures capabilities beyond the limitations of individual economic agents. This latter aspect of regional specialization, that is, the accumulation of collective capabilities leading to their progressive institutionalization, we have described elsewhere as local 'institutional thickness' (Amin and Thrift, 1994), to stress the commonly available and institutionalized nature of assets which enhance firm-level capabilities. These assets might include hard institutions such as business support agencies and widespread interest representation. They might also include softer institutional forms such as the build-up of regional strategic vision and broad economic governance capability, through activities serving to build a strong sense of place and belonging, through high levels of public participation in civil and political life, through the development of an independent political identity that provides influence in arenas beyond the locality. In short, 'institutional thickness' also refers, at least potentially, to locally generated cultures serving as a base for economic activity and economic innovation.

Michael Storper (1994) has suggested that in those fortunate places in which globalization is consistent with localization of economic activity, a distinctive territorial aspect is that of locality as a nexus of 'relational assets' or 'untraded interdependencies'. This specifies our concept of 'institutional thickness' further, by emphasizing territories as geographical points of encounter between the 'intricate web of external, inter-firm transactions, or internal, infra-firm transactions' (Storper, 1994:15) that characterize most production systems. Storper (ibid). explains how such transactions might become the relational assets of a region, taking on a 'life' beyond their original corporate or industrial context:

> Many such [transactions] – buyer–supplier . . . relations, or R&D–producer relations, or firm–labour market relations – come to be structured in ways that are highly specific to a given, initially geographically bounded transactional context. . . . Over time, they become more and more specific as unwritten rules of the game (conventions), formal institutions, and customary forms of knowledge are built up.

Against a background of globally available mobile factors of the sort described at the start of the chapter, and the consequent intensification of international competition, it could be argued that the competitive significance of territorially rooted immobile assets is increasing. Dunning (1993: 28) argues, for instance, that 'the globalising economy is unearthing a new importance to concepts such as trust, forbearance and reciprocity; and of informal rather than formal, organisational forms in affecting national competitiveness, and, hence, the disposition of resources and capabilities.' For firms, such assets – which we have argued are in part at least territorial assets – can make the difference between growth and decline not only because of their increased importance in defining competitive advantage but, more significantly, because they are offered as part of the general supply-side milieu offered by governments and communities that firms use to improve efficiency without imposing significant cost burdens on individual actors.

But the case for the continuing salience of proximity can be made in less functionalist terms, following on from socio-economics itself, with its emphasis on the decisive influence on economic activity of interpersonal relations, habits, conventions, institutions and culture. These are influences, as Storper hints in the earlier citation, which are geographically constituted, by organizations and people in specific places. Indeed, for some observers (Maskell and Malmberg, 1995; Becattini and Rullani, 1993), proximity, or better, bounded cultures might play a positively unique role in supplying *informally constituted* assets. For instance, Maskell and Malmberg argue that tacit forms of information and knowledge are better consolidated through face-to-face contact, not only because of the transactional advantages of proximity, but also because of their requirement for a 'high degree of mutual trust and understanding', which is often constructed around shared values and culture. Similarly, Becattini and Rullani distinguish between codified knowledge, such as technical and scientific knowledge, as a feature of translocal networks, and non-codified knowledge, such as workplace skills and practical conventions, as aspects locked into the 'industrial atmosphere' of individual places.

To recognize the significance of territorial assets is not in any way to argue that all places can flourish equally in the global economy. Globalization represents a tying in of nations and regions to the common set of global forces discussed earlier, and indeed in ways which suggest that the stakes for survival are getting higher. As such, it might represent an accentuated version of uneven development based around inclusion and exclusion whereby only those regions and nations which can mobilize assets for local advantage are rewarded. Nations and regions with assets that are crushed or ignored by the forces of globalization are likely to face punishing consequences as they fail to establish a foothold in the global economy. To judge by the increasing tendency of growth to become centralized in a shrinking number of 'asset-rich' nations such as Japan and Germany, regions such as California, and global cities such as Tokyo, London and Singapore, the circle of exclusion might be enlarging to a disturbing size.

## CONCLUSION: PRACTICAL ORIENTATIONS, OR, IT'S THE GEOGRAPHY, STUPID!

The argument in this chapter has been that there is still much that nation-states can do. We have argued that in at least three areas on the supply side – corporate governance, innovation and labour markets – the scope for national policy action remains considerable, and likely to have a major influence on economic performance. In addition, our attempt to connect a socio-economic perspective on globalization with issues of territoriality has uncovered the enduring significance of place-bound institutional and cultural assets. The relevant practical question that follows, therefore, is not *whether* globalization allows scope for national or local action, but *what kind* of action is necessary for positive engagement within the global economy.

The literatures on socio-economics, learning and place-specific assets suggest two areas we wish to discuss in bringing this chapter to a close. They refer to externalities which remain highly territorialized (and therefore amenable to policy or societal grasp) core conditions for competitive advantage. The first area is the building and sustaining of learning capability, and the second is the need to build diffuse and interactive institutional networks to support such capability. In the discussion which follows, we outline the arenas in which action is desirable, rather than putting forward specific policy proposals.

It is widely recognized that a key condition for economic success in the vagaries of today's international economy will be the capacity to generate and sustain knowledge and innovation (both technical and social). Globalization, together with the evolution of markets towards intensified competition, product obsolescence and consumer discernibility, is placing a greater premium than before on factors which can be grouped together under the label 'learning capability'. We have argued elsewhere (Amin and Thrift, 1994) that economic success will be consolidated in the hands of those who can develop and internalize a capacity to renew products; raise productivity through technological and organizational upgrading; dominate or master information and communications networks; generate and disseminate discourses and stories about latest advances; establish coalitions, maintain trust and develop rules of behaviour; and develop strategic capability.

The benefits of learning-based competitiveness, building on formal and informal means of operating innovation, information and knowledge, are obvious. As Storper (1994: 4) argues:

> Those firms, sectors, regions and nations that can learn faster or better . . . become competitive because their knowledge is scarce and therefore cannot be immediately imitated by new entrants or transferred, via codified and formal channels, to competitor firms, regions, or nations.

As inimitable 'brain trusts', which always manage to keep one step ahead of the game, the centres of *pooling* of learning capability, be they nations, regions or networks of association, offer the added advantage of density and, consequently, the magnet-like quality of attracting further investment, employment and growth. Thus, pooled learning capability becomes a condition for both dominating the relevant global economic networks and securing the cumulative industrial development of the 'home base', by attracting and supporting the best-quality domestic and overseas firms.

The difficult policy question is how to build learning capability and whose responsibility it should be to do so. At the simplest – and normally most discussed – level, it is possible to identify a number of closely related supply side measures, where much of the responsibility for action can lie with the state or other public policy networks. The most general of these is the need to gear up the supply-side, especially in ways that relate to the circulation of information (e.g. advanced communications programmes, the upgrading of information standards and technologies) and the generation of skills and expertise (e.g. investment in higher education, training programmes, vocational qualifications). Following on, there is the need to ensure the accumulation and transfer of knowledge (e.g. investment in research activities and establishments, strengthening higher education – industry links, efforts to integrate supply chains between firms). Another measure is a need for enterprise support systems, such as technology centres or service centres, which can help keep networks of firms innovative. Finally, there is the need for measures to ensure continuity of learning within networks (e.g. periodic retraining exercises, exposure to other cultures and international 'best practice' routines).

These are arenas of action where *formal* programmes can be launched with relative ease, involving the state, learning institutions such as universities and schools, research centres, technology centres and actors in the economy including firms, labour organizations, chambers of commerce, and so on. And here, successful intervention is likely to be related to the scale of resources, continuity of effort, level of collaboration and degree of strategic vision that are on offer, together with the depth of translation of such institutional support into the entrepreneurial system.

It has to be recognized, however, that many other aspects of the socio-economy are beyond the direct influences of formal policy initiative, which is why we have argued that economic success must also be tied to aspects such as the propensity to co-operate within and between networks, the ability to mobilize or change embedded cultures within organizations, the unlocking of tacit and informal sources of knowledge that are bound into habits, routines and conventions, the ability to generate new conversations and languages of dialogue, and so on. These are the products of social and institutional evolution and historical sedimentation, framed in the attitudes, behaviours and dispositions of individuals, institutions and communities. They are the economic life-stream from which, as O'Donnell (1994: 22) notes, 'traditional levers of economic management, macroeconomic policy, directive industrial policy and bureaucratic welfare policies are . . . too remote'. Public policy intervention cannot be expected to generate these conditions. The implication, however, does not have to be that the state should therefore continue to restrict its role to supply-side or market interventions of a formal nature.

What the state can do is to *underwrite* the building-blocks of the learning economy. A first necessary step might be, to use Dunning's (1993) phrase, a 'restructuring of government involving a fundamental redefinition of the state as 'an enabler or facilitator' (p. 31). The need is to have a state that learns from the leading networks of the socio-economy to be 'flexible and anticipatory of change' (p. 33). This is a fundamental reformulation of how the state sees and organizes itself, in a direction away from a simple choice between command and market forms of governance, towards a facilitative role which internalizes the principles of interaction and dialogue within the learning economy. The state itself has to become permeable, learning-orientated, strategic and enabling as well as sanctioning in its actions (Amin and Thrift, 1995; Amin, 1996).

This shift in attitude and behaviour, in turn, necessitates a reformulation of how the state sees itself in relation to other institutions. The central tenets of the learning economy are interaction and negotiation, or, to use Hirschman's famous phrase, '*voice*'. It is a voice that involves interactive communication as a principal logic of governance, as well as acceptance of plural authority structures, thus offering the combined benefit of decentralized innovation without loss of strategic direction and co-ordination.

In such an 'associationist' (Hirst, 1994) or 'negotiated' (Hausner, 1995) institutional framework, the state has to become one strategic node of the socio-economy, working in *partnership* with interest groups and voluntary organizations, rather than standing apart from them. The 'interactive' state, like the 'managed' or 'market' state, continues to formulate universal rules and laws, provide collective services and establish a framework of incentives, rewards and penalties, but significantly it also places a much greater premium on devolution of responsibilities and co-ordination through voice. As Hausner (1995: 250) explains:

> In this method, the central authority . . . becomes a participant and treats the other participants as independent agents. The behaviour of each participant can only change as a consequence of mutual interaction. The task of the central authority . . . is not to lay down the new systemic rules and force the participants to respect them; it is rather to stimulate the process by which these rules are formulated and defined; thus allowing the participants to satisfy their needs and realise their interests.

Governance of this sort, however, begs the question of where the boundaries of decentralization should be drawn. The learning economy, in our view, presupposes institutional pluralism and the activation of many voices as a basis for formulating strategies. To extend the analogy into areas of economic governance, a likely implication is that alongside state and other major institutions of the socio-economy (unions, employer federations, trade associations, etc.), civic associations, too, need to be drawn in and empowered. Indeed, Robert Putnam (1993) has gone so far as to suggest that both state and economic efficiency suffer without the institutional 'infill' of voluntary organizations, community groups, weaker economic groupings, and so on. Such a build-up of 'social capital' can be seen not only as a means of spreading responsibility or establishing checks on the state, but also as a basis for economic innovation (e.g. giving life to third-sector and community initiatives, alternative forms of work, exchange networks, etc.).

Interactive governance therefore requires the development of and entitlement to voice *across* the whole spectrum of interests in civil society, economy and politics, and to such a degree that this mode of social regulation becomes normal and, indeed, habitual (Amin and Thrift, 1995). A part-explanation of why such voice-based systems can then produce economic dynamism is offered by Bianchi and Miller (1994: 7, 8, 12):

> the more the social regulation of the group is closed and internally rigid, the greater the risk that innovative behaviour by an individual will either become monopolistic behaviour . . . or it will provoke a reaction by . . . potential losers who . . . will form a coalition which applies sanctions to the innovator in order to preserve the existing structure . . . when voice is present, a collective is stable enough to welcome change, even at the institutional level, without jeopardising the system itself . . . alternatives are suggested and the project for dynamic evolution remains, as does the right of entry.

To conclude, the emerging literature on socio-economics makes it clear that territorially based strategies remain feasible in the global economy. But it also reveals that the terms for centre-stage participation, based around developing and retaining learning capability and institutional adaptability, are very demanding. Nothing short of state and societal reform in the direction of decentralization, democratization and participation seems to be called for, thus moving beyond the call for the construction of networks of association that maximize learning and adaptation. These are lofty goals to aim for and ones which are easily sacrificed in a contemporary policy culture driven by formal and quick-fix solutions. In the main, public policy is troubled by arguments which appeal to long-term, holistic, evolutionary and processual discourses. Yet it is becoming increasingly clear that these are the discourses of the real economy. Thus, imagining the parameters of a dialogic policy rationality is the next challenge, one that ought to be taken up to get us out of the cul-de-sac of contemporary debates on whether, in an era of globalization, the state matters and at what spatial scale.

## ACKNOWLEDGEMENT

This chapter is a reworked and extended version of an article published in *Nordisk Samhallsgeografisk Tidskrift* (1995) 20, 3–16.

# CHAPTER TWELVE

# UNPACKING THE GLOBAL

PETER DICKEN, JAMIE PECK AND ADAM TICKELL

## APPROACHING GLOBALIZATION

Although the concept of globalization has become widely diffused only since the 1980s, particularly following Levitt's (1983) paper on the globalization of markets, the underlying concept can be traced back to at least 1960 when McLuhan (1960) coined the term 'global village' to capture the impact of new communications technologies on social and cultural life. Subsequently, the idea suffused broad reaches of popular communication as well as of academe. Yet while Chesnais' claim that 'globalization is largely a term coined by journalists and politicians . . . [which] has been thrust on the academic community' (1993: 12) is difficult to sustain in the face of the voluminous academic literature on the subject, much of this material is stereotypical and exaggerated. Quite often, globalization is represented not so much as a historical tendency or a complex process, but as an outcome: a 'new order'. How often do we read that, in the face of the unstoppable juggernaut of globalization, nation-states will inexorably disappear (if they have not done so already), that homogenization of markets, consumer tastes and cultures is inevitable (if it has not happened already), that a small number of gigantic global firms will, or do, control the global economy through an internalized global production system? Part of the problem is that too much of the current debate proceeds on the basis of very loose, often contradictory and self-serving, definitions. Most writers either do not bother to define globalization at all, or carelessly use the term interchangeably with 'internationalization'. But they are not the same thing (Dicken, 1992). Definitions are not mere semantic peccadilloes; they remain crucial not least because they caution against the kind of caricaturing of globalization which has become all too common.

Globalization as caricature, however, is a powerful and highly politicized act. It is no coincidence that many of the prophets of globalization are business gurus and/or neo-liberals. Caricature serves their interests only too clearly. For example, in an open letter to the *Atlantic Monthly* in June 1993, Akio Morita, chair and founder of the Sony Corporation, implored global leaders at the impending G7 meeting in Tokyo to find 'the means of lowering *all* economic barriers between North America, Europe and Japan – trade, investment, legal, and so forth – in order to begin creating the nucleus of a new world economic order that would include a harmonized world business system with agreed rules and procedures that transcend national boundaries' (quoted in Korton, 1995: 122). Looking at this from the perspective of the nation-state, globalization stories pertain to the widely held view, critically summarized by Amin and Tomaney, that there is increasing

> pressure on states, both weak and strong, to accommodate and perhaps also give way to other powerful global economic institutions . . . whose interest is to secure the adoption of rules and policies of economic governance which facilitate accumulation on a global rather than purely national scale. . . . It would appear that a new era is unfolding in which governments and global firms bargain with one another on the world stage for a favourable position within the international division of labour, rather than in pursuit of national economic interests. . . . This stands in sharp contrast with the once dominant priority of states to place the national economic system in a good trading position within a world economy which was composed of competing national economies. (Amin and Tomaney, 1995b: 172).

Globalization rhetorics are consequently being deployed *prescriptively*, by both political leaders and business strategists (see Reich, 1992). For example, the leader of the UK's then opposition Labour Party, Tony Blair, recently declared that 'globalization is changing the nature of the nation-state as power becomes more diffuse and borders more porous. Technological change is already reducing the power and capacity of government to control its domestic economy free from external influence' (quoted in Rees-Mogg, 1995: 18). Significantly, these claims were made not at some mass rally in the northern industrial heartlands of Britain, but at a conference on a privately owned Pacific Island convened by the global media magnate Rupert Murdoch! While there is something to be said for taking what politicians say with a pinch of salt, the fact that they are saying them *is* important. Such interpretations have the effect of naturalizing the global; of treating globalization as some sort of relentless and inevitable process, driven by the twin imperatives of capitalist competition and technological change. A corollary is that the nation-state is presented as an eviscerated shell, as (some forms of) political intervention are rendered effectively redundant or outmoded. While breeding political defeatism in some quarters, it also serves to legitimate and rationalize certain neo-liberal strategies. There are, then, real political dynamics caught up in the globalization debate which it would be dangerous to mistake for mere rhetoric (Peck and Tickell, 1994a). So, for Piven (1995: 108), the concept of globalization 'itself has become a political force, helping to create the institutional realities it purportedly merely describes' (*see also* Bienefeld, 1996).

Rather like the discourse of deregulation during the 1980s, globalization discourses speak in politically loaded ways about phenomena and processes which are far from clearly understood. Some of the problems stem from the very elasticity of the term 'globalization' itself, others from the premature polarization and foreclosure of the debate. Here, analyses have rapidly polarized between, on the one hand, a 'booster' line in which globalization tendencies are seen as being all-encompassing, all-powerful and – literally – everywhere and, on the other, a 'hypercritical' line in which the very existence of these same tendencies is questioned or denied, and their historical significance trivialized. Whereas one camp tends to attribute all contemporary developments to globalization, as if it were some kind of universal causal agent, the other insists equally vigorously that the process is little more than a mirage. While both sides of the debate have made useful, and often valid, contributions, they have at the same time severely distorted and mis-specified the meaning of the term and the nature of the process. There is in fact little to be gained – from a theoretical, a political or a policy perspective – from either overstating or understating the extent and nature of globalization, but these seem to be the only positions currently on offer. It is certainly necessary to be sceptical about caricatures of globalization, and to be aware of the political contexts in which these are being deployed, but this need not entail denying the very existence of the process.

This chapter steers an alternative course to that defined by the 'boosters' and the 'hypercritics'. It represents not some flabby 'middle way' but a serious attempt to define globalization *as a set of interrelated tendencies*. The effects of these, it will be argued, are significant, important and in some senses novel (contrary to the claims of the 'hypercritics'), but they are not universal and homogeneous (contrary to the claims of the 'boosters'). Fundamental changes *are* occurring, but by way of a complex of *globalization processes*. We can claim no unique or privileged insight into how these might function, given that a great deal of work remains to be done once the terms of analytical engagement are clarified. With this objective in mind, we can provisionally state that globalization processes are intrinsically uneven – heterogeneous rather than homogeneous – in both their form and their effects. They involve highly intricate interactions between a whole variety of social, political and economic institutions operating across a spectrum of geographical scales. They involve processes of both collaboration and competition and of differentiated power relationships. Globalization is, in the basic sense of the term, *dialectical*. The relationships between its underlying processes and outward forms are mutually transformative; the same process can be associated with a range of concrete forms. Paradoxically, these processes are as complex as the rhetorics of globalization are simple.

There is a need for clarification in both the terms and the terminology of the globalization debate. As Cox (1992: 428) has put it, 'the questions of globalization, hypermobility of capital and their [respective] effects on the balance of political forces . . . need to be confronted on theoretical grounds. For a start, there is a lot of conceptual unpacking that needs to be done.' We seek to begin this process here by unpacking the globalization debate itself, mapping out some of the lines of agreement and disagreement, and dissecting the theoretical premisses of the protagonists. This is followed by an attempt to 'repack' – to clarify and to reconceptualize – the concept of globalization. A central message of the chapter is that globalization involves *qualitative*, and not just quantitative, changes. Globalization is not just about one scale becoming more important than the rest; it is also about changes in the very nature of the relationships between scales. We argue that there is a need for a more nuanced and iterative analysis which conceptualizes the global within the local and the local within the global. Ironically, much of the debate around globalization has been profoundly *ageographical*, as structures and processes of spatially uneven development have been underplayed, or even – in several renderings of the globalization thesis – *counterpoised to* the globalization process. In contradistinction, we argue here that a more *geographical* understanding of the process of globalization is required, one which is sensitive to the changing nature of (cultural, economic, political, institutional) relationships across and between scales.

## UNPACKING GLOBALIZATION

The diagnostic features of a fully globalized economy and society – the idealized world of the 'global dreamers' – have been summarized by Korton (1995: 131) in the following way.

- The world's money, technology and markets are controlled and managed by gigantic global corporations.
- A common consumer culture unifies all people in a shared quest for material gratification.
- There is perfect global competition among workers and localities to offer their services to investors at the most advantageous terms.
- Corporations are free to act solely on the basis of profitability without regard to national or local consequences.
- Relationships, both individual and corporate, are defined entirely by the market.
- There are no loyalties to place and community.

In this global scenario, capital would be infinitely mobile and completely footloose, shaking off all forms of local and national allegiance or dependence; the principal agents of change would be the all-powerful transnational corporations, the epitome of 'placeless' capital. Unregulated market forces would become dominant, effortlessly sidestepping all attempts at regulation, containment or political direction. As a consequence, the nation-state would erode to the point of insignificance, both as a unit of analysis and as a political agent. Global homogenization of social, political and economic conditions would erase regional differences and uneven development.

Many politicians and businesspeople implicitly or explicitly accept the essence of this scenario. Recently, however, the globalization debate has been pushed forward by a group of critics (mostly on the left) concerned to challenge the empirical and theoretical bases of the globalization thesis, questioning its extent, novelty and implications (*see*, for example, Gordon, 1988; Cox, 1992; Ferguson, 1992; Glyn and Sutcliffe, 1992; Hirst and Thompson, 1992, 1995, 1996; Campanella, 1993; Ramesh, 1995; Boyer and Drache, 1996). These critiques, which seek to reinsert political agency and the necessity of struggle into the debate, provide a useful corrective to the uncritical eulogies to globalization, but, for the most part, they go for the easiest targets. They tend to focus on the most exaggerated claims of the boosters, typically citing their least restrained uses of 'globaloney' in order to cast doubt on their entire project. The objective seems to be to debunk the 'myth of globalization' rather than to qualify and quantify those developments which *can* be observed. Even three of the most thorough analyses – those by Glyn and Sutcliffe, Hirst and Thompson, and Gordon – are guilty of this charge in their attempt to challenge globalization as a political-economic project and to reclaim the terrain of (domestic) political action. Their critiques are detailed, dealing with both the substantive empirical claims and the underlying conceptual foundations of the globalization thesis.

The three papers separately maintain that advocates of globalization have mis-specified the nature of the world economy, partly because they have failed to identify what globalization *as a condition* would actually mean. This represents the nub of the globalization debate, cutting to the heart also of how we understand the present conjuncture. In effect, each argues that we are not in a *new phase* of globalization. By focusing on the long-run history of capitalism, rather than simply on the period after 1945, they set out to show that economic interdependence and large-scale trade were commonplace prior to 1914 (*see also* Zevin, 1992; Obstfeld, 1995). The aberration, therefore, is not the present period of global instability and fluidity, but the protectionism that developed during the 1920s and 1930s and which was progressively undermined, first by the Bretton Woods settlement, and second by the liberalization of (some) markets after its collapse in the early 1970s.

In criticizing the exaggerated claims which are made in some parts of the globalization literature, and in more closely specifying what globalization would entail, Gordon, Glyn and Sutcliffe, and Hirst and Thompson have made an important contribution to the globalization debate. However, they have themselves underestimated the degree to which a *qualitative*, as well as a quantitative, series of changes has occurred. Globalization tendencies can be at work without this resulting in the all-encompassing end-state, the *globalized economy*, in which all unevenness and difference is ironed out, market forces are rampant and uncontrollable, and the nation-state merely passive and supine. Gordon, for example, makes the argument rather easy to win by insisting that globalization must involve 'the construction of a *fundamentally new and enduring* system of production and exchange' (1988: 54; emphasis added), while Hirst and Thompson review current trends only to conclude that 'If one calls this *outcome* "globalization", so be it, but it *does not conform to the ideal type* elaborated above' (1992: 370; emphasis added).

The result is a caricature to which few serious analysts of globalization should adhere. Although critics of globalization would insist that a strong case against some of the more audacious claims had to be made, there is also a sense in which they distort the explanatory structures of the globalization thesis, rendering tendencies as outcomes (and therefore finding it easier to dismiss them), and connecting together the most injudicious claims of (parts of) the globalization literature (in order to render the end-state of globalization practically unattainable). In effect, the ideal-type of globalization deployed by the critics rests on five interrelated claims (viz. the infinite mobility of capital; the prevalence of unregulated market forces; the attainment of absolute power by transnational corporations; the demise of the nation-state; and homogenization in social, political and economic conditions). Trawling the vast globalization literature, it is not difficult to find evidence of such claims (e.g. Korton, 1995; Reich, 1992). It is, however, the next step that these critics take which is the crucial one: implicitly, they link each of the five claims together into chains of mutual causal necessity. So, there is a need only to dislodge one claim – say, debunking the notion of global homogenization – for the whole artificially created edifice to come crashing down.

In our view, the process of globalization is best understood as a *complex of interrelated tendencies*, not as a rigid set of mutually dependent and absolute causal claims. Because these are globalization *tendencies* they will interact with one another in unpredictable ways and they will result in uneven effects. They do not imply the 'automatic' generation of a set of guaranteed outcomes – as suggested in the critics' interpretation of the global*ized* economy. Our preferred approach to globalization, as a set of contingently related tendencies, is developed further in the following section, in which we seek to 'repack' the notion of globalization.

## REPACKING GLOBALIZATION

In contradistinction to the rigid and formulaic positions of both the boosters and the critics, we propose a preliminary conceptualization of globalization as *process*. In the absence of an adequate understanding of globalization, it is premature to engage in the kinds of undertheorized empirical claim and counter-claim which have characterized aspects of the debate. The study of globalization and spatial restructuring represents more than a mere focus of study. As critics of globalization imply, it is the site, more fundamentally, of a series of methodological, political and conceptual dilemmas (Lipietz, 1993b), and its mis-specification eviscerates our methodological, political and conceptual responses to the process and its outcomes.

Little progress is likely to be made in defining globalization, as Hirst and Thompson do, as a 'negative ideal-type' (1992: 394) which is practically unattainable because, by definition, it *requires* the complete capitulation of the nation-state and the dominance of immutable and unregulated global 'market forces'. This would imply, in effect, that the 'real' global economy had come to resemble the caricature of globalization presented in the 'global dreamer' literature. We prefer a process-based approach which requires that a conceptual distinction be drawn between the process of internationalization and the process of globalization. This distinction can be established in a number of ways. With respect to the organization of *economic relationships*, internationalization refers to the *simple extension* of economic activities across national boundaries while globalization implies that *functional integration* has occurred among geographically dispersed activities (Dicken, 1994a). With respect to the organization of *political relationships*, internationalization denotes a situation in which the primary agents of change and sites of struggle are nation-states, while globalization implies a deprivileging of this pre-eminent role, as both the strategies of nation-states and the parameters or terms of national political struggles are increasingly constrained and conditioned, *though not determined*, by extra-national forces (R. Cox, 1993). Similarly, in terms of *cultural relationships*, whereas internationalization suggests that the primary interchanges are between and around nation-states, globalization implies a complex rearticulation of cultural forms and channels at a range of scales (Featherstone, 1990).

The distinction, therefore, between internationalization and globalization is one which must be made on *qualitative* grounds. That the two terms refer to different *processes* seems to have been lost on those critiques of the globalization thesis which seek to reduce the debate to questions of quantification and outcomes. It is not simply the case that in a globalizing economy there is more international trade (though there may be), that transnational corporations have extended their global reach and significance (though they may have), that local and national economies are increasingly exposed to global competition (though they may be), or that nation-state powers and capacities have been eroded (though they may have). Such issues have a role to play in debates around globalization, but they represent only part of the picture. Equally important are those rather more qualitative and slippery problems such as the spatial (re)constitution of power relationships, the changing form and purchase of regulatory and governance structures, the reorganization of state-strategic capacities and so on. It is on this terrain, first and foremost, that the globalization debate should be conducted.

One way of illustrating this perspective is with respect to perhaps the most controversial aspect of the globalization debate, the changing structure and influence of the nation-state. To reiterate, in our process-based conception of globalization it is not necessary for nation-states to disappear, to lose their place as a key site of struggle and regulatory practice, or to become insignificant or marginal actors. Rather, globalization is associated with a *qualitative reorganization* of the structural capacities and strategic emphases of the nation-state. So, in Held's view, the capacities of the nation-state have paradoxically 'been both curtailed and expanded, allowing it to continue to perform a range of functions, which cannot be sustained any longer in isolation from global or regional relations and processes' (1991a: 208). More formally, Jessop develops a similar argument with respect to Poulantzas' distinction between *generic* (or 'global') and *particular* state functions, maintaining that

> the erosion of one form of national state should not be mistaken for its general retreat. It may be, on the contrary, that, as the frontiers of the KWNS [Keynesian welfare national state] . . . are rolled back, the boundaries of the national state are being rolled forward in other respects and/or that other forms of politics are becoming more significant. . . . But this does not mean that the national state has lost its key position in securing the 'global' [or generic] political function of the state. For *the national state remains the primary site for this crucial generic function*. . . . It is still the most significant site of struggle among competing global, triadic, supra-national, national, regional, and local forces. (Jessop, 1995a: 22; emphasis in original)

In this respect, the impact of the process of globalization is measured not in the crude terms of whether there is 'more' or 'less' of the nation-state, but in its changing structure and orientation. While some of the nation-state's capacities are most certainly being eroded (particularly with respect to aspects of macro-economic policy), this is a complex *process*, not one, in R. Cox's (1993: 260) words, in which nation-states are reduced to passive 'transmission belts from the global into the national economic sphere'. Emphatically, the nation-state remains a key actor in the global economy; it is more than a pawn in some totalizing and all-determining globalization process (*see* Ramesh, 1995; Piven, 1995). Johnson *et al.* (1994: 9) strike a more appropriate analytical balance when they insist that 'nation-states are much less prisoners of circumstance than is implied in some rather deterministic interpretations of economic globalization. . . . Economic globalization obviously matters. However, ideologically conditioned policy responses to globalization also matter'. Some of these responses, moreover, involve the rolling forward of new institutions, regulatory projects and policy programmes, as nation-states have been drawn more heavily, for example, into the sphere of the micro-economic regulation of labour markets (in the wholly misleading name of 'deregulation') as the scope for macro-economic management has narrowed (Peck, 1996).

Only the most perverse and exaggerated analysis could, on the basis of the continued existence and importance of nation-states, deny that a process of globalization is under way. So, it is necessary simultaneously to reject Hirst and Thompson's audacious claim that 'globalization has not taken place and is unlikely to do so' (1992: 393), *and* to recognize that they are correct to argue that while 'national governments may no longer be "sovereign" economic regulators in the traditional sense, they remain political communities with extensive powers to influence and sustain economic actors within their territories' (p. 371). This highlights the paradox of globalization that, like a constitutional monarch, nation-states remain both sovereign *and* subject. The failure to recognize this contradictory nature of globalization stems, in many cases, from the adoption of an absolutist and untenable definition of globalization. Contrariwise, those approaches which stress the qualitative nature of the globalization process, and its uneven and variable effects, need not labour with a self-imposed imperative to define the nation-state out of existence. Thus, Jessop insists that the

> national state retains crucial generic political functions despite the transfer of other activities to other levels of political organization. In particular, the national state has a continuing role in managing the political linkages across different territorial scales and its legitimacy depends on its doing so in the perceived interests of its social base. . . . In short, there remains a central political role for the national state. But it is a role which is redefined as a result of the more general re-articulation of the local, regional, national and supra-national levels of economic and political organization. . . . [The] national state will remain a key political factor as the highest instance [at present] of democratic political accountability. How it fulfils this role will depend not only on the changing institutional matrix [but also] on shifts in the balance of forces as globalization, triadization, regionalization, and the resurgence of local governance proceed apace. (1995a: 22–23)

Certainly, there is a sense in which the current conjuncture is associated with a rescaling of economic governance functions. Whether this represents a 'new order' or continuing (*after*-Fordist) 'disorder' is an issue which remains open to debate (Kotz *et al.*, 1994; Peck and Tickell, 1994b; Gough, 1996). What is becoming clear, though, is that just as the spatial reorganization of economic governance capacities continues to reveal the structural limitations of forms of economic management and intervention, so also it is opening up new possibilities (Hirst and Thompson, 1992; Jessop, 1995a). In the process, state capacities and structures will be reorganized in complex and subtle ways, one of the outcomes of which will be a restructuring of the institutional apparatus and economic orientation of the *nation*-state.

We can look at what is happening to transnational corporations (TNCs) in an analogous way, while recognizing that the structures and capacities of TNCs and nation-states remain different. The issue is not whether all TNCs are, or are becoming, 'global firms'; the real point is to recognize the diversity and complexity of the processes and structures involved as firms increasingly operate across national boundaries. The empirical evidence for the global firm in some 'pure' sense is scanty to say the very least. It tends to rely overwhelmingly on a small number of cases in which it is especially difficult to get behind the corporate hype. From a firm-based perspective, 'globalization is best interpreted as a strategic objective rather than as an accomplished reality. Put differently, a firm may hope to realize "globalization" yet fail' (Ruigrok and van Tulder, 1995). There are very few, if any, firms which meet the criteria for *pure* globalization, which would imply the fulfilment of the following conditions: the globally optimal location of each element in the firm's production chain; both the intention and the ability to switch and reswitch activities between alternative locations at a global scale; and the severance of the ties that bind firms to their home base, making them indifferent, in other than a purely economic sense, to alternative locations (in other words, global firms would be 'placeless').

Although there has been a very substantial degree of geographical restructuring of production chains (Gereffi and Korzeniewicz, 1994; Dicken, 1995), only to a limited extent can that restructuring be regarded as being 'truly global'. The sinking of costs (Clark and Wrigley, 1995) undoubtedly inhibits the rapid reswitching of production between alternative locations, as do the political constraints within which firms continue to operate. There are, indeed 'barriers to exit'. Finally, even the most globalized firms, when closely examined, turn out to have a substantial, and continuing, degree of embeddedness in their home base (Dicken, 1994a; Dicken *et al.*, 1994; Hu, 1992; Ruigrok and van Tulder, 1995; Stopford and Strange, 1991).

However, none of this is to argue that there are not global*izing* processes at work within and between TNCs. There certainly are. But they are highly uneven in both space and time. There is no single globalization strategy. Rather, there are many variants on the theme of globalization strategies being pursued by firms, short of the extreme case of a global intra-firm division of labour as implied in the analyses stimulated by the work of Porter (1986) and others. In that sense, it is perhaps better to talk in terms of *quasi-global strategies* which certainly contain elements of 'pure' globalization but fall short of complete implementation. The most obvious example, for which there is substantial empirical evidence, is the regional integration of production at the scale of the major triadic regions of North America, Europe and East Asia. Whereas few, if any, of even the leading firms in the world are genuinely globalized in their activities, virtually all the major firms have become '*globally regionalized*'; that is, they have constructed (or are constructing) integrated production systems within each of the major regions. Such regional systems vary greatly in their degree of self-containment, particularly because firms show a very strong propensity to retain home-country control of the highest-order functions. They also vary in the extent to which their production system in one region is connected functionally to that in other regions, apart from the firm's home region.

To most writers in the management sphere, such regionally orientated production is regarded as being locally responsive; as a manifestation of *glocalization*, where the loci of power and activity have moved from the nation-state to supra- and subnational state levels (*see* Swyngedouw, 1992a). To geographers, in particular, the use of the term 'local' at such a broad spatial scale seems somewhat perverse. However, there is an important issue here which concerns the extent to which firms need to remain locally responsive in the larger-scale sense of the term. As long as local, regional and national states continue to exert a regulatory influence – as they certainly do – then firms operating internationally need to remain responsive to 'local' conditions. This represents just one of the ways in which the (unevenly developed) strategies and structures of the state interpenetrate with the (unevenly developed) strategies and structures of TNCs. Here, again, it is apparent that the particularities of *place* continue to exert a powerful influence on firms despite – or rather, because of – the alleged annihilation of space through time. Firms with the capability of locating some of their operations almost anywhere are undoubtedly more discriminating in choosing between specific places (Harvey and Scott, 1988).

In summary, the contemporary production system is undoubtedly subject to *globalizing processes* but it is far from being global*ized*. The major driving force in such processes is the transnational corporation, although this institution must be defined in terms which go beyond simple criteria and focus upon the power to co-ordinate organizationally – and geographically – dispersed activities within and between production chains. The production system is best conceived of as a series of interlocking dynamic networks, articulated through complex sets of intra- and inter-firm relationships whose size and shape are produced and reproduced through place-specific processes of embeddedness. The geographical extent of such networks has undoubtedly been increasing. At the same time, their functional complexity and degree of integration has also increased, but in a very uneven manner both spatially and temporally. However, what these processes have certainly created is an intensification of the transition from an internationalized world economy characterized by the kind of 'shallow' integration based upon flows of trade and of international capital movements to the 'deep' integration of globalized production (UNCTAD, 1993). Furthermore, such processes have occurred simultaneously with the restructuring of markets, which have a far greater global reach than in earlier eras.

The restructuring of both the nation-state and the transnational corporation are connected in a complex dialectical fashion to the processes of globalization. These processes are articulated *through* both firms and states operating in complex interaction. Globalization does not exist as a free-floating structure, unrelated to the economic and institutional context in which it arises. It is constituted through those very practices which it subsequently transforms. Analysis, therefore, needs to extend beyond 'studies of how "general" processes and structures are modified in particular contexts, [to recognize] that the general structures do not float above particular contexts but are "reproduced" within them' (Sayer, 1989a: 255). As Thrift (1990a: 81) points out, the global is not some *deus ex machina*; the 'local and the global intermesh, running into one another in all manner of ways'. While there will be those who continue to analyse globalization as a, if not *the*, metaprocess, all such processes are in fact embedded in other relations and structures. This may be messy, but that is the way the world is.

To sum up, how does the conception of globalization advanced here differ from what might be understood as internationalization? How far, in other words, is it appropriate to use the term 'globalization' under present conditions? First, under globalization, the degree of (functional) integration between national economic spaces is much deeper, implying a high degree of mutual interdependence between national economies and an attendant reorientation of the economic capacities, strategic orientations and institutional apparatuses of the state. Here, the global economy is associated with imperatives which are not transcendental and immutable, but do exert a strong discipline upon economic actors (including both corporations and states). Through their very actions, of course, these actors are actively engaged in the reproduction of a system of global political-economic relations through which these disciplines are enforced (*see* Cochrane *et al.*, 1996). This system is beyond the control not only of individual cities, nation-states or corporations, but also of any putative global hegemon by virtue of its 'more-than-the-sum-of-the-parts' nature. Globalization denotes a complex of political-economic relations which, in this sense, have ceased to be merely international but which – in slipping the leash of hegemonic control – have become *extra*national. A defining feature of globalization, then, is this form of 'deep integration', one which has important implications for economic management, corporate restructuring and political action.

Second, it follows that while the nation-state will remain a primary site for the articulation and reproduction of globalization tendencies, and notwithstanding its discursive 'self-abasement' (Denis, 1995), it will no longer constitute the only channel of mediation for these processes. Again, another way in which the global system has taken on extra-national features stems from the fact that dynamics of globalization are worked out not solely through and between nation-states (as realists presuppose) but through a range of sites and channels. The list certainly includes nation-states but extends also to transnational corporations, institutions of global and supranational governance, regional and local states, and so on. Such institutions are both responding to *and* reproducing globalization tendencies. Nation-states are certainly no longer – if they ever were – the sole 'authors' of this process.

These claims add up to a different conception of globalization from that proposed by the global dreamers and their critics. Our approach is rooted in the following six principles.

- Globalization is a complex of processes, not an end-state or a 'new order'.
- Globalization is a contradictory process, not an unbending force or unidirectional trend.
- Globalization will proceed hand in hand with uneven spatial development; it is not the opposite to it.
- Globalization processes, just like any other, do not float in the air, but are realized in specific institutionally, historically and geographically specific sites.
- Globalization implies qualitative as well as quantitative change, in the sense that there are changes in the relationships between scales, social structures and agents.
- Globalization involves the complex diffusion, rearticulation and reconstitution of power relationships, not simply a zero-sum redistribution among nation-states and TNCs.

This conceptualization provides a distinct alternative to the established positions of rendering all effects as globalization effects (as do the 'boosters') or of defining globalization out of existence (as do the 'hypercritics'). It offers a way of moving forward in the globalization debate, which has become needlessly polarized.

## CONCLUSION

The debate around globalization has usefully focused attention on the spatial reconstitution of political-economic relations, both contemporarily and historically. The debate has rapidly polarized, in part, perhaps, because current conditions (and the way in which they are being characterized by some) seem to be calling for a reconsideration of existing theoretical categories and constructions. As Lash and Urry have argued, for example,

> The task for social science is to make sense of [the] new kinds of global–local relations and to examine their implications for what we used to examine, namely national societies and economies. . . . [It] is increasingly difficult to think of the nation-state as still the appropriate power-container of important economic and social relationships. (1994: 312, 279)

Similarly, sceptics such as Cox concede that notions of globalization are 'provocative and stimulating. No attempt to explore the spatiality of contemporary politics can afford to ignore them' (1992: 427). The critical questions, however, concern how we make sense of what is happening in the global economy and what, in turn, this means for accepted theoretical-political positions. It is possible, we maintain, to be profoundly sceptical about the readings of globalization fostered by corporate boosters or neo-liberal ideologues, and yet recognize that there is something going on 'out there'. One can study globalization without succumbing to the excesses of 'globaloney'.

In this chapter we have argued that an adequate conception of globalization need not be all-encompassing, but should focus on globalization as process. This means that, in contrast to the hypercritics of the thesis, we argue that globalization *tendencies* exist under which the global system – which remains, of course, organized on fundamentally capitalist principles – is being reworked in the context of a particular set of institutional, political and cultural conditions. These shifts we prefer to capture here in the language of reorganization and restructuring. In this sense, there have been, and no doubt there will continue to be, important changes, as globalization tendencies work themselves out. But this is no final, ultimate or terminal state of capitalist development, because the situation remains contradictory and unstable (*see* Swyngedouw, 1992a; Peck and Tickell, 1994a). There is no law which says that tendencies cease to operate during periods of disorder, or that social systems cannot be reorganized during such periods. On the contrary, phases of crisis are often periods of rapid and fundamental readjustment. Thus there is a need to trace the ways in which globalization tendencies are feeding, and then being fed by, the current global disorder. Certainly, there is a danger in perpetuating the kind of globalization rhetoric that breeds political fatalism, but at the same time it is imperative that the causes and consequences of globalization are carefully specified, if for no other reason than because this must be a prerequisite for decisive, effective and correctly formulated political action *at whatever scale*.

There is a need, then, to be clear about terms and terminology, not to misrepresent emerging structures and tendencies. Rather than respond to the exaggerated claims of (some of) the globalization literature by way of an equally exaggerated critique, we have chosen here to steer a different course, defining globalization not as some (practically unattainable) end-state but as *a complex of interrelated processes*. As with all processes, globalization is realized unevenly, interacts in unpredictable ways with counter- and co-functioning processes and melds in a complex and contingent fashion with extant (historically and geographically specific) institutional and economic structures. By emphasizing the distinctiveness of the process of globalization *vis-à-vis* that of internationalization, we have also tried to draw attention to some of the qualitatively different effects of the globalization process. More than anything, though, our approach has been to define globalization, not define it out of existence. If this provides a start, then it is just that.

## ACKNOWLEDGEMENTS

Thanks to Martin Jones, Steve Quilley and Kevin Ward for useful discussions around the ideas developed in this chapter and also to Kevin Cox, Roger Lee, Jane Wills and Henry Yeung for their comments on an earlier draft. Adam Tickell would like to acknowledge the support provided by his ESRC Research Fellowship, 'Regulating finance: the political geography of financial services' (award number H2427001394). The usual disclaimers apply.

# CHAPTER THIRTEEN

# EXCLUDING THE OTHER: THE PRODUCTION OF SCALE AND SCALED POLITICS

ERIK SWYNGEDOUW

The two extremes, local and global, are much less interesting than the intermediary arrangements that we are calling networks. (B. Latour, 1993: 122)

## PREAMBLE

A few months ago, I spent an afternoon with my children in the Africa Museum of Tervuren, near Brussels. Originally established as Belgium's colonial museum to celebrate the country's imperialist mission in Central Africa and to display a domesticated Congolese social ecology which had been tamed, exported and packaged to serve the twin purpose of educating Belgium's working classes about distant places and cultures on the one hand and of mustering a sense of national identity and celebrating the grandeur enacted in the colonial enterprise on the other. The impressive neo-classical museum houses the domesticated exotica of African socio-ecological life. The display of the main social, ethnographic and ecological niches of Central Africa as well as its economic modernization and contributions to world manufacturing and trade combines with the penetrating odour of mothballs and the nose prickling generated by decades of accumulated dust. The odours of the place recall, it seemed to me, both the nostalgia of lost imperial glory and the current political and economic disintegration of much of Africa.

A stuffed elephant separates the ethnographic section that stages the arts and artefacts of Congo's main 'tribes' from the vast section of the museum devoted to biogeography. Thousands of stuffed animals, skulls, furs, a bewildering variety of pinned-up butterflies, beetles, spiders and scorpions celebrate a domesticated wilderness that has become part of everyday life in Brussels. Eva, our oldest daughter, tried to figure out why female animals, invariably situated next to – but one step behind – their sole male companion and their head moving gently downwards in a protective and loving gesture toward one or two younger siblings sporting around, seemed always smaller than the proudly erect figures of the male variant. Our younger boys revelled in the anxiety of being confronted with the to them threatening exhibition of snakes, crocodiles and scorpions. When we then moved to displays of the evolution of Zambia's copper production output and the advances brought to Africa by the modernizing impacts of imported manufacturing production technologies, we decided it was time to get a Coke in the nearby cafeteria. The depressing figures, together with the mothballed display, were too much to take. And surely our children got tired of listening to my futile attempts to 'deconstruct' the display and offer narratives that tried to weave the thread from here to there, from nature to society, from the global economic to the gendered body, from Brussels to Central Africa in ways different from the story intended by the museum curators.

When I reflected on this experience, Latour's (1993) networks came readily to mind. Had our experience been local, national or global? Had we peered into the Heart of Darkness or did we remain trapped in the concrete cage of the museum space and the rationality imposed by imperial museum builders? Had our encounter been real or imagined, material or discursive? Surely it had been all of the above, but first and foremost it had been a remarkable experience of time and space, history and geography, of places and nations, of nature and society, of ecology and culture, of economies and sexualities, of bodies and continents, of power symbolized, iconized and crystallized along the hazy corridors of an underlit (post-)colonial display. The stratigraphic deposits of layer after layer of historical and geographical processes, of the amalgamation of different environments, habitats and spatial scales, seemingly randomly overlain in a maelstrom of imagined, suggested or explicit relations and interactions, hammer home not so much how the global becomes localized and the local globalized, but rather how a variety of geographical scales (the body, family, building, city, nation, ecological niches, communities, international trade and economic relations) become condensed and embodied in commodified display.

The threads that bind it all together through time and space and jump from scale to scale in a mesmerizing collage of interpenetrating materials, ideas, geographies, visions and representations open up the problematic that I wish to explore further in this chapter. In particular, I intend to focus on the production of scale and scaled geographies. As the museum experience suggests, everyday life operates through the internalization of continuously shifting articulations of complex interwoven sets of particular but socially produced geographical scales through which action, meaning, sense and explanation are constructed. In recent years, the problem of scale has become increasingly important, both academically and politically, as the contemporary whirlpool of social and cultural change and economic transformation is accompanied by transgressions of scale boundaries, the production of new scales and the restructuring of others. The hypermodern pulverization of time and space, the transformations of everyday life and the still accelerating globalization of commodification and commodified relations junk geographical scales we have taken too long for granted as fixed, stable and frozen moments, as static containers that organize and regulate life, and render our existence apparently transparent and intelligible. The boundaries of the body, the city, the nation are rapidly redefined in ways often enabling and emancipatory for some but often also in ways deeply disempowering for others. The intention of this contribution, therefore, is to contribute to the formulation of a sounder foundation for theorizing scale and to bring out the political importance of the process of rescaling of life and the contested production of a new 'gestalt of scale' (Smith, 1993) in terms of strategies of empowerment and disempowerment, of repression and emancipation.

In a first section, the problem of geographical scale will be explored in some greater detail. In a subsequent section, key features of recent rescalings and new scale articulations in the political economy of everyday life will be discussed. None of these new scale formations is socially neutral. Scales of social regulation/reproduction and scales of production have changed, but while social regulation tended to move to the individual, the private or the bodily, scales of production have become supernational if not global. The third section comments on the contested and empowering/disempowering processes associated with the production of such new 'gestalt of scale'. I shall argue that these transformations of scale do not go unchallenged and social movements organize around a variety of issues to challenge the 'jumping of scales' of dominant groups. However, recognition of the importance of the 'politics of scale' has not yet resulted in formulating adequate scaled strategies. The final part concludes with considerations of the role and importance of the politics of scale in emancipatory and progressive strategies. Indeed, resisting the disempowering effects of the new 'gestalt of scale' necessitates reaching out from local and particular identities and issues to find the threads that enable solidarity and extend lines of power for those that remain otherwise trapped in place.

# RESCALING LIFE

## Scales and Lines of Power

All social life is necessarily 'placed' or 'situated', and engaging place is fundamental to maintaining the process of life itself. Engaging place(s) is inevitably a transformatory process and often implies some sort of 'creative destruction' or 'destructive creation' of nature/place. This 'engagement' is always an already social *process:* it is a metabolic transformation that *takes place* in association with others and extends over a certain geographical space. Lefebvre (1974), Harvey (1985d; 1996b) and Massey (1994), among others, explore in a variety of ways this socio-spatiality of everyday life and its expression in the 'the production of space(s)'. The process character of socio-spatial relations means that life is in a state of perpetual change, transformation, reconfiguration (*see* Ollman, 1993; Harvey, 1996b). These social relations are always constituted through temporal and spatial relations of power with respect to the social and physical ecology that is being transformed.

That is what Massey (1992, 1993a) refers to as 'the geometry of power'; that is, the multiple relations of domination/subordination and participation/exclusion through which social and physical nature are changed. These social relations are 'grounded' in the sense that they regulate (but in highly contested or contestable ways) control over and access to transformed nature/place, but these relations also extend over a certain material/social/discursive space and operate over a certain distance. It is here that the issue of geographical scale emerges centrally. Scaled places, then, become the embodiment of social relations of empowerment and disempowerment and the arena through and in which they operate. This scaling of the everyday, as Smith (1993) insists, is expressed in bodily, community, urban, regional, national, supernational and global configurations whose content and relations are fluid, contested and perpetually transgressed.

Starting any geographical analysis from a given geographical scale (local, regional, national) is deeply antagonistic to apprehending the world in a dynamic, process-based manner (Howith, 1993). Scalar spatial configurations, whether physical, ecological, in terms of regulatory order(s) or as discursive representations, are always already a result, an outcome of the perpetual movement of the flux of socio-spatial dynamics. The theoretical and political priority, therefore, never resides in a particular geographical scale, but rather in the process through which particular scales become (re)constituted. A process-based approach focuses attention on the mechanisms of scale transformation through social conflict and struggle.

## Scalar Metamorphoses

Spatial 'scale' has to be theorized as something that is 'produced'; a process that is always deeply heterogeneous, conflictory and contested. 'Scale' and 'scale articulations' become one of the arenas and moments where socio-spatial power relations are contested and compromises are negotiated and regulated. If the capacity to appropriate place is predicated upon controlling space, then the scale over which command lines extend will strongly influence the capacity to appropriate place. More importantly, as the power to appropriate place is always contested and struggled over, then the alliances that social groups or classes forge over a certain spatial scale will shape the conditions of appropriation and control over place and have a decisive influence over relative socio-spatial power positions. For example, the present struggle over whether the scale of social, labour, environmental and monetary regulation within the European Union (EU) should be local, national or European indicates how particular geographical scales of regulation are perpetually contested and transformed. Consider how the UK's earlier opt-out from the Social Chapter of the Maastricht Treaty left a whole range of social regulatory issues in the hands of a decidedly conservative English national elite.

In a context of heterogeneous social and ecological regulations, organized at the corporeal, local, regional, national or international level, mobile people, goods and capital and hypermobile information flows permeate and transgress these scales in ways that can be deeply exclusive and disempowering for those operating at other scale levels (Smith, 1988a,1988). As Massey (1996: 112) put it:

> The power to move, and – the real point – to move more than others, is of huge social significance. Moreover, that it is *relative* mobility which is at issue is underlined . . . by the need of some, the relatively mobile/powerful, to stabilise the identities of others in part by tying them down in place.

Geographical configurations as a set of interacting and nested scales (the 'gestalt of scale') become produced as temporary stand-offs in a perpetual transformative, and on occasion transgressive, socio-spatial power struggle. These struggles change the importance and role of certain geographical scales, reassert the importance of others, and sometimes create entirely new significant scales, but – most importantly – these scale redefinitions alter and express changes in the geometry of social power by strengthening power and control of some while disempowering others (*see also* Swyngedouw, 1993, 1996a). This is the process that Smith (1993) refers to as the 'jumping of scales', a process that signals how politics are spatialized by mechanisms of stretching and contracting objects across space:

> This [stretching process] is a process driven by class, ethnic, gender and cultural struggles. On the one hand, domineering organizations attempt to control the dominated by confining the latter and their organizations to a manageable scale. On the other hand, subordinated groups attempt to liberate themselves from these imposed scale constraints by harnessing power and instrumentalities at other scales. In the process, scale is actively produced. (Jonas, 1994: 258)

These scales are, of course, operating not hierarchically, but simultaneously, and the relationships between different scales are 'nested' (Jonas, 1994: 261; N. Smith, 1984, 1993). Clearly, social power along gender, class, ethnic or ecological lines shapes and is shaped by the scale capabilities of individuals and social groups. As power shifts, scale configurations change both in terms of their nesting and interrelations and in terms of their spatial extent.

The historical geography of capitalism exemplifies this process of territorial 'scalar' construction of space and the contested production of scale. Engels (1968 [1845]) long ago suggested how the power of the labour movement, for example, depends on the place where and the scale over which it operates, and labour organizers have always combined strategies of controlling place(s) with building territorial alliances that extend over a certain spatial scale. Capitalists have usually also been very sensitive to and have skilfully strategized – often much better than labour or other movements – around issues associated with the geographical scale of their operations and paid careful attention to the importance of controlling greater spaces in their continuous power struggle with labour and with other capitalists. Scale emerges as one of the sites for control and domination, but also as the arena where co-operation and competition find a fragile standoff. For example, national unions are formed through alliances and co-operation from lower-scale movements, and a fine balance needs to be perpetually maintained between the promise of power yielded from national organization and the competitive struggle that derives from local loyalties and inter-local struggle. At the same time, the thorny issue of maintaining or consolidating local power versus the danger of incorporation and compromise at a higher scale remains an eternal quandary for social movements. Similarly, co-operation and competition among capitals is also deeply scaled (Herod, 1991b; Smith and Dennis, 1987). Tendencies toward cartel formation or strategic alliance formation point in that direction (Cooke *et al.*, 1992). Clearly, these processes of alliance formation are cut through by all manner of fragmenting, dividing and differentiating processes (nationalism, localism, class differentiation, competition and so forth). In sum, as Smith (1984, 1993) points out, scale mediates between co-operation and competition, between homogenization and differentiation, between empowerment and disempowerment.

As tensions and conflicts generated by the restless transformation of modern life proliferate, spatial scale constitutes one of the moments and arenas where compromise is struck and power relations become routinized. Indeed, scalar configurations contain and regulate social relations and enable social reproduction in the face of conflicts and tensions until the latter burst out of their cocoon imposed by scale boundaries. The modernist fallacy, which denounced spatial scales as remnants of a pre-modern past that needed to be shattered in order for emancipation and freedom to become total and universal, proved to be a mirage in the face of the often brutal repression of those trapped in place by those who command ever larger spatial scales. Inevitably, spatial scales were shattered and new scaled configurations emerged as boundaries were transgressed and new frontiers erected. During periods of great social, economic, cultural, political and ecological turmoil and disorder, when temporal/geographical routines are questioned, broken down and reconfigured, important processes of geographical rescaling take place that interrogate existing power lines while constructing new ones. These changes in scales of production/reproduction can go either upwards or downwards, but will always express new power relations and shift the balance more to one side than another. Over the past decades, it has been mainly capital that 'jumped' upwards, while in many cases (and with varying degrees of resistance) the regulation of labour moved downwards. In the next section, some of the recent transfigurations of scale and the emerging new 'gestalt of scale' will be documented.

## RESCALING THE ECONOMY, RESCALING THE STATE

During the past quarter-century or so, a profound shake-up of established geographical scales, their content and their relations with other scales has taken place. In particular, regulatory codes, norms and institutions as well as economic processes are spatially jumping from one scale to another. The overall pattern is one that I have termed elsewhere 'glocalization' (Swyngedouw, 1992a, 1992b; *see also* Luke, 1994, 1995; Robertson, 1995) and refers to (a) the contested restructuring of the institutional, regulatory level (the level of social reproduction) from the national scale both upwards to supranational and/or global scales and downwards to the scale of the individual body, the local, the urban or regional configurations (Swyngedouw, 1996b), and (b) the strategies of global localization of key forms of industrial, service and financial capital (*see* Cooke *et al.*, 1992; Swyngedouw, 1992a).

Local or regional production *filières* and firm networks, deeply rooted in local/regional institutional, political and cultural environments, co-operating locally but competing globally (Swyngedouw, 1991; Amin and Thrift, 1994), have become central to a reinvigorated – but often very vulnerable and volatile – local, regional or urban economy. A variety of terms have been associated with such territorial economies such as learning regions (Maskell and Malmberg, 1995), intelligent regions (Cooke and Morgan, 1991), *milieux innovateurs* (Aydalot, 1986), reflexive economies, competitive cities (Philo and Kearns, 1993), etc., while new organizational strategies have been identified (the 'embedded' firm (Grabher, 1993a), vertical disintegration (Scott, 1988b), strategic alliances, and so forth). Similar processes can be identified in the service sector (Moulaert and Djellal, 1994). Such territorial production systems are articulated with national, supranational and global processes. In fact, intensifying competition on an ever-expanding scale is paralleled exactly by the emergence of locally or regionally sensitive production milieux. Yet these localized or regionalized production complexes are organizationally and in terms of trade and other networks highly internationalized and globalized. The insertion of networked global companies into the particularities of regional production milieux is part and parcel of a strategy of globalization and global integration. In fact, the 'forces of globalization' and the 'demand of global competitiveness' prove powerful vehicles for the economic elites to shape local conditions in their desired image: high productivity, low direct and indirect wages and an absentee state (Group of Lisbon, 1994). Companies are simultaneously intensely local *and* intensely global (Smith, 1992).

These 'glocalizing' production processes and inter-firm networks cannot be separated from 'glocalizing' levels of governance. The rescaling of the regulation of wage and working conditions or the denationalization/privatization of important companies *and* public services throughout Europe, for example, simultaneously opens up international competition and necessitates a greater sensitivity to subnational conditions. The bureaucratic regulation of the wage nexus at the scale of the national state (something that the labour movement struggled hard for throughout most of the century) became more problematic as a significant part of the production system became supernationalized. The lowering of the scales of regulation of work and of social reproduction coincided with an increasing scale in the organization of the economy and the forces of production. This is just one of many possible examples of the growing separation between the scales of production and the scales of regulating reproduction.

Perhaps the most pervasive process of 'glocalization' and redefinition of scales operates through the financial system (Swyngedouw, 1996b). When the Bretton Woods agreement broke down in 1972 as a result of the tensions and frictions associated with differential scalings of regulating money on the one hand and the expanding scale of production and trade on the other, the global financial order was shattered. In the interstices of this mosaic, new global–local arrangements, new money flows and new geographical configurations would emerge. As Jeelof (1989) pointed out, the volatility in the money markets made production planning extremely risky and uncertain. The internationalization of production and world planning of production chains and input–output flows which characterized much of the post-war international division of labour became a high-risk strategy. Different locations of production as well as sites of production and sites of commercialization were located in different currency zones and subject to often rapid and dramatic relative exchange rate fluctuations. This made a shambles of long-term corporate strategic locational planning. Globalizing companies, trying to deal with rapidly changing relative locational conditions, launched a series of strategies which I have defined above as 'glocalization' strategies. These allow for rapid temporal and spatial adjustments in production, distribution and marketing arrangements (Cooke *et al.*, 1992).

More importantly, perhaps, the liberated money markets and the volatility of the international money markets created a new market environment. Buying and selling currencies and speculating on exchange rate fluctuations allowed for the development and rapid growth of a speculative foreign exchange (forex) and, from the mid-1980s, a burgeoning derivatives market (*see* Swyngedouw, 1996b). Interestingly enough, making money by buying and selling money and speculating on future (however near this future may be) currency values became a prime vehicle for accumulation. Speculating on future values and the buying of time proceeded through the creation of new spaces and spatial relations. For example, the forex market has grown from a modest US$15 billion in 1970, when most deals were directly related to settling trade, to well over a trillion dollars today. The bulk is driven by constant hedging, arbitrage and speculative position-taking in the international financial markets. Almost all deals involve spatial transfers of money as well as changes in the relative positions of national currency values (which, in turn, influence national interest rates, buying capacity, regional competitive positions, trade flows, monetary and fiscal policy and so forth). This volatility enables speculative gain, while the flows of money further contribute to reaffirming these fluctuations.

The bumpy history of the European Monetary Union, for example, illustrates how the confrontation of national demands and global financial integration and strategies results in perpetual tensions and continuous friction (Gros and Thygesen, 1992; Leyshon and Thrift, 1992). The uncertain road to the Euro (the proposed European common currency) is an example of how a particular and hotly contested politics of scale is inserted in this emerging new scalar gestalt of money. This will be discussed in greater detail below.

In the midst of this glocalizing economy, the role and position of the 'traditional' nation-state are revamped. While the state constituted, although by no means uniquely, the pivotal scale for the regulation and contestation of a whole series of class practices in the post-war period, its relative position and importance is shifting in decisive ways (Jessop, 1994a, 1994b; Swyngedouw, 1992a, 1997). In a context in which the capital–labour nexus was nationally regulated, but the circulation of capital spiralled out to encompass ever larger spatial scales, there was a concerted attempt to make the 'market imperative' the ideologically and politically hegemonic legitimation of institutional reform. This took shape through a variety of processes which combined (a)) the 'hollowing out' of the national state, both downwards and upwards, with (b)) more authoritarian and often softly but sometimes openly repressive political regimes. Of course, the key question, in short, is not whether the state is globalizing or localizing, but rather what kind of struggles are waged by whom and how the rescaling of the state towards the 'glocal' produces and reflects shifts in relative socio-spatial power geometries. In the following section, I shall briefly explore some examples of this contested rescaling and how processes of inclusion and exclusion operate through the reconfiguration of the 'gestalt of scale'.

## THE CONTESTED POLITICS OF RESCALING: SCALES OF INCLUSION/SCALES OF EXCLUSION

This section explores the intricate relationship between recent changes in the role and the position of the national state and the formation of new and differently 'scaled' institutional forms on the one hand and spatial restructuring processes on the other. I shall discuss the ways in which the drive to produce competitive regional spaces in an increasingly globalizing and intensely competitive world economy coincides with a more prominent position of both local and supranational state forms or sites of governance. This 'rescaling' of the state takes place through the formation of new elite coalitions on the one hand and the systematic exclusion or further disempowerment of social groups already weaker, politically and/or economically, on the other (Jessop, 1993, 1994a, 1994b; Peck and Jones, 1994). A particularly telling example is the process of 'European integration' and the whirlpool of contested rescalings it implies.

As Lefebvre (1976, 1978) has argued on a number of occasions, the state (at whatever scale – local, national, international) is always spatially organized as its interventions are profoundly spatial strategies to regulate social and physical relations. Spatialized interventions are, consequently, one of the strategies where state tactics to control and mediate social relations among individuals, classes, class fractions and social groups in the context of the maelstrom of perpetual shifts in the global economy are played out. The scale of governance, then, becomes an integral part of these tactics. The rescaling of the state and the production of new articulations between scales of governance in turn redefines and reworks the relationship between state and civil society or between state power and the citizen (Swyngedouw, 1996a, 1997). One of the remarkable institutional-political tendencies over the past decade or so has been the simultaneous internationalization and decentralization/devolution of key policy/regulatory/economic issues. The interventionism of the state in the economy, for example, is rescaled, either downwards to the level of the city or the region, where public–private partnerships shape an entrepreneurial practice and ideology needed to successfully engage in an intensified process of inter-urban competition (Harvey, 1989b), or upwards. Upward rescaling is manifested in the albeit highly contested and still rather limited attempt to create a supernational Keynesian interventionist state at the level of the EU.

In a different sort of way, institutions such as NAFTA (the North American Free Trade Agreement), GATT and others are testimony to similar processes of upscaling of the state. Furthermore, a host of informal global or quasi-global political arenas have been formed. OPEC may have been among the first and most publicized of quasi-state organizations, but other examples abound: the G-7 meetings, the Group of 77, the Club of Paris and other 'informal' gatherings of 'world' leaders who attempt to regulate the global political economy. Of course, competitive rivalries among these 'partners' prevent some form of effective co-operation that could otherwise ultimately lead to a frightening global, authoritarian state-form.

In addition to the socially deeply uneven, socio-spatially polarizing and selectively disempowering effects of the 'jumping of scales' that exemplifies this 'glocalization' of the state or of other forms of governance, they also take place through disturbingly undemocratic procedures. The double rearticulation of political scales (downward *and* upward – as well as outward to private capital) leads to political exclusion, a narrowing of democratic control, and, consequently, a redefinition (or rather, a limitation of) of citizenship. Ironically, while the creation of local institutions is often defended and legitimated on the basis of a potentially enhanced democratic control from locally rooted organizations, the evidence suggests a tendency toward a loss of democratic control (Saunders, 1985). In addition, the privatization of governance (the displacement of state power outwards) increases the control and power of (inter)national or regional business elites which take centre-stage in promoting a 'boosterist' entrepreneurial development vision (Cox and Mair, 1989, 1991).

The rescaling of the state does not suggest, therefore, a diminishing role of the state apparatus. In fact, these new global/local institutions, in close co-operation with private capital, launch the redevelopment largely on the basis of public funds and state capital (Peck, 1995; Peck and Tickell, 1995a). However, the power and control over social capital is increasingly diverted to a small elite who shape the urban and regional fabric in their own image and fashion, and define the very content of the restructuring process. In short, the 'glocalization' or rescaling of institutional forms leads to more autocratic, undemocratic and authoritarian (quasi-)state apparatuses (Morgan and Roberts, 1993). These new institutional forms are riven with all manner of conflict and tension. First, this double rearticulation of the scalings of the state is highly contested, particularly by those who become marginalized in or excluded from these new institutions. Second, the new alliances that are forged and their need to affirm their legitimacy accentuate the need from the part of the boosters to try to create a new hegemony of vision (Zukin, 1996), particularly through the 'spectacularization' of both development perspectives and political programmes which take away the focus from the substantive, on-the-ground transformations of the urban-regional socio-economic fabric (Debord, 1990).

In a European context, the process of 'jumping scales', the stretching of scales and the contested nature of the new 'gestalt of scale' is most vividly expressed in the debate over European integration and the unquestionable rearticulation of scales that accompanies this process, which is arguably the most radical political-economic and cultural transformation of space in Western Europe since the French Revolution.

There is, of course, an intense discursive invocation of the politics of scale in most national ideological and political debates, in which the European scale is often presented as the scapegoat to cover up or displace internal conflict and struggle. For example, when conservative prime minister Juppé in France attempted to introduce a radical national austerity programme in December 1995, the Maastricht Treaty and the provisional agreement to reduce annual state deficits to below 3 per cent were invoked as the 'external' legitimation to push through a major overhaul of the social security system in France. The French workers pierced through this veil, mounted the greatest mobilization in France since 1968, and successfully resisted the state. When Britain chose to opt out of the Social Charter under the guise of protecting national sovereignty, it curtailed the rights of women and men in the workplace, maintained a stance of social dumping to protect the short-term interest of capital located in Britain (whatever its national origin), withheld a movement towards participation of workers in company management, and allowed companies to maximize absolute surplus value production by not accepting a ceiling to the maximum working week. The present Dickensian socio-economic conditions of the poor in the UK, resulting, for the first time since at least the Second World War, in poverty-in-employment for those who manage to escape from the dole queue, also accelerate delocalization and increasing unemployment in other member states. Beggar-thy-neighbour politics dig the graveyard of social protection and social security, not only at home but abroad as well.

In addition to the rhetoric of the politics of scale, there is an important substantive process of rescaling of authority taking place. Such politics of scale are central to almost all issues associated with European integration:

(1) Money and finance: the contested introduction of a single currency transcends the debate on national sovereignty and control. Under conditions in which global financial speculation and mass transfers of currencies amount to over a trillion dollars a day, the room to manoeuvre in monetary matters for national governments is rather limited anyway. The conflict over the introduction of a single currency is leading to the formation of a strange new set of alliances which cut through all manner of traditional cleavages and fractures. The UK financial sector, worried about maintaining the pre-eminent position of the City of London (where US banks have become the dominant players any-

way), British nationalists who see the currency as the ultimate expression of national identity and difference, speculators who welcome currency and interest rate fluctuations (on which most of the speculative derivatives markets are based) and a conservative ideological bloc that has traditionally pursued aggressive devaluationary policies to maintain competitiveness by transferring the failure of domestic capital to maintain a competitive edge on to both domestic and foreign workers, join ranks in pursuing a politics of scale that revolves around maintaining a national currency in the face of the demands by industrialists, international merchants and some fractions of the working class that see a common currency as a tool to break with the infernal cycle of competitive devaluations and speculative financial movements, and to align monetary policy more closely to the interests of the 'real' economy.

(2) Social issues are perhaps the most visible and problematic aspect of the European integration process and its associated redefinition of scale. While the British opt-out represents the most radical and regressive position, the Danish and Norwegian scepticism about European integration revolves rather around the fear of the potential dismantling of the social welfare network that has characterized the 'Scandinavian model' for almost a century now. From the Treaty of Rome of 1958 onwards, the production of a new 'European' scale of regulation and governance was inspired by predominantly economic considerations and centred on the introduction of a free trade zone that would allow the unhindered movement of capital, commodities and labour. However, this liberalization of European economic space took place under conditions of different national social regimes existing side by side. In order to maintain a level playing field, social and other rules and regulations needed to be streamlined at the new scale of the economic free zone. With a virtually absentee welfare state at the level of the EU, the tendency has been to align national institutional frameworks on the basis of the lowest common denominator. This, in turn, threatens the cohesion and continuation of the welfare regimes built up over decades of intense working-class struggle and hard-won victories. Perhaps the most important change associated with this rescaling process is the absence of structural redistribution programmes at the European scale (with the possible exception of agriculture and, to a far lesser extent, the structural funds). During the post-war period, national welfare regimes were undoubtedly the most important structural redistribution mechanisms that alleviated socio-spatial differences by means of long-term inter-regional transfer payments. The disappearance of inner-city slums and the relatively small inter-regional differences within states is largely the result of systematic socio-spatial redistribution schemes. With the withdrawal or slimming down of these social security transfer payments, socio-spatial differentiation is again accentuated. The European geographical project has only the bare bones of such spatial redistribution mechanism while growing economic homogenization and international competitive pressures fed the call for a reduction in national regulatory systems. The upscaling of the economy in a context of trimmed-down national redistributive mechanisms and intensified inter-place and inter-regional competition contributed to an acceleration of processes of exclusion and marginalization, and deepened social polarization in ways that tie down a growing part of the European population in unemployment, poverty and reduced citizenship rights. The two-or-three speed Europe is not one linked to a geographical core and periphery in terms of their determination to accelerate integration, but is rather an internal differentiation between those who revel in and benefit from greater command over space on the one hand and those who remain trapped in the doldrums of persistent marginalization and exclusion on the other.

(3) Gender relations equally revolve around an intense politics of scale, from rights to abortion, more egalitarian divorce laws, the regulation of non-heterosexual relations, social security rights to workplace discrimination, etc. Each of these extend from the scale of the body to the regulation of these corporeal rights at higher scales. These scales are again the arena of conflict and struggle in which different groups pursue scale strategically to maintain or alter existing power relations. Consider, for example, how the decision of the UK to opt out of the Social Charter undermines women's already precarious legal and other rights. Although many of the persistent discriminatory practices have been successfully brought to the European Court (notably on the equal treatment of part-time workers), these had to be fought for through lengthy, time-consuming and expensive legal procedures available only to the rich or to those who have sufficient backing from grassroots organizations.

(4) The regulation of capital–labour relations tended to devolve from some kind of national collective bargaining to highly localized forms of negotiating wages and working conditions. The UK, for example, has moved a long way towards this, and continuous pressure is exercised to make unions and workers accept 'local' pay deals. Similar movements have been documented elsewhere (*see* Cox and Mair, 1991; Ohmae, 1995a), but, depending on particular political configurations, resistance

to these movements towards downscaling has been more successful in some countries, such as Sweden and Germany, than in others. At the same time, attempts are made to supernationalize some of the issues related to the capital–labour divide. In particular, but not exclusively, socialists and ecologists are fighting for an upscaling of respectively capital/labour and environmental regulations to higher scale levels. A key example is the struggle over workers' participation, consultation and involvement in corporate decisions and negotiations. Collective consultation, which is part of many national regulatory frameworks (although all of them firmly maintain the final powers in the hands of capital), is hotly contested at the European level. In an environment in which competitive relocations (the Hoover relocation from Dijon to Scotland is a notorious case) are taking place, workers' involvement and co-decision-making at the same level as the economic space is central to a more democratic participation of the workers in economic processes.

(5) Migration and citizenship: the most problematic area of the new 'Gestalt of scale' in Europe is the political (as well as economic/cultural) exclusion of immigrants. While procedures to be allowed into the EU are tightened and a new 'European wall' is built to keep the undesired Others out, the growing immigrant community remains without basic political rights in most EU member states in terms of its participation in the political process. Its exclusion in the construction of the new hierarchy of scales of governance intensifies immigrants' cultural and political-economic marginalization and feeds the already acute tensions between different ethnic communities.

## MOBILIZING SCALE POLITICS

Engaging place, restructuring places, occupying places, metabolizing physical and social nature take place through conflicting socio-spatial processes. The transformative continuation of socio-spatial relations that operate through deeply empowering or disempowering mechanisms produces a set of related and nested spatial scales which define arenas of struggle, where conflict is mediated and regulated and compromises settled. Socio-spatial struggle and political strategizing, therefore, often revolve around scale issues, and shifting balances of power are often associated with a profound rearticulation of scales or the production of an altogether new 'gestalt of scale'. The socio-spatial transformations that have characterized the past two decades or so are testimony to such scale restructurings through which older power relations are transformed. The disturbing effects of these recent 'glocalization' processes suggest that the spaces of the circulation of capital have been upscaled, while regulation of the production–consumption nexus has been downscaled, shifting the balance of power in important polarizing or often plainly exclusive ways. The rescaling of the state and the production of new articulations between scales of governance in turn redefines and reworks the relationship between state/governance and civil society or between state power and the citizen.

While the rich and powerful revel in their freedom and ability to overcome space by commanding scale, the poor and powerless are trapped in place (Harvey, 1973). It remains deeply disturbing to find that the power of money and a homogenizing imperialist culture take control of ever larger scales, while very often the 'politics of resistance' seem to revel in some sort of 'militant particularism' (*see* Harvey, 1994) in which local loyalties, identity politics and celebrating the unique self from the different Others attest to the impotence to embrace an emancipatory and empowering politics of scale. Certainly, local loyalties are central in any emancipatory politics, but solidarity, inter-place bonding and collective resistance demand a decidedly scaled politics that can challenge the totalizing powers of money and commodified culture and provide a credible alternative. What is disturbing in contemporary politics of resistance is not that the paramount importance of scale is not recognized, but rather that oppositional groups have failed to transcend these confines of a 'militant particularism' or 'particular localism'. The angst for negating the voice of the Other has overtaken the resistance to the totalizing powers of money or capital. Ironically, the retreat from collaboration and coalition formation out of fear of perverting the Other's identity, of assuming some sort of homogenized identity and of annihilating difference, swings the leverages of power, of marginalization and exclusion decidely in the direction of the totalizing and homogenizing forces of global commodification and repressive competition, controlled by a band of 'glocal' elites.

While the power of the (sometimes new) elites becomes consolidated and is often reinforced, new social movements (sometimes in alliance with the politically and socio-economically excluded) begin to challenge the new elite programmes and question the legitimacy of the institutional framework from which they are excluded (Mayer, 1994). Such strategies of resistance can take a variety of forms, ranging from the rise of deeply anti-statist forms of anti-politics which feed the electoral support for extreme right-wing – exclusive and nationalist – political parties to active contestation of the development vision by all sorts of groups, from discontented – since excluded – local businesspeople to the green movement and immigrant groups. In addition, as the state or other forms of governance become increasingly authoritarian, the excluded target extra-state objectives to voice their discontent or to launch (often direct) actions and strategies of contestation (Mayer, 1989, 1993) which have become one of the few avenues open to contest and express citizenship rights. In sum, the new 'gestalt of scale' is accompanied by new forms of social movements and new sorts of voicing and mediating social conflict.

Oppositional groups, whether organized around working-class, gender, environmental or other politics, are usually much better and empowering in their strategies to organize in place, but often disempowered and fragmented when it comes to building alliances and organizing collaboration over space (Harvey and Swyngedouw, 1993). Empowering strategies in the face of the global control of money flows and competitive whirlwinds of 'glocal' industrial, financial, cultural and political corporations and institutions demand co-ordinated action, cross-spatial alliances and effective solidarity. Strategizing around the politics of scale necessitates negotiating through difference and similarity, to formulate collective strategies without sacrificing local loyalties and militant particularisms. New and old progressive social movements and, in particular, progressive ecologists, feminists and socialists attempt to struggle through the difficult process of formulating cross-space strategies that do not silence the Other, exclude the different or assume the particular within a totalizing vision and political project. For example, when Shell, with the full approval of the British government, planned to dump the *Brent Spar* oil platform in the ocean, it was the German consumer, led by Greenpeace, who brought Shell to its knees on this matter. Both the national state and a global company were successfully challenged by an alliance-based organization that cuts through all sorts of traditional cleavages (production/consumption, male/female, labour/capital) and strategically mobilizes on the basis of particular local, regional or national sensitivities in the context of broader and global objectives. When France was testing nuclear weapons around Mururoa, French wine producers were hurt because of consumer boycotts in some of the most affluent parts of the world. The demand of local and national unions in some countries to at least complement (if not replace) the GATT agreement with a social 'pact' to guarantee minimum social and work standards (often in an effort to save jobs, locally or nationally, in the face of social dumping practices elsewhere) brings out the issue of scaled politics and the articulation of scales (with all its tensions and contradictions). The combination of self-interest and solidarity with social struggles elsewhere can construct a platform with which to resist the totalizing powers of an unqualified free trade ideology.

Negotiating through local interests, socio-spatial differences and constructing alliances over space is a difficult, contested and always fragile process, riddled with tension and conflict. There is no easy answer or simple solution to the tensions generated by particular scalar configurations. Nevertheless, these examples show how scaled politics weave together in new and, to some, rather surprising ways, and illustrate how recent scale metamorphoses recombine the political and economic, the cultural and the ecological, in new and hitherto rarely, if ever, explored ways. The politics of scale are surely messy, but ought to take centre-stage in any successful emancipatory political strategy.

# Chapter Fourteen

# GLOBALIZATION AND GEOGRAPHIES OF WORKERS' STRUGGLE IN THE LATE TWENTIETH CENTURY

KEVIN R. COX

## INTRODUCTION

Today, it would seem, a point of departure for *any* discussion of working-class struggle and its prospects is what is referred to as 'globalization'. 'Globalization' is a vague term that can be defined in many different ways. But regardless of its diverse concrete expressions, it is to the fore in discussions of the recent and massive demobilization of the labour movement. Declines in unionization and static or declining real wages for large segments of the workforce are placed at the door of a superior bargaining power that capital has supposedly achieved by its creation and subsequent exploitation of more all-encompassing arenas for its own mobility. The essential complement to this argument is the failure of workers to colonize through their organizations or their own mobility those same arenas. These conceptions have been of particular interest to geographers and to those others in planning, urban and regional studies who typically prioritize space in their discussions.

The arguments have assumed two major and separable forms. The first dwells on changes in the geography of production: a language of cores and peripheries, of deindustrialization and rust belts, of deskilled workforces in small towns and free trade zones, and of changes in the spatial divisions of labour of firms. As a result of disinvestment and plant closures, capital has been able to turn resistance to wage demands around into a request for worker givebacks. Living with the unions has given way to the active encouragement of decertification elections, all justified by claims of an intensified international competition and the attractions of relocating to a free trade zone.[1] Globalization has also become a central part of growth coalition rhetoric in efforts to build popular support for programmes aimed at improving 'the business climate' – a shorthand for the rewriting of labour law and a shifting of tax and regulatory burdens at the expense of labour. In brief, globalization has become the new form in which claims to the naturalness of capital as a social relation are asserted.

A second argument has focused on the emasculation of macro-economic policy – a retreat that has served to further sap worker resistance. Policies have become markedly less expansionary and now give priority to the defence of the national currency. Subsequent deflationary tendencies reduce the demand for labour and intensify the competitive pressures between firms. These changes too are related to the overall context of globalization. Some have emphasized the greatly enhanced magnitude of international financial transactions. For others the changes in the geography of production referred to above have been the crucial determinants, along with subsequent shifts in trade and trade dependence.

In this chapter I want to subject these arguments to a critical scrutiny. It is my view not only that they seriously overgeneralize but that through their rhetorical power they serve to mystify and so, in their own way, undermine the collective power of the working class. Spatial fetishism is an old accusation in critical human geography and one that seems rarely to be used any more. But there are good reasons to argue that that is precisely what is at issue in the politics of globalization, both as an object of academic scrutiny and as it is practised on a day-to-day basis by corporate interests. Moreover, both academics and capital seem at one on this interpretation of the world and the subsequent inevitability of a shift in bargaining power. The only difference seems to be that for most of social science it is something to regret.

In addressing the nature of worker resistance, the terms of the chapter are quite narrow, and this reflects the bulk of the globalization literature. It emphasizes changes in the bargaining situation confronted by workers, both individually and collectively, in the leverage they have with respect to employers, and has little to say about forms of worker organization and discourse. It also has little to say about the relation between workers, political parties and party political competition. In a more comprehensive critique of the literature this would surely have to be addressed. The whole rhetoric of co-operation with employers in confronting competition 'elsewhere' underlines the significance of the meaning systems that workers subscribe to and the importance of an independent organizational base for the construction of those meanings. Too often workers have been persuaded by a discourse of globalization where a deeper understanding of capital, and the pressures and temptations to which firms are subject, might have led to a better bargain for labour: wage concessions, perhaps, but so the firm can finance its way out of that particular industry? (Cohen, 1991). The importance of a counter-hegemonic discourse, and indeed of the geographies that facilitate it, cannot be overemphasized, but it falls outside the scope of the current chapter.

My discussion is divided into two parts. First I address those globalization literatures which foreground changes in the geography of production. I then turn and consider the arguments that link globalization in some of its quite diverse expressions to the decline of full employment as a policy goal and the increased popularity of macro-economic regimes that are deflationary in character.

## CHANGES IN THE GEOGRAPHY OF PRODUCTION

The arguments here are fairly well known. Changed production conditions have facilitated the extension of industrial production away from the old heartlands of the advanced capitalist societies. A New International Division of Labour has sprung up in which less-skilled work is decanted to Third World sites while headquarters functions, R&D and more-skilled manufacturing activities remain behind in the First World. Similar patterns emerge within the First World countries themselves as less-developed national peripheries, small towns formerly catering to dispersed rural populations, sunbelts, are colonized by waves of branch plants employing deskilled, often female, labour in branch plants. Behind them these diffusion processes leave behind a landscape of rustbelts, or, in Dick Walker's evocative words, a 'lumpengeography of capital' (1978: 34).

In accounts of this 'shift', two changes in production conditions are typically stressed: the deskilling of labour processes; and changes in transportation and communication. To a large degree these processes work their effects in tandem. The deskilling of labour processes liberates employers from dependence on scarce skills and the high wages accordingly demanded. Major gains in wage reductions are to be had if the deskilled processes can be moved out of major metropolitan centres and old industrial heartlands into small towns and lower-wage areas around the world where, for a variety of reasons – living costs, unemployment – wages are lower. But in order to do this, in order to create new firm spatial divisions of labour in which the deskilled work processes are moved out to more distant locations, easier, cheaper, speedier transportation and communication are required.

This is not to confine the changes to the investment patterns of First World firms. The experience of the newly industrializing countries (NICs) can be interpreted in terms of the same two conditions: deskilled labour processes that have allowed them a toehold on the ladder to more advanced forms of technology through a process of industrial learning, and the changes in transportation – the container revolution, air cargo in some cases – which have given them access to First World markets.

The implications of this for labour in the older industrial centres of the First World and for organized labour more generally are seen as universally detrimental. The changing geographic configuration of demand for labour, it is argued, is severely weakening labour's bargaining power. Instead of fighting for increased wages, benefits and enhanced job security, their representative organizations now have to contest plant closures and requests for worker givebacks. Likewise, in a rhetorical climate that appeals to these changing geographies, even campaigns for union certification encounter serious difficulty. Instead of making things easier for workers, unions are seen as a threat to the continued existence of employment. Union membership in the advanced industrial societies has decreased – in some cases, like that of the USA, quite drastically – and much of this is laid at the door of so-called 'globalization'.[2]

There are, however, some quite serious problems with these arguments. An immediate difficulty is the focus on cost minimization as *the* strategy of profit maximization for firms. For there is no reason why competitive strategy should be confined in this way. Certainly, the tendency to cost minimization is a necessary one under capitalism: this is the whole thrust of the law of value. But capitalism revolutionizes not just the forces of production – how things are produced – but also the things that are produced. To the extent that the forces of production are embodied in new machines and technologies that are products for some firms at least, this goes without saying: cost minimization strategies and product development strategies are necessary complements. But in addition, firms seek out new products, new product attributes, which are themselves products for final consumption, and not for productive consumption.

Through the development of new products in these ways firms can strive for a level of profit above the average – a quasi-rent. Patent laws and other strategies for protecting proprietary knowledge allow some period of time in which that advantage can be preserved – whereas a cost-minimizing strategy is one that is quickly generalized to other firms, so that any initial advantage quickly evaporates. Beyond that, of course, the variety of competitive strategies reminds us that the purpose of capitalists is *not* cost minimization but profit. And as far as the individual capitalist is concerned, it does not matter how that profit is gained, whether through seeking out cheaper labour as per the globalization thesis, or through the discovery, development and exploitation of new products.

But beyond this, there are important geographical conditions for product development and innovation that need to be taken into account – conditions which contradict easy resort to an overriding globalizing tendency in contemporary economic geographies. Storper has emphasized this in his discussion of the conditions for what he calls 'product based technological learning' (PBTL) (1992). For firms, the significance of PBTL is that it facilitates the achievement of competitive edges through continual innovation. It tends to be concentrated in what Storper calls technological districts. Technological districts afford access to a highly disintegrated social division of labour. This allows firms to avoid the technological 'lock-in' that comes from more vertically integrated forms of organization.

In a development of this sort of argument, others have emphasized the importance of close spatial relations between the producers of technology and their users (for a discussion, *see* Gertler, 1997a; *see also* Fagerberg, 1995). At another stage in the development of new products, their actual launching on the market, a spatial concentration again becomes important, owing to the intense and frequent personal communications and rapid decision-making necessary at that stage (Patel, 1995: 152). Recent work on the introduction of new financial products exemplifies this point (Pryke and Lee, 1995).

Even on its own terms of a cost-minimizing competitive logic, the globalization thesis is far from secure. Cost minimization can in no way be reduced to firm strategies of dispersion and the creation of far-flung spatial divisions of labour tapping into low-wage peripheries. The work of Allen Scott (1985 for a good general statement) has been exemplary in its emphasis on the virtues, in terms of external economies, of vertical and horizontal disintegration under conditions of locational convergence. The advantages range from the creation of new roles in the social division of labour, through the pooling of labour reserves, to the achievement of thresholds for collectively consumed items of infrastructure, social and physical. In addition to bringing cost minimization, accessibility can facilitate the collaborative interaction of those at different positions on the same input–output chain: collaboration on the development of new models, or on the elimination of bottlenecks or of quality control problems, for example. The agglomerative implications of just-in-time as a quality control programme underline the significance of some of these effects (Sayer, 1986).

On the other hand, the conditions for social learning and for the development of the division of labour can by no means be reduced to some ideal type of small, vertically and horizontally disintegrated firms. They may, rather, be within-plant and within-firm as much as they are between-plant and between-firm.[3] The occurrence of extremely dynamic firms, like Boeing, Cummins or Pilkingtons, dominating local economies but with little local synergy is exemplary.

The point is that such locational convergence, once achieved, and whether between firms or among the employees of the same firm, is difficult to reconstitute elsewhere and is therefore resistant to the sort of globalizing logic inherent in current arguments. Even when we turn and examine those conditions at the heart of globalizing claims – deskilling and changes in transportation and communication – the case for a logic of globalization falls short of convincing. Deskilling is far from necessary as a process and there is no good reason why it should be (Cohen, 1987). If indeed, as per cost minimization strategies, the interest of the capitalist is in reducing costs, then that cannot be equated to reducing any *particular* cost. Some processes are more resistant to deskilling as opposed to change in other directions. There is a lot of technical change that develops the productivity of the worker but which has nothing to do with deskilling. Rather, it may have more to do with the nature of the materials being processed, the nature of the machinery that those materials facilitate or both. To take an obvious case, the food processing industry is now able to consider new technical possibilities as a result of the homogenization of raw material inputs opened up by biotechnology.

Furthermore, even when deskilling *does* occur it is by no means inevitable that it will result in relocation. Services like printing have been deskilled, making former typographic skills obsolete and reducing printers to button-pushers. But a lot of printing – job printing for small businesses, newspaper printing – remains resolutely where it has always been: close to its market (Goss, 1987). And this is true of many so-called service industries, particularly those requiring personal contact with the customer. Likewise, there has been an efflorescence of new job categories that require little in the way of training but are, again, highly market-orientated: the armies of cleaners in downtown office buildings and hotels, and the rapidly expanding corps of private security guards and parking garage attendants, to take cases in point that will appeal to advocates of that variant on globalization claims, the world city hypothesis (Friedmann and Wolff, 1982). Moreover, and market arguments aside, once a work process *is* deskilled there is no reason why the new workforce that is tapped should be peripheral or small-town in its location. 'First World' or 'core' labour conceals a bundle of differences, vulnerabilities and potentials for the construction of new, marginalized workforces, as has been discussed in work on back-office location (Nelson, 1986).

Neither do arguments about changing modes of transportation and communication and their implications for locational calculi withstand careful, critical scrutiny. Just as not all labour processes are subject to deskilling, not all firms are able to take, or even want to take, advantage of lower transportation costs. Small-volume producers are obviously at a disadvantage. Where quality control is an issue, as with components or subassemblies, bulk transport on the scale that allows economies to be reaped has disadvantages. This is because in order to achieve the necessary thresholds, deliveries may have to be large and infrequent with subsequent delays in remedying any deficiencies in the manufacturing process.

Reduced transportation costs are contradictory in their effects. Just as they can facilitate industrial dispersion, so too can they foster the growth of existing agglomerations. As per Adam Smith's dictum, the division of labour in existing metropolitan areas can undergo further development, throwing out new external economies, as the market for the area's products undergoes expansion. Within those same metropolitan areas, transportation change in the form of the widespread use of the automobile facilitates the integration of urban labour markets and on expanding scales, bringing about a pooling of labour reserves and a lowering of housing costs. Similarly, it is the universalization in the advanced capitalist societies of the two-car household that has been a necessary condition for the construction of housewives as a new labour force for suburbanizing back offices, though this is by no means to identify it as the sole social change on which that construction depended.

I would conclude, therefore, that the contingent effects of deskilling and transportation change on geographies of production are either overstated, contradictory, or both. This means that labour in the heartlands of Western industrialization is not necessarily placed at a disadvantage, and the arguments about new industrial spaces, world cities, even the continued vitality of many old industrial spaces and single-industry towns, whether characterized by vertical distintegration or not, lend support to this. Obviously there are numerous cases which confirm the direst predictions of the orthodoxy. There *is* competition between localities on the basis almost purely of labour cost; plant closures *are* averted by worker givebacks; firms move in the direction of non-union labour while the unionized are left jobless, with consequent implications for the future of union membership. But what I am arguing here is that the world is much more complicated than this and the contemporary discourse of globalization represents the tightest of procrustean beds into which to force that variety. And, I would add, given what I have argued at the level of first principles about capitalism and its relation to space, this is a necessary variety.

Despite the intense cost competition stressed by the orthodoxy and the subsequent propulsion of firms in the direction of national and international peripheries, quasi-rents can still be achieved in those areas that are, according to this logic, being 'hollowed out'. This can occur through product innovation or through the external economies central to Scott's model of the development of the social division of labour in metropolitan areas. Furthermore, the way in which many of the firms affected are, in effect, locked into these locations by the difficulties of reconstituting their webs of exchange relations and their other forms of transaction elsewhere suggests that workers may enjoy some leverage over wages and work conditions. This does not mean to say that this will be exercised collectively. Rather, it may be through more individualized forms, as in the case of internal labour markets. But even less skilled workers can share in the flow of value through the social relations of a local economy, to the extent, that is, that they organize; and the metropolitan character of the labour market they participate in often means that they have some leverage – even, say, as security guards, janitors, or fast-food workers.[4] The difficulty in those instances may be less the invulnerability of employers to worker claims and more the problems that labour unions have had in organizing such dispersed workforces. That does not mean to say that those difficulties are insuperable. More likely is the failure of the union movement to adjust to changing labour processes.

## THE QUESTION OF MACRO-ECONOMIC POLICY

During the 1980s, and presaged by some of the changes occurring towards the end of the previous decade, macro-economic policy in the advanced capitalist societies underwent a dramatic change. Instead of the prioritization of full employment as the policy goal, fighting inflation became the emphasis. Keynesian demand management through deficit spending lost its attractions, and the orthodoxy came to be dominated by monetarism: the management of the money supply in the interest of maintaining stable prices. To the extent that full employment remained a concern, it was to be handled by micro-economic policies: the so-called supply side, of which there have been different versions from both ends of the political spectrum. This ushered in policies of what Albo (1994) has dubbed 'competitive austerity': 'competitive' because in an attempt to make themselves credible to investors, for whom the world, apparently, had become their stage, states saw themselves as being evaluated relative to each other in terms of such indicators as government deficits, wage pressure, strikes, profitability and inflation; and 'austere' because of the effects of such policies on employment levels. The failure of the Mitterrand experiment at the beginning of the 1980s seemed to confirm this new conventional wisdom.

These policies had important effects on the internal economic geographies of countries since while the general tendency was to reduce the bargaining power of workers, this was quite uneven. For a start, the decline in easy money pressed more marginal firms and more marginal regions to the wall. As governments strove to rectify their deficits, this differentiation was intensified by general reductions in the social safety net – reductions which obviously hit hardest those regions suffering most from unemployment. Typically, the effect of centrally funded social services and income transfers had been to provide automatic stabilizers, mediating inter-regional transfers from winners to losers. As government expenditures were reduced, so this transfer became less effective because it was diminished in scope.[5]

The micro-economic policies that were supposed to enhance national competitiveness and increase the demand for labour had similar regionally differentiating effects. Tax reforms curtailed income transfers even more than was required by simply bringing the budget into balance. Privatization and deregulation were uneven in their effects since some areas were more dependent on to-be-privatized or to-be-deregulated activities than others: steel towns in the British case, small towns dependent on marginal railroad branches or cross-subsidized trucking and airline services in the USA provide examples.

In some cases, divergence in regional fortunes had the perverse effect of immobilizing people and making an adjustment through migration difficult. For on top of divergence in employment records and incomes went diverse pressures on housing markets. Home-owners in depressed areas found themselves owning assets which, if they sold them, not only might not realize what they had paid for them, in the case of a town like Youngstown or Liverpool, for example; but would certainly not convert into the sort of capital sum needed for home-ownership in the hothouse housing markets of the growth areas like Silicon Valley or south-east England (for the British case, *see* Muellbauer, 1990).

From the standpoint of the argument of this chapter, however, what is most interesting about this policy shift is the way in which it has been attributed to that somewhat vague congerie of effects that has come to be known as 'globalization'. This is a common argument in both lay and academic circles. It is useful, however, to distinguish between two separate versions. The first roots itself in changing global patterns of industrial location, investment and trade. The second foregrounds what is regarded as a dramatic upsurge in the magnitude of international financial transactions.

In other words, there is an argument which leans on the rise of the NICs, the emergence of a New Industrial Division of Labour, and an overall growth of import penetration of national economies in the First World – the sorts of changes discussed earlier in the chapter. These changes have, it is suggested, altered in a qualitative way the economic environment with which macro-economic policy must cope. Several related effects are commonly identified. For a start, the possibility of effective expansionary policy has been vitiated by the increased openness of national economies.

Expansion through demand management has limited relevance for firms for whom major markets are overseas or for branch plants keyed into the demands of plants in other countries. More critically, expansionary policies, unless all countries are engaging in the same sorts of policy at the same time, may simply boost imports from overseas rather than domestic production.

An interesting wrinkle on these arguments has come from regulation theorists, or at least those influenced by that particular problematic. The argument here is that the crisis of labour in the First World countries is a crisis of the mode of regulation and the collapse of Fordism. Fordism, the basis of labour's 'golden age' during the 1950s and 1960s relied on a close articulation of domestic production and domestic consumption and on industries that were relatively oligopolistic and immune to foreign competition. Globalization has destroyed this (*see* Schoenberger 1988, for a good statement). Relatively stable price levels encouraged the heavy fixed investment characteristic of Fordist mass production and served to maintain the demand for labour. Wage increases subsequent to improving productivity allowed the realization of the product. Now, however, price stability has been disrupted by the international spread of Fordism. And since the NICs rely on external markets rather than on domestic consumption, the imbalance in First World countries between production and consumption is aggravated further. As Schoenberger (1988) points out, Fordism always had an international component but this did not throw it into disarray since multinational corporations went largely to other Fordist countries with rapidly expanding mass consumption bases.

There is, however, another argument. This too draws on ideas of globalization, but this time with respect to financial flows. The emphasis here is on the circumscribing of the macro-economic discretion of states as a result of an unparalleled surge in international financial transactions (*see*, for example, Stewart 1983, Webb; 1991). Wealth holders, it is believed, are using this power to shift their investments in bonds, securities, bank deposits and the like to other countries in order, in effect, to exercise a new control over the macro-economic objectives of states. Accordingly, maintaining the value of the currency against speculative shifts of capital overseas has become a major concern of states. This policy stance conflicts with measures aimed at reflating the economy – monetary or fiscal stimulation, for example – so as to alleviate the unemployment that has become so prevalent, especially in Western Europe. Policies of high interest rates and stable currencies have become *de rigueur*.

In my view, neither of these arguments can withstand careful scrutiny. They both assign a necessary role to the globalization of economic relationships that on closer examination is quite suspect. The first argument is suspect largely on empirical grounds. Growth in trade is less than dramatic. For the OECD countries of Europe, exports as a percentage of GDP were 19.5 per cent in 1950 and 26.4 per cent in 1991–2. It is true that there has been increased import penetration in manufactured goods, but, as Glyn has argued:

> greater competition within manufacturing has been offset (in terms of its effect on average import propensities) by the declining importance of manufacturing in the OECD's output and employment structure (27.4% of OECD value added in 1973, 22.2% in 1990) . . . the 'sheltered' sector of the economy (notably government and personal services) is probably growing. Such relatively modest increases in import and export shares could hardly constitute an independent explanation for slow growth in the world economy. If countries had no reason to hold back from Keynesian policies other than the effect on their payments balances, it is hard to see why coordinated macroeconomic expansion (at the European level for example) would not happen. (Glyn, 1995: 48)

The recent growth in international financial transactions, on the other hand, is something else. It *has* been quantitatively dramatic. But two qualifiers should be immediately inserted. The first is that there is nothing new about the particular constellation of events with which it has become associated. Prior to the 1930s and the introduction of policies aimed at reviving depressed national economies there was a similarly high level of international financial transactions. This exercised the same sort of external constraint over macro-economic policy via the Gold Standard (Notermans, 1997). But, and second, this does not mean to say that capital flight is the only possible response to concerns about the deteriorating value of a national currency. When faced with unpromising macro-economic environments, investors can also retreat out of money and into debt or speculative real assets. So, and *pace* contemporary thinking about the role of the globalization of trade in financial instruments, an isolated economy would not be free of the sorts of pressures experienced by economies undergoing serious inflationary pressures.

Rather, I would suggest that the crucial condition to be taken into account is not globalization, either of financial transactions or of production and trade, but the heightened levels of class struggle that developed during the 1970s[6]. This was a period of intense distributional conflict characterized by growing labour unrest, wage concessions that could not be justified in terms of rates of productivity increase, a profit squeeze, pressure on governments to pre-empt unemployment by credit expansion and, to the extent that governments yielded to that temptation, inflation.

Given such a policy context, there are two ways out: a tightening of credit sufficient to provoke unemployment and re-establish wage discipline; or the successful reassertion of a prices and incomes policy. The earlier success of what has come to be known as Fordism in preventing overheating of the economy resided precisely in the latter. This was achieved through various means ranging from legislative fiat to state-orchestrated corporatism. With tight labour markets, however, and the leverage they afforded to labour, these disciplines tended eventually to break down.[7] This meant that unemployment was to be the means of reasserting capitalist control and pushing back the threat to what Glyn (1995) has described as 'its prerogatives'. But unemployment policies are unpopular, particularly if exercised by a state that had defined full employment as a policy goal. As Notermans (1993, 1997) suggests, however, appeal to the high level of cross-border financial transactions, a level which those states themselves constructed through making currencies convertible, provided a legitimation – in effect by appealing to a renaturalization of the laws of motion of capital in the form of supposedly uncontrollable financial flows to which government policy had to adapt. Without appeal to some external pressure, policies that produce unemployment will be unpopular. It *looks* as if it is the level of international transactions which is causally determinant when what it really is is the desire to rein in inflationary pressures which have their origins within countries and, though Notermans does not make this point himself, in the desire to re-establish capitalist domination of the accumulation process.

The danger then, though, is that economies will enter into a spiral of a different kind, a deflationary one. As prices fall, capital shifts out of real investment into money. The demand for labour falls, and to the extent that labour accepts lower wages this aggravates the situation since it reduces the demand on the basis of which a reflation of the economy might conceivably take place. As business conditions worsen, banks have trouble collecting on loans and a run on them by fearful depositors becomes a real possibility. This, Notermans believes, if not in quite such stark terms, is the situation with which the advanced capitalist states are currently flirting. It is, moreover, precisely what led to the depression of the late 1920s and 1930s.

These sorts of policy will persist until failures to reflate the economy are regarded as more serious than the possibility of renewed inflation. At such moments in economic history, countries resort to restrictions on the convertibility of currencies, retreats from free trade and a search for floors to wage and price declines: minimum wages, agricultural price supports, enhanced welfare supports for the unemployed – precisely the sorts of thing that laid the basis for Fordism. But as Glyn has affirmed, that point has yet to be reached.[8]

# CONCLUDING COMMENTS

The politics of globalization elevates space to a key role. Changing space relations, in turn embedded in changing technologies, both of transportation and communication and of the immediate labour process, result in an enhanced mobility of capital. In a context of immobile workers this gives it an increased leverage over the terms of employment. The arguments regarding finance capital are similar, the frenetic level of international transactions impeding the implementation of expansionary macro-economic policies on behalf of particular national and immobile constituencies.

What seems to be at issue is what I will call here the overspatialization of social relations. This is the tendency to mis-specify the (contingent) nature of *some* spatial relations, at least, as necessary aspects of the social relations with which they are correlative: that, for instance, a search for cost-minimizing will necessarily lead manufacturers in the direction of Third World locations. In other words, while space is a necessary aspect of social relations, it does not dictate particular strategies, either for labour or for business. At their most abstract, social relations define certain necessities and opportunities: the need to make a profit, to earn a wage, the possibility of shifting employment, etc. But there is a considerable range of concrete socio-spatial practices and strategies through which those social relations can be realized. In many cases these are practices for which the necessary conditions in the form of changes in technology or organization have yet to be developed. Accordingly, there is always a variety of geographies that can come about. Globalization is one possibility among several; another is the industrial district. Some technologies can be deskilled, affording the possibility of relocation to the Third World or to some First World periphery. But many cannot be. And even if they can be, there may be compelling reasons for keeping the industry close to core locations of the First World, perhaps by constructing a low-wage labour force *in situ* out of suburban housewives or moonlighting high-schoolers.

Part of the reason for this overemphasis can be identified by distinguishing between a reality and a discourse aimed at becoming a reality. Globalization arguments are a form of discourse which is part of the contemporary class struggle. As Piven has suggested:

> The key fact of our historical moment is said to be the globalization of national economies which, together with 'post-Fordist' domestic restructuring, has had shattering consequences for the economic well-being of the working class, and especially for the power of the working class. I don't think this explanation is entirely wrong but it is deployed so sweepingly as to be misleading. And right or wrong, the explanation itself has become a political force, helping to create the institutional realities it purportedly merely describes. (1995: 108)

And as she adds later on in the same paper: 'Put another way, capital is pyramiding the leverage gained by expanded exit opportunities, or perhaps the leverage gained merely by the spectre of expanded exit opportunities, in a series of vigorous political campaigns' (p. 110).

In other words, and from a more general standpoint, what seems to be at issue here is a particular construction, a particular representation, of space. For one group of globalization theorists space is differentiated by pools of (immobile) labour of varying cost and between which, as a result of the deskilling of labour processes and changes in transportation and communication, capitalists enjoy powers of perfect substitutability. For another group, those more concerned with recent changes in macro-economic policy, space is differentiated according to the different rates of inflation that attach to different countries and between which another faction of capital, money capitalists, is able to enjoy similar powers of substitution as they move their money from one country to another. To paraphrase Piven, these representations have become political forces, helping to create the reality they are supposed merely to describe.

This is clearly not a representation for everybody. Capital, or at least important fractions of it, has tended to gain. In this way, it has been able to take control once more of an accumulation process which seemed to be eluding its grasp during the distributional conflicts of the 1970s. For it, the particular concepts of space represented by globalization were the right concepts at the right time. On the other hand, and as I have been at pains to point out, this representation stands in need of correction in light of defensible first principles regarding the pressures and opportunities which characterize capitalist space economies. And while not all those opportunities are opportunities for labour, neither do they run so uniformly to the advantage of capital. In material terms the bargaining context for labour is not nearly as one-sided as contemporary argument would have us, and most importantly the working class, believe.

## NOTES

1. As Cohen (1991) points out, this process has gone much further in the USA than in the UK. On the other hand, for evidence of how far down this road the UK has progressed, *see* Holloway (1987). For the way it has been struggled over, *see* Foster and Wolfson (1989).
2. Gill and Law (1989) write that 'the widening of the scope of the market in the 1980s and probably during the 1990s, along with certain changes in technology and communications, contributes to the rising structural power of internationally mobile capital' (p. 480). And: 'With regard to the structural power of capital, the key contrast at the international level is the relative mobility of capital and the relative immobility of labor in most sectors of activity' (p. 487). And according to Burawoy (1985: 150), 'The new despotism is the "rational" tyranny of capital mobility over the *collective* worker. The reproduction of labor power is bound anew to the production process, but, rather than via the individual, the binding occurs at the level of the firm, region or even nation-state. The fear of being fired is replaced by the fear of capital flight, plant closure, transfer of operations, and plant disinvestment' (p. 150).
3. For an example, *see* S.K. Smith's (1995) discussion of minimills.
4. For the case of organizing janitors, *see* the discussion of the Justice for Janitors campaign, whose strategy is to organize janitors in a single metropolitan area, in Hurd and Rouse (1989) and in Howley (1990).
5. This point has been made quite vigorously by Dunford and Perrons (1994). The work of Mackay (1994) on the British case, however, suggests that some caution might be required since the regional stabilizer effect may have been more durable than they argue.
6. I am not arguing, however, that more Fordist types of arguments, such as those of Schoenberger (1988) and Dunford and Perrons (1994), are oblivious to these types of claim. They do indeed tie the development of a New International Division of Labour and subsequent shifts in trading relations to failures of productivity relative to wages in the First World. What I *am* arguing is that subsequent shifts in location and trade, as per above, cannot account for the shift in macro-economic priorities from full employment to stable prices and exchange rates.

7. Not surprisingly, the reassertion of some form of prices and incomes policy or concertation of the different demands of labour and capital is now being stressed by more radical observers. In his discussion of the contemporary urge to maintain 'credibility' with international investors in financial instruments, Glyn writes, 'But maintaining such credibility only rules out the expansion of employment if there are no means other than unemployment for regulating conflicting claims over distribution and control. Viable policies for expanding employment entail costs which must be explicitly counted and willingly shouldered by the mass of wage and salary earners' (1995). This echoes Hirst and Thompson (1992: 373) regarding the importance of constructing a distributional coalition. In this case, though, they are more accepting of the arguments linking globalization to the emasculation of expansionary policies and see a new corporatism as underpinning the micro-economic changes necessary for employment growth. More specifically, they talk about the need for the state to construct a distributional coalition that wins acceptance of key economic actors and the organized interests representing them for a sustainable distribution of national income and expenditure which promotes competitive manufacturing performance.
8. Glyn has difficulty explaining the failure of business investment to revive, particularly given the fact that distributional conflict clearly died down during the 1980s. His conclusion is that it stems from 'Anxiety that a renewed period of high employment would lead to recurring inflationary pressures and threats to profitability', and that 'this underpins both the hesitancy of employers to invest and of government to expand' (1995: 47).

# CHAPTER FIFTEEN

# NOTES ON A SPATIALIZED LABOUR POLITICS: SCALE AND THE POLITICAL GEOGRAPHY OF DUAL UNIONISM IN THE US LONGSHORE INDUSTRY

ANDREW HEROD

Although they are not often recognized as such by many, labour relations are explicitly geographical in nature. Not only are industries, trade unions and labour markets organized spatially, but the question of where the locus of decision-making capabilities within them should lie (nationally or locally, for instance) is fundamentally one of the geographic scale at which power is located and exercised. A number of commentators have suggested that the recent trend in many advanced industrial economies, together with those economies in transition in Eastern Europe, has been one of labour relations' decentralization, with the focus of activity being shifted away from central or national employer, trade union and government bodies to regional or local institutions (*see* Martin *et al.*, 1994, for a good review of these arguments). Such decentralization is evidenced in the break-up of national agreements, efforts to outlaw secondary boycotts (so hindering national solidarity actions by workers), the passing to state and local governments of responsibility for regulating workplaces (*see* Herod, 1997a for more on this in the US context), and the development of collective bargaining agreements tailored to specific plant conditions. This process has led some to assume that employers prefer decentralized systems of labour relations because they allow management greater flexibility to develop individualized systems of local-level bargaining and contracts, and thus more easily to play workers off against each other. The corollary of this assumption is that workers logically should prefer to negotiate at regional or national levels as a means of preventing employers from whipsawing plants in a never-ending spiral of concessions. Other observers, in contrast, while arguing that employers prefer decentralized systems of labour relations, have assumed that rather than fighting to maintain a national organization and structure, logic instead demands that workers' organization follow that of their employers. Consequently, they suggest, trade unions would do best to decentralize their own organization to match that of their employers and the transfer of central state power to the local state.

It is not my purpose in this chapter to argue for normative strategies of trade union organization with regard to the issue of centralization versus decentralization, although I would maintain that despite the attention it has received, decentralization is in fact only one element of the current restructuring of labour relations occurring in the advanced capitalist world and that a second element, one much less explored in the industrial relations literature, is that of the movement towards the superconcentration and/or 'translocalization' of some elements of labour relations. (Such translocalization is evidenced by growing cross-border worker organization (*see* Herod, 1995); the implementation of internationally imposed labour and workplace standards by entities such as the European Union (the recent Social Clause and the European Works Councils initiative being examples); efforts to develop co-ordinated transnational bargaining strategies and practices in response to globalization (*see* Herod, 1997b for more on this); and the international activities of trade unions in, for instance, helping to reconstruct the trade union movements of the Eastern European economies (*see* Herod, 1998, for an example).) Instead, in this chapter I wish to examine through a case study an example of the actual construction of scales of labour relations and to suggest that simply assuming, rather than demonstrating, that employers or workers prefer to negotiate at particular geographical scales as a result of some internal organizational logic is problematic for a number of reasons related to geography.

First, such assumptions forget that whereas the development of national contracts by various trade unions may provide minimum standards below which the weakest local unions will not be permitted to fall and that consequently the destruction of such agreements opens the field to much greater variability of conditions throughout an industry, such centralized agreements may also restrain more powerful or militant local unions within the confines of concessionary national agreements. Indeed, the development of national agreements may actually be favoured by some employers in those regions where labour is stronger as a way precisely of limiting the local power of workers, while destruction of such agreements actually allows more powerful local unions greater freedom to negotiate contracts without the constraints of nationally imposed conditions. Likewise, those employers with higher wage costs may favour national agreements as a way of encumbering their competitors with the same high costs while low-cost producers may prefer local systems of collective bargaining to maintain their competitive advantage (*see* Herod, 1997c, for an example of this). Such examples raise the question of how contingent factors, which vary geographically, may shape pressures towards centralization or decentralization, national or local control.

Second, such arguments fail to see the connection between different scales and the fact that decentralization, for instance, may sometimes be achieved only through the centralization of authority. In the USA, although the process whereby local unions may form is a highly decentralized one (Clark, 1988), it is a process in fact established and regulated by federal, that is to say national, labour law. Equally, the right of individual states to pass anti-union Right to Work laws is enshrined in a federal piece of labour law, the 1947 Labor-Management Relations ('Taft–Hartley') Act. Similarly, current efforts to undermine the federal Occupational Safety and Health Administration's power to regulate workplaces from Washington, DC, are involving dramatically increased Congressional oversight to overburden the system so completely with red tape that regulation becomes, in effect, impossible and must be monitored, instead, by in-house committees in individual workplaces (*see* Herod, 1997a).

In what follows, then, I present a case study of conflicts over the construction of the scale of representation and collective bargaining in the US longshore industry as a means of making some general observations about the central problematic of geographical scale in labour relations. In particular, I wish to move the debate away from whether or not there is some internal logic of geographic scale which best suits workers' or employers' organization and, instead, towards the issue of how such scales are produced as a central part of the politics and geography of labour relations. Efforts to decentralize or to centralize decision-making processes are designed to remake the geographical scale of labour relations and reflect the changing contexts within which employers, government and workers make decisions. Indeed, geography is crucial to understanding labour relations. While local economies may be booming, national ones may be in recession. While national political conditions may favour employers, local political conditions may be more favourable to workers. Political tensions between local and national union leaderships (or, as we shall see below, between different unions) may cause each to want different scales of bargaining. Such contingent geographical variation fundamentally shapes the political practice of employers, workers and their organizations. In the USA, unions have typically sought to come to terms with such geographical variability by developing multi-scalar contracts in which some conditions are determined by a national or regional 'master' contract and others are determined by a local contract, a structure which allows the incorporation of certain minimum standards yet also allows for some flexibility to respond to local issues. However, even within such a multi-scalar structure, the question of what gets negotiated as a national or a local issue is politically contested and informed by geography.

Frequently, workers are seen as rather passive elements in determining the geography of labour relations. Decentralization, for example, is often portrayed as the result of corporations' efforts to adopt more flexible systems of production and employment relations, and of the state's effort to dismantle national regulations believed to limit the ability of capital to take advantage of local opportunities. This portrayal, perhaps, is a result of the dominant view within economic geography (held by many neo-classical and even more critical writers) that labour is little more than a 'factor' in the calculus of economic location and industrial organization, and that the primary shaper of the economic landscape is capital (*see* Herod 1994, 1997d, for more on this). In this chapter, in contrast, I provide a case study to illustrate the significance of workers' struggles to construct geographic scale as part of a generalized spatial praxis. In particular, I examine the political conflicts to represent waterfront workers waged between two rival dockers' unions – one the dockers' traditional representative, the other a new union – in East Coast ports of the USA between 1953 and 1959. In the following narrative, I show how the two unions sought to construct radically different geographies of representation and collective bargaining as part of their political struggle. Indeed, the conflicts between the two unions centred precisely on the issue of the geographical scale at which representation of workers and bargaining with employers would take place; that is to say, whether they would continue to be conducted on the traditional port-by-port basis (as the new union wanted) or on a new regional or national basis (as preferred by the old union). The outcome of these conflicts would dramatically change the economic and political geography of the industry. As the example rendered below illustrates, not only is geographic scale socially made in a very real sense but the business of making such scale is a deadly serious business indeed.

# FORGING SCALE: THE SPATIAL PRAXIS OF TWO LABOUR UNIONS

In August 1953 the American Federation of Labor (AFL) expelled from its ranks the International Longshoremen's Association (ILA), which represents dockers in East Coast, Great Lakes and Pacific North-West ports (*see* Barnes, 1915; Larrowe, 1955; Hoffman, 1966; Russell, 1966; Kimeldorf, 1988, for more on the union's history and geographical organization). The primary reason for this action was that the union's leadership had engaged in a host of corrupt practices which, the AFL argued, warranted expulsion from the Federation. Having expelled the old ILA (which now began calling itself the ILA-Independent (ILA-IND)), the AFL sponsored a new union to replace it, the AFL-ILA (later also referred to as the International Brotherhood of Longshoremen (IBL)) (for more on this, *see* Rosenbaum, 1954; Larrowe, 1955; Hutchinson, 1970; Robinson, 1981). These two acts began a period of bloody labour warfare on many waterfronts throughout the East Coast but particularly in New York, where recent government investigations had revealed gangsterism on the part of many of the old ILA's officials (*see* New York State, 1953; Dewey Commission, 1953). However, the expulsion from the Federation of the old ILA did not automatically give the newly formed AFL-ILA the right to represent waterfront workers in contract negotiations with the employers. To replace the ILA-IND as the dockers' bargaining agent, the AFL union would first have to win a National Labor Relations Board (NLRB) representation election designed to indicate waterfront workers' union preference. Yet this was not a straightforward matter. Because of their conflicting political agendas, the two rival unions lobbied to have geographically quite different units designated by the NLRB as appropriate for union representation and collective bargaining purposes. As a result of these conflicting pressures brought to bear on it, before any representation election could take place the Board first had to determine which waterfront workers were even entitled to vote in such a contest.

## Debates Concerning the Geographical Extent of the Representation Unit

In the midst of the political turmoil between old and new unions in New York, two issues dominated the struggle to construct the geography of participation and exclusion in future waterfront collective bargaining negotiations. The first was the question of which waterfront job classifications were to be included in the representation unit (for more on this, *see* Herod, 1992). In particular, the ILA-IND preferred a broader definition that included many categories of waterfront workers whereas the AFL-ILA argued for a more narrowly defined unit. The second issue, and what is the focus of this chapter, concerned the geographical scope of such a unit and whether it should be confined to New York or, instead, constituted on a broader, multi-port basis. Both the ILA-IND and the AFL-ILA recognized that primarily the contest to represent waterfront workers nationally would be won and lost in New York. New York was the nation's busiest port, the heart of the old guard's organization. AFL officials calculated that if the new union could capture New York, then the job of persuading dockers in other ports to leave the ILA-IND would be made that much easier. The ILA-IND also realized the strategic importance of holding on to New York, both from the political perspective that it would lose thousands of its members if the port's dockers chose to be represented by the AFL union, and because of the psychological impact this would have along the coast. The employers' association (the New York Steamship Association (NYSA)), too, preferred a unit confined to New York, for this would allow it to continue negotiating contracts locally. Whereas the AFL's political and geographical strategy sought to split the ILA-IND spatially by driving a wedge between New York and other ports, thereby allowing the old guard to be confronted in piecemeal fashion, the ILA-IND determined to ensure that New York remained integrally connected to the union's larger coast-wide organization. Hence, in the subsequent political struggle between the rival unions and between the unions and their employers to convince the NLRB as to what was the most appropriate bargaining unit, the AFL would continually stress New York's insularity while the ILA-IND's strategy hinged on maintaining that union–management relations had historically been coast-wide in nature.[1]

Understanding why the issue of whether or not labour relations should be conducted on a single- or multi-port basis became so central to struggles in the industry requires an initial examination of the contested tradition of bargaining between the old ILA and the NYSA. Historically, bargaining over the contract for the Port of New York had been conducted between the ILA's Wage Scale Committee and the NYSA's Wage Scale Conference Committee.[2] The ILA Wage Scale Committee was composed of approximately 120 delegates from all ILA locals along the coast from Maine to Virginia. This constituted the US contingent of the ILA's Atlantic District (although the district stretches into Canada, the Canadian ports negotiate with employers on a different basis from their US counterparts).[3] ILA committee members would first hold preliminary discussions among themselves concerning what gains they hoped to secure from the employers in the upcoming contract negotiations. Once a proposal had been agreed upon by the delegates, it would be presented to the NYSA. Some kind of compromise agreement would then be thrashed out. This would then be formally voted on by the Wage Scale Committee and, if accepted, presented to the *entire* US membership of the Atlantic District for ratification or rejection (Connolly, 1956).

The NYSA's Wage Scale Conference Committee generally consisted of 15 individuals (12 members representing steamship lines, supplemented by 3 contracting stevedores). Unlike the ILA's deputation, which was drawn from ports throughout the North Atlantic, all the NYSA representatives were based in New York. Herein lay the source of the conflict. Seeking to reinforce its claim to multi-port bargaining, the ILA-IND argued that because its Wage Scale Committee was made up of union delegates from throughout the US ports in the Atlantic District, all of whom had equal say with the New York representatives when it came to negotiating the details of the contract, any contract negotiated for New York was equally applicable to other North Atlantic ports (ILA-IND, 1956). Even the union's own constitution (Article XXI, Section 5), officials pointed out, prevented it from reaching any agreement with the NYSA unless a majority of its Atlantic District membership voted in favour of the agreement hammered out in New York. ILA-IND officials maintained that although some issues pertinent only to particular ports were discussed on a local level, the contract's core elements – dealing with such basic issues as wage rates, hours to be worked and the general terms of the contract – had been negotiated on a multi-port basis at least since the NYSA's inception in 1932.

Furthermore, the union submitted that increases and adjustments agreed upon in the North Atlantic Coast negotiations were also invariably embodied in the South Atlantic and Gulf Coast agreements, with employers in these ports 'wait[ing] for New York or the [North] Atlantic Coast district to finish their negotiations, and then [offering their own workers] the same thing' (Connolly 1956: 249; *see also* the statements by Gayle, 1963). Pointing out that NYSA members controlled some 85 per cent of the Atlantic and Gulf Coast shipping business, ILA-IND officials insisted that the New York employers were in a position to speak for the whole shipping industry with regard to contract negotiations. Clearly, they argued, such evidence compelled no less an NLRB finding than 'that the appropriate bargaining unit embrac[e] all the ports from Portland, Maine to Brownsville, Texas, or, alternatively, from Portland . . . to Hampton Roads, Virginia' (ILA-IND, 1956: 12).

The NYSA, opposed to the ILA-IND's attempt to claim for itself a contract covering at a minimum the North Atlantic coast from Maine to Virginia and potentially the entire East Coast, vehemently contested this version of the longshore industry's collective bargaining history. In particular, the NYSA did not wish to concede to the union the prize of multi-port bargaining, a concession which would have prevented employers from taking advantage of different wage rates and working conditions operating in different ports, or of shipping through other ports in the event of a strike in New York (since a strike over a multi-port contract was more likely to be enforced in other ports than would a strike over a contract applicable only to New York). Whereas the ILA-IND contended that the inclusion of delegates from all the Atlantic District ports on its Wage Scale Committee sustained its position concerning multi-port bargaining, the NYSA (1953: 8–9) argued that such delegates were admitted to the negotiations only as 'a matter of courtesy' and that therefore 'the mere presence of Union representatives from the other ports is . . . immaterial'. NYSA executives maintained that although they may occasionally have sought advice from employers' associations in other ports, at no time did the NYSA's Conference Committee negotiating body 'have any representatives of any employers from any port except the Port of Greater New York'. Nor did the NYSA accept the ILA-IND's contention that the wage scale agreed upon for the Port of New York was automatically accepted by employers in other ports. Indeed, executives maintained that although employers may have given similar wage increases in different ports, each port along the coast was in fact covered by an *individual* contract agreement signed between that port's local unions and the appropriate employers' associations, associations which bore 'no relationship to [the] NYSA'. The NYSA argued that the weight of the evidence clearly demonstrated that it had never bargained on a multi-port basis.

The AFL-ILA, which was also seeking to confine the bargaining unit to New York as a means to geographically split the old ILA and so facilitate its own challenge, echoed this position (IBL, 1956). AFL-ILA officials emphasized the distinctiveness of the agreements covering the different ports, pointing out that employer representatives from New York did not go to Philadelphia, for example, to negotiate over local conditions, but that instead such negotiations were conducted locally by the appropriate employer association in the port. The AFL union sustained that the evidence produced by the ILA-IND did not suggest a history of co-ordinated multi-port bargaining. Rather, it represented little more than a system of pattern bargaining in which similar agreements were sometimes executed in other ports after the negotiations had been concluded in New York.

Clearly, whether implicitly or explicitly recognized as such by the participants, this was a struggle to shape the geographical scale of union representation and hence contract bargaining. All three parties – the ILA-IND, the AFL-ILA and the NYSA – sought to present the particular history of bargaining in the industry in such a manner as to bolster their own arguments before the NLRB, the body which would ultimately decide the representation matter based on the evidence presented. The resolution of this issue took on crucial significance. Were the ILA-IND to prevail, not only could the old guard use its national organization to shake off the AFL-ILA challenge more easily, but it would also be able to negotiate as a single entity with the various employers' associations along the coast, something which might lay the groundwork for ultimately developing an industry-wide contract. Furthermore, a coast-wide representation unit might give International officers in New York a mechanism to control rebellious union factions in other ports. If, on the other hand, the AFL-ILA and the NYSA prevailed and the NLRB ruled against a coast-wide representation unit, the ILA-IND would face a tougher representational fight with its AFL rival. Additionally, rejection of the coast-wide unit would hamper the union's efforts to equalize conditions throughout the industry by requiring each port's employers' association to adopt the same basic contract terms.

## The NLRB Decides

On 16 December 1953 the NLRB issued its decision. With regard to job classifications, the Board ruled in favour of a more broadly constituted unit. With regard to the unit's geographical scope, however, the Board found that 'the record shows that the [NYSA] has not bargained for employers in any port other than the Port of New York' (*New York Shipping Association*, 107 NLRB 364 [1953], p. 368). Even though the wage scale arrived at in the New York negotiations may have set a pattern for other ports, the Board determined, this did not alter the fact that bargaining between the NYSA and the ILA-IND had been conducted on a Port of New York basis only. As a result, the NLRB confined the unit to the Port of New York and vicinity. With the unit issue seemingly resolved, the Board ordered a representation election to be held on 23 and 24 December 1953 for the Port of New York, an election designed to settle the question of which union – ILA-IND or AFL-ILA – would represent the port's dockers.

The New York waterfront had remained tense ever since the summer. In some sections of the port, ILA-IND pickets refused to work with AFL dockers and closed down piers, both to pressure the NLRB to rule in its favour and to demonstrate that the rival union's inability to keep piers working was a sign of how little support it really had in the port. Countering assertions that the AFL union lacked support, in October 1953 Federation President George Meany had claimed that some 9 000 New York dockers (out of a total of about 40 000) had joined the new union (*Journal of Commerce*, 1953b). This war of words soon erupted into physical confrontations. In the struggles to control the piers, scuffles between the two rivals became commonplace. Whereas the ILA-IND sought to shut piers down, AFL-ILA dockers on the other hand fought to keep them working, hoping thereby to show that the AFL had sufficient support to break the old union. On the actual days of the election, three AFL-ILA organizers were stabbed in Brooklyn, while several reputed waterfront mobsters were seen loitering around the ballot-boxes wearing large ILA-IND buttons, well within the 'no electioneering' area. Although the ILA-IND received the largest number of votes, the AFL quickly petitioned the NLRB to overturn the results because of the violence surrounding the election.[4] This the Board did, ordering that a second election be held in May 1954.

## Of Mergers and Affiliations: Towards a New Geography of Union Representation?

When the AFL-ILA had first challenged the old union power structure, it seemed that dockers might potentially be split between two separate unions, with some ports remaining loyal to the ILA-IND while others elected to rejoin the AFL fold. However, as the ILA-IND and the AFL-ILA each sought out allies to buttress their own positions, a new dynamic began to shape the political geography of union representation and bargaining in the longshore industry. Strapped financially and isolated from other AFL unions, ILA-IND officials looked to explore the possibility of a merger with John L. Lewis' United Mine Workers of America (UMWA), who themselves appeared to have waterfront ambitions, to build a political (and geographical) alliance capable of defeating the AFL-ILA challenge. If brought to fruition, affiliation with the miners' union would dramatically change the geography of representation on the East Coast and significantly affect the future course of the ILA-IND's attempts to secure a coast-wide collective bargaining agreement.

The first indication that such a merger might be afoot had come in late November 1953 when the ILA-IND announced that its top officials had held preliminary talks concerning a potential amalgamation with the catch-all District 50 of the UMWA. On 21 December ILA-IND President William Bradley stated that he would recommend to the union membership that it vote to affiliate with the UMWA (*see Journal of Commerce*, 1953a). There was even talk that the ILA-IND and the UMWA (which had itself split from its own governing body, the Congress of Industrial Organizations) would join forces to form a new labour federation financed largely by Mine Workers' money (*Journal of Commerce*, 1954). Such speculation took on added meaning when, in late January 1954, UMWA organizers arrived on the New York waterfront.

The geographical significance of the UMWA's entrance to the waterfront quickly became apparent. The ILA-IND had originally turned to the UMWA to provide political backing (and a $50 000 loan) which it hoped would help frustrate the AFL's ambitions and ensure that a single union (the ILA-IND) continued to represent all East Coast dockers. But with jurisdictional battles continuing in New York, many ILA-IND officials began to worry that South Atlantic and Gulf Coast locals might decide to abandon New York and the North Atlantic District, thereby splitting the old organization in two. Ensuring that the southern ports did not desert the union thus became a top priority for the ILA-IND hierarchy. As part of their strategy to maintain the union's coast-wide organization, International officers in New York sought to persuade locals in the South Atlantic and Gulf ports to follow their lead and join with the UMWA (*New York Times*, 1954). However, southern dockers' fears that their ports would be tied up by the continuing jurisdictional battles being waged in New York, combined with attempts by the AFL to woo them back to the Federation, left many unwilling to contemplate such a merger. Increasingly it seemed that the North Atlantic ports would affiliate with the UMWA while many South Atlantic and Gulf ports were leaning more towards the AFL.

But even in the North Atlantic trouble loomed. In its effort to outmanoeuvre the AFL, the ILA-IND leadership appeared to have fallen foul of its own plans. In January 1954 ILA-IND officials had decided to allow dockers in the other North Atlantic ports to break with tradition and conclude their contract negotiations before New York. This, they hoped, would preserve unity by placating outport dockers' fears that their settlements would be held up by the jurisdictional questions in New York. By the end of March, however, there were indications that having already settled their own contracts, these dockers would simply abandon New York and form an independent national longshore union if the representational matter were not solved soon (Burke, 1954). Whereas the ILA-IND leadership had originally considered affiliation with the UMWA as a means to consolidate a single longshore union from Maine to Texas, it appeared that the resolution of this political struggle could perhaps now leave *three* longshore unions: the AFL-ILA in the South, the ILA-IND (affiliated to the UMWA) in New York, and an independent North Atlantic union (minus New York). Not only might this geographical fragmentation spark intensified wage competition, but it would also hinder the three unions' ability (should they so wish) to develop co-ordinated national agreements.

There was clearly a certain geographical irony to these developments. The ILA-IND had turned to Lewis and the UWMA to provide the financial and logistical support necessary to defeat the AFL and prevent the industry's geographical division between the two rival unions, and yet this very strategy might actually leave dockers splintered between three unions. For a short time confusion reigned. Only in late March 1954 did representatives of 103 South Atlantic and Gulf Coast locals (with a combined membership of 17000) finally reject the AFL's overtures and elect to stay with the ILA-IND. Although rumours about the potential break-up of the union continued until the ILA-IND's victory in the second NLRB election (held in New York on 26 May 1954), the decision of these southern dockers to reject the AFL and remain loyal to the old ILA was enough to persuade the other North Atlantic ports not to abandon New York.[5] Despite the help provided by Lewis and his Mine Workers' union, the ILA-IND's success in the second NLRB election was enough to dissipate the immediate threat posed to the union and obviate the political and geographical urgency of merging with the UMWA.

## Multi-port Bargaining and the End of the Threat of Dual Unionism

While seeking to affiliate its locals along the East Coast with the UMWA had been one (spatial) strategy the ILA-IND considered as a means to outmanoeuvre its AFL challenger, a second geographic strategy involved developing a national multi-port contract to replace the port-by-port system of bargaining which had traditionally been the norm for the industry (for more on this, *see* Herod, 1997c). In particular, the ILA-IND argued that while the NLRB's 1953 decision might have prevented the union from seeking a coast-wide *representation* unit, it did not preclude it from continuing to press for multi-port *bargaining*. Although in the 1953–4 negotiations the ILA-IND was unsuccessful in this regard, during the 1956 contract negotiations the union again sought such an agreement as a means to achieve two goals. First, the union's success in developing a regional or industry-wide contract would reduce the ability of employers to play workers in different ports against each other on the basis of different wage rates and working conditions. Second, and more directly related to the continued challenge it was facing from the AFL-ILA, if the union could negotiate such a multi-port agreement it would greatly improve its political position *vis-à-vis* its rival because for the immediate future dockers along the coast would be covered by the ILA-IND contract. Given the NLRB's usual policy of waiting for a contract to expire before allowing a rival union to petition for a representation election, a multi-port contract would severely undermine the AFL-ILA's chances of being able to force representation elections throughout the industry. Thus, attempting to build a regional or national contract was a consciously thought out spatial strategy designed to eliminate the political challenges the ILA-IND faced from the new AFL union.

Despite its defeat in the May 1954 representation election, the rival AFL union (now generally referred to as the International Brotherhood of Longshoremen (IBL)) had remained a force on the New York waterfront and in July 1956 filed with the NLRB for a third representation election to be held in the port. In a replay of its 1953 stance the ILA-IND once again tried to show the Board and the employers that it had historically negotiated on a coast-wide basis and that therefore a multi-port representation unit including all ports from Maine to Texas (or at the very least to Virginia) was the most appropriate for the industry (ILA-IND, 1956). The AFL union, on the other hand, maintained that the election should be restricted to the Port of New York, the representation unit mandated by the NLRB's earlier decision (IBL, 1956). After hearing opposing arguments, the Board soon reaffirmed its 1953 decision concerning the geographical extent of the bargaining unit and ordered a third representation election be held in New York, an election which the ILA-IND won handily.[6] While the NLRB's decision meant that the ILA-IND could not hope to gain a multi-port *representation* unit, union President Bradley asserted (*Journal of Commerce*, 1956c) that the substantial margin was an overwhelming 'mandate from [the] membership to seek only [a collective bargaining] agreement that will include all ports from Portland . . . to Brownsville'. This the union proceeded to do by ordering dockers from Maine to Texas to strike in mid-November 1956.

At first, the employers' associations beyond New York refused to concede to the ILA-IND's demands for coast-wide bargaining for fear this would legally tie them into the historically more generous agreements negotiated in New York. The NYSA, too, resisted what Chairman Alexander Chopin called the union's demands 'to coerce the employers into accepting national bargaining', arguing that it had no authority to bargain outside New York (quoted in *Journal of Commerce* 1956b). However, in mid-September ILA-IND officials announced that they had granted charters to five Great Lakes and inland waterways locals. Recognizing that the St Lawrence Seaway (scheduled for opening in 1959) would play a crucial role in future freight movements in and out of the Midwest, and fearing that shippers might divert business from the

North Atlantic to ports on the Great Lakes, where wages were lower, ILA-IND officials demanded that any contract negotiated for New York not only include all ports on the East Coast but now be extended to those on the Lakes (Gleason, 1955: 824; *Journal of Commerce*, 1956d). Not only would this significantly increase the scope of any future multi-port agreement, but it represented an incursion into the heartland of IBL support where many old ILA locals had affiliated with the AFL union.[7] Coming so quickly on the heels of the ILA-IND's recent announcement that it was seeking an agreement with the West Coast International Longshoremen's and Warehousemen's Union (ILWU), which could possibly lead to a common contract expiration date (and hence simultaneous strikes) on both coasts, ILA-IND leaders had significantly changed the geographical parameters of the political struggle over multi-port bargaining and with the IBL. The political stakes were now higher than ever.

Concerned that any strike over multi-port bargaining would be costly, the NYSA finally agreed on 11 January 1957 to offer the union a contract which would cover the ports from Maine to Virginia. The union, on the other hand, continued to push for an agreement covering the entire coast to Texas. Only after several months of continued haggling did it become obvious that while the union had the power to secure a multi-port contract in the North Atlantic, for the moment it did not have the political power to force South Atlantic and Gulf employers to agree to a national contract; that would have to wait until later (*see* Herod, 1997c, for more on the union's securing of such a national contract after several more years of struggle). Finally, the ILA-IND agreed to accept the new North Atlantic multi-port contract covering five items, and the master agreement was ultimately put into place in December 1957.[8]

Securing the master contract and constructing a new scale of bargaining in the industry was a significant coup for the ILA-IND, for it enabled the union to do two things. First, it was effectively able to shut out its rival. Dockers in ports that were signatories to the master agreement now all had ILA-IND contracts, as did those in South Atlantic and Gulf ports, which, while not actually parties to the master contract, nevertheless adopted its provisions and also signed their own ILA-IND contracts. Indeed, having failed to win the votes of New York dockers in three representation elections and unlikely now to be able to raise the representation issue anywhere along the East Coast for another three years until the ILA-IND contracts expired, the IBL held its final convention in Milwaukee in October 1959, voting to disband itself and merge with the old ILA (which had been re-admitted to the newly formed AFL-CIO in August 1959). Second, through the master contract the ILA-IND was able to begin reducing some of the inequalities in conditions between dockers' wages, benefits and conditions of work in the various East Coast ports. Consequently, the union could now limit the employers' ability to play dockers in different ports off against each other on the basis of wages, etc., thereby removing this geographical strategy from their arsenal. Furthermore, the fact that the union had negotiated a legally enforceable master contract in the North Atlantic meant that dockers in one port could now strike in support of dockers in other North Atlantic ports without fear of injunctions under the secondary boycott provisions of the federal Taft–Hartley Act.[9] Evidently, the production of a new scale of bargaining in the industry represented a significant political and economic achievement on the part of thousands of dockers, an achievement which would have very real consequences for their wages, conditions of work and political power *vis-à-vis* employers.

## IMPLICATIONS FOR A SPATIALIZED LABOUR POLITICS

Although it may seem a somewhat strange concept, there are several reasons why recognizing that geographic scale is socially constructed is crucial for articulating any kind of spatial praxis and spatialized politics. First, as Neil Smith (1984) has argued, it is scale which gives coherence to notions of uneven development, by allowing us to delineate, for instance, between 'more developed' and 'less developed' or 'underdeveloped' parts of the economic landscape. Scale is writ large in the landscape as it divides space into its more and less developed parts. Understanding how scale is produced thus provides important insight into the uneven development of capitalism. Second, spatial ideologies are often expressed through the creation of particular scales of feeling or identification. Notions of localism, nationalism, xenophobia or anti-outsiderness all rely on the ability to delineate the landscape and/or groups of people between an 'us' and a 'them', a process which itself is often based in the ability to construct an identification with some spaces which must be clearly delineated from other spaces. Hence, Mitchell (1998) shows how agribusiness interests in 1930s California sought to create a localist ideology in which they portrayed themselves as legitimate 'local' operators (despite the fact that many were organized nationally if not internationally) whereas union organizers among farm workers were portrayed as 'outsiders' or even 'non-local' (despite the fact that many actually came from the very communities which were being organized). Third, by constructing scale in particular ways, social actors can constrain enemies and facilitate the activities of friends. For instance, recent struggles in South Africa between the African National Congress, which has a wide geographic base of support and which favours a strong central government, and Gatsha Buthelezi's Inkatha Freedom Party, whose support is more geographically concentrated and which wants a system with strong regional government, have been precisely struggles over the scale at which political power should be articulated. Likewise, workers' efforts to expand the scale of conflict in industrial disputes are invariably designed to allow them to tap into a larger resource base and make important political connections with workers living in different communities or even different countries. As we can see, when viewed this way the making of geographic scale suddenly takes on grave political significance.

The expulsion of the ILA from the AFL and the Federation's sponsorship of a rival union spawned myriad spatial struggles over the geography of representation and collective bargaining in the longshore industry. These were evidenced in many ways: as struggles to construct the geographical scale of representation (coast-wide versus port-by-port units); through changed geographical strategies (such as the ILA-IND's January 1954 reversal of the traditional order of contract bargaining in which New York led the pack); the ILA-IND's efforts to develop a multi-port regional and/or national contract; and with the potential geographical splintering of dockers into at least three separate unions. However, these were not simply political struggles that just happened to have a particular geography to them which was impacted by the struggles themselves in some secondary manner. The changing spatial structure of the industry was not simply an innocent outcome or a naively given consequence of the political struggles between the two unions and between the unions and the employers. Rather, these political and economic struggles were at their very core *geographical* struggles. They were efforts to restructure the very geography of the industry. They were efforts to make the geography of contract bargaining in the industry in particular ways so as to facilitate the pursuit of particular political and economic goals, these being in the ILA-IND's case the elimination of competition from a rival union and the reducing of inequalities in wages and working conditions experienced by dockers in different ports, in the AFL-ILA's case facilitating its challenge to the corrupt old guard, and in the case of the employers preventing the union from developing a multi-port contract which would limit shipowners' abilities to play different ports off against each other.

The fact that these struggles were at heart *simultaneously* both political and geographical raises some interesting questions concerning workers' spatial praxis. Certainly, it suggests that economic geographers should pay much greater attention to the spatial practices of workers and how these practices shape the geography of capitalism. But more than that, it suggests also a need to theorize more intimately the role played by the production of space in the political praxis of workers. Workers have a vested interest in making space in certain ways. They must ensure that the landscape is made in such ways that they are able to reproduce themselves as social beings both on a day-by-day basis and on a generational basis – as a landscape of work and not of unemployment or as one of union power rather than impotence, for example.

Conceiving that workers have a vested interest in making the landscape in some ways and not in others allows us to begin to consider workers' political practices under capitalism as both intimately geographic and crucial to understanding the making of the geography of capitalism. It allows us to begin to talk about labour's spatial praxis and the efforts by workers to make what we might call, after Harvey, 'labour's spatial fix'. It is, in other words, a recognition that workers strive (not always successfully) to make space in their own image just as capital seeks to represent 'itself in the form of a physical landscape created in its own image' (Harvey, 1978: 124). Clearly, this is not to say, however, that both labour and capital are equally capable of making landscapes as they wish, for this would be to ignore the contingent realities of power between the two which fluctuate historically and geographically. But, on the other hand, it is to say that we should not simply assume that the economic geography of capitalism reflects the whims of capitalists. As the dockers on the East Coast have shown, it is possible to take on capital politically and geographically – and to win. This fact has significant implications for theorizing the politics of the production of space.

Equally, we should also not assume that all workers have a vested interest in making space in the same way simply because they are workers in a particular class relation to capital. Trade union organization is a messy process, in no small part because of the geographic variability which workers and union officials face in attempting to implement particular strategies and structures. The conflicts between the ILA-IND and the AFL-ILA were manifested largely in their different spatial visions for the industry and the geography of representation and collective bargaining in it. This itself raises questions about how geography is implicated in class practices, and how the production of space both shapes and is shaped by class forces. We must think of class formation and class processes as geographical in the first rather than the last instance. Hence, for example, some workers' geographical fixity in space may lead them to engage in civic boosterism or give contract concessions to defend their jobs in particular places even while this may serve to divide workers across space (cf. Cox and Mair, 1988; Herod, 1991a). Likewise, processes of building solidarity between workers are inherently geographical, involving creating links between workers in different regions or even countries (cf. Herod, 1995) and the spatial translation of union culture and tradition across the landscape (Wills, 1995).

Evidently, the ability to make the economic landscape in some ways and not in others not only is highly contested, but confers on the makers potentially great social power. In recognizing this fact the landscape must be thought of as an active social entity in the playing out of the political geography of capitalism and class relations. Labour struggles do not simply occur *in* or *over* space. Rather, *they actively shape that very space*. Such a realization requires that we see workers' activities not simply in terms of the labour *histories* which they create, but also in terms of their labour *geographies*.

## NOTES

1. When it challenged the ILA-IND in New York in 1953, the AFL-ILA did not file any formal representation petitions in the other East Coast ports, although some of its members were active in these ports. The new union instead preferred to focus its resources first and foremost on the representation struggle in New York (Board of Inquiry, 1953).
2. The NYSA was formed in 1932 as the bargaining agent for New York waterfront employers. In the early 1950s it was made up of about 160 members who included steamship lines, stevedores (direct hirers of dock labour), and cargo repair, checking, clerking and maintenance firms. During this period the steamship lines dominated the organization. However, in 1971 the NYSA was reorganized and the stevedoring interests gained control (*see* Herod, 1992).
3. The ILA was historically divided into four districts. These were the Atlantic District (covering ports from Canada to Virginia); the South Atlantic and Gulf District (covering ports from North Carolina to Texas); the Great Lakes District; and the Pacific Coast District. In the industry the terms 'industry-wide', 'national', and 'East Coast' are often used interchangeably, whereas the term 'coast-wide' is usually used to refer to the area covered by each of these four districts. However, it is also sometimes used to refer to the entire East Coast. For sake of clarity, unless referring to proper names or direct quotes where they are used otherwise, in this chapter I use the terms 'industry-wide', 'national', 'coast-wide', and 'East Coast' interchangeably. I also refer to ports in the Atlantic District as 'North Atlantic ports' so as to distinguish more clearly which part of the coast I am referring to. On the West Coast, ports had originally been organized by the ILA. However, during the 1930s the more militant West Coast dockers became increasingly disillusioned with the ILA leadership in New York and finally left the union in 1937 to form the International Longshoremen's and Warehousemen's Union (*see* Kimeldorf, 1988, for a comparison of waterfront unionism on the East and West Coasts).
4. The results of the 23–24 December 1953 election were: 9060 votes for the ILA-IND; 7568 for the AFL-ILA; 95 for no union; and 4405 challenged.
5. The results of the second election were: 9110 votes for the ILA-IND; 8 791 votes for the AFL-ILA; 51 votes for neither union; 49 void votes; and 1797 challenged votes. For more on these elections, *see* Jensen (1974: 125–35).
6. The vote count was ILA-IND 11327 votes to the IBL's 7428.
7. The Great Lakes formed the backbone of the IBL's support, providing about 8000 members for the new union. Within 18 months of the union's formation, all ILA Great Lakes locals (with the exception of those which had earlier amalgamated with UMWA District 50) had joined the IBL (Larrowe, 1959).
8. Under the terms of the newly negotiated master contract the NYSA was authorized by employer associations in Boston, Philadelphia, Baltimore and Hampton Roads to negotiate on their behalf with the Atlantic District of the ILA-IND on issues of basic wages, hours of work, length of contract, and employer contributions to welfare and pension funds (but not the benefits). Once the master contract terms had been settled, each individual port would then negotiate on a local level over such items as holidays, vacations, working conditions, gang sizes, and pension and welfare benefits.
9. Although they had often ignored such injunctions, after 1947 when the Taft–Hartley Act went into effect, ILA-IND locals in separate ports had been unable legally to strike on each other's behalf because they were considered separate bargaining units by the NLRB and so such strikes represented

illegal secondary boycotts under the terms of the Act. With the negotiation of the master contract, however, all ILA-IND local unions in the North Atlantic were now signatories of the same regional contract and so could legally strike over issues related to it without fear of injunction, although sympathy strikes over 'local' contract issues were still considered illegal.

## CHAPTER SIXTEEN

# LOCAL FOOD/GLOBAL FOOD: GLOBALIZATION AND LOCAL RESTRUCTURING

**ROBERT FAGAN**

In recent years, the concept of 'globalization' has become one of the most powerful characterizations of contemporary economic restructuring. Competing theses about globalization have arisen since the early 1980s and these contain within them different ways of conceptualizing relations between economy, society and place. Theories among those most influential in economic geography have pointed to the emergence of an international economy underpinned by information technology – a new 'informational capitalism' (Luke, 1994: 619) involving 'space–time compression' (Harvey, 1989a). According to these theories, economies have become organized increasingly around a genuinely global financial system, and have experienced rapid increases in the volume of both world trade and international flows of money and information. There have also been major changes in the behaviour of transnational corporations (TNCs), causing a thoroughly internationalized division of labour to emerge. This powerful representation of the global economy has been summarized recently by Barnet and Cavanagh (1995), who continue to portray 'imperial corporations' as using the lever of global competition between workers and national governments to retain their control over key sectors of world production and, increasingly, consumption.

There is now general consensus that there is greater integration of economies at the global scale. By contrast, it has proved exceptionally difficult to handle the concept of globalization in either theoretical or empirical research into contemporary social and economic change. Indeed, globalization is now such a loose term (Robertson, 1992) that the usefulness it promised to geographers and sociologists for much of the 1980s is in danger (Fagan and Le Heron, 1994). Partly as a result, a paradoxical alternative vision of economic restructuring has elevated 'localization' to centre-stage. New forms of flexible production, emerging from the breakdown of Fordist regimes of accumulation and modes of social regulation which dominated developed capitalist countries after 1950, have given ascendancy to new industrial spaces at regional and local scales which must be negotiated by governments and TNCs alike (Scott, 1988b; Storper, 1992).

Research into restructuring at these local scales has commonly privileged labour markets as entry points for understanding widely different experiences of global change faced by people according to their gender, age, ethnicity, social class or geographical location. Since the mid-1980s, attempts to reveal such locally bounded stories have been affected strongly by developments in feminist theory, the rise of post-structuralism and critical cultural studies. These alternative discourses have had major impacts on each other but have also mounted a strong challenge to economic geography and its attempts to theorize global restructuring. In feminist analysis of relations between gender, work and place, Hanson and Pratt make only passing reference to globalization in pointing out that 'attention [given] to the globalization of culture and economy should not allow us to lose sight of the rootedness of local lives' (1995: 22).

This chapter seeks to extend these well-known debates about global and local by tracing relationships between globalization and recent experiences of industrial restructuring in Australia's food processing industries. Since the mid-1980s, much attention has been paid in economic geography, political economy and rural sociology to 'the development of a worldwide agribusiness system [which] is in the process of bringing about deep-seated and lasting changes in the conditions governing the production and consumption of food, on a global scale' (Leopold, 1985: 315). Part of the reason for the upsurge of research is a relative neglect of food production as the restructuring paradigm took hold, especially compared with earlier attention paid to industrial sectors such as clothing, electronics, motor vehicles and steel. Yet a focus on the globalization of food commodity chains and the emergence of an increasingly globalized food regime (Friedmann and McMichael, 1989; Le Heron, 1994; McMichael, 1994b; Whatmore, 1995: 39) has drawn attention to the complexity of relations between global and local scales. These commodity chains run from food farming, through agritechnology suppliers, food manufacturers and retailers to final consumers. Hence, the food industries draw attention to the importance of markets and consumption in the globalization process, indeed to the commodification of culture, in contrast to the dominant structuralist emphasis on production. In addition, contemporary agrifood systems increasingly demonstrate how 'subnational spaces . . . are woven into, or excluded from, global production and trade networks' (Bonanno *et al.*, 1994: 2).

Understanding global and local scales simultaneously, requires avoiding the deafening silences about the importance of these 'subnational spaces' and place-specific processes in Barnet and Cavanagh's analysis of global capitalism (1995) while extending the rich examination of localism offered by Hanson and Pratt (1995) so that it can engage with impacts of globalization. The chapter argues that one key to such avoidance and extension in contemporary economic geography is to recognize globalization not only as a complex set of economic and social processes but also as a powerful political-ideological discourse conducted at national and local scales (Koc, 1994; Marcuse, 1995). Processes of globalization can be traced through different sectors and places as they bring about locally uneven impacts of technological change and greater global economic integration. Yet a political discourse of globalization and its imperatives is used by powerful stakeholders in an attempt to conceal this unevenness and the social conflicts which underlie it.

Hence, the global–local debate in academia has strong parallels in the political sphere. Globalization has become one keystone of a dominant discourse within the powerful policy-making cultures ascendant in most industrialized countries since the early 1980s. Global imperatives have become part of the rationale for complex processes which underlie major shifts away from regulation strategies implemented by the state during the so-called Fordist long boom – although not necessarily towards deregulation as such. Yet in their industry and employment policies, nation-states have been driven to a greater extent by a particular ideology of globalization than by measurable impacts of global change on their local industries and industrial places. Increasingly, the discourse of globalization has also marked a revival of interest by the state in specific regional and local approaches to labour market 'adjustment'. These are made necessary by the increasingly uneven political impacts at regional and local scales of state strategies promoting engagement with the global economy (O'Neill and Fagan, 1995).

This chapter first considers attempts to theorize globalization against the background of these powerful ideologies, and reviews attempts by both academics and technocrats to connect global and local scales. The framework is then applied to the Australian food industry, which is now widely held to be rapidly globalizing. The chapter shows that recent changes have been shaped by domestic as much as 'global' process and, especially, by local impacts of a national political discourse about globalization. The Australian food processing industry provides a good example of the importance of national and local processes, not only in responding to global change but also in bringing about the processes now commonly described both within the food industry and among Australian policy-makers as globalization.

## THEORIES OF GLOBAL AND NATIONAL RESTRUCTURING

In Australia, orthodox economics academies have been dominated for at least two decades by a neo-liberal view equating globalization with a new framework for 'international competitiveness' – a response to 'the spread of liberalisation, privatisation and de-regulation throughout the world' (Simon, 1995: xv). Leaving aside major problems of cause and effect here, neo-liberalism assumes the nation-state unproblematically as the basic economic unit of analysis. Imperatives subsequently arising from this globalizing economy are then advanced as a rationale for nations to abandon their regulatory structures of the long boom while sharply reducing unsustainable growth rates of expenditure on welfare and social infrastructure. In this economic orthodoxy, newly industrializing countries (NICs) around the Pacific Rim are often held to be examples of the triumph of market forces over state intervention. This standpoint was epitomized by official pronouncements emanating from the Asia Pacific Economic Cooperation (APEC) summit in November 1994, despite the fact that even the World Bank had recently recognized the crucial role played by state institutions and government intervention in Pacific Rim development since 1980 (Mathews and Ravenhill, 1996).

The vast majority of research into global restructuring by economic geographers, urban sociologists and planners since about 1980, however, has been situated within an alternative discourse about the internationalization of capital (*see* Waters, 1995: 65–95). The dominant economic thesis has been that, to solve accumulation crises precipitating, then accompanying, the collapse of Fordism, capital moved production to new sites around the world offering cost advantages such as cheap labour. Imports from these NICs further undermined production in the world's major mass consumption markets. Far from reflecting the interplay of international market forces, these places of growth and decline are linked directly through TNCs and the emergence of a competition between places for their investments and between workers for employment opportunities. This competitive framework, also a key to the neo-liberal thesis, would reduce costs, help restore profits and compel both organized labour and the state in developed countries into acquiescing to comprehensive changes in the social relations and distributional arrangements developed under Keynesian-Fordism.

This thesis argues that restructuring has involved purposeful strategies by capital, state and organized labour (Clark *et al.*, 1992) and is primarily about power relations and reconstructing capital accumulation, not markets. The most sophisticated versions of the globalization thesis still have notions of a new international division of labour at their core but see a global economy with five central features. First, economic transactions transcend nation-states through global integration of the circuits of production, realization (trade) and reproduction (finance capital) (*see* Fagan and Le Heron, 1994). Second, globalization is underpinned by information technology and its ability to compress dramatically both space and time. Third, older industrialized countries continue to experience automation and restructuring of production *in situ*, processes with far greater net impact on industrial employment than import competition from NICs, but these processes are increasingly 'bench-marked' against world markets which exchange both high-productivity, knowledge-intensive production from the 'triad' markets of North America, Western Europe and Japan, and high-productivity, low-cost production from NICs. Fourth, rapidly changing global flows of money have had dramatic impacts on production since the mid-1980s, permitted by the global financial system and widespread abandonment by the state of financial regulation.

Finally, industrialized countries have experienced continued casualization of their industrial workforces – a *de facto* deregulation of labour markets – sometimes leveraged with the threat of offshore production in NICs but commonly also reflecting supply-side changes. These include the loss of bargaining power by workers in regions where production has been rationalized severely; entry of increasing numbers of women into waged work; and international labour migrations, for example from Latin America, the Middle East and, increasingly, Pacific Asian countries. This has produced a new multiculturalism in urban labour markets of the United States, Western Europe and Australia, and new low-cost labour reserves providing capital with increased numerical flexibility in their labour forces.

The theoretical and empirical turn towards globalization during the 1980s coincided with, and partly created, a proliferation of global metaphors: global factory, global market, global village, global workplace and 'the borderless world' (Ohmae, 1990). Yet all of these are high-level abstractions which can bury the continued importance of national, regional and local scales of analysis. They often exaggerate both the degree of economic integration and determination of the system by 'hypermobile' forces coming from the top down. The essential importance of national and local scales is obscured by regarding globalization not as process but as outcome (McMichael, 1994a: 278). This is the 'steamroller' model of globalization; changes have taken place 'out there', for example in the Pacific Rim, and must be accommodated 'in here' – for example, Australia's cities, manufacturing and exporting regions. The metaphors have become as important as measurable impacts of global change in the globalization discourse.

This leaves at least two major issues unresolved. First, the role of the nation-state in the face of globalization is controversial. It remains difficult to conceptualize global change and deal satisfactorily with the impact of processes operating beyond (outside?) nation-states. This has led inevitably to variants of the thesis that globalization has 'hollowed out' the power of nation-states as 'national governments lack both the strategic vision and the managerial tools to play an integrative role in the societies that they are elected to govern because to a considerable extent they have lost control of the levers of economic change' (Barnet and Cavanagh, 1995: 340).

Yet intervention by nation-states remains integral to the complex processes of capital accumulation, social distribution and maintenance of political legitimacy. State agencies and instrumentalities have contradictory, and continuously fractured, roles which remain integral to the construction of markets for both products and factors of production rather than somehow being detachable (O'Neill, 1996). The role of the state might have become qualitatively different in the era of globalization, but both the neo-liberal discourse of deregulation and the hollowing out thesis can become smoke-screens behind which lie complex patterns of policy abandonment, policy reconstruction and re-regulation in most industrialized countries since 1980. The policy changes themselves have played a central role in bringing about the complex changes which constitute the globalization process (Christopherson, 1993; Gertler, 1995a).

Since the early 1980s, for example, the Australian state has moved towards a new accumulation strategy (P.M. O'Neill, 1994) involving financial deregulation, substantial tariff reduction opening the economy somewhat to global market forces, strategies to encourage exports of goods and services in which Australia should have international competitive advantage, and (limited) moves to deregulate the labour market. The imperatives of globalization have been employed as a rationale for most of this change. Yet policies designed to encourage new patterns of connection with globalizing accumulation have resulted from political processes at national and local levels within Australia. The so-called deregulation which has resulted is just as much an intervention by the state as was the policy regime it has been designed to replace but often on behalf of different fractions of capital, and encouraging different connections with different firms and trading partners (Fagan and Le Heron, 1994). This has been carried out in the context of dramatic changes in the Pacific Asian region and, especially, Australia's relationship with Japan. Yet the USA and the European Union (EU), the other two corners of the 'triad', also remain crucial to this global context, especially for Australian food exporters, mining and manufacturing firms, and financial institutions. Trade and production links between the markets of this 'triad' still dominate the global economy rather than the other way around.

The second unresolved issue concerns the search for connections between global and local scales. Often simply assumed both in academic research and in policy formulation, they remain poorly understood despite empirical work on both global restructuring and localities. Much geographical literature assumes a simple dichotomy between the global economy and changes occurring within specific nation-states; global and local become 'a meaningful opposition of different scales and contraposed sites . . . two ends of a geographic continuum divided by, but defined through, the national' (Luke, 1994: 617). Yet challenges to this representation come from recent research into global–local connections.

Capital moves through circuits of production, realization and reproduction in the accumulation process. Throughout the history of capitalist economies, internationalization of these production, trade and investment circuits has played an integral part in reproducing the system. The more integrated global economy emerging since the mid-1970s, however, is qualitatively different from the internationalization of earlier regimes of capital accumulation (*see also* Dicken, Peck and Tickell, Chapter 12 of this volume). Globalization marks greater integration of all three international circuits of capital across a multitude of specific sites around the world rather than the relations of dominance and subordination between circuits which existed previously (Fagan and Le Heron, 1994; Fagan and Webber, 1994). Hence, privileged status can be accorded no longer either to international trade (as in neo-liberal models of competitive advantage) or to internationalized production through TNCs (as in theories of the new international division of labour).

Fig. 16.1 divides total capital within an economy according to different ways in which individual capitals become inserted into global accumulation through their patterns of production, trade and investment. Firms in the national fraction, for example, produce, sell their commodities, raise capital and reinvest largely within the territorial space of their home nation. By contrast, a global fraction can be recognized in which local production takes place within globalized networks of TNC branch plants; production is sold primarily on international markets (at world market prices); and finance capital is obtained from both onshore and offshore profits or by raising loans from transnational banks.

Following Bryan (1987), Fagan and Le Heron (1994) recognize two other fractions of capital. Investment-constrained firms produce and reinvest locally but sell a significant proportion of their output on world markets. These include large domestic exporters (other than TNCs) whose integration with the global economy is constrained by their producing and investing largely inside a particular nation. By contrast, firms in the market-constrained fraction obtain finance capital outside the nation-state and, in addition, often invest their domestic surpluses internationally or repatriate profits. The integration of these firms with the global economy is constrained because most of their output is sold to the national markets in which they are located. This fraction typically contains foreign-owned branch plants of TNCs serving markets which may be protected by tariff barriers.

| FRACTION | INTERNATIONAL CIRCUIT OF CAPITAL | | | LINKAGES WITH GLOBAL ECONOMY |
|---|---|---|---|---|
| | Production | Realization | Reproduction | |
| National | N | N | N | **Direct**: imported means of production; licences and franchises **Indirect**: local competition with multidomestics and global firms; deregulated financial system |
| *Investment-constrained* | N | I | N | **Direct**: exports; imported means of production **Indirect**: competition from TNCs for export markets; export franchising; deregulated financial system |
| *Market-constrained (inc. 'Multidomestics')* | N (G) | N | G | **Direct**: TNC branch plants status; imported means of production; offshore finance **Indirect**: competition with firms in global fraction |
| Global | N (G) | G | G | **Direct**: part of TNC networks; global competition with other TNCs; imported means of production; offshore finance |

**Fig. 16.1** Globalization and circuits of capital. N = nationally bound circuit; I = international circuit; G = global circuit (intra-corporate); TNC = transnational corporation. Source: after Fagan and Le Heron (1994)

In recent years, TNCs primarily adopting this strategy have been called 'multidomestics' and remain common in food and clothing industries where strategies of global localization are important for securing market share (Morita *et al.*, 1986).

In this conception, restructuring has seen globalizing corporations arising next to the multidomestics, which grew rapidly during the long boom. Internationalization of production was an integral feature of Fordism, but global firms do not simply produce, market and invest in many nations; they seek to integrate the three circuits of capital inside global corporate structures. Yet even firms in the national division are now connected to the global economy through relationships with international banks (even if they borrow money 'locally'); through technology imports, franchising and the payment of patent royalties; through subcontracting relationships with global firms; and through increased competition for local markets from TNC branch plants or imports. Globalization of accumulation encompasses all the activities of firms in all four fractions rather than just the strategies of global firms.

Firms 'move' between the fractions shown in Fig. 16.1 according to contingent interactions between capital, labour and state in specific places. This can occur in either direction over time, not simply from national to global. Further, capital, labour and state interact at local scales where local representatives of 'global' firms (including large domestically owned TNCs) interact with national (and local) firms, governments, workers and consumers. These national and local relationships are not subordinate to globalized accumulation but simultaneously create its architecture. The idea of a two-way flow between local and global undermines the 'steamroller' metaphor that there is no alternative for nation-states but accommodation to global change. Finally, the state cannot intervene on behalf of all fractions at once. Patterns of intervention are determined in both formal and informal political spheres and may be resisted strongly by some stakeholders while supported (pressured?) by others.

While the concepts underlying Fig. 16.1 show that theories of globalization need not be deterministic, such metatheoretical frameworks for analysis have been seriously challenged by post-structuralist and feminist discourses which reject static binaries such as centre/margin and global/local, replacing them with fluid, non-deterministic conceptions. Post-structuralist discourse theory recognizes that social sciences, in common with the technocracies, have been constructed by power relations and the political and economic apparatuses which sustain them. Meaning and signification within the dominant restructuring discourse about economic change has excluded other discourses and meanings, for example those relating to gender and race. Hence, one task of feminist theory and critical cultural studies has been to uncover subjugated discourses about the restructuring and economic change experienced in places and shaped by characteristics of place. Their critique opens up scale as social and political construction and metaphor (Barnes, 1996; Jonas, 1994; Massey, 1992). Scales such as global and local should be seen as purposefully constructed starting-points for narratives about social change. The powerful discourse of globalization constructs metaphors of the scales from which dominant processes emanate. The exclusion of other discourses and meanings is sometimes done in ignorance but more often for economic or political purposes.

## INTEGRATING THE GLOBAL AND LOCAL

Progress in integrating global and local in both theoretical and empirical work requires a non-deterministic and political reading of scale. Four arguments seem crucial in attempting to reshape the concept of globalization along these lines (Fagan, 1995). First, it is a mistake to represent 'global' as the scale most obviously ruled by general (economic) laws of capitalist society while 'local' is seen as unique, chaotic, parochial and the realm of culture and human agency. For policy-making in the areas of industry, employment, urban and regional development, the widely employed geographical metaphor of the hypermobility of transnational capital has made matters worse. There can be no argument about the mobility of finance capital in an era of financial deregulation, global banking and information technology. Yet productive capital remains much less mobile because of the sunk costs associated with plant closure. These are often determined locally and can include problems of realizing value from plant, equipment and productive sites, large-scale redundancy payments to workers, and political problems (including assiduously cultivated corporate image) associated with breaking contracts or agreements with national, regional or local governments. Similarly, local-scale processes always reflect the national and global contexts in which places are embedded (Massey, 1993b).

Second, local agents of global capital are often crucial in developing corporate strategies for change. Such agents often represent their local strategies as reflecting global imperatives, such as the need to internationalize the Australian economy or respond to the North-East Asian ascendancy (Garnaut, 1989). Yet these corporate representations are often aimed at controlling bargaining with non-local head offices, with national and local tiers of the state, and local communities which will bear the social and economic costs of change. Similarly, national or global processes can be constructed as local (*see* Luke, 1993) – for example, recent policies of the Australian government which encourage localities and regions to take responsibility for their own employment development strategies – to suit the purposes of the most powerful stakeholders in change (O'Neill and Fagan, 1995).

Third, coming to grips adequately with global and local scales mutually determining each other is difficult, but post-structuralism challenges political economy and economic geography for adopting a mechanistic view of global–local links. The dominant discourse about globalization has implied hegemony of the global whereas recent critical theory would reject a fixed and hierarchical notion of scale. Finally, it follows that globalization is always an uneven and contested (and thus contestable) process. There is no single path to restructuring either globally or locally because multiple paths are constructed contingently by stakeholders. Further, these various agents construct images of both past and future to sell their contemporary restructuring strategies to others. Corporate managers, bureaucrats, politicians and academics all employ such image-making. From this, it is easy to see how the discourse of globalization can be used to legitimize state deregulation and cut-backs in state investment, or a business push for greater 'flexibility' in markets or industrial relations. In this way, globalization can exert a major influence over local change. Yet it is also clear how the 'global leviathan' imagery of most globalization theses can be locally or regionally disempowering.

## RESTRUCTURING OF THE AUSTRALIAN FOOD INDUSTRIES: GLOBAL AND LOCAL

The Australian food industries illustrate well the flaws inherent in hierarchical conceptions of global-local relations. They also demonstrate basic problems arising from limited conceptions of globalization in formulating public policy about industrial development. In Australia, as elsewhere, globalization has gained a narrow political expression as the requirement that domestic economic activity meet international standards of profitability and adopt 'international best practice'. Globalization has been posed as a constraint on domestic activity at the macro-economic level, epitomized by a focus since the mid-1980s, in both politics and the media, on the balance of payments deficit as a critical barrier to Australia's economic progress. Economic policy has been aimed at increasing exports at a rate faster than the growth of imports, usually by focusing on industries with demonstrable competitive advantage, a construction influenced strongly by the ideas of Porter (1990), which were endorsed by industry policy think-tanks close to the federal Labor government. These industries include resource-based manufacturing, tourism and, rather problematically, skill-intensive services.

Since the early 1980s the neo-liberal discourse about 'levelling the playing field' to promote restructuring (Stewart, 1994) has coalesced with two others to provide the Australian state with a powerful basis for formulating and legitimating its new accumulation and distribution strategies. The first hinges around the central role of Pacific Asia in Australia's experience of globalization since 1980. In addition to the importance attached to Japan as Australia's principal trading partner, and a major buyer of foodstuffs, the experience of the Asian NICs has captivated influential policy advisers. Australia has little choice, they argue, but to seek closer integration of its production, trade and financial systems with this dynamic Pacific Rim to secure the long-term future of the Australian economy. A second powerful discourse is based on assertions about culture. Notions of social hierarchy, authoritarianism and 'leadership' are held to provide new cultural underpinnings to economic success and international competitiveness – an 'Asian way'. Such notions have been encouraged by some academic think-tanks, bureaucrats and influential sections of the financial press (Byrnes, 1994). Australians have little alternative but to get their acts together, so the argument runs, or get left behind. Yet ideas about these so-called cultural and political differences between Asia and the West have been locally constructed by some political leaders in Pacific Asian countries, notably those from Malaysia and Singapore, often for domestic political purposes (for further discussion, *see* Robison, 1996).

Finally, the discourse of globalization itself has intersected powerfully with neo-liberalism and arguments about the dynamism of East Asia. These three policy mind-sets have mutually reinforced the steamroller model of globalization – a set of imperatives to internationalize the Australian economy in particular ways. Neo-liberalism, integration with East Asia and globalization have become conflated within Australia's dominant political discourse since the mid-1980s. The global economy remains 'out there', to be defended against, or brought into fruitful partnership through extending exports and maximizing domestic import replacement at prices reflecting world markets. The two-way relationship between globalization and domestic restructuring is ignored.

Major studies commissioned by the federal government (for example, Australia: Department of Industry, Technology and Commerce, 1991; Australia: East Asia Analytical Unit, 1994) have singled out processed foods (including beverages such as soft drinks, wines and beer) as one of Australia's most promising sectors for increasing the value-added to exports of unprocessed raw materials. This is based on conventional wisdom about the country's large agricultural production capabilities and biophysical resources, and perceptions of a huge future East Asian market for commodified food. Food processing constitutes the largest (remaining) sector in Australian manufacturing employing, about 180 000 people in 1995, 17 per cent of the national manufacturing workforce. Of more relevance to this chapter is its export orientation and links with globalizing agri-food commodity chains. While the mineral booms of the 1970s radically restructured Australian exports and completed their reorientation to the Pacific Asian region, foodstuffs (both raw and processed) constituted between one-fifth and one-quarter of total exports throughout the 1980s. By 1990 four-fifths of food exports were traded as unprocessed commodities, leaving one-fifth as processed foods. Hence, Australian government policy has been to encourage higher value-added food exports since, by the 1990s, most still received only simple processing such as slaughtering, freezing or canning. Forms of non-tariff protection have been important in some food sectors – for example, quarantine regulations which effectively close Australia to chicken-meat imports but effective rates of protection of the food industry have been among the lowest in manufacturing.

While a large number of small businesses manufacture food and beverages in Australia, and there has been much discussion about the rise of boutique producers cultivating niche markets, all key food sectors are dominated by large companies, often subsidiaries of TNCs. Most of these arrived in Australia during the 1960s as multidomestics, attracted to rapid population growth and relative affluence of the market. The 20 largest food manufacturers in 1992 accounted for nearly 70 per cent of total industry turnover and the vast proportion of Australia's processed food exports. By 1992, 11 of these 20 largest producers were subsidiaries of foreign-owned TNCs, including six of the largest 10 firms. Hence, food processing was seen not only as an industry with potential competitive advantage but one already linked to a globalizing agrifood system.

Yet while the annual value of processed food exports from Australia did rise after 1980, the growth did not keep pace with rates achieved in other manufacturing sectors, a matter of great disappointment to policy-makers. Since 1980, Australian corporate food exporters have played a very small part in the dramatic growth of world exports of fresh and processed foods, including those targeted at Japan and new Pacific Asian markets. Exports of Australian manufactured food peaked at nearly one-third of production during the late 1970s, fell during the 1980s and were below 20 per cent in 1990 for the first time since the early long boom. The absolute growth in exports has been significant in some sectors, while volumes exported have increased between 1992 and 1995. This has been most noticeable in the wine industry, where exports are targeted at the EU and North America. Further, increasing quantities of processed cereals and vegetables have been sold recently in Asian markets. While these trends give some encouragement to the hopes of both state and labour for benefits from a globalized food industry, the export performance has been disappointing considering the quantities and qualities of Australian foodstuffs potentially available for export and the government's strong embrace of the concept of competitive advantage. Meanwhile, there has been a significant growth in imports, not only of traditional 'exotic' foods or higher-value, internationally branded products (such as European wines and cheeses), but most notably of orange juice, canned vegetables and even potato products.

The behaviour of the corporate food groups in relation to the state's internationalization strategy provides a key to understanding food restructuring. The sector was slow to begin restructuring in the 1970s but, by the mid-1980s, was changing more quickly than many other manufacturing sectors. Total employment in food processing stabilized after 1990 but this disguises major job-shedding from larger plants in key sectors supplying both domestic and export markets. Significantly, the growth of processed food imports in the 1990s has followed, not preceded, restructuring. Import competition has not been a principal cause of restructuring in this industry – perhaps least of all, in the 1980s, that from countries in Asia where labour costs are lower (Fagan, 1996). The reasons for restructuring must be found elsewhere.

## Global and Local in Corporate Strategies

In the 1980s there was an epidemic of mergers and take-overs in the food sector and major offshore investments by Australian firms. In the 1990s, this has led to an upsurge in Australian activity by food TNCs. The categories employed in Fig. 16.1 can be used to show how this came about. Four overall types of corporate strategy dominated the 1980s (Fagan and Rich, 1990). First, in the national fraction, fierce struggles for market power developed in brewing and the production of wine, meat products, confectionery and biscuits, often constructing national markets to replace what had been regional food markets for most of the century. Second, established food corporations in slow-growth sectors in both national and investment-constrained fractions, such as sugar refining, brewing and commodity exporting, attempted to diversify away from food, often with poor results. Third, there was a massive corporate shift into food production during a frantic restructuring period after 1984, encouraged by the Australian government's new deregulated financial environment and based largely on debt-raising from transnational banks. Leveraged bids from debt-financed corporate raiders such as Bond Corporation and Adelaide Steamship Co. stimulated defensive strategies among vulnerable local firms, often also financed by overseas borrowing (*see* Fagan, 1990). Food processing became attractive to highly geared globalizing companies because of steady cash flows which arise from controlling a dominant domestic market share for products such as beer, biscuits and processed foods.

Finally, some Australian companies attempted to reconstruct themselves as food TNCs, largely through expensive take-overs in the major markets of North America and at vantage points inside the European Community. Investments in Asian NICs grew slowly in the 1980s; indeed, some of Australia's largest agrifood organizations reduced their direct investments in the Pacific Asian region between 1985 and 1990. Far from chasing low-cost production sites, these large Australian companies joined the market-constrained/multi-domestic fractions of countries in Europe and North America. Often this was done through acquiring local brand names, although Elders IXL Ltd, Australia's largest agribusiness conglomerate and controller of its largest brewer, attempted to find new local vehicles, especially in the UK and Canada, for what it hoped would become a global brand name: Foster's Lager. Yet there was no inexorable shift of food firms towards the global fraction. Multidomestic TNCs continued to compete for domestic market share with the largest Australian-owned food manufacturers, which were, increasingly, new corporate conglomerates. While some of these Australian groups commenced off-shore production, most became linked to globalized accumulation through debt-financing. Very little of the huge investment in Australian food processing after the mid-1980s was related to new technology, innovative product development or the exploring of new export markets in Pacific Asia.

The debt burdens being carried by the globalizing agrifood enterprises, such as Elders IXL Ltd, and by corporate raiders such as Adelaide Steamship and Bond, brought about major changes in the 1990s. Many of the globalization strategies were abandoned because of 'debt overhang' (P.M. O'Neill, 1994). Some of the shake-out was a return towards securing power within particular segments of local food markets, but virtually all of it was financially driven as companies struggled to restructure overseas debts by divesting food assets, some of which had been controlled for relatively short periods. Often priced for rapid sale, or broken into specific commodity parcels, these assets became targets for the global agrifood enterprises and, after 1990, foreign control of Australia's food manufacturing increased quickly. The divestments coincided with new strategies of the global food corporations, now emphasizing the power of nationally branded food products in widely differentiated markets (Arce and Marsden, 1993) but increasing their global sourcing of processed foodstuffs to market under these 'local' brand names. Securing production bases and strategic market shares in Australia and Pacific Asian countries has become part of their strategy.

While much of the food restructuring was financially driven, impacts on local production, trade and employment have been significant. Larger firms have attempted to develop simpler structures based on fewer commodities, often returning to the national fraction (Fig. 16.1). This is epitomized by Elders IXL divesting its agribusiness divisions after 1991, renaming itself Fosters Brewing around one of Australia's cultural icons, and ultimately withdrawing from production in the UK and North America, leaving Fosters to be brewed overseas under licence. The sale of the Elders agrifood divisions in 1992 brought ConAgra Inc. of the USA, one of the world's largest food conglomerates, into a controlling position in Australia's export beef industry. By 1992, all five of Australia's largest beef exporters were foreign-owned and able to source exports to the USA and Japanese markets from multiple production sites around the world. By 1996, Australian producers were bemoaning the fact that beef sourced from the USA had overtaken Australian market share in Japan.

Patterns of restructuring among firms in the national fraction have varied sharply between food commodity chains and within sectors, even from plant to plant within firms. With the financial collapse of Adelaide Steamship, whose path towards globalization was blocked by unsustainable foreign debt and an audit for unpaid tax by the Australian Taxation Office, new national firms were created from divestments. An example is the aptly named National Foods Ltd, which has become a major player in a fierce competitive struggle for control of the Australian milk and dairy products market. Yet the restructuring of Adelaide Steamship in 1991 also brought into frozen foods, canned vegetables and ice-cream Pacific Dunlop Ltd, one of the largest Australian-based TNCs with experience of an aggressive global rationalization strategy in its clothing, textiles and rubber products divisions. The newly acquired food division was quickly rationalized in Australia with plant closures and centralization of remaining processing in key agricultural areas.

The rationale for Pacific Dunlop's strategy, at least for public consumption, was to bring Australian production costs in line with international best practice with a view to increasing exports to Pacific Asia in future. The domestic market remained the crucial base for this plan, however, and the company began to import vegetables as part of the cost-cutting exercise, thus increasing the pressure on contract farmers and process workers in the remaining Australian supply regions. By 1995, facing increasing impatience among its major shareholders waiting for the food division to show the returns obtained elsewhere within the conglomerate, Pacific Dunlop abandoned its market-leading positions in the Australian food industry. The frozen and canned vegetable enterprises were sold to the American agribusiness firm J.R. Simplot, while Nestlé of Switzerland purchased the dairy products division to consolidate its position in the local industry.

While some food production was hived off from disintegrating conglomerates, market concentration has continued to increase in other Australian sectors. In chicken production, for example, restructuring has caused something resembling Fordist production to emerge for the first time with mass production of a mass high-protein and standardized commodity. By contrast, rapid growth of 'boutique' producers occurred in wine production, brewing and specialized foods. Yet problems in raising finance locally for small enterprises (ironic, given the volumes of capital moving in and out of the Australian food industry), and the costs of maintaining market share against the scale economies of corporate producers, caused several 'success stories' to become absorbed into the large-firm sector by 1995, notably in brewing and wine-making. As food markets have become locally differentiated, new technologies of product design, transport and warehousing have allowed production from TNCs exploiting scale economies to be targeted at local niche markets. Boutique-style beers and premium wines can be placed into these markets at prices cross-subsidized by their bulk production lines. This makes survival difficult for the genuine niche producers beloved of flexible specialization theorists.

## Global and Local in Food Consumption

Flexibility in food consumption has been more prevalent than in production. 'Localized' marketing has become crucial in food company strategies. Style and cultural values are elevated as Australian food processors promote food as entertainment or healthy lifestyle. Some of the key changes include rapid growth in production of fast foods which supposedly add value (including household convenience related directly to changing social divisions of labour and the feminization of the workforce); shifts to forms of protein other than red meat; a shift towards fresh and packaged frozen foods away from canned and processed foods; so-called 'green' foods, where marketing emphasis is placed on nutritional value and hygiene; and the growing health food and 'real' food market, especially among the new urban service classes in the capital cities. Here, value is added by 'naturalizing' rather than homogenizing the food product, but among the global agrifood branch plants, the naturalized and localized product is a cultural and industrial construct.

Faced with a renewed struggle for market share, large firms have resorted increasingly to brand consciousness. Brand names allow companies to cash in on the cultural icon status of many food products. Indeed, the local construction of these icons has become a major strategy for globalizing food companies, almost always involving images of a past, easier time when food was 'home-style' and its nutritional value less problematic. Brand names help these firms to create barriers to entry and protect market share. Sometimes, companies raised cash by disposing of brands as they withdrew from certain products or markets following the leveraged buy-outs or divestments. They have also become crucial in the growing struggle between Australian retailers and manufacturers for a share of profits from domestic food markets where branded products compete with the supermarkets' generic brands.

Finally, brand name loyalties have helped companies to obscure product origins. This helps construct a barrier between the final consumer and the place and conditions under which the food was produced, a further demonstration of the localization of globalizing products. An example is Pacific Dunlop's market-leading brand of canned vegetables. When the company began to import tomatoes from low-cost sources in the 1990s, ironically from subsidized production in the USA and Italy, the wording on its top-brand labels was changed to state that when home-grown tomatoes were not available because of 'adverse weather conditions', the company looked after local consumers' interests by searching the world. While consumers might have been suspicious of this claim given that the canning of vegetables is designed partly to overcome seasonal shortages, they would probably not know that Pacific Dunlop's facilities to process Australian-grown tomatoes had been closed (Pritchard, 1995: 322). This is a good illustration of how globalization has been hidden under a cloak of local interest.

## Globalization and the State Strategy

Hence, the most important link with globalizing accumulation for many Australian food firms in the 1980s was debt-raising from transnational banks. With the prevalence of financially motivated restructuring, and the instability of the 1990s, it is not surprising that development of new food export markets in Pacific Asia has been unspectacular and that export orientation has declined overall. At the level of local food processing plants, the financial and management uncertainties commonly caused productive investment strategies to become stalled. By the 1990s, levels of investment in research and development in Australian food processing were among the lowest in Australian manufacturing (Scott-Kemis, 1990). During this period, the largest TNC branch plants showed little interest in exporting Australian processed foods (except for the meat-packers and wine producers). Indeed, by 1993 the highest export orientation had been achieved by Australia's two largest milk product exporters, both co-operatives owned by dairy farmers in Australia's most productive dairying state, and by the rice-growers' co-operative.

The globalization of the Australian food industry has been a far from smooth process and is not captured by deterministic models or hierarchical notions of scale. While foreign ownership increased sharply in the 1990s, much of this resulted from spectacularly unsuccessful attempts by Australian conglomerates to globalize in the 1980s. Both of these outcomes, however, have been encouraged by two state policies; first, financial deregulation and second, a virtual 'hands-off' attitude to foreign take-over, through Australia's Foreign Investment Review Board (FIRB), to encourage construction of globalizing food firms with a critical mass to compete on world markets. The Australian government's public stances in approving the largest food take-overs since 1990 have demonstrated the confusion inherent in its position on deregulating the economy to foster both domestic export-orientated manufacturing and further economic integration with Pacific Asia. Some policies, however, have made globalization more difficult for some firms, notably quarantine and other food quality regulations. Absorption of food assets into global corporations during the 1990s has often been contested strongly by local firms, shareholders and sections of the labour movement. Yet national and local policy changes have been constructed as global imperatives, from which the food industry was expected to benefit, while large firms have constructed as local their globally sourced products. These processes cannot be decoupled from the national and local spaces in which Australian and Pacific Rim food consumption and marketing, production and food agriculture take place.

## CONCLUSIONS

The case study in this chapter exposes major flaws in the prevailing state discourse about international competitiveness. By the mid-1990s, the food industry's decade of sustained globalization had not led the large Australian food processors, mostly now TNC branch plants, to capture positions in new Pacific Asian export markets despite their considerable rhetoric, especially during FIRB inquiries into take-overs. There has been a rise in food imports from Asia, however, often reflecting offshore sourcing by Australia's leading supermarket chains to provide price competition against locally produced branded products of the TNCs or to source their own generic brands. Finally, in the 1990s some Australian companies have established food processing plants in China, Indonesia and Vietnam to serve local markets in these countries. This has little to do with the quality of Australian raw materials, the productivity of its agricultural sector or the variety of biophysical environments in which Australians can grow high-quality food. In reality, food exporting has more to do with changing corporate structures, driven partly by global finance capital; with bilateral trade negotiations and other macro-economic policies of the state; with levels of R&D and product innovation, especially in the 1990s among small firms; and with the global sourcing strategies of the largest food corporations. All these things have contributed to increased integration of the Australian economy with Pacific Asia but not necessarily towards locally revitalized, export-orientated food manufacturing.

Neo-liberalism and the determination to further open the Australian economy to the dynamism of the Pacific Rim remain powerful among the Australian government's principal policy advisers and showed no signs of diminishing with the election of a conservative federal government in 1996. It has become a matter of urgency, however, that debate about economic policy and Australia's relationships with Pacific Asia be lifted from the cul-de-sac engineered by the political discourse about globalization so that the direction of public policies relating to both accumulation and social redistribution can be re-engaged. Simply attempting to level the playing field, to the exclusion of understanding local corporate restructuring behaviour and its relationships with globalizing production, trade and finance, seems to have put Australian production at a disadvantage relative to key players controlling both triad markets and the NICs. The example of the food industry shows that a revitalized food manufacturing sector will not result from simply joining the alchemy of competitive advantage with the magic of Asian dynamism.

In economic geography, a non-deterministic and explicitly political reading of scale constructs a research agenda sensitive to post-structuralist and feminist critiques of economistic and mechanistic readings of global–local relations. Identifying multiple pathways involved in globalization processes, such as those found in the Australian food industry, offers an important research agenda focused on the simultaneous construction of global and local spaces both materially and discursively. Yet there should not be a conflation of the importance of the local in determining processes of economic and social change and its importance as a scale at which such changes can be contested. Theories of power structures which shape post-industrial society, albeit partial, and of connections between global, national and local scales, seem crucial to understanding the environment in which local strategies have to operate and identifying ways in which processes reinforcing local marginalization can be contested at all scales from local to global.

## ACKNOWLEDGEMENTS

This chapter develops a preliminary version presented as the 1995 Griffith Taylor Memorial Lecture, University of New South Wales (Fagan, 1995). The author is grateful to Sherrie Cross for assistance in researching the Australian food industry. The author is also grateful to Katherine Gibson, Richard Howitt, Richard Le Heron and Phillip O'Neill for constructively critical debates over a lengthy period about globalization and what it might mean.

# Chapter Seventeen

# GLOBALIZATION OF R&D IN THE ELECTRONICS INDUSTRY: THE RECENT EXPERIENCE OF JAPAN

**DAVID P. ANGEL AND LYDIA SAVAGE**

## INTRODUCTION

In this chapter we examine the extent to which manufacturing firms are adopting a global approach to research and technology development. If the concept of a new international division of labour was once a useful shorthand to describe the changing geography of manufacturing worldwide, the internationalization of core R&D functions suggests the advent of a new phase in the ongoing globalization of manufacturing systems. We take as our point of departure in this regard what Hirst and Thompson (1996) characterize as the 'strong' thesis of globalization. In other words, we draw a strong distinction between the continuing internationalization of economic processes and a structural shift towards a global economy. While the two ideal types of internationalization and globalization are multi-faceted, we place particular emphasis upon the difference between a multinational and a transnational firm, and upon the organization and geography of the firm. It is commonly asserted that manufacturing firms with global reach have shed their allegiance to any home country and are now beginning to make investment decisions for R&D and other core business operations on a global scale. And yet the evidence for a truly global, as opposed to an international, approach to R&D remains limited (Angel and Savage, 1996; Casson and Singh, 1993).

The prominence of globalization no doubt derives in part from a widespread sense that a transformation in regimes of accumulation is currently under way within advanced industrial economies. But the rhetoric of globalization has specific material effects and we should not be innocent to its promotion as a strategic and policy tool by different interests. Nowhere are the policy and political implications of globalization more evident than with respect to R&D and manufacturing investment by foreign firms in the USA. From the late 1970s onwards, the level of Japanese and European investment in R&D laboratories, technology alliances and start-up high-technology firms in the USA increased substantially. Many analysts in the USA suggested that the growing internationalization of R&D should be resisted as it would lead to an accelerated transfer of technology from US firms to foreign competitors (*see*, for example, National Research Council, 1992; Prestowitz, 1988; Reich and Mankin, 1986). In the ensuing policy debate, this 'techno-nationalism' was countered by a vision of a global or transnational economy in which the nationality of the firm was of secondary significance to the location of the investment (Reich, 1991).

To what extent does the changing organization and geography of R&D support the dominant rhetoric of globalization? We explore this question in the context of changing technology development practices in the Japanese electronics industry. With sales in excess of $200 billion, the Japanese electronics industry warrants attention on the basis of its size alone. Two additional factors, however, underlie our selection of this industry for study. First, precisely because of the rapid increase in international R&D investments by Japanese firms, and attendant fears of declining competitiveness of US high-technology firms, the character of the internationalization – or globalization – of Japanese electronics R&D has been a central policy concern in the USA. Second, the Japanese electronics industry is of interest because a new round of restructuring of technology development practices is under way, motivated in part by the limitations of the initial phases of internationalization of R&D. A survey conducted by the Japanese Science and Technology Agency (1994) found that 47.3 per cent of private firms had over the past three years implemented or created plans to implement a major reform of their R&D organization and practices. Our research indicates that one of the main targets of this restructuring is the international organization of R&D, many aspects of which have failed to meet the expectations of corporate leadership in Japan.

Thus our central research question concerns the contribution of foreign R&D and international alliances to the shifting fortunes of Japanese electronics firms. A full evaluation of the sources of competitive advantage of Japanese electronics manufacturers is beyond the scope of this analysis. Even the internationalization of technology development is a complex multi-dimensional process involving (a) foreign R&D laboratories; (b) international technology partnerships; (c) acquisition of foreign firms; (d) minority investment in foreign firms; (e) participation in universities and other research institutes, and (f) technology development at affiliated production facilities, as well as various forms of informal communication and information scanning. We focus here on two important elements of this overall structure of international technology development, namely, the contribution of foreign R&D facilities and of international technology development partnerships involving Japanese and US electronics firms.

In the next section, we review briefly the increasing internationalization of the Japanese electronics industry and examine levels of R&D and production investment by Japanese electronics firms in the USA. In subsequent sections of the chapter we draw upon the results of questionnaire survey and interview data to consider the role of US R&D laboratories and international technology partnerships in the manufacturing operations of Japanese electronics firms. Two research findings are emphasized. First, the increasing number of Japanese-owned R&D laboratories in the USA notwithstanding, Japan remains the locus of decision-making for technology development. There is little evidence of US facilities constituting a semi-autonomous technology development platform for Japanese electronics firms. Second, the technology development partnerships identified as being of greatest commercial significance by Japanese electronics firms have in most cases worked to the mutual benefit of both Japanese and US partners.

What these results and a growing body of other research suggest is that much of the current restructuring of technology development practices is more accurately described as a process of internationalization rather than globalization of R&D. Tendencies towards globalization are at best partial and constrained, and are often inconsistent across different dimensions of the technology development process (e.g. foreign R&D laboratories versus international technology development alliances). More generally, the experience of Japan points to distinctly national models of, or pathways to, globalization. As Boyer and Drache (1996: 14) suggest, 'The very process of internationalization reveals the persistence of national systems of innovation which are deeply embedded in a web of interconnected political, educational and financial institutions which cannot easily be copied or adopted.'

## INTERNATIONALIZATION OF JAPAN'S R&D CAPABILITY

Much previous research has emphasized the strength of Japanese firms in technology development (for a recent review, *see* Westney, 1994). Relative to international competitors, Japanese firms have been identified as having shorter product development times (Mansfield, 1988), better integration of technology development with overall corporate strategy (Roberts, 1993), more effective acquisition of external technology (Florida and Kenney, 1990; Kenney and Florida, 1992a), strong capability in incremental innovation, and effective integration of innovation and manufacturing (Bowonder and Miyake, 1992; Odagiri and Goto, 1993; Wakasugi, 1992). While historically the vast majority of this technology development activity has taken place in Japan, the level of foreign R&D and numbers of technology development partnerships involving foreign firms increased dramatically from the early 1980s onwards. This shift constituted part of a more general trend within advanced industrial economies towards the internationalization of R&D (Casson, 1991; Howells, 1990; Westney, 1993).

Perhaps the most visible aspect of the internationalization of technology development capability on the part of Japanese electronics firms has been the establishment of free-standing R&D laboratories in the USA, Europe and elsewhere. Several studies have documented the growth in Japanese R&D investments in the USA, and the motivations behind this investment strategy (Florida and Kenney, 1993; Herbert, 1989; Peters, 1991; Serapio, 1994). Dalton and Serapio (1993) note that Japanese corporations operated 155 stand-alone R&D facilities in the USA in 1992 and spent $1.2 billion on US-based R&D in 1990. Of these 155 R&D

facilities, more than 70 per cent were established between 1986 and 1992.

Electronics, including computers, communications and semiconductors, constitutes the lion's share of this R&D activity. Estimates of the actual numbers of Japanese electronics R&D facilities in the USA vary depending on the precise definition of R&D and of ownership status. For the purpose of this study, we developed a database comprising the identity and location of 73 Japanese-owned electronics R&D laboratories in the USA. This total includes R&D laboratories involved in computers, communications, semiconductors and other areas of electronics manufacturing. Fig. 17.1 shows the location of these R&D laboratories. Japanese electronics R&D is overwhelmingly concentrated in California, especially in the Silicon Valley region. Subsidiary clusters of laboratories are observed in southern California and other centres of high-technology manufacturing.

In addition to free-standing R&D laboratories, much technology development activity takes place at affiliated manufacturing facilities in the USA. The numbers of such affiliated manufacturing facilities increased dramatically during the 1980s both through the establishment of new branch plants and through the acquisition of a majority or minority holding in existing facilities in the USA. Indeed, the acquisition of existing US companies by Japanese manufacturers has been one of the most contentious aspects of Japan's increasing involvement in the US electronics industry. Of particular concern to US policy-makers has been the acquisition of innovative start-up companies in semiconductors, semiconductor equipment and computers (Mowery and Teece, 1993; Sun, 1989).

Drawing upon a database maintained by the Japan Economic Institute, Genther and Dalton (1992) identified over 360 manufacturing plants operated by Japanese electronics companies in the USA in 1990, up from 291 plants in 1989. This total includes facilities in which Japanese firms have majority ownership as well as facilities in which they hold minority ownership of no less than 10 per cent. The total is based upon a relatively liberal definition of electronics that includes, for example, firms manufacturing ultrasound machines and other medical equipment. Acquisitions accounted for approximately 51 per cent of the 360 manufacturing facilities. The major concentrations of facilities are in electronic components, computers and semiconductors.

**Fig 17.1** Location of Japanese electronics R&D facilities in the USA

The Electronic Industries Association of Japan identifies 148 wholly owned production facilities operated by its members in the USA in 1993, of which 88 are involved in manufacturing electronic components and parts of one type or another. In our own research, we identified a total of 163 wholly owned Japanese production facilities in the USA. Fig. 17.2 shows the location of these facilities. Once again, there is a heavy concentration of production facilities in California. In contrast to the geography of R&D laboratories (*see* Fig. 17.1), there are also large numbers of plants in the north-eastern USA, as well as in the South. The majority of the latter facilities are high-volume production plants for consumer electronics systems and components, such as televisions and video-cassette recorders.

The third major dimension of the internationalization of R&D by Japanese electronics firms comprises technology development partnerships established with foreign firms. This is in practice the least well documented aspect of the internationalization process. The total number of international alliances of all types in electronics has certainly increased during the 1980s (Hagedoorn, 1993; Hagedoorn and Schakenraad, 1990). By one account, more than 1000 international alliances involving Japanese firms were announced during the 1980s in semiconductors alone (National Research Council, 1992). Hitachi Semiconductor, for example, currently maintains 11 technology alliances of one type or another, including major alliances with Texas Instruments in DRAM technology and VLSI Technology in application-specific integrated circuits.

A database developed by the Department of Commerce identified 450 US–Japan alliances in high-technology industries during the period 1989–90 (Dalton and Genther, 1991). Of these alliances, approximately 38 per cent were marketing agreements and 14 per cent were R&D and production joint ventures. The greatest concentration of alliances was in the computer industry (87 alliances), computer software (91 alliances), semiconductors (105 alliances) and semiconductor manufacturing equipment (38 alliances). As already mentioned, of particular concern for policy-makers have been asymmetrical alliances between large Japanese corporations and small US start-up firms, alliances that typically provide Japanese manufacturers with access to US technology in exchange for capital investment or guaranteed production capacity (National Research Council, 1992).

Fig 17.2 Location of Japanese-owned manufacturing facilities in the USA

## JAPANESE ELECTRONICS R&D IN THE USA: INTERNATIONAL BUT NOT YET GLOBAL?

Electronics is in practice a diverse sector of production that involves a variety of products of different levels of technological complexity and different market dynamics. Certain electronics products, such as printed circuit boards and computer keyboards, are relatively low-technology goods; manufacture of these products is now dominated by low-cost producers in Taiwan, South Korea and elsewhere. Other segments of the electronics industry involve immensely complex production technologies, as exemplified by advanced semiconductors and thin-film-transistor flat-panel displays. These are technologies in which Japanese firms, with their close integration of R&D and manufacturing and their emphasis upon incremental innovation in the production process, have excelled. The targets of much recent Japanese R&D activity in the USA – design-intensive semiconductors, computer servers and workstations, software, and so on – are generally characterized by high levels of technological sophistication *and* rapidly changing market demand and application opportunities.

As the recent experience of IBM, DEC and other companies has shown, it has proved difficult for many large bureaucratic organizations to achieve the twin goals of sustaining both technological advantage and speed-to-market in these segments of the electronics industry. Most successful US electronics firms have over the past five years extensively reorganized their manufacturing operations in an effort to reduce the development time for product and process technologies (Angel and Engstrom, 1995). Japanese firms attempting to expand market share in the US electronics industry face the additional challenge of managing the technology development process on an international scale. Various models of the global corporation have been articulated by Japanese electronics firms to meet the challenge of short product development cycles and rapidly changing market opportunities on a global scale (most notably the concept of global localization). In practice, however, there is now growing evidence that the ways in which Japanese R&D laboratories in the USA have been integrated into the overall technology development structure of their parent firms have undermined the ability of these firms to respond rapidly to changing market demand and technological opportunities in the USA. The large number of Japanese R&D laboratories in the USA is certainly an indication of a substantial internationalization of R&D, but not a fully developed globalization of technology development capability.

Partial documentation of the pattern of industrial organization is provided by the results of a mailed questionnaire survey of Japanese R&D laboratories in the USA. The survey was mailed in 1993 to all free-standing Japanese R&D laboratories in the USA. We report here only on the responses provided by 25 participant laboratories involved in the electronics industry (the survey response rate for electronics firms was 34.2 per cent). Results for automobiles and other industries, as well as detailed discussion of the survey itself, are provided in Angel and Savage (1996). Respondents were asked to provide information on the organization of their R&D activities, the pattern of contact with manufacturing facilities, and the structure of decision-making at their US laboratories.

Drawing upon data collected in the questionnaire survey, Table 17.1 shows the relations maintained by US R&D laboratories with various other units of the parent Japanese electronics firm (the corporate headquarters in Japan, the main research laboratory in Japan, and so forth). Respondents were asked to rate the importance of each of the different units of the firm on a five-point scale (1 indicating not important, 5 indicating extremely important). The two most important sources for product ideas at the sample R&D laboratories were customers (mean score 3.48) and central R&D laboratories in Japan (mean score 3.60), and the corporate headquarters in Japan (mean score 3.28). When reporting the results of research, the primary destinations for research results are the corporate headquarters in Japan (mean score 3.88) and the main research laboratory in Japan (mean score 3.46). What is striking here is that the primary intra-organizational axis is between the US laboratory and the central headquarters and main research facility in Japan. The sample R&D laboratories have little contact with the US subsidiary headquarters, or with manufacturing facilities in Japan or the USA. As we might expect, the sample laboratories have considerable autonomy in day-to-day decision-making over the direction and budgets of research. Beyond the laboratory itself, the corporate headquarters in Japan is the unit with the most substantial involvement in decision-making.

Table 17.2 shows complementary information on the structure of contact, by mail, phone, fax and in person, between the sample laboratories and other units of the parent corporation. The highest frequency of contacts by mail, phone or fax are with the corporate headquarters in Japan (mean score

**Table 17.1** Intra-organizational linkages of Japanese electronics R&D laboratories in the USA

| | US subsid. | Japan HQ | Japan R&D lab. | US prod. | Japan prod. |
|---|---|---|---|---|---|
| Importance as source for project ideas | 2.24 | 3.28 | 3.60 | 1.64 | 2.16 |
| Frequency with which you report results | 2.64 | 3.88 | 3.46 | 1.48 | 2.00 |
| Decisions on the direction of R&D | 2.16 | 3.88 | 3.67 | 1.52 | 1.92 |
| Decisions on the budget for R&D | 2.21 | 4.00 | 3.12 | 1.40 | 1.80 |

*Source*: questionnaire survey.
*Notes*: Variables are measured on a five-point scale (1 = low, 5 = high). Number of cases: 25.

**Table 17.2** Contact patterns of Japanese electronics R&D laboratories in the USA

| | US subsid. | Japan HQ | Japan R&D lab. | US prod. | Japan prod. |
|---|---|---|---|---|---|
| Contact by phone, fax or e-mail | 2.92 | 3.96 | 3.68 | 1.72 | 2.20 |
| In-person visits to facility | 2.52 | 2.92 | 2.88 | 1.68 | 2.00 |
| In-person visits from facility | 2.48 | 3.00 | 3.00 | 1.56 | 2.00 |

*Source*: questionnaire survey.
*Notes*: Variables are measured on a five-point scale (1 = low, 5 = high). Number of cases: 25.

3.96) and the main research laboratory in Japan (mean score 3.68). Broadly the same pattern is observed with respect to in-person visits, though in this case visits to and from the US subsidiary headquarters are similar in frequency to visits to Japan. Once again, the frequency of contact with manufacturing facilities is low. Even at this low level, however, the sample laboratories are more likely to be in contact with manufacturing facilities in Japan than in the USA.

Rather than serving as a semi-autonomous platform for technology development, the majority of R&D laboratories operated by Japanese electronics firms in the USA function as specialized satellite operations, or listening posts, that feed technological knowledge and information back to central R&D laboratories in Japan. The locus of technology development for global markets is in Japan.[1] Typically, initial production of new products also takes place in Japan; production is shifted to high-volume facilities in the USA once the manufacturing process has been fully stabilized. As a result, there is little contact between Japanese R&D laboratories and affiliated production facilities in the USA.

These results, based on a small sample of laboratories, are supported by the conclusions of other studies on the operation of Japanese electronics R&D laboratories in the USA. Thus Serapio (1994) found that Japanese R&D laboratories in the USA were subject to a greater degree of centralized decision-making than was the case for US-owned laboratories in Japan. About 60 per cent of the Japanese R&D laboratories in the USA studied by Serapio (1994) followed a centralized mode of decision-making (i.e. the research agenda is set by the parent company), as compared to only 30 per cent of US R&D laboratories in Japan. Over 80 per cent of the Japanese R&D laboratories and about 40 per cent of US-owned R&D laboratories have their budgets determined by the parent company, rather than by the local subsidiary organization (Serapio, 1994). Westney (1993: 178) draws the following conclusion from her case study analysis of Japanese-owned electronics R&D facilities in the USA:

> There is great variation across the other corporate and divisional-level facilities in size, technology mandate, and management systems. But one commonality is their relatively high specialization and focus: each lab tends to work in one technology or product area, and to lack the capacity to generate a complete product. In consequence, they must interact closely with the Japanese parent technology organization.

No doubt this centralized organizational structure has many advantages, including the close integration of innovation and production for which Japanese firms are renowned, as well as the ability to integrate effectively different kinds of technologies (what Kodima (1992) calls technology fusion). It is interesting to speculate, however, as to whether this organizational structure also has detrimental impacts upon the ability of Japanese firms to respond quickly to changing market opportunities in the USA. Certainly the chain of decision-making and of response is extended in organizational and geographical terms by the relatively low level of autonomous technology development authority accorded to Japanese R&D laboratories in the USA. As one observer has noted,

> 'despite their massive investments in U.S. sales and marketing, manufacturing, and even research and development (R&D facilities), few of these (Japanese) companies possess the full range of institutional skills needed for globalization. The necessary approaches to planning, measuring, rewarding, communicating, and day-to-day decision making all fly in the face of the centralized functionally driven style of most Japanese MNCs [multinational corporations]. (DeNero, 1993: 169)

Significantly, many Japanese electronics firms are now re-examining the organizational structure of their international technology development activities in an effort to accelerate the pace of decision-making and innovation. Sony and NEC among other companies have begun to incorporate organizational forms now common in large US electronics firms, such as the creation of semi-independent internal companies, and greater decentralization of decision-making from central to divisional units. Commenting on changes at Sony, one manager notes, 'We are having to deal less with the corporate headquarters since most of the decision making is done within the [in-house] company . . . . I was always aware of the size of the company getting in the way of new business decisions' (*Financial Times*, 1994). Ristelhueber (1994: 58) reports that Fujitsu is preparing to assign financial and operating responsibility to operating units outside of Japan: 'Under that arrangement, the logic products division, based in San Jose (California), would have the responsibility for defining, designing, and marketing logic chips like the SPARC microprocessor; the same setup would apply in Europe for communications circuits.' Hitachi, Mitsubishi and NEC all have recently reorganized their R&D units in the USA. These changes suggest that at least some Japanese electronics firms are identifying problems associated with the existing international technology development structure.

## ALLIANCES OF THE STRONG

A second key element of the internationalization of R&D by Japanese electronics firms is the proliferation of technology partnerships with foreign firms. Most of the policy debate concerning US–Japan technology partnerships has focused upon alliances between large Japanese corporations and small start-up companies in the USA. In many cases these partnerships involve the licensing of the US firm's technology to the Japanese partner in exchange for capital investment, guaranteed access to advanced production capacity, and assistance in marketing products in Japan. A number of observers have suggested that such alliances involve a trade-off of short-term gain against long-term disadvantage. While US firms may benefit from an infusion of capital or by increased market access, the longer-term effect is to strengthen the technological capability of Japanese manufacturers relative to US firms. In the semiconductor industry, for example, the partnership between Hitachi Semiconductor (Japan) and VLSI Technology has been crucial in allowing Hitachi to gain an important position in the market for application-specific integrated circuits. Partnerships with HAL Computer and with Sun Microsystems have been a key part of Fujitsu's attempts to diversify out of mainframe computers into the manufacture of computer workstations. In the semiconductor industry, alliances between large Japanese firms and small US firms make up approximately two-thirds of all US–Japan alliances (National Research Council, 1992).

But are these highly contested asymmetrical alliances representative of technology partnerships between US and Japanese firms? Preliminary results from survey research suggests that while partnerships with small start-up firms may be numerically dominant, Japanese electronics manufacturers in most cases ascribe greatest importance to partnerships with large foreign firms. In April 1994 we conducted a questionnaire survey of the 100 largest electronics manufacturers in Japan, seeking to gather information on the experience of these firms regarding international technology partnerships. Of the 100 manufacturers contacted in the survey, 38 firms completed the questionnaire and a further 9 firms returned the questionnaire indicating that they currently had no international technology partnerships. The questionnaire covered a range of issues concerning the organization of technology partnerships, obstacles encountered in managing alliances, success in achieving partnership goals, and so forth. The results of the questionnaire were supplemented by interviews with senior management officials at eight of the largest electronics firms.

We report here on just one interesting result from the questionnaire survey, namely the importance ascribed by Japanese firms to partnerships with large companies, as opposed to small start-up firms. In the questionnaire survey, Japanese electronics manufacturers were asked to identify the one international technology development partnership that was judged to be of greatest technological and commercial significance to the firm. For 30 (78.9 per cent) of the 38 respondent firms, that partnership was with a large firm with sales in excess of $500 million. In 28 (73.6 per cent) of 38 cases the partnership was with a US firm; the remaining partnerships were with companies based in Germany, South Korea, Taiwan, Singapore and Australia. Most of the alliances are relatively recent in origin, with 24 (63.1 per cent) established during the 1990s.

Table 17.3 shows the importance ascribed by the survey respondents to different reasons for entering into these alliances. Respondents were asked to rate the importance of each reason on a five-point scale (1 indicating not important, 5 indicating very important). As we might expect, the primary reasons for the Japanese firms entering into the alliances are accelerating the pace of technology development (mean score 4.53), accessing the technology of the partner firm (mean score 3.91) and sharing the cost of technology development (mean score 3.70). What is of particular interest here is that in the vast majority (81.5 per cent) of cases, these goals were to be pursued through *joint* research, rather through a licensing agreement, or an arrangement that provides access to technology in exchange for production or marketing assistance. In other words, these are partnerships in which predominantly large firms in Japan and the USA have agreed to co-operate towards the goal of reducing the cost and accelerating the pace of technology development. Typically such 'alliances of the strong' are entered into not from a position of weakness (e.g. lack of capital), but from a desire to leverage existing strengths for further growth and accelerated technology development.

**Table 17.3** Reasons for Japanese electronics firms entering into technology partnership

| | |
|---|---|
| Share cost of technology development | 3.70 |
| Accelerate technology development | 4.53 |
| Access technology of partner firm | 3.91 |
| Access to production facilities | 2.35 |
| Access to foreign markets | 3.50 |

*Notes*: Reasons are scored on a five-point scale (1 = not important, 5 = very important). Sample size: 38 firms.

Partnerships between Texas Instruments and Hitachi and between ATT and NEC are illustrative of these alliances involving large US and Japanese firms.[2] Of the 11 technology development partnerships in which Hitachi Semiconductor is currently involved, the company identified its partnership with the large US electronics firm Texas Instruments as being of greatest significance to its technology and market development activities. This alliance, initiated in 1989, focuses upon the development of DRAM process technology. The primary motivation for entering the alliance concerned the need to share the rapidly rising costs of developing next-generation process technology. In selecting Texas Instruments as a partner, Hitachi specifically sought out a firm of comparable technological strength in semiconductor technology. The two firms currently maintain a joint process technology development centre in suburban Tokyo, staffed by engineers from both Hitachi and Texas Instruments. The partnership is widely viewed by both partners as a success in strengthening their technological capability (for a discussion of the partnership from the perspective of Texas Instruments, *see* Angel, 1994). The initial partnership has been renewed on three separate occasions, progressively moving to greater levels of co-operation between the two partner firms. One senior Hitachi engineer (Dr Kazuya Kadota) compared the progress of the alliance to a dinner party: 'In 1989 the two companies joined together around the dinner table; now they do the cooking together. In addition to sharing the costs of process technology development, Texas Instruments has benefited from Hitachi's experience in high-volume integrated circuit manufacturing.'

While not the primary cause of the recent strong performance of US firms in semiconductors and other electronics industries, many technology alliances between US and Japanese firms have worked to the mutual benefit of both partners. Interestingly, in another study Bleeke and Ernst (1993: 2) draw similar conclusions in reporting that among a sample of 49 global business alliances studied, two-thirds of those involving balanced partners (partnerships of the strong) succeeded whereas two-thirds of the alliances involving unequal partners failed.

What are the implications of these findings for our understanding of globalization? Two issues are of particular importance. First, it is clear that the process of globalization remains in flux and it is important to resist premature closure of debates as to the precise ways in which the internationalization and globalization of R&D are taking place. As with other aspects of industrial organization, firms learn by doing and modify practice based upon experience. Second, the different aspects of technology development practice can reveal alternative models of globalization. Thus in the area of technology alliances, there is growing evidence of a 'network' model of globalization that diverges from the dominant rhetoric of global localization.

## CONCLUSION: CONSTRAINED GLOBALISM?

Internationalization of manufacturing has been a central feature of the Japanese electronics industry for much of the past decade. Driven on by a rising yen and recession in Japan, processes of internationalization have intensified during the 1990s in what amounts to a remarkable shift in the geography of Japanese electronics production. The amount of overseas production of colour televisions carried out by Japanese firms, for example, is now more than twice that of domestic production. In addition to the large-scale shift of production out of Japan, Japanese electronics manufacturers have also engaged in a substantial internationalization of their technology development capability. Japanese manufacturers have for many years sought to access foreign technology through patent and licensing agreements. US and European electronics firms typically licensed Japanese manufacturers as additional sources of supply for products and production processes. During the 1980s, Japanese firms expanded beyond this focus upon patents and licensing agreements as their primary mechanism for accessing foreign technology, opening large numbers of R&D laboratories in the USA and elsewhere, and establishing large numbers of technology development partnerships of one type or another with foreign firms. By the end of the 1980s, total R&D performed by US affiliates of Japanese electronics firms exceeded $1.2 billion.

In this chapter, we have highlighted two features of the international technology development practice that contradict conventional rhetoric concerning Japanese manufacturing firms. The first feature comprises the way in which R&D laboratories located in the USA have been incorporated into the overall technology development structure of Japanese electronics firms. The dominant rhetoric is one of global localization; that is, Japanese firms would be expected to build an integrated manufacturing presence, comprising R&D, production and sales, in each of the major world markets of Asia, Europe and North America. In practice, Japanese R&D facilities in the USA have tended to operate as satellites of the main technology development presence in Japan, maintaining only tenuous links with production facilities in the USA. The second feature concerns the nature of US–Japan technology partnerships. The policy debate here focuses on asymmetrical alliances between large Japanese firms and small US start-ups, typically involving a trade of technology for cash or market access. Our own research, by contrast, indicates that Japanese electronics manufacturers judge so-called alliances of the strong as being of greatest importance to their technology development activities. These are alliances in which both firms typically enter the partnership from a position of strength, and if the alliance is successful, both derive benefits from the relationship.

More generally, our research suggests that the globalization of technology development capability in Japanese electronics firms is a partial process that remains in flux, still heavily constrained by a dominant core in Japan. It is interesting to speculate as to what contribution this constrained globalism has made to the relative market performance of Japanese and US electronics firms during the 1990s. This issue must be treated with considerable caution; the full effects of Japanese investment in the US electronics industry (e.g. in basic research) will not emerge for some time. Moreover, whatever the organizational characteristics, the sheer volume of Japanese R&D investment is likely to have a substantial impact upon the technological capability of participant firms. Nevertheless, the fact that Japanese manufacturers are beginning to restructure their international technology development activities suggests that the current organizational and geographical form is to some degree constraining the ability of Japanese firms to develop and deploy technologies rapidly on a global scale.

Much recent work has stressed the existence of distinctly national systems of innovation in Japan, the USA and elsewhere (Nelson, 1993). The growing internationalization and globalization of economies has stimulated debate as to whether such differences are likely to disappear as economies converge around global 'best practice' (*see* Berger and Dore, 1996). Our own research suggests paradoxically the presence of a series of national models of internationalization within advanced industrial economies.

## ACKNOWLEDGEMENTS

This research is supported in part by an Abe Fellowship awarded by the Center for Global Partnership in conjunction with the Social Science Research Council and the American Council of Learned Societies.

## NOTES

1. Sony Corporation, for example, has two corporate laboratories and 11 divisional laboratories in the USA. Nevertheless, approximately 90 per cent of total R&D investments are still concentrated in home office operations in Japan.
2. This section of the chapter is based upon interviews conducted in Tokyo with Mr Shigeki Matsue, Vice-President of NEC Semiconductors on 20 May 1994, and with Mr Koichi Nagasawa and Kazuya Kadota, department managers at Hitachi Semiconductor on 28 June 1994.

# CHAPTER EIGHTEEN

# FALLING OUT OF THE WORLD ECONOMY? THEORIZING 'AFRICA' IN WORLD TRADE

JOHN AGNEW AND RICHARD GRANT

## INTRODUCTION

In recent years 'globalization' has become a key term for organizing our thinking about contemporary changes in world economic geography. Processes of globalization are seen as leading to a profound geographical reorganization of capitalism and the role that particular geographic units, such as states or regions, play within the system. To date the processes of globalization have been studied mainly as transformations affecting developed countries and their highly specialized economic activities such as trade, finance and communications. In our view both the scale and scope of globalization have been exaggerated, and vast areas of the world, such as Africa, are either being excluded from globalization of production and finance or marginally affected by them. Africa's continued experience with underdevelopment represents a countervailing trend to the spread of economic growth associated with versions of the globalization thesis.

Africa has been largely ignored in contemporary economic geography research. Economic geographers have been preoccupied with 'success stories' in the world economy such as the NICs, silicon valleys and Asia-Pacific dynamism. Less attention has been directed to more marginal areas of the world economy such as Africa. Despite the fact that Africa has over 450 million inhabitants, comprises one-quarter of all the states of the world, and occupies a landmass three times the size of the USA, Africa is largely overlooked. Still worse, Africa is often represented as a homogeneous, undifferentiated unit and the region has come to connote a 'coming of anarchy' (Kaplan, 1994), decay, crisis, war, etc. (Watts, 1991). Huge continent-wide generalizations are often made on the basis of single-country or subregional studies (Agnew and Grant, 1996). As a result, African experiences are simplified and generalized to such an extent that they may be meaningless.

In this chapter we examine theories that explain the position of Africa in world trade. A widely accepted view about present world trade is that the African continent is increasingly marginal to it. Indeed, a number of authors have gone so far as to claim that 'Africa' is falling out of the world economy (e.g. Bayart, 1992; Krueger, 1992; Terlouw, 1992). Authors differ, however, as to why they think Africa is in such dire straits. Three general groupings of theories – neo-classical, dependency and state-centred – are most frequently offered as explanations for why Africa has been marginalized in world trade. After briefly reviewing these types of theory, and their shared stereotypes about 'Africa', we propose that each theoretical perspective only captures part of the 'African' experience. But each does so not as part of an eclectic 'super-theory' but in relation to a particular epoch of Africa's evolving relationship with the world trading system. The world trading system has changed over time and the appropriateness of particular explanations for Africa's place within it likewise varies temporally. Each of the theories best captures the dominant trading practices of one of three time-periods and should be deployed accordingly. Some elements identified by each theory – for example, the mix of commodities traded or the continuing impact over time of biases in infrastructure – have persisting effects over time that are even acknowledged as such by the proponents of theories that do not focus on them. This is why the theories can make sense 'outside' their time-periods and have a persisting popular appeal. But the overall appropriateness of the theories is not set for all time; the 'power' of a theory varies depending on its fit to historical-geographical circumstances.

By way of example, dependency theory captures in large part the nature of the colonial trading system and its imperial trading preferences (1930s to 1950s). State-centric explanations fit best the period of national development in the 1960s and 1970s when import substitution and other interventionist policies were widely promoted throughout Africa. Neo-classical theory bears a close relationship to the paths followed in response to IMF/World Bank structural stabilization programmes from the early 1980s onwards that are an aspect of the emerging hegemony of transnational liberalism within the world trading system. During different historical epochs, different geographies of trade have prevailed in which 'Africa' has had different roles and in which different geographical scales have characterized dominant trading relationships: imperial/local in the first, international/national in the second, and global/national/local in the third. Existing theoretical accounts overreach both historically and geographically. Using a range of empirical evidence, this chapter proposes a historicization of theory to match the prevalent geographies of trade. The topic of 'Africa' in world trade, therefore, speaks to an important feature of contemporary social theory: the need to place theories in historical-geographical context.

## 'FALLING-OUT' THEORIES AND IMAGES OF AFRICA

Writing on Africa and world trade is filled with comments about the marginality of the continent in the contemporary world economy. To Bayart (1992: 208), for example, Africa is 'being erased from the map of world capitalism'. To Krueger (1992: 467), sub-Saharan Africa's trade performance shows it as 'the world's "problem area"'. For Terlouw (1992: 88), commenting on the trajectory of Africa's involvement with the world economy, 'almost the whole of Africa had less structural ties with the world-system after 1972 than before'. This leads him to conclude that the recent experience of Africa is of 'excorporation' rather than continuing 'incorporation' into the modern world economy (Terlouw, 1992: 106). In 1988 Africa as a whole, a region with 450 million inhabitants, had export revenues below those of Singapore, a city-state of 2.5 million people (Svedberg, 1991: 549). What is clear is that after growing at a generally rapid pace in the century from 1870 to 1970, Africa's participation in world trade has recently collapsed. In the 1970s export growth subsided and in the 1980s there was a huge decline (*see*, for example, Svedberg, 1993) .

Three different approaches to theorizing about Africa in world trade animate the fateful prognosis about 'Africa's' marginality. The first is that of 'dependency theory'. With respect to trade, its major tenet is that resource-export earnings (Africa's main offering in world trade) are an unsatisfactory basis for economic development. It is dependence on the export of basic commodities to the old colonial masters that has inhibited the growth of trade and the development of more diversified economies. The reproduction of the colonial economy – if now in the form of institutional relationships to the European Union as expressed in the Lomé Convention and GSP (Generalized Systems of Preferences) schemes – is at the root of Africa's continuing marginality in the world economy (Mahler, 1994).

The second perspective draws attention to the divergent export performance of different African countries, suggesting that government policies and state–society relations are fundamental to the declining aggregate performance of Africa in world trade. The outcome of bargaining with multilateral institutions on the timing and sequencing of trade reforms is a specific aspect of relative performance (Booth, 1991). World market conditions for Africa's basic commodities have been no worse than those for a wider range of goods produced in other world regions. So it is circumstances internal to states that are responsible for trade performance. In particular, political barriers to rational economic strategies (favouring urban food consumers over farmers in pricing policies, subsidizing some producers over others on the basis of ethnic identity, etc.) restrict the growth of efficient trading sectors. Moreover, the tendency of political elites to blame their economic woes entirely on 'the North' undermines their capacity to address national and local barriers to change.

The third, and most dominant, perspective in the literature on African trade is that of mainstream (neo-classical) economics. In this perspective, growth in trade is the result of countries (or other spatial units) establishing their own niche within the world economy and then exchanging goods and services with others specializing in different ones. The liberalization strategies mandated by multilateral agencies such as the IMF and the World Bank can serve to expand the scope of comparative advantage in export production by encouraging African governments to remove the policies that limit specialization and the growth of trade. Government intervention (and politics), therefore, are the 'shackles' that have limited the growth of African trade.

Each of the types of theory has assorted facts to back it up. Each also emphasizes different empirical measures, making theoretical comparison problematic. For example, there is evidence that illiberal trade policies have contributed over the years to Africa's overall trade deficit (theory number 3); the composition of Africa's trade supports theory number 1; and the evident importance of politics in determining trade policies validates theory number 2. But each piece of evidence reflects factors that emerged in different historical epochs and have had different causal trajectories over the years. Each theory, however, presents 'Africa' as a self-evident entity that can be fitted into a specific universalist understanding of trade as either a global (dependency) or state-level (neoclassical and state-mediator) phenomenon about which singular generalizations can be made.

All of these theoretical perspectives are informed, therefore, by a specific image of 'Africa' as a world region (Grant and Agnew, 1996). One aspect of this is an implicit neglect of the size and complexity of Africa. Geographical generalizations are made on the basis of studies of one or few countries that are then interpreted as 'typical' of Africa as a whole. There is an associated temporal 'telescoping' of African trade experience such that, irrespective of the character of the world economy at any time, Africa can be understood in terms that stay the same. The same 'suspects' are always lined up to account for 'Africa's failure', be it bad economic policy-making (theory number 3), colonialism (theory number 1), or rent-seeking and corruption by ethnic groups (theory number 2). Africa's niche in the world economy is still understood as that laid out in the late nineteenth century. The terms that could be used then to explain its position in world trade continue to apply today. The representation of the 'African state' adds a final component to the image of pervasive and persisting economic incompetence. In the context of world trade it glosses over the unique ways in which different countries were incorporated into the world economy. Typically, an archetypal African state (or 'quasi-state') is compared to an ideal-typical Western state against which it is seen as a failure. Though actual Western states are characterized by plenty of rent-seeking behaviour and systematic criminality, this tends not to figure in comparisons. African states are pictured as singularly venal.

Though Africa's position in world trade has worsened, therefore, dominant explanations do not account for this so much as project already deployed concepts on to the purported present-day situation. The causes of falling out, therefore, are the same as the causes of Africa's previous participation in world trade. Nothing need be changed in accounting for the contemporary position of Africa in the world economy.

## SYNCRETIC SOLUTIONS TO THE THEORETICAL IMPASSE

One solution to the 'deadlock' between theories is to develop alternative accounts of African trade that either stress the 'contingency' of all the theories (a theory identifies 'necessary relations' that are often vitiated in practice by 'local' conditions which other theories may identify; e.g. Miyoshi, 1993) or propose a 'super-theory' that combines elements of all of them (using the concept of 'contingency' as a means of relating the distinctive theoretical elements; e.g. Anderson and Bone, 1995). The first strategy has become increasingly popular in development studies. In this construction, the overall thrust can be maintained – dependency theory, perhaps – but local conditions – a particularly 'corrupt' state, perhaps, and other features indicating 'human agency' – are included as a contingency. Many world-system studies are of this type; responding to critiques of the lack of attention to 'the state' in the main statements of world-system theory by bringing it back in as a 'local' contingency. The most recent example of an approach that would 'rescue' dependency theory yet allow for local differences as an integral part of the theory is 'post-colonial' theory. Common largely in literary studies, this perspective claims that once brought into the linear development framework of the 'secular West' ('chronopolitics'), 'colonised space cannot reclaim autonomy and seclusion' (Miyoshi; 1993: 730). Though alert to the local differences that colonialism always strove to erase, by its own global framing the perspective tends to replace a timeless Africa (or wherever) with a timeless world divided geographically into two hostile and mutually unintelligible blocs. Africa is still distant and passive, but now as part of a larger disenchanted sphere. In a trade context, this perspective emphasizes that the trade strategies of particular states are now determined largely by the actions of an international corporate elite.

The second strategy is more sophisticated in that it recognizes that there may be a mutuality of interests between global business, on the one hand, and states at different stages of development, on the other (Corbridge, 1990). From this perspective, what is critical is the need to combine a global with a national or a local 'level' of theory. So-called regulation theory is offered as one approach that can do so; at least to a certain extent. This combines the idea of a global 'regime of accumulation' with that of a national or local 'mode of social regulation'. For most of the twentieth century a Fordist regime of accumulation is held to have prevailed globally. This has recently disintegrated and is being replaced by a regime of 'flexible' accumulation. The mode of social regulation refers to the ways in which economic transactions are institutionalized and managed in different geographical areas. The coupling of a mode of regulation with a regime of accumulation is seen as defining a 'mode of development': the local and national process of growth generation. Together they define a 'functional' whole: a system of relationships that are mutually supporting and persist through time and over space. But this system maintenance is also created through the activities of key groups and individuals; so a regulation approach should be supplemented through attention to local entrepreneurial classes and/or 'grassroots' development activists (e.g. Anderson and Bone, 1995) as well as customs, social norms, enforceable laws and local state organization (e.g. Peck and Tickell, 1992).

Apart from regulation theory's relatively crude distinction between a historical epoch in which a 'Fordist' regime of accumulation prevailed globally and the recent trend towards flexible accumulation, neither strategy of theorizing would pay any attention to the historical course of Africa's (or any other world region's) interaction with the world economy. The attempt is still to a offer an aspatial, (largely) timeless, structural rendering of the course of economic development (and trade). More specifically, both of the syncretic approaches fail to address trade as a central issue, implicitly render world trade as exchange between two undifferentiated (but empirically fuzzy) global regions – core and periphery – and offer no implications as to what course of action 'Africa' might best follow in world trade. In a more philosophical vein, both strategies indicate a certain cast of mind. Faced with contradictory evidence but wanting to preserve a commitment to timeless theory, syncretic solutions propose bringing everything in. But this is to avoid rather than resolve theoretical disputes. In the memorable words of the historian Marc Bloch (1950: 3), 'If your neighbour on the left says two times two equals four, and the one on the right says it is five, do not conclude that it is four and a half.'

## THEORIES OF DOMINANT PRACTICES

Part of the problem in evaluating types of theory involves the lack of consensus about what theories are supposed to do. From a positivist point of view, they are supposed to account (fairly) completely for all instances of a phenomenon (be it trade flows or whatever) at all times and in all places. An alternative perspective, however, regards theories as heuristic or 'thought-producing' devices for identifying the most important causes of a phenomenon relative to scope conditions which set limits (historically and geographically) to the generalizations that can validly be drawn (Schrag, 1975). In the social sciences, given that attention is focused on the activities of knowing historical subjects, this point of view translates into seeing theories as capsulating accounts of dominant practices situated historically and geographically. In this understanding, trade flows are best explained in terms of those theories that capture the main features of practices in different historical-geographical epochs as they affect trade flows. It is at the 'level' of epistemology (how we think), therefore, that theory must be defined historically-geographically (Livingstone, 1995). Simply adding 'space' and 'time' (in the form of case studies of regions at different time-periods) to timeless or spaceless theories is radically insufficient, and misses the point.

In this light, the three theories sketched earlier take on a new significance. Dependency theory fits best the historical-geographical conditions under which African trade took place during the colonial and early post-colonial periods. State-as-mediator theories account best for the period of 'national development' in the 1960s and 1970s. The neo-classical theory parallels closely the practices of the 'structural stabilization regime' that has affected most African states since the mid-1980s. These periods fit closely those identified by recent work in critical international relations theory (Cox, 1987; Agnew and Corbridge, 1995). The period from the 1870s until 1945 was one of inter-imperial rivalry in which competing empires attempted to dominate the world territorially. From 1945 until the 1970s a Cold War geopolitical order prevailed in which the USA and the former USSR vied for influence in the Third World (including Africa) by sponsoring competing models of political-economic development. Since the 1970s a global neo-liberalism has been ascendant, in which multilateral institutions such as the IMF and the World Bank have sponsored a model of development based on specialization within a worldwide division of labour. Trade practices and Africa's position within them, therefore, correlate with different periods of geopolitical order. Different theories are needed to explain what has happened in each period.

## THE IMPERIAL TRADING SYSTEM

From the late nineteenth century until independence in the 1950s and early 1960s, most of Africa was involved in a trading system based on the exchange of goods within tightly defined imperial networks. Initially, this was a straightforward exchange of primary commodities from specific African colonies with a (limited) export of manufactured goods from specific European imperial centres (particularly the UK and France). Regional specialization within a territorial empire was the economic motif of this period. Single colonies specialized in the production of specific raw materials, food crops or beverages (such as tea, coffee and cocoa). 'Choice' of commodity was usually imposed through either force or market domination. Some colonies, without initial advantages in raw materials, climate, social organization or accessibility, became providers of labour to the primary commodity economies (e.g. the Sahelian colonies to the north of the coastal ones in West Africa).

At the global scale, the imperial trading system was bound together by a network of steamship and communication routes and 'home' markets in Europe. Within African colonies, railway and road connections often converged on a limited number of coastal ports. These became privileged locations, acquiring administrative as well as economic functions for the entire colony; examples include Lagos, Dakar and Cape Town. Inter-colony (and intra-colony) ties were weak, principally because different European empires ruled contiguous colonies and exporters seldom looked for adjacent markets. A long-run external bias in communications followed from this early pattern of transportation orientation.

Although always less important in its contribution to world trade than Asia and Latin America, Africa contributed a steadily increasing proportion of world exports and imports throughout this period; from 3.7 per cent of world exports in 1913 to 5.3 per cent in 1937 and from 3.6 per cent of world imports in 1913 to 6.2 per cent in 1937 (Yates, 1959: 32–3). It did so as part of a multilateral system of trade, even though its own trade was overwhelmingly with the European colonial countries. The industrializing countries of Europe and North America bought large amounts of African raw materials. The UK, as a result of its free trade policy (operative down until the 1920s), ran substantial trade deficits as a result of importing manufactured goods from such industrializing countries as the USA and Germany. In turn, the UK financed its deficits through the export of its manufactured goods to its colonies, such as those in Africa. Thus the circle of international trade and dependence was closed.

In the late 1920s this system suffered a number of blows from which it never recovered. One was the overall decline in trade as the world economy experienced a major depression. The increased efficiency of colonial production had led to falling prices and the subsequent collapse of demand for manufactured goods. At the same time, the UK faced increased competition in manufactured goods from Japan, India and China in its traditional colonial markets in Asia and Africa. The second was the decline in the UK's relative position as the linchpin of the imperial trading system as a result of increasing inter-imperial rivalry. The UK was left by the early 1930s with an 'increasingly asymmetric bargain' (Stein, 1984: 375). Other states devalued their currencies and imposed import tariffs in the belief that the UK would remain 'open'. Yet in order to respond to the UK's own economic collapse, British governments chose to close off the empire from a deteriorating world economy.

From the 1850s until the 1930s the UK and France were the largest and joint second-largest traders, respectively, in the world economy. France shared second place with Germany and the USA for most of this period (Kuznets, 1966: table 6.3). Between 25 and 35 per cent of the UK's trade was with its colonies and between 10 and 18 per cent of France's trade was with its colonies. There were some minor imperial tariff preferences before the 1930s, but the Depression produced a radical increase in their incidence and intensity. In the British case the preferential margin on within-empire trade was increased by around 10 per cent on British exports and around 12 per cent on British imports between 1929 and 1937 (MacDougall and Hutt, 1954: 237). This gave rise, as did similar preferences within the French and other European empires, to a significant jump in imperial trade shares. From 1928 to 1948 (the UK) and 1955 (France), the UK's and France's trade with their colonies went from 38 to a 48 per cent share and from 16 to a 27 per cent share, respectively (share = annual average of merchandise exports and imports; Anderson and Norheim, 1993: 92–5). The disintegration of the multilateral imperial trading system produced, therefore, an intensification of the trade dependence between the European imperial powers and their colonies (in Africa and elsewhere).

This pattern was not to last. Even though imperial trade preferences represented a commitment to an image of empire that imperial ideologues such as Joseph Chamberlain and Halford Mackinder in Britain (and others elsewhere) had proposed at the turn of the century, their implementation had been a response to an economic emergency. The dismantling of the European territorial empires under American pressure (and as a result of successful colonial independence movements) led to a reduction of tariff preferences and a decline of imperial trade. British trade within its (former) empire declined from 48 per cent of total trade (average of all exports and imports) in 1948 to 10 per cent in 1989. Likewise French trade within its (former) empire went from 27 per cent of the total in 1955 to 3 per cent in 1989 (Anderson and Norheim, 1993: 92).

Trade with Africa persisted at somewhat higher levels than with elsewhere in the two former empires, largely because most African countries enjoy some preferential access to European markets under the Lomé Convention. At the same time, trade between African countries has expanded considerably, perhaps as result of worldwide declines in trade barriers since the 1960s (Anderson and Norheim, 1993: 95). The average number of trading partners for African countries has undergone a significant expansion (Fig. 18.1), indicating an erosion of the dependence on single colonial markets. What is even more vital theoretically is that trade between Africa and Europe no longer serves the development of Europe at Africa's expense, the major tenet of dependency theory.

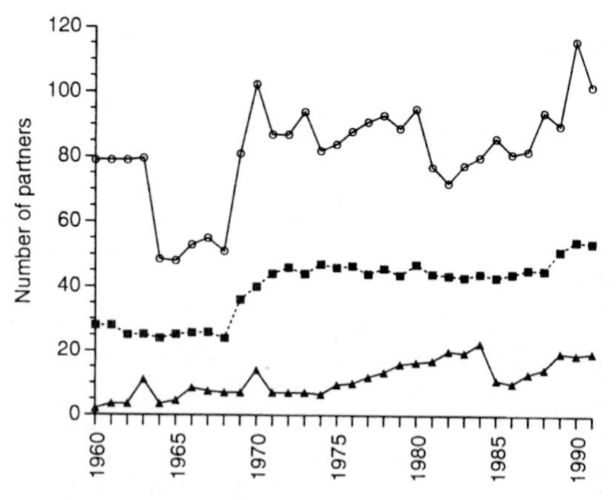


**Fig. 18.1** The average number of trading partners for African countries, 1960–91

Trade within Europe and with other industrialized world regions (such as North America and East Asia) is now at the centre of the world trading system, involving considerable intra-industry (and intra-firm) trade rather than the 'classic' exchange of raw materials for manufactured goods that at one time made Africa important to Europe's economic growth (*see* Grant, 1994). Even though under the Lomé Convention two-thirds of Africa's exports to Europe are *still* primary exports, total exports from Africa are now also part of the developing trend of intra-firm trade. For example, Page (1994: 128) estimates that 21.7 per cent of total African imports into the USA come from the subsidiaries of USA firms in Africa. Needless to say, however, these imports are not vital to American economic development.

## NATIONAL DEVELOPMENT

The example of the USA and its institutionalization in the UN and Bretton Woods systems after the Second World War suggested that industrial economies did not need empires in order to prosper. This was obvious to all but the most obdurate Leninists by the late 1950s. The arms race between the USA and the former USSR and, above all, mass consumption stimulated economic growth. The major difficulties of the industrial economies were no longer underconsumption and excessive saving (*pace* Lenin) but their opposites and their offspring: inflation. This change had profound effects on Europe's former African colonies. Previously, European industrial prosperity had depended on favourable terms of trade for raw materials and agricultural products. Now, however, 'the demand that once depended on prosperity in the nonindustrial world was being stimulated at home' (Calleo, 1987: 148).

Political independence meant that the colonies achieved a degree of autonomy in a global situation in which powerful states had to compete for influence and markets. In the absence of European political domination African states had to 'define' themselves and engage in their own policy-making. They could now pursue policies that might lead to economic diversification and national economic development. At first the prospects seemed limited. State boundaries had been drawn by colonial powers without reference to ethnicity, so the inhabitants lacked any common language, history or culture; national identity had to compete with prior claims to ethnic identity and (for some in the elites) a sense of 'African' identity. Nevertheless, the political map of Africa was firmly established; elites, particularly the military, have generally stood against secession and boundary corrections. External recognition of state sovereignty has been crucial to this process, by limiting foreign support for secessionist and expansionist movements (Strang, 1991). At the same time, political elites embarked on 'modernization' projects, designed to diversify their economies and attract foreign aid by playing off the global superpowers against one another and using the appeal of historic (colonial) ties to Europe. In mediating between the world economy, on the one hand, and the various local economies within their boundaries, on the other, states inserted themselves into a process that had hitherto been one with limited political mediation.

It was in this context that ethnic politics, rent-seeking activity by various economic and status groups, and bureaucratic pay-offs took root. National industrialization was given an especially high priority. This was largely because the declining terms of trade for most agricultural commodities in the 1950s and 1960s suggested that export of primary commodities was not a good long-run development strategy. But it was also because of an intellectual bias in the development economics of the time which saw draining agriculture to fuel industrialization as the best approach to economic development (because of industry's presumed multipliers effects) and the 'prestige' and patronage possibilities of industrial projects for political leaders (*see* Knox and Agnew, 1994: 332–5).

National development proved a limited success. The global situation changed dramatically by the 1970s such that the terms of trade for primary commodities had improved enormously. Yet African producers proved incapable of taking advantage (Table 18.1). Their shares of world markets in many commodities (except coffee and tea) slipped precipitously from the 1960s to the 1980s. Paralleling this decline in export performance was a dramatic increase in manufacturing imports without any compensating growth in manufacturing exports (Figs 18.2 and 18.3). The anti-trade ideology of political elites (Sender and Smith, 1986: 127), the failure of most planned attempts at industrialization, frequent military *coups d'état* and associated Cold War conflicts, and the reduced importance of Africa within the world trading system conspired to produce economic stagnation. Even as they traded less (relative to an overall expansion in world trade), most African countries were unable to substitute for this through growth of their domestic economies (Fig. 18.4) (Harris, 1990: 244–85). Attempts at regional economic integration in Africa also proved unsuccessful. Insufficient transport and communication links, a multiplicity of currencies outside the (French) franc zone, and ethnic, linguistic and political enmities are among the most cited barriers (Foroutan and Pritchett, 1993).

**Table 18.1** Sub-Saharan African exports as a percentage of total world exports of selected commodities, 1961–3, 1969–71, 1984-5

| | 1961–3 | 1969–71 | 1984–5 |
|---|---|---|---|
| Coffee | 25.6 | 29.3 | 21.8 |
| Tea | 8.7 | 14.4 | 15.7 |
| Groundnut oil | 53.8 | 57.6 | 30.6 |
| Groundnuts | 85.5 | 69.1 | 7.2 |
| Palm kernel oil | 55.2 | 54.8 | 7.5 |
| Palm oil | 55.0 | 16.4 | 1.8 |
| Bananas | 10.9 | 6.5 | 2.7 |
| Cotton | 10.8 | 15.5 | 11.9 |
| Rubber | 6.8 | 6.8 | 5.3 |
| Tobacco | 12.1 | 8.2 | 9.6 |

*Source*: UNCTAD (1960–90).

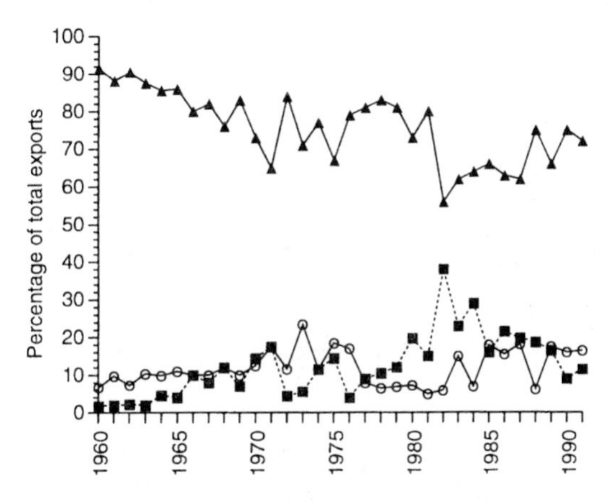


**Fig 18.2.** Africa's exports by type, 1960–91

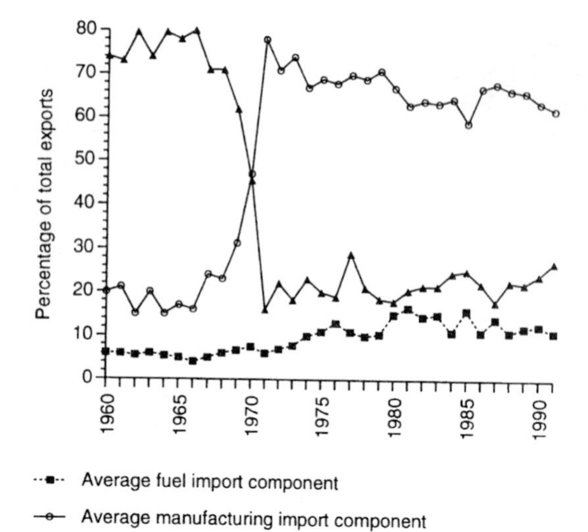


**Fig. 18.3** Africa's imports by type, 1960–91

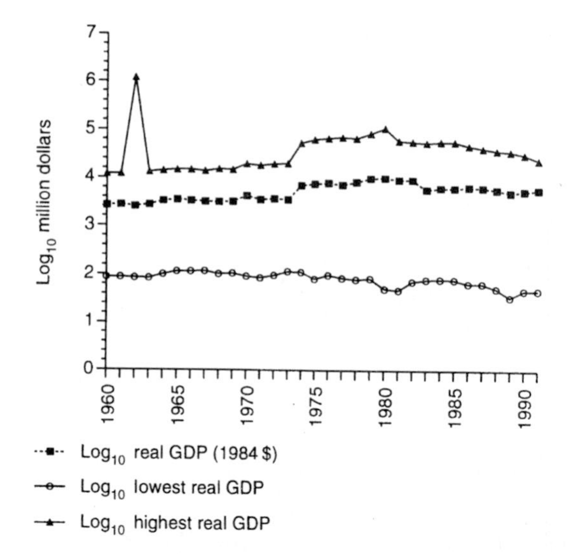


**Fig. 18.4** Africa's GDP by highest case, average, and lowest case, 1960–91

## NEO-LIBERALISM AND THE STRUCTURAL ADJUSTMENT REGIME

By the mid-1980s, many African states had become increasingly repressive as their over-regulated (and politicized) economies unravelled and their capacity to govern declined (e.g. Barratt Brown, 1995a: 269–86). Households and local community groups made up for the absence of the state by the expansion of the 'informal' economy and through new demands for political liberalization and improved state accountability. Associational life (civil society) emerged to compensate for the decline of the state and the collapse of official markets (Harbeson *et al.*, 1994). In some extreme cases, such as Somalia, Liberia, Rwanda and Sierra Leone, states effectively unravelled, as clans, ethnic groups, army units and social classes used violence rather than politics or associational life to gain access to resources (de Waal, 1994; Keen, 1995). At the same time, however, and more positively, new social movements and new ideologies critical of the character of development as the mere growth in leading economic indicators have emerged in some countries to give voice to those heretofore excluded from the benefits of state-centred development programmes (Corbridge, 1993; Routledge, 1995). The 'hegemony' of the state as an inheritance from the colonial period has been opened to question in ways not possible previously (Engels and Marks, 1994). More generally, the question of the relationship between the indicators used to measure economic growth and the actual condition of significant numbers of people has become a political and intellectual issue in 'rich' countries such as the USA, given the apparent increase in income polarization in these countries in recent years (*see*, for example, Cobb *et al.*, 1995).

This disenchantment with the state coincided with three trends in the world economy that presaged an abandonment of state-centred development strategy all over the world. One was the increased openness of the world economy, manifested in such phenomena as the globalization of production and the emergence of a truly global financial system. Even erstwhile 'centres' (such as the UK and the USA) now had to worry about the impact of trade on incomes and their inability to stabilize their economies by means of fiscal and monetary policies (e.g. Harris, 1986; Abowd and Freeman, 1990; Richardson, 1995). This trend has also involved an increasing degree of differentiation and competition between regions and localities within states. 'Commodity chains' – flows of goods and components from one stage of production or processing to another – face fewer state-level barriers than was the case until very recently (Gereffi, 1989). Small areas and their populations are now directly connected to the world economy with less effective protection and regulation on the part of the states of which they are part (Agnew and Corbridge, 1995: 164–205). A second was the rise of a dominant neo-liberal discourse about economic development that displaced state-centred doctrines with ideas about 'comparative advantage' and 'export orientation' (e.g. Biersteker, 1993; Richardson, 1994). These intellectual currents drew inspiration (correctly or not) from the upward economic mobility of the so-called newly industrializing countries (NICs), particularly those in East Asia such as Taiwan and South Korea. But it was their sponsorship by multilateral lending institutions that was decisive. This guaranteed that the ideas would become dominant practices in a context where these institutions had tremendous leverage over many states. Third, the end of the Cold War meant an end also of the ability of small states, such as those in Africa, to extort aid and support out of the global superpowers. That this coincided with Europe's 'turning inward' around an expanded European Union made the difficulties for Africa only more dramatic.

Falling commodity prices in the 1980s were compounded by the heavy foreign debt loads that African states had taken on in the 1970s to finance their efforts at industrialization. Servicing the debt accounted for an increasing part of export earnings. Those countries which sought financial assistance from the main international financial institutions, the World Bank and the IMF, were forced to adhere to strict rules reducing public expenditures and encouraging privatization of state assets.

> At the same time, they were encouraged to expand their commodity exports to repay their debts. The result could have been predicted. Since all producers were encouraged to expand their commodity exports simultaneously, stocks built up and prices fell even more sharply, with catastrophic effects on producers' incomes and on government revenues. (Barratt Brown, 1995a: 276)

Structural adjustment policies, therefore, premissed on the long-run tenets of neo-classical economics – removal of state mediation except for 'minimal' regulation, free trade, comparative advantage, etc. – and enforced through the near-bankruptcy of most African states, have come to exercise a major influence over the contemporary situation of Africa in the world economy. The main effect so far has been to encourage continued specialization in primary commodities and, as a result, dependence on shrinking markets (Fig. 18.5). Africa's position in the global division of labour is likely to become even more marginalized, therefore, unless by a miracle of comparative advantage Africa can follow the path trodden by the NICs. For this to happen, low wages are not enough. There would have to be substantial improvements in productivity, either through the application of technology or by means of a mix of factors of production that would make African products competitive in world markets. If low wages automatically meant low costs, the world's poorest countries would dominate world trade. They do not, because low wages go with low productivity and this reduces their net

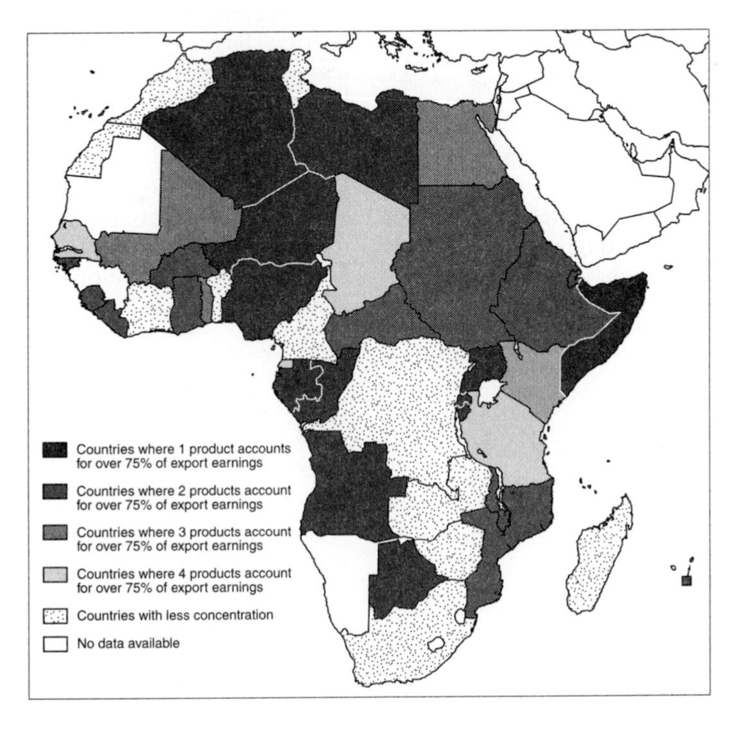


**Fig. 18.5** Map of export dependence, 1992

comparative advantage, particularly in capital-intensive sectors (Golub, 1995).

Adjustment policies began initially as a way of dealing with the debt crisis of the early 1980s. Only after 1985 did they become more general in focus, orientated towards total economic reform rather than debt repayment. Ironically, the structural adjustment programmes mandated by the IMF and World Bank (on behalf of the industrialized countries that control them) require a 'denationalized state' that can intervene effectively within its boundaries. Unfortunately, the reduction of state expenditures undermines the legitimacy of the state's attempts at doing so. As this happens, the outside experts from the international agencies are likely to become a *de facto* government within the shell of the African state. As Ould-Mey (1994: 329) puts this for the case of Mauritania, 'SAPs [structural adjustment programmes] dismantled the national state and replaced it with a multilateral state where the government shares power with international institutions.' Though this could be called a 'multilateral imperialism' (as it is by Ould-Mey), it is a long way from the classic imperialism of the colonial period. It is based, after all, on the globalization of capital rather than its containment within the boundaries of empire or state as occurred in the past. We should find theoretical space for these transformations by means of historical-geographical analysis such as that presented here and not keep squeezing new practices into old theoretical pigeon-holes.

## CONCLUSION

Wide agreement about 'Africa's' fate within contemporary world trade (even though this representation of the continent as a whole is itself subject to challenge; *see* Grant and Agnew, 1996) does not reflect theoretical consensus about what has brought it about. Three theories have jostled for greatest popularity in terms of explanatory power. Each offers some purchase on the available empirical evidence. However, their advocates insist that their theories are total accounts. If a different view of theories is adopted, one which stresses their heuristic character, then each of the three theories can serve as a guide to the dominant trading practices and their geography in three different time-periods. This historicist as opposed to positivist view of theory offers a way of acknowledging the insights of the three theories without the hyper-eclecticism that insists on bringing everything in and in the process undermining the whole point of theorizing: identifying dominant practices and their consequences.

This 'take' on theory is not one that positivists – only one theory can ever do – or post-modernists – no theory can ever do – will find satisfactory. For those interested in the historical-geographical specificity of theory there is no alternative. That this has taken so long to acknowledge shows, on the one hand, the power of positivist thinking and the strong urge to transcendental/universalist knowing that few 'scientists' would want to reject, and on the other hand, the easy attraction of rejecting all theorizing simply because no one theory offers extended historical reach and extensive geographical scope.

The conclusion on Africa's 'falling out of the world economy' is equally radical. Accepting that 'Africa' has an existence that is not entirely intellectual, its 'problem' with respect to trade is that it has lost the niche in the world economy that it had under colonialism. Within a discourse of 'development', this means, of course, that Africa's problem is not that it is exploited more than anywhere else or that it has peculiarly bad governments. It is that it has only a limited role in the emerging global economy. For development, as conventionally defined, it needs a larger one.

# SECTION THREE

# NEW GEOGRAPHIES OF UNEVEN DEVELOPMENT

# THEORIES OF ACCUMULATION AND REGULATION: BRINGING LIFE BACK INTO ECONOMIC GEOGRAPHY INTRODUCTION TO SECTION THREE

**TREVOR J. BARNES**

The Mecca of the economist lies in economic biology rather than economic mechanics. But biological conceptions are more complex than those of mechanics; a Volume on Foundations must therefore give relatively large place to mechanical analogies, and frequent use is made of the term 'equilibrium' which suggests something of a static analogy. (Marshall, 1961: xiv)

Alfred Marshall, the late nineteenth- and early twentieth-century Cambridge economist, is probably best known to contemporary economic geographers as an adjective. In his work on the new industrial spaces of post-Fordism, Allen Scott (1988b) popularized within economic geography the idea of the 'Marshallian' industrial district, a dense network of interlinked, vertically disintegrated firms set within a relatively small geographical area. As a result, Marshall emerges, if not a hero, at least a seer of a new type of economic geographical theory (*see also* Amin and Thrift's (1992) similar invocation of Marshall).

It is true that Marshall was a seer of a new type of economic theory, but it was neither the neo-Marxist one that Scott later sought to develop, nor even the embryonic institutional one outlined by Amin and Thrift. Marshall's main claim to fame was in codifying neo-classical economics in his tome *Principles of economics* (first published in 1890); it was economic orthodoxy that he saw.

That said, Marshall's relationship to neo-classical economics was never straightforward (Joan Robinson (1964) thought him 'a sly fox'). So although taking a second wrangler in mathematics as an undergraduate at Cambridge, he remained suspicious of the use of formalism in economic theory, and in one hyperbolic moment even suggested 'burn[ing] the mathematics' (Keynes, 1951: 157–60; Mirowski, 1989a: 264). Then, for an economist, there was his unlikely interest in the nitty-gritty geographical details of economic life exemplified by the case studies found in his *Elements of economics of industry* (1910), which formed the basis of the idea of the industrial district. Finally, there was his improbable concern with metaphor, which in turn was bound up with his worries about 'the element of time', which he saw as the 'chief difficulty of almost every economic problem' (and illustrated by this chapter's epigraph; Marshall, 1961: vii).

It is this last issue of metaphor and time that is the starting-point for this introductory essay. Marshall half knew that there was something wrong with the neo-classical treatment of time, and which was a result of the mechanistic metaphors that were deployed. He also half knew that biological metaphors did not face the same kinds of problems, but as a basis of an economic method they suffered from being messier, less determinate and aesthetically less pleasing than their mechanistic counterparts. It was for these latter reasons that it was necessary for him 'to give a relatively large place to mechanical analogies' (Marshall, 1961: xiv). For example, he famously likened the supply and demand framework that is the pivot of his neo-classical scheme to the intersecting blades of a pair of scissors.

Marshall's penchant for the mechanistic over the biological became both more widely accepted and more exaggerated over the course of the twentieth century as subsequent neo-classical economists, such as Paul Samuelson, became in effect putative physicists, besotted by mechanistic analogies (Mirowski, 1989b). The same thing, of course, also happened in economic geography, albeit later. Spatial science was defined by its use of mechanistic analogues, either imported directly from the physical sciences or borrowed second-hand through economics (Barnes, 1996: ch. 4).

The metaphorical tide seems to be turning, however, certainly in economic geography, and possibly even in economics, and there is now an interest in biological analogies rather than the mechanistic kind. In this sense, some economic geographers and economists may be arriving at Marshall's Mecca.

It is certainly a destination of many of the subsequent chapters contained in this section of the book. Each of these pieces is substantively concerned with the issue of geographical accumulation and regulation, where by accumulation is meant investment, often, but not exclusively, carried out by private capital, and by regulation the set of formal and informal institutions, rules and norms that help govern its maintenance, size, spatial distribution and type. Rather than relying on Marshall's mechanistic metaphors as an explanation of accumulation and regulation across time and space, the authors in this part of the collection frequently rest their argument, although often only implicitly, on biological ones; for example, on metaphors of reproduction, or evolution, or on organic wholeness. The purpose of this chapter is to provide a general review of theories of accumulation and regulation inspired by biology rather than mechanics. Three broad theories are discussed, all of which fall into the political economy tradition:[1] Marxist, including French regulationist, theories, which begin with the metaphor of reproduction; evolutionary and institutional theories, which make use of notions of acquired characteristics and inheritance; and post-Keynesianism, which emphasizes holism and organicism.

## ON METAPHOR

It might seem odd that a chapter about theories of accumulation and regulation would begin with metaphorical talk from a long-dead economist. Accumulation, in particular, seems about as concrete and non-metaphorical as anything could. A distinction needs to be made, however, between, as Richard Rorty (1985) puts it, texts and lumps, between dollops of brute reality and their interpretation. While the things that make up accumulation and regulation have a brute materiality – iron and steel foundries, optic fibre cable lines, classroom complexes for labour training – how they are theorized requires interpretation. It is with interpretation that metaphorical talk comes to the fore.

In brief, the argument developed over the past three decades, particularly in the history, sociology and philosophy of science, is that objectivism, the idea that theories mirror reality, is not sustainable (Bernstein, 1983; Woolgar, 1988). Both the Duhem–Quine thesis, which claims that by necessity the validity of a theory is always underdetermined by the facts used to verify it (Harding, 1976), and Kuhn's (1970) conclusions about the value-ladenness of empirical enquiry, make the task of hooking theory to the 'real world', whether it be theories of quantum physics or accumulation and regulation, inherently problematic. With this realization, attention has switched increasingly to how theories are formulated and developed, and away from issues of strict empirical validation. In undertaking that former task a number of people have turned to metaphor.

Although there is much controversy over the meaning of a metaphor, most agree that it involves asserting a similarity between two or more different things; for example, that supply and demand is like the intersecting blades of a pair of scissors. Typically we tend to be most aware of the small metaphors that pepper individual texts, like the metaphor 'pepper' used in this sentence. But there are also big metaphors, which, while we may be less aware of them, are potentially even more important, structuring whole research paradigms, and when first introduced producing 'revolutionary science' (Barnes, 1996: ch. 5). Richard Rorty (1979: 12) thinks so, at least, writing that 'it is pictures rather than propositions, metaphors rather than statements, which determine most of our philosophical convictions'. Certainly, the recent history of economic geography is the history of different metaphors: places are points of mass that interact according to Newton's gravity formulation (Stewart and Warntz, 1958), places are mental maps that capitalists store in their heads (Gould and White, 1974), places are deposits of dead labour time born from the reproduction of capitalism (Harvey, 1982), places are geological strata laid down by the weight of historically changing divisions of labour (Massey, 1984), and places are flexible poles of investment (Storper and Scott, 1988). In each case it is a central set of metaphors that carries forward the research programme (Barnes, 1996: chs 4 and 5).

Typically, a characteristic of these big metaphors is that their users are frequently unaware of all the intellectual freight that they carry. For metaphors are never innocent: they represent a particular slant on the world, albeit often multi-layered and submerged. For this reason it is necessary to inspect critically the metaphors that form the basis of the enquiry, checking them for their implicit politics, hidden assumptions, and their logical coherence, consistency and compatibility. Doing so also involves scrutinizing the historical and material origins of the original metaphor which shaped their meaning, and in this sense, taking metaphor seriously implies taking the world seriously too.

Metaphors require 'worlding' (Gregory, 1994), indicating again the complex relationship between theories and things.

Thinking of theories of accumulation and regulation along metaphorical lines is, therefore, not so odd as it might first appear. All our theories are metaphors of some kind. The critical question is what kind. Marshall was starting to answer just that question in the Preface to his *Principles*, recognizing as he does the two metaphors of mechanics and biology. By way of a bench-mark for the subsequent discussion, let me start with the mechanistic type, and their deployment in understanding the geography of accumulation and regulation.

## METAPHORS, MECHANICS AND MRS THATCHER

Much has been written about mechanistic analogies and their use by neo-classical economists (Lowe, 1951; Sebba, 1953; Samuelson, 1972; Thoben, 1982; Mirowski, 1989a). By mechanistic analogies I mean metaphors taken from classical physics, a form of enquiry that begins with Galileo, includes Descartes and Newton, and continues through to the end of the nineteenth century when it is finally superseded by the beginning of quantum mechanics and the theory of relativity. Such literature suggests that the influence of classical physics on neo-classicism was twofold: first, through a set of associated metaphysical doctrines (Lowe, 1951; Sebba, 1953; Thoben, 1982); and second, through using one of physics' specialized branches, 'energetics', as a theoretical template (Mirowski, 1984a, 1989a). Together, both shaped the methodological agenda of neo-classicism in at least four important respects; these are the 'hidden assumptions' of the metaphor.

There is *determinacy*, and the critical role played by mathematics. Mirowski (1989a: 65) calls this presumption 'the Laplacian Dream', the belief that it is possible to find 'a single mathematical formula that described the entire world' (Mirowski, 1989a: 28). In this view the universe is a clockwork mechanism, and mathematics describes it best because, as Galileo put it, 'mathematics is nature's own language'. Neo-classical economics is certainly suffused in the language of mathematical determinacy. The University of Chicago economist Gary Becker (1976) in his Nobel prize-winning work portrays all aspects of human life, literally from birth to death, as determined and representable by the calculus of variation. The important corollary is that just as in classical physics and energetics, points of mass have no choice but to behave in accordance with physical laws such as the law of least effort, so too must neo-classicism similarly treat economic agents. With only a 'photograph' of an agent's tastes (Pareto, 1971: 120), the analyst determines (using constrained maximization techniques, first developed in energetics) a person's every move. The problem with this perspective, though, is that it misses out much that seems to be part of the human predicament, and which enters into such events as accumulation and regulation. Absent are deliberation, mistakes, misunderstandings, blind passions, surprises and even human agency itself. In fact, Georgescu-Roegen (1971: 343) argues that neo-classicism is not really about human agents at all, 'because there is no economic process. There is only the jigsaw puzzle of fitting given means to given ends, which requires a computer and not an agent.'

There is a *reductionist* methodological strategy, the attempt to explain the world by reducing its complexity to elemental components. In classical physics, for example, everything is explained by beginning with individual particles, and their interaction. Known also as atomism, the counterpart in the social sciences is methodological individualism (Elster, 1983: 20–4), the idea that an adequate explanation of society must refer to only the properties of, and relations among, the individuals who compose it. In neo-classical economics this reductionist position is manifest in the centrality accorded to rational individuals, who are treated as the mainspring of all events, and the backstop of any explanation. But what emerges from this metaphorical projection of a physicalist universe on to a social one is a world of isolated, self-contained agents, where there are no wider social institutions that bind or regulate. As Margaret Thatcher famously put it, 'there is no such thing as society'. Society is simply the sum of the rational individuals who compose it. In such a conception, the notion of social regulation is utterly meaningless because the social is evacuated of content.

There is an *ahistoricism* that relates directly to Marshall's anxiety about time. In classical physics there is only 'motion, . . . [which] is completely reversible and in no way gives rise to any qualitative changes' (Thoben, 1982: 293). There is no arrow of time because in the world of classical physics anything that goes forwards just as easily goes backwards, thus always ensuring the possibility of equilibrium. As Marshall suggests in the epigraph, by adopting an energetics metaphor neo-classical economists also necessarily committed themselves to deriving static equilibrium conditions. But those conditions, because of the metaphor from which they were taken, define a time in which there is no time, where yesterday is like today and today is like tomorrow (Robinson, 1973a). Such a conception, as Joan Robinson (1979) called it, represents logical, not historical, time. The past never matters (path dependence is irrelevant), and uncertainty never arises.

Finally, there was often a *teleology* implied by classical physicists, which in turn is linked to a type of naturalism: what is natural is by definition good. This emerges, for example, in energetics, in the presumption that there is an absolute efficiency of particle movement which it is the purpose of the wider system to fulfil (Mirowski, 1987: 84). The same type of teleology is present in neo-classical economics: the purpose of the system is to optimize over everybody's 'objective' function, and because in this case what people want is naturally good, then optimization occurs. There is a clearly a paradox here. Science has long struggled to rid itself of, in particular, religious teleological conceptions, but they none the less make a return through naturalism. The same holds for neo-classical economics. By beginning with rational individuals the aspiration is of a non-teleological explanation of human behaviour, but teleology returns in the form of giving individuals a purpose: the meaning of life is to maximize. As Mirowski (1984b: 473; emphasis in original) puts it:

> Ultimately, the only reason for praising the maximum principles in economics is the belief on the part of the theorist that the players *should* and *do* maximize some quantum which they deem as their ultimate goal. The conception endows the human drama with a scope and a purpose which it has not had since the intelligentsia broke away from the theological institutions which earlier had performed that function.

## Mechanistic Metaphors and Economic Geography

If these are the mechanistic metaphors that moulded neo-classicism, how did they influence the portrayal of accumulation and regulation by economic geographers? Often entering economic geography during the 1950s and 1960s through the side-door of regional science (Barnes, 1996: ch 4), versions of neo-classical regional growth theory became for a period a disciplinary mainstay, especially for pedagogical purposes. Perhaps the best-known example is a version put forward by Borts (1960).

Very briefly, making use of constrained maximization techniques, Borts' theory demonstrates that in a two-region world, where each region initially possesses different amounts of the internally homogeneous resources, capital and labour, factors of production will move between the regions until equilibrium is attained. By rationally responding to price signals (that is, by maximizing), workers and capitalists in both regions will eventually earn respectively equal wage and profit rates, thereby erasing any spatial inequality.

Neo-classicism's four hidden assumptions are easily revealed here. Determinacy is represented by a set of mathematical equations that set equilibrium prices of labour and capital. Given the initial conditions of rational agents and resource distribution, capital and labour move like a well-oiled clockwork mechanism to equilibrium. Second, reductionism is embodied in the system's movers and shakers; that is, those 'homogeneous globule[s] of desire', as Thorstein Veblen once referred to rational economic agents, each of which 'is an isolated, definitive human datum' (Veblen, 1919: 73). As a consequence, missing in the depiction is any recognizable society, or form of social regulation. Third, also missing is any sense of history. The kind of troubled, restless and unsettled landscapes of late twentieth-century capitalism, which are the focus of many of the essays that follow, are irreconcilable with supply and demand diagrams that posit equilibrium through backward and forward movements on a two-dimensional page. For when historical time is admitted, lying as it does 'at right angles to the plane on which the diagram is drawn' (Robinson, 1979: 52), such equilibrating movement is impossible. Finally, there is a suffocating sense of purpose to the scheme. The economy will always work in such a way as to ensure that all is best in the best of all possible worlds.

Hopefully this is enough to show that the mechanistic metaphors of neo-classicism are in so many different ways anathema to the useful representation of accumulation and regulation. As many of the subsequent chapters illustrate, accumulation and regulation do not operate like clockwork, but are haphazard and contingent; they are intimately bound with social institutions and the collective cultivation of trust, responsibility and understanding, and not reducible to individual rational calculations; they are embedded in a historical time where the past cannot be undone and the future cannot be known; and they are directed towards no end other than that of endless possibilities.

That these features of accumulation and regulation were not recognized in the standard neo-classical model, I would argue, is because of its mechanistic metaphorical origins which necessarily occluded them and silenced their alternative political implications. For neo-classicism's veneration of mechanistic metaphors was as much political as intellectual. Fortunately, compared to neo-classical economics, economic geography never had the same either intellectual or political vested interests in, or length of institutional memory of, mechanistic metaphors. As a result, their grip on the theoretical throat of the discipline was never life-threatening, and after a decade or so, certainly by the mid-1970s, it was relaxed sufficiently to allow for other kinds of traditions and other kinds of metaphors. Prime among them was political economy – the theories of which, especially those of accumulation and regulation, were frequently based on biological analogies rather than the mechanical kind.[2] Consequently, life was brought back into economic geography.

# BIOLOGY AND THEORIES OF ACCUMULATION AND REGULATION

But what kind of life? To answer this question I intend first to advance a set of general arguments about the advantages of employing biological metaphors. I will suggest that for each of the four problematic 'hidden assumptions' of mechanical metaphors there is for users of biological metaphors a potentially opposite and favourable counterpart. In making this argument, I realize that not all biological theories are equally apt as metaphors, and there are some that take on the same complexion as classical physics. Likewise, the uses to which biological metaphors have been put have not always been politically salutary: *lebensraum* is one such case, Spencer's social Darwinism is another (and incidentally was favoured by Marshall), and the connection forged between climatic regimes and racial types by nineteenth- and early twentieth-century environmental determinists is yet a third. While biological analogies are not necessarily exemplary – there is always need for critical scrutiny – I will nevertheless argue that they are more appropriate as a model for understanding accumulation and regulation than any mechanical kind. That general argument is fleshed out in the second part of this section, where the usefulness of biological metaphors is discussed by reviewing the three different political economic theories of accumulation and regulation: Marxists, institutional and post-Keynesian.

## Biological Metaphors

Paralleling in reverse the hidden assumptions of neo-classicism's mechanical metaphors are four methodological characteristics of biological, and in particular, evolutionary, theory.

There is a rejection of the Laplacian dream of full mathematical determinacy, which in turn implies quite a different view of the agent. By its very nature, evolutionary theory supposes randomness and determinacy, chance and necessity. As a consequence the Laplacian dream as applied to the biological world is just that, a dream. The combination of determinacy and randomness makes relationships too complex and unamenable to general mathematical solutions (reflected, for example, in some biologists' use of chaos theory, which admits no determinate mathematical relationship between initial conditions and final outcomes). Once metaphorically extended to the world of economics, biological theories accordingly give quite a different view of the agent. It is of someone embedded in a very complex ongoing system, which is both stochastic and determined, and as a result can never be fully known or precisely predicted. Perturbations arise, mistakes occur, there are continual surprises, and processes are always changing.

Rather than striving for reductionism, biological theory often seeks a holism or organicism where the emphasis is on emergent properties; that is, properties generated by the interaction among parts of the whole system the nature of which cannot be known in advance. The positions of holism and emergence both stem partly from the presumed complexity of the biological system and its mathematical intractability, and partly from the belief that biological systems operate quite differently from mechanical ones. For example, Mayr (1985: 57–8) writes, 'Nowhere in the inanimate world can one find a system, even a complex system, that has the ordered internal cohesion and coadaption of even the simplest of biological systems.' Mayr means here that biological systems consist of a series of internally cohesive subsystems or levels, which then interact in complex ways through adapting one to another, and in the process producing emergent properties. In this view, there is no elemental component to which everything is reduced: the system must be examined all of a piece. As Mayr (1985: 58) states, 'by the time we have dissected an organism down to atoms and elementary particles we have lost everything that is characteristic of a living system'. The theoretical corollary is that there is unlikely to be a single general theory that explains everything. Instead, there is the need for a pluralist mix of different theories. When transposed to the social world, the imperative of holism and emergence is to emphasize internal interrelations among parts, rather than separating them off, and reducing them to presumed basic entities such as individual rational choice.

Instead of timeless equilibrium, biological theories accentuate historical process and change. As Hodgson (1993: 32) writes, biological theories are concerned with 'irreversible and on-going processes in time . . . with variation and diversity, with equilibrium and non-equilibrium situations, and with the possibility of persistent and systematic error-making and thereby non-optimizing behaviour'. There are at least two points here. Because of irreversibility, time path-dependence or chreodic development is endemic; that is, events in the past necessarily influence the course of the future. To use David's (1985) economic example, once the QWERTY arrangement of keys on a typewriter was developed, there was no going back in spite of its inefficiencies for users of word processors. This example goes to the second point: that path dependence does not imply optimality. Initial accidents, such as the QWERTY system, may well culminate in suboptimal and eccentric paths of evolution. As a result, it is not possible to believe that history is always progressing as a series of discrete steps leading to optimality. The best that can be hoped for is a temporary equilibrium, one arrived at by a series of often accidental path-dependent processes, and which is not stable (a 'punctuated equilibrium'; Gould and Eldredge, 1977).

Finally, at least in contemporary biological theory, there is no teleology; no sense that the wider system is necessarily improving or pursuing some ultimate goal of perfectibility. Hodgson (1993: 200–1) writes:

> [N]atural selection does not lead to the superlative fittest, only the tolerably fit . . . evolution is not necessarily a grand or natural road leading generally towards perfection. Change can be idiosyncratic, error can be reproduced and imitated, and a path to improvement can be missed.

Stephen J. Gould (1987: 14) concurs, suggesting that by necessity there must be mistakes, for 'imperfections are the primary proofs that evolution has occurred, since optimal designs erase all signposts of history'. The counter-argument is that evolutionary theory is teleological because it equates survival to success: provided DNA characteristics are passed on to the next generation, success is achieved. Such an argument is not compelling, though, because of its circularity. Given the premiss that success is equivalent to survival, the conclusion that only the successful survive tautologically follows. The trick is not to fall into the circular argument in the first place, which means thinking of evolutionary theory in a non-teleological way.

It is a way of thinking particularly appropriate to accumulation and regulation. There is no predetermined end-point to which economies are drawn, no final perfectibility that is realized. There is only the continual hustle and bustle of change, albeit shaped by capitalist competition and conflict, sometimes contradictory objectives and countervailing policies, and unintended consequences and mislaid plans. It is not that 'mere anarchy is loosed upon the world'; there are still institutional norms and mores guiding behaviour, but they change quickly, are not universally shared, and clash, producing unpredictable results. Likewise, the associated geography will be complex, restless, and fickle. There is still a need for 'envisioning capital' (Buck-Morss, 1995), but it can no longer be done using the geometrical certainties of circles and regular polygons. Their use implies a final spatial order which, while compatible with the mechanistic metaphors of spatial science, is, as will be now argued, irreconcilable with the biological ones of political economy which continually distort and corrupt the clean lines of smooth geometries.

## Theories of Accumulation and Regulation

### MARXISM

Marx first outlined his scheme of accumulation, called simple reproduction, in volume 2 of *Capital* (Fig. 1). There are two departments (sectors), I and II, each of which produce different kinds of goods: department I manufactures capital goods, and department II consumption goods, which includes both wage goods for workers and luxury goods for capitalists. Marx shows that, in order for production to continue, output levels in each department must be such that they exactly match the input needs of their own department and that of the other. That condition, though, is difficult to meet, and becomes even harder when the more common case of expanded reproduction is admitted; that is, when the capital and consumption goods sectors are themselves growing (Desai, 1979). For Marx, the difficulty of achieving balanced reproduction – that is, for outputs of one period to meet exactly the input needs of the next – was yet another reason why capitalism is beset by crisis (defined as an interruption in the cycle of reproduction).

The idea of reproduction that underlies the Marxist theory of accumulation clearly draws upon a biological analogy. It is the assertion that an economic system reproduces itself like a living organism. In both cases, there is a cyclical process where an entity or relationship in one time-period is transformed in such a way as to allow for the possibility of producing similar entities or relationships in the future. It was the French physiocratic economist and physician to the court of Louis XIV, François Quesnay, who first conceived the economy in this form by explicitly basing his work on William Harvey's seventeenth–century discoveries about the pulmonary circulation system (found in Quesnay's *Tableau économique*; Gudeman, 1986; Buck-Morss, 1995). Oxygenated blood circulates around the body, and as it does so it is transformed, but not before providing the necessary wherewithal for the body to inhale more oxygen and begin the process again. Similarly, the English classical economist David Ricardo makes use of the same general metaphor by setting his work within the context of a 'corn model' of the economy (Gudeman, 1986). Here corn seed is planted, nurtured and harvested in one period, thereby producing more corn seed for the process to be renewed in the next time-period. In both these cases, should the cycle of reproduction be interrupted in some way – say, by a blockage in a blood vessel or by a crop disease – then by the logic of the metaphor, crisis necessarily follows.

This basic reproduction analogy has been elaborated upon in a number of different ways by subsequent political economists, but I will examine just two which are especially germane to the subsequent essays: the first is the theory of uneven development, and specifically a version prosecuted by David Harvey (1985a, 1989a) and Neil Smith (1984); and the second is the theory of the French regulationists.

## HARVEY'S AND SMITH'S THEORY OF UNEVEN DEVELOPMENT

Both Harvey and Smith begin with the yardstick of simple reproduction where the cycle of accumulation is undisturbed. What is special about their presentation is that reproduction is occurring not only in time but also across space and in place (termed a rational landscape by Harvey (1985a: 190) and an equilibrium one by Smith (1984)). At least temporarily, regions such as the US manufacturing belt hold their position as pre-eminent industrial sites. As Marxists, however, Harvey and Smith argue that the cycle of reproduction never lasts, and that sooner rather than later something will happen to disturb it. For them, a key disabling factor is the very quest by capitalists for surplus value, itself a central imperative of capitalism. Their argument is that one means to increase surplus value is to speed up the turnover time of capital by altering the spatial configuration of production, which is made possible, for example, by new transportation and communication technologies. Such a strategy is even given the imprimatur of Marx, who talks about such a process as the 'annihilation of space by time'. The problem, though, is that by altering the spatial arrangement of production in order to raise profits, capitalists necessarily undo what they have already done; that is, to dismantle existing economic sites so as to allow the formation of new ones. Reconfiguring the landscape in this

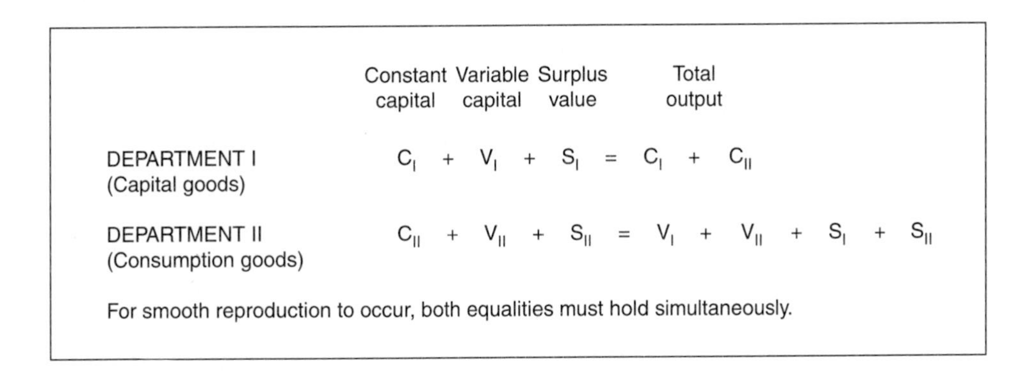

Constant capital
Variable capital
Surplus value
Total output
DEPARTMENT I
(Capital goods)
$C_I + V_I + S_I = C_I + C_{II}$
DEPARTMENT II
(Consumption goods)
$C_{II} + V_{II} + S_{II} = V_I + V_{II} + S_I + S_{II}$
For smooth reproduction to occur, both equalities must hold simultaneously.


**Fig. 1** Marx's scheme of simple reproduction

way, though, may well produce the crisis that such a reconfiguration was in part designed to avoid. For in the process of annihilating space by time, large amounts of fixed and sunk capital will be scrapped and large numbers of workers laid off. Places like Buffalo, Gary and Youngstown become part of the Rustbelt, as capital moves off to create a new geographical landscape that is more profitable. But again sooner rather than later, even this new locational configuration will itself become a spatial barrier to accumulation, thereby requiring annihilation like its predecessor. As Harvey (1985a: 55) writes:

> Capitalist development has to negotiate a knife-edge path between preserving the values of past capitalist investments embodied in the land and destroying them in order to open up fresh geographical space for accumulation. A perpetual struggle ensues in which physical landscapes appropriate to capitalism's requirements are produced at a particular moment in time only to be disrupted and destroyed . . . at a subsequent point in time.

Internal factors within capitalism itself, then, are responsible for the restlessness of its landscape. But the details of that restlessness are not predictable. Regional class alliances may stall devaluation, or through luring liquescent capital reverse it altogether (Harvey, 1989b); the state at any number of different geographical scales may intervene either beneficially or detrimentally; and new technologies and working methods may precipitate an entirely different economic regime with its own distinctive geography. The only certainty found on this gyrating landscape is that of uncertainty.

If we go back to the original reproduction metaphor, Harvey and Smith are suggesting that there is something about the very process of reproduction within a given environment that eventually becomes obstructionist, thus necessitating change. It is like saying that the circulation of blood produces clogging of the arteries, or the growing of corn year after year on the same field produces soil infertility. The only means to avert this internal crisis is to change the environment while also keeping life-giving reproduction going; for example, by performing bypass surgery, or switching fields, or in Harvey and Smith's case by moving to a different landscape, say, from the manufacturing belt to the Sunbelt. In this reading, uneven development occurs when reproduction is switched from one environment to another.

Note that because of its origins in a biological metaphor this version of uneven development has none of the mechanical hidden assumptions of its neo-classical counterparts. Final determinations are impossible (Sheppard and Barnes, 1990: ch. 9), emergent properties unpredictably emerge as various collective entities and institutions interact, equilibrium is at best temporary, and absolute end-points are never reached because of continual flux and change. Again, this is not 'mere anarchy', because there are strong and powerful forces at work that give some shape to events, but none are so powerful that they always have their way (for example, in Harvey's rendering, unintentional consequences are particularly important in creating disorder and discord; Sheppard and Barnes, 1990: 220–2).

One of the problems with this reproduction metaphor, though, is that it can sometimes shade into a functionalist methodological position, one in which events are explained by simply pointing to their beneficial consequences. For example, in Harvey's and Smith's works there is sometimes the argument that uneven development, and even crisis itself, exists because it is beneficial (functional) for the reproduction of capitalism. Smith (1984: 127) writes, 'for no matter how disruptive and dysfunctional, crises can also be acutely functional for capital. The mergers, takeovers and bankruptcies . . . that accompany crisis also prepare the ground for a new phase of capitalist development.' However, there are a number of problems with functionalism as a method (*see* Elster, 1985, and also my own critique of specifically Harvey's and Smith's work: Barnes, 1992). There are other ways, though, to use the basic reproduction metaphor, that explicitly attempt to circumscribe the criticisms of functionalism. French regulationism is one of them (and one which interestingly Harvey (1989a) seemingly has now taken up).

## FRENCH REGULATIONISTS

The French regulationists draw more directly on Marx's scheme of reproduction than Harvey and Smith (there are copious reviews of the regulationist school, of which a small sample are Dunford, 1990; Tickell and Peck, 1992; Benko, 1995; Peck and Tickell, 1995a). For the regulationists, departments I and II are the basis of what they call the regime of accumulation, a term designating a stable proportion of capital and consumption goods production in the economy. In addition to the regime, there is the mode of regulation. The mode consists of things like state forms, social practices and institutional norms and mores that bear upon the regime. Such an addition is necessary, the regulationists argue, because it helps explain why capitalism survives despite its seeming instability: the mode of regulation ensures that capitalism stays on the rails. However, once the mode and regime become decoupled then a crash of some kind occurs. (Here there are clear similarities in reasoning with Marx's invocation of the disjuncture between the forces and the social relations of production as the 'engine of history'.)

Regulationism is not just an abstract scheme, but has been brought to bear empirically on the recent history of both Western and Northern European and North American capitalism (and exemplified by many of the chapters in this section). In crude terms, the thirty years following the end of the Second World War were dominated by a Fordist regime of accumulation that combined mass production with mass consumption, and was able to maintain stability because of an appropriate mode of regulation, a Keynesian welfare state that both ensured full employment and guaranteed high levels of consumption. By the late 1960s and early 1970s, however, the prevailing Fordist regime of accumulation and associated mode of production became increasingly unhinged, creating a sustained period of crisis. Because, as the argument went, the crisis was a result of rigidities in the regime of accumulation, it was necessary for capitalists to become more flexible by switching to new technologies, work practices, products and markets, that collectively crystallized in a fresh regime of accumulation, post-Fordism. While the shape of that new regime has become well-defined in the literature, its corresponding mode of social regulation is still only faintly drawn. Some initially linked the new regime with the rise of the political neo-conservatism of Margaret Thatcher and Ronald Reagan (Scott, 1988c), and later to what Jessop (1993) has called a 'Schumpeterian workfare state'. But that seems less clear now (Tickell and Peck, 1992, 1995; Peck and Tickell, 1992, 1995b).

In terms of their metaphorical practice, the regulationists take the initial reproduction metaphor, which is couched in terms of the whole system, but graft on to it another that turns on chance variation and the inheritance of acquired characteristics. The possibility of adding this extra metaphor is a result precisely of making the distinction between the regime and the mode (in contrast, in Harvey's and Smith's scheme, where no such distinction exists, the kind of role played by the mode is absent). Under the pure functionalist position, reproduction occurs because by the very logic of that methodology capitalism ensures that its characteristics are such that maintenance is guaranteed. However, once one moves away from functionalism, which is one of the objectives of regulationism, then it is necessary to provide a different kind of explanation for capitalism's continuance. It is here that the second metaphor of chance variation and acquired characteristics is crucial: it is the basis of that alternative explanation.

Suppose by chance a viable regime of accumulation is established, and by chance some individuals, but especially institutions, act in such ways as to reinforce that pattern. The word 'chance' here does not denote improbability or unlikeliness; there are many potential feasible combinations of regimes and modes, and chance means here only that there is no common single cause that determines which of those combination is selected. However the selection process happens, though, once it occurs, and provided it is a feasible combination, a process of positive feedback results. The very success of the regime or mode is an inducement for those institutions associated with it to repeat their actions, which are then later passed on to others through habit and convention. In this sense, institutions act like genes in conveying information about appropriate behaviour to future generations. In fact, de Bernis, the originator of the term 'regulationism', explicitly made the analogy between genetic codes and institutions, thus allowing him 'to explain why, in certain periods . . . , a certain coherence . . . can be maintained . . . [provided] adequate institutions are established' (Benko, 1995: 197–8).

Because in this scheme regimes and modes are not predestined for one another, but are the result of chance discovery, two other consequences follow. First, there is scope for considerable national and even local variation, which, in turn, provides the beginning of an explanation of uneven development. Fordism, for example, which is defined by the same basic regime of mass production and consumption, is associated with a series of feasible but quite different national modes of regulation, the combination of which creates a specific kind of mutation. 'Classical Fordism' of the USA, for example, is quite different from the neighbouring 'permeable Fordism' of Canada (Jenson, 1989). So while different regimes and modes may be feasible, they are not necessarily all alike in terms of the level of wealth they generate or in its form of social distribution. As a result, the specific combination of mode and regime, say the one in the UK, may make that country's version of Fordism the laggard of Western Europe, or West Germany's the envy of the world. Note also that regulationists emphasize national variations rather than those at any other scale because it is national institutions, and particularly the state, that are the most important in regulating because of the state's financial and legislative power. This is certainly a theme picked up by many of the chapters in this section, although some authors are also keen to stress that regulation can occur at both sub- and supranational scales, and be non-state driven (*see also* Peck and Tickell, 1992, 1995b).

Second, there is a vanquishing of the devil of functionalism. Whereas under classical Marxism the character of the social is always such to optimize the development of the forces of production, in the regulationist scheme the social takes on a variety of different forms. Of course, for stability to occur the mode must be compatible with the regime, but it need not be the best, nor need its feasibility be predicted beforehand. There is what Lipietz (1986: 20) calls an 'a posteriori functionalism': only after the event of chance discovery is it known whether a regime and mode are consistent. A priori one cannot know whether a regime and mode will hook up to one another, or the consequences of their so doing. As Lipietz (ibid.: 21) puts it, 'In reality certain compatible relations become combined with each other, and that is all. Had it happened with other kinds of relations, the story would have been different.'

In sum, by relying on two different but related sets of biological metaphors, the regulationists are able to tap into some desirable methodological features that further their political economic analysis. By relying on the chance discovery of a feasible combination of mode and regime, the regulationists shun the Laplacian dream and any talk of final determinations (even in the last instance). Rather than being reduced to the rational calculus of maximizing individuals, the social takes on a semi-autonomous role. Defined by a rich mixture of institutions and social classes, beliefs and norms, it interacts in complex ways with the economic, producing a set of emergent characteristics not predictable before interaction. Historical change and transformation is the lifeblood running through the regulationist account, but it is a history that pauses in periods of stability. Finally, there is no sense that history is going anywhere; that is, takes on a purpose. For Marx, and possibly Harvey (1984), capitalism will end with a bang, whereas for the regulationists there is only the continuing process of chance discovery which, while certainly driven by macro-economic imperatives, need not finish at the end-point of a socialist state. Indeed, the socialist state may, rather than being an end-point, be just another beginning.

## INSTITUTIONAL AND EVOLUTIONARY ECONOMICS

Institutional and evolutionary economics is primarily an early twentieth-century American invention, and infused by a variety of biological analogies and approaches. For much of this century, however, institutionalism has been consigned to 'the underworld of economics', as Keynes once put it, associated with sometimes shady characters – of which its founder, Thorstein Veblen, is perhaps the best known (Diggins, 1978) – and sometimes shoddy work. Over the past 15 years such a reputation is on the way to being repaired, a result, in part, of two bodies of work that follow institutional and especially evolutionary precepts and which at least implicitly apply them to issues of regulation and accumulation: an evolutionary theory of the firm and technological change following Nelson and Winter's lead, and writings on the so-called new growth theory. Before outlining both these bodies of work, let me begin by reviewing Veblen's main theses about accumulation and regulation, and their relation to a biological sensibility.

### Veblen

Given Veblen's (1919: 68) antipathy to 'symmetry and system making', it is not surprising that he leaves no grand synthesis of the Marshallian type. Instead, his forte is brilliant one-liners, and memorable pithy phrases ('conspicuous consumption', 'leisure class', 'idle curiosity'). That said, throughout all his works there is a consistent call for an evolutionary approach, a 'post-Darwinian' economics as he called it. It has at least two foci that both bear upon accumulation and regulation.

There is an argument about institutional change. For Veblen the mainspring of human action is certain instincts, which he initially attributed to biological drives but later ascribed to the influence of culture and society. Those instincts, in the form of habits, conventions and customs, crystallize in the form of institutions defined as 'settled habits of thought' (Veblen, 1919: 239). It is the settled nature of the institutions – that is, their stability and durability – that is so important here. That quality allows Veblen to treat institutions metaphorically as individual organisms within a wider species. Just as in Darwin's scheme only those individuals that are most fit for their environment will replicate in sufficient numbers (that is, pass on their DNA characteristics), so too for institutions in Veblen's scheme. Specifically, given the variety of institutions that exist, and because of an environment that is changing, there will be a continual process of institutional selection. While some institutions fall by the wayside, others thrive, and through passing on their good habits, the 'genotype' of surviving institutions multiplies. The upshot is that all human activities, including those around accumulation and regulation, will be resolutely social, shaped by a set of insitutional norms and expectations.

The other, related focus of Veblen's work is cumulative causation. By that he means 'a continuity of cause and effect. It is a scheme of blindly cumulative causation in which there is no trend, no final term, no consummation' (Veblen, 1919: 436). A post-Darwinian economics, then, implies no ultimate equilibrium or purpose, just change. But it is change that is structured by the past, a consequence of Veblen's theory of institutions: 'The situation of today shapes the institutions of tomorrow through a selective, coercive process, by acting upon men's habitual view of things, and so altering or fortifying a point of view or mental attitude handed down from the past' (Veblen, 1953: 190–1). So while there is change, there are always limits imposed on it by historic conventions, or, as Veblen puts it in another passage, 'a cumulative sequence of economic institutions [should be] stated in terms of the process itself'.

## Nelson and Winter

Nelson and Winter's (1982) work is not directly based on Veblen's writings, but they prosecute a very similar argument to the one that he makes about institutional evolution. Before I set out their argument, though, it is helpful to review briefly three central features of Darwinian natural selection. There must be *variation* among individuals within a species with respect to some significant selective trait; if every entity were the same, there would be nothing on which natural selection could gain purchase. At least for the period of investigation, individuals must remain stable, and be able to transmit characteristics to future generations (*inheritance*). Finally, there must be a *selection* mechanism such that some individuals are better suited to their environment than others, and therefore increase their relative significance compared to inferior forms. One final distinction also worth making here is between Darwinian and Lamarckian forms of natural selection. While the Lamarckian form recognizes that certain character traits of individuals can be learned as a response to the environment and then passed on to their progeny, the Darwinian form does not.

Each of these three features is employed by Nelson and Winter in developing their own evolutionary theory. Variation takes the form of each firm possessing a different kind of routine for selecting a particular production technology which is the basis of profitability. Specifically, in their simplified model, each firm faces a set of linear production functions among which it must choose, where the choice is based upon a firm-specific stable set of decision rules or norms. Provided the firm can meet its own threshold level of positive profitability (firms are satisficers), the same routine can be repeated. If its chosen profitability levels cannot be met, then it must mutate or go bankrupt. Mutation means engaging in an active search for alternative techniques either through imitation of other successful firms' strategies, or by initiating its own R&D programme. In the sense that Nelson and Winter suggest that firms can adapt to their environment they are proferring a Lamarckian view of evolution. As they say themselves, 'our theory is unabashedly Lamarckian: it contemplates both the "inheritance" of acquired characteristics and the timely appearance of variation under the stimulation of adversity' (Nelson and Winter, 1982: 11).

Inheritance follows from the very stability of firms as institutions. Provided that they meet their own level of profitability, firms will continue to practise the same routines and strategies. Furthermore, in so far as new firms or even old ones imitate successful firms there is a form of inheritance.

Finally, there is selection. Key here is profitability. Those firms that choose techniques that are most profitable accrue the greatest resources for future investment, and hence increasingly take a larger market share. As Hanappi (1994: 108) writes:

> At any time, a firm's past results determine its productive opportunities. Firms need money, as a generalized resource to enlarge their scale of operation in much the same way that organisms need energy as a generalized resource, to be reproductively successful.

Because of the explicitly biological metaphor, the Nelson and Winter model takes on all the methodological characteristics already discussed. It has also spurred a number of other authors to adopt an evolutionary approach specifically to growth and technological innovation (for examples, *see* Dosi *et al.*, 1988; and Metcalfe, 1988; provides a good overview). Their argument is that one of the greatest sources of firm variety is technological innovation, which, in turn, is the principal catalyst of accumulation (Metcalfe, 1988: 83). However, because they are also institutional and evolutionary economists, they add to this insight the recognition that innovation is bound up with particular kinds of institutional practices: that technical change is embodied within a specific set of routines and habits.

There is some work in this tradition by geographers. Webber *et al.* (1992) examine the different components that bear upon technological innovation and firm mutation (*see also* Sheppard and Barnes, 1990: chs 9, 13 and 14). And Storper and Walker (1989) also in part draw upon the same tradition in trying to understand the 'inconstant geography of capitalism'. Arguing that technological change is localized because the institutions and firms associated with it are themselves place-bound, they suggest that regional economic change must be 'understood, like all industrial development, as an evolutionary path in which each step moves one way from a past that cannot be recovered and that limits future directions' (Storper and Walker, 1989: 113). In this reading, regional growth, a result of technological innovation, is path-dependent, although there may well be events, even small ones, that alter that trajectory. This last conclusion is bolstered by writings in the second body of work, the new growth theory.

## The New Growth Theory

The new growth theory first arose in the early 1980s, although its intellectual antecedents go back at least as far as Marshall (indeed, versions of it are very close to neo-classical economics – classic statements are Romer, 1986; Arthur, 1989; and Krugman, 1991). Central to all the versions is the supposition of increasing returns, which provide for a positive feedback mechanism: any advantage is made more advantageous over time. One of the recent theoretical inspirations here is the work on complex systems pioneered by von Bertalanffy, Weiss and more recently Prigogine. Their general argument is that coherent, complex systems through a process of positive feedback can evolve out of initial chaotic conditions. However, the end-point is not stable equilibrium growth, because sooner or later there is an abrupt morphogenic change triggered by either endogenous or exogenous factors. Furthermore, at the point at which that change occurs – known as the point of bifurcation (Laszlo, 1987) – even a small variation of one variable can be enough to produce massive alterations in the system trajectory (for a geographical take on this work, *see* Stern, 1992; Sunley, 1993: 11–13).

As an illustration, let me use the work of Arthur (1988a, 1988b, 1989), who is concerned, particularly with the evolution of different technologies as they compete with one another. His argument is that the one that wins out is not necessarily the best, but rather the one that most possesses 'increasing returns to adoption.' That is, as more people adopt the technology the more favoured it becomes. This is because of such factors as scale economies in production, a greater likelihood of improvement and development, and the technologies probably being better known (Arthur, 1988a: 591). Whatever the specific cause of increasing returns to adoption, once the balance is tipped in favour of one technology there is a process of cumulative causation leading to its dominance. At that point there is, in Arthur's terms, technological 'lock-in'. Once a technology dominates there is no going back.

There are other interesting features of Arthur's model that tie into the evolving complex system's metaphor.

1. There is path dependency; that is, outcomes depend upon the way in which adoptions build up. For example, the outcome of a process in which rival technologies were for a long time neck-and-neck in terms of adoptees (for example, Apple versus Microsoft operating systems) would be quite different from a case in which right from the outset there was a clear-ahead winner (gasoline- versus steam-powered motor carriages).
2. There is the possibility of a suboptimal result. In Arthur's scheme, factors that lead to adoption are modelled as random and incidental to any intrinsic beneficial features of the technology. Arthur (1988a: 596) cites as examples the utilization of light water nuclear reactors chosen over the superior gas-cooled type, and the use of QWERTY keyboards. Dominant technologies dominate because of chance and happenstance, not because of the ineluctable unfolding of some superior rationality.
3. There is the possibility that small changes in variables will produce massive consequences. Because adoption is modelled as a random process, it 'is inherently unstable, and it can be swayed by the culmination of small "historic events", or small heterogeneities, or small differences in timing' (Arthur, 1988a: 595). As a result, it is 'the small events of history [that] become important' (Arthur, 1989: 127).
4. There is the emergence of a self-regulating system. For 'what we have in this simple model is "order" (the eventual adoption-share outcome) emerging from "fluctuation" (the inherent randomness in the arrival sequence)' (Arthur, 1988a: 595).

In comparison to Nelson and Winter's approach, Arthur's work is focused on technology selection rather than the routines of firms themselves. His argument is that if there are increasing returns to adoption, one technology exercises 'competitive exclusion' over others, thus creating technological 'lock-in'. In terms of a spatial economy the corollary would be where because of proximity effects a given geographical area would lock in to a given technology, but it might well be different from another area. As Arthur (1988a: 604) says, 'here geographical clusters of localities locked in to different technologies might emerge, with long run-adoption structure depending crucially on the particular spatial increasing-returns mechanism at work'. Such a vision is at least congruent with Storper and Walker's (1989) work, which speaks of 'pathways of industrial change,' and highlights the irreversibility of technical choice and its uneven spatial consequences.

## POST-KEYNESIANISM

Post-Keynesianism is the least practised within economic geography of the three approaches (the most explicit review is Sunley, 1992, and scattered elements are found in Clark *et al.*, 1986 and Sheppard and Barnes, 1990). Arising in the 1950s primarily in Cambridge, England, and associated particularly with the work of Nicholas Kaldor, Luigi Pasinetti and Joan Robinson, post-Keynesianism is characterized by both a substantive and a methodological eclecticism. Combining Marx's reproduction scheme, the Polish economist Michael Kalecki's ideas about pricing and oligopolistic competition, and Keynes's concerns about uncertainty, post-Keynesianism is more formal than regulationist theory, not as politically charged as Marxism, but much more interested in theorizing issues of class and income distribution than institutional and evolutionary economics.

Holding all these pieces together is the biological metaphor of organicism. The economy is conceived as a changing complex system of many parts and levels, the relationships among which are internal or necessary rather than external or contingent. By external or contingent is meant the classic Humean conception of cause which is of a constant conjunction between two independent entities: if A then B; if a heart beats then blood flows to other organs. In contrast, and characterizing organicism, internal or necessary relationships are defined such that there is something about the very definition of one element or a set of elements that necessarily connects it to others. For example, there is something about the very definition of a heart that requires blood to flow to other organs once it pumps. A heart, therefore, gains its meaning from the different necessary relationships that it takes on with other parts of the body. If it did not have those relationships it simply would not be a heart, but something else. Specific examples will be given below, but once applied to the economy, the imperative of the organicist metaphor is to examine the economy as a whole, with a focus on the multiplicitous and variegated necessary relations that exist among its parts, and which define each. As such, this organicist position clearly relates to notions of emergence, and therefore a non-reductionist methodological stance.

Hodgson (1993: 11) argues that in large part the metaphor of organicism entered post-Keynesianism through Alfred Whitehead's influence on Keynes while the two were at Cambridge (*see also* Winslow, 1989). Specifically, Whitehead propounded a non-Cartesian philosophy that held that the body and mind, the physical and mental, should be kept together because both are interrelated such that 'lower forms of life, such as vegetation and lower animal types, . . . touch upon human mentality at its highest, and upon inorganic nature at its lowest' (Whitehead, 1938: 204–5; quoted in Hodgson, 1993: 13). The world in all its variety, then, needs to be seen as an interconnected whole. But post-Keynesianism is not a pure derivative of Keynes's work. It has been influenced by Marxism and institutional economics, which also come with organicist intellectual baggage, adding to the mix.

Whatever its precise origins, the organicist metaphor has shaped the central characteristics of post-Keynesianism. Perhaps the clearest link is with respect to uncertainty. Because of the complexity of the (organic) system, and its evolving character, agents (and theorists) are necessarily uncertain about the future. Useful here is a distinction made by Keynes. Whereas risk can be assessed by means of probability calculations, uncertainty cannot. As Keynes (1973: 113–14) writes, 'By uncertain knowledge . . . I do not mean merely to distinguish what is known for certain from what is only probable. . . . About these matters there is no scientific basis to form any calculable probability at all. We simply do not know.' Much subsequent post-Keynesian work is, then, about working out the consequences of not knowing. One of those consequences, as Kaldor (1985) puts it, is an 'economics without equilibrium', an economics that 'looks upon the economy as a continually evolving system whose path cannot be predicted any more than the evolution of an ecological system in biology'. For this reason, post-Keynesians are also not interested in defining economic optima. Even talk of satisficing makes little sense, because one must still need to know how much effort is required to maximize even if one chooses not to. In an uncertain world not even that can be known.

Another of the foci that follows from the organicist metaphor is post-Keynesianism's stress on macro-economics, and an examination of the interrelationship among the system's parts (Keynes's work, for example, was behind the establishment of an office of national accounts in the UK). Moreover, it is not just that the economy is composed of irreducible but interconnected subcomponents, but that it is itself only one component of an even wider system. For example, in the post-

Keynesian analysis the economy is directly related to social class and conflict through the distribution of income, which is a key factor in determining saving rates. Certainly 'the principal variable of post-Keynesian analysis is not the isolated agent so beloved of orthodoxy' (Harcourt, 1985: 134), but broader social agents, such as institutions, which interact in complex patterns. This emphasis on institutions within the post-Keynesian analysis also follows from the base supposition of uncertainty. As Sunley (1992: 60) writes, 'institutions not only shape the formation of expectations but also . . . exist to reduce uncertainty and to make decision-making possible'.

Finally, along with post-Keynesianism's recognition of the openness of the organic system of which the economy is part goes a theoretical openness: a recognition that one theory will not explain everything, and so there is a need to adapt to the circumstances. Dow (1990) calls this a 'Babylonian' sensibility, and Harcourt (1985: 138) uses the phrase 'a horses for courses approach', by which he means the awareness that 'a number of outcomes are possible, depending upon concrete situations and circumstances, rather than simple, slightly dogmatic views which follow from the universalist nature of much modern neoclassical theory'.

Showing how these different elements of the organicist metaphor are translated into formal post-Keynesian theory is beyond the scope of this chapter (for reviews of post-Keynesianism, *see* Sawyer, 1989; Arestis, 1992, 1996). Instead, my focus is on post-Keynesianism's theory of growth, and, in particular, Kaldor's work, including his use of Verdoorn's law, which he applies to regional economies.

A central tenet of the Keynesian revolution was the realization that effective demand, rather than supply, is the critical economic variable in generating employment. As a result, Keynes made the marginal propensities to consume and save key variables because of their effect on aggregate demand. However, he connected neither to income distribution, nor to long-term growth. This was Kaldor's (1956) contribution. He demonstrated that the share of profit in income and its rate depended positively upon the level of capitalist investment and negatively on the capitalist propensity to save. While this framework was later modified by Pasinetti (1962) to include both savings by workers and profit on any capital they owned, Kaldor's basic result held. This was not a strictly one-way relationship, however. Following Keynes' ideas about expectations, and Kalecki's about planned investment, Kaldor argued that the very realization of higher profits, a result of a high investment and low savings, led to even higher capitalist investment and even lower savings, producing yet higher profits still. The result is a 'virtuous circle' or its converse, the 'vicious circle'. Higher (lower) profits in the past lead to higher (lower) levels of investment, which increases (reduces) profits even more. This was Kaldor's first cut at a theory of 'cumulative causation', but he was to develop it further by adding an explicit analysis of technological change.[3]

Unlike the neo-classical conception, in which technical change is exogenous, Kaldor and Mirlees (1962) argue that it is endogenous, driven by the very rate of investment itself. Their notion of embodied technical progress presumes that technological change is incremental, and occurs on the job – 'learning by doing', as they label it. It follows that those capitalists that invest the greatest amounts are also the ones that experience the greatest improvement in technological progress. That technological progress, in turn, improves productivity, raises profits, which spurs further investment, thereby improving productivity even more. It was this relationship that Kaldor (1966) called Verdoorn's law, the idea that the growth of productivity is dependent upon the growth of output itself. More generally, Verdoorn's law is an example of a dynamic scale economy which simply means that unit costs are continually falling (productivity rising) because of consistent increases over time in the output of a firm, industrial sector, or regional or national economy.

Such scale economies, of course, are well known to economic geographers, partly through Kaldor's work, but also through that of Perroux, Hirschman and Myrdal. Clearly, once they are applied to a geographical economy, as Kaldor (1972) did, they imply uneven development: fast-growth regions grow ever more faster and slow ones ever more slowly. From the perspective of this chapter, though, the interesting point about Kaldor's post-Keynesian theory is its undergirding by an organicist metaphor. There are the mutual interrelations among the different parts of the system: class relations bear upon investment levels which bear upon technological change which bear upon income levels which bear upon class relations. There is the high level of complexity, which results in persistent uncertainty, and manifest as disequilibrium and suboptimality. There is a concern with necessary relationships among the parts, rather than conjunctural ones. In Kaldor's scheme there is something about the very process of investment under capitalism that necessarily precipitates productivity improvement. And there is the focus on the capitalist system as a totality. As Joan Robinson (1973b: 267) puts it, the post-Keynesian question is not Marshall's little one, 'why does an egg cost more than a cup of tea?', and addressed by his mechanical metaphors, but Ricardo's and Marx's big one, 'what is the level of "output as a whole?"', requiring analysis by the biological metaphors of political economy.

## DISCUSSION OF THE FOLLOWING CHAPTERS

The nine chapters that form this section of the book each to varying degrees theoretically draw upon, or empirically apply, one or more of the theories reviewed above, and hence also rely upon some type of biological metaphor.

Michael Storper's chapter is perhaps the most explicit about metaphor. Even 'heterodox regional economics', Storper (p. 000) writes, 'continues to be controlled by the metaphor of economic systems as machines, with hard inputs and outputs, where the physics and geometry of those inputs and outputs can be understood in a complete and determinate way'. Because of the kinds of problems already discussed with these physical metaphors, Storper argues for the adoption of a new metaphor, 'the economy as relations' (p. 000). The rest of Storper's chapter is then concerned with unpacking that analogy, but in so far as it is concerned with conventions and institutions, complexity and uncertainty, and reflexive individuals embedded within a wider social and cultural matrix, it is inspired in part by institutional economics, and notions of evolution and path dependence.

In contrast, Michael Dunford's contribution sits squarely within the regulationist approach, although he offers a much more nuanced and complex presentation of the mode than was given above. In part he does this to help make his substantive argument, that an incongruent set of institutions, or in come cases the absence of them altogether, has created deep regional inequalities in the UK. The interesting point here is that the institutional lacuna is a direct result of the neoliberal policies of Mrs Thatcher. In terms of the argument made above, it was in effect Mrs Thatcher's application of neo-classicism's physical metaphors that caused the problem.

Martin and Sunley's chapter is an argument for continuing to accept at least the spirit of a Keynesian analysis, if not all its substantive conclusions. Their thesis is that because of uncertainty, large institutions and especially the state are still required for economic success. This still remained true for countries such as the UK, which, while its Tory government continued to use the rhetoric of privatization, maintained a heavy reliance on state intervention. Moreover, *a fortiori* it applies to such success stories as Japan, the newly industrializing countries and regions such as the Third Italy. So while their's is not an explicitly post-Keynesian analysis, by emphasizing the centrality of economic uncertainty and the ameliorating role of the state Martin and Sunley head towards post-Keynesian policy implications (set out in Arestis, 1992: ch. 10).

O'Neill concurs with Martin and Sunley that the state remains a pivotal institution within the economy. His argument, though, is that it is often not theorized explicitly, and instead is treated as a black box. This is true even for the regulationists, who should perhaps know better. Drawing upon a number of theorists, O'Neill urges that the state should be viewed as a necessary part of economy and society, and not left dangling as if it were an afterthought. While never using the biological metaphor explicitly – except for the occasional reference to the body or to suturing – O'Neill's view can be interpreted as an organic one: the state is necessarily related to other parts of economy and society and, by excluding it, theorists are omitting a vital organ.

Ray Hudson's and David Sadler's respective chapters can usefully be discussed together. Both rest upon a regulationist theoretical reading but elaborate upon it in different ways, both of which turn on geography. In Hudson's case, the argument is that post-Fordist production – high-value production, in his vocabulary – has transformed the Fordist stalwart of the mass collective worker. Large numbers of workers, and even their labour organizations, are still needed but, under a regime of high-value production they are deployed now on capital's terms. There is another consequence: because of the undermining of the mass collective worker, capitalists have much greater flexibility with respect to locational choice, resulting in increasingly complex patterns of spatial inequalities. That greater geographical complexity is also a key feature in Sadler's chapter, which looks at glocalization by examining the process of Japanese automobile subcontracting in Europe, and specifically Nissan in north-east England. Particularly interesting is the effect of institutional inertia on this firm. More generally, Japanese auto companies have tried to replicate their previously successful domestic strategy of subcontracting in another geographical context, Europe. In terms of the earlier concepts, this is a case of path dependence, but, as was also discussed, this does not imply that it is the best path to follow.

In some ways Dina Vaiou's chapter is complementary to Hudson's. While Hudson regrets the breaking up of the traditional form of the mass collective worker, Vaiou regrets that it has never appeared at all in Greece, especially among women. Although there is no explicit theoretical template in the chapter, there is the sense of organic wholeness to the economy. Different parts of economy and society – gender, family life, processes of urbanization, small-firm formation, and definitions of the economy – necessarily relate in complex ways.

In reacting against free-marketers such as Jeffrey Sachs, who say that the shock therapy of the pure market is enough to kick-start the former Soviet states into economic take-off, Adrian Smith argues from an institutionalist (and also regulationist) perspective that this must be wrong becuase the market is never pure; it must always be buttressed by the right kind of institutions if it is to evolve and reproduce successfully. That has been the problem in most of the post-Soviet states. There are either the wrong kind of institutions or no kind. The result is a mutant kind of capitalism that does very few people any good.

California is, in many senses, a long way from Eastern and Central Europe. And yet, for Dick Walker, the state has also been following the wrong path, especially for the past five years. In a damning account, Walker details the various components that have turned California's dream into a nightmare. Again, there is no explicit theoretical template, but what is impressive about Walker's narrative is its organic quality: economy, polity, race and gender are seamlessly integrated within his political-economic approach.

## CONCLUSION

My purpose was to introduce the chapters in this section of the book, and in so doing highlight a fruitful line of theorizing about accumulation and regulation around biological metaphors. My argument was that because of their shunning of determinancy, embracing of organicism and emergence, recognition of the primacy of historical change and adoption of a non-teleological position, biological metaphors are a more useful point of departure for representing accumulation and regulation than the mechanistic kind. Indeed, three of the leading political economic theories of accumulation and regulation are already significantly infused by biological metaphors, and in so far as economic geographers have drawn upon them their work benefits too (as is illustrated by subsequent chapters).

But it is not just a one-way flow. The various authors whose chapters follow add to this tradition by their clear geographical sensibility, one that is often couched in terms of uneven spatial development. The specifics are worked out in different ways by the different authors, but the general outcome is similar. Because of the very differences among places and across spaces – whether it be, for example, in terms of fixed capital, modes of regulation, institutional norms or technological change – inequalities emerge, become chronic and frequently self-reinforcing. The biological metaphors that underpin the theories of accumulation and regulation employed by the authors do not create the uneven development, but they allow it to be seen. In contrast, this was not possible under the mechanistic metaphors of spatial science: uneven development was removed from view by such assumptions as isotropic plains, distance minimizers and general spatial equilibrium.

That said, clearly not all biological metaphors are equally useful, and as with the mechanistic kind there is a need to be alert to the limitations of their hidden assumptions. Also, in urging the consideration of biological metaphors I am not calling for a one-to-one mapping of biological theory on to a economic geographical phenomenon: in any case, that might be impossible (*see* Sunley, 1993, for elaboration). Rather, the role of metaphor is as a catalyst in thinking new thoughts; it is as a jolt, or a *frisson*, to doing things in a different way. So in urging the application of biological metaphors I am urging an open mind; a willingness to experiment and be novel. Certainly, capitalism has been experimental and novel in its forms of accumulation and regulation, and as capitalism has been so must we.

## ACKNOWLEDGEMENTS

The title of this chapter is an economic geographical take on Hodgson's (1993) book, which in many ways was the inspiration for this chapter. I would also like to thank Roger Lee and Jane Wills for both their very constructive comments and their Job-like patience.

## NOTES

1. Political economy is defined here as the study of the process of producing, distributing and accumulating economic surplus in class-divided societies (Barnes, 1995a).
2. Certainly, the three approaches to accumulation and regulation reviewed in this introduction are part of the political economy tradition, and explicitly rest upon biological analogies. This suggests that there may be something about the kinds of questions that political economists pose – questions about historical processes of change, conflict and transformation – that are homologous to those asked by biologists. Interestingly, here Darwin was influenced by the work of political economists, and in particular the writings of Robert Malthus (Hodgson, 1993: ch. 4). In this sense, there is perhaps a metaphorical 'spiral', as Mirowski (1994: 15) calls it, with biology originally inspired by political economy, which now in turn draws upon biology.
3. I use the phase 'cumulative causation' because there is a link between Kaldor's theory and Veblen's work. Kaldor was a student of Allyn Young, who in 1928 wrote a well-known paper on increasing returns. Young, in turn, was both a student of the institutional economist Wesley Mitchell and an avid admirer of Veblen.

# CHAPTER NINETEEN

# REGIONAL ECONOMIES AS RELATIONAL ASSETS

MICHAEL STORPER

## THE 'HOLY TRINITY' OF REGIONAL ECONOMICS

Confronted with such dazzlingly complex and important phenomena as the globalization of production systems, deindustrialization and reindustrialization, new economic spaces, global/multinational regions, and enormous planetary flows of goods, capital and labour, regional economics and economic geography, like much of economics as a whole, has in recent years seen a heterodox paradigm emerge in its midst. Where the orthodox paradigm remains fundamentally concerned with prices and quantities in a rather abstracted way, the heterodox paradigm breaks the problem of economic development in regions, nations and at a global level into a series of substantive empirical and theoretical domains, and attempts to build up a multilayered explanation for it. The heterodox approach involves what we might call a new 'holy trinity' by which it analyses heterogeneous labour and capital: technologies–organization–territories.

*Technology* and *technological change* are now recognized as among the principal motors of changing territorial patterns of economic development; the rise and fall of new products and production processes takes place in territories, and depends to a great extent on their capacities for specific types of innovation. *Organizations*, most importantly firms and groups or networks of firms tied together into production systems, not only are inscribed into and in some cases dependent on territorial contexts of physical and intangible inputs, but have greater or lesser relationships of proximity to each other. *Territories*, whether peripheral regions or agglomerations, may be characterized by either strong or weak local interactions and spillovers between factors, organizations or technologies.

The heterodox paradigm integrates the significant theoretical advances that have been made in each part of the holy trinity in recent years. Its emphasis on heterogeneous labour and capital in all of them is fundamentally correct. Heterodox regional economics, like economics in general, however, continues to be controlled by the metaphor of economic systems as machines, with hard inputs and outputs, where the physics and geometry of those inputs and outputs can be understood in a complete and determinate way. This focus on the mechanics of economic development must now be complemented by another focus, where the guiding metaphor is the *economy as relations*, the *economic process as conversation and co-ordination*,[1] the subjects of the process not as factors but as *reflexive human actors*, both individual and collective, and the nature of economic accumulation as not only material assets, but *relational assets*. Regional economies in particular, and integrated territorial economies in general, will be redefined here as *stocks of relational assets*, and regional development as the *evolution of regionally specific relational assets*.

This shift in guiding metaphors reflects new content for each of the elements of regional economics' holy trinity, content which goes beyond what is found even in the heterodox paradigm. Technology involves not just the tension between scale and variety, but that between the codifiability or non-codifiability of knowledge; its substantive domain is learning, not just diffusion. Organizations are knit together, their boundaries defined and changed, and their relations to each other accomplished not simply as input–output relations or linkages, but as untraded interdependencies subject to a high degree of reflexivity. In a globalizing world economy, territorial economies are created not only by proximity in input–output relations, but more so by proximity in the untraded or relational dimensions of organizations and technologies. The principal assets of territories – because scarce and slow to create and imitate – are no longer material, but relational.

## REFLEXIVITY AS THE CENTRAL CHARACTERISTIC OF CONTEMPORARY CAPITALISM

The economic capabilities of capitalism have undergone great expansion and deep qualitative change in the past twenty years. Among the new 'metacapacities' of modern capitalism, several are most important. First, the technological revolution in production, information and communication technologies permits vast expansion of the *nature and spheres of control* of firms, markets and institutions, involving deeper and more immediate feedbacks from one part of these complex structures to others than ever before, dramatic cheapening of many forms of material production, and great increases in the variety of material and intangible inputs and outputs. Second, there has been a vast *spatial extension and social deepening* of the logic of market relations, in part facilitated by the technological leap (especially through the cheapening of telecommunications and media as vehicles of market relations, and through the extension of physical infrastructure); greater percentages of the population, and greater percentages of their relations than ever before, are inscribed in a process of markets and commodities, and these are more and more tied into far-away places than ever before. This is, in one sense, a continuation of long-term processes of 'modernization'; in another sense, it involves the crossing of a qualitative threshold in terms of extent and depth. Third, and combining the effects of the first two processes, there has been a *generalization* of the 'grid' of modern organizational methods, bureaucratic rule and communicational processes to more dimensions of economic and non-economic life than ever before. This does not mean the extension of a single, hierarchically administered regime to all peoples, but the sharing of certain general ways of life which are common to contemporary industrial-market society (Giddens, 1994; Beck, 1992; Beck *et al.*, 1994).

The qualitative consequences of these metacapacities are more novel than the mere quantitative expansion of the capitalist market system. In the most general terms, they may be summed up as an enormous leap in *economic reflexivity*. The term 'reflexive' refers to the possibility for groups of actors in the various institutional spheres of modern capitalism – firms, markets, states, households and others – to influence the course of economic evolution as a result of their own critical distance from the traditional functions of these spheres, this distance itself facilitated by contemporary technologies and communicational practices. The actions thus made possible can no longer be described as 'rational action within established parameters', which is the keystone concept of modern social and economic theory. The procedural rationality of contemporary economic and social actors considers the parameters themselves. The temporal rhythms, evolutionary pathways and role of positive feedbacks in social and economic dynamics today render them radically different from those which the social sciences have attempted to understand for much of the twentieth century.

This in no way implies that such reflexivity is free from constraint. Instead, the old debate in the social sciences between determinism and free will, structure and agency, has become largely irrelevant, because it has been empirically left behind by the course of real socio-economic evolution, in which the two sides of these traditional oppositions have become inextricably produced by each other.

Translated into more narrowly economic terms, the evolution of relations of production and the physical and technological infrastructures of production have multiplied the possibilities for economies of variety in the organization of firms, markets and other institutional domains of economic life, this variety being the expression of reflexive action in all these realms. When we speak of a growth in variety, it is with respect to the preceding industrial system, that of mass production, and not with respect to artisanal economies of yore. Even today, small artisanal producers, those who remain, are threatened by the extension and deepening of market logics.

The possibilities for product and process variety in the new industrial system, once tried out, are then subject to the effects of competitive or imperfect competition. But the competitive dynamic is more endogenous than ever, generating new constraints. The variety of one becomes risk for the other (economic, ecological, social, psychological, personal), to which the other must respond. In the economic sphere, these risks are expressed through the redefinition of competition: what it takes to win and how it is possible to lose. Winning has become a much more complex target, because the conditions which a firm, region or production system must now satisfy in order to win are manufactured and remanufactured more thoroughly and more rapidly than ever before, creating a moving target for success and a shifting minefield of risks of failure. This is directly a consequence of the increase in the reflexivity of economic activity in the context of a generalized market system.

Theories of competitiveness have struggled to capture these phenomena over the past twenty years, developing many descriptive monikers for the new economy: post-industrialism, the information economy, the knowledge-based economy, flexible specialization, and post-Fordism (Cohen and Zysman, 1984; Castells, 1989; Piore and Sabel, 1984; Boyer, 1992). Though each of these labels helps in understanding some dimensions of the contemporary economic process, the deepest and most general way to describe the logic of the most advanced forms of economic competition is that of 'learning' (Lundvall and Johnson, 1992; Arrow, 1962; Rosenberg, 1982). Learning is the competitive outcome of heightened reflexivity. Those firms, sectors, regions and nations which can learn faster or better (higher quality or cheaper for a given quality) become competitive because their knowledge is scarce and therefore cannot be immediately imitated by new entrants or transferred, via codified and formal channels, to competitor firms, regions or nations. The price–cost margin of products generated in this way can rise, while market shares increase; the resulting knowledge or technology rents alleviate downward wage or profit pressure. Learning-based activities are not immune to relocation or substitution by competitors. Once they are imitated or their outputs standardized, then there are downward wage and employment pressures. Firms or territorial economies must therefore be equipped to keep outrunning the powerful forces of imitation in the world economy. They must become moving targets by continuing to learn. *The learning economy is therefore an ensemble of competitive possibilities, reflexive in nature, engendered by capitalism's new metacapacities, as well as the risks or constraints manufactured by the reflexive learning of others.*

The dimensions of the new economic reflexivity are therefore principal concerns of any kind of economic analysis interested in developmental processes. These dimensions may be seized, at least in a preliminary way, by such keywords as 'action', 'created rules', 'action frameworks' and 'routines'. Substantively, their study requires that we focus on how individual and collective reflexivity operate in the contemporary economy, through cognitive, dialogic and interpretative processes, with the substantive goal of understanding how *relations of coordination* between reflexive agents and organizations are established.

## THE RELATIONAL TURN IN ECONOMIC ANALYSIS: TECHNOLOGIES, ORGANIZATIONS, TERRITORIES

In the field of regional economics and territorial development, the developments described above mean that the content of the theoretical holy trinity – technologies, organizations, territories – must be redefined, from a series of machines to a set of relations (Asanuma, 1989) and their constituent reflexive processes.

### Technology

Technological change is no longer the 'black box' it used to be. The heterodox paradigm has adapted the discoveries of the economics of technology to analyse the effects of technological change on the geography of production, distribution and transport. In the geography of production, we know that activities based on standardized technologies permitting scale economies to be realized inside the firm can delocalize, while those based on non-standardized technologies, and generally having higher levels of variety and flexibility, tend to agglomerate. The heterodox paradigm has, therefore, better understood the spatiality of the input–output machine (technology + division of labour) of the modern economy, and in so doing it has revolutionized the theory of agglomeration.

The limits of this paradigm can be found in the analysis of the *causes* of technological change, and the geography of innovation and learning, which are the heart of economic competition today. Its model of the process of agglomeration–diffusion is upstream of the motor forces of technological change. Its image of spatial dynamics is that of the emergence of new technologies or industries in 'centres', followed by their 'diffusion' towards 'peripheral' regions. This image corresponds closely to the linear model of technological innovation, 'research–invention–innovation/deployment', where technological progress is said to be embodied in the teleology from innovation to standardization and scale economies (Mansfield, 1972).

Yet it now appears that development, at least in wealthy countries and regions, comes about through *destandardization* and the *generation of variety*. A new style and rhythm of technological learning have radically expanded the variety of technologies, in two ways: at any given time, the specialization and numbers of technologies for given kinds of uses; and, more importantly, over time, the pace of their modification and replacement. This forces a complete reconceptualization of the technological innovation process in economic development: it now involves not only the gigantic formal organizations of research in laboratories, universities and multinational firms, which correspond to our image of the process as hierarchical and linear, but the proliferation and dramatic complexification of relations among those institutions and between them and other elements of the economic environment. Paradoxically, the rise of bigger and bigger science and R&D has been accompanied not by its increasing isolation upstream, but by its increasing integration into a host of other economic and social processes. Within 'big' R&D, for example, there are now more complex feedbacks between science and *savoir-faire* in the high- and medium-technology industries than ever before (Nelson, 1993; Griliches, 1991; von Hippel, 1987, 1988; Jaffe, 1986, 1989; Jaffe *et al.*, 1993; Antonelli, 1995), while in many medium- or low-technology sectors, *savoir-faire* is now subject to deliberate reflection, attempts at systematization, and appropriation of the results of science and engineering (Lundvall, 1990). Research on technological change has documented the importance of user–producer relations (inter-firm, inter-industry and consumer–producer); science–production relations; inter-firm relations in technologically cognate areas; and firm–government–university relations in technological innovation. It has also shown, importantly, that these relations are increasingly organized as non-hierarchical, networked, complex and substance-filled communication and action processes (Hakansson, 1987, 1989; Johansen and Mattson, 1987; Cohendet and Llerena, 1989; Callon, 1992). Research on the proliferation of 'flexibly specialized' industrial districts has shown, in addition, that the capitalisms of a number of very wealthy regions and countries are built around practical forms of technological innovation, involving relatively small or indirect roles for formal science or R&D, where complex relational feedbacks in the production systems are responsible for successful innovative performance (Russo, 1986; Storper, 1997: ch. 1).

The technological enterprise which is so central to contemporary capitalism seems to involve a set of circular processes today. The increasing density and complexity of relations is the means to new forms of collective reflexivity, leading to a quantum leap in the possibility for generating technological variety; that is, to learning. This variety has two principal consequences. On the one hand, it sets off traditional cycles of codification, standardization, imitation and diffusion of knowledge. On the other, at any given time, there are innumerable 'islands' of non-cosmopolitan (Rip, 1991) knowledge in this variety-centred economy, where only those actors who are inscribed in the relations required to get access to the knowledge – and, perhaps even more importantly, the relations required to *understand, interpret, and effectively use* the knowledge – will be able to deploy it in economically useful ways. In turn, these nodes of relationally linked actors may 'spin off' new standardization, decodification processes, but they may also regenerate variety within their field of endeavour, sustaining the viability of non-cosmopolitan nodes of interaction. This is but one of the many new dynamics of an economy of reflexivity and its manufactured opportunities and risks.

For regional and territorial economics, this means a reorientation of the central issues posed by technological change: from standardization to *destandardization and variety* as the central competitive process; from diffusion to *the creation of asymmetric knowledge* as the central motor force; and from codification and cosmopolitanization of knowledge to the organizational and geographical dimensions of *non-codified and non-cosmopolitan knowledge*.

## Organizations

The second element of the holy trinity is organizations, by which I mean, principally, firms and production systems.[2] In the post-war period, organizations have figured prominently in economics generally, and regional and industrial economics in particular. The theory of the firm – stemming from Coase and developed by transactions cost economics – has defined, as its core subject, the functional boundaries of the firm, the division of labour between firm and market, and the relations or transactions between firms (Coase, 1937; Williamson, 1985; Dosi and Salvatore, 1992). The theory of production systems received a major push in the late 1940s and early 1950s, with the Perrouxian notion of economic spaces and industrial complexes, and was given greater generality and empirical-analytical power in Leontief's development of input–output models of the entire economy (Perroux, 1950a, 1950b; Leontief, 1953). Regional economists made major efforts to use input–output theory and techniques in the modelling of regional economies (Richardson, 1973).

Transaction cost economics, as developed by Williamson, provided a more precise understanding of the cost drivers for input–output structures, thus bringing the theory of the firm and that of the production system closer together.[3] In turn, the theory of industrial complexes and agglomeration was given new dimensions by consideration of the geographical dimensions of transacting. On the one hand, it was shown that geography figures in transactions costs in general, and hence influences the boundaries of the firm and production system (i.e. geography influences the degree of internalization or externalization in the production system) (Scott, 1988a). On the other, it was shown that the geography of transactions costs helps explain agglomeration and spatial divisions of labour. Much of this work on spatial divisions of labour shared similar concerns with research on multi-locational or multinational enterprise, with the first approaching the problem from the side of geography, and the second from the side of the firm, meeting around the subject of locational dynamics of complex production systems (Dunning, 1979).[4] In addition, transaction cost theory was extended to output markets and to labour markets on the input side, and both were integrated into geographical transaction cost theory and modelling. Finally, innovation theory, in many guises, has attempted to understand the transactional context for technological change, and geographers and regionalists have claimed that this context has strong territorial dimensions; though still at an early stage (Camagni, 1991; Malecki, 1984; Maillat *et al.*, 1990, 1993; Russo, 1986; Bellandi, 1986, 1989, 1995; Djellal and Gallouj, 1995), it is an active area of work today, and its goal is nothing less than an integrated theory of economic space, consisting of the interrelations between organizational, technological and geographical space.

It can be seen that great theoretical progress has been made in the past half-century towards understanding economic organization, and its extension to location, and the geography of production systems. The fundamental concerns of theory and modelling, however, are focused almost entirely on the traded relations between firms and places (factor markets, institutions), on traded relations between firms (inter-firm trade), or on exchanges between production units of big firms (intra-firm trade). The mechanism which accounts for organizational and geographical outcomes is the prices, quantities and qualities of these *traded interdependencies*. Geographical analysis concentrates on the prices, quantities and qualities of these interdependencies.

The notion that such relations among economic actors ultimately are expressed in terms of direct and traded interdependencies, however, can no longer be sustained. For network production systems – whether they take the form of industrial districts or of big multinational firms and their suppliers – the geometry of such linkages is only the tip of the iceberg. The visible structure of these networks – traded linkages – is underpinned by a complex structure of interrelations between firms, between firms and labour markets, and between firms and institutions – which allows learning to take place. These interrelations depend on the conventions and relations underneath transactional linkages, as well as certain relations between organizations which are not 'hard' at all, which are *untraded interdependencies*. We shall return to them shortly.

Learning as a form of reflexivity is fundamentally a dynamic process, where the parameters of interaction must be unstable if learning is to take place. A high degree of uncertainty (what we earlier called 'risk') is an endogenous property of reflexivity. This uncertainty has to do with two 'critical situations' of reflexive action. On the one hand, all productive activity depends on the pragmatic necessity for actors to co-ordinate with each other. But there are very few situations where the uncertainty over what the other might do can be eliminated by cognitive structures, norms or knowledge which are completely codified, or by structures of bureaucratic authority or incentives sufficiently powerful to eliminate the possibility that the actor could choose to deviate from established norms. In addition, the exercise of bureaucratic authority is increasingly incompatible with the goal of the economic process, learning, because it would substitute hierarchy for reflexivity.

On the other hand, learning is frequently associated with informational or practical problems whose cognitive content cannot be codified or routinized. The corollary of this is cognitive uncertainty. It follows that transactions between actors will be more and more dependent on the means of *interpreting* information and – above all – on establishing confidence that the interpretations which are used (by partners or by commercial relations) are correct, or at least on having confidence with respect to their intentions.

In these two situations, the problem for the actor is how to *co-ordinate* with other actors, in the absence of bureaucratic or authority-based means of doing so, and in the presence of cognitive ambiguity and complexity. These situations are not at all rare; they are more and more typical of the capitalist economy at the end of the twentieth century, especially of its value-intensive functions, those which emerge from technological learning.

This co-ordination is made possible by transactions between actors which are strongly shaped by relations and conventions. In the former, personal contacts, knowledge of the other, and reputation generate the required confidence. In the latter and more frequent case, transactions are less idiosyncratic than in the first, because they are organized by conventions which permit economic agents to absorb, interpret and utilize information (especially non-cosmopolitan and non-codified) such that the uncertainty with respect to the other is abated. Collective action can then take place. Conventions involve mutually coherent routines and expectations. They are a kind of halfway house between idiosyncratic relations and bureaucratic and impersonal rules.

Conventional or relational transactions (henceforth C-R) affect many dimensions of production systems, but the nature and functions of such conventions differ from industry to industry, according to the nature of the product, the economic fluctuations associated with its markets and production processes, and the type of learning which is possible.[5] C-R transactions may be found in at least five principal domains: (a) inter-firm 'hard' transactions, as in buyer–seller relations that involve market imperfections; (b) inter-firm 'soft' transactions, as in the diffusion of non-traded information about the environment or about learning, for example through circulation of personnel through the same external labour market or through contact between producers; (c) in hard and soft intra-firm relations, as the bases for the functioning of large firms which are 'internally externalized' in the way we noted above; (d) in factor markets, especially labour markets, which involve skills that are not entirely substitutable on an inter-industry or inter-regional basis, i.e. where there are industry- or region-specific dimensions to workers' skills; and (e) in economy–formal institution relationships, where universities, governments, industry associations and firms are able to communicate and co-ordinate their interactions only by using channels with a strong relational-conventional content.

In every critical action situation, the ensemble of relations and conventions defines an *action framework* by which actors come to co-ordinate with each other in spite of radical uncertainty. The specific qualities of the resulting co-ordination will vary according to the sphere of pragmatic action in question: the product, its technologies, the nature of the market, other historical and cultural factors and, above all, the specific evolutionary trajectory of the actors involved, as defined by their existing relations and conventions (Storper and Salais, 1997).

It should be remembered that the conventional-relational foundations of economic co-ordination do not refer to a stark contrast between internal ownership and externalization of production systems, or on hierarchies versus markets or external-embedded networks, but rather on the notion that manufactured opportunities and risks (respectively, learning or the competitive challenge of others' learning), carried out through organizational reflexivity, are becoming pervasive in contemporary capitalism. Every kind of production system has to cope with some form of fluctuations in markets, product design, available technology and prices, which make difficult the full cognitive routinization of relations between firms, their environments and employees.

Hence, a major additional focus in the analysis of organizations – firms and production systems – is now required. It has three principal components: attention to untraded interdependencies and not simply traded transactions as the corner-stones of the organizational question; the conventional and relational qualities of such untraded interdependencies; and the ways that conventions and relations organize and make possible many of the *traded* transactions of the contemporary economy.

## Territories

Most social science has traditionally considered regional economies, or – more generally – territorial economies at any subnational geographical scale to be derivative reflections of the more 'basic' forces of technologies and organizations. Today, even national economies are being demoted, by many analysts, to the same secondary status traditionally assigned regions, owing to the increasing reach of global technologies and global organizations. Thus, in the standard view, two elements of the holy trinity generate a set of outcomes in the form of the third, territory.

In contrast to this view, the apparent resurgence of regional economies and the growth of economic differentiation between major world trading economies has stimulated the notion that territories are levels of economic action in their own right, with defining contributions to, and feedback effects on, technologies and organizations. Moreover, some branches of contemporary innovation theory, as noted above, propose a set of dynamic interrelations between technological, organizational and geographical spaces. In these views, territory is a basic and not a secondary element of the holy trinity.

This said, and in spite of its innovative conceptualization of the role of territory in economic development, the heterodox paradigm has not succeeded in developing an analytical framework sufficient to realize its ambitions. The common way in which economic analysis deals with geographical proximity and distance is by analysing the geography of economic transactions: exchanges of goods, information and human resources over geographical distance. Economic geography considers the price dimensions of transacting activity, to identify circumstances where geographical concentration is necessary to efficient transacting, and those where geographical dispersion of firms, consumers, workers and institutions is consistent with it. In some analyses, agglomeration is the means to realization of superior pecuniary efficiencies of each transactor (i.e. firm).[6]

Geographical proximity is strongly probable in the presence of uncertainty; for example, a high rate of technological change along a given technological frontier in a high-technology sector, differentiation of products (i.e. rapid turnover) via deployment of traditional knowledge, and highly unpredictable market structures are all circumstances which make the environment uncertain. Even transactions that are very cost-efficient, owing to scale economies or low transport costs, can be swept away by these circumstances. In these cases, firms must try to avoid locking-in their transactional relations in the medium run, by being always ready to modify them. It follows that they reduce their risks through geographical proximity, which allows them maximum access to other potential transactors, or by vertical integration, but this latter generates other kinds of risks (lock-in, high overheads, etc.).

As with transactions in general, proximity-based transactions are underpinned by relations and conventions. It is important to understand to what extent this idea goes against the grain of modern economic analysis. As we have seen, agglomeration is seen by it as a machine for risk dispersion: in the face of uncertainty, producers agglomerate in order to have access, in case of need, to other transactors. All of this occurs in a context of rational action, where everyone seeks to reduce their costs and, above all, everyone is ready, at any moment, to break their relations to others if they find it advantageous to do so, including promises made earlier. This is the environment of 'moral hazard' which can be found in Williamson (1985) as well as many other contemporary institutionalist analyses in economics. Agglomeration thus becomes the result of efforts to reduce the risks of morally hazardous, opportunistic behaviour on the part of other economic actors on whom we depend.

In our analysis, by contrast, there is not one single moral environment, one single universal rationality of transactional relations, but a great diversity and heterogeneity of these behaviours. They are shaped by relations and conventions, not given by a supposedly universal rationality. As a result, the degree, the nature and the effects of proximity manifest a great diversity from one territory to another. In addition, conventions and relations which develop in association with a given industry or *filière*, in a given territory, can have long-term effects on the evolution of technologies and organization in that industry. These are dimensions of economic life which reflect territorial specificities through co-evolutionary processes. They are not necessarily subject to a single universal law of development or to one best practice, because contemporary capitalism, globalized as it is, depends on the flourishing of variety, in the form of multiple equilibria even within a given single industry. Beyond this, the ensemble of relations and conventions which exists in a territory can have spillover effects on multiple industries which are located there, conferring advantages or disadvantages which are regionally specific to a group of interrelated industries in that territory. It is because of the durability of conventions and relations that geographical proximity is often much more long-lasting than would be necessary merely to minimize transactions costs of an input–output system.

The theoretical status of territory is, in this way, radically different from that in the heterodox paradigm. Conventions and relations of production systems have always been essential to the functioning of the latter, even though little examined by economics. In the era of contemporary capitalism, however, they are all the more important, because they are the essential vehicles of the collective reflexivity of the economic system, the heart of its competitive dynamic and the way in which wealth is accumulated unevenly within the system. Conventions and relations are not 'obstacles' to modernization or to the 'perfect' functioning of markets. They constitute veritable economic assets, which are specific to the territories and organizations caught up in them.

In sum, the territorial element of the holy trinity needs refocusing, from the geography of input–output relations – industrial complexes and spatial divisions of labour – and the economics of proximity in traded linkages, to the geography of untraded interdependencies and the technology and economics of proximity and distance in them. This, in turn, is necessarily bound up with the geography of conventions and relations, which have cognitive, informational and psychological and cultural foundations. Throughout all of this, there must be simultaneous consideration of territory and region as derived outcomes of technology and organizations, and as the locales of differentiated conventions and relations – action frameworks – which often cut across specific technologies and organizations and affect their evolution.

## THE WORLDS THAT MAKE REGIONS, AND REGIONS AS WORLDS

It remains now to begin reconstructing concrete areas of enquiry and explanation in the field of territorial economic development, economic geography and regional economics. Our field can be reconstructed as a series of intentional, collective human projects, fields where pragmatic actions search for some kind of effectivity. The holy trinity – as reconceptualized – supplies some basic building-blocks, in that technologies, organizations and regions are pragmatic domains of intentional human activity. But they are not equal in power and importance. Territories and regions are not, in the current era, the principal pragmatic action spaces of capitalism. People do act to save regions and they act consciously to develop and promote them, in some countries more than others. Regional societies have strong regionalist sentiments in some places, weak ones in others (Markusen, 1985). It remains none the less the case that regionalist pragmatics are subservient to other pragmatic action networks today: this is because capitalism is increasingly based on geographically extensive product markets, firms and factor markets. As a result, *markets*[7] have become the principal arbiters of what is *legitimate* collective action in contemporary capitalism: other groupings, such as regions, nations, families and firms, must submit themselves to the test of the market, and they are more and more subject to political regimes which require proof that such groupings are not erected in opposition to markets.[8] Markets – in conjunction with contemporary technological capacities – make certain kinds of action spaces very important. To begin with, there is the *product*, the essential focus of markets. Product markets involve two principal elements of the holy trinity: technologies (of products and processes) and organizations (especially firms, but also the organizations that support firms, such as schools and states). Factor markets involve mostly organizations (firms, but also those of collective social reproduction, such as the state, schools and the public R&D organizations). These two elements of the holy trinity are the principal vehicles of the primary intentional projects of economic action today. The deployment of these actions principally 'makes' regional economies today,[9] when they are situated or subdivided into locations.

Through complex locational structures and patterns, however, such activities may come into close proximity in the restricted geographical spaces of regions, where they are constituted as *territorial economies*. In turn, these activities may develop various forms of regional coherence, spillovers and feedbacks; when this occurs, it is because regional economic actors have developed conventions and relations which enable such regionally centred co-evolutionary processes between organizations and technologies to unfold. Both the physical and the relational assets of production *become* – to some degree – regionally specific assets. In other words, *regional worlds of production* can emerge out of the *technological and organizational worlds that make regions*. But this occurs in only some cases; in many others, the regional economy remains – for the most part – a mere locational repository of organizational and technological worlds or artefacts, exogenously driven, exhibiting little regional co-evolution, or what regionalists have traditionally labelled 'disarticulated' or 'peripheral'.

The modern economy can therefore be conceived as a complex organizational puzzle, consisting of multiple and partially overlapping worlds in which reflexive collective action unfolds. For any given domain of economic analysis, the task is to understand the functional nature of the action spaces involved, and the substantive content of the conventions-relations – the world of action – by which actors co-ordinate and give shape to their concrete, functioning activities in that domain.[10]

In operational terms, these domains which have strong influence on the evolution of regional economies when they become co-ordinated worlds of action are different 'cuts' at regional analysis. Four such domains, which are complex interactions within the holy trinity, may be defined as priorities for theory and research:

### Technologies and Organizations

Technologies and organizations are the principal generators of the 'production possibilities' of capitalism. The first defines the envelope of physical and intellectual possibilities, while the second defines the institutional possibilities for deploying the first in an economically feasible manner. As we have noted, each of these elements of the holy trinity has been revolutionized in recent years, by the reflexive turn. In combination, they generate complex co-ordination possibilities and problems, two sorts of which are of greatest importance. The first is *products*, which are the results of co-ordinated reflexive action, against a background of technological and organizational constraints and possibilities; products are the results of different, conventionally relationally founded action frameworks, or 'worlds of production'. The second *is systems of innovation* (Nelson, 1993), which are based on action frameworks through which physical-intellectual capabilities are developed and evolved; these are 'worlds of innovation' (Storper, 1996).

## Organizations and Territories

Organizations, especially firms, 'make' regions through their locational behaviour, but it is also the case that organizations such as firms are the products of the institutional environments of their locations. This is most obviously true for single-location firms, but a case can be made that even the biggest multi-locational firms are, in some ways, strongly influenced by the localities in which they situate certain of their activities (Pavitt and Patel, 1991). For other sorts of organizations, such as schools, government institutions and politically or culturally defined institutional 'environments' (formal and informal rules for governance of the economy), the relationship to place is a great deal more direct. As we noted above, territorial economies may involve transversal effects between their different activities, through technologies (localized knowledge spillovers); through organizations (localized input–output linkages); or through aspects of the local action frameworks by which multiple sectors of the economy are co-ordinated and resources mobilized. These localized conventional-relational environments are *regional worlds of production*.

## Technologies and Territories

The development of knowledge and know-how is subject to a complex dialectic of codification/economic diffusion, and innovation/tacitness. While the first tends to lead to geographical diffusion, the second may – in some, but not all, cases – emerge from restricted geographical contexts and impede, at least for a certain time, easy geographical diffusion. The role of localization in technological innovation and deployment is made all the more potent because certain forms of innovation emerge from inter-activity knowledge and know-how spillovers, which themselves sometimes occur in restricted geographical spaces, as well as defined organizational spaces. One of the major issues for students of economic development in the age of the reflexive learning economy of contemporary capitalism is, therefore, the geography of knowledge and know-how development; that is, the geography of innovation. Accompanying the geography of innovation is the question of how this exceedingly complex form of collective action emerges and is co-ordinated in particular contexts. Paralleling enquiries into worlds of innovation in general, then, we must examine how localization of knowledge and learning comes about in the form of *regional worlds of innovation* (Storper, 1997: ch. 6).

## Technologies, Organizations and Territories

When all the elements of the holy trinity are considered equally and simultaneously, there is no theoretical 'bracketing' for the purpose of simplification. As a result, only the most complex and concrete problems of economic development can be considered. But we can build up to them using insights gained through rigorous theorizing of the individual elements of the trinity, and the limited combinations identified above.

# CONCLUSION

The approach to territorial economic development outlined here has little to say about standard problems of 'spatial economics' or 'location theory', staples of the literature on the geography of economic development, but it has much to say about the territorial differentiation of economic development, performance and institutions. Its principal contribution to the spatial disciplines is to analyse the role of territorial proximity in the formation of conventions; the role of conventions in defining the 'action capacities' of economic agents, and hence the economic identities of territories and regions; the economic status of regional conventions of production as a type of regionally specific collective asset of the economy; the status of conventions as untraded interdependencies in economic systems; why it is so difficult for some places to imitate or borrow conventions and institutions from other places; why agglomerated economic activity comes into being and why it persists even when the costs of covering distance are of relatively little importance to the activities at hand; and why there is so much heterogeneity and diversity within a capitalist economy which is more integrated and extensive than ever. As a research programme, they would vastly enhance the explanatory power of regionalist social science, bringing it closer to the principal subjects of many other contemporary social sciences while making distinctive new contributions to those debates.

## ACKNOWLEDGEMENTS

This is a much-abridged version of chapter 2 in Storper, M., 1997: *The regional world: territorial development in a global economy*. New York: Guilford Press. Earlier versions of this chapter were presented at the conference 'Industrial Economics, Spatial Economics' held by the French-language regional science association, Toulouse, August 1995; and to the 'Local Development Conference' held by the University of Florence and the Prato local development association, Artimino (Tuscany), September 1996; as well as to the Spanish regional science association, Madrid, October 1996. The author thanks all the participants at those meetings for their stimulating comments and criticisms of earlier drafts, as well as colleagues in the LATTS laboratory in Paris.

## NOTES

1. I am referring here to what has become known as the 'cognitive turn' in economics, and to 'constructivist' approaches to economic life, mostly from sociology. *See* Rip (1991) on the cognitive turn and Storper and Salais (1997) more generally.
2. NB: I am choosing to use the term 'organizations' to refer to firms and production systems, rather than 'institutions' which is the term favoured by institutional (transactions cost) economics. This is because I want to reserve the use of the term 'institution' for routines, practices and formal 'non-private' organizations, such as governments, trade associations, and so on. It is also a way to link organizations (plural) to the subject of economic organization in general.
3. As did Stiglerian development of the scale-division of labour analysis, and some neo-Sraffans (Stigler, 1951).
4. *See also* the discussion of the 'geography of enterprise' in Sayer and Walker (1992).
5. For extensive discussion of this point, *see* Storper and Salais (1997), and Storper (1996).
6. There is much ambiguity about external economies in both the geographical and the economic literatures. The basic question is whether agglomeration is simply an additive effect of individual, optimizing producers, where there are no truly collective goods with spillovers involved in the transacting system, in which case no real externalities exist. If, on the other hand, agglomeration is a site of such spillovers and feedbacks to the production system, i.e. where proximity opens up possibilities for production organization and development which would not otherwise exist, then true externalities exist. In the literature, two suggestions have been made along these lines: one is that there are intimate feedback effects between proximity and specialization within the division of labour (the work of Scott suggests this. The other is that agglomerations are sites of transaction-dependent technological innovation. In both cases, agglomeration is not merely a static 'Stiglerian–Smithian' effect, but a dynamic 'Youngian' effect. In this respect, I agree with Krugman (1995), that external scale economies are a theoretical key to economic geography, but not with his emphasis on the pecuniary and static efficiencies thereof.
7. This does not imply that markets are perfect, but rather refers to markets as a general institutional format for the organization of legitimate interactions in contemporary capitalism. There are innumerable concrete variations on the theme of the market.
8. I have said little about the links between pragmatic action and the 'justification' and 'legitimacy' of action undertaken. But suffice it to say that all pragmatic action – especially in so far as it targets reciprocity by other actors – rests on some notion of legitimacy, some form of justification, whether implicit or explicit, which must be shared by the actors caught up in the collective action. These issues have been extensively explored in Boltanski and Thevenot (1989). In the case of economic models of products, Salais and I discuss different principles of justification for different possible worlds of economic action (Storper and Salais, 1997).
9. Even admitting that there is much 'drag' from the past, and feedback from the present, to existing regional economies.
10. It cannot be overemphasized, however, that the

functional domains of action are not predefined, whether by a Parsonian functionalist logic of social organization or even by any higher capitalist structure. The whole point of the theory of pragmatics is that structure and action unfold and redefine each other simultaneously. We can model the basic functional domains that appear to us now, but these are indicative, not in any way causal.

# CHAPTER TWENTY

# DIVERGENCE, INSTABILITY AND EXCLUSION: REGIONAL DYNAMICS IN GREAT BRITAIN

MICHAEL DUNFORD

## INTRODUCTION

The past twenty years have been difficult ones for most advanced capitalist countries. Among the symptoms of these difficulties, three trends warrant particular attention. First, average rates of economic growth and of productivity growth have slowed down. The slow growth of productivity is particularly remarkable as it coincides with developments in information and communications technologies that amount to a technological revolution and that seem to offer the possibility of dramatic increases in the productive potential of human societies. Clearly, advanced societies have so far failed to realize this potential of new technologies, as the American economist Solow (1987) indicated when he made the remark that 'I can see the computer age everywhere but in the productivity statistics.'[1]

Second, slow growth has coincided with increases in unemployment. With the arrival of the trough of each successive economic cycle, the level of unemployment increased. In the 1993 recession 18 million people in the European Union (EU) were unemployed. As unemployment increased, the size of the inactive population expanded as 'discouraged workers' dropped out of the workforce altogether. As a consequence, in 1993 less than 60 per cent of the EU's population of working age was in work. Of those in work, many had had to accept part-time, insecure or casual employment.

Third, social and spatial inequality have increased. Job loss and growing unemployment, along with the fact that the large majority of women who have entered employment came from 'job-rich' households with at least one other income earner, have led to an increase in the share of 'job-poor' households and of the population dependent on state welfare benefits. At the same time an increasing dispersion or polarization of employment structures and earnings has led to an increase in gross wage differentials and social divisions between those with secure and well-paid jobs and those whose jobs are less secure and lower paid. This divide is particularly marked in the case of male wage-earners. In Great Britain, for example, the wages of the top 10 per cent of male wage-earners increased from 1.67 times the median wage in 1979 to twice the median in 1993, while the wages of the poorest-paid 10 per cent declined from 68.5 to 58.2 per cent of the median (Gregg and Machin, 1994).[2]

These divergent trends are often viewed as indicators of different degrees of success of different companies, social groups and places in their adaptation to a new post-Fordist era. In this literature the emphasis is on the importance of innovation and technology transfer, training and education, the economics and sociology of localized learning, untraded interdependencies and external economies, the degree of co-operation and trust, and the richness and performance of institutions (*see*, for example, Sabel, 1992; Amin and Thrift, 1993; Cooke *et al.*, 1995; Storper, 1995b). In this chapter I shall suggest that these supply-side interpretations, which dominate the literature on industrial districts and 'successful' regions, are insufficient fundamentally because they fail to recognize that training and competitiveness strategies do not, in a context of competitive modes of social regulation, lead to a re-employment of resources released as a result of structural change. What I shall argue is, first, that divergence and greater inequality are symptoms of an unresolved crisis. Second, deregulation and greater reliance on market mechanisms are not elements of a sustainable post-Fordist future but defensive reactions to the crisis of an earlier order and, increasingly, a cause of the unresolved crisis in that their deflationary consequences are important contributors to unemployment, non-employment, exclusion and inequality. Third, the current situation of widespread unemployment and exclusion represents a new state of equilibrium, around which fluctuations occur. The roots of this cul-de-sac, in which all localities must encourage changes in the productive order and their resource endowments to survive, but in which human resources are not re-employed, lie in the confinement of attention to supply-side issues and a failure to recognize that only a new social compromise and new rules of the game can resolve contemporary problems of non-employment, exclusion and inequality.

To make this case I shall develop two sets of arguments. In the first section I shall outline some of the evidence for regional divergence and for greater instability in the performance of regional economies in Great Britain and I shall present some of the trends in productivity, the distribution of earnings and income, employment and social exclusion that lie behind them. In the second section I shall identify the roles of a number of structural, cyclical and conjunctural mechanisms that help explain these phenomena. Attention will be paid to the reasons why competitive modes of social regulation do not lead to a re-employment of resources released as a result of structural change, why supply-side strategies are not a sufficient response to contemporary problems of unemployment, exclusion and inequality, and why a way out of the current cul-de-sac depends on the political construction of a new regulatory order.

## THE DYNAMICS OF REGIONAL AND URBAN ECONOMIES IN CONTEMPORARY GREAT BRITAIN

There are a number of ways in which the structure and dynamics of regional and urban economies can be analysed. In economic geography, much of the emphasis is on the dynamics of the market sector and the geographies of the production of goods and services. To understand the dynamics of advanced economies and societies, however it, is essential to bear in mind a more complex model which recognizes the structure, role and interactions of the three fundamental orders of contemporary mixed economies: the economic, political and domestic orders (*see* Théret, 1994 and Fig. 20.1). Each of these orders involves a set of practices, structured by a set of social relations, and characterized by a different and contradictory logic. The economic order in its widest sense is the set of practices through

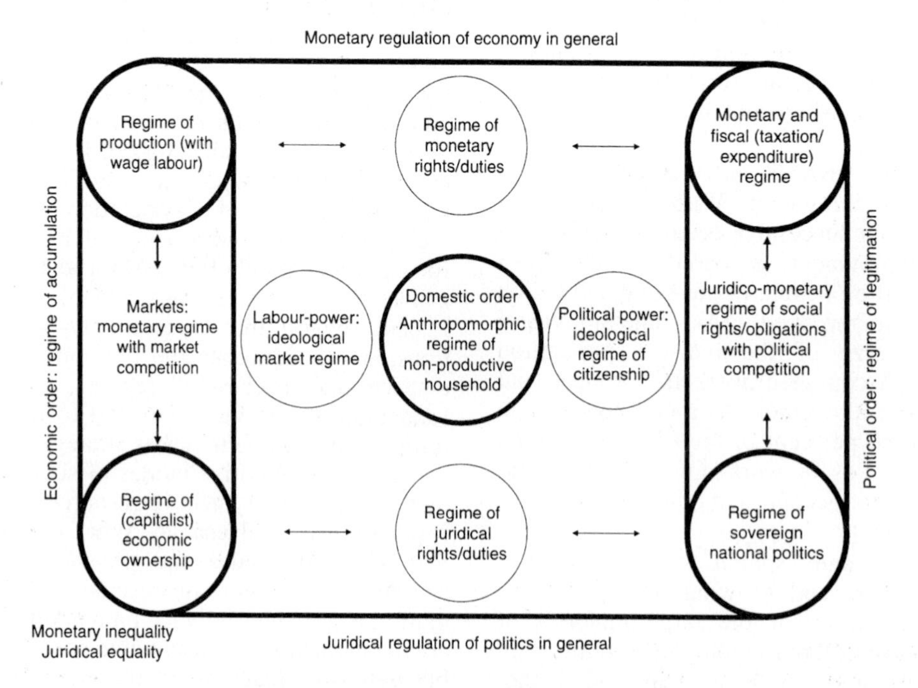


**Fig. 20.1** Conceptualizing the socio-economic order of mixed economies (after Théret, 1994)

which human beings use the material resources they derive from inanimate and animate nature to meet human needs. In capitalist societies a more restricted conception of the economic order limits it to those economic practices orientated towards the accumulation of material wealth, where material wealth is understood as material and immaterial things, and signs that represent or command things, of which the most important are monetary titles. In order to regulate their internal contradictions and improve their mutual correspondence, these orders are mutually regulated and articulated via a set of monetary, legal and ideological/discursive (audio-visual, literary, statistical, linguistic, accounting, etc.) systems of mediation.

This conception of a social order, which derives from developments in regulation theory, and which rests on a recognition of each suborder as a complex of economic, political and cultural elements, neither ignores those dimensions of social life (the organization and economic impact of the domestic and political/welfare orders) so frequently overlooked in economic geography, nor forgets that the economy is a complex combination of activities of production, distribution, consumption and exchange, nor overlooks (as is common in cultural geography) the crucial role of economic mechanisms in the reproduction of human societies and human life. As a step in the direction of the use of a wider conception of the economy, the economic order is conceived, in this chapter, as

- first, a nexus that organizes processes of labour and the production of goods and services, not just in in the market sector but also in the welfare state;
- second, a nexus that organizes the distribution of income to the agents who own productive assets or are employed as wage earners; and
- third, a nexus that embraces the expenditure of wages and other incomes on commodities and the role of consumption in the reproduction of individuals as wage-earners and citizens in the domestic order.

(To these three elements one should add the processes of exchange that articulate these nexuses one with another.) This concept of the economic order involves, therefore, a consideration of its articulation with the political and domestic orders, and, as I shall argue, it also leads to a concern with systems of regulation and mediation and with what I shall call the rules of the game.

## Growth and Change: from Convergence to Divergence

Just how dynamic and successful have regional and urban economies been in recent years? Are differences in their relative economic performance increasing or diminishing and are spatial inequalities in the 'creation' of wealth increasing or declining? To answer these questions I shall concentrate on variations in gross domestic product, which is a measure of the value of all the market and collective-sector goods and services produced in an area or produced by those inhabitants of an area who are 'employed'. What it measures is, in fact, the capacity of the economic activities in an area or involving an area's inhabitants to create or appropriate wealth.[3]

As Fig. 20.2 shows, there was a turnround in trends in spatial inequality in Great Britain in the mid-1970s. In the years up to 1976, regional variations in GDP per inhabitant diminished. In 1976–89 there was a dramatic increase in spatial inequalities. As Fig. 20.3 shows, divergence was a result of the relative growth of the South-East, South-West and East Anglia and the relative decline of much of the rest of the country. With the onset of recession in the early 1990s, the divide declined. In 1977–91 as a whole there was nevertheless a very substantial relative growth of GDP per head in a cluster of counties in the Greater South-East (Surrey 25.9 per cent, Buckinghamshire 24.1, Wiltshire 12.4, Berkshire 11.0, Oxfordshire 10.9 and West Sussex 10.6), though similar cumulative increases were recorded in a number of other counties (Warwickshire 27.6 per cent, Grampian 21.1, Clwyd 18, Cumbria 16 and North Yorkshire 10.6). At the

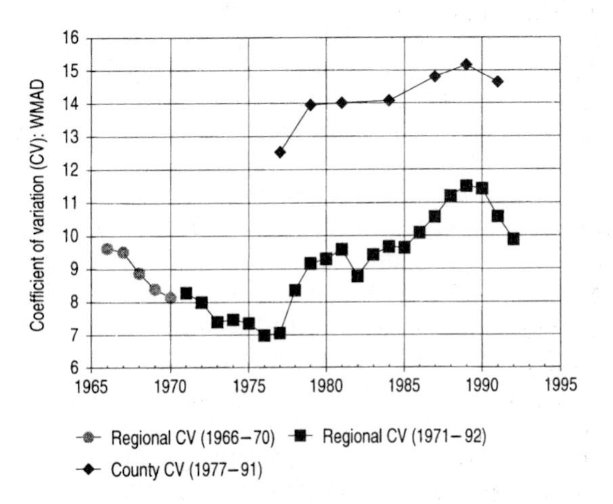


**Fig. 20.2** Trends in regional inequality in Great Britain 1966–92

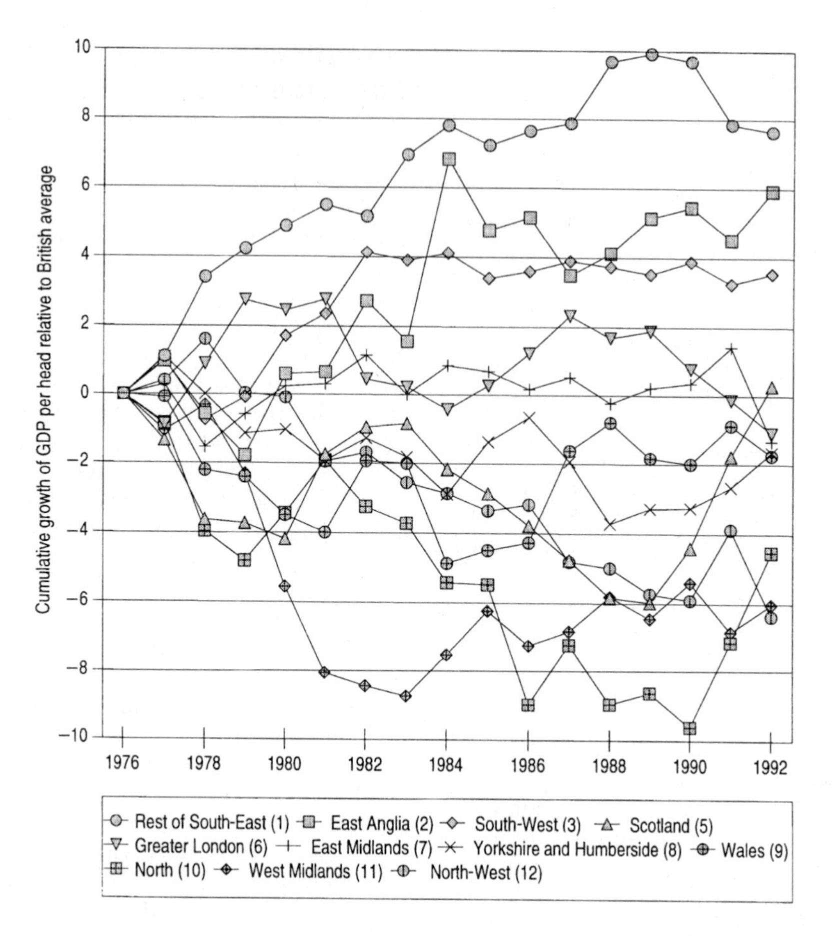


**Fig. 20.3** Cumulative growth in regional GDP per inhabitant relative to the national average 1976–92

other end of the spectrum there were very sharp drops in GDP per head in a number of northern industrial cities (Merseyside −23.4 per cent, Cleveland −21.1 and South Yorkshire −20.0). (It is important to note that there is a difference between regional and county GDP series. In the regional GDP series income and output are attributed to place of residence, whereas in the county series commuters' incomes are included in their county of work rather than in their county of residence. In the case of cities where commuting is substantial, the consequences are dramatic: in 1991, for example, according to the county data GDP per head in Greater London was 145 per cent of the national average, while according to the regional series it stood at 124.4 per cent of the national average.)

Differences in GDP per inhabitant can be divided into two elements: differences in productivity, which themselves reflect differences in physical productivity, prices and earnings, and differences the share of the population in employment. More formally:

$$\frac{\text{Gross Domestic Product}}{\text{Resident Population}} \equiv \frac{\text{Gross Domestic Product}}{\text{Employed Population}} \times \frac{\text{Employed Population}}{\text{Resident Population}}$$

This disaggregation produces a number of insights into the determinants of geographical disparities in development. As a first step, data for 1981 and 1991 are analysed. In Fig. 20.4, regional disparities in GDP per inhabitant in Great Britain are recorded as percentage deviations from the national average. In 1991 three regions had above-average levels of GDP per head: Greater London, the South-East and East Anglia. All the rest were below average, with Wales in the lowest position. In 1981–91 the South-East, East Anglia, the South-West, the East Midlands, the West Midlands and Wales improved their relative positions. All other regions lost ground.

Fig. 20.4 also indicates the respective roles of labour force mobilization and productivity growth in 1981 and 1991. In 1981 the dominance of Greater London and to a lesser extent of the

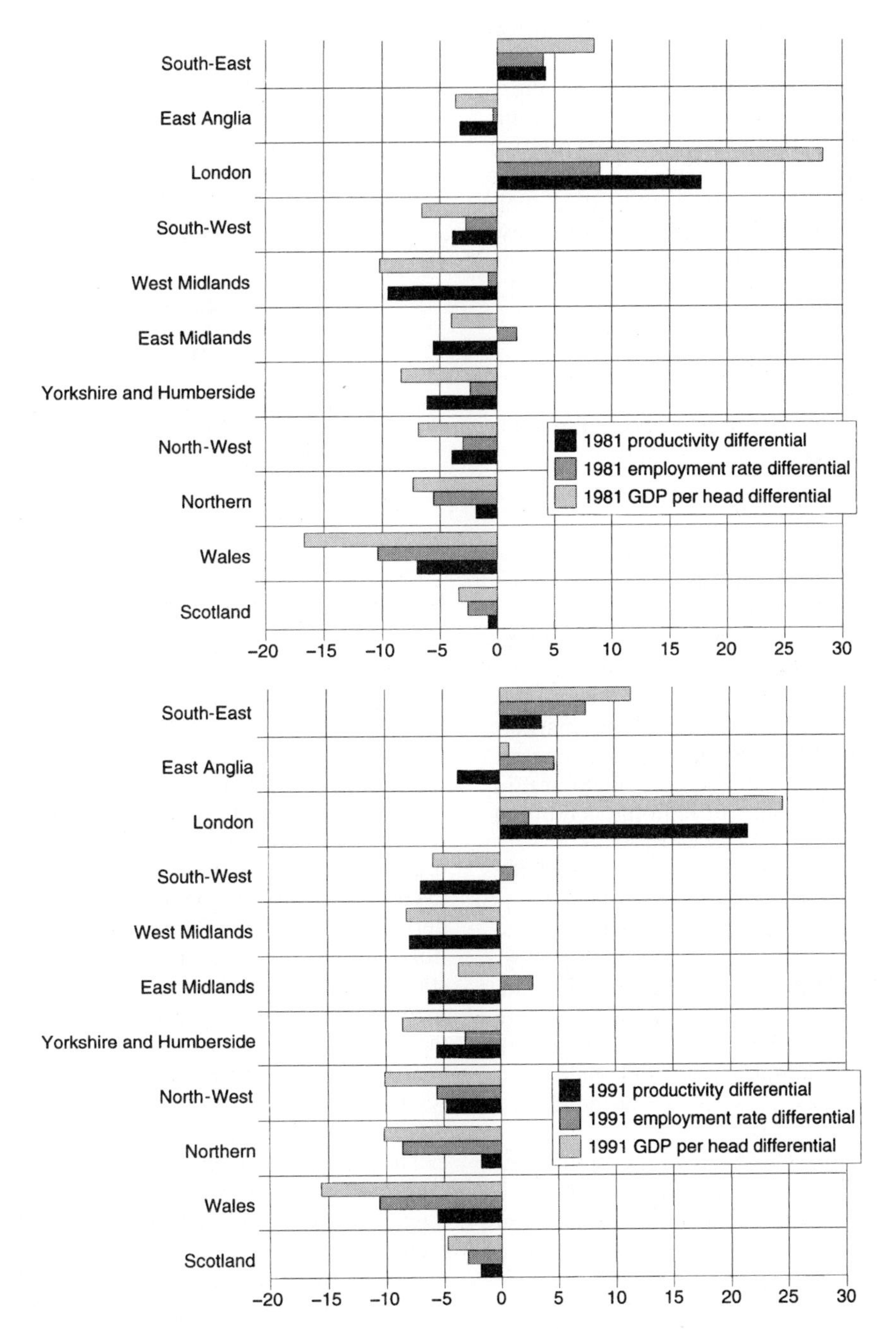


**Fig. 20.4** Differential trends in regional economic performance 1981–91 (regional figure as a percentage of the Great Britain average)

South-East rested on above-average productivity levels and employment rates. In Wales and the Northern region the major determinant of less than average GDP per head was low employment rates, whereas in most other areas a productivity divide predominated.

In the 10 years from 1981 to 1991 several changes occurred. First, the employment rates deteriorated quite sharply in a number of areas: included were Greater London, Yorkshire and Humberside, the North-West, the Northern region and Wales. The position of Greater London, which was eroded over this period, depended to a much greater extent than in the past on its productivity lead. This productivity lead increased in the 1980s. This increase was in all probability a result of changes in the structure of employment, with an increase in the relative importance of jobs for executives, and in the relative earnings of those employed in high-status jobs in the capital. A number of areas around the capital saw a relative improvement in their employment rates but their productivity differentials declined, indicating their dependence on the expansion of relatively low-productivity jobs: included in this group were the South-East, East Anglia, the South-West and the East Midlands.

## The Dynamics of Unemployment and Non-employment

The division of differences in GDP per head into productivity and employment rate elements shows that it is not just differentials and trends in productivity and therefore production system dynamics that determine the map of development and inequality but also differences in employment rates and therefore in the dynamics of labour markets and the articulation of the productive and domestic order. Disparities in the rate of employment are, in short, important determinants of regional and urban development differentials.

Differences in the degree of mobilization of the human potential of a society are important determinants not just of geographical disparities but also of the general economic dynamism and performance of a country, region or city. In Great Britain, as in other advanced countries, the recent increase in disparities has gone hand in hand with a growth slowdown. Greater unemployment and male non-employment are causes, as well as consequences, of this slowdown.

Data on the dynamics of British unemployment are plotted in Fig. 20.5. Two features of this graph warrant particular attention. First, unemployment varies with the economic cycle, as the cyclical movement of the line plotted in Fig. 20.5 shows. As that line also shows, however, the point around which unemployment revolves has increased after each of the major shocks since the end of the post-war 'golden age' of full employment and comparatively high growth. Market societies do not automatically restore full employment. Instead; unemployment tends to oscillate around an 'attraction point' or an equilibrium which reflects the institutional characteristics and the nature of the social compromise in each society. Second, in Great Britain there is a much greater degree of instability than in other European countries (*see* Dunford, 1996). Compared with those of other

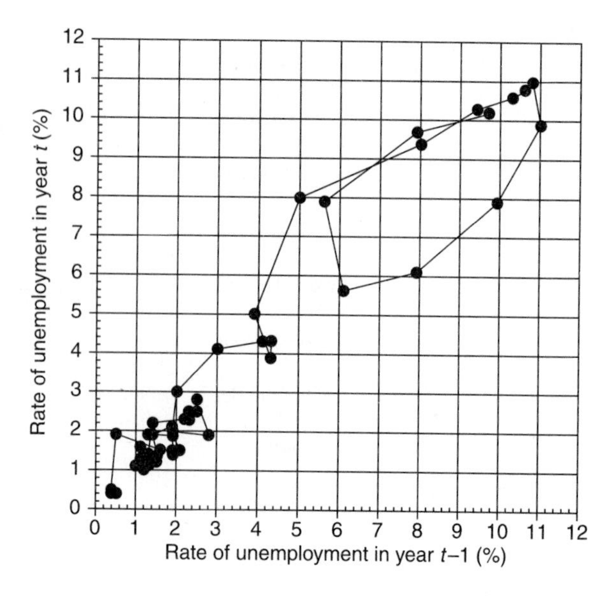


**Fig. 20.5** Unemployment in Great Britain 1943–93

European countries the amplitudes of the cycles in Britain are relatively large. The volatility of economic cycles is a consequence and a cause of the greater emphasis on market-led models of adjustment: in the job market, the insecurity of employment, short notice periods and comparatively small redundancy payments as well as short-term profit pressures amplify employment and unemployment cycles as, in recessions, comparatively large numbers of workers are laid off. The costs of reproduction of a part of the workforce are transferred to the state, though the consequent loss of disposable income amplifies the drop in demand, as do the tax increases or reductions in public expenditure required to offset increases in the size of the public-sector deficit as a result of lost tax revenues and increased social security payments.[4]

The adverse movement of employment rates is, however, not just a consequence of increases in unemployment. The UK definition of unemployment, which is subject to almost constant change, measures the number of people eligible for unemployment benefit. This definition overlooks people who are available and actively looking for work but who are not eligible for unemployment benefit and people who do not have a job but have given up the search for employment. Census data (*see* Table 20.1) show that between 1981 and 1991 the share of the population aged 16 and over in employment declined slightly, though there were quite significant declines in a number of regions (−2.6 per cent in Greater London, −2.0 per cent in the Northern region and −1.6 per cent in the North-West). This aggregate figure conceals, however, marked gender differences: while there was a 3.7 per cent increase

**Table 20.1** Non-employment in Great Britain 1981–91

| | Employment rate in 1991[1] | | | Change in employment rate[2] | | |
|---|---|---|---|---|---|---|
| | All | Male | Female | All | Male | Female |
| South-East | 58.5 | 58.5 | 58.5 | 1.3 | 1.3 | 1.3 |
| East Anglia | 56.8 | 67.5 | 46.8 | 2.0 | −3.1 | 6.8 |
| London | 55.6 | 64.5 | 47.6 | −2.6 | −6.4 | 0.8 |
| South-West | 54.4 | 64.6 | 45.1 | 1.7 | −3.7 | 6.6 |
| West Midlands | 54.9 | 64.5 | 45.9 | −0.4 | −4.0 | 3.1 |
| East Midlands | 56.2 | 65.6 | 47.3 | −0.2 | −5.0 | 4.4 |
| Yorkshire and Humberside | 53.1 | 62.0 | 44.9 | −0.9 | −5.7 | 3.5 |
| North-West | 52.1 | 60.7 | 44.4 | −1.6 | −5.6 | 2.0 |
| Northern | 50.0 | 58.0 | 42.7 | −2.0 | −6.7 | 2.4 |
| Wales | 48.9 | 57.6 | 41.1 | −0.4 | −5.7 | 4.5 |
| Scotland | 53.1 | 62.3 | 44.9 | −1.1 | −5.2 | 2.7 |
| Great Britain | 54.6 | 64.1 | 45.9 | −0.4 | −4.8 | 3.7 |

*Source*: Censuses of population.
*Notes*: 1. Percentage of people aged 16 and over in employment in 1991.
2. Change in percentage of people aged 16 and over in employment between 1981 and 1991.

in the female employment rate, there was a 4.8 per cent fall in the male rate. The male figure is indicative of a sharp increase in male non-employment (men of working age who are unemployed, on government schemes, permanently sick or in early retirement), which itself is correlated with crime and a wide range of other indicators of social breakdown. An increase in male non-employment occurred in all regions except the South-East (that is, the South-East excluding Greater London), with particularly sharp increases in the Northern region, Greater London, Wales, Yorkshire and Humberside, and the North-West.

## The Articulation of Employment and Exclusion with the Domestic Order

In the last section I presented some evidence relating to the scale of, and regional and temporal variations in, the exclusion of sections of the population from employment. Earlier disparities in the rate of employment were identified as an important determinant of disparities in development.

The data used in the last section referred to the number of people with jobs and the number of people excluded from employment. To understand the dynamics of urban and regional economies, this evidence should be related to the structure of households and of what was earlier called the domestic sector. There is other evidence that suggests that a large share of the people counted in the data on employment are members of the same households, as are a large percentage of the non-employed. In particular, a large majority of women who have entered employment came from households with at least one other income-earner. At the same time, the share of households and of the population dependent on state welfare benefits has increased. What is emerging is a strong distinction between work-rich and work-poor households. As Table 20.2 shows, 35.7 per cent of households have two or more earners, while 35.6 per cent have none. More astonishingly, 17.4 per cent of dependent children were being brought up in households with no earners. In Greater London that figure reached 22.9 per cent, while in the North-West it reached 21.3 per cent.

The implication, with reference to Fig. 20.1, is that analyses of regional and urban dynamics should endeavour to explain the processes of domestic income formation through an analysis of the differential integration of households into the worlds of work and employment through the labour market but also through the system of state transfers, which themselves reflect the nature of politically defined citizenship rights.

A second phenomenon that warrants attention is the trend for people to do more than one paid job. (The scale of undeclared jobs is more difficult to assess, but also of considerable importance.) To get some indication of the scale of dual job-holding, which itself is in part a consequence of the struggle to earn sufficient income in an era in which the number of ill-paid jobs has exploded, different employment estimates can be compared. Between 1981 and 1991 there was an increase in employment. According to the 1981 and 1991 censuses, the number of people employed increased from

**Table 20.2** Earners and dependent children in 1991

| | Percentage of households with the following number of adults in employment | | | Percentage of dependent children in households with no earners |
|---|---|---|---|---|
| | None | One | Two or more | |
| South-East | 31.3 | 29.5 | 39.2 | 11.7 |
| East Anglia | 33.7 | 28.6 | 37.7 | 11.8 |
| London | 34.0 | 33.4 | 32.6 | 22.9 |
| South-West | 36.1 | 27.9 | 36.0 | 13.0 |
| West Midlands | 35.0 | 27.6 | 37.4 | 18.0 |
| East Midlands | 34.1 | 27.7 | 38.2 | 14.7 |
| Yorkshire and Humberside | 37.8 | 27.0 | 35.2 | 18.8 |
| North-West | 38.6 | 27.1 | 34.3 | 21.3 |
| Northern | 40.4 | 26.8 | 32.8 | 21.0 |
| Wales | 40.2 | 27.3 | 32.5 | 19.9 |
| Scotland | 37.7 | 28.5 | 33.8 | 19.0 |
| Great Britain | 35.6 | 28.7 | 35.7 | 17.4 |

*Source*: Census of Population, 1991.

22 880 922 to 23 940 549. This figure is, however, much less than the figure for the civilian workforce in employment (25 035 000 in 1991) produced by the Department of Employment. The reason why is that the census counts people, whereas other Department of Employment series count jobs. In a series that counts jobs, anyone with two jobs is counted twice. The two series can therefore be compared to estimate the number of jobs that are 'second' jobs. What this comparison shows is that in 1991 1 096 451 jobs (1 619 830 if the Census Special Workplace Statistics are used) were second, third or fourth jobs. In 1981 the comparable figure was smaller (708 078).

## Employment Dynamics in Great Britain

In wage-earning societies, employment is a critical determinant of the quality of life and social cohesion as the existence of a regular earned income is a presupposition of an acceptable degree of access to the resources on which extra-economic activities and social reproduction depend. A resolution of the problems of exclusion and of insufficient mobilization of a society's human potential accordingly depends upon either greater employment growth or a more equitable distribution of employment. At the same time, the growth of income and output per head, or the rate at which free time can be created, depend on the productiveness of work. Compared with other advanced economies, Great Britain lags in productivity and income growth and is characterized by relatively low levels of output per head. In 1989 Britain lay almost 20 per cent below the EU average and 30–45 per cent below the levels of the other members of the G7 (OECD, 1991). What is more, compared with its own record up to the crisis of the mid-1970s, Britain's growth and productivity record deteriorated: GDP growth fell from 3.4 per cent per year in 1968–73 to 1.5, 2.3 and −0.9 per cent in 1973–9, 1979–89 and 1989–91 respectively, while its rate of productivity growth dropped from 3.2 per cent in 1968–73 to 1.3, 1.7 and 0.1 per cent in 1973–9, 1979–89 and 1989–91.

Within Britain, the growth that did occur was very unequal with sharp regional, inter-urban and intra-urban contrasts. The data in Fig. 20.6 depict cumulative trends in regional employment growth relative to the British average. The areas that gained most in employment terms were East Anglia, the South-West and to a lesser extent the East Midlands. In the case of the South-East, no distinction is made between Greater London and the rest of the region. As a whole, the South-East gained until the second half of the 1980s and then lost ground. The indication is that it gained less in employment than in GDP per head, suggesting the existence of higher rates of productivity growth in the South-East than in other parts of the south and of a selective concentration of higher value-added activities in the South-East.

The map of regional inequality is none the less a product of a more complex mosaic of development and underdevelopment of a set of city regions: in

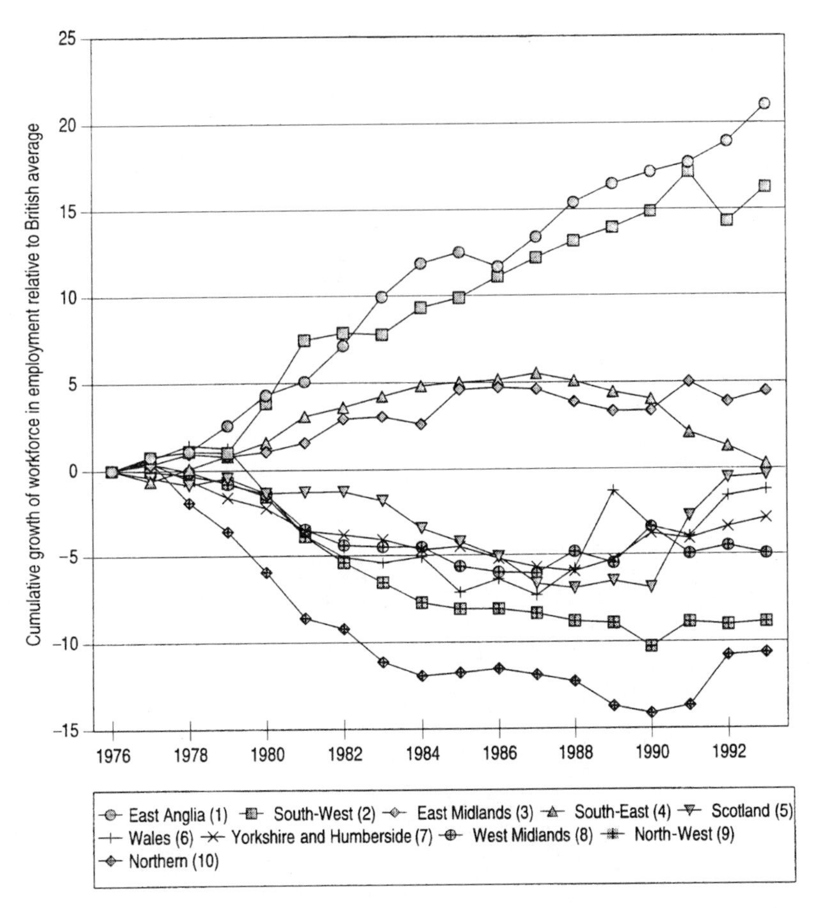


**Fig. 20.6** Cumulative percentage employment growth relative to the British average by region in 1976–93

1991 employment in Britain was concentrated in a number of clusters of travel to work areas (TTWAs), with the 42 largest accounting for almost 58.8 per cent of all jobs and Greater London alone accounting for 13.4 per cent. As the differential dynamics of the regions was a result of the dynamics of employment and output of these city regions and of the evolution of their productive systems, attention will be paid to some of the main trends in the location and composition of employment at this scale.

To identify the main changes in the geography of employment between 1981 and 1991, Fig. 20.7 plots the absolute change in employment – the number of jobs rather than the number of people with jobs – by TTWA in 1981–91 (*see also* Dunford and Fielding, 1997). As Fig. 20.7 shows, there were significant losses of jobs in the London and Heathrow TTWAs and in most of the large conurbations of the north and west. At the same time, however, there was very striking growth in a whole series of TTWAs in an arc that stretches around Greater London (with increases of 31 988 in Milton Keynes, 26 174 in Crawley, 24 713 in Bristol further to the west, 22 514 in Swindon, 21 522 in Guildford, 19 946 in Cambridge, 19 863 in Reading, 19 233 in Aylesbury and 16 293 in Basingstoke).

At a national level, the only sectors in which employment increased were services: 23 per cent of the increase was in distribution, 39 per cent in banking and 38 per cent in other services (*see* Fig. 20.8). At the same time there was a dramatic change in the gender composition of employment and in the role of part-time work: 1.3 million male full-time jobs were lost. Jobs for women accounted for 83 per cent of the increases in employment and part-time jobs for 71 per cent.

As Table 20.3 shows, in the 10 areas with the greatest increases in employment, 24 per cent of the increases were in distribution, 43 per cent in financial services and 23 per cent in other services. Just 7 per cent were in high-tech activities. As far as the gender composition and status of employment were concerned, 23 per cent of the increases were made up of male full-time jobs, 39 per cent of female full-time jobs and 29 per cent of female part-time jobs. In most of these areas female jobs

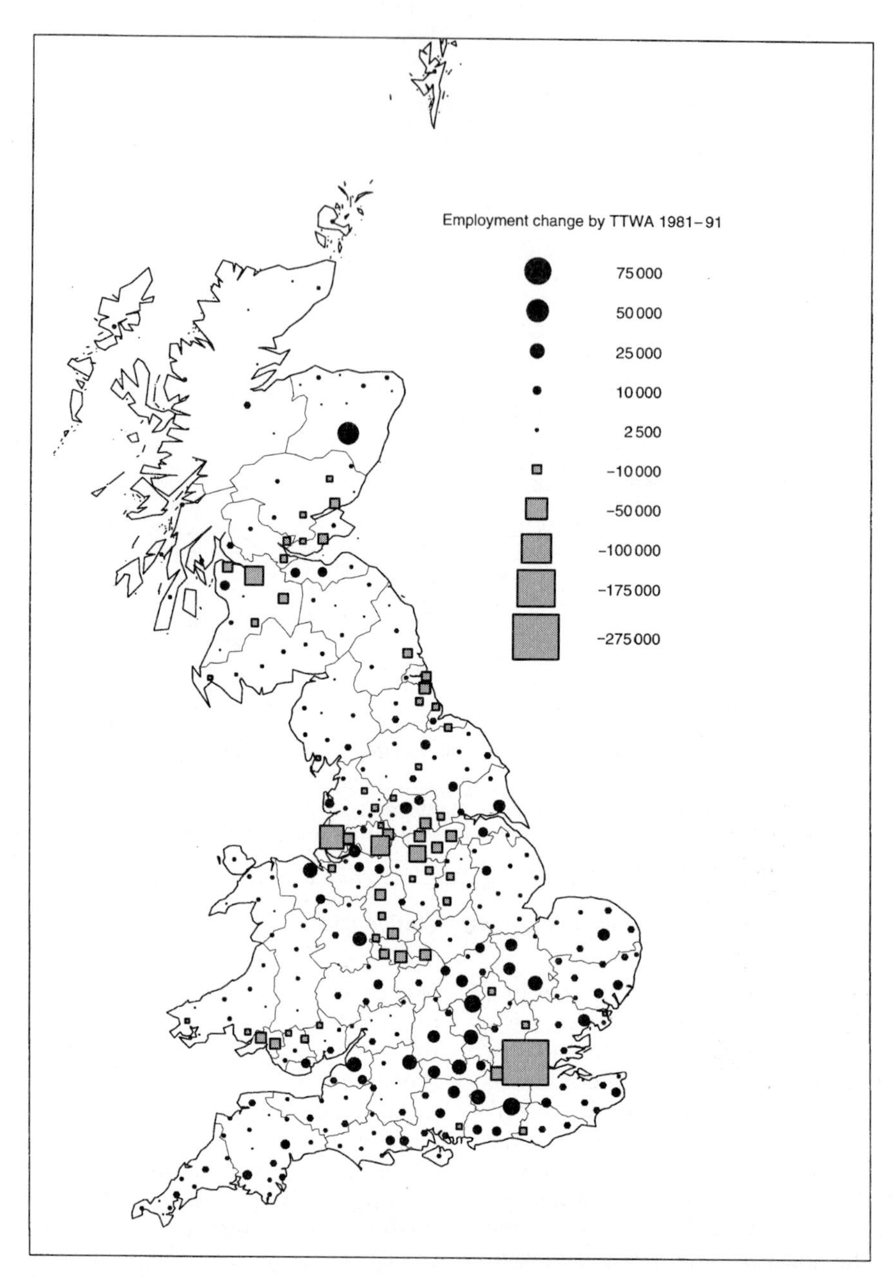


**Fig. 20.7** Employment change by travel to work area (TTWA) in Great Britain 1981–91

counted for a very high share of additional jobs: 70 per cent in Crawley, 85 in Bristol, 64 in Swindon, 77 in Guildford, 70 in Cambridge, 86 in Aylesbury, 87 in Peterborough and Northampton, and 88 in Oxford. Also important were jobs that were part-time: of the increases that occurred, 36 per cent were made up of part-time work in Crawley, 51 in Bristol, 31 in Swindon, 60 in Guildford, 38 in Cambridge, 64 in Aylesbury and 76 in Norwich and Hull.

At the other end of the spectrum, in the TTWAs that lost most jobs, 76 per cent of the increases that occurred were in finance and 24 per cent in other services. In terms of gender and status, 38 per cent of the growth was in male part-time work, 24 per cent in female full-time work and 37 per cent in female part-time work. In some of these places (Liverpool, Glasgow, Wigan and Barnsley), the only categories of jobs that increased in number were part-time. The significance of increases in female employment varied: it was important in Manchester (72 per cent), Heathrow (67), Birmingham (84) and Sunderland (67) but was insignificant in TTWAs such as London, Liverpool (where it

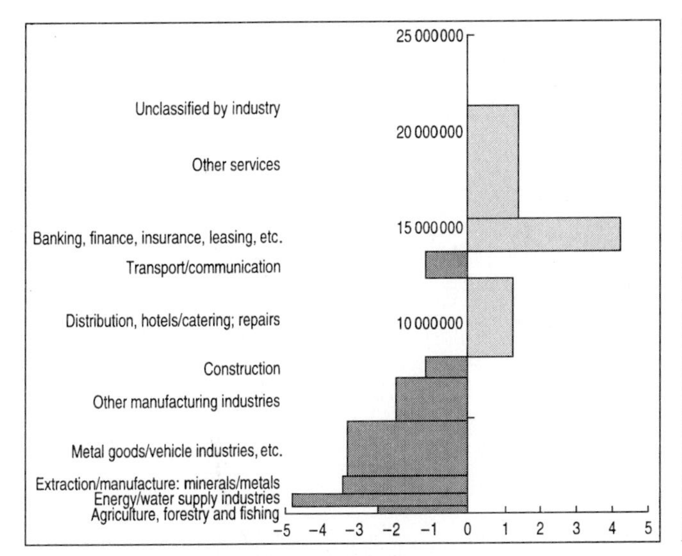


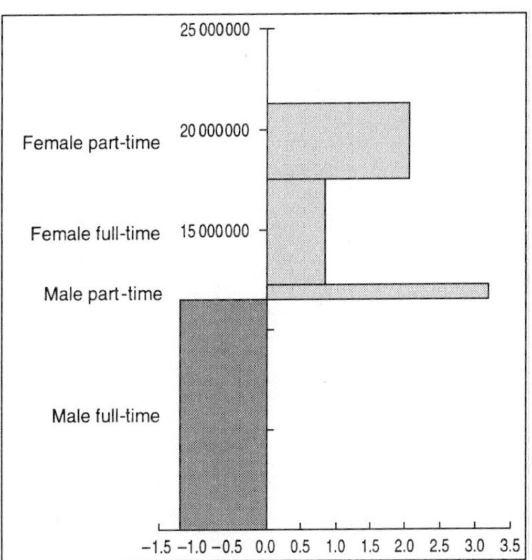


**Fig. 20.8** Employment change by sector, gender and status in Great Britain in 1981–91

declined), Wigan and especially Barnsley, owing to the loss of female full-time jobs.

The growth in employment (and in output) was, in short, to a large extent service sector led. What is also clear is that there was a large wave of inward investment, though most of this investment involved acquisitions. Many of the investments that occurred in the arc around Greater London were in logistic and marketing activities, whose growth was fuelled by a consumer-led boom, asset sales and a redistribution of wealth and income to the rich and affluent, though there was also a significant growth of white-collar jobs in industry and of producer services. At the same time, the expansion of incomes led to substantial growth of a wide range of subordinate consumer service and low-level manufacturing activities characterized by part-time employment, short-term contracts, informal work and low productivity. For a few years rapid growth occurred. As the dramatic recession of the past few years showed, however, this unbalanced model of development was quite unsustainable.

## INSTABILITY, EXCLUSION AND THE RULES OF THE GAME

So far, attention has been paid to the nature of trends in spatial inequality and economic performance. Trends in spatial development were shown to depend on the dynamics of the productive sector and on trends in labour force mobilization. Two conclusions emerged. The first is that recent trends in inequality, and in particular the growth slowdown, are to a significant extent a product of a regionally differentiated failure to re-employ resources released as a result of structural change, to provide sufficient jobs and to ensure an equitable distribution of employment. The second is that job growth has occurred not in the industrial/productive sectors on which economic geographers tend to concentrate but in services. The aim of this section of this chapter is to identify some of the mechanisms that explain these trends.

Trends in spatial development are a result of a number of sets of mechanisms: secular trends that depend on the mode of functioning of a particular social order; cyclical trends that depend on the successive institutional orders established to regulate social contradictions; and conjunctural trends which reflect day-to-day political and economic developments (*see* Dunford and Fielding, 1997). In this chapter I have focused on developments since the middle of the 1970s, as this period represented a turning-point in the cyclical development of Great Britain and of other advanced countries. In the case of Great Britain, the crisis of the mid-1970s saw the collapse of the compromise that had underpinned the post-war order and the emergence of a new neo-liberal agenda that involved a removal of constraints on the operation of market forces, deregulation and privatization.

The change of direction of the early 1970s was a response to the breakdown of the post-war model of development. As a response to a slowdown in rates of output and productivity growth, the neo-liberal

**Table 20.3** Employment change by sector, gender and status in the 10 travel to work areas (TTWAs) with largest and smallest absolute increases in employment between 1981 and 1991

| | Agriculture, forestry and fishing | Energy and water supply | Manu-facturing | Construction | Distribution, hotels and catering, repairs | Transport and communi-cation | Banking, finance, insurance, leasing, etc. | Other services | Male full-time | Male part-time | Female full-time | Female part-time | Total | High-tech | High-tech location quotient |
|---|---|---|---|---|---|---|---|---|---|---|---|---|---|---|---|
| Aberdeen | −1 312 | 12 282 | 1 273 | 3 040 | 6 991 | 3 134 | 10 748 | 12 078 | 23 813 | 3 862 | 9 034 | 11 525 | 48 234 | −548 | 0.4 |
| Milton Keynes | −351 | 41 | 1 629 | 1 161 | 11 661 | 2 279 | 10 093 | 5 478 | 14 609 | −1 842 | 13 687 | 5 537 | 31 991 | 2 153 | 1.3 |
| Crawley | −1 404 | 94 | −10 561 | −779 | 8 010 | 7 079 | 14 689 | 9 044 | 4 950 | 2 858 | 11 851 | 6 513 | 26 172 | −1 946 | 1.5 |
| Bristol | 340 | −936 | −18 208 | −1 902 | 7 383 | −2 293 | 26 999 | 13 329 | −6 910 | 4 740 | 15 426 | 11 456 | 24 712 | 3 237 | 1.7 |
| Swindon | −598 | 1 815 | 512 | −941 | 5 126 | 1 367 | 9 032 | 6 202 | 6 322 | 1 817 | 9 101 | 5 275 | 22 515 | 2 720 | 1.7 |
| Telford and Bridgnorth | −311 | 6 | 8 584 | −314 | 6 544 | 277 | 2 073 | 5 561 | 8 842 | 836 | 7 969 | 4 773 | 22 420 | 2 865 | 0.9 |
| Guildford and Aldershot | −724 | −332 | −4 487 | −583 | 8 998 | 691 | 15 995 | 1 965 | 1 758 | 3 166 | 6 820 | 9 779 | 21 523 | 5 587 | 2.2 |
| Cambridge | −2 000 | 182 | 1 789 | −1 308 | 2 883 | 1 088 | 9 616 | 7 696 | 3 847 | 2 164 | 8 532 | 5 403 | 19 946 | 5 044 | 2.3 |
| Reading | −461 | 855 | −7 277 | −1 338 | 8 640 | 2 999 | 14 045 | 2 401 | 4 018 | 1 980 | 9 441 | 4 425 | 19 864 | 2 793 | 1.5 |
| Aylesbury and Wycombe | −784 | 111 | −9 055 | −716 | 6 324 | 1 142 | 16 819 | 5 391 | −1 208 | 2 936 | 7 384 | 10 120 | 19 232 | 842 | 1.2 |
| Barnsley | −93 | −10 893 | −2 702 | −1 298 | 2 247 | 150 | 1 447 | 805 | −11 113 | 510 | −2 659 | 2 925 | −10 337 | −830 | 0.1 |
| Sunderland | 33 | −11 302 | −4 253 | 947 | −3 285 | −230 | 3 368 | 2 846 | −16 163 | 1 417 | 58 | 2 812 | −11 876 | −838 | 0.5 |
| Birmingham | −212 | −4 258 | −72 382 | −590 | 11 125 | 1 939 | 27 677 | 22 736 | −48 694 | 5 389 | 11 295 | 18 045 | −13 965 | −3 718 | 0.8 |
| Wigan and St Helens | −214 | −7 865 | −16 289 | −259 | 3 985 | 220 | 3 551 | 1 078 | −19 821 | 1 573 | −2 745 | 5 200 | −15 793 | −767 | 0.5 |
| Heathrow | −744 | 622 | −75 917 | −9 831 | 16 039 | 6 059 | 43 363 | 427 | −40 074 | 4 997 | 17 512 | −2 417 | −19 982 | −13 840 | 1.5 |
| Sheffield | −97 | −5 451 | −36 528 | −2 108 | 1 519 | −708 | 7 059 | 6 877 | −34 775 | 2 514 | 1 320 | 1 504 | −29 437 | 628 | 0.7 |
| Manchester | −179 | −4 825 | −67 881 | −3 616 | 2 328 | −2 630 | 27 770 | 14 561 | −55 868 | 5 960 | 2 460 | 12 976 | −34 472 | −6 594 | 1.0 |
| Glasgow | −354 | −2 385 | −50 989 | −7 622 | −3 438 | −7 189 | 14 778 | 20 871 | −44 179 | 3 810 | −5 869 | 9 910 | −36 328 | −2 529 | 0.8 |
| Liverpool | 67 | −2 148 | −47 082 | −4 749 | −2 855 | −15 447 | 7 520 | 5 104 | −57 249 | 4 611 | −8 261 | 1 309 | −59 590 | −7 982 | 0.7 |
| London | −177 | −15 310 | −271 510 | −34 691 | −42 467 | −64 446 | 143 054 | 11 117 | −306 515 | 18 511 | 17 855 | −4 281 | −274 430 | −51 589 | 0.9 |

*Source*: Censuses of Employment and NOMIS.

agenda has proved a complete failure, particularly in comparison with the order it sought to replace: rates of growth have remained at less than one-half of those of the post-war golden age, while earlier tendencies for inequalities to diminish were thrown into reverse gear. As indicated in the last section, mass unemployment and widespread social exclusion returned, insecurity and inequality increased and the performance of regional economies diverged.

In this new environment, some places were comparatively successful – increasing their competitive advantage and increasing their market shares in the 1980s and 1990s – and much effort has gone into the identification of the determinants of success and the possibilities of transferring experiences from one region to another. Most attention has been paid to supply-side factors: the transformation of the productive system, the creation of a framework for co-operative industrial relations, the development of transport and telecommunications infrastructures, the establishment of synergies between public research and industry, the identification of strategies for technology transfer and investment in education and skills. A preoccupation with supply-side adjustments is understandable: each organization and each area wants to do as much as it can for itself. Supply-side approaches are also consistent with the view that the crisis of the golden-age model essentially involved a set of supply-side difficulties associated with the crisis of its characteristic (Fordist) productive order and the saturation of solvent markets. This interpretation is in part correct. There is, however, a second side to the problem: globalization added a set of demand-side problems (of effective demand in particular, which will be considered in a subsequent section) as a result of the destruction of regulatory orders put in place to manage the contradictions of market societies. The implication of this view is that it is in fact questionable whether the current development model crisis would end if all regions modelled themselves on regions that were successful: a productive-system change is not enough to end a crisis of accumulation.

An emphasis on the transformation of the supply side of successful regions is also problematic for a second reason: much of this work underemphasizes both the extent to which the growth in employment in the 1980s was dependent on services and the unsustainability of expansion due to the predominance of rent-seeking and speculative motives. The significance of an innovation-orientated and needs-determined productivist logic is conversely exaggerated.

A new model of development would indeed involve the development of a new productive order. An emphasis on productive-order change often occurs, however, at the expense of an analysis of the growth slowdown and of the growth of non-employment and inequality. A recognition of the existence of the latter would suggest that what contemporary societies confront is not a new order but an unresolved crisis in which current changes in production and the growth of speculative services are consequences and causes. In this section I shall develop this view and consider just three of the reasons why the current institutional order cannot resolve and is perhaps exacerbating some of the contradictions of market societies. First, I shall argue that greater uncertainty created a strong preoccupation with short-term considerations and increased the attractiveness of liquid ways of holding wealth, that the rise of neo-liberalism was closely correlated with the consequent ascent of the ideology and the influence of money capital and that speculation and the growth of rent-seeking played a significant part in the expansion of pseudo-services. Second, I shall suggest that in market societies, the human resources released as a result of major structural change will frequently not find alternative employment: a shortage of solvent demand is an inherent feature of competitive regulation. Third, I shall argue that a recognition of the essential role of state intervention in any strategy for full employment is no longer sufficient and that if cities and regions are not to engage in a process of competition that will in the end prove self-defeating, new supranational rules of the game are required.

## Economic Instability, the Dominance of Commercial and Financial Gain, the Growth of Pseudo-Productive Services and the Restructuring of the Productive Order

The economic slowdown coincided with much greater uncertainty and financial instability, and marked and rapid changes in the volume and composition of demand. The ascendancy of private over public financial markets and their globalization reduced the economic autonomy of nation-states and reinforced economic instability and a politics of austerity. These new market conditions created severe difficulties for large (rigid) Fordist firms which could not adapt quickly to a rapidly changing economic environment and for anyone whose wealth was tied up in fixed assets committed to particular ends. A desire to raise profit rates and to re-establish profit as the central indicator of economic performance went hand in hand with the emergence of a situation in which productivist strategies whose goal is the creation of wealth seemed increasingly risky. Conversely, *rentier*-type strategies of appropriation of wealth seemed increasingly attractive. These two strategies are inconsistent with one another, as Keynes clearly indicated in the inter-war years: productivist strategies are long-term and depend on low interest rates and economic stability, whereas commercial and financial strategies are often short-term, speculative and centred on the acquisition and disposal of assets, dealing in land and property, and trading in currencies, stocks and shares, debt and commodities. A major characteristic of the market-led strategies of the 1980s was the diversion of investment from productive to financial activities and the leading role of rent-led programmes centred on speculation and on the quest for commercial and financial gain.

To make this claim is not to fail to recognize that in market societies, commercial, logistic and financial actors play an indispensable role: commercial and logistic activities play a crucial role as intermediaries between producers and between the owners of resources and the consumers of the goods and services made with them, while financial institutions perform a critical function as providers of credit money. Nor does it overlook the fact that commercial and financial activities are, in the last instance, dependent on the capacity of the productive sector – domestic or overseas – to create wealth. In the period since the mid-1970s, however, the scale, influence, scope and relative significance of commercial and financial activities have increased at the expense of productive activities.

The significance of the growth of commercial, financial and speculative activities lies in the fact that they do not create wealth. Instead their remuneration depends on their capacity to secure a part of the wealth created elsewhere. As Palloix (1993) argues, such redistributions of created wealth occur wherever intermediaries can control and levy charges for access to markets. A crucial way in which wealth is at present redistributed is, in his view, through the sale of what he calls pseudo-services: services which must be purchased to get access to markets but which create no new value and whose remuneration implies a compression of the margins of the producers of the goods and services in question. Commercial and financial gains are made, however, not just through the appropriation of wealth from the productive sector but also from other spheres of life. Gains are made from the exploitation of the domestic economy, the informal economy, drug dealing, corruption, crime, long-distance trade in agricultural surpluses and the scramble for the spoils of the fragile economies bequeathed by the collapse of communism in East-Central Europe.

Palloix (1993) goes as far as the suggestion that this increase in the relative importance of pseudo-productive financial and commercial capital, which he dates from the early 1970s, is so marked as to warrant the claim that there has been a historical reversal of the relationship between productive and pseudo-productive capital. Whereas in the industrial era commercial gain was subordinated to productive gain, in recent years productive interests have been subordinated to commercial and financial interests. Palloix (1993) identifies three aspects of this shift. First, pseudo-productive capital has used its control over the access of productive capital to markets to shift decisively the distribution of value added in its favour. Second, the squeeze it has applied to the productive sector has increased competition, reduced wages and prompted a strong drive for productivity growth. As productivity growth occurs without output growth, the consequence is the exclusion of a large and growing section of the population from the world of paid work and from the world of consumption of commodities. Finally, through its dominant role in the restructuring of global cities and metropolitan economies, pseudo-productive capital has created new hierarchical urban systems and new orderings of geographical space which raise its status and profile at the expense of the productive sector.

Neo-liberalism was not just the ideology of the *rentier*. Another of its characteristics was the critical role of redistribution: wealth was redistributed from poor to rich and, through the growth of indebtedness, from future generations to the current generation. This redistribution of wealth reinforced the polarization of incomes, which itself stimulated the demand for – through the growth of elite incomes and of work-rich households – and the supply of – through the expansion of a low-wage sector – a large range of low-status consumer services.

The implication is that the expansion of services was in part dependent on the redistribution of assets, increases in inequality and greater indebtedness on the one hand and speculation and rent-seeking on the other. Growth of this type is not sustainable in the long term, as the depth of the late 1980s and early 1990s recession showed.

Greater uncertainty and increased volatility in the volume and composition of demand also played a major role in the transformation of the productive system. Widespread use was made of new information and communications technologies, and the organization of productive work was transformed to allow the mass production of quality goods at low cost. At the centre of this transformation was an attempt to increase the ease and speed of reaction of firms to changes in their external environment. First, owing to the saturation of markets and the absence of new products capable of renewing the consumption norm, strategies were developed to accelerate obsolescence and to follow shifts in fashion in order to speed up replacement. Second, attempts were made to increase the variety of goods and services offered to consumers, to improve the quality, price and performance of normalized products and to improve the responsiveness of manufacture and assembly to variations in the volume of demand in order to gain market share at the expense of rivals. These two solutions require more production flexibility, which was achieved in a variety of ways: greater co-ordination and integration of R&D and product manufacture; integration of marketing, design, manufacture and management; creation of networks and integration of activities in 'extended firms'; and an increase in the skills, learning capabilities, functional flexibility and involvement of the workforce in the struggle for improved productivity and quality.

Changes in the organization of production are, however, just one part of the story. At the same time there were changes in the character of the wage relation and new meanings were given to profit. Profit was interpreted not just as the return on the shareholders' capital but also as the investment required for survival of the firm. An emphasis on the second interpretation meant that all wage demands and all attempts to maintain the purchasing power of the wage could be presented as a threat to jobs and to the firm's survival. Accordingly, after 1977–80, wages in most countries were seen largely as a determinant of competitiveness and not as a determinant of the level of final demand for the output of domestic firms. This redefinition of profit was accompanied by a deregulation of the wage relation and the industrial relations system and the development of Japanese methods of employee involvement. Often there was an erosion of collective agreements, a squeeze on real wages, more individualized negotiation of wage rates, individualized assessment of career progress, increases in the intensity of work and greater economic insecurity.

## Market Adjustment and Full Employment

In the last section I argued that instability and uncertainty were critical determinants of the redeployment of resources towards commercial and financial activities, the restructuring of production and the increase in the economic insecurity of wage-earners. A situation of instability and insecurity is, however, not conducive to sustained productivist models of development. On the contrary, a large and sustained volume of investment in fixed assets requires an absence of fear of a sudden devaluation of assets that itself requires either that fixed assets are adaptable or that investors can have a substantial degree of confidence about the future.

A climate of rapid and unpredictable change in a market environment is also problematic because it is questionable whether market-dominated societies can ensure that the resources released as a result of structural change are re-employed. Almost all regional economists assume that markets will lead to a full employment of resources, and if full employment does not occur the reasons why lie in the existence of monopolies, minimum-wage legislation or other institutional constraints which prevent the adjustment of prices. If structural change releases resources that do not find alternative employment, the policy recommendation is, accordingly, a removal of restrictions, and in particular reductions in real wage rates. (In a neo-classical world, what exists are two hypothetical vectors of non-negative prices and interest rates which, if established, would result in full employment, while movement from one to another simply involves changes in relative prices and a reallocation of factors of production from one activity or region to another.)

The curious fact is that the endurance of unemployment in the inter-war years led Keynes to reject neo-classical concepts of market adjustment and to advocate wage rigidity as a policy for combating unemployment. What Keynes argued was that the initial response to a decline in demand was not a price adjustment but a quantity adjustment. Keynes did not deny the existence of a hypothetical vector of non-negative prices and interest rates which, if established, would result in full resource utilization. What concerned Keynes were the difficulties of reaching the market clearing vector in economies in which (a) all the information needed to ensure the perfect co-ordination of all the current and future activities of all traders does not exist and is not provided free of charge by an all-knowing Walrasian auctioneer, and (b) money is the medium of exchange.

The questioning of these assumptions had two implications. First, if a resource is made unemployed owing to a change in demand, traders do not have perfect information about the new market clearing price. The seller will set a reservation price in the light of past experience and a knowledge of the current price of comparable services, and search for the highest bidder, adapting the reservation price as the search unfolds. Adjustment is not therefore instantaneous, and in this state of disequilibrium trading occurs while the resource itself remains unemployed.

To this argument it is necessary to add a second. In this state of disequilibrium the loss of income from the services of the unemployed resource will impose a constraint on the owner's effective demand for other goods and services. This constraint on money expenditure and on effective demand provides the rationale of the multiplier analysis of competitive markets, and is, as Marx and Keynes demonstrated, a defining characteristic of a money economy. As Clower (1969a, 1969b) has pointed out, in a money economy it is essential to make a clear distinction between activities that are often conflated. First, it is essential to make a distinction between offers to exchange money for goods (purchase offers) and offers to sell goods for money (sale offers). Second, it is important to make a related distinction between planned transactions (what an economic agent intends to do) and realized transactions (what an economic agent actually does). According to Say's law, no economic agent plans to spend money on the purchase of goods and services without at the same time planning to earn the money to pay for them whether from profit receipts or the proceeds of the sale of other commodities. If resources are unemployed, however, realized current receipts will fall short of planned receipts. In these circumstances actual consumption expenditure, as expressed in effective market offers to purchase goods and services, will fall short of desired consumption expenditure. As Keynes showed, therefore, current income constrains current expenditure: an individual who is forced by a lack of buyers to sell less of a factor than he or she wishes to sell is also forced by a lack of money income to spend less than he or she wants to spend. To make sense of the mode of functioning of a decentralized money economy, the Walrasian *tâtonnement* process, which lies at the heart of neo-classical conceptions of price adjustment, should at the very least be redefined to ensure that no purchase order is accepted unless the purchaser already has sufficient income to pay for the transaction as a result of the completion of a previous sale order. (This analysis abstracts from the fact that an economic agent may be able to borrow money.)

The implications of this conception of market adjustment are profound. In orthodox general equilibrium theory it is assumed that the sum of excess demands – the difference between the quantity demanded and the quantity supplied of each good and service – is zero. A consequence of this proposition – which is called Walras' law – is that the existence of excess supply of any factors of production necessarily implies the simultaneous existence of excess demand for some goods and services. In the Walrasian universe the coexistence, in any disequilibrium situation, of an excess supply of some goods with excess demand for others will lead to movements in the structure of relative prices that will cause the economic system to converge on an equilibrium. (It is this argument that underpins the orthodox view that a reduction in involuntary unemployment simply requires that prices change and in particular that real wages fall and the prices of the goods that the unemployed wish to consume increase.) If, however, allowance is made for the mediation of money in all exchanges, it is clear that Walras's law is valid only in situations of full employment. If full employment does not prevail, there will be an excess supply of some factors of production, yet this situation of excess supply will not correspond to an effective excess demand for other goods and services. Wherever goods can be exchanged for money and money for goods but goods cannot be exchanged directly for other goods, realized current income is in short an independent constraint on effective demand. Excess supply in the labour market – involuntary unemployment – diminishes effective excess demand elsewhere. As Clower shows, with unemployed resources there is an excess supply of factors of production and there is a notional excess demand for goods, but adjustment will not occur as the effective excess demand for goods is zero owing to the fact that the demand for goods is constrained by the lack of money income that stems from unemployment.

What this account of market adjustment shows is that unregulated market economies do not tend towards full employment and that involuntary unemployment is a result not of market imperfections but of the mediating role of money and the costs and time involved in the acquisition of information. As soon as this point is understood, it is possible to understand why the neo-liberal agenda has had such devastating deflationary consequences. What this argument also emphasizes is that the regulatory order established after the Second World War was a response to fundamental contradictions in market societies and in particular to the unemployment of the inter-war years. To restore competitive market adjustment is to allow these contradictions to resurface and mass unemployment to reappear.

### Full Employment in a Post-Fordist Order

The implication of the arguments in the last section is that a renewal of demand management policies is a pre-condition for creating a set of rules which will allow all regions to win. The (partial) erosion of the demand side of the post-war order was a consequence, however, of globalization. Any re-establishment of demand-side measures would therefore have to occur at a supranational scale. In this sense, international Keynesianism is a necessary feature of a new order capable of restoring full employment.

Although necessary, international Keynesianism is, however, insufficient. Admittedly there are, in advanced countries, great social needs to be satisfied, and an investment of resources to meet these needs would significantly reduce unemployment and non-employment. What this view takes for granted, however, is the possibility of a renewal of the consumption norm on which a sustained renewal of growth would depend (*see* Fig. 20.9). In fact, the scope for a renewal of the consumption norm is questionable for two reasons. First, the Fordist era ended in part because of the saturation of many mass markets. Second, the new information and communications technologies have not yet led to the creation of new products and new needs of sufficient value to relaunch the virtuous spiral of mass consumption and mass production. Nor have information technologies yet led to major increases in the productivity of unproductive labour. There is therefore a structural contradiction between the renewal of productivity growth and the non-renewal of consumption norms.

A characterization of a post-Fordist order would therefore have to encompass not just an analysis of a productive system change as a response to the difficulties of the classic mass production model, nor just also the development of a new (liberal or neo-Keynesian) international economic order adapted to the globalization of economic and political life but also a new compromise designed to reconcile potentially fast productivity growth and (in advanced countries) slow output growth. In essence there is a choice of worlds. The first is a world that is in the making in which the developed world experiences slow growth in divided societies characterized by wide disparities in income and employment, an erosion of the welfare state and all the social ills that go with wide disparities, while at an international scale there is a sharp functional division of labour between developed and less developed countries. The second is a world in which in advanced countries there is a reduction of labour time organized at a supranational scale and the creation of more cohesive societies in which work is shared, everyone has more free time and extra-economic activities expand, while on a wider global scale there is a renewal of consumer-led growth – whose merits are questionable on environmental grounds – as a result of co-ordinated increases in income, consumption and production in less developed countries. Of these choices it is the second that offers the greatest prospects of a decline in disparities and a constant dynamic recreation of full employment in places where, at present, the rate of productivity growth exceeds the rate of output growth.

## CONCLUSION: TOWARDS A NEW REGULATORY ORDER

The essential argument of this chapter is simple. In countries such as Great Britain the neo-liberal turn has resulted in the creation of a divided society with growing insecurity and social exclusion and has achieved rates of economic growth that are worse than those in the post-war golden age. Deregulation has reinforced speculation and eroded the institutional structure put in place to regulate the contradictions of market societies. In this new environment the competitive struggle has intensified: all cities and regions must seek to improve their competitiveness and secure greater market share at the expense of others. This game is one in which not everyone can win, as the non-re-employment of resources shows. Any economic geographer must clearly be concerned with ways in which places can improve their competitiveness. But economic geographers must also pay attention to the rules of the game. In this chapter I have suggested that new rules may make it possible to create a situation in which everyone can play a role and everyone can win. The essence of regulation theoretic approaches to the Fordist era is the demonstration that an institutional order could regulate social contradictions and create the foundations for generalized increases in prosperity. A challenge for the present is to see whether economic geographers can contribute to the design of a new set of rules for the re-creation of a cohesive and sustainable model of regional and urban development. What I have argued is that such a new set of rules would involve the creation of a new neo-Keynesian international order in which substantial increases in output and consumption occur in less

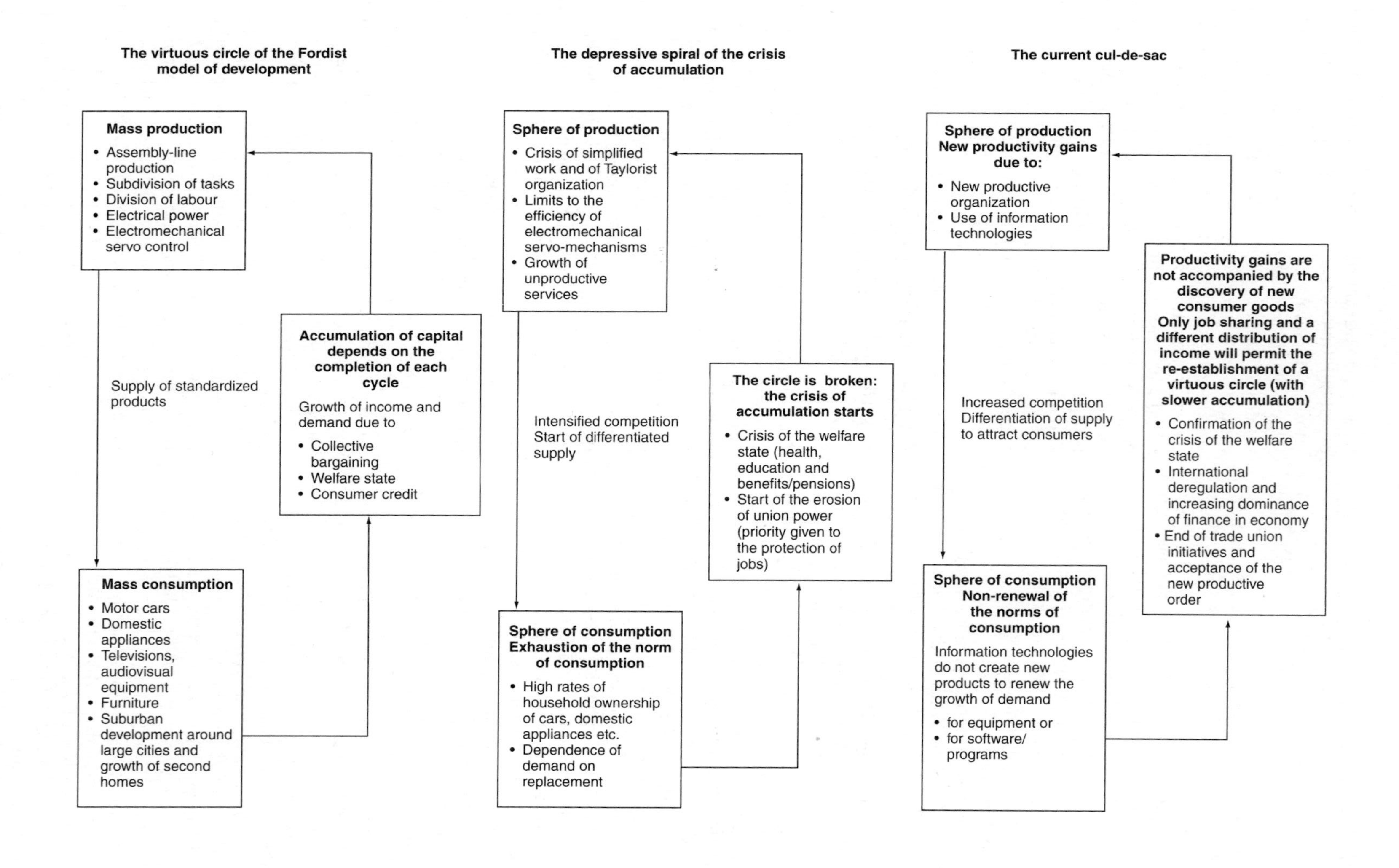


**Fig. 20.9** Typologies of the Fordist model of development and the post-Fordist crisis (after Boyer and Durand, 1993: 84–86)

developed countries, and in which in advanced countries employment is shared, working time is reduced and more free time is created. These choices are of course political, though what I have argued is that measures of this kind are a pre-condition for ensuring that constant structural adjustment carries with it the possibility of a dynamic creation of new jobs to replace those that are lost and the participation of all in the world of paid work. As a design for the future this model is far from complete. To it should be added the ideas about the development of a Third Sector (*see*, for example, Lipietz, 1992). What is clear is that the future nature of the places in which we all live will depend on our capacity to shape this wider institutional environment and these wider rules as well as on the development of vibrant and dynamic local productive systems.

## NOTES

1. Aggregate data on productivity suggest that, in manufacturing, the productivity pay-off of the new information and communications technologies has been limited and very slow to appear, while in the service sector productivity seems to have declined. There are a number of explanations of these phenomena. The decline in service–sector productivity is due in part to the expansion of low-productivity services, while in some of those areas in which computers and telecommunications do add significantly to human capabilities the productivity gains associated with automation were offset by increases in the amount of data to analyse, store and disseminate. According to a 1990 OECD report cited in an essay by Sweeney (1994), the problem is that 'the social environment has not been sufficiently adaptive to, and supportive of, a rapid diffusion of these technologies. In particular the reluctance of firms to change their work organisation, labour relations, decision-making structures and management styles means that the spread of information technologies is currently slower than it could be, or is not achieving the full potential productivity gains. . . . The full exploitation of the new technologies implies a shift away from the Taylorist model of production typical of the previous phase of industrialisation.'
2. In the 1950s, 1960s and early 1970s the situation was very different. In what is increasingly seen as an economic 'golden age', growth rates were double those of recent decades, there was near full employment and social and spatial inequalities declined. In that era, in other words, dynamic growth went hand in hand with economic convergence and increasing economic and social cohesion. For a discussion of the causes and implications of this contrast, *see* Dunford and Perrons (1994) and Dunford (1994).
3. GDP measures of wealth differ from measures of the income of residents. Differences in the two result from the inclusion in the income measure of incomes paid to economically inactive individuals such as pensioners and the unemployed and from the inclusion of profits in GDP and investment income in household income. In this chapter I shall not consider these transfers and the redistribution of wealth that stems from them.
4. In other European countries, firms are slower in shedding employees in a recession and slower in recruiting unemployed workers in the subsequent upturn as fewer were laid off in the recession. In Germany, for example, the *Kurzarbeit* system financed out of a 2.5 per cent levy on gross wages allows employees to be put on short time at full wages for up to 2 years. (This system was put in place as a result of the devastating consequences of the instability of a market-led model in the inter-war period.) German firms therefore suffer less from the loss of skilled staff, gain from less volatile demand and can expand output more quickly in an upturn (Bischof, 1994).

# CHAPTER TWENTY-ONE

# THE POST-KEYNESIAN STATE AND THE SPACE ECONOMY

RON MARTIN AND PETER SUNLEY

## INTRODUCTION: BRINGING THE STATE INTO ECONOMIC GEOGRAPHY

Traditionally, all of the mainstream schools of economics – classical, neo-classical, Marxist and Keynesian – have had difficulty coping with the role of the state in the capitalist economy, and have tended to treat the state in a subsidiary and reductive way (Schott, 1984; Caporaso and Levine, 1992).[1] States may help clear the way for accumulation, and may help alleviate some of the shortcomings and undesirable consequences of the accumulation process, but they do not decisively influence the *forms* of economic organization and co-ordination of economic activities. Recently, a number of new 'state-centred' approaches to political economy have appeared (*see* Caporaso and Levine, 1992), which at last recognize that states may interfere in the economy not simply to readjust it or to compensate for its undesirable externalities or to manage its social tensions, but with the explicit political-ideological intention of shaping its very functioning and organization.[2] However, these still tend to see the state and the market as regulatory institutions that are necessarily antithetical to one another, so that when one increases the other correspondingly declines. As a result, our understanding of the *forms of capitalism* that are likely to be promoted by state intervention remains limited.

Likewise, economic geographers have been remarkably reluctant to integrate the state into their theorizations and analyses of the space economy. To the extent that the state is considered at all (and in some recent key works, it is ignored almost completely), it tends to be viewed simply as a regulatory *deus ex machina* to be lowered on to the economic landscape to resolve this or that specific aspect of uneven development (typically via 'regional policies'). Consequently, we still know little about whether different spatial configurations of production, accumulation and welfare are likely to be associated with different forms of state–economy relations. The precise nature of this association will, of course, depend on various contingent factors, including the state's *capacities*, the particular economic and political *strategies* (both domestic and international) it pursues, its institutional-territorial *organization*, and the prevailing material spatio-economic tendencies and constraints (both domestic and international) that it faces. Nevertheless, it is clear that the state and the space economy are inextricably interrelated: state intervention helps to *constitute* the spatial structure of the economy, and that spatial structure in turn influences the state's economic policy actions and their outcomes.

However, ironically, this argument to bring the state into economic geography comes precisely at a time when there is a rising chorus of opinion that the capitalist state is in fact 'withering away', that its role and influence are being undermined by new socio-economic forces from both within and without. The emerging view is that as we move into a new phase of capitalist accumulation and development that is at once both increasingly more localized and fragmented on the one hand, and more market-driven and globalized on the other, so the state is no longer an appropriate or effective agent of economic regulation and co-ordination, and is itself undergoing fundamental restructuring. The implication of this view would seem to be that the state's influence on the space-economy is evaporating and that the locus of regulation and intervention is shifting both downwards to local and regional institutions and upwards to supranational bodies and organizations.

Our aim in this chapter is to evaluate and move beyond these claims about the 'withering away' and 'retreat' of the state. We begin, in the next section, by outlining the influence of the post-war Keynesian welfare state form on the economic landscape, as a backcloth for then critically examining the nature, extent and spatial implications of the changes in state intervention and regulation that are allegedly being driven by new social-economic-political pressures and forces. We suggest that the argument that the state is being undermined by globalization is overdrawn. Likewise, the extent of the alleged shift from the welfare to the 'workfare' state is argued to be much more problematic and piecemeal than has been claimed. The state may have ceded or lost its powers of intervention in some spheres of socio-economic life, but it has extended or reinforced them in others. A change in the *mode* of state intervention is occurring rather than the withdrawal or weakening of the state. We suggest that future debate surrounding the future form of state–economy relations is likely to revolve around two organizational-regulatory models: a trend towards decentralized local micro-social and economic regulation on the one hand, and the rise of the co-ordinatory or network state on the other. While these models may appear to mark alternative trajectories for the future of state–economy relations, each with different implications for the processes and patterns of regional development, we argue that they should be seen as complementary. Without central orchestration, support and co-ordination, local micro-regulation is likely to accentuate, and ultimately flounder on, problems of regional instability and imbalance. Conversely, central state regulation on its own is unlikely to create the indigenous socio-economic dynamism and flexibility which are needed at local and regional levels. Hence the changing character of state–economy relations is fundamental to the economic geography of contemporary capitalism and vice versa.

## THE SPACE ECONOMY UNDER THE KEYNESIAN WELFARE STATE

From its origins in the 1930s (in the UK, the USA and Sweden), the Keynesian welfare state project emerged as the dominant post-war model of economic regulation among the advanced industrialized nations. Its twin goals were the stabilization of the inherent cyclical trend of capitalist growth, and the construction of mass societal support and harmony through the maintenance of full employment and the provision of a public welfare system. Different capitalist states pursued different variants of this model. According to the particular balance of socio-economic forces and historical political-institutional legacies in each country, the specific policies and forms of intervention employed differed, as did the degree of incorporation of capital and labour interests into the policy-making process.[3] Notwithstanding this diversity, however, there were also sufficient common features to permit some generalization as to the implications of the post-war Keynesian welfare mode of state intervention for the patterns and processes of uneven regional development within nations during this period.

In the Keynesian welfare model, the *national* economic space is the essential geographical unit of economic organization, accumulation and regulation over which the state is sovereign actor. As Radice (1984: 116) points out, 'The national economy is privileged in Keynesian theory for the purely practical reason that the nation-state system defines geopolitical space with the necessary features convenient for the theory: a common currency, common laws, and shared institutions.' Indeed, the Keynesian welfare project required a high degree of *closure* of the national economic space in order for the state's domestic policy measures to have their desired effects. Of course, extensive flows of money, capital and goods took place across the borders of the national space-economy, but under the post-war Keynesian regime these flows were controlled by and negotiated between nation-states, through such international bodies as the IMF, World Bank and the General Agreement on Tariffs and Trade, precisely in order to guarantee the stability of each *national* economy.

At the same time, as an accumulation strategy the Keynesian welfarist mode of intervention necessarily involved a high degree of *spatial centralization* of political regulation of the domestic economy. The management of aggregate national demand – the key innovation of Keynesianism – itself required new, centralized powers of co-ordination and manipulation of basic fiscal and monetary measures, and of concertation with the representatives of large national organizational interests, especially of capital and labour (Regini, 1995). Likewise, the use of a wide range of regulatory and legislative controls, on markets, prices, corporate organization, wages, labour, unions and the like, all part of the panoply of *extensive* intervention that characterized the Keynesian state, brought regions and localities within the economy under much greater central state control and dependence. To be sure, the degree of spatial centralization of political power over the economy varied from country to country, reflecting the spatio-organizational structure of different state systems (especially whether federal or unitary), although in most nations the administration of certain aspects of taxation and welfare spending was decentralized to subnational regional or local levels, as was responsibility for other aspects of allocation. But in every instance, the regions were subordinated to the macro-economic and macro-redistributive imperatives of the centre.

The counterpart of this centralized political regulation was *spatial socio-economic integration*. One of the direct consequences of the development of the post-war welfare state has been the dramatic growth in state spending, on production, allocation and redistribution (Cochrane and Clarke, 1993; Cerny, 1990).[4] Whereas in 1937 government spending averaged only 21 per cent of GDP among the industrialized countries, by 1980 this had grown to 43 per cent (Tanzi and Schuknecht, 1996). In all of the advanced economies, public spending increased on utilities, social and physical infrastructures, and various collective goods, especially housing and nation-wide education, health care and social benefit systems. The effect of these programmes was to foster consistent standards of social welfare and social infrastructure provision across regions and localities, thereby incorporating them into an increasingly *collective* or *public* space-economy, which in some countries extended to large-scale state ownership and management of key industries, in addition to utilities and other collective goods.

Finally, the Keynesian welfare state model of economic intervention and regulation was a *spatially redistributive and stabilizing* one. A defining feature of the post-war Keynesian welfare state – though again one that varied in intensity between countries – has been its redistributive role. Progressive tax regimes combined with benefit programmes aimed at low-income groups had the effect of redistributing income from rich sections of society to the poorer sections.[5] Similarly, the scale and nature of public finance under the Keynesian state has involved substantial fiscal transfers between regions and localities. Most of these transfers are 'automatic', in the sense that they derive from the way in which national (and federal) tax and benefit systems operate, and from the public expenditure stabilizers that are automatically activated by the economic cycle (especially unemployment and related social security payments). In addition, various other forms of expenditure, for example on infrastructure, industrial support and military and other procurement, also involve inter-regional fiscal transfers. The same is true of the explicit regional and urban policies adopted by most Western advanced countries over the post-war period, aimed at reducing spatial income and employment inequalities. The limited evidence that exists on inter-regional fiscal transfers suggests that they have played a critical role in stabilizing regional economies. For example, one of the most comprehensive studies, the McDougall Report (Commission of the European Communities, 1977), found that inter-regional fiscal transfers reduced inter-regional per capita income differentials by an average of 46 per cent in unitary states like the UK, France and Italy, and by 35 per cent in federal systems such as the USA, West Germany and Canada (*see* MacKay, 1995).[6] These regional redistributive-stabilizing effects were probably one of the most significant spatial consequences of the growth of the post-war Keynesian welfare state, the more so because they have largely resulted from the workings of public finance without any explicit government intent.

While it is difficult to isolate the precise regional impacts of post-war Keynesian policies in different individual countries, especially in the absence of any counterfactual evidence as to what would have happened in the absence of this form of state intervention and the state-led post-war economic boom, there can be little doubt that the direct and indirect impact has been substantial. Keynesianism helped to *underwrite* regional economies in a variety of ways. By stimulating and maintaining mass consumer demand, it helped to support those manufacturing regions specializing in the mass production industries of the period.[7] By intervening in or acquiring direct ownership of key industries, such as coal, steel, power, aerospace, shipbuilding, even motor vehicle manufacture, it influenced the fortunes of those local communities built around such sectors. And by promoting the substantial expansion of public-sector employment, it introduced a whole new layer to the spatial division of labour. Further, in those countries where Keynesianism became entwined with large-scale state spending on military production and research, as in the USA and UK, this spending itself produced specific geographies of investment and jobs, in some cases boosting economically backward areas but in others reinforcing pre-existing spatial inequalities (*see*, for example, Markusen *et al.*, 1991).[8] In many countries, Keynesian welfare policies became allied with more explicit attempts at reshaping the space-economy through various 'regional policies' aimed at directing the location of industry and jobs (as in the UK), or through more strategic spatial economic planning (as in Japan). More generally, post-war state policies and expenditures on housing, transportation and public utilities encouraged massive investments in particular spatial patterns of work, consumption and residence. In the USA and UK, for example, one expression of this process was a wave of suburbanization, which in turn helped to maintain aggregate mass demand for consumer durables. Once in place, these spatial configurations of industry, employment, infrastructure and population shaped the alternatives available to policy-makers on a range of fiscal, welfare and regulatory issues. On balance, as a result of these various policies and mechanisms, in most countries post-war Keynesian interventionism was a key factor behind the steady process of *regional convergence* in per capita incomes that characterized most advanced capitalist nations until the late 1970s.[9]

Over the past two decades, this model of the regulated and managed space economy has been undergoing radical change. The Keynesian welfare state has been undermined from without and from within, and is now widely viewed as obsolete, a project no longer relevant to the changing imperatives and contours of socio-economic development, and perhaps inherently flawed even during its apparent heyday (Janicke, 1990). On the domestic front, the inability of the Keynesian state to resolve supply-side economic rigidities as evidenced by inflation, unemployment and 'lame-duck' industries, its endemic 'fiscal crisis' of rising costs of social welfare and public resistance to higher taxation, and its inability to control the demands of organized labour had culminated by the end of the 1970s in a crisis both of economic management and social legitimation.[10] In addition, the very bases of national Keynesian interventionism have, it is widely argued, been completely undermined by changes in the international economic context, in particular the collapse of the international regulatory regime (Bretton Woods) that underpinned national financial stability and, more recently, the 'globalization' of economic activities and economic spaces. These internal and external challenges have thus raised the fundamental issue of whether and in what sense there is a future economic role for the state. The political response to this question, as embodied in the widespread shift to New Right neo-liberal state models throughout much of the advanced capitalist world during the 1980s and early 1990s, epitomized by Thatcherism in the UK and Reaganism in the USA (*see* King, 1987; Thompson, 1990), has been to roll back national systems of regulation, intervention and welfare support in an attempt to give national economies the 'flexibility' needed to compete in today's global markets (Regini, 1995). This reorientation of state–economy relations has in turn had equally profound implications for the space-economy.

## GLOBALIZATION VERSUS THE STATE: THE END OF NATIONAL ECONOMIC SPACE?

The idea that *globalization* is undermining the economic sovereignty of the nation-state is now widespread. The term 'globalization' itself is a 'chaotic concept'. For some it conveys the notion of increasing time–space *compression* of economic activity, symbolized by the almost instantaneous hypermobility of money, capital and information around the globe without reference to national boundaries (what Ohmae, 1990, labels the 'borderless world'). For others, it denotes an increasing process of socio-economic *integration* and *convergence*, in the sense of the replacement of nationally distinct products, firms, services and markets by truly global products, markets, trade and corporations. In yet other accounts, the term 'globalization' is a process of increasing time–space *distanciation*; that is, 'stretching' and interdependency of socio-economic events and processes across the globe irrespective of geographical separation. Driven by the conjoint forces of competitive deregulation by nation-states themselves, the market-driven global strategies of firms, internationally convergent consumer tastes, and by the thrust of new revolutionary, supranational information technologies, globalization, it is claimed, renders conventional notions of 'national' economic space, 'national' money, 'national' firms and 'national' technologies increasingly irrelevant (*see* Reich, 1995a, 1995b; Ohmae, 1995a, 1995b).

According to this view, then, globalization is 'decentring' national economic space, thereby undermining one of the basic tenets of Keynesian interventionism, indeed of all the mainstream schools of political economy. National economic boundaries are now so 'porous' that attempts to manipulate domestic demand by fiscal or monetary means are almost certain to be futile, even counter-productive. In particular, states have lost their economic sovereignty, their control over their exchange rates, money supplies and currencies to 'stateless' financial institutions and global markets, and these have no overriding obligations of national interest (Allen, 1994; Banuri and Schor, 1992; Camilleri and Falk, 1992; O'Brien, 1992; Martin, 1994b).[11] Likewise, states are no longer able to exercise control over the investment, employment and location decisions of firms, an increasing proportion of which are no longer committed to their home nations, but see themselves as global, and willing to switch production, jobs and investment between countries in response not only to market opportunities and cost advantages but also in search of less regulated environments. Nor, in today's world of global corporations and global information flows, can states easily confine the external economies associated with domestic R&D activity within their national boundaries (Comor, 1994). Under these conditions, it is claimed, states have no option but to withdraw from extensive regulation and intervention in their domestic economic spaces, to remove all barriers to the free flow of money, goods and capital, and to try to reduce their levels of corporate and individual taxes in order to create the low-tax 'enterprise' spaces required by the competitive forces of the new global economy (*see* Tanzi, 1995). The New Right response to globalization has thus been a sort of transnational economic liberalism or neo-classicism, in which the need for states to cede economic power to global markets and corporations is seen not only as inescapable but optimally efficient.[12] For extreme exponents of this view, nation-states are dinosaurs waiting to die, a 'cartographic illusion'; once efficient engines of wealth creation, nation-states today have lost that role and have become reduced to inefficient engines of wealth (re)distribution (Ohmae, 1995a).

The flip-side of this loss of economic sovereignty and power by nation-states is that the individual regions within nations have been exposed to the intense competition and uncertainties associated with globalization and the global economy. In a sense the national economic space is becoming a 'glocalized' composite, and individual regions within it increasingly linked into and integrated with diffuse webs of overseas markets, suppliers, technology and competitors, and less and less with 'domestic' ones. This makes individual regions and localities more prone to idiosyncratic demand and technology shocks, as well as to externally originating decisions with respect to investment and disinvestment, employment and production. Some states have actively encouraged this increased exposure of their regions as part of a 'shock therapy' to force their economies to restructure away from rigid, old, inefficient industries and activities towards new and more flexible ones capable of competing in world markets. The running down of regional aid, the deregulation of industry and labour markets, legislative attacks on unions, and the privatization and marketization of public-sector activities – these and other strategies pursued by free-market conservative governments since the early 1980s, especially in the UK and USA, but also elsewhere – have all been directed at increasing the flexibility of, and reducing central support for, the space-economy. While this state-sponsored retreat from the regions has certainly heightened the 'shock' felt by localities, whether and in what sense it has been 'therapeutic' remains debatable. If the Keynesian state was concerned to integrate its constituent regional and local economies and to cushion them from economic instability, the approach of its successor, the neo-liberal conservative state, has been to *dismantle* and *fragment* those systems of central support in deference to the restructuring forces of global competition, destabilizing its regions in the process.

Both because of this withdrawal of central state intervention in the space-economy, and because of the local impact of globalization itself, individual regions and localities have increasingly sought to assume responsibility over their own economic destinies. *Indigenous growth* and *local economic governance* are the new buzz-words in regional economics and politics. As economic geographers and other regional analysts endeavour to unravel the structural, technological and socio-economic foundations of 'indigenously generated' localized growth and adjustment within a globalized economy, regional states and authorities themselves are busy trying to pursue their own local enterprise strategies, whether these be based on local deregulation, local technology parks, or local small-business initiatives. In some accounts, the contemporary counterpart of the demise of the national economy is the rebirth of 'regional economies'; the emergence of a global network of specialized local industrial districts, varying in the type of activity involved but sharing the key trait of being based on localized networks of small, flexible firms (Amin and Thrift, 1994). Some commentators, again led by Ohmae (1995a), take this vision of current trends one step further, and see new *region-states* as the emerging successors to nation-states.[14] According to Ohmae, these 'region-states', defined by economic activity rather than by political borders, are emerging as the growth regions of the global economy precisely because they embrace the very characteristics demanded by the logic of that economy: deregulated, flexible and multinationalized accumulation, and little of the centralized and collectivized interventionist baggage of the nation-state.

However, compelling though these arguments and accounts may seem, their validity and generality are questionable. The 'death of the nation-state' by globalization has been much exaggerated.[15] While globalization has indeed weakened the economic sovereignty and power of nation-states in certain spheres – most notably in the realm of monetary policy – individual states still exercise substantial independence and authority in the regulation and management of their domestic political economies (Porter, 1990; Pooley, 1991; Hirst and Thompson, 1992, 1996). States still possess a large measure of autonomy over their fiscal policy, still control large sections of industry and services, still set much of the regulatory framework governing economic markets, and still exert considerable influence through their public spending programmes. Indeed, some policy areas, such as public spending, have been characterized by intensified central state regulation. Likewise, although regions and localities within nations are now subject to considerable competitive pressures and internal disarticulation from global economic forces, their economic fortunes and prosperity still depend in fundamental ways on the economic policies and expenditures of their central states. In fact, despite the political rhetoric of monetarist supply-side governments of the 1980s that 'Keynesianism was dead', political reality has been somewhat different: Reagan's expansion of the US budget deficit, especially through increased military spending, and Thatcher's tax cuts in the UK both had the classic Keynesian effect of boosting national consumer demand. And in these and other industrialized economies, programmes of privatization and restructuring of public sector activities and post-Cold War demilitarization are having highly significant impacts on regions and localities.

There is little doubt that the globalization process is changing the nature and meaning of 'national economies', in the sense that this concept has traditionally been understood, but it would be wrong to assume that state intervention is rendered obselete in a world of post-Keynesian globalized capitalism (*see also* Drache and Gertler, 1991). Rather, the challenge is one of rethinking conventional notions of 'economic nationalism', 'national' economic policy and the division between the 'private' and the 'public' in ways that are more appropriate to new economic realities. The New Right political response, of seeking to cut the size of the state, to squeeze and privatize the public sector, and to reduce taxes and workers' real wages, all so as to compete with low-wage, low-welfare newly industrializing countries, is one that threatens to lock the advanced nations into a vicious circle of progressively declining living standards (*see* Wood, 1994). Globalization does not justify less state intervention, but a *redirection* of that intervention. There is a growing connection between the kind of investment and spending that the state sector undertakes and the capacity of a country to attract worldwide capital and technology. Although states may now exert less influence over economic development through conventional fiscal and monetary policies, their role in moulding the socio-institutional embeddedness of that development is, if anything, becoming more important (Amin and Tomaney, 1995a). For it is the specific nature of the socio-institutional framework that in large part accounts for the differences between 'national capitalisms' and for the differences in national growth performance within them (*see* Hutton, 1995a; Berger and Dore, 1996). In each case the economic, political and social interlock in different, distinctive ways.

Thus even if money and technology have 'gone global', different nation-states continue to differ in their relationship between finance and industry (Cox, 1986), and the generation of new technologies (Archibugi and Michie, 1995). Perhaps above all, as the advanced economies become global, each state's most important competitive asset becomes the skills and cumulative learning of its workforce and the quality and efficiency of its public infrastructure. Unlike capital, technology, raw materials and information, all of which have become much more mobile, even 'borderless', the workforce and public infrastructure are unique to a state, and are vital to the prosperity of its constituent regions.[16] The extent to which different states invest in and upgrade their workforces and their social and infrastructural capital is likely to be a decisive factor in determining the ability of their regions and localities to compete in the global market-place. The state is in a pivotal position to promote these investments in education, training, R&D and in all the infrastructure that moves people and goods and facilitates communication. These are the investments that distinguish one state's economy from another: they are the relatively non-mobile factors in global competition, and key determinants of the local outcomes of that competition.

## FROM THE WELFARE STATE TO GEOGRAPHIES OF WORKFARE?

The New Right attempt to refashion the state has been designed not only as a response to perceived globalization but also as an attack on the whole ideology of welfarism and state support. This has involved a displacement in the focal point of politics from the maximization of general welfare to the promotion of enterprise, innovation and profitability in both public and private sectors (Cerny,

1993). Jessop (1994a, 1994b) argues that the transition from Fordism to post-Fordism, the rise of new technologies and the trend to continental regionalism have all acted to hasten the succession of the Keynesian welfare state (KWS) by the Schumpeterian workfare state (SWS). In prioritizing innovation and competitiveness, the SWS furthers the 'hollowing out' of the nation-state as powers and responsibilities are transferred to smaller regional and local governments whose intervention is closer to the sites of competitiveness (*see also* Goodwin *et al.*, 1993). In this view, social policy is becoming subordinate to the needs of labour market flexibility (Geddes, 1994) so that welfare systems are being replaced by policies designed to increase the skills and flexibility of those already in work (hence the label 'workfare').[17] Others see this shift towards a competition focus not as an inevitable institutional logic caused by the crisis of Fordism but as a reflection of deliberate policy choices by neo-liberal governments (Peck and Tickell, 1994b; Amin and Tomaney, 1995b).

There is already some consensus on the geographical implications of this reorientation from welfare to competition and enterprise. The disengagement of the state from its former spatially distributive role has stark implications for less favoured regions as it removes an important mainstay of their socio-economic welfare and weakens their protective framework. Thus according to Amin and Tomaney,

> Most obviously, the logic implies a reduction in the level of direct state support for industry in the less favoured regions, as the commitment to regional incentives is reduced, as the state disengages from and restructures industry under its ownership, as it deregulates public utilities and services, and as it ceases to direct public procurement contracts to firms located in the less favoured regions. (1995a: 39)

Furthermore, any retrenchment of welfare expenditure would particularly threaten the less favoured regions and poorer localities in which low-income and marginalized groups are disproportionately concentrated. The geographical danger is that the state will increasingly, perhaps inadvertently, support the most competitive regions and cities and that the less prosperous will be left to the market.[18] It could be argued that this general move to a type of neo-liberal workfare or competition state will undo the integrative and stabilizing effects associated with the Keynesian welfare state, with the result that geographical inequalities will widen and social cohesion will fracture.

This picture of radical discontinuity in state–economy linkages, however, should be viewed with circumspection. It relies on a selective emphasis and tends to impart too much coherence and logical neatness to contemporary experiments with new policies.[19] Studies which have examined the restructuring of welfare states have found that they have proved surprisingly resilient. Both in the UK and the USA, the state has experienced difficulty in retrenching the welfare system (Pierson, 1994). In the UK, the expenditure effects of cuts in programmes such as housing, the freezing of real benefit levels and wider means-testing have been cancelled out by the increased costs of persistent mass unemployment and population ageing, so that total welfare expenditure has been broadly static. While the Keynesian goal of full employment has undoubtedly been abandoned by most states, the Beveridgian aim of maintaining a minimum standard of living has proved more robust (Mishra, 1990). Welfare state retrenchment is difficult as welfare institutions have become fixed over a *longue durée* and exert considerable inertial power.[20] Moreover, welfare state reform has been restrained by its political sensitivity and unpopularity: governments are reluctant to alienate their social and regional support, and maintaining some socio-spatial coherence is still vital.[21] States' responses to international economic conditions will therefore be conditioned by political struggles and their need to sustain the social bases of their autonomy (Gourevitch, 1986), and in this respect universal benefits, in particular, are harder to cut back.

The argument that welfare is being subordinated to labour market flexibility is also problematic in that it portrays welfare states and labour market flexibility as incompatible alternatives. This seems to concede too much to the view that there is an inevitable trade-off between equity and efficiency. In fact, there are a large number of micro-economic flexibilities and efficiencies which are secured by the operation of welfare systems (Barr, 1993; Esping-Andersen, 1990).[22] Indeed, this may be one of the reasons why there is no easy cross-national correlation between the size of states' expenditures on welfare and their relative economic performances (Morley and Schmid, 1993; Pfaller *et al.*, 1991). It also needs to be emphasized that welfare reform has operated from very different starting-points and, despite its widespread incidence, these differences have not by any means been erased.[23] A great deal of confusion and uncertainty over the future direction of welfare remains, and it would be premature to assume that all states are converging on a post-welfare enterprise model (Cochrane and Clarke, 1993).[24] In general, despite the rhetoric of the 1980s, public spending has continued to increase, to an average of 47 per cent of GDP in the industrial countries by 1994 (Tanzi and Schuknecht, 1996). Furthermore, if there is a new convergence in welfare provision, then it seems more likely to centre on the use of '*quasi-markets*' through which social services are centrally financed but provided by contractors, quangos and private agencies (Taylor Gooby and Lawson, 1993). The creation of more decentralized and pluralistic systems may be in tune with demands for greater consumer responsiveness but its implications for resources are unclear. The geography of service provision will play an increasingly important role in determining the outcomes for public spending, for where suppliers are able to exploit a local monopoly the costs of provision may increase.[25] It is also likely to exacerbate differences between localities in the resourcing and provision of services (*see*, for example, Mohan, 1995).

It is important, then, not to exaggerate the logical and practical coherence of the 'post-welfare' agenda as it is not easy to explain precisely what the new focus of state policy is to be. According to Jessop (1994b), for example, 'The growing importance of structural competitiveness is the mechanism which leads me to believe that we will witness the continuing consolidation of the "hollowed-out" Schumpeterian workfare state in successful capitalist economies' (p. 36). The key problem here is that *competitiveness* is a vague and in some ways obfuscatory idea. As Krugman (1994a, 1994b) argues, the notion is often misleading as it implies that international trade is a zero-sum game in which one country's gain is always at the expense of another's loss.[26] Competitiveness cannot simply be reduced to balances of trade and, to be meaningful, it must refer to comparative levels of productivity and living standards (Porter, 1990). But it is then difficult to argue that contemporary state policies are coalescing on raising productivity and living standards. Raising productivity undoubtedly depends on significant investment in public infrastructure and on a well-motivated and trained workforce, both of which are contradicted by neo-liberalism's determination to reduce public expenditure and lower real wages. It is clear that there are different ways in which states can try to construct competitiveness, including both social democratic and neo-liberal policy models (Garrett and Lange, 1991). Moreover, conflictual processes of social learning within institutions are as important as economic pressures in determining which model is adopted.

This means that the geographical implications of the current changes in state–economy relations are graduated and complex. On the one hand, the departure from Keynesianism has encouraged the widening of socio-spatial inequalities. The reluctance to cushion the impact of economic adversity has exposed regions to international trade and a greater risk of instability and decline. Simultaneously, the neo-liberal shift from a fiscal reliance on direct taxation to one on indirect taxation has acted in a spatially regressive way by falling hardest on areas containing relatively poor tax-payers. Conversely, policies of tax cuts for higher earners have no doubt acted to boost leading regions. On the other hand, many of the geographical consequences of recent shifts in state policy are indeterminate and difficult to predict in an abstract and generalized fashion. For instance, allowing regions to specialize to a greater extent on the basis of international trade also increases the risks faced by hitherto successful regions.[27] More generally, if some policies do come to focus more effectively on supply-side improvements then there is no inherent reason why these policies should be of less benefit to less favoured regions and cities. Finally, in the context of persistent mass unemployment, automatic fiscal transfers to poorer regions have remained substantial. MacKay (1995), for example, shows that in the UK since the early 1980s these transfers have come to account for a greater proportion of the total income of the peripheral regions. While some foresee the decline of fiscal transfers to poorer regions under post-Fordism (Dunford and Perrons, 1994), there is little evidence of this as yet.

We should also treat the claim that economic powers are being devolved to regional-level governments with caution. Experiences in Germany, Italy and Spain demonstrate that while regional governments may well foster dynamic economic efficiency, their autonomy is constrained by the operation of nation-wide redistributional mechanisms (Newlands, 1995; Zimmerman, 1990). The contradiction typically faced by regional institutions is that while their economies are increasingly vulnerable to depressions and shocks, their ability to respond is typically strongly constrained by limited resources and national regulatory and legislative structures (Anderson, 1992). The German *Länder*, for example, are often seen as models of the new economic regionalism, but have been subject to greater central controls and fiscal redistribution over the post-war period (Newlands, 1995; Amin and Tomaney, 1995a). The technological and industrial policies developed by the *Länder* have not been an *alternative* to national policies, for the latter have also been expanded (Esser, 1989). Indeed, it appears that regional-level initiatives work best when there is an effective integration of regional and national policies (Gertler, 1992; Weiss, 1989). Moreover, there is a great deal of difference between genuine local economic capability and the local implementation of central state directives, as the case of the British 'quango boom' (there are now over 5000 of these bodies) testifies (Holtham and Kay, 1994).[28]

While there are aspects of state regulation which have undoubtedly changed radically, others have been more resistant to reform, producing a complex assemblage of new and old institutions, pressures and experiments. Rather than all industrialized states shifting to a single post-Keynesian model of intervention in the space economy, different types of policy innovation and experimentation are being pursued in different states. The austerity and marketization associated with neo-liberal policies over the past two decades, and the trend to greater spatial inequities brought in their wake, are undoubtedly common experiences, but it is misleading to represent this primarily as a transition to a locally based, supply-side and micro-interventionist state. For example, privatization programmes have been witnessed in a wide range of countries,[29] but, in many cases, this has not meant a corresponding decline in state intervention but rather a shift to *intensive* regulation in response to the need to protect public interests in the privatized provision of public goods (Thompson, 1990; Helm, 1994; Parker, 1993). Within different national frameworks and models of intervention, there appears to be a growing use of a *contract style* of administration and service supply, separating finance from service provision, and this may well fracture the national uniformity associated with Keynesian welfarism and exacerbate spatial differences in regulatory environments and service provision.[30]

## CONCLUSION: MICRO-REGULATION AND THE NETWORK STATE

It is clear, then, that nation-states continue to exert a profound influence on their space-economies and it is also clear that the question of the scale of state intervention and regulation will be fundamental to its future economic role. There appear to be two emerging models of the future of state–economy relations. The first refers to local social regulation modelled on the Italian experience. Both Regini (1995) and Locke (1995), for example, argue that the economic success of Italy during the 1980s was due not to a central state strategy but rather to the 'vibrancy' of local associational networks which encourage entrepreneurialism. The most successful regions are distinguished by their polycentric networks of egalitarian associationalism, interest group organization, and local institutions which facilitate the pooling of information and scarce resources, mediate conflict, and generate trust among local economic actors. Hence micro social regulation provides collective goods such as co-operative industrial relations, the co-ordination of wage dynamics, training and the development of human resources. In this view, what has happened in Italy is indicative of more general trends sweeping across all advanced industrial nations and, in an era when governments are losing macro-economic control and divesting themselves of micro-economic control through deregulation, established national and *étatist* models are of little help (*see also* Warren, 1994; Schmidt, 1996). This Italian model is proving to have a very wide impact, not only through the industrial districts literature but also by stimulating debate on the wider regional possibilities of associational networks in civil society (Amin and Thrift, 1995). What these accounts imply, however, is a much more pronounced and dramatic experience of regional disparities as those regions and cities with effective local regulatory orders outperform those which lack 'vibrant' micro-regulatory structures.

The second emerging model of state–economy relations is predicated on the view that both the wealthiest states and the fastest-growing economies (particularly the East Asian newly industrialized countries) owe their success, in part, to what might be called an *institutionalized and co-ordinatory* type of state intervention. States such as Germany, Sweden, Austria, Switzerland and Japan have employed co-ordinatory policies at both macro and micro scales (Soskice, 1990; Matzner and Streeck, 1991). At the macro scale the development of co-operative institutions between industry, finance and labour has restrained real wage growth and provided relatively cheap capital, while at the micro scale, competition has been balanced by long-term co-ordination centred on high-skilled labour forces and co-operative, long-term relationships with suppliers. As Holtham and Kay (1994) have argued, 'In these successful capitalist economies it is not simply that the rough edges of capitalism are disguised by a veneer of collective activity. It is that such *collective action is essential to making capitalism work*' (p. 2; emphasis added). At the same time, explanations of the phenomenal growth of the East Asian 'late industrializers' have stressed the *co-ordinated model* of state involvement in the economy. In South Korea and Taiwan, for example, the dense networks which exist between industrial groups and the state have facilitated the rapid development of new sectors and the promotion of export-led industrialization (Amsden, 1989; Wade, 1990). Similarly, Japan has also been described as a 'network state' (Sheridan, 1993; Wilks and Wright, 1991). In the view of many, this type of co-ordination is not so much bureaucratic as based on formal and informal policy networks which connect public, intermediate and private actors. In this model, state capacity is reconceived as the ability to sustain a dense structure of *policy networks*,[31] which allows the state to work through and in co-operation with other organizations (Hall and Ikenberry, 1989; Weiss and Hobson, 1995). Moreover, such institutionalized networks are seen as increasingly important for economic growth (*see*, for example, Lazonick, 1991).

However, the view which we have tried to develop in this chapter is that, contrary to the arguments of some, these two models of public regulation and co-ordination are not necessarily alternatives. Rather, the need for *state-organized* networks to provide the facilitatory infrastructure, co-ordination and redistributed resources in which more local and regional types of social actor co-operation and micro-regulation can flourish is likely to continue.[32] Without central orchestration, micro-regulation is likely to flounder on the problems caused by intensified regional instability and imbalance. Conversely, central state regulation, on its own, is likely to prove incapable of creating and making use of the wide range of external economies and increasing returns which operate at a regional level. One result of the Anglo-American dominance of economic geography, however, is that we know very little about the ways in which macro- and micro-co-ordinatory networks can be made to work symbiotically and to reinforce one another. It appears that this fusion can not only benefit leading industrial regions but can also be strikingly effective in managing the decline of older industries (for example, *see* Young, 1991) and may well increase the adaptability of declining industrial regions.[33] Yet we also know little about the geography of national and local effective policy networks: the extent to which they can be cultivated in states currently dominated by neo-liberalism and whether and how these co-ordinatory networks can be created in differing legal and administrative cultures and within different regional social and political traditions. In short, the whole question of the changing character of state–economy relations remains central to understanding the economic geography of the post-Keynesian era.

## NOTES

1. Even regulationist political economy, which stresses the role of the state as a key component and mechanism of the capitalist 'mode of regulation' (Boyer, 1990) fails to escape this limitation (Goodwin *et al.*, 1993; Jessop, 1990b). Most regulationists simply adopt an already available account of the state to fill out their model of political economy. And despite their efforts to the contrary, their treatment of state intervention is close to being a functionalist one, in which state regulation serves the needs of the accumulation regime; hence the regulationists' idea of the 'Fordist' state, and now its successor the 'post-Fordist' state.
2. For example, 'transformational state' theories of economic governance (*see*, for example, Lindberg and Campbell, 1991), Jessop's (1990b) strategic-relational theories of the state and 'state projects', and the new, 'historical institutionalist', comparative political economy (Steinmo *et al.*, 1992). Though these various new approaches differ in their specifics, all share what has been termed a 'relative autonomy' or autonomous view of the state.
3. There was, in fact, a (largely unresearched) international geography to post-war Keynesian interventionism, involving such versions as 'military' Keynesianism in the USA, consensual Keynesianism or 'Butskellism' in the UK, the 'social market' Keynesianism of Germany, the 'regulatory' Keynesianism of France, 'pragmatic' Keynesianism in Italy, and 'social democratic' Keynesianism in Sweden (*see* Hall, 1989, for detailed accounts of these different national forms).
4. While there is nothing inherent in the theory or practice of Keynesian macro-economic management that requires an accompanying welfare state, the two are certainly complementary. By providing a social wage, the welfare system helps to support aggregate demand and mass consumption, and through its education, housing and health components it helps in the reproduction of the labour force needed for full-employment production. By the same token, the maintenance of full employment through demand management ensures the tax take required to fund the welfare state. An extensive welfare state, then, is predicated on the economy being run at or near full employment.
5. To the extent that some social benefits have been universal, and indirect taxes have been regressive, critics of the post-war welfare state argue that contrary to what has been intended, the poor have not always been the main beneficiaries.
6. As Krugman (1993) argues, such 'non-market' fiscal stabilizers are as important as market forces in limiting the impact of regional economic downturns and crises (*see also* Sala-i-Martin and Sachs, 1991). Sala-i-Martin and Sachs (1991) have shown for the USA that declining federal tax payments and rising

federal transfers provide roughly a one-third offset to a regional-specific decline in economic activity.

7. In this sense the Keynesian state played a key role in the growth of Fordist industry in the 1950s to 1970s.
8. Keynes himself only made one reference to the applicability of his policies to the regional development question. Interestingly, this was when he advocated directing the British Government's rearmament programme in the late 1930s to the depressed regions of the north of the country, where surplus labour and capital could be utilized without encouraging national inflationary pressures.
9. For empirical evidence on the convergence of regional incomes in the USA, European countries and Japan, *see* Barro and Sala-i-Martin (1995). Although those authors tend to explain this convergence in terms of market forces and technological diffusion, they admit that government policy is likely to have been important. Our argument here is that Keynesianism was indeed of fundamental importance.
10. The Keynesian welfare state was based on two assumptions concerning both the *level* and the *composition* of employment: that the economy was kept at or near full employment, and that the typical worker was a male who normally earned sufficient to support hinself, his wife and a family of at least one child. In the labour market of the 1980s and 1990s, however, not only is high unemployment endemic but employment has become much less secure, less full-time and increasingly feminized. The implications of these changes for the social security system are profound.
11. It is significant in this context that Keynes himself feared the 'globalization' of finance and the rise of 'stateless monies'. Of all aspects of economic and social life, he argued, 'let finance be primarily national' (Keynes, 1933: 758). In his view, 'economic internationalisation embracing the free movement of capital and loanable funds as well as of traded goods may condemn [a] country to a much lower level of material prosperity than could be attained under a different system' (ibid.: 762–3). The contemporary loss of financial autonomy by the leading industrialized states is, in large part, of their own doing, the result of the wave of competitive financial deregulation that they embarked upon during the 1980s.
12. Indeed, some see the dichotomy between nation-states and global capitalism as ultimately unsustainable and as necessitating a shift from a system of national regulation to one of transnational regulation (McMichael and Myhre, 1991).
13. Ohmae has fallen victim to a common complaint among management 'gurus' and 'airport economists', namely gross over-simplification and exaggeration. Ohmae's work is certainly long on sensationalism, but it is much shorter on detailed analytical exegesis and empirical evidence.
14. He cites numerous instances, for example: northern Italy, Baden-Württemburg, Hong Kong/southern China, the Silicon Valley–Bay Area in California, Fukuoka–Kitakyushu in the north of the Japanese island of Kyushu, the growth triangle of Singapore–Johore–Riau, Triangle Park in North Carolina, and the Rhône-Alps region of France.
15. It can equally be argued that the challenge to the nation-state and state intervention has more to do with the expanding trend towards *regional integration* than with globalization; that is, the formation of continental regional free-trade blocs and customs unions (*see* Anderson and Blackhurst, 1993; De Melo and Panagariya, 1993; Cable and Henderson, 1994; Hirst and Thompson, 1996). Although the European Union represents the most advanced example of this new regionalism, there are more than 30 of these regional initiatives worldwide, one of the most recent being NAFTA, the North American Free Trade Agreement. As the case of the EU illustrates only too clearly, regional integration agreements raise critical questions concerning the scope for and nature of economic intervention and regulation by the nation-states that are signatories to such agreements, and concerning the need to ensure that individual regions and localities adjust to the economic shocks and changes that integration agreements bring. Although some believe that integration leads to a reduction of spatial economic disparities within the regional bloc, most observers expect such disparities to be exacerbated as a result of locational shifts in investment and jobs towards the core growth regions. Thus it is often argued that a necessary corollary of the advanced stages of economic and monetary union is the establishment of a corresponding integrated system of spatial fiscal transfers and stabilizers. It is arguably within these blocs that Ohmae's 'region-state' economies assume their real significance.
16. Critics of the globalization thesis have repeatedly emphasized that labour was more internationally mobile during the period of mass migration in the nineteenth century.
17. Jessop argues that there are neo-liberal, neo-statist and neo-corporatist forms of the SWS but insists that we will witness its continuing consolidation in successful capitalist economies. The reason is that this new state form could help to resolve the crises of Fordism, so that 'Thus it seems that the hollowed out Schumpeterian workfare state could prove structurally congruent and functionally adequate to post-Fordist accumulation regimes' (1994b: 27).
18. This conforms with the vision of writers on globalization who also argue that the weakened governance capability of the nation-state leads to new inequalities and dualisms between those who can compete in the global economy and those who cannot (Sassen, 1994).
19. This exaggerated coherence may reflect the influence of regulation theory on these ideas; the regulationist notion of a coherent mode of regulation may be misleading (Painter and Goodwin, 1995).

20. For example, contrasting labour market policies have been shaped and constrained by patterns of institutions fixed early in the twentieth century (King, 1995).
21. While maintaining social consensus may have been given lower priority in recent years, it remains one of the key imperatives of modern states. In this context, the dramatic ageing of the population structures of many OECD poses a twin problem for welfare reform, for while it will escalate costs, it will also endow moves to privatize pension and health systems with enormous political sensitivity.
22. As Barr (1993) argues, 'The welfare state is much more than a saftey net; it is justified not simply by a redistributive aim one may (or may not) have, but because it does things which private markets for technical reasons either would not do at all, or would do inefficiently. We need a welfare state of some sort for efficiency reasons, and would do so even if all distributional problems had been solved' (p. 433).
23. For example, the liberal, social assistance model characteristic of the UK and the USA, the conservative, corporatist regimes of Germany, France and Austria and the social-democratic regimes of the Scandinavian countries (Esping-Andersen, 1990).
24. For instance, the dramatic reduction of state spending under New Zealand's 'Rogernomics' from 46 per cent of GDP in 1988 to 36 per cent in 1994 is clearly different from Australia's combination of stringency with increases in benefit levels for low-income families. Thus those like Bennett (1990) who talk of a 'post-welfare' agenda are thus guilty of *pars pro toto*.
25. This argument is supported by the example of the welfare system in The Netherlands, which is highly decentralized and pluralistic but relatively costly (Glennerster and Le Grand, 1994).
26. Krugman (1994a) also argues that, because of the domestic importance of service industries, raising productivity in these 'non-traded' sectors would have an enormous effect on standards of living, but this has little to do with international competitiveness.
27. For example, Buck *et al's* (1992) comparison of London and New York argues, 'A significant decline in activity in the international financial system, or a competitive loss to other cities . . . would now pose as severe a threat for the two cities as the decline of manufacturing and goods circulation did in the past' (p. 103).
28. While some regulation theorists have distinguished between local government and local governance, regulation theory has failed to explain why these different local political trends prevail in different states (Hay, 1995).
29. These include not only the UK and USA, but also France, Japan, Italy, New Zealand and Sweden.
30. For example, Clark (1992) describes a move in the USA towards an administrative state based on regulatory agencies and argues that this type of regulation is inherently geographical as it is enforced by, and enmeshed in, a legal environment shaped by local precedents and cultural practice. His argument clearly reflects the legalism of the US regulatory environment, and we need to know much more about the different national forms, the local consequences and the incomplete geographies of regulation of this administrative state model.
31. Policy networks in this view represent the linking processes within policy communities and are constituted by complexes of organizations and actors connected to each other by resource dependencies.
32. Regini (1995) himself notes that 'To what extent micro-regulation processes will become an effective alternative to macro-regulation in the production of other public goods is, as I have said, difficult to predict. The boundaries between micro and macro, between their respective ranges of action, are still uncertain. But if macro-political regulation continues to be scaled down in favour of micro-social regulation, this will not necessarily mean the increasing irrelevance of institutions and the greater weight of the deregulated market in determining economic outcomes' (p. 145).
33. A key issue for research is how, and under what conditions, dense policy networks become obstacles to, rather than facilitators of, structural economic change (Katzenstein, 1987; *see also* Grabher, 1993c).

# CHAPTER TWENTY-TWO

# BRINGING THE QUALITATIVE STATE INTO ECONOMIC GEOGRAPHY

**PHILLIP M. O'NEILL**

## INTRODUCTION

It is commonly argued that there has been substantial erosion of state power during the past two decades. This is seen to have occurred as a direct result of changes in the nature of capitalist accumulation which, in turn, are perceived to be driven by enhanced and extended circuits of capital. This chapter will advance three arguments to illustrate how this view of the state is inadequate. First, it is shown that the conception of the state as a totalized entity with centralized structures and purposes is erroneous. It is a conception that fits neatly with periodized views of capitalism (monopoly capitalism, Fordism, post-Fordism, and so on) but which provides little assistance in the analysis of contemporary economic change, at any scale. It is argued in this chapter that a more powerful understanding of contemporary economic change can be derived from depicting the state as a domain where a complex and heterogeneous state apparatus is engaged in constant interplay with non-state institutions and agents, including those from other nations, in an irresolvable contest over accumulation and distributional goals.

Second, the chapter challenges the politically charged discourse that markets are capable of a separate, private existence beyond the actions of the state's apparatus. Rather, a *qualitative* view of the state is preferred. In this view, the state is seen to play an indispensable role in the creation, governance and conduct of markets, including at the international scale. Consequently, arguments about the *extent* of state intervention are seen as being feeble. Because the state is always involved in the operation of markets, the salient debate should be about the nature, purpose and consequences of the *form* of state action, rather than about questions of magnitude of intervention.

Third, the chapter shows the depowering outcomes of arguments that the macro-economic powers of the state are being eroded. At the heart of the neo-liberalist project, for example, is a discourse that promotes a form of capitalism, *laissez-faire*, which thrives under conditions where economic transactions are conducted outside the realm of state action. In this view, it is the very absence of state action that produces the best conditions for the efficient allocation of productive resources and the most desirable distributions of production's rewards. Similarly, accounts of contemporary economic change such as globalism and certain variations of regulation theory argue that there is a relentless hollowing of nation-state structures and powers. In short, capitalists are seen as being able to accumulate how and where they wish and, thereafter, states intervene to alter distributional outcomes. With the alleged collapse of nationally organized regimes of accumulation, however, opportunities for state-led redistributions are thought to be fading as state effort is redirected into shoring up the conditions for successful accumulation by national capitals in the face of growing international competition. In this contest, extant distributional positions are seen as being bargained away in competition for increasingly mobile investment projects. A *qualitative* construction of the state can avoid this immobilizing view of distributional possibilities. Since capitalism is incapable of operation outside the realm of the state, and since each and every step in the accumulation process involves distributions, then the state is *ever* involved in the distributional processes of capitalism.

These arguments are developed through the four sections of this chapter. The first section points to the need for a review of the way that the state is perceived in economic geography by referring to three theoretical contexts in which formulations of the state are advanced. These contexts are neo-liberalism, globalization and regulation theory. The second section of the chapter demonstrates that by decentring the way in which we portray the state and by concentrating on the interactions between the state's apparatus and capitalist processes, the distinction between state and market is broken down and issues of accumulation and distribution become inseparable. The third section of the chapter applies these arguments in an examination of the role of the state at new scales in the circumstances of rapid growth in international transactions. The final section examines distributional implications and new opportunities for intervention to achieve more desirable distributional outcomes.

## VIEWS OF THE RELATIONSHIP BETWEEN THE STATE AND THE ECONOMY

One of the major difficulties in undertaking an analysis of the role of the state in the construction of economic geographies is finding common answers to the question, What is the state doing? (O'Neill 1997). On the one hand, many economic geographers argue that, as a result of liberalized international financial flows over the past two decades, a *global* economic sphere has emerged which is beyond nation-state control (*see* Martin and Sunley in the previous chapter). On the other, a small number of empirical analyses show how the nation-state plays a major role in both *managing* and *promoting* international financial flows (e.g. Leyshon, 1994b; Martin, 1994; Hutton, 1995; O'Neill, 1997). It seems, then, that the major problem with attempts to generalize about the changing role of the modern state on the basis of empirical study is that there is almost always a case study of national economic change available somewhere in the world to match a writer's particular view of the state. Compare, for instance, the Hudson and Williams (1995) view of an increasingly restricted British state to the active and purposeful state roles depicted by Doraisami (1995) for Malaysia, Webber (1994) in the North-East Asian economies, and Enderwick (1995) for New Zealand state, irrespective of political hue. The range of state responses to the economic crises of the past two decades is summarized in Hutton (1995a, especially ch. 10), where attention is drawn to the historical, institutional and cultural differences within which state actions are inevitably situated. In the main, economic geographers have largely *observed* rather than *theorized* state action, concentrating chiefly on the rise of two forces, *neo-liberalism* and *globalism*. In turn, explanations for these phenomena have relied heavily on the *regulation theory* template.

### The Rise of Neo-liberalism

Neo-liberalists argue that freely operating private markets are the most capable instrument for maximizing output (technical efficiency), welfare (allocative efficiency) and ease of adjustment (dynamic efficiency). Neo-liberalists see the state as an outsider to the processes of real production and believe that it should constrain private activity to promote public interest only in situations of market failure – where desired goods and services are not supplied at acceptable prices. Yet, even in these circumstances, neo-liberalists point to a dilemma in placing the responsibility for corrective action in the hands of the state apparatus. They argue that the motivation of public regulators is not to maximize public interest but rather they will seek out exclusive benefits (or *rents*) for themselves and the groups which have secured their patronage. This view is encapsulated in *public choice theory* (Stigler, 1971) and represents the attempt by neo-liberalists to impugn the principle of public service and the existence of public goods (Ernst and Webber, 1996). Neo-liberalist discourse frustrates the development of state distributive actions by arguing, first, that state intervention is best minimized and, second, that it will be corrupted in any event.

It is axiomatic, according to neo-liberalism, that the absence of state intervention *is* the market, that market failures are never failures of the market *per se* and, therefore, they can only ever be failures of the state (drawing on Hayek, 1948). The political consequence of this view is the drive to deregulate, since modern states are thought to be incapable of managing either the social or the private economies – only through deregulation and privatization can the preconditions of economic growth be re-established (Francis, 1993: 33). This apparent displacement of a state's distributional functions has meant that not only does the discourse of neo-liberalism drive the accumulation strategies of many nations but, *ipso facto*, it has became the dominant distributional tool. The important point to be made here is that the neo-liberalist vision of 'less state' is entirely illusory. Neo-liberalism is a self-contradicting theory of the state. The geographies of product, finance and labour markets that it seeks to construct require *qualitatively* different, not less, state action. Neo-liberalism is a political discourse which impels rather than reduces state action (Hirsch, 1991: 73; Jessop *et al.*, 1990: 86–9[1], Tickell and Peck, 1995).

## Globalism

Internationalization and more recently its descendant, globalism, have been portrayed by economic geographers, and others, as all-powerful tendencies of capitalism which have had major impacts on the role of the state. During the 1970s and 1980s, international capital was commonly seen as controlled by stateless multinational corporations which aimed to penetrate the capital and consumer markets of every country and every region. Analysis drew heavily on the theory of the *international division of labour* (Fröbel *et al.*, 1980), and privilege was accorded to those events seen to operate at the global scale (e.g. Thrift, 1986; Dicken, 1986; Chase-Dunn, 1989). In an extreme version of this tale, Ross and Trachte (1990; *see also* Ross, 1995) advise that global capitalism dominates all economic spheres, leading to the crippling of nation-states, whose meagre resources shift from the pursuit of legitimation goals (especially social welfare) to the accumulation goals of once-national capitals in their global ventures.

Common to the inexorable-rise-of-global-capitalism accounts of contemporary crisis is the portrayal of the state in classical Marxist form: an organizational unit which serves the dominant classes which, in the contemporary scene, are the controllers of global fractions of capital. Two rebuttals of this position can be made. The first acknowledges that states, acting through autonomous state agents and apparatus, maintain sufficient economic authority and purpose to produce 'discontinuous' economic and political spaces (Dicken, 1992: 149–50; 1994b). The state remains capable of producing 'location-specific' supply and demand conditions *within* nations – and these conditions are more than the passive response to globally mobile fractions of capital; they are conditions which *drive* foreign investment itself (Dicken, 1992: 223). Certainly, there is ample evidence that Keynesian macro-economic management powers are eroded by globalization processes, but one cannot conclude that a powerless nation-state has been left as residue. Inevitably, states are developing new capacities and structures to exert new forms of political and economic power, even across the territories of different states (Dicken, Peck and Tickell, in Chapter 12 of this book).

The second rebuttal of the inexorable-rise-of-global-capitalism-and-decline-of-the-nation-state story is based on distinguishing between globalization as discourse and evidence (or lack of evidence) for the existence of globalization as a general trend (Cox, 1992; Dicken *et al.*, this volume). There is growing understanding that the rise of the vocabularies of globalism has been based on their use for coercive purposes in framing national economic strategies, in reformulating workplace practices and in garnering community support for national and local economic reforms. These vocabularies necessarily portray the state as a depowered entity – for obvious reasons. Yet there is a difference between a *loss of* state role contained in the globalism story and a *shift in* state role which a qualitative view of the state would assert. However, if indeed the shift in state role is qualitative, then what is changing? What new state capacities and structures have emerged? And what guides their formulation and reformulation? The following section demonstrates the inadequacy of one version of regulation theory in answering these questions.

## Regulation Theory and Jessop's Hollowed-out State

One of the major impacts of recent regulation theory on thinking about the state has been the idealization of the post-war Keynesian welfare state and its positioning as a historical yardstick against which subsequent state roles are measured. In particular, regulation theory assigns a nanny role to the post-war state: alleviating crisis in the Fordist economy and nurturing capitalism's drop-outs. Peck and Tickell (1994b), for example, argue that the decline of the Keynesian state is a fundamental cause of contemporary economic crisis, a crisis which persists because a stable replacement regulatory order has yet to emerge. In this context, the nation-state is simultaneously coerced by international economic forces and driven by the accumulation needs of its domestic economy (Tickell and Peck, 1995: 359). Similarly, Jessop (1994a) argues that the 'hollowing-out' of the nation-state is a feature of the contemporary period. Jessop maintains that, at the same time that nation-states have abandoned demand management in favour of supply-side initiatives, supranational state apparatuses have emerged as the new regulators. Activities at the nation-state level are seen to be driven by global economic trends (Jessop, 1994a: 263). These include the development of new legal forms to support cross-national co-operation and strategic alliances, the reform of currency and credit systems, rules for technology transfer, trade governance, intellectual property negotiations and the regulation of international labour migration. At the same time, according to Jessop, investment policies have been devolved to the local state, resulting in new cross-national groupings of regions. A residual nation-state thereby is left to engage in the development of work practices and other measures to promote international competitiveness, a situation which Jessop labels as the Schumpeterian workfare state.[2]

Jessop's hollowed-out metaphor, then, is an account of *shift in* rather than *diminution of* state role with complex new relationships emerging between state apparatuses at different scales. Yet, irrespective of the original conception, it has become commonplace for the hollowed-out metaphor to be used in the literature to represent a universal condition of state depowerment. For example, the metaphor's signification that the nation-state has been gutted is advanced by Thrift (e.g. 1990b, 1992) and in many places in an *Environment and Planning A* (1995) special issue on changes to the welfare state. The metaphor has come to represent an extreme paralysis of the nation-state, different from the shift process described by Jessop.

Putting aside these misinterpretations of regulation theory, to what extent does regulation theory provide an enduring explanation of state process and behaviour? Regulationists Tickell and Peck (1995) are ambivalent. Certainly, they use regulation theory effectively to explain the role of the state in the construction of social regulation as a stable form of governance during the Fordist period. Yet they acknowledge that regulation theory offers little explanation for the key processes of transition (Tickell and Peck, 1995: 361) which, surely, have been the key concern of nation-states for at least the past two decades! Regulation theory has insufficient to say about the crucial role of the state in moving societies from one period of stable economic conditions to the next, beyond describing them as intervals of crisis.

Much of this inadequacy stems from regulation theory's attempt to totalize the economy in its form, its history and its methods of governance. Further, accumulation crisis is portrayed by regulation theory as a singular, totalized event. There is no possibility in regulation theory for a multitude of unrelated economic events, in different cycles of growth and prosperity, under different forms of governance. There is no allowance for incremental, strategic, state-driven economic restructuring and transition such as has been the hallmark of East Asian economic change since the 1960s; and there is an underlying denial that conflict and tension in the operations of state apparatuses may be normal events.

Further, as a result of presenting an idealized (and nostalgic) view of the social and economic outcomes of Keynesian-Fordist state management, regulation theorists have a tendency to conflate successful accumulation with successful distributional outcomes. The theory assumes that the successful management of national accumulation processes is the most important coercive tool in securing state legitimacy. This view is almost the opposite of the view of many state theorists, such as Claus Offe (*see* Offe, 1984), and is denied by considerable empirical evidence. Bakshi *et al.* (1995), for example, point to the racialized and gendered history of the British state in the context of successful national accumulation during the Keynesian-Fordist years. Social compliance was never earned through long periods of economic growth, and fair and reasonable distributions of national income were not automatic for each social group. Bakshi *et al.* point to those large sections of the population whose economic seclusion was maintained by a hierarchy of oppression within or through the compliance of the state apparatus.

It is argued, then, that there is an absence of a theory of the state in regulation theory[3] stemming, first, from the way regulation theory idealizes the *form* and *functions* of the contemporary state and, second, from a heavy reliance on observations of the UK experience – an experience which Hutton (1995a) admits is unusual. Regulation theory fails to acknowledge that national economic strategies must be continuously managed and renegotiated since, *per se*, they are continuously involved with both accumulation *and* distribution questions.[4] Better abstractions of state process are needed. These need to draw on a multitude of state experiences and situate the realm of the state *within* accumulation processes. Further, there needs to be greater understanding of the ongoing nature of distributional processes, thereby avoiding the crippling placement of the state in a position where its capacities are limited to *ex post* income transfers through welfare assistance. In summary, what is needed is a theory of the state which says something about (a) the *way* the state functions; (b) how it stabilizes *and* transforms regimes of accumulation; (c) how it operates through geographic scales – not simply how it is constrained by scale; and (d) how it is, and should be, involved with *redistribution.*

# THE QUALITATIVE STATE

This chapter advances the paradigm of the *qualitative state* as a way towards a theoretical position which rejects the possibility of a privately constructed realm of freely operating markets and which asserts the indispensable role of the state in providing the means (*inter alia*) for privately performed production and consumption. It is a view that, as yet, has not found prominence in economic geography with the notable exceptions of Drache and Gertler (1991), Clark (1992), Christopherson (1993) and Marsden and Wrigley (1995). Block (1994) notes that a qualitative view of the state rejects the assessment or measurement of state role by *degree of intervention*. Wherever, and whenever, commercial transactions occur, the state plays a key role. Thus Block proclaims the arrival of a new state paradigm based on the concept of the qualitative state:

> The new state paradigm begins by rejecting the idea of state intervention in the economy. It insists instead that state action *always* plays a major role in constituting economies, so that it is not useful to posit states as lying outside of economic activity. (Block, 1994: 696; emphasis in original)

Developments in political theory which assert the qualitative role of the state draw strongly from the historical works of Karl Polanyi (especially *The great transformation*, 1957b). Polanyi demonstrated how, on the one hand, the rise of industrial capitalism could never have been a purely private process and, on the other, that economic institutions are inevitably political creations. He exposed the myth of spontaneous emergence of free-market relationships in post-feudal society. Instead, the emerging modern state is shown to have established crucial conditions for the operation of capitalist relations including exclusive property rights, a legal system based on the inviolability of contracts, the establishment of large national markets (through administrative structures, monetary systems and common standards such as in food purity, weights and measures), and the means for the penetration of other national markets, especially through imperialism (Block, 1990a, 1990b, 1991; *see also* Cerny, 1990: 206). The chief message that state theorists have drawn from Polanyi is that markets simply cannot operate in a *laissez-faire* environment.[5]

Block (1994) presents a detailed summary of views of the state which have led to the emergence of the qualitative state paradigm. The article is the culmination of a series of arguments about state role which establish four major tenets. First, *economy* is necessarily a combination of three events: markets, state action and state regulation (Block, 1990a, 1990b). A corollary of this constitution is that there is an infinite number of ways in which an economy can be organized. Second, although economic efficiency is dependent on markets, markets are state-constrained and state-regulated and thereby incapable of operating in a *laissez-faire* environment (Block, 1992). Third, neither capital nor the state is capable of achieving its goals simultaneously nor independently. Finally, it should be recognized that any coherence that exists about the idea of *economy* derives essentially from our cultural beliefs, which (in Anglo cultures at least) have led to constructions of economy being overlain with the dichotomy of *planned versus market*, which, in turn, has had the effect of denying the existence of multiple forms of economy (Block, 1993). This is an extension of the Marxist recognition of the power of economic ideologies to make particular economic arrangements appear as natural and inevitable.

These tenets stand opposed to the basic assumptions of what Block (1994) terms the 'old state paradigm'. Various forms of the old paradigm construct the state as occupying a position (*a priori*) external to the main economy. The idealized (or normative) *public goods state*, for example, describes the state as having the duty to provide the goods and services, such as blood or policing, which private markets are incapable of supplying efficiently and universally. The *macro-economic stabilizing* form of the state *intervenes* to adjust market aggregates, especially consumer demand, in order to move equilibrium (or market-clearing) positions of private markets closer to full employment.

Certainly, markets remain the best available device for aggregating individuals' commercial preferences. At the same time, state involvement is inevitable so that markets can be formed and operate efficiently. A common set of roles can be drawn up for all states in all economies. These include the maintenance of a regime of property rights; the management of territorial boundaries; the establishing and administering of legal frameworks to ensure economic co-operation; the provision of basic infrastructure; the creation and governance of financial markets and product markets; the ensuring of the production and reproduction of labour; the controlling of macro-economic trends; and the conduction of legitimation activities to guarantee the transition of the economic system through time and within a wider system of social events. Table 22.1 details the role of the qualitative state in the construction and maintenance of

**Table 22.1** Roles of the qualitative state in a modern economy

**A Maintenance of a regime of property rights**

i. maintenance of private property rights
ii. recognition of institutional property rights
iii. basic rules for the ownership and use of productive assets
iv. basic rules for the exploitation of natural resources
v. rules for the transfer of property rights (between individuals, households, institutions and generations)

**B Management of territorial boundaries**

i. provision of military force
ii. economic protection through manipulation of:
- money flows
- goods flows
- services flows
- labour flows
- flows of intangibles

iii. quarantine protection

**C Legal frameworks to maximize economic co-operation**

i. establishment of partnerships and corporations
ii. protection of intellectual property rights
iii. the governance of recurring economic relations between
- family members
- employers and workers
- landlords and tenants
- buyers and sellers

**D Projects to ensure social co-operation**

i. maintenance of law and order
ii. undertake national image-making processes
iii. other coercive strategies

**E Provision of basic infrastructure**

i. Provision or organisation of:
- transportation and communications systems
- energy and water supply
- waste disposal systems

ii. assembly and conduct of communications media
iii. assembly and dissemination of public information
iv. land use planning and regulation

**F Creation and governance of financial markets**

i. rules for the establishment and operation of financial institutions
ii. designation of the means of economic payment
iii. rules for the use of credit
iv. maintenance of the lender of last resort

**G Creation and governance of product markets**

i. regulation of the market power of firms
ii. the selection and regulation of natural monopolies
iii. the promotion and maintenance of strategic industries
iv. the provision of public goods
v. the provision of goods unlikely to be supplied fairly

**H Production and reproduction of labour**

i. demographic planning and governance
ii. provision of universal education and training
iii. governance of workplace conditions
iv. governance of returns for work
vi. social wage provision
vii. supply and governance of childcare
v. provision or governance of retirement incomes

**I Control of macro-economic trends**

i. fiscal policy
ii. monetary policy
iii. external viability

**J Other legitimation activities**

i. elimination of poverty
ii. maintenance of public health
iii. citizenship rights
iv. income and wealth redistribution
v. urban and regional development
vi. cultural development
vii. socialization
viii. enhancement of the environment

modern economies and demonstrates the vast and complex operations common to all Western states.

The analysis so far has been concerned with what the qualitative state *does*. Consideration must also be given to the *structures and mechanisms* of the qualitative state which enable the performance of the roles identified. Cerny (1990) notes the inadequacies in the many analyses of the state which assume that it has a hierarchical and centralized character (in contrast to freely operating markets, which are seen as pluralistic and atomistic). In contrast to these analyses, Claus Offe's work provides valuable insights into the *processes* by which the state engages with capitalism (for a review of Offe's view of state function, *see* O'Neill, 1996). There are two major thrusts in Offe's work. The first is his search for a useful theoretical analysis of economic crisis which incorporates the general questions of state authority and legitimacy as well as the functional problem of how states actually achieve their fiscal and welfare goals (*see* Offe, 1984: 65–6). The second is Offe's concern with the question of whether the state is capable of producing the means of overcoming the contradictions of capitalist production (*see* Offe, 1975). For Offe, state involvement is more than the actions of public institutions *on* various societal groupings. Rather, the state participates directly *in* the domains of other institutions and associations such as political parties, trade unions and corporations (Offe, 1976: 397), and in the processes by which social and economic interests are represented to government (Offe and Ronge, 1975; Offe, 1976). Not surprisingly, therefore, social turbulence and political resistance – threats to both capital accumulation and state legitimacy – are seen to be continuously internalized within the state apparatus (Offe, 1984) as it seeks to manage and distribute resources in ways that contribute not just to the achievement of economic growth but also to prevailing notions of justice. Offe's point is that the state is neither an arbiter nor a regulator nor an uncritical supporter of capitalism, but is 'enmeshed' in its contradictions (Held, 1989: 71). Capitalism is anarchic, requiring the state to sustain the processes of accumulation and protect the private appropriation of resources. The social processes necessary for the reproduction of labour, private ownership and commodity exchange, then, are regulated and sustained by *permanent* political intervention (Offe, 1976: 394). In other words, there is no possibility that the economic system can exist as a 'pre-political substrate' (Offe, 1976: 413).

Following Offe, Cerny (1990) also notes contradiction of state role: on the one hand, having to maintain legitimacy through the organization of collective activities and the pursuit of common goals and, on the other, relying on capitalism for financial viability. Like Block (1994), Cerny seeks a new paradigm for state theory – but being new or different by way of its treatment of state *mechanism*, not just role. And like Clark (1992), Evans *et al.*, (1985: viii) and Skocpol (1985: 27–8), Cerny insists that state theory should give major consideration to the structure and behaviour of the state's apparatus – a response to the lack of explanation for the behaviour of the administrative apparatus in normative, hierarchical and centred models of the state.[6] So Cerny emphasizes the importance of the organizational characteristics of the state and the ways these are manipulated by groups which seek (often competitively) to dominate the state's agenda. Thus, the state adopts two complementary roles. First, it provides the organizational structures through which groups can pursue their distinctive goals. Second, through its own organization and actions, the state influences the behaviours of other groups and classes in society. This is a view that accords the state a relatively autonomous structure as accumulation and distributional struggles are played out (a view similar to 'the middle way' argued for by Evans *et al.*, 1985, and Clark and Dear, 1984), even though the autonomy of the capitalist state is outstripped, from time to time, by the contradictions of capitalism (Cerny, 1990: 17).

Cerny's key contribution to a new qualitative state theory, then, is his insistence that the state is a decentred, non-totalized entity. The state's *apparatus* is the key vehicle for state functioning (Cerny, 1990: 43ff.), at one level underpinning the conduct of market transactions (Rueschemeyer and Evans, 1985: 46) and, at a deeper level, being the vessel for the complex political processes of negotiating accumulation and distributional pathways. Here, tension and conflict is normal as groups with different and competing logics seek to be included in, or control, state action.

So the state is more than just government. The state is constituted by continuous administrative, legal, bureaucratic and coercive systems that not only build relationships *between* the state and other groups in society, but also heavily influence relationships *within and between* these groups (Stepan, 1978, referred to in Skocpol, 1985: 7). Further, it should not be surprising that different states have different levels of power. The point is that these differences are produced less by extant economic conditions and more by the capacities of states to create or strengthen their organizations, to employ enough appropriate personnel, to co-opt political support, especially through programmes to assist economic enterprises, and to facilitate social programmes (Skocpol, 1985: 17). Further, these capacities are in no small part due to historical attitudes to governance and state role. That is, *qualitative* differences in states arise and are sustained by prevailing and historical structures and conditions influencing the state's apparatus.[7] Differences are not simply partial outcomes of the inexorable rise of the global financial system.

## THE STATE'S RESPONSE TO SUPRA-NATIONAL SCALE ISSUES

This section explores the impact on the state of the growing proportion of market transactions which are international. It is argued that international transactions reinforce the structure and importance of the nation-state, rather than diminish them (*see* Cerny, 1990). As Pooley (1991) demonstrates, there is considerable overlap between the trajectory of capital and the operations of the nation-state. First, the site of the production and reproduction of labour remains nation-bound, with the state playing the major role in determining the physical, ideological and cultural spaces in which investments are located. Second, because production always occurs at specific, solitary sites, its managers must negotiate with the state for the enforcement of behavioural norms for both capital and labour, and for the provision of basic infrastructure and so on. Third, the state maintains the structures and conditions in which capital circulates, including monetary systems, legal jurisdictions and the prevailing monetary and fiscal policy frameworks. Fourth, the state has a critical influence on realization conditions, including through financial and institutional stability, the level and distribution of aggregate demand, and through taxation and transfers policies. In other words, the roles of the state described in Table 22.1 require fulfilment irrespective of the nature or origin of an accumulation process.

Yet the emergence of global market-places has transformed states' role, and there is ample evidence that global market-places themselves are transformed by the involvement of nation-states in the creation and operation of supranational governance regimes and structures. Examples include nation-state involvement in the prudential supervision of international financial transactions through the Bank of International Settlements (BIS), the supervision of quality and safety in traded products through the operations of the International Standards Organization (ISO) and, of course, the governance of access to domestic product markets by the World Trade Organization (WTO). Thus the rise of a more prominent supranational tier of governance has required the *increased* involvement of nation-states, including, in many circumstances, *new* participation by apparatuses of the state at the level of local and regional governments, development agencies, sectoral-industrial and financial instrumentalities and so on. Paradoxically, then, pressures arising from an increasingly integrated world market-place reinforce and reconstruct state role rather than usurp it.[8]

The nation-state has also played a key role in the trend towards international regionalism as an avenue for successful international accumulation. Hay (1995) notes that this trend has been accompanied by the emergence of supranational state structures such as the North American Free Trade Agreement, the General Agreement on Tariffs and Trade and the European Union. In contrast to the Jessop argument that the emergence of supranational state structures is part of the hollowing-out process of the nation-state, Hay argues that tendencies towards strengthening supranational power structures are constrained by 'the fact that the inter-state bargaining required . . . is driven by the exigencies of maintaining *national* legitimacy bases' (Hay, 1995: 403; emphasis in original). In other words, national state agents pursue national rather than international interests in the global political arena. The irony, according to Hay, is that supranational bodies are incapable of intervening in the circulation of capital unless they are constituted by vibrant national state power. Moreover, global capital circulations *require* state structures in order for accumulation to proceed without persistent chaos. Hirst and Thompson (1992) make this point in arguing that the internationalization of financial markets and product movements has enhanced, not diminished, competition between national capitals whose competitive characters stem from distinctively state-structured national economies. Hay summarizes:

> To understand global political dynamics and their impact upon supra-national political agencies, therefore, we need to consider the *national* constructions of crisis arising out of *nationally* specific experiences of the exhaustion of modes of economic growth that in fact condition the interests reflected (differentially) in policy arenas at levels above the national. (Hay, 1995: 403; emphasis in original)

To assist their supranational operations, nation-states are devising methods which strongly contest both the depowering images of globalism (*see* Dicken *et al.*, this volume) and the growing contestability of domestic markets. Commonly, the nation-state moves to strengthen what its populations see as 'national'. It is involved increasingly in coercive strategies seeking societal approval of national economic change and to legitimize the adoption or reconstruction of a national accumulation strategy. These strategies build on pre-existing national identities which are historical products of myriad state policies dealing with immigration, foreign investment, sport, the arts, school curriculum, telecommunications, and so on. And, not surprisingly, many strategies are erroneous, for they involve the performance of relatively new roles to address new problems in new operational domains. Further, new state actions are not necessarily designed to produce more acceptable distributional outcomes. The argument here is simply that increasingly open and integrated national markets do not so much threaten or undermine the operation and effectiveness of state apparatuses as require that they undergo qualitative change. Critically, this qualitative shift is not optional, for it is fundamental for the continuation of successful accumulation processes and, in the absence of oppression, for the production of distributional outcomes that maintain legitimacy. Along the way, the process of adjustment produces problems which few state managers relish and which require much experimentation in ideas, management structures and cultures, and policies (Cerny, 1990: 5). Some discernible adjustment trends in the current period include a preference for micro- rather than macro-interventions; a shift from the protection of selected industries in specific market segments to the construction of internationally competitive conditions across markets; the adoption of enterprise cultures which promote innovation and competition, including in the public sector; and a shift in state expenditures towards the maximization of economic outcomes rather than the maximization of social welfare. Not surprisingly, there is considerable discontent among progressive groups in all nations with the distributional consequences of these trends.

Two consequences of thinking about the supranational scale in these ways emerge. The first concerns the role of the nation-state in translating activities from the supranational domain to production and consumption activities which take place within national boundaries. Not only does the state create the basic competitive conditions essential for successful accumulation, including trade rules, property rights and exchange rate stability, it also 'sutures' (Hirst and Thompson, 1995) or blends these conditions with emerging apparatuses from the supranational scale. Critically, though, the successful blending of domestic and supranational domains to produce stable national and international circuits of capital with desirable distributional outcomes depends fundamentally on nation-states combining to agree on common objectives and implement common regulations and standards (Hirst and Thompson, 1995: 426).

The second consequence is pointed to by Cerny (1990), who notes that national economic openness in trade, finance, information flows and communications produces an 'overloaded state'. This overload consists of problems which arise from the international transmission of recession, the incorporation of both private and public economic goals into an international context, and from struggles to maintain the political legitimacy necessary for national economic management when the traditional tools are found wanting. For example, a common constraint to the maintenance of legitimacy is the persistence of chronic public funding deficits which absorb national savings and exacerbate current-account imbalances. Another constraint is the incompatibility between centralized labour regulation and actions by firms to reduce real unit labour costs through labour shedding, new shift patterns and outsourcing. A third constraint stems from demands for state assistance in maintaining or restructuring unprofitable economic sectors especially through direct subsidy and micro-economic reforms. Cerny concludes that alongside growing internationalization, 'the total amount of state intervention will tend to *increase*, for the state will be enmeshed in the promotion, support, and maintenance of an ever-widening range of social and economic activities' (Cerny, 1990: 230; emphasis added). However, Cerny adds that 'The domestic redistribution of wealth and power, which is at the heart of the social democratic welfare state, will become more difficult and complex to achieve' (ibid.: 231).

Continued crisis in capitalism, then, maximizes, rather than reduces, the demands for state intervention. Predominantly, however, state intervention seems to favour the types of accumulation practices which involve distributions of incomes in favour of capital – an outcome at the heart of neo-liberalist discourse. The 'overloaded state' is faced with addressing distributional problems through targeted welfare assistance from a distressed public budget. In contrast, a reconstructed discourse which takes the view that the state and its apparatus are inherent to the processes of production and consumption, and not lying outside them as detached overseer, regulator or undesirable intruder, offers opportunities for the identification and manipulation of distributional outcomes at the point of the accumulation process.

# PROPOSITIONS ABOUT DISTRIBUTION

Because capitalism is incapable of an existence outside the realm of state action and because capitalist processes involve distributive processes *per se*, then the state is always involved in redistribution activities. Hence, the argument that the internationalization of the world economy is a natural tendency of modern capitalism resulting in depowerment of the nation-state is simply a restatement of a coercive discourse designed to defer or deny the benefits of restructuring to particular groups. This section of the chapter advances this argument through three propositions about the distributional opportunities of the paradigm of the qualitative state.

The first proposition is that the state and its apparatus constitute an arena for the struggle over distribution (*see* Offe, 1976, 1984, 1985; Offe and Ronge, 1975). Importantly, though, at the same time as being enmeshed in capitalist production and exchange, the state is driven by the need to preserve its own autonomy and power as the arbiter of class conflict and the sustainer of decommodified social production. Accordingly, it is not because of its being a servant of capitalism that makes the state interested in successful accumulation; successful accumulation is critical for the sustenance of the state's own interests, not the least reason being the state's reliance on economic growth for the provision of taxation revenues particularly for redistribution purposes (O'Connor, 1973). A struggle then ensues. The state seeks to fund fiscal actions promised to the electorate. Capital resists regulations which inhibit capacities to secure advantage over competitors, to extract surplus value from labour and to minimize its distribution to consumers. Thus the state's dilemma is the maintenance of the accumulation process (which ideally seeks *minimal* state intervention) while successfully pursuing legitimization goals (which ideally require *maximum* state intervention). In other words, the state has to engage simultaneously in commodification and decommodification.

It can be concluded from this reasoning that, while only the state has the power and apparatus to organize economic spheres of action, the state's redistributive actions inside market-places are continuously opposed by agents of capital which claim that economic spheres should be held as 'natural and inviolable' (Offe, 1976: 395). On the other hand, the state's 'natural' domain is seen as being within social systems such as education, health and welfare which render the 'ingredients of a "decent" life' (Offe, 1975: 256). The agents of capital contend that the state's functions should be funded by the state's appropriation of revenue *after* the redistribution processes inherent to accumulation have occurred. Company taxation, for example, is levied on the basis of *net* income flows *following* a financial year of economic activity. Accordingly, when the state accepts a position of *ex post* distributor, it will always suffer fiscal crisis during downturns in the economic cycle when there are increased distributional demands and falling revenues. Hence, the successful simultaneous performance of state functions is impossible for any length of time. Instead, the state is forced to make a constant attempt to reconcile their contradictions with its own internal structures and modes of operation (Offe, 1975: 144).

A second proposition that must be inserted into thinking about the qualitative state and its inherent distributive functions is that the state plays a critical role as a coercive instrument during periods of economic restructuring. Webber *et al.* (1991) and O'Neill (1997) show how restructuring stories gain political hegemony within and through the apparatus and emerge from time to time to drive economic policy towards particular accumulation and distribution outcomes. Clark *et al.* (1992) argue that the nation-state is the only organization capable of producing the societal co-operation necessary for a successful transition through economic restructuring. The authors identify a range of state coercive strategies. The most subtle forms are *subordination* and *idealization*. These involve gaining the co-operation of workers and citizens through public arguments which extol the virtues of a particular restructuring pathway and warn of otherwise unpalatable consequences. At the other extreme, workers' and citizens' compliance with change can be achieved through *intimidation*, such as by threatening industry-wide closures, and *domination*, involving the use of direct force such as military or police action to defeat extant opposition.

Less systematically, Hirst and Thompson (1995) point to globalization as presenting an attractive image or story to conservative politicians especially in the parts that argue that local labour needs to acquiesce to the demands of international capital and world competitive pressures. This type of discourse displaces traditional national social democratic strategies and active macro-economic policies. The point is that the construction of the idea of a market economy that is rational, self-constituting, self-regulating and independent of the political sphere is a *normative prescription*. The idea, then, is available for those who demand that only a minimalist, non-interventionist, night-watchman state can accompany a maximum performance economy. Thus, the state is at the centre of discourses of economic restructuring which, in turn, generate distributional outcomes.

A third proposition picks up from Jessop (1994a) and argues that since internationalizing processes involve an expansion of nation-state organization, then internationalization presents greater, not fewer, opportunities for state intervention into distributional outcomes. As has been argued above, there is growing evidence that global circulations of capital, goods and services, ideas and people require supranational state structures and apparatuses. In turn, these depend on the vibrancy of nation-states for legitimacy and power. Only nation-states have territorial authority to deal with the social outcomes, including conflict, which inevitably follow internationalization processes. Obviously, the discourse of internationalization used by nation-states is critical here. Finally, local and regional authorities have little chance of pursuing their international interests unless nation-states 'suture' the supranational domain to the national territory. Greater realization of the role of the nation-state in internationalization processes can lead to improved distributional outcomes in regions which have often suffered from the timid explanation that events causing local economic devastation arise from an uncontrollable global capitalism.

## CONCLUSION

Thinking about the *qualitative* state offers new opportunities for invigorated state action for more desirable distributional outcomes. It involves accepting the autonomy of the state; accepting the crucial role of the state in the governance of private markets; accepting that the state is not a homogeneous unit but exists as a contested domain continuously interacting with society; and accepting that internationalization is not a singular logic of capitalism as investment (allegedly) flees collapsing Keynesian-Fordist national economic spaces. Thinking about the qualitative state also involves rejecting the notion of a hollowed-out state in the sense that internationalization has made the nation-state redundant as a macro-economic manager; and rejecting the logic that redistribution is an act which follows accumulation processes, its extent dependent on their success. Finally, the idea of the qualitative state is not a question of bringing the state back in. Close empirical work shows that the state never departed. The crucial question concerns the ways in which we have represented the state during the recent decades of economic crisis when commercial transactions have become increasingly internationalized. New discourses about the qualitative state have the potential to enhance opportunities for intervention into economic processes and make them more successful – especially when judged by their distributive outcomes.

## ACKNOWLEDGEMENTS

I would like to thank Rod Francis and Kris Olds for referring me to key articles and the editors for their very helpful comments on an earlier draft.

## NOTES

1. Perplexingly, Jessop claims that neo-liberalism creates a 'political vacuum' with state intervention 'operating indirectly and at a distance' (Jessop *et al.*, 1990: 94). Jessop misses the key point that neo-liberalism is only ever the illusion of 'hands-off' in order to deflect distributional concerns to other scales such as the local region and to other organizations such as voluntary agencies. However, Jessop acknowledges the contradiction between neo-liberalist discourse and action in accounting for pathways of reconstruction in the Eastern European economies (*see* Jessop, 1995b: 675).
2. At the same time, however, Jessop argues that the nation-state is 'best placed' to deal with redistributive policies (e.g. Jessop, 1994a: 263), acknowledging Offe's assertion that 'Capitalism cannot coexist with, neither can it exist without, the welfare state' (Offe, 1984: 153, cited in Jessop, 1991: 104). He also qualifies the hollowed-out metaphor by acknowledging that the restructuring pathway for the welfare state is far from clear and, in any event, involves contested processes and 'continual changes in the articulation of government and governance' (Jessop, 1995c: 1624).
3. This is an omission conceded by Hay (1995), Hay and Jessop (1995), and Weiss and Hobson (1995).
4. It is noteworthy that the process of continuous negotiation of economic crisis was central to Aglietta's (1979) original account of US Fordism.
5. Block and Somers (1984) and Cerny (1990) provide excellent reviews of Polanyi's works and discussions of the contemporary applications of his arguments. In addition, Goodman and Honeyman (1988) provide a detailed account of the rise of industrial Europe with close parallels to Polanyi's arguments on the role of the state in building international trade

pathways during the seventeenth and eighteenth centuries.

6. An important qualification to the view of an active state apparatus as forming the basis of state actions is provided by Rueschemeyer and Evans (1985), who note the 'functionalist trap of assuming that because the state is "necessary" it will therefore have the inclination and capacity to fill the required role' (p. 46).
7. A variation from this literature on state role in constructing economy is the concept of governance. This extends the state as a domain of control to include private and other non-state actors, institutions and practices (Hirst and Thompson, 1995). Jonas (1996), for instance, points to the role of non-state agents in local labour market governance. Yet acknowledging the potential for governance to be extended to non-state instrumentalities should not be seen as a diminution of state power. Rather than being supplanted, the state should be seen as *not* home alone!
8. Picciotto (1991) claims that transformations in state role are historic phenomena which pre-date the contemporary period of international capitalism. He rejects any characterization of a correlation between national capital and the national state, and international capital and the international state, pointing instead to the historic development of 'loose and overlapping jurisdictions' (1991: 47) across national territories including an international hierarchy of state regulations, the annexation and colonization of new territory, mutual legislation, and the co-ordination of regulatory frameworks to assist realization and accumulation on an international scale. In illustration, Piciotto cites the harmonization of intellectual property laws in the mid-nineteenth century in the context of a wave of scientific innovation, similarity in the construction of labour laws prior to 1915 in the face of new forms of labour organization, and anti-trust laws associated with the international transfer of the thrust of New Deal policies in the post-Depression USA.

# CHAPTER TWENTY-THREE

# THE END OF MASS PRODUCTION AND OF THE MASS COLLECTIVE WORKER? EXPERIMENTING WITH PRODUCTION AND EMPLOYMENT

RAY HUDSON

## INTRODUCTION

There has been growing recognition of the changing character of work in the advanced capitalist countries (Beynon, 1995), and in the decline of industrial employment there as it expands in newly industrializing countries (Dicken, 1992). These changes in the volume, character and spatial distribution of employment are one facet of profound changes in the character of contemporary capitalism, manifestations of 'the crisis of Fordism' and the search for post-Fordist successors to it. This is the case both at the micro level of a particular form of organizing production within the workplace and at the macro level of a model of societal political and economic development. The meaning and status of the changes continue to be hotly debated. Broadly speaking, there are competing alternative interpretations, with different implications for labour and the spatial organization of the economy. For some, a shift of epochal significance has occurred. Fordism's successor is already known, although the terminology used to describe it varies. One implication of this is that the mass collective worker, characteristic of the big urban factories of Fordism, is a thing of the past, disempowered in and by the new economy of decentralized small flexible firms (Murray, 1983). Others contest this, arguing that the death of Fordism may have been prematurely announced, based on a partial reading of the evidence. Seen from this perspective, large-scale production in big factories is far from being a thing of the past. The search for Fordism's putative successor(s) remains an ongoing process and may or may not represent some epochal shift. Claims to the contrary are premature and potentially damaging, theoretically and politically. The shape of the future remains to be determined, but its developmental trajectory will be more complex than a neat, clean break from a Fordist past to a post-Fordist future. This is so for two different sorts of reasons. First, even at the high point of Fordism as a macro-scale development model, only a minority of labour processes were organized on Fordist lines (Pollert, 1988). 'Fordism' was constituted as an uneven mosaic of places, production and labour processes. Second, as Fordism reached its limits in the core countries of capitalism, companies began to search elsewhere for locations in which Fordist production would remain economically viable. This intersected with the desires of governments in peripheral countries (initially those which became characterized as the first generation of newly industrializing countries and more latterly others such as China) to promote industrialization as the route to development. The crisis of Fordist production in the core of advanced capitalism thus became the proximate cause of changes in its location. It continues, but nowadays in new locations in the peripheries of the global economy. Consequently, the mass collective worker has not necessarily simply become a subject of history.

As these are ongoing changes in the character of a *capitalist* economy, understanding them requires a political-economy approach that acknowledges this and also that there are limits to capital (Harvey, 1982). The ways in which, and the perspectives from which, we represent the world are certainly important. These are, however, competing accounts of a material world, socially produced according to 'rules of the game' in which people make historical geographies of employment and production. The economy remains subject to the structural class relations and boundaries that define capitalism. Consequently, class relations, especially those between capital and labour in the labour market, in the wage relation and at the point of production, remain of central significance. Class relations in the workplace can be reconfigured but they cannot be erased. It is for this reason, and because such relations are treated in a one-sided and idealized fashion which seeks to deny their class character in much of the literature on new 'post-Fordist' forms of work, that they are the focus of attention here. Companies must be able successfully to purchase labour power and then organize and deploy it to ensure the production of surplus value and the possibility of profitable production. This emphasis on the central importance of the class relation between capital and labour, of value analysis in emphasizing the asymmetrical but mutually defining power relations between capital and labour, may smack to some of essentialism (Barnes, 1996). But insistence upon the pivotal significance of capital–labour relations in a capitalist economy does not deny the significance of other processes and relationships which are deeply involved in the reproduction of this class relationship; nor does it reduce them to mere reflections of it (cf. Massey, 1995a). Capitalist societies are not *simply* divided along class lines. There are also divisions along dimensions such as gender and ethnicity, which are intertwined with those of class in a variety of ways. Attempts to understand the ongoing restructuring of production and work that fail to acknowledge the class basis of these changes are, however, at best partial, theoretically impoverished and politically dangerous.

Capitalist societies and the conditions that allow production to take place within them are not, however, automatically reproduced. The constitution and representation of the social relations of capitalism remain contested and a focus of struggle. Accounts which deny this, or gloss over its implications, do not do so innocently. It is, therefore, important to acknowledge the processes and institutions through which the class basis of production and wage labour is reproduced and regulated and through which contradictory interests are kept within limits that permit profitable production. While regulation involves more than the activities of states, these are integral to the institutions and processes that, in different ways, help make production possible. Social relationships within households and the institutions of civil societies are, however, also central to the reproduction of labour power and are bound up with the (re)formation of class relations, as the spheres of production and social reproduction are reciprocally, but not necessarily equally, determining. State policies help define the shifting boundaries and articulations between commodified and non-commodified social relationships. One implication of this shifting articulation, as well as of the different forms that class relations can take, is that the broader social relations of capitalism can be cast in varied moulds within these structural limits.

Given the continued salience of class relations in production, the focus in this chapter is therefore upon changing forms of work and competing interpretations of the implications for labour of capital's attempts to experiment with and find new methods of high-volume production (HVP) in manufacturing. Other aspects of contemporary economic restructuring and employment change such as the search for alternatives to mass production,[1] the impacts of human resource management practices on employment in parts of the services sector, or indeed the Taylorization of large swathes of service sector employment (Beynon, 1995), will be considered only in passing. This is not because they are unimportant but rather because of the continuing significance of large-scale industrial production and the leading-edge role that this has occupied in debates and discourses about the redefinition of work. The focus is micro- rather than macro-scale, although some consideration will be given to those macro-scale conditions that make micro-scale changes both necessary and possible. The remainder of the chapter is organized as follows. First, the broad lineaments of HVP will be outlined. Second, the implication of HVP for the character of work and geographies of employment will be discussed. Third, the reconfiguration of the collective worker and the implications of new forms of work and industrial relations practices for labour will be explored. Finally, some conclusions will be drawn.

## EXPERIMENTING WITH NEW MODELS OF HIGH-VOLUME PRODUCTION

The term 'high-volume production' (HVP) denotes approaches to production that seek to combine the benefits of economies of scope and greater flexibility in responding to consumer demand which are characteristic of small batch production with those of economies of scale characteristic of mass production. They include approaches designated as just-in-time (Sayer, 1986), 'lean' production (Womack *et al.*, 1990), dynamic flexibility (Coriat, 1991), flexible automation (Veltz, 1991) and mass customization (Pine, 1993), which, with its ambition of batch sizes of one – uniquely customized commodities assembled from mass-produced components – epitomizes the goal of HVP. While typically presented as distinctive, all these different HVP approaches emphasize the emergence of new structures of relations between companies and of new forms of work, work organization and industrial relations practices – in part because they typically refer to the same set of exemplar companies and industries. Since one of their shared defining features is a blurring of the distinction between mass and craft methods of production, HVP approaches also show considerable continuities with, as well as differences from, existing approaches such as 'just-in-case' Fordist production and small batch production. In practice they combine elements of 'old' methods of production with new production concepts and practices, both within and between firms. These innovations in production methods have often been associated with changes in corporate anatomy. Growing company size as a result of acquisitions and mergers and an increasing prevalence of strategic alliances among these bigger companies in search of economies of scale (in R&D, product development and so on) and scope has often been a prelude to experiments with HVP, with the how and where of production (Rainnie, 1993).

These new approaches have typically involved a reorganization of production in new factories which achieve enhanced labour productivity via some combination of new fixed capital investment in more automated production technologies, more intensive ways of organizing work and the labour process (so that employment declines), and more efficient ways of processing material inputs. These *are*, however, forms of *high-volume* production, incorporating variations around the basic mass production theme. Consequently, there are strict limits, in terms of the material and social requirements of commodity production, that define the range of industries and products in which HVP approaches *can* be applied. Companies such as Dell and Motorola may provide commodities such as pagers, PCs and workstations on a mass customized basis (Pine, 1993). Other major companies, such as Ford, are seeking to move more towards mass customization as an automobile production strategy. It is difficult to see how soap powder or screws could be profitably produced in this way, however, and they will continue to be mass-produced. Conversely, other commodities, major items of fixed capital equipment such as power-stations, will continue to be produced on a 'one off' basis while exclusive fashion goods will continue to be produced in small quantities in small production units. Consequently, there is not *necessarily* a complete divide between small batch production, mass production and various forms of HVP.

Equally, while these new HVP approaches share characteristics in common, there are also significant differences between them. Companies thus face a choice in deciding the what, how and where of production. This has definite implications for geographies of employment, the amounts and types of waged work on offer, and forms of organization of the labour process. Furthermore, the economic viability of such approaches assumes that certain labour and product market conditions will be fulfilled. The choice of a particular HVP approach involves seeking a balance between responding to more differentiated product markets via economies of scope and the need to retain scale economies. For example, especially in those approaches, such as flexible automation, that encompass very highly automated production processes, there are (usually tacit) assumptions as to very high levels of demand and of capacity utilization to allow fixed capital investment to be depreciated sufficiently quickly. In addition, the introduction of such automated approaches has sometimes been in response to tight labour market conditions (for example, as prevailed in the Japanese auto industry in the 1980s) in circumstances in which relocating in order to maintain the competitiveness of existing production technologies was infeasible. As a consequence, the introduction of robots to replace human labour in automated production processes has been very uneven, sectorally (concentrated in automobiles and also electronics) and spatially (with the greatest concentrations per worker in Japan, Singapore, Sweden, Italy and Germany; United Nations Economic Commission for Europe and International Robotics Federation, 1994). Conversely, there is evidence of companies switching to rather different automated HVP technologies in the 1990s, with robots as an adjunct to rather than a replacement for human labour at the point of production. These offer greater scope for combining flexibility with profitability as aggregate levels of demand have declined and/or as labour market conditions have allowed the requirements of such HVP approaches for very specific types of worker to be satisfied (*see* Hudson and Schamp, 1995).

## WORK, WORKERS AND HVP AND ITS GEOGRAPHIES

The literature on HVP has paid considerable attention to new forms of relations between companies (Crewe and Davenport, 1992; Hudson, 1994). Restructuring the links between companies impacts upon relations between workers, upon those between workers and managers, and upon forms and conditions of work. These changing forms of work and of capital–labour relations reveal both breaks from and continuities with existing forms of work and the ways in which it is organized.

From one point of view, HVP approaches are represented as abolishing the mass collective worker and class conflict at the point of production, incorporating more satisfying, individualized forms of work. From another perspective, they are seen as transforming the forms of, rather than abolishing, class conflict at the point of production and reshaping rather than removing the collective worker. These competing claims about the character of employment in HVP approaches can illuminated by comparing it with the character of employment under Fordism. Fordism is characterized by a deep technical division of labour, a sharp distinction between occupations requiring mental and manual labour, informed by Taylorist views of scientific management within companies. It is marked by a strong vertical hierarchy of control; individual production line workers are confined to single, specialized – often deskilled – tasks within a finely disaggregated detail division of labour (Braverman, 1974). Management relies upon increasing line speed and intensifying the labour process in this way as the route to greater labour productivity. This leads to extremely monotonous jobs, which fail to capture the knowledge that workers develop through doing the job. This can result in problems – for capital – of alienation, lack of motivation, and resistance from workers, disrupting production via strikes and other forms of industrial action, resulting in decreased productivity and profitability (*see*, for example, Beynon, 1973). This is especially so in large factories, locationally concentrated in major urban areas, in labour market conditions of 'full employment' – in short, in circumstances in which the capacities of the mass collective worker spontaneously to organize and resist the demands of capital are favourable. Such conditions are not, however, automatically realized, and capital deploys a variety of strategies, in terms of the how and where of production, precisely to prevent them from being brought into existence. The invention of new methods of HVP and associated ways of working constitutes one element in this repertoire of tactics.

Alternative forms of HVP require workers to perform a wider range of tasks than on the Taylorist line; in that sense the technical division of labour is not so deeply inscribed. This is partly because of the incorporation of principles of 'just-in-time' production into HVP approaches, but there are other reasons associated with claims about producing better and more satisfying forms of work. There is considerable debate as to the implications of this change for those actually carrying out the tasks of production. There are those who see workers as multi-skilled and much more creatively involved in the process of production. Companies (and their intellectual supporters, such as proponents of human resource management (HRM)) emphasize that these new forms of work provide, from the perspective of workers, better jobs in an empowering environment, built around the themes of flexibility, quality and teamwork (*see* Wickens, 1986). There are strong claims, which allude to an alleged golden age of craft production, about the re-emergence of multi-skilled polyvalent workers, employed in jobs which recombine the mental and the manual which Taylorism had torn asunder. Florida (1995: 168), for example, writes, approvingly, of the emergence of 'high-performance manufacturing' in sectors such as automobiles and consumer electronics in the USA. He associates this with a shift to a more knowledge-intensive economy in which the keys to success are harnessing the ideas of all workers from the R&D laboratory to the factory floor to create the high-quality, state-of-the-art products that the world's consumers want to buy.[2] Under this new form of organization, the factory itself is said to be becoming more like a laboratory, with knowledge workers, advanced high-technology equipment and cleanroom conditions free of dirt and grime. This does indeed powerfully suggest that the Taylorist distinctions between manual and mental labour are being swept away to the advantage of all workers.

Others dispute this. They stress not the qualitative differences from the old Taylorist model but rather the continuities between the old and the new. In part, this is because of the inconsistencies between the claimed advantages of the new forms of work and the realities of the labour market and the labour process (Blyton and Turnbull, 1992; Peck, 1994; Pollard, 1995). Critics contend that what is involved is not multi-skilling but multitasking, part of a search for new ways of intensifying the labour process. The production line keeps running all the time but not *necessarily* at its maximum possible overall speed. Indeed, the aim is not to maximize line speed but to minimize the number of workers needed for a given speed, as dictated by the implementation of just-in-time and 'lean' production principles. The emphasis in lean production on 'halving the human effort' has drastically reduced the number of jobs available. In this way is the labour process intensified and, in terms of its direct inputs of labour power, production becomes 'lean'.

There are therefore serious doubts as to whether the jobs on offer in these factories are actually any better than those on Taylorized mass production lines. In contrast, critics claim that the new jobs display very clear parallels with the discredited employment forms of Fordism, and indeed can be thought of as an intensified extension of them (for example, see Garrahan and Stewart, 1992: 125). From the point of view of labour, therefore, HVP approaches involve greater intensification of work and greater stress than before (*see* Okamura and Kawahito, 1990). There are fewer jobs on offer and greater competition for them. The notion of 'lifetime employment' has ceased to be meaningful for the majority of employees, not least because intensification of work leads to a physical incapacity to meet productivity norms with increasing age, and the average age of workforces has fallen sharply. As there are fewer jobs and so greater competition for them, especially in locations blighted by high unemployment, firms can be extremely selective about whom they recruit, and about the terms and conditions on which they offer employment. Against a background of high unemployment, recruitment is typically based at least as much on 'appropriate attitudes' and expressed commitment to the company, and age, personal and family circumstances and physical fitness, as it is on requisite technical knowledge and skills. As well as changing the terms and conditions on which full-time jobs are offered, there is a tendency to replace full-time with part-time and casualized jobs, as a further stage in customizing labour supply in relation to fluctuating demands in product markets. If the new forms of work and HRM approaches within HVP are based upon fostering loyalty to the company, then this is a loyalty born of fear of unemployment in circumstances where individual jobs are no longer guaranteed. This suggests a commitment that may be shallowly based and very much dependent upon the context of labour market conditions.

Enhanced selectivity in recruitment in turn allows radical and regressive changes in the organization of the labour process. Work is organized within a disempowering regime of subordination, characterized by control, exploitation and surveillance, bound together through team working (Garrahan and Stewart, 1992: 109–11). There is, however, a tension between an emphasis on team work and increased individual competition for jobs, promotion and pay. There are strong pressures for workers to regard themselves as competitive individuals, dealing with the company individually rather than collectively over wages and terms and conditions of employment. At the same time, through the rhetoric of team work, workers discipline themselves and their colleagues (identified as their 'customers' further up, or 'suppliers' further down, the line). This in itself both increases stress and changes the nature of the mode of regulation of the labour process. No longer is it 'us' versus 'them'; 'them' are now part of 'us'. Considerable ambiguities and uncertainties follow from this change of identities, not least in terms of forms of organization and representation of workers' interests. In so far as there is collective representation of workers' interests, there are pressures for this to be via works councils rather than trade unions – although the latter may still be permitted. The net result is that jobs within HVP approaches are actually worse jobs than those offered within Taylorized mass production approaches.

Thus, to summarize the argument so far, these new forms of HVP must be understood both as a response to a profound crisis of mass production and as enabled by it. As part of an attempt to restore profitability, they have to respect and so reproduce the defining structural limits and parameters of a capitalist economy. This implies the structuring of work and the labour process so that profitability criteria are met. In the process they redefine the social relations of production. Whatever the rhetoric about empowerment of workers, production has to be organized in such a way that it is sufficiently profitable. In order to introduce new ways of working, there have been corresponding changes involving greater selectivity in recruitment, putting more emphasis on appropriate attitudes, making sure companies hire the 'right' people on the labour market – workers who will accept and adapt to new ways of working. At this point, links to broader macro-economic and geographical contexts become crucial. High unemployment, spatially concentrated, both in deindustrialized old industrial areas in which the power of the mass collective worker can be broken, and in non-industrialized areas in which it never existed, provides the context in which the texts about employment practices can be rewritten.

A crucial pre-condition for companies being able to introduce drastic changes in the organization of the labour process therefore has been either to find locations untainted by an earlier history of capitalist industrialization or to break the power of existing forms of workers' organizations in areas with such a history. The historical geography of Fordism clearly demonstrates these twin tendencies, as mass production of consumer goods was established in new locations, and companies subsequently sought a series of spatial fixes to find viable new locations for such production once it became problematic there. Similar points can be made about other forms of production organization. The defeat of the militant Japanese trade unions in the 1950s allowed the introduction of innovatory new forms of HVP by automobile producers there, for example. The selective location of experimental new HVP plants in parts of the European periphery in the 1980s and 1990s, often directly or indirectly associated with inward investment from South-East Asia, is another example (Hudson, 1994; Hudson and Schamp, 1995). The creation of locationally concentrated mass unemployment has often been a crucial precondition for attracting such investment via erosion of the power of established trade unions. This again emphasizes the significance of relationships between state policies, macro-economic conditions and feasible factory regimes of production (Burawoy, 1985). In particular, it highlights the importance of the replacement of Keynesian welfarism by neo-liberal workfarism in establishing labour market conditions which made the experiments with new forms of HVP possible over much of the advanced capitalist world.

The geographies of employment associated with the introduction of HVP show that spatial variation in labour market conditions is clearly a critical consideration in decisions as to whether, and where, to introduce new forms of HVP. Localized high unemployment has been equally important in permitting established firms with more 'traditional' forms of work and labour organization to restructure and introduce new ways of working, informed by the precepts of HRM, often in response to the competitive challenge of HVP. The literature on geographies of HVP tends to emphasize the importance of new forms of relations between companies rather than forms of capital–labour relations, however. It initially seemed that just-in-time approaches and close inter-firm linkages would lead to a regional reconcentration of production, offering growth opportunities and a considerable increase of new sorts of industrial jobs in at least some peripheral regions. This could be seen as heralding an opportunity for the resurgence of the mass collective worker. While acknowledging the emergence of new forms of relations between companies, such an interpretation underestimates the significance of locationally concentrated high unemployment in attracting such clusters of factories in the first place. Their production strategies are predicated on the ability selectively to recruit particular types of employee and prevent the reconstruction of a militant mass collective worker.

There is no 'obvious' geography to other new HVP methods such as lean production, flexible automation or mass customization in the way that there initially seemed to be with just-in-time. The relationship between location and labour market conditions is indeterminate and variable, though it may well involve a sophisticated use of spatial differentiation. As the strategies of companies such as Motorola and Dell reveal, there is considerable flexibility over choice of production location, enabled by advances in information technologies in communication and production (Pine, 1993). As a concern with economies of scale remains central in many industries, lean production almost certainly means fewer factories. There will be intensified place market competition for them, though *where* these factories might be located remains an open question. From one point of view, one might expect heavily automated assembly plants, requiring only relatively small inputs of living labour, to be drawn to 'core' regions, near to the main markets of contemporary capitalism. The greater weight attached to the role of R&D and the increasing emphasis placed upon post-sales services as part of the product could add to the attractions of core locations for mass customized production. On the other hand, the ready availability of labour in peripheral locations, facilitating continuous shift working and new working practices, and the availability of substantial state financial subsidies make *these* attractive locations (Conti and Enrietti, 1995; Schamp, 1995). Furthermore, labour-intensive production, organized on classic Taylorist principles, will continue to find the cheap labour peripheries an attractive destination.

## SO IS THIS THE END OF THE MASS COLLECTIVE WORKER?

Fordism was created as a way of breaking the power of the craft worker to control the production process in the 'traditional' heartlands of industrial capitalism. This it did, at least in certain industries, times and places, exploiting existing spatial variations to create its own distinctive geographies of employment. At the same time, however, in concentrating production in massive factories in major conurbations, Fordism created the conditions for the emergence of a militant mass collective worker that in due course threatened its viability as a production strategy in what had become its 'traditional' heartlands in the advanced capitalist world. Consequently, from the moment that the mass collective worker emerges as a threat to it, capital has striven to achieve '*the destruction of the spontaneous organization* of the mass worker on a collective basis' (Murray, 1983: 93; emphasis added). Capital remains, of necessity, committed to this objective of destroying challenges to its power to organize the production process on its terms. Capital is not, however, necessarily opposed to the existence of the collective worker *per se*. Many companies continue to require large numbers of workers organized into individual workplaces, and there are limits to the extent to which they can deal with them and organize their work individually, as opposed to on some sort of collective basis. There are tendencies to underestimate the extent to which production must remain a collective enterprise, both in the HRM literature and in some of the critiques of it. One implication of the continuing collective character of production is that spatial variations in labour market conditions remain of great significance to companies in their search to produce a compliant collective worker.

One expression of capital's continuing requirements for large amounts of malleable labour is the growing shift of routine mass production to branch plants in parts of the peripheries of the First World and into some parts of the Third World. There has been a succession of such fixes, first intranationally within the advanced capitalist world, then internationally, as companies have sought to contain crises of profitability via decanting routine production into peripheral branch plants (Hudson, 1988). Capital's latest search for a spatial fix to preserve the viability of mass production and re-create the mass collective worker involves investment in locations such as the Special Economic Zones of China. With labour costs in industries such as clothing and textiles in China a mere 2 per cent of those in Germany, and those in other parts of the Far East less than 4 per cent of German levels (according to Coats Vyella: *see* Rich, 1996), there is considerable scope to preserve mass production strategies on the basis of the availability of very large masses of very cheap labour. While such shifts in production challenged the powers of trade unions both in their 'traditional' manufacturing heartlands and in those peripheral locations in which earlier branch plants were located, they also open up potential opportunities, at least for a while, in new locations. Such potential has often remained latent, however, as a consequence of broader macro-economic and political conditions in the new destination areas.

At the same time, within the First World, different strategies are being followed. The emergence of new forms of HVP in the First World involves the reconfiguration and re-formation, in slimmed-down form, rather than the death of, the collective worker. While the current 'round' of HVP factories require less labour than their Fordist predecessors of thirty or more years ago, many still require a lot of labour. While companies can be much more selective as to whom they recruit and retain as compared to the decades of 'full employment', workers still need to be organized in ways that conform to capital's needs for profitable production. The rhetoric may emphasize individually more satisfying work, but producing profitably necessarily remains a collective and social process that has to be regulated to conform to the disciplines of commodity production. Companies therefore still need workers with *particular* collective forms of labour organization and activity that align the interests of workers with, rather than against, those of the company. Therefore companies searching for viable models of HVP of necessity have sought out new ways of disorganizing and then reorganizing labour, typically with the assistance of national states and their restructuring of regulatory regimes. Consequently, HVP approaches both require and permit the shattering of old forms of trade unionism and the institutions of labour and their recasting in new no-union or one-union moulds. There is a much greater emphasis on plant-level or local rather than national-level bargaining, in part linked to competition between places for jobs as well as between unions for members. Trade unions have often been willing to trade off sole bargaining rights for various 'sweetheart' deals as one way of combating their own falling memberships. Furthermore, 'selling places' often involves emphasizing the passivity, flexibility and malleability of their workers as compelling attractions to mobile capital. These new deals are thus grounded in a conception of capital–labour relations very different from that previously dominant, as the already asymmetrical power relations between capital and labour have swung sharply in favour of the former but in a context in which production none the less remains a collective and social process.

This situation is not, therefore, without dangers for capital and potential opportunities for labour. On the one hand, the new models of production are undeniably predicated on there being *no* return to 'full employment'. Such labour market conditions would at a stroke destroy capital's capacity to be so selective in recruiting labour and deciding which individuals will become incorporated as part of the collective worker. Former trade union strategies based on organizing workers in large industrial plants on a national basis in a 'full-employment' economy have undeniably been weakened. Companies are nevertheless acutely vulnerable to interruptions to production in 'lean' production strategies built around vertical disintegration, just-in-time principles, sole-supplier deals and minimal stock levels. These characteristics can lead to production being compromised by rapid labour turnover and to companies being particularly susceptible to the effects of industrial action. This was sharply demonstrated, for example, by strikes at Ford's UK factories in 1988, which soon disrupted production in Belgium and Germany, and at Renault's Cleon plant in the autumn of 1991. The Cleon plant supplied a very high proportion of Renault's engines and gearboxes, and Renault had recently adopted just-in-time production principles.

Potentially, these new forms of HVP could strengthen the position of labour more generally, subject to two important caveats.

The first is that aggregate labour market conditions become more favourable, possibly as a direct result of alternatives to neo-liberal regulatory regimes being put in place. Other regulatory regimes, embedded in a different conception of economic development policy, which sought to lower aggregate unemployment, reduce spatial variability in labour market conditions, improve working conditions, and create better and more humane working environments would clearly generate a more favourable terrain for organized labour. Such regulatory regimes would need to be constructed internationally; the Social Chapter of the European Union's Maastricht Treaty and European legislation on minimum wage levels and working conditions represent small but none the less invaluable steps in this direction. At a minimum, they serve as a reminder that there are alternatives to the free-market rhetoric of neo-liberalism.

Secondly, trade unions need to devise new forms of international organization that recognize the increasingly sophisticated ways in which companies use spatial differentiation in labour market conditions. This could, for example, involve trade unions in strategies of collaboration rather than competition for transnational investment, possibly seeking to build new forms of combines between plants that are dispersed globally but linked within the production networks of companies, or seeking to build alliances across sectors and industries and national boundaries. Workers in different places are increasingly bound in dense webs of interdependencies woven through corporate global strategies that offer the potential for co-operation. Trade unions could thus seek to use the economic linkages between companies as the basis for their own forms of industrial and political organization and strategy. There is evidence that they are beginning to do so. Organizations such as the International Conference of Free Trades Unions are increasingly sensitive to the challenges posed by globalization, for example, and are seeking to develop new ways of co-operating across national boundaries in response to them. Such strategies will, however, also need to be sensitive to the politics of place, the varying conditions of local labour markets with different industrial histories and traditions of labour organization, and the extent to which production is necessarily embedded in particular places. These variations in the geographies of labour markets and production will influence the possibilities for the sorts of strategies that trade unions will be able to pursue in particular places within a framework which recognizes the collective but spatially segmented character of production.

## CONCLUSIONS AND REFLECTIONS

The crisis of mass production and the introduction of new methods of HVP has led to a sharp reduction in industrial employment, profound changes in the character of work and continuing changes in the geography of industrial employment. HVP has not, however, replaced mass production, any more than mass production replaced craft production, but has come to coexist alongside it. It is therefore unlikely that HVP will provide a general long-term solution to capital's problems of profitability, any more than did the once revolutionary innovations of Taylorism and Fordism. This is because, as with mass production methods, there are strict limits as to the conditions under which HVP strategies can be profitably deployed. Some companies will undoubtedly successfully restructure. Some will do so because of the continuing potential to discover spatial fixes to preserve mass production. Others will do so in part precisely because of the persistence of conditions within a global labour market that allow them to introduce new models of HVP. The enhanced organizational and technological capacities of major companies to exploit differences within an increasingly differentiated global production space have led to new geographies of employment in a global labour market that is simultaneously more deeply integrated and more sharply segmented. There are increasingly marked differences in labour market conditions between and within countries, regions and cities – *a fortiori*, within the so-called world cities (Sassen, 1991). There are enormous qualitative and quantitative locational differences in the terms and conditions on which companies can purchase and deploy labour power as a part of their routine repertoire of tactics in search of profits. In many areas companies find no difficulty in recruiting workforces that will accept the terms and conditions which the new models of production necessitate – or that allow old production technologies to be preserved (depending on whether companies are pursuing strategies of 'strong' or 'weak' competition; Storper and Walker, 1988).

Labour's position within this emergent new order would seem, from this point of view, to be necessarily worse in terms of number and types of available jobs. The restructuring of employment is disabling and dissecting, and then reshaping and reforming, the collective worker as part of a more general process of restructuring of capitalism. This process is unfolding on terms that are increasingly disadvantageous to labour and which outflank its 'traditional' forms of sectoral and territorial organization. The capacity of the mass worker to organize spontaneously and collectively to challenge the imperatives of capital has been seriously eroded. Corporate restructuring and inter-place competition will continue to pose acute problems for community organizations and trade unions seeking to come to terms with a shifting labour market terrain, not least because they have been active, albeit unwilling, subjects in producing these changes. The mass collective worker may remain in the peripheries of the global economy but in a weakened position. In the heartlands of industrial production, however, for the foreseeable future at least, the collective worker will exist in emaciated form, no longer possessing the capacity for spontaneous autonomous action that challenges the imperatives of capital.

On the other hand, the introduction of HVP approaches enhances the possible vulnerability of the production process to disruption by industrial action, and in important ways this potentially enhances the power of organized labour, subject to some important broader conditions being met. For realizing this potential requires that trade unions evolve new forms of co-operation between companies and across national borders so that they can take positive advantage of the new forms of relations between companies and between companies and their workers. In short, a renewed recognition of the collective character of production must become much more prominent in framing trades union strategies.

Second, and most importantly, it requires the construction of new regulatory regimes centred around greater social and economic equality, perhaps linked to notions of greater environmental sustainability (Weaver and Hudson, 1995). This presupposes, at a minimum, that the dominance of neo-liberal discourse in economic policy formation is broken. There are clear signs that it is creaking under the weight of its own internal contradiction and that the necessary is becoming impossible, the impossible necessary, in reproducing this particular mode of regulation. There is no doubt that the neo-liberal regulatory regimes and their representation of the character, constraints and opportunities of contemporary capitalism, which have dominated over the past decade or so, have had enormous influence. Not least, they suggested that a pattern of changes which was inimical to the interests of workers, their families and communities was desirable, necessary and unavoidable. It was none of these things. The trajectory of changes that has been set in motion is not, however, sustainable, economically, socially or environmentally. It is highly improbable that a new stable macro-scale model of growth and mode of regulation will be discovered that does not break sharply with its legacy. There are undoubtedly alternative approaches to production and regulation that place more priority on the interests of workers, their families and communities and which offer greater possibilities for the future. The future for labour is not therefore necessarily one of a continuation of the debilitating trends of the past two decades, but remains to be fought and struggled over as part of a process of redefining the political and regulatory terrain of contemporary capitalism.

## ACKNOWLEDGEMENTS

Roger Lee, David Sadler and Jane Wills all kindly and constructively commented on an earlier draft of this chapter. Their assistance is gratefully acknowledged but the usual disclaimers apply.

## NOTES

1. There is now a very extensive literature dealing with these alternatives to high-volume production, notably the proliferation of small firms producing in small quantities and, more specifically, flexibly specialized production within networks of small firms organised into industrial districts (for example, *see* Asheim, 1996).
2. Thus such a view attributes considerable weight to the notion of consumer sovereignty. It ignores the extent to which companies shape consumer tastes and structure markets via their advertising and marketing strategies.

# CHAPTER TWENTY-FOUR

# THE ROLE OF SUPPLY CHAIN MANAGEMENT STRATEGIES IN THE 'EUROPEANIZATION' OF THE AUTOMOBILE PRODUCTION SYSTEM

**DAVID SADLER**

## INTRODUCTION

Recent years have witnessed a lively debate over economic restructuring processes and the changing geography of production. While much attention has focused on the proclaimed resurgence of regional economies (for a review, *see* Storper, 1995b), there has also been analysis of the continued salience of internationalization, and of the significance of connections between global and local processes (Amin and Thrift, 1994). For some writers there have been epoch-making changes in the organization of production, constituting a decisive break with past practices (e.g. Scott, 1988b); others, less programmatically, stress complexity and indeterminacy (e.g. Amin and Malmberg, 1992; Hudson, 1989a). The concept of the production system has also been elaborated, as a shorthand notion encompassing recognition of the complex interaction between processes of transformation, distribution, circulation and regulation. In Europe, such debates have been given heightened relevance by two further developments: moves towards a single internal market within the enlarged European Union, and the dramatic political-economic reforms set in motion in Eastern Europe after 1989. Both these developments fundamentally altered the terrain on and through which production systems are constructed, but altered it in as yet uncharted ways; indeed, it is likely that their full implications will only become evident over the next ten to twenty years.

This chapter addresses one aspect of that wide-ranging agenda. It investigates the extent to which recent changes have been associated with an intensified *Europeanization* of production systems. That is to say, the chapter explores whether or not there has been a reorganization of production within Europe such that the spatial frame of reference within which decisions take place, and the way in which these decision-making processes are spatially configured, have become qualitatively different from those spatial frames and organizational structures which characterized an earlier era. There remains particular uncertainty on this question, not least because the Europeanization of production systems is a multi-faceted process. It might encompass changes in terms of corporate ownership, the inter-firm division of labour in activities such as R&D and logistics, and in regulatory mechanisms, as well as questions to do with production in the more narrow sense, including labour management and component sourcing strategies.

Dicken and Oberg (1996), for example, argued that most, if not all, European industries were becoming increasingly regionalized as firms sought to create organizational networks orientated to Europe as a whole, rather than to individual national markets. They also suggested that the most adept corporations in this process of Europeanization would be non-European by origin, either because they were less encumbered with existing nationally focused systems, or because they had initiated a move towards Europeanization relatively early. In partial contrast, Nilsson and Schamp (1996) stressed differences in the way that activities are organized, co-ordinated and spatially configured both between sectors across European states, and even within industries in each country. They focused on the intense uncertainty facing key actors, and the need to consider corporate strategies as adaptive and learning responses, in which Europeanization remained just one from a series of options.

It is reasonable to allow that there may be differences between industries and firms in the precise combination of these processes of change. The account that follows is, however, limited to just one branch of manufacturing activity, that of automobile production. While this sector has frequently been taken unquestioningly as an exemplar of wider changes (and this has led to some debate over the extent to which this emphasis is merited), it remains the case that through its absolute size and demonstrative effect the auto industry is still close to, if not actually at, the leading edge of production reorganization. This is not to claim that trends for all industries can be 'read off' from the auto sector, but rather to suggest that indicative questions can be raised for further investigation.

The auto industry has been the focus of considerable attention in terms of its changing geographies of production in Europe. In the 1970s and early 1980s the search for low-cost locations – for a spatial fix to the twin issues of high levels of labour organization at longer-established plants and intensified inter-firm competition – focused on the southern periphery of Europe (Lagendijk, 1993; Ferrão and Vale, 1995). More recently, the attractions to producers of such options have been rivalled by those of Eastern Europe (Sadler *et al.*, 1993; Sadler and Swain, 1994; Swain, 1996). These processes led Hudson and Schamp (1995) to conjecture that a new map of auto production was being established in Europe, though they stressed the continued salience of existing national and regional differences (*see* de Banville and Chanaron, 1991). There has also been investigation into the impact of just-in-time (JIT) supply arrangements in the auto industry (Hudson and Sadler, 1992). Contrary to some expectations, JIT does not appear to have been a proximate cause of wholesale geographical relocation, at least not for more than a few products. For instance, Schamp (1991) found only limited evidence of spatial polarization around assembly plants in Germany; instead, there was a reduction in the density of local interconnectedness as global sourcing substituted for local linkages. Wells and Rawlinson (1992) similarly found only limited spatial concentration in the patterns of component supply for the Ford Escort produced in Europe.

The analysis that follows, however, focuses upon the geographical implications of another and subtly different issue, that of supply chain management. This has increasingly been recognized as being of central importance to the competitiveness of individual firms. Popularized by the influential 'lean production' thesis (Womack *et al.*, 1990), a growth in out-sourcing has been accompanied by a quest for more cost-effective component purchasing solutions. These solutions are tangibly associated with decisions to do with both co-ordination and configuration, the spatial location of chosen suppliers. Supply chain management is therefore highly significant in the context of debates about the barriers to integration of formerly discrete national systems, and about the connections between processes of globalization and localization. As such, component sourcing offers an important 'window' on to more general questions.

The chapter is structured as follows. It first considers the extent of Europeanization of the auto industry in Europe, exploring debates about component sourcing strategies and evidence from a sample of assemblers and component producers. Then a case study of supply chain management issues is deployed. By focusing on a major Japanese-owned plant, the account examines the ways in which one of the most recent investors in Europe, less burdened by legacies of the past than European or US firms, has adapted to and helped to shape the internationalization of activity within the continent. The case is the Nissan plant at Sunderland in north-east England, the largest single Japanese manufacturing investment anywhere in Europe. As the first fully integrated Japanese vehicle assembly facility there, it is at the centre of a complex and sophisticated web of component sourcing. The lessons of this case study, in the light of the discussion of industry-wide trends, form a basis for the chapter's concluding remarks.

## EUROPEANIZATION AND SUPPLY CHAIN MANAGEMENT IN THE AUTOMOBILE INDUSTRY

Europe is but one constituent part of a broader global system of automobile production and trade, and the role of European firms within this worldwide competitive struggle should not be overestimated, particularly given their relatively limited achievements in Japanese and North American markets (Commission of the European Communities, 1994a: 1–2). None the less, the European market-place is the biggest of these three major sales areas and Europe has figured centrally in the global strategies of more car producers than any other region. This latter point is as valid for long-established notions such as the quest for the 'world car' which could be produced at different locations across the globe (Beynon, 1984) as it is for more recent conceptualizations which stress the interconnections between global strategies and those cast within specific regional markets.

There have been many variations of such accounts. For instance, van Tulder and Ruigrok (1993) distinguished globalization – aimed at a worldwide intra-firm division of labour – from glocalization, intended to establish a geographically concentrated inter-firm division of labour in each of the three major trading blocks. On this basis, glocalization entailed enhanced interaction with component suppliers in each region. Lung (1992) referred to strategies of organization across three different areas (Europe, North America and Asia) which were sensitive to the (very real) differences between them as 'transregional', while Mair (1994) described a new model of production organization (characterized by Honda) as 'global-local'. Above all, it is important to recognize that there have been many differences between firms in the auto industry in their quest for the optimum geographical configuration (Bélis-Bergouignan *et al.*, 1996); and that by virtue of its tremendous range of national and local conditions, Europe offers a particularly challenging set of questions for firms on this search path.

Many accounts, however, take processes of restructuring within the component sector for granted, if they address such questions at all (Sadler, 1994). Even though there has been considerable concentration in the component sector (partly in response to the new demands posed by assemblers, but also derived from the competitive struggle between these firms), the supply chain remains markedly less concentrated than the assembly industry. The top ten assemblers accounted for around three-quarters of vehicles produced worldwide in the early 1990s, while the top 30 parts suppliers accounted for only one-third of independent components production (OECD, 1992). As yet, very few components firms command the same global reach as the leading assemblers. None the less, the largest component firms have increasingly become significant corporations in their own right, particularly within Europe (Sadler and Amin, 1995). For instance, Wells and Rawlinson (1994) charted patterns of supply chain organization in two key sectors: pressed steel parts, and friction materials for braking and clutch systems. They concluded that globalization in vehicle assembly had been accompanied by changed 'procurement regimes', but that these changes were very uneven both across different assemblers and according to the components involved; they also stressed the key role played by a small number of leading independent component producers.

Taken together, these developments – the enhanced significance of supply chain management, the struggles over competitive advantage based on the precise combination of component sourcing strategies adopted worldwide, and changes within the component sector – point to the growing centrality of questions to do with the *location* of supply chains within, across and even outside Europe. Bordenave and Lung (1996) identified four possible configurations for the European automotive industry in 2005, based on the location of assembly plants (either concentrated within the European core, or decentralized to the peripheries) and of supplier firms (either dispersed or polarized): agglomerated unipolar, dispersed unipolar, multipolar, or continental integration. They concluded that the last of these – entailing decentralization and dispersal, and leading to the disintegration of national industries with the emergence of a vertical division of labour within Europe – was the most probable, provided that certain preconditions were met. These included monetary union, improved transport infrastructures, further liberalization in Eastern Europe, and homogenization of demand. Such an outcome was, however, recognized as being far from inevitable.

Similarly, Lagendijk (1997) argued that the European auto industry in the mid-1990s remained dominated by a small number of national champions – the assembly firms – which were strongly dependent on nationally focused supply chains. He suggested that a 'post-national' *filière* had not yet been achieved, but that there were pressures leading to an unlocking of this situation. These included the empowerment by assemblers of their in-house component manufacturing divisions (such as Peugeot's ECIA and Fiat's Magneti Marelli) so that they could act as free-standing business units; new forms of collaboration between assembler and supplier; and the internationalization of suppliers. There is some evidence that nationally focused systems had in fact already begun to diminish in significance even before this. For instance, Bélis-Bergouignan *et al.* (1993) reported that from 1986 to 1992 the proportion of components sourced in Germany by Ford's assembly plants there fell from 80 to 60 per cent. The same scale of reduction took place in terms of Ford's purchases in Spain from its factories there, while Ford's UK purchases from its UK assembly plants fell from 77 to 52 per cent.

One alternative strategic option for the industry is that of parts sourcing outside Europe for assembly within the continent, leading to a 'hollowing-out' of the European auto industry through decentralisation to low-cost locations beyond its borders (*see* for instance, Lamming, 1990: 682). Belzer and Dankbaar (1993) suggested that by the year 2010 this process would be partly responsible for a 25 per cent reduction in employment in the European auto industry. It is an option for assemblers which needs to be set alongside their prospective further development of supply chains within Europe.

One way of exploring these questions with respect to part of the auto industry production system in more depth is through a systematic evaluation of the extent of Europeanization of firms in the assembly and component sectors. Table 24.1 therefore presents evidence on the geographical distribution of turnover by destination for a selected sample of major automotive assembly and components firms (*see also* Sadler, 1996). While evidence on turnover alone is not sufficient to draw comprehensive conclusions with respect to processes of Europeanization in the automotive industry, it is none the less significant given the extent of interconnections within the production system. Three points are particularly evident.

In the first place, over the period from 1990 to 1994 the five leading European-owned assemblers became less dependent on sales in their respective domestic markets. The proportion of their turnover derived from the rest of Europe grew slightly, although the proportion of turnover realized from the rest of the world increased even more markedly. These trends were most in evidence at Fiat and Renault, while BMW was the major exception, as the proportion of sales in the rest of Europe outside its base in Germany increased markedly, partly as a result of its acquisition of Rover.

Second, part of the same picture was evident among the leading component firms. Their mean proportionate dependence on domestic sales fell slightly but in this case there was no generalized shift towards sales elsewhere in Europe. Rather, their most significant increase in proportion of turnover was in the rest of the world. This trend was apparent at six of the eight firms, the only exceptions being GKN (whose increased dependence on its domestic UK market was a consequence of expansion through acquisition in a different activity altogether, the defence sector) and Bosch, which saw only very little movement in its overall distribution of sales.

One combined consequence of these changes from 1990 to 1994 was that, third, leading European component firms remained more internationalized (in the sense of sales realized outside Europe) than European assembly firms. By 1994 component firms gained roughly one-third of their turnover in each of their respective domestic markets, the rest of Europe, and the rest of the world. For assembly firms these proportions were 40, 40 and 20 per cent respectively. Among component companies the extent of internationalization was

**Table 24.1** European automotive assembly and component firms: turnover by destination, 1990 and 1994

| | 1990 | | | 1994 | | |
|---|---|---|---|---|---|---|
| | Home | RoE | RoW | Home | RoE | RoW |
| BMW | 38 | 32 | 30 | 31 | 42 | 27 |
| Fiat | 56 | 35 | 9 | 35 | 38 | 27 |
| Peugeot | 47 | 47 | 6 | 45 | 47 | 8 |
| Renault | 59 | 35 | 6 | 48 | 38 | 14 |
| Volkswagen | 40 | 39 | 21 | 41 | 34 | 25 |
| Mean ($n$ = 5) | 48 | 38 | 14 | 40 | 40 | 20 |
| BBA | 22 | 32 | 46 | 18 | 33 | 49 |
| Bosch | 49 | 34 | 17 | 46 | 36 | 18 |
| GKN | 33 | 40 | 27 | 39 | 36 | 25 |
| Lucas | 33 | 38 | 29 | 31 | 36 | 33 |
| Pirelli | 23 | 44 | 33 | 15 | 40 | 45 |
| T&N | 29 | 35 | 36 | 18 | 38 | 44 |
| Valeo | 44 | 39 | 17 | 40 | 37 | 23 |
| ZF | 45 | 36 | 19 | 45 | 33 | 22 |
| Mean ($n$ = 8) | 35 | 37 | 28 | 32 | 36 | 32 |

*Source*: Company annual reports.
*Notes*: Home: country of origin;
RoE: rest of Europe;
RoW: rest of world.

particularly marked at BBA, Pirelli and T&N; for these suppliers, respective domestic markets accounted for less than one-fifth of turnover. Such evidence reinforces the significance of component firms within an emergent *global* supply chain.

Thus there are some indications that for first-tier component suppliers at least, globalization was at least as much a strategic priority as Europeanization during the first half of the 1990s. Such aggregate statistics, however, cannot fully capture the processes through which inter-firm relationships between assembler and supplier are being and have been restructured. In order therefore to explore in more depth the ways in which particular component sourcing strategies have evolved, and to relate these to patterns of change within the production system, the account that follows turns to a case study. This should not be regarded as an attempt to capture all the key changes taking place in the organization of the production system. Rather, as a new investor, and one seen as being at the forefront of organizational innovation in Europe, Nissan represents an indication of some of the ways in which one firm repositioned its operations. The case study explores these changes and highlights some of their implications.

## COMPONENT SOURCING POLICIES IN WESTERN EUROPE: THE CASE OF NISSAN IN NORTH-EAST ENGLAND

Some of the restructuring which took place among European and American assembly firms in Europe during the late 1980s and early 1990s was in response to the impact of Japanese assemblers within the continent. Strong political pressure was placed on Japanese assembly companies in Europe, encouraging them rapidly to attain high levels of 'local' (that is, European) content before their output could be freely exported across the European Union (*see* Sadler, 1992). Additionally, the relatively slow build-up of production volumes meant that there was less incentive for Japanese component companies to invest in Europe than there had been in North America (Kenney and Florida, 1992b; Mair *et al.*, 1988; Reid, 1990). These factors brought Japanese assembly firms into close contact with the European automotive components industry. While Nissan began producing cars at Sunderland in 1986, its rivals took slightly longer to decide on and implement their plans, and most Japanese companies did not begin production in Europe in earnest until 1992. This meant that Nissan was something of a pioneer in the development of qualitatively different forms of labour organization and supplier recruitment (on the former, *see* Garrahan and Stewart, 1992).

Nissan made a conscious decision to work with an existing European components base, despite early reservations about quality. The first vehicle produced at Sunderland, the Bluebird, initially incorporated parts from 31 European suppliers (of which 27 were from the UK and just 6 were in north-east England), accounting for 20 per cent of the car's content by value. 'Localization' of this vehicle to meet UK government-imposed targets expanded the European supply base to 126 companies by 1988, accounting for 70 per cent of the car's content. At this point, emphasis switched to component sourcing and development for a replacement model. The Bluebird was superseded by the Primera range (which had a local content of over 80 per cent) in 1990, by when the list of suppliers had grown to 180. In 1992, with the addition of a second model range (the Micra), Nissan had a European supply base of 195 companies, of which 129 were based in the UK and 27 were located in north-east England. The remainder came from Germany (29), France (13), Spain (11), Belgium, Ireland and Italy (3 each) and Austria, Portugal, The Netherlands and Switzerland (1 each). The only components not sourced by Nissan from Europe in 1992 were transmissions, engine blocks, diesel engines and fasteners. Output was originally scheduled to increase to 270000 vehicles in 1993, when it was forecast that the plant would purchase components worth £850 million within Europe, of which £655 million (or 77 per cent) would be spent in the UK, and £280 million (or 43 per cent of the UK total) in north-east England. Purchases from Japan were estimated to amount to £240 million. In the event, Nissan's Sunderland operation was eventually affected by declining demand in the European market, and output reached only 246000 in 1993, with external components purchases of £790 million. Both output and the volume of purchases subsequently fell in 1994 (*see* Table 24.2).

Much of this increased local purchasing came about *not* through a wholesale expansion of Japanese-owned facilities, but rather through Nissan's commitment to its selected European supply base. Just 10 of the 195 suppliers in 1992 were Japanese, with a further 12 joint ventures involving Japanese participation. The emphasis was on partnership with suppliers in a search for continuous improvement, on single-sourcing (of particular components

**Table 24.2 Nissan at Sunderland: production and suppliers, 1986–94**

| | Production ('000 vehicles) | Number of suppliers (year end) | Components purchases (£m) |
|---|---|---|---|
| 1986 | 5 | 46 | 4 |
| 1987 | 29 | 62 | 28 |
| 1988 | 57 | 126 | 90 |
| 1989 | 77 | 126 | 170 |
| 1990 | 76 | 177 | 220 |
| 1991 | 125 | 180 | 400 |
| 1992 | 179 | 195 | 575 |
| 1993 | 246 | 197 | 790 |
| 1994 | 205 | 202 | 670 |

*Source*: Nissan Motor Manufacturing (UK) Ltd.

rather than generic groups), and on cost reduction rather than price, entailing complex 'open-book' discussions with component manufacturers. While this might superficially resemble the way in which the company did business in Japan, Nissan's UK management also recognized the differences in terms of social and institutional context. Managing director Ian Gibson, for instance, commented that:

> We will never have the same relationship with our suppliers in Europe as we do in Japan. Such a system can only work in Japan. Our challenge here is not to look at how it's done in Japan, but to find a way of achieving the same results within a European environment. (Quoted in *International Management*, May 1992, p. 42)

A large part of the benefits of this system to Nissan depended upon working with a limited number of carefully selected suppliers. This carried several advantages: simplified communication, easier adoption of closer working relationships, and delivery of a greater number of more nearly complete subsystems (simplifying the task of final assembly). For component companies, the attraction lay in a relatively stable business relationship, which became a high-volume one: by 1992 external purchases amounted to 75 per cent of costs, a level high even by Japanese standards.

From 1990 onwards Nissan developed a highly detailed monitoring system to evaluate the performance of its European suppliers in five areas: quality, cost, design capability, delivery and management. Delivery schedules were driven by Nissan's desire to minimize stock at the Sunderland plant, both to cut costs directly and to encourage greater process discipline from its workforce. In late 1992 Nissan carried an average inventory for European-produced parts of just 1.6 days, five times better than the next best in Europe and well ahead of the industry average of 20 days. The target for 1993 was to get this figure down to one day. With the Bluebird model, orders were placed on a monthly basis; the launch of the Primera in 1990 was accompanied by a switch to weekly ordering of parts, with deliveries expected at least on a daily basis. The 1993 goal entailed full Electronic Data Interchange linkage with all suppliers, and a move to daily ordering. For many suppliers (particularly those from mainland Europe), meeting these standards involved the use of external warehousing facilities close to the plant for storage of components, which could then be 'called off' against production schedules. In 1991 Nissan also established a collection system from 30 suppliers based in the Midlands, simplifying the logistics of delivery of some components. For others, 'synchronous' production was established, with order times for seats and carpets down to less than one hour.

Quality was a further significant point at issue. Nissan rated quality standards in the UK in 1990 as 65–70 per cent of the levels of Japanese suppliers in Japan (NEDO, 1991), and a little below the European average of 80 per cent. During 1991 reject rates fell to less than 1000 parts per million (ppm), against 200 ppm for parts delivered to Nissan plants in Japan. The target for 1992 at Sunderland was 200 ppm, compared with one of 10 ppm in Japan – a measure both of the extent of differences between the two continents, and of the never-ending search for improvement. UK companies were, however, attaining higher standards faster than the European average rate of improvement, as Nissan's managing director Ian Gibson explained:

> What has been really impressive is the extent to which UK components makers have been working with us as an opportunity to leapfrog their continental competition. They have got hold of techniques and technologies which previously they'd only seen arriving as Japanese-built cars. (Quoted in *Financial Times*, 17 January 1992)

By 1995, 47 per cent of Nissan's European suppliers had achieved a quality standard of fewer than 10 defective parts per million. This was inferior to the performance of its suppliers in Japan (where 70 per cent had met this target), but a big improvement on the 17 per cent of two years previously. Such 'leapfrogging' rested upon the kind of close technical collaboration between assembler and some suppliers that was prevalent in Japan.

In other words, the significance of Japanese transplant investment with new systems of production organization extended into long-term issues to do with the design and development of new and existing products. The first vehicle made at

Sunderland, the Bluebird, was already nearing the end of its model life when it went into production, so that questions of supplier involvement in product development were not particularly relevant. With the Primera, local design capacity became significant, for the model was Nissan's first created specifically for European customers (*see* Nonaka and Takeuchi, 1995). The decision to build the car was taken in 1986. It was intended as a 'global car', built primarily for Europe, where high performance was identified as a key marketing factor, but with lower volumes sold also in North America and Japan. In recognition of the specificities of the European environment, 1500 staff were sent from Japan to Europe during the first three years of the project, to experience local driving conditions and to deepen their understanding of local consumer preferences. This was co-ordinated by the firm's European Technology Liaison Office in Brussels. A key design concept became 'sure, fast and comfortable on the Autobahn', indicative of the expectations that designers were targeting. Certain key components were refined in Japan to provide enhanced long-distance high-speed ride comfort, including the engine and the suspension system. The issue was not just one of tailoring vehicles to particular markets, but also of involving suppliers in the design process from the earliest stages, maximizing cost reduction. A four-channel braking system was developed by Lucas which was quite different from the Japanese domestic market specification, while other items such as the steering wheel were also of wholly new design as compared with the original Japanese conception. By and large, however, much of the core development work was undertaken in Japan, so that production of the vehicle in Europe was initiated six months later than in Japan, and the firm sent 300 engineers from Japan to Sunderland in order to ease the introduction of the new model.

Nissan had established a European Technology Centre (NETC) at Sunderland in 1988, with the tasks of trim and body design for the European market, the collection and analysis of European market data, and the collection of technical data for medium- and long-term planning. In 1991 – after the launch of the Primera – the company brought its investment in R&D facilities in Europe to over £60 million through the establishment of a new site at Cranfield in the UK, with a broader range of functions which included new vehicle design. The new site was selected partly because of its closer proximity to certain key component suppliers in the West Midlands. The Sunderland base retained just the roles of mid-life product enhancement and vehicle testing (*see* Charles and Feng, 1994: 14, 21). Nissan's target was to establish a world-class European supply base, and to achieve 100 per cent simultaneous engineering – production engineers working with vehicle designers from the outset of a project. This was achieved to the extent of 40 per cent for the Primera, and 70 per cent for the Micra. It estimated that more than half of the design work on the European version of the Micra (sold in Japan as the March) had been done in Europe by Nissan and its suppliers.

Much remained to be done, however, in terms of full localization of design activity. Whilst the Primera had initially been very successful in Europe, with sales well over an initial annual target of 100000, demand for the car fell sharply during the early 1990s recession. Criticism focused on the *styling* of the model rather than its technical performance. As Nissan launched a new-series Primera in 1996 – still largely developed in Japan – attention switched to the prospect of a third model being introduced at Sunderland, which might be wholly designed in Europe. This option was particularly strongly favoured by European design staff, in order to combat the problems which had been identified with the Primera in European markets. Suppliers in Europe were already working on possible components well in advance of their involvement with previous models.

One indication of Nissan's commitment to working with selected suppliers lay in its Supplier Development Team (SDT). Such activity began in Japan in 1971 and initially grew only slowly, involving just 18 of the company's suppliers there within the first three years and only 38 by 1986. It subsequently expanded rapidly to embrace all suppliers in Japan by 1990. Quality improvement was the main motivation in the 1970s, supplemented by cost reduction from the mid-1980s onwards. Discussions over the prospect of introducing such activity in the UK – strongly supported by Sunderland management – began in 1987, and after initial reluctance from Japan had been overcome, a UK-based SDT was established in 1988. This began with one Japanese and two British engineers, the latter trained specifically in Japan with Nissan's SDT there.

The emphasis in SDT activity from the outset in the UK was on total cost reduction and continued quality improvement at supplier firms. Much of the team's work involved relatively simple refinements to the manufacturing process, eliminating waste and unnecessary complexity and making production easier. At one plant making windscreen wipers, for instance, Nissan engineers devised a new, low-cost and safer way of keeping plastic tubing supple (and making it easier to install) by replacing a pan of boiling water with a rig of light bulbs and protective netting. From small beginnings (where only 3 of 12 suppliers contacted agreed to participate), the range and breadth of involvement by Nissan's SDT steadily increased. The team itself grew to encompass seven engineers by 1992, working with 73 companies in 1991 and 62 in the first half of 1992 alone.

Initial contact frequently involved process-level improvement, such as the introduction of *poka yoke* (mistake- or fool-proofing devices), which rendered it difficult, if not quite impossible, to manufacture a defective part. One member of the team, for example, described how he tried for half an hour without success to get a manufacturing cell at a supplier in Japan, protected by many *poka yoke* devices, to make an unsatisfactory motorized window winder. A nearby visual display proclaimed that it was over 500 days since the machine in question last produced a substandard component. Over three years, SDT activity gradually broadened to encompass a series of factory-level improvements, such as the linking of machines within a given cell. Dramatic changes were made over this period: productivity improvements over 90 per cent, defect reduction over 80 per cent and space savings of greater than 50 per cent were achieved in some instances. The first reaction of many suppliers in the late 1980s was either suspicion or disbelief (particularly in the light of such figures). Gradually, SDT activity came to seem less revolutionary and more evolutionary, as it encompassed (in some cases) ideas about purchasing, logistics and long-term strategy. In 1992 Nissan even began to work with companies in its 'second tier' of suppliers, and to encourage 30 of its suppliers to start their own SDT-style operations with their (Nissan's second-tier) suppliers. To further this end, local 'self-help' Nissan supplier groups were established in south Wales, the Midlands and north-east England.

Nissan's supplier development efforts focused on the full range of operations at a given company, not just those cells or lines producing Nissan parts (although companies with a high percentage of their turnover with Nissan were targeted relatively early in the exercise). This wide-ranging and long-term approach was in contrast to the very particular and limited ways in which some assemblers in Europe worked with their suppliers. Emphasis was initially placed on the mutual pursuit of reduced costs, and the sharing of benefits from such cost reduction efforts. Such concerns gradually took the SDT closer into the areas of supplier monitoring and evaluation.

Monitoring and evaluation were closely connected with appraisal against Nissan standards and those of actual and potential competitors in the supply chain. The supplier evaluation system established in 1990 embraced aspects of quality, cost, design, delivery and (subsequently) management. Each supplier was evaluated quarterly against these variables and informed of its score, its position in relation to the distribution of other scores, and its performance in relation to the general trend. During 1991, for instance, the mean score increased from 67 per cent to 75 per cent, with a tail of poor performances removed; the lowest score increased from 16 per cent to 56 per cent. In that year, too, a supplier award scheme was instigated, with the best performers publicly congratulated in full-page newspaper advertisements. Display boards were also established at the Sunderland plant's final assembly area, listing the ten best- and the ten worst-performing suppliers every month. The SDT was drawn into this evaluation procedure because of the range of its involvement with the supply chain, and in recognition of the need for a more co-ordinated approach to supplier relations as production built up to higher volumes, with fewer and fewer parts imported from Japan.

While one objective of evaluation and public praise of high performers was to encourage suppliers and to help them win business from other assemblers, there was also an important aspect of control from Nissan's point of view. Although conscious of suppliers' need to avoid undue dependence upon just one source of orders, Nissan was also very aware – as any business in its position would have been – of the ways in which partnership translated, frequently via open-book discussion on costs, into very real issues of profitability. In 1992 60 per cent of suppliers shared their full cost structure in dealing with Nissan, well above the proportion at that time at other assemblers in Europe. In the long term, such bargains or negotiations were likely to be increasingly crucial, both for Nissan and for the evolution of the European automotive components industry.

As its role shifted, Nissan's SDT in Sunderland acquired new objectives, embodying changed dealings with component firms. 'Globalization' of parts sourcing was established as a strategic goal for Nissan in Japan in 1989, and the European production system gradually adapted to that end. During 1992 engineers from Nissan Motor Iberica trained in Sunderland, and formed the first SDT in Barcelona in 1993. Efforts were made to begin to develop a common European supply chain for the two factories. Additionally, attention focused on defining 'world standards' to adopt or adjust the supplier evaluation system so that performance could be improved across different continents. In this way, global sourcing – based on accurate and up-to-date comparison – could be transferred from being an abstract goal to becoming concrete reality. In this, too, Nissan was following a path which most other assembly companies pursued in the early 1990s. For Nissan's European suppliers, this entailed the challenging target of achieving Japanese standards by 1996.

Nissan's operationalization of a 'partnership' philosophy via supplier development, monitoring and evaluation embodied many of the far-reaching changes to the relationship between assembler and supplier which took place in Western Europe during the early 1990s. The concluding section relates the implications of this case study back to the earlier discussion of processes of Europeanization.

## CONCLUDING COMMENTS

The case of Nissan's supply chain management strategy provides an emblematic illustration of the way in which relations between assembler and component supplier have been restructured in Europe in the past decade. Key elements of such changes (for recent inward investors in particular) were a search for suppliers on a Europe-wide basis, and deep involvement with these carefully selected suppliers from across the continent. For assemblers which were longer-established in Europe, such changes – and the competitive advantage which they endowed – represented a major challenge, one that was particularly marked given their legacies of nationally focused production systems. Such differences between assembly firm procurement strategies are important. Each firm's response to new competitive conditions – in the auto sector and more generally – has unfolded as part of a series of specific strategic decisions, contingent upon prior development paths and worked out as part of an ongoing restructuring process. This process is not determined by technical change, nor is it inevitable; it is instead highly context-dependent and can only be fully understood as such.

This implies that there will be no single, unidimensional geography of new supply chains in European industry. Resulting patterns of production and employment are likely to be complex and indeterminate. It is, however, demonstrably the case that the establishment of closer inter-firm relationships is inextricably linked to Europeanization in the production system; that, to use van Tulder and Ruigrok's (1993) word, some form of 'glocalization' is taking place, at least in the automotive sector and arguably more generally. On present trends, existing centres of production within Europe are likely to continue to prosper – as evidenced by the focus of Nissan's purchasing pattern in Germany, France and the UK – but the attractions of lower-cost production environments to the south and east may also weigh heavily in the locational decisions of leading component suppliers.

On the other hand, supply chain management in Europe presently rests on a delicate balance between political and economic pressures towards *Europeanization* (that is to say, co-ordination of activities across the continent), and strategic priorities towards *globalization* (the integration of operations in two or three different major market areas, including Europe, North America and East Asia). Both involve the construction of complex networks of two-way flows between different units of a polycentric system, and of webs of inter-firm connections (as opposed to the unilateral flows and restricted inter- and intra-firm divisions of labour of earlier modes of internationalization). They differ, however, in their geographical pre-conditions and implications. The precise outcome of these different tendencies will vary between assembler and across products, in a fashion partly dependent upon the interaction with existing nationally focused systems. The relatively limited extent of globalization of European-owned assemblers may be a serious weakness in the future in this context.

As assemblers seek to put together a combination of preferred global suppliers – amounting to perhaps as few as 600 worldwide – it is increasingly likely that firms making components in Europe will also make the same or similar products for the same customers, elsewhere in the world. Leading European component firms have already proceeded further than assemblers in internationalization beyond Europe, at least in terms of sales. This in turn raises another set of challenging questions both for companies and for policy-makers, to do with the prospect of a hollowed-out European automotive assembly industry. If component firms continue to globalize more rapidly than assemblers, then production and employment in the sector in Europe may well undergo significant change regardless of the future of European-owned assemblers. This is not to pass a value judgement on the desirability or otherwise of such changes, but to point out that they represent a policy issue which should be addressed.

The evidence concerning Nissan's links to its supply chain is particularly significant too for what it says about learning, trust and power in inter-firm relationships. The ways in which suppliers were both encouraged and tacitly expected to participate in development activities through the work of the Supplier Development Team and of NETC provide an object lesson in the importance of densely structured inter-firm communications in the establishment of embedded production structures. In this process it should be remembered that these are fundamentally *power* relations at work. The way in which Nissan's SDT gradually evolved and shifted emphasis in its dealings with component suppliers from development to monitoring is one indication of this power balance. These relationships are not just confined to inter-firm dealings; they also affect intra-firm decisions, as evidenced in internal debates within Nissan over the geographical location of responsibility for new vehicle design projects. Whether (and if so, how) such relationships of trust and understanding can be socially constructed between and within organiszations across major market areas (such as Europe, North America and/or East Asia) will probably have a pronounced impact on the future evolution of European production systems in the auto industry and more generally.

## ACKNOWLEDGEMENTS

Arnoud Lagendijk, Roger Lee, Yannick Lung, Anders Malmberg and Jane Wills read and made helpful comments on an earlier draft of this chapter. A version of it was presented at the European Science Foundation workshop on 'Learning and Embeddedness: Evolving Transnational Firm Strategies in Europe', held at the University of Durham in 1996.

CHAPTER TWENTY-FIVE

# INFORMAL CITIES? WOMEN'S WORK AND INFORMAL ACTIVITIES ON THE MARGINS OF THE EUROPEAN UNION

DINA VAIOU

The informal sector/economy/activities, after a long association with Third World development studies since the 1970s, has gained currency in industrialized countries as a tool for understanding industrial restructuring and changes in the labour market. A mass of publications has appeared, from the early 1980s onwards, prioritizing various aspects of informal activities, with a predominantly urban focus. The European Commission commissioned a report entitled *Underground economy and irregular forms of employment (travail au noir)* which gave an overview of the 'underground economy' in member states, emphasizing aspects of employment conditions (Barthelemy *et al.*, 1988). In 1995, a new report was compiled on the informal sector, thus acknowledging its (growing?) importance in member states (Mingione and Magatti, 1995). The new report is a follow-up of the White Paper *Growth, competitiveness and employment*, where flexibility and restructuring of employment relations are given a high priority in the process of European integration. In European Union (EU) documents, the term 'atypical' has also been introduced, overlapping but not identical with 'informal' since it refers to 'non-standard' forms of employment, particularly with regard to women's involvement in paid work (Meulders and Plasman, 1992).

The heterogeneity of activities and processes that have been called informal has contributed to a conceptual confusion around the term, rendering it a catch-all category whose usefulness has at times been questioned. Yet its use persists, since it captures, even if imperfectly, trends that would otherwise be overlooked (Roberts, 1994). The content and uses of the informal have changed with its introduction in 'First World' debates. In the EU in particular, the debates are coloured by the dramatic changes of the post-war development patterns, the growth of unemployment, urban poverty and social exclusion, as well as by the terms imposed by policy documents of the Commission, to which states and particular places have to adapt.

Informal activities and forms of work are not, in the majority of cases, a new phenomenon, as some of the literature seems to suggest. In Southern Europe, the informal is part of the development history of many places and has occupied a more or less prominent position in the study of uneven development as well as in political debate, although at times it has been treated as a 'deviation' from a dominant pattern of development characteristic of the North.[1] However, processes of restructuring since the 1970s (which other chapters of this book examine), a passage from Fordism to more flexible forms of production and the rediscovery of 'other' forms of work in the North have contributed to the challenging of old certainties about the 'economic' and have problematized the meaning and content of work and employment in uneven development.

This Chapter aims to contribute to this European part of the debate on the informal sector/economy/activities. My emphasis is on informal activities and forms of work, rather than a distinct sector or economy, and I draw from research in Greek cities – places on the margins of the EU, both geographically and in terms of the political and academic debates. I briefly discuss the relevance of the formal/informal dichotomy and I focus on three interrelated themes – the diffusion of economic activity and employment into a large number of very small firms, scattered in the urban area; the multiple functions of the family; and women's work – as important aspects in understanding the geography of production and the organization of urban space.

# INFORMAL ACTIVITIES AND FORMS OF WORK

The growing literature includes much controversy over and confusion in the use of terms, but also different theoretical and disciplinary perspectives and goals – which, to a certain extent, are reflected in the terms. 'Economy', 'sector', 'employment' (or 'work') or 'activities' are variably characterized as 'informal', while 'informal' often alternates with 'black', 'irregular', 'underground' or 'shadow'. Economic activities and forms of work included in the informal (as part of its definition) vary with the intentions and perspectives of particular writers. It is possible, though, to discern some differences in emphasis and content between literature originating from the North and that originating from the South of Europe. It is beyond the scope of this chapter to review this vast literature. Some points are summarized here, somewhat schematically, to help put in context the discussion that follows.

## Definitions and Dichotomies

Definitions of the informal have so far been couched in negative terms, by juxtaposition to the formal. This definition in negative terms and in the context of dichotomy (formal/informal, typical/atypical) is a hierarchical one: the second part of the dichotomy is secondary, less important, defined in terms of lack. By the same token, those involved in informal activities and the places where such activities and forms of work predominate are also defined as less important. A discussion of definitions is therefore necessary, in order to highlight how activities/workers/places are *in*formal only from the point of view of dominant debates and interpretations of uneven development. The latter are based on experiences particular to North-Western Europe, but are less helpful in understanding *different* histories of capitalist development.

As has frequently been pointed out, the informal and the formal are parts of the same economy, not simply in close relation but in an explicit domination-subordination relation (Roldan, 1987; Lawson, 1992; Roberts, 1994). The informal does not exist in a vacuum, but in specific contexts of formal regulatory systems; thus the same activity or practice may be perfectly regular and formal in a certain place and time but informal, irregular or illegal in another (Hadjimichalis and Vaiou, 1990a).[2] It is therefore important to consider the historical and spatial origins of what we call informal at any given time and place, as well as the relative nature of the informal and its close connection with place-specific regulatory regimes. Despite a possible lack of clarity, the informal does signify and bring to the fore patterns of activity and forms of work which are important for the analysis and understanding of uneven development.

## Northern and Southern European Variations

The literature originating from the north of Europe emphasizes, and includes in the informal, those activities that violate the rules of welfare systems; for example, unemployed people on benefits carrying out work without the permission of the social insurance authorities, social security fraud, evasion of social security payments. Also included in the informal are self-employment and the productive activity of small firms – which, however, are prone to be unregulated (for a review, *see* Boer, 1990; Williams and Windebank, 1995). Such activities, along with tax evasion, prostitution and criminal activities, are counterposed to a formal, post-Second World War model of production, based on large firms, developed welfare systems, regular employment relations with full social security rights and a strongly unionized workforce (the 'norm'). As far as forms of employment are concerned, informal is almost identified with atypical;[3] but atypical is, in this context, declared and regulated (e.g. part-time employment).

In the case of literature from Southern Europe, the emphasis is on tax evasion, criminal activities and the quality of employment in 'other' forms of work – which are much more widespread (*see*, among many, Mingione, 1988; Recio *et al.*, 1988; Benton, 1990). The limits between formal and informal are less clear, since the formal model has been only partially applied. In a context where the productive structure is characterized by small firms and where welfare systems are weak, the dichotomies are not self-evident or easy to support. One is rather faced with a continuum of activities ranging from purely formal to purely informal – with all intermediate categories present (for classifications according to various criteria, *see*, for example, Mingione, 1985; Hadjimichalis and Vaiou, 1990a). Table 25.1 is an attempt to summarize this continuum, based on characteristics of work.

**Table 25.1 Formal and informal forms of work**

| | Informal ⟶ | Formal |
|---|---|---|
| Working time | Casual, fragmented or unsocial, seasonal, part-time | Full-time |
| Ways of payment | Unpaid, block payment, payment by piece, hourly rate, with salary derogations | Salary or wage |
| 'Occupational status' | Domestic labour, family helpers, apprentices and trainees, self-employed, subcontractors | Salary or wage earners |
| Social security, benefits | Rarely or never, paid by the worker | Paid by the employer |
| Place of work | Worker's home, small workshop, employer's premises | Employer's premises |
| Destination of the product of work | Self-consumption, insertion in the productive cycle of a firm, informal or formal market for goods and services | Formal market |

## The Family and Women's Work

Women have, to a large extent, undertaken 'other' forms of paid work – whose expansion is now part of the restructuring of the formal 'norm' – and have been involved in unpaid work of various types in the context of families. In the writings on the informal sector, particularly in Southern Europe, women's involvement is hidden in the family, as an institution facilitating the proliferation of such practices and the availability of individuals for work. But an exclusive emphasis on the family obscures the gender and age divisions of labour and power within it and their reproduction through, among other processes, informal activities.

Work based on the diverging experiences of women caught between two worlds of work – 'non-standard', although not necessarily informal – work has been a steady focus of feminist research and theory-making. Thus, there has been brought into the forefront of analysis a wealth of experiences of work, and their specificities in time and space, beyond those of the collective mass worker (for a review of arguments, *see* Leacock and Safa, 1986; Bradley, 1989; Vaiou and Stratigaki, 1989). Combinations of different types and relations of work in women's everyday lives – an all-inclusive definition of work, and not only formal, full-time, paid work – are examined in this context as a set of social practices where gender identities and gender relations are constructed, at least in part.

## The Specificity of Place and Urban Studies

Informal activities and forms of work are quite diverse and differ from one place to another. But they are closely linked to particular places, as they form part of increasingly segmented local labour markets and depend on local networks of contacts and information, and sometimes also on local markets for products and services (*see also* Roberts, 1994). In urban studies, crisis and restructuring of the post-Second World War pattern of production and work has given rise to a renewed interest in the informal. The dramatic increase of unemployment, particularly in cities, and the deterioration of the protective net of the welfare state – which undermined mechanisms of social integration more generally – have opened relevant areas of research (Jazouli, 1992; Lianzu, 1991). The new urban poverty, increasing inequalities and social polarization, and the survival strategies of different groups of urban dwellers are some of the themes through which informal activities have come back on to urban agendas.[4]

Informal activities, far from being sectorally and spatially homogeneous, are associated with such conditions, which form part of the specificity of particular cities and places. Traditional and modern forms of production, which are part of urban productive stuctures, create different demand for informal labour. Place-specific compositions of employment and unemployment lead particular groups of people to informal activities for longer or shorter periods of time, as a first job or in the context of multiple employment, while patterns of migration may bring still others to informal activities. Traditions of gender relations place women and men in different power positions in formal/informal mixes of activities and work – positions which are in turn reproduced through gender divisions of labour in families and in the labour market. The particularities of systems of regulation and of their enforcement leave room to devise ways of avoiding legislation and regulation. The relative affluence of certain sections of the population and the development of markets for the products contribute to further expansion of informal activities.

# WORK AND PRODUCTION IN GREEK CITIES

A prominent feature of the productive structure of Greek cities is a vast sector of very small (dwarf) firms, low activity rates and high proportions of self-employed people. In this context, informal activities find room to develop. A regulatory framework which is complex and full of gaps creates openings and permits a continuity and mutual support of formal and informal activities. Thus informal activities are not exactly illegal and do not develop in the sphere of social anomy. But, despite the widespread consensus that informal activities, in Greece as in other parts of Southern Europe, form part of the development histories of cities and particular places, research into their spatial aspects is limited (*see*, for example, Vaiou *et al.*, 1991; Tsilenis and Hadjimichalis, 1991; Leontidou, 1993). Historic research, on the other hand, has testifed to the constant presence of what we now call informal activities in Greek cities (Pizanias, 1993).

## Informal Activities

In the Greek debate on the informal, the terms are often set by government attempts, at least at the level of rhetoric, to 'catch' tax-evading activities and thereby reduce budget deficits. The discussion has thus been dominated by economists and has concentrated on the problem of non-registered income, and hence tax evasion, in the context of the so-called *paraeconomia* (Pavlopoulos, 1987; Vavouras, 1990). One of the major concerns is the quantification of this income as part of the GNP. The increasingly obvious signs of urban crisis, including homelessness, poverty and high unemployment, albeit at much less acute levels than in other parts of Europe, have contributed to shift the emphasis to informal forms of employment and to the terms and conditions of work. Such approaches do not ignore tax evasion as a result of unregistered activity, but prioritize the effects of such practices on workers (Kravaritou, 1988; Mouriki, 1991).

On the part of the state, there has not been any significant attempt either to penalize or to regularize informal practices. On the contrary, tolerance is the most common 'policy', as well as periodic arrangement of pending tax affairs to the benefit of evaders. There is widespread agreement on the fact that a substantial part of urban productive activities are informal, totally or in part, while the urban economy and the livelihood of many households depend, again totally or in part, on such activities. The trade unions, trapped as they are in conceptions of employment which include primarily full-time workers in large firms, only from time to time 'discover' and include in their arguments, and more seldom in their demands and policies, the large numbers of people working on the margins of legality.

Quantification of informal economic activity and workers is an almost impossible task, and only approximate estimates are produced from time to time which have to be treated cautiously. According to such estimates, informal activities made up 18–30 per cent of GNP in the late 1980s (Barthelemy *et al.*, 1988). The trade unions of clothing and leather-workers estimated in 1987 that, along with the 96000 workers registered by the National Statistical Service (NSS), there were at least another 200000–250000 informal workers in those branches. Sixty-two per cent of the registered workers and over 85 per cent of the informal ones were women. If such an estimate is taken on board, it changes both the weight of industrial employment in total employment and the gender composition of industrial workers in those branches.

An indirect indication of the presence of informal activities is the low proportion of 'economically active', and the even lower of 'employed', in the total population aged over 14 (OECD indicator, cited in Waldinger and Lapp, 1993). The activity rate in Greece, as in other Southern European countries, is quite low, less than 50 per cent, and has diminished since the mid-1980s. It is even lower in urban areas, with the lowest in Greater Athens. The same is true for the proportion of employed in the total population aged over 14 (Table 25.2). Other indirect indicators may be found in the consumption patterns in particular cities and social groups, as they are identified in local studies: expensive cars and houses, commercial streets full of luxury goods in shop windows, money gambled in casinos – all testify to available incomes much higher than

**Table 25.2 Economically active, employed and unemployed in total population over 14 years old (%)**

| | Greece | Greater Athens | Greater Thessaloniki | Other urban |
|---|---|---|---|---|
| *1984* | | | | |
| Econ. active | 50.7 | 46.6 | 46.3 | 48.8 |
| men | 70.0 | 66.4 | 66.4 | 70.8 |
| women | 33.4 | 29.5 | 28.6 | 28.8 |
| Employed | 46.6 | 40.9 | 41.1 | 43.6 |
| Unemployed | 8.1 | 12.1 | 11.3 | 10.6 |
| men | 6.0 | 8.9 | 7.7 | 7.3 |
| women | 12.1 | 18.1 | 18.6 | 18.0 |
| *1987* | | | | |
| Econ. active | 49.6 | 45.3 | 44.8 | 47.7 |
| men | 66.6 | 62.6 | 62.5 | 67.8 |
| women | 30.4 | 30.4 | 29.6 | 29.8 |
| Employed | 45.9 | 40.8 | 40.6 | 43.6 |
| Unemployed | 7.4 | 10.0 | 9.4 | 9.0 |
| men | 5.1 | 6.7 | 6.8 | 3.7 |
| women | 11.4 | 15.9 | 14.5 | 16.5 |
| *1991* | | | | |
| Econ. active | 47.4 | 45.8 | 46.1 | 46.3 |
| men | 63.5 | 61.0 | 61.4 | 64.7 |
| women | 32.6 | 32.2 | 32.6 | 29.4 |
| Employed | 43.8 | 41.2 | 42.2 | 42.1 |
| Unemployed | 7.6 | 10.0 | 8.4 | 9.2 |
| men | 4.8 | 3.8 | 5.8 | 5.2 |
| women | 12.8 | 16.4 | 12.7 | 17.2 |

*Source*: National Statistical Service, *Labour Force Surveys*, various dates.

those declared and taxed (for a more detailed discussion, *see* Hadjimichalis and Vaiou, 1990a, 1990b). General figures indicate the importance of informal activities, but it has to be kept in mind that there are pronounced variations across sectors, branches and particular places, and much controversy concerning methods of calculation (Lolos, 1989).

Informal activities which support urban development in Greece, as in other parts of Southern Europe, are not a homogeneous whole, nor do they entirely comply with images of backwardness and marginality. They are linked to a continuation of old forms of organizing production, as well as with dynamic firms and restructuring processes. As far as those who work informally are concerned, they can be found in all sectors and branches of economic activity, with incomes ranging from very low to very high. Examples include owners of small firms, self-employed professionals, unpaid family workers, homeworkers, many seasonal workers in tourist businesses, domestic helpers and child-minders, teachers of private courses at home, construction workers, but also those who engage in smuggling, drug trafficking, arms trading, prostitution, illegal fishing, illegal building, illicit dealing in antiquities and other illegal or socially undesirable activities. An increasing number of foreign migrants, both legal and illegal, as well as 'non-active' people such as students, housewives, pensioners and minors, work informally on a regular or *ad hoc* basis.

Through such activities, individuals and firms use an inefficient and centralized bureaucracy for personal favours, can evade taxation and avoid an in any case inadequate inspection system meant to enforce, for instance, labour or land use legislation. But it makes a lot of difference whether they engage in them as a way to supplement income from other sources or as the only source: a public employee who keeps the accounting books of several small firms and individual professionals in his spare time is in a completely different position from the stitcher of uppers of shoes who struggles to survive on ever-lower piece rates and combine paid with domestic work. It also makes a lot of difference whether there are prospects and stability of employment or complete precariousness and insecurity: the owner of a, usually undeclared, disco on a tourist island or the professional doctor or lawyer who receives higher fees than he declares is in a different position from the Albanian assistant on a construction site or the Polish or Greek cleaning lady. In the former case informal activities may multiply opportunities, while in the latter they may be a step towards marginalization (for similar observations about Italy, *see* Paci, 1995; Mingione and Magatti, 1995; *see also* Mantouvalou *et al.*, 1993).

## Small Firms

Economic activity in Greek cities is diffused into a vast number of very small firms – 'small' here meaning with fewer than 10 employees per firm. These firms are scattered in the urban area and are in part documented in the censuses conducted by the NSS. They are, however, seldom considered in more detailed statistical surveys and/or in academic research which is based exclusively on those surveys. These refer almost exclusively to 'big' industry; that is, firms employing more than 20 people (*see*, for example, Vergopoulos, 1986; Yannitsis, 1988). However, 97 per cent of firms employ fewer than 10 people and provide 57 per cent of total employment. The average size of firm is 3.1 employees and has been decreasing since 1969 (NSS, Census of Industrial and Commercial Estabishments, various dates). The size of firms is not very different in different major urban centres such as Greater Athens and Greater Thessaloniki (Table 25.3).

In terms of officially registered employment, it is men more than women who work in the very small firms: 46 per cent of employed men as compared with 27 per cent of employed women work in firms with fewer than 10 employees. Conversely, 29 per cent of employed men as compared with 35 per cent of employed women work in firms with more than 100 employees – where 20 per cent of total employment is concentrated. Furthermore, 54 per cent of the firms surveyed by the NSS employ 2 or fewer than 2 people – which in practice corresponds to a large proportion of self-employed. In fact, 84 per cent of employers and self-employed people work in firms with fewer than 5 employees; approximately 9 out of 10 of them are men and 1 in 10 is a woman (NSS, *Census of Industrial and Commercial Estabishments*, various dates).

This vast sector of small firms contains many different situations and patterns of operation. As in other Southern European countries, of which Italy is best known and documented, a large number of micro-firms are marginal and precarious, a 'refuge' for people who find a way of getting by in a tight labour market; small neighbourhood shops and manufacturing workshops are a case in point here. But there are also firms which form part of subcontracting chains or networks of firms working for the local, national or international market. Some of these are technologically advanced, flexible, dynamic, inserted in particular niches of the market, while others use older technology and defensive strategies of reducing production costs (Vaiou *et al.*, 1996).

Some micro-firms are completely 'formal' (they are registered, they follow regulations, etc.). There are others, part of whose activity is informal and still others which operate in complete illegality. The informal part of their operation includes undeclared activity and income; violation of labour law, in terms of payments and/or conditions of work, and firing and hiring practices; avoiding social security payments, totally or in part; non-observance of locational and environmental regulations; and profiting from the gaps in the formal regulatory system. Through such practices, combined in original ways, micro-firms manage to contain their

**Table 25.3 Firms and employment by size of firm, 1988**

| | | 0–9.9 | 10–49.9 | 50 and over |
|---|---|---|---|---|
| Greece | % firms | 96.8 | 2.7 | 0.5 |
| | % workers | 56.9 | 16.5 | 26.6 |
| Greater Athens | % firms | 95.8 | 3.6 | 0.6 |
| | % workers | 55.4 | 19.2 | 25.4 |
| Greater Thessaloniki | % firms | 96.0 | 3.6 | 0.4 |
| | % workers | 62.3 | 21.8 | 16.3 |

*Source*: National Statistical Service, 1988, *Census of Industrial and Commercial Establishments.*

cost of operation and survive for longer or shorter periods of time.

In such a complex universe, one cannot generalize about small firms and their position in the productive structure of cities. In particular, it is not possible to classify them into an undifferentiated whole of 'traditional' or 'backward' informal activity nor, conversely, into a 'flexible' and vertically integrated one (for an elaboration on the Italian case, *see*, for example, *La piccola impresa*, 1990). In the same line, it is not possible to place all types of work in small firms in one category of underpayment, bad conditions and lack of alternatives. There are at the same time some positive connotations, linked to self-employment, personal achievement and high earnings.

## Families and Urban Development

The patterns of activity discussed in the previous sections of this chapter rely heavily on the family, as an institution which facilitates their proliferation and the availability of individuals for work. Unlike in the North, in Southern Europe, and in Greece in particular, the family has retained its importance as a productive unit (*see also* Benigno, 1989). The multitude of small firms and family ventures which support the dynamics of urban development are a case in point here. But the family includes and organizes a variety of functions. Families pull together income from many formal and informal sources and make it available to their members when they want to start a business, study or look for a job. Through traditions of owner-occupation,[5] housing is secured for its members, but also, family wealth is increased through illegal building and property exploitation (for a discussion of the patterns, *see* Mantouvalou *et al.*, 1993). Family networks are the main mechanism of finding a job, while families provide security and assistance to their members at times of unemployment, study and sickness, and render services to children, to the old, the sick and the disabled which are either not provided or poorly provided by the state and/or the market.

Assistance of the family in all these areas substitutes, to a certain extent, for a poorly developed welfare state. But its 'protective net' entails a large number of conditions for the help and services rendered. It perpetuates prescriptive behaviours and divisions of labour and power, for it is not families but specific members within them who perform different functions (*see also* Papataxiarchis, 1992). The interaction of family and productive system presupposes that the family includes a full-time person, the housewife – who is often trapped in her professionalized role (for a discussion concerning the Italian situation, *see*, among many, Paci, 1992). Men engage in family activities as employers or heads of micro firms, while women and youngsters are 'family workers', quite often unpaid and unregistered, but whose labour is essential for the survival of small family businesses. Caring is done, as a rule, by women of different ages, since men's identities do not include involvement in domestic and caring labour. But such demanding and time-consuming labour ties women to their home and restricts opportunities to participate in paid work and in public life. It is no surprise that they are 'available' for homeworking or for regular or casual 'help' in family businesses.

As employment opportunities decrease, young people have no other choice but to engage in low-paid, precarious jobs which do not lead to independence from the family. In fact, employed members of a family are under more pressure to work to support the unemployed, usually young, members. As already mentioned, it is the employed who are better placed to engage in informal work since they are part of relevant networks and possess the necessary skills (*see also* Mingione and Morlicchio, 1993). Elderly people also remain tied to the family since pensions and other benefits, even when they exist, are very low. In all its capacities, extending to production and work, property relations, caring and support, the family is key to understanding urban development processes, in which the 'not informal' is more an exception than a rule.

## Women's Work

An important feature of the past 15 years has been the increasing presence of women in the (formal) labour market, with activity rates reaching 29–32 per cent in urban areas (Table 25.2). But there is also growing evidence, from branch or local surveys and from qualitative research, that large numbers of women have always been in the labour market for longer or shorter periods of time (for a review of the literature, *see* Thanopoulou, 1992). Among working-class families and the urban poor, women's paid work has always been a necessity, while being able to afford to 'maintain' a housewife has been a status symbol for men (Varikas, 1987).

The amount and quality of women's unacknowledged labour have supported some of the most important patterns of urban development in Greece, most notably diffused industrial and commercial activity in small family firms, service provision and informal paid work, both of which have led to rising standards of living throughout the post-war period; it also makes it possible to inhabit the remote and under-served neighbourhoods which resulted from urbanization of the urban periphery through illegal building. Some forms of paid work in which women are involved in different cities are summarized here from a number of research projects (Vaiou and Hadjimichalis, 1997).

## PROCESSING FISH IN KAVALLA

Kavalla is an industrial port in the north of Greece, with a concentration of major chemical plants of the country which employ almost exclusively men on stable contracts, with high wages and other benefits. Fishing and processing fish for the internal market and for export is a major port activity. The work in fish-processing consists of packing fish in tins after removing the heads and entrails. It is performed in shacks where humidity is high and workers stay in the water up to their knees. By the end of the working day their clothes are stinking, wet and full of scale. The work is low-paid, seasonal and unstable, as its volume depends on the day's catch, and it is exclusively done by women, mainly pensioners with very low pensions, seasonal workers in agriculture, mothers with many children.

## WORKING FOR SUBCONTRACTORS OF GERMAN DISTRIBUTORS IN DRAMA

Drama is a city in the north of Greece where many large clothing firms (with more than 100 workers per firm) located after the mid-1970s, profiting from regional incentives. These firms are subcontractors of mainly German distributors and department stores and, in their turn, put out work to smaller firms. Nine out of 10 of the workers are women (who do the sewing) and 1 in 10 is a man, mainly working as cutter, ironing staff or foreman. Work starts very early, an hour before any of the nurseries in town open; it is performed in bad conditions, particularly in the summer when winter clothes are sewn, and payment follows the national collective sectoral agreement, with no major derogations. In the small firms, employers avoid paying social security and many workers work without contracts; wages are lower and many women take up second jobs as cleaners or child minders in order to survive.

## INDUSTRIAL HOMEWORKING IN THESSALONIKI

Thessaloniki is the second largest city and the second largest concentration of manufacturing in Greece. Along with the registered (formal?) enterprises for which data are regularly collected, there is the diffused and 'invisible' activity of small workshops and homes which is to a great extent informal but very important in terms of output, value and employment. In some of its neighbourhoods, there are homeworkers in approximately 10 per cent of the houses. Homeworkers are predominantly women who have started to work 'temporarily' in order to cover some family need; but needs keep cropping up and many end up doing industrial homework until they are quite old. Their working day is over 8 hours and, in peak periods, over 12 hours, with no social security or other benefits. The 'housewives' of employment statistics solder automation devices for elevators, assemble TV games and electric equipment, attach semi-precious stones to jewellery, stitch uppers of shoes, handbags, belts and other leather articles, make toys and paper flowers, pack food, sew clothes.

## YOUNG WOMEN IN ATHENS: EDUCATED AND UNEMPLOYED

The younger generation of women mostly have much higher levels of education and training. For them, paid work is not necessarily associated with dire need; it is rather part of forging an identity. Also part of this identity is the drastic fall in birth rates in a context of renegotiation of gender contracts. Integration in the urban labour market, however, is on unfavourable terms and unemployment is at the top of the agenda. It is seldom noted that women's (registered) unemployment rates are double those of men (Table 25.2), and the differences are even more acute when the data are disaggregated by age. Young women in Athens are educated and unemployed but not inactive. While searching for a job, they undertake precarious and temporary jobs for varying durations and levels of pay. And many oscillate between unemployment and underemployment, 'formal' and informal forms of work.

Analyses of the urban economy and its restructuring concentrate on large firms and formal male employment, and pay scant attention to women's informal work, despite its importance for, among other things, urban development and patterns of survival and everyday life in cities. A formal/informal dichotomy, combined with gender divisions, determines a scale of importance in theoretical and political terms which continuously constructs and reproduces 'others'.

## INFORMAL CITIES?

In my discussion of urban development in Greece, I have emphasized the part of processes that relates to informal activities, small firms and women's work. This is not to deny that big employers and formally regulated work relations are not also part of the story. In state institutions, banks and big private enterprises, the situation is more or less close to what is generally meant by the term 'formal'. However, my intention has been to draw attention to patterns of economic activity and work which, in mainstream debates, are often not considered or are seen as residuals bound to disappear. These patterns are crucial for understanding urban development in Greece and the way in which people's livelihoods are organized.

The complex patterns of formal and informal activities, diffused in small firms all over the urban area, affect local employment and incomes and contribute to the dynamics of urban development – but are also part of its crisis. Along with the patterns of ownership of housing, they have prevented, to a great extent, extreme cases of social segregation and conflict. Informal activities are not an outcome of social division but rather a means of social integration – albeit often on unfavourable terms, particularly for women, along with youth and migrants. On the other hand, diffusion of economic activity and lack of controls have resulted in a patchwork of land use which accounts for the liveliness and density of activity throughout the urban area.

A key to understanding these patterns of urban development is the family in all its capacities as production and reproduction unit. In the absence of its functions and of the different labours of its members, some of those patterns would be unsustainable. However, the family is also an institution in which power relations and hierarchies are constituted and reproduced, not least by concealing the contributions of particular members. Women's paid work is a case in point. Its extent and quality remains largely undervalued; it is often classified as non-work, hidden in family relations and in the gender contracts negotiated therein. The prospects open to young and better-educated women are not very promising, caught as they are between unemployment and informal jobs with limited prospects.

The patterns of paid work in which women are involved, in my examples as in many others, are part of the experience of work in Greece and an important part of the development histories of cities. In these patterns of work, the limits of 'formal' and 'informal' activity are not always clear. In definitions of the informal in terms of a formal/informal dichotomy, a hierarchy is established in which all that is classified as 'other' (idiosyncratic or bound to disappear) is undervalued, in the process itself of defining the 'norm' (formal). In this process analytical tools have been developed that are appropriate for the study of an increasingly homogenized, predominantly male, mass worker in the era of expanding capitalism, and then have generated 'facts' upon which this image of modern society was based (*see also* G. Smith, 1994).

If such standard definitions are accepted, then, in the cities of Greece, we are faced with islands of 'regularity' and formality, in a non-regular and informal ocean. If, however, informal is what does *not* form part of the socio-economic patterns that predominate in a given place and time, then an interesting inversion has to be considered: what has until now been called informal is predominant, while the presumed 'formal' is 'other'. In this context, the activities and forms of work encountered in Greek cities are 'informal' only from the point of view of those who define them as 'other', by projecting on them experiences from different contexts. It is probably the first post-war decades in the north of Europe that are an atypical period as far as forms of work and patterns of urban development are concerned (see also Pahl, 1984). But a full understanding and explanation of urban processes require a longer historical and a broader geographical perspective.

## NOTES

1. 'North' and 'South' are used to counterpose broadly different development patterns and are in no way viewed as uniform entities, given the important differences between countries and places.
2. A widely accepted and general definition of the informal comes from the International Labour Organisation: 'professional activity, whether as a sole or secondary occupation, exercised gainfully and non-occasionally on the limits of, or outside, legal, regulatory or contractual obligations' (cited in Mingione and Magatti, 1995).
3. 'Atypical employment' is a term launched by the EC Network of Experts on Women in the Labour Market and includes 'all those forms of employment which are distinguished from traditional occupations

by characteristics as diverse as the number and distribution of hours worked, the organisation and localisation of production, wage determination and statutory regulation and conventions' (Meulders and Plasman, 1992: 1).

4. Many relevant journals have produced special issues or have devoted part of their pages to these topics; *see*, for example, *International Journal of Urban and Regional Research* 17 (3) and 18 (1).
5. Owner occupation in Greece reaches 70 per cent. The percentage is higher in rural areas (92 per cent) and lower in cities, diminishing as the size of the urban area increases: 54 per cent for Greater Athens, 58 per cent for Greater Thessaloniki, 64 per cent for the rest of urban areas (NSS, census data). In urban areas, such high levels of owner occupation have been achieved through successive rounds of semi-squatting of the urban periphery and, later, when properties became legalized, building densely on every plot (Mantouvalou and Mavridou, 1993; Leontidou, 1990).

# CHAPTER TWENTY-SIX

# BREAKING THE OLD AND CONSTRUCTING THE NEW? GEOGRAPHIES OF UNEVEN DEVELOPMENT IN CENTRAL AND EASTERN EUROPE

ADRIAN SMITH

## INTRODUCTION

The crisis and collapse of Central and East European (CEE) state socialisms has provided a space for one of the world's most daring economic experiments: the construction of a capitalist system on what was seen as the deathbed of communism. The attempt to construct a new regulatory fix has resulted in the ascendancy of a policy agenda and ideological discourse orientated towards the creation of globalized market economies through trade and price liberalization, privatization, the attraction of foreign investment, the development of small and medium enterprises, and the restructuring of the state towards a neo-liberal order (Amsden *et al.*, 1994; Gowan, 1995). Six or seven years into the much celebrated 'transition to capitalism', however, we have become more sober about the prospects for a rapid transformation of political-economic life in the region.[1] This is perhaps not surprising; those who reject the various one-way transition positions of the likes of Francis Fukuyama (1989), Jeffrey Sachs (1990) and the International Monetary Fund (IMF) (1990) have argued that capitalism cannot be constructed in the belief that the collapse of state socialism left a *tabula rasa* in CEE. While transformations in the region have seen the emergence of new forms of political-economic organization and integration, it is the contention of this chapter that the 'breaking of the old and the construction of the new' is complex, unevenly developed, multi-determined and 'embedded' in the nature of the state socialist system, the struggles that arose out of that system, and its current transformation. This argument suggests that the social relations of production, exchange and consumption are not merely replaced with the fall of state socialism, but are being reworked in complex ways as state socialism meets the capitalist law of value. First, I consider some of the alternative theoretical claims regarding 'transformation' and 'restructuring' rather than 'transition'. Following this, the chapter discusses some of the very material consequences of the new economies of CEE and the former USSR. It also examines the complexity and uneven development of geo-economic change in the region and highlights the connections between old and new material practices.

## THE 'EMBEDDEDNESS' OF TRANSFORMATION

Neo-liberal and modernizationist approaches to the transition (what Michael Burawoy (1992) has called 'transitology') argue for a stock set of policy measures which can discipline the macro-economic situation of CEE economies as the optimum method for providing the conditions for capitalist accumulation: fiscal austerity, reduction in state expenditure, privatization, etc. Such claims encompass a 'weak theory' of restructuring with debates revolving around the pace of change ('shock therapy' versus 'gradualism') (Sachs, 1990; United Nations Economic Commission for Europe, 1993), the timing of the implementation of policies (whether the macro-economic environment can be restructured before any substantive change in the micro-economic context) (Gowan, 1995), and the method of privatization (voucher, foreign investment, joint ventures, worker–management buy-outs) (Stark, 1992). However, this remains *weak theory* because it fails to take into account the social relations of 'transition' and focuses, rather, on policy positions. Consequently, a number of counter-arguments to 'transition' have emerged which stress the evolutionary and embedded nature of economic and societal *transformation* rather than straightforward *transition*, and focus on the institutionalization of transformation rather than market and price mechanisms. They suggest that we need to rethink the way in which 'the economic' is conceptualized away from the formal nature of macro-economic dynamics towards locally institutionalized practices of economic life emerging from the social relations structuring surplus extraction, distribution and consumption.

Three key areas of work can be identified within this broad institutionalist agenda. First, there is work from evolutionary and institutional economics. This, in itself, is a diverse group linked through the concern to understand the dynamics of CEE from a perspective emphasizing *continuities* and the slow emergence of the institutionalized practices of the market economy. Poznanski (1995), for example, has argued that the current economic dislocations (the 'post-communist recession') arise primarily from economic agents regressing to previous habits when unpopular and unfamiliar policies are implemented. Poznanski emphasizes the dissonance between the importation of 'core' (Western) institutional structures and policies and the 'relative peripherality' of CEE. Institutional arrangements imported from the core not only 'belong to a different – and more mature – stage of development but, when imitated, they are often deformed and thus made even less suitable' (Poznanski, 1995: 21). Similarly, Stark (1993: 15; 1996) has argued that instead of focusing upon the teleological project of transition we should look instead 'for patterns of *transformation* in which new elements are typically combinations, adaptations, rearrangements, permutations, and reconfigurations of existing organizational forms' (emphasis in original). The consequence of this approach is to recognize the diversity of organizational and institutional formations emerging in the region and to argue that this 'bricolage' is a rational product of the actions of key agents in the transformation of state socialist relations.[2]

This project has been taken further by a second group which focuses upon the 'embeddedness' of economic reproduction (Grabher, 1994a, 1994b; McDermott, 1994; Smith and Swain, 1997). This approach stresses the way in which economic activity is 'bedded down' in local and regional contexts – through understanding the integration of economic life via both formal and informal networks and institutional formations. It considers not only institutional arrangements but also the dynamic relations of various forms of economic integration through networks, joint ventures, supplier integration and so on. McDermott's (1994) discussion of Czech privatization, for example, has found that firms remain embedded in the social relationships inherited from state socialist economic organization which stressed the devolution of control to enterprises, the importance of local alliances which 'became more or less institutionalized frameworks to define and renegotiate claims to assets and production flows' (McDermott, 1994: 2). The consequence is that, even within the context of relatively rapid privatization, enterprises still rely upon and are 'embedded within' networks of social relations which are utilized for survival:

> The central problem for the renovation of industries is not, as is typically argued, how to assign rights so as to avoid conflicts of interest, but rather how to distribute rights amongst the interested groups and create institutions for the consultation among them to mediate conflicts and reach compromises that facilitate the reorganization of production. (McDermott, 1994: 33)

There is also a third approach which examines economic and social transformation as the institutionalization of social relations (Altvater, 1993; A. Smith, 1994b, 1995; Smith and Swain, 1997). Derived from regulation theory (Aglietta, 1979; Dunford, 1990; Jessop, 1990a; Tickell and Peck, 1992), this approach takes a broader 'cut' at transformation by looking at the inherent contradictions of particular social formations, how these contradictions are regulated and come into crisis, and the likelihood of new regulatory fixes being established. The key argument is that no clear, coherent regulatory fix is emerging to regulate the contradictions of the transformation in CEE. While the establishment of a neo-liberal order has been attempted by the World Bank, the IMF and other key actors (Gowan, 1995), the reality is that there are a variety of strategies pursued at local, national and cross-national scales to consolidate particular regulatory dynamics, none of which is becoming dominant. The results are divergent realities throughout the region and a mixing of 'old' and 'new'. In the following section I turn to some of the key dynamics which are restructuring economic life in the region. It highlights the diversity of 'fixes', and some of their contradictions, along with the profound nature of uneven development.

# THE NEW GEO-ECONOMIES OF CENTRAL AND EASTERN EUROPE

## Economic and Geographical Change: Divergent Realities and Uneven Development

While it is very difficult to generalize about the experience of restructuring in CEE, it is clear that it has been divergent from the expected transition-to-capitalism thesis and from the expectations of many of those who 'made' the 'revolutions' of 1989. The transition-to-capitalism thesis, most cogently expressed through the work of Jeffrey Sachs (1990)[3] and implemented through G7, the IMF, the World Bank and the European Bank for Reconstruction and Development policies, argues that rapid liberalization through marketization, stabilization and privatization is the key mechanism to ensure the rebirth of capitalism in the East. Furthermore, intra-CEE trade and regional economic integration should be rejected and replaced by what Gowan (1995: 6) calls a 'hub and spoke' structure of East–West relations in which each eastern state is competitively aligned to the West to ensure that it follows the requirements of the Western 'hub': trade liberalization, enlargement of the European Union, access for foreign investors.[4] Only if these measures are implemented, it is argued, will capitalism succeed in the East, and for Sachs (1990), experiments in 'market socialism' or more evolutionary forms of economic change will doom the region to further impoverishment.

The recent shift to positive economic growth rates in the region has been seen as evidence that 'the prospects for an extended period of strong growth are very good' (European Bank for Reconstruction and Development, 1995: 5) and that shock therapy has been a success.[5] The predominant experience in the first few years of the 1990s has, however, been economic collapse (Fig. 26.1). While positive growth is now occurring in several economies, the depth of the collapse between 1990 and 1994 has meant that economic rejuvenation is at best a long way off. Indeed, we are beginning to witness the emergence of complex patterns of international uneven development. Even in the economies where positive growth rates have been registered, there is still a long way to go before output returns to pre-1989 levels.[6] Rollo and Stern (1992) have estimated, for example, that it could take until at least 2000 before per capita GNP levels return to those of 1988 in most 'transition economies'.

Modernizationist arguments over transition posit that economic flexibilization and reconstruction must occur through enterprises restructuring by shedding labour. Indeed, in the absence of developed capital markets, the prime mechanism for enterprise restructuring has not been through borrowing for new investment in plant and production but through 'downsizing' by shedding labour and reducing its price (Gowan, 1995). Consequently, employment loss has been a universal, yet uneven, experience (Table 26.1). Bulgaria, Hungary, Poland and Slovakia stand out as the worst examples. The Czech Republic and Russia seem to have fared better. Yet what lies behind the relatively low levels of decline is complex. In the case of the Czech Republic labour has been retained (up until 1992) in industrial enterprises, suggesting that the perceived success of the Czech model was *initially* based upon limited intra-enterprise restructuring which has only recently begun to make an impact.[7] Furthermore, relatively low levels of unemployment and labour-force reductions have been achieved through encouraging large-scale retirement of post-retirement age workers, notably women (Paukert, 1995). In Russia, the continued and clear commitment to the enterprise 'labour collective' is still in operation (Clarke, 1993). A recent International Labour Office report on Russia, for example, argued that some 35 per cent of employees had no job to do, thus producing high levels of 'hidden unemployment' waiting to be 'shed' (Williams, 1994).

The key result of employment change has been the growth of unemployment, often within the context of only limited social security provisioning and in societies in which commitments to full employment were a fundamental part of the model of development up until 1989. Official unemployment has been particularly severe in Albania, Bulgaria, Croatia, Poland, and Slovenia where levels reached over 15 per cent in 1994 (United Nations Economic Commission for Europe, 1994: 86). Furthermore, female unemployment rates have tended to be higher than those for males as the commitment to 'equalizing the sexes' has given way to mass lay-offs, differentially affecting women and also, although figures do not exist, ethnic minorities, particularly Roma (Ladanyi, 1993).

Together, then, the experience of 'transition' after five or six years has been one of economic collapse, labour shedding, rationalization and an onslaught on labour, and social and political disorientation (collapsing birth rates and increasing

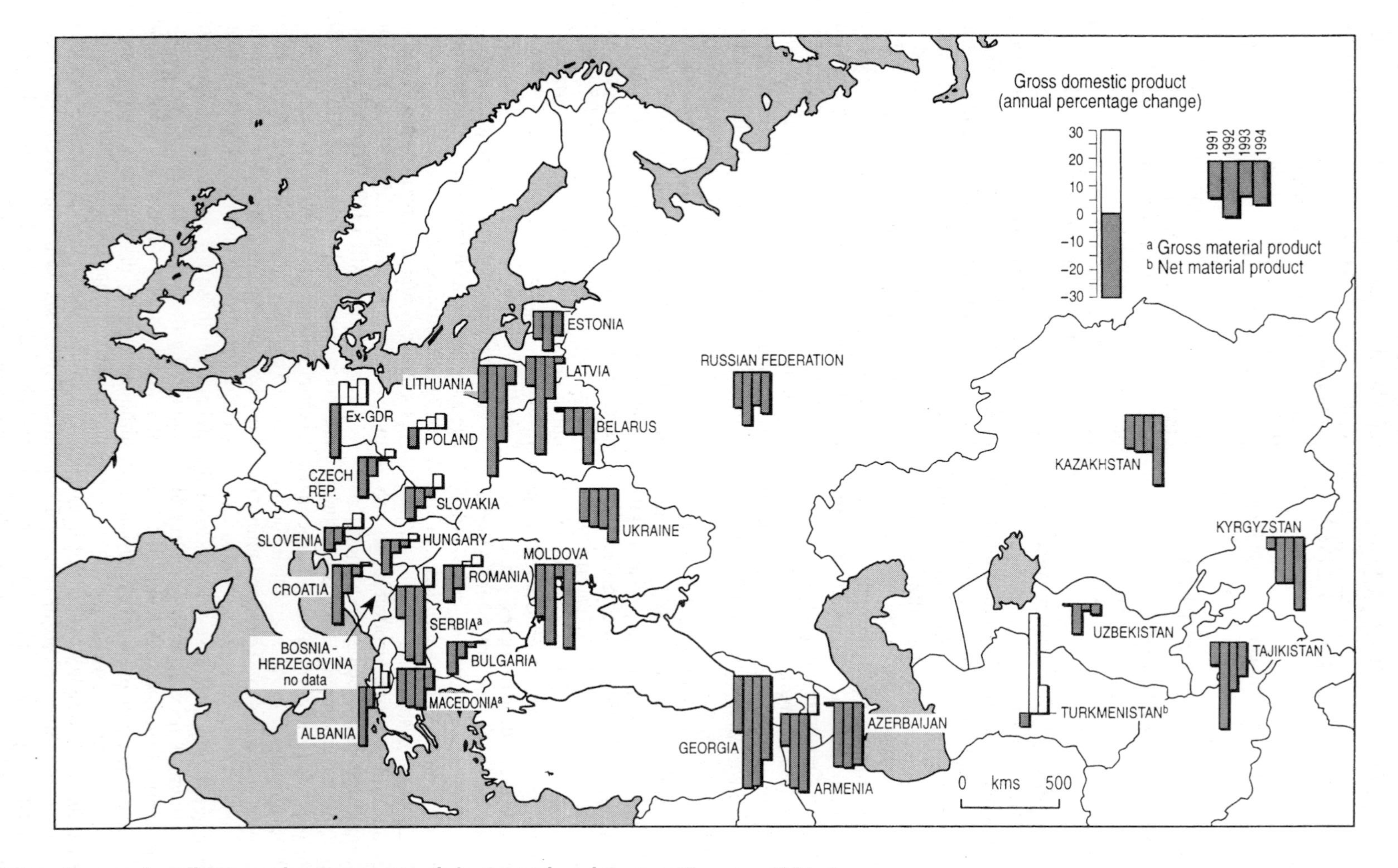


**Fig 26.1 Economic collapse and uneven growth in Central and Eastern Europe, 1991–4**
*Source*: United Nations Economic Commission for Europe, 1995, Table 3.1.1.

**Table 26.1 Percentage employment change in selected Central and East European countries, 1990–2**

| | Total | Agriculture | Industry | Services |
|---|---|---|---|---|
| Bulgaria | −28.7 | −31.2 | −37.1 | −16.8 |
| Czech Republic | −8.8 | −29.5 | −14.2 | 2.7 |
| Hungary | −12.1 | −34.3 | −17.7 | −0.5 |
| Poland | −12.6 | −13.3 | −20.4 | −5.7 |
| Slovakia | −13.5 | −27.5 | −22.6 | 1.8 |
| Russia | −4.2 | −3.7 | −6.9 | −1.7 |

*Source*: UNECE, 1994: Table 3.4.2, p. 85.

death rates, suggesting a deep-seated social and psychological crisis (Ellman, 1994)), while at the same time establishing the ability of some to prosper on the backs of others. The result has been a profound increase in poverty and inequality.[8] Milanovic (1994), for example, argues that poverty now affects some 58 million people in CEE, or 18 per cent of the region's population. Real wages have dropped dramatically throughout the region, and in situations where wage growth has been recorded for two consecutive years (Slovenia, the Czech Republic, Latvia and possibly Estonia (United Nations Economic Commission for Europe, 1994, 1995)) these increases may be based more upon wage differentiation (Večerník, 1995) than any kind of general increase in levels of living, with large expansion in a small group of high-income-earning 'new professionals' and those involved in speculation and illicit activities. Alongside increased inequality have been the rise of homelessness,[9] declining health levels and other social problems associated with polarization. Furthermore, the emergence of poverty and inequality has been a breeding-ground for two forms of activity. The first is the increased use of household survival strategies (Piirainen, 1994a) such as the exchange of household production including food and other basic items between friends and in networks established in the workplace which have led to a burgeoning of the informal economy. The second is the rise of illegal and semi-legal activities, which I will return to.

While increasing inequality and uneven development have been hallmarks of the transition, it is also the contention of this chapter that the transformation of the new economies of CEE will represent something more akin to a mixing of economic formations created under state socialism and those forged by the current processes of both real and pseudo-marketization and globalization. New forms of capital accumulation emerge unevenly and bring with them new class relations, but these represent the contingent implementation of capitalist and pseudo-capitalist practices and social relations rather than the creation of 'shimmering landscapes of capitalism'. We have already seen how economic collapse and uneven development have dominated the first six years of the transition. It is perhaps no surprise that in such economic conditions survival strategies develop which build upon past capacities and relations as a way of mediating emerging markets. In what follows I attempt to pull apart and piece together some of the new geo-economies in CEE in the light of a coupling of 'old and new'. This treatment does not claim to be exhaustive. No account could be. Rather, it aims to highlight some of the key dynamics of the 'transition to capitalism' which suggest divergence from the claims of transitology.

## Deindustrialization

The economic crisis that has hit CEE has resulted in the decline of industrial output and employment throughout the region. Only three economies have sustained positive growth in industrial output for the period between 1993 and 1994 (Hungary, Poland and Romania) (United Nations Economic Commission for Europe, 1995: Table 3.1.1, p. 70). Consequently, a report for the European Commission argued that '[t]he closure of particular industrial plants, or the decline of production, with redundancies as a natural consequence, are likely to cause some of the most serious regional problems' (Commission of the European Communities, 1992: 108). Deindustrialization has affected industries and regions which have only a weak ability to respond to rapid liberalization. The old regional dependencies, with enterprises dominating local economies cut off from the operation of the law of value, have been devastated. In particular, areas dominated by one-sided industrial structures – the classic model of forced state socialist industrialization (A. Smith, 1994b) – and areas with concentrations of armaments, heavy engineering, mining, steel and chemicals sectors have fared badly. Clear, comparable data on the extent of deindustrialization are limited. Yet work on particular countries has highlighted the extent to which industrial collapse is one of the key experiences of 'restructuring'.

Data published by the Russian government, for example, suggest that between 1991 and the end of 1994 industrial production declined by some 41 per cent (*Russian Economic Trends*, 1994: 63–4), while employment in industry declined from 21.5 million in 1992 to 19.1 million in mid-1994 (ibid.: 91). In Slovakia, research has shown that deindustrialization has been particularly apparent in areas with high concentrations of armaments production, in regions more recently industrialized through branch plants where the system of integration through vertically integrated industrial associations has collapsed, and in peripheral regions where industrial expansion came late in the mid-1980s but had only a limited transformative impact upon local social relations (A. Smith, 1994b, 1995, 1996a). Plant closures, however, have been relatively limited, and the experience has been characterized by one of large-scale employment and output decline and shifts to unpaid holidays, with clear implications for local economic life as personal consumption declines.

A similar situation has been experienced in Hungary, where, for example, the crisis of iron and steel production in the north-east (Research Institute of Industrial Economics, 1992) has led to the re-emergence of profound east–west inequalities in the Hungarian space-economy with the continued polarization of development in Budapest and the northern Transdanubian region (Cséfalvay, 1994; Nemes Nagy, 1994). In Ozd during 1991, one of the key centres of ferrous metallurgy,[10] there were 7 vacancies for every 1000 unemployed persons, compared to the worst regional situation in the UK where there were 42 jobs per 1000 unemployed in the North and 39 per 1000 in Northern Ireland (Central Statistical Office, 1992).

In the case of East Germany, Grabher (1993b) has pointed to the re-emergence of north–south regional disparities as deindustrialization is emerging in regions where relatively low levels of new investment have meant the loss of some 70 per cent of the total jobs lost in East Germany. Of particular significance is the loss of some 80 per cent of R&D jobs between 1989 and 1992 alone, which has tremendous long-term development impacts through the loss of innovative potential (Grabher, 1993b; *see also* Grabher, 1992).

Similarly, in Bulgaria, Pickles (1995) has found that the almost complete dislocation of the national economy has particularly affected the most marginal areas.

> The regional allocation of production under central planning has given way since 1991 to widespread branch plant closure as state industries seek to maintain benefits and labor in core plants. The core plants of the formerly multi-plant state enterprises have closed branch plants, labor has been shed, and the semi-autonomous enterprises have withdrawn from the workshop economy of peripheral, low-wage, often female based factories. . . . The result has been devastating in the peripheral localities and regions of the branch plant economies. (Pickles, 1995: 8–9)

In some of the most peripheral areas, unemployment rates have soared to a staggering 90–95 per cent. Core–periphery competition has therefore emerged, leading to struggles between managers, labour and communities, all dependent upon large plants for survival. Thus, the regional industrial structures of large enterprises dominating local economies – what Illner (1992) has called 'industrial paternalism' – have not given way to new forms of dynamic, productive, capitalist practices. The intersection of the 'old' local dependencies with the 'new' law of value has meant that the 'other side' of capitalist practice (economic decline and closure (Martin and Rowthorn, 1986; Hudson, 1989b)) has been all too prevalent as the key experience of 'transition'. Indeed, what lies behind the experience of uneven development discussed above is the eradication of economic capacities and livelihoods as 'new' economic relations devastate institutional structures weakened by economic decline. What, however, have some of the new landscapes of capitalist relations constructed after 1989 begun to look like, and to what extent do they draw upon the past?

## Speculative and Merchant Economies

While deindustrialization represents the destruction of space-economies in the face of the international law of value, some of the major new forms of economic life in CEE have been 'speculative economies'. By 'speculative economies' I mean a diverse set of semi-formalized, institutionalized structures through which capital is recycled through semi-legal and legal activities revolving around the basic tenet of 'fast money'.[11] There are four main elements constituting the new speculative economies: formalized capital and money markets; 'political' and merchant capital; 'kiosk economies' and small-firm services; and the rise of protection, extortion and the mafia. Each is characterized by money recycling through legal, semi-legal and illegal activity, yet each has its own dynamics and relationship with the past.

## CAPITAL, STOCK MARKETS AND MONEY MARKET SPECULATION

The most formalized example of speculative accumulation is through the recently developed capital markets and stock-markets of the region. While these markets have limited strengths in the new economies of CEE, they suggest the establishment of formal institutional structures of capitalist regulation – money and capital markets (Altvater, 1993). The monetary and deflationary shocks of the early transition period created a severe shortage of capital throughout the region (Poznanski, 1995) which deepened the economic crisis and constrained the availability of funds for new investment. The new capital markets aim to rectify this situation, but the 'fast' nature of their speculative growth may set limits to their long-term investment potentials.

The most dynamic markets are in Budapest, Prague and Warsaw (Hutton *et al.*, 1994). For example, in 1994 the Budapest, Prague and Warsaw stock-markets closed the year respectively some 29 per cent, 96 per cent and 787 per cent higher (Commission of the European Communities, 1994b).[12] However, Herr (1994: 163) has argued that these markets are 'deformed monetary economies' in which disorders exist because the economies 'do not possess a national currency that is able to adequately perform all money functions in the eyes of domestic and foreign economic agents'. Thus, relative isolation from the world economy established under state socialism continues to have important implications for the nature of capital market development. First, capital flight is encouraged as domestic currencies are readily exchanged for hard currencies, producing a collapse of national investment activity. Second, the domestic power of hard currencies produces a situation in which hard-currency speculation becomes an important mechanism for accumulating capital. Third, capital market development has been associated with the growth of a host of financial and producer services, suggesting the emergence of metropolitan areas potentially similar to the global cities of the advanced capitalist world. Furthermore, the increasing synchronization of trading between capital markets in the Visegrad countries (the Czech Republic, Hungary, Poland and Slovakia) has led to a rapid 'speed-up' in capital flows, increasing the volatility of these markets (Balaž, 1995b), which may undermine any attempt to construct long-term sustainability – a kind of regionalized compression of space and time (Harvey, 1989a).[13] For now, however, the growth of metropolitan markets has been part of the expansion of large urban economies in the region that helps account for some of the uneven development discussed earlier.

Foreign exchange markets have also provided new forms of accumulation. However, much of the activity that occurs constitutes part of the informal economy of survival strategies as individuals engage in foreign currency speculation as part of a bundle of strategies. For example, a study of informal economic activity in the former Czechoslovakia and Bulgaria found that 29 per cent of households in the former and 18 per cent in the latter engaged in foreign currency dealing (cited in Piirainen, 1994a). Such 'black market' practices represent the continuation of survival strategies already established in the hard-currency-scarce environment of state socialism. Today, however, currency dealing has become one of the paramount methods through which personal economic survival can be guaranteed, at least in the short term.

Balaž (1995a: 8) has argued that '[h]ealthy banks, liquid capital markets and well governed companies do not spring automatically from centrally planned economies', suggesting that capital markets are only marginally embedded in many of the economies of CEE. While they have provided a dynamic for metropolitan growth, the rapid development of speculative capital and foreign exchange markets suggests that the process is unsustainable in the medium to long term, and riddled with corruption.

## POLITICAL AND MERCHANT CAPITAL

'Political capitalism', Staniszkis (1991) argues, is a hybridization of the economy in which divergent forms of new productive relations emerge on the deathbed of the state socialist system and represent an intersection of existing relations and new dynamics of power. 'Political capitalism' closely resembles the recombinant property identified by Stark (1993, 1996) in Hungary. Political capitalism represents a new relationship between political power and command over capital (Staniszkis, 1991).[14] First, power in industry and the state has become linked through the role of the bureaucracy in private companies. Second, *nomenklatura* consumption levels are maintained through the close supplier linkages between 'private' and state firms. Third, 'profits are derived from the exclusive access to attractive markets, information, and supply [with] . . . such access [being] . . . made possible by the dual status of the *nomenklatura* owners' (ibid.: 137). Political capitalism has operated unevenly throughout CEE, yet it has become central to economic reproduction – through segmenting markets into 'formalized' and 'informalized' it enables the shifting of costs of accumulation from the 'pseudo-private' sector to the state sector, thereby enhancing the 'pseudo-private' basis for accumulation. Political capitalism therefore has an embedded rationality in 'economies in limbo' with poorly defined regulatory structures – an environment in which established practices and networks derived from the past are utilized in the new circumstances in the now legitimized rush to accumulate capital.

A closely related form of economic organization that builds upon past practices of bargaining is that of *merchant capital*, in which economic transformation is 'confined to the realm of exchange', while 'economic transactions . . . increasingly governed by the pursuit of profit though trade . . . leave production more or less unchanged' (Burawoy and Krotov, 1993a: 52). Thus, Burawoy and Krotov (1993b: 64–5) have posited that merchant capital[15] involves a set of partially new relations based around (a) the increased importance of self-generating, lateral linkages of inter-enterprise integration, primarily barter exchange; (b) the fusing of the economic and the political as 'parastatals' take on an increased importance involving an expansion in the power of regional monopolies through the maximization of profit rather than the satisfaction of social need; (c) the fact that 'the source of profits is based on trade, speculation, or even extortion rather than on the transformation of production'; and (d) an increase in worker control over production as large and small monopolies emerge in the form of trading companies 'putting out' production to worker groups within enterprises. Burawoy and Krotov (1993a: n. 6, p. 54) have, however, distanced themselves from Staniszkis' (1991) position on political capitalism:

> we do not believe that there has been as much change in the arena of production as in the arena of redistribution . . . we stress the subordination of production to trade . . . [and] the idea of 'political capitalism' still exaggerates the 'political' and overlooks the autonomous dynamics of the economic.

A. Smith (1994b), in an investigation of the armaments industry in Slovakia, has used part of the merchant capital thesis to argue that state ownership, or forms of quasi-state ownership, remain dominant as continued subsidization is received from the state. The collapse of systems of inter-enterprise integration, and the need to seek out new markets as old ones disappear, have resulted in the transformation of the 'local embeddedness' of firms. These new systems of integration are based around 'spin-off' and newly established trade companies which fulfil a mercantilist role through the effective 'putting out' of production to the main enterprise. Many of these mercantilist firms have been more readily able to secure inputs and market opportunities for such enterprises, partly because of the increasing liquidity crisis in core enterprises, and partly because of the need to develop new contacts and markets, which large state enterprises have found difficult to achieve. Payments, however, are made through a barter system which represents a partial continuation of the systems of integration through reciprocity established under state socialism (Grabher, 1994b), resulting in the enterprise being a producer of goods which are sold by mercantilist firms for hard currency on global markets. The result is a new form of local dependency arising from a system of local integration connected to global markets. Indeed, in the context of high levels of inter-enterprise debt, barter has become one of the standard forms of exchange in CEE. The Association of Russian Banks, for example, estimates that 65 per cent of all industrialists, banks and businesspeople use the shadow economy to avoid taxes, primarily through barter (Prokhvatilov, 1994), and recent reports have highlighted the persistence of barter relations, where, in the case of the Siberian Barnaul match factory, employees are paid in matches which they sell in local markets to gain a 'wage' (Freeland, 1995).

Similarly, barter relations have structured the survival strategies of some enterprises in Bulgaria during what Pickles (1995) calls the 'actually occurring transition',[16] which represents the reworking of 'old' experiences under central planning to secure budgets, subsidies and supplies. Thus, 'far from transforming the industrial workplace and rationalizing production systems and the labor process, this form of "modernization" is actually resulting in both mass unemployment . . . and . . . continued collusion between management and workers to protect their budgets, supplies, and wages' (Pickles, 1995: 12). Rent-seeking strategies thus represent an entirely rational response to the introduction of competitive conditions without, as in the past, the oversight of the state as regulator of the planned economy.

## THE 'KIOSK ECONOMY' AND NEW CONSUMPTION SPACES

While the rise of mercantilist accumulation has been documented for the industrial economy, very little work has focused upon the dynamics of change in the small-scale mercantilist or 'kiosk economy'. This sector has burgeoned in importance and has transformed the morphological landscapes of cities and towns. The small-scale trade economy has resulted in large part from two processes: the privatization of retail services formerly run by state firms, and the establishment of low-overhead cost, kiosk-style retail outlets in central urban areas and in outlying housing estates. In Russia, for example, this material transformation of urban space and the rise of the small-scale merchant economy has been characterized as follows:

> Buying, selling, trading, haggling, lowering the price or raising it to adjust to the customer, Russians have taken to the streets, not to protest but to indulge in a huge, nationwide rummage sale. . . . A woman hawks a bottle of vodka, another a pair of shoes . . . a third is selling six months' worth of birth-control pills. (quoted in Varese, 1994: 239)[17]

The emergent kiosk economy thus suggests a new site for consumption, retailing and the production of new urban spaces that builds upon existing survival strategies.[18]

However, while the street and kiosk economies remain small-scale sites of consumption, other forms have become internationalized and have taken on great significance in some new spaces of consumption. For example, at the Dziesieciolecia football stadium in Warsaw, Russians and others from the FSU buy a whole range of goods, largely Western consumer items, for sale back at home. Recent estimates put the number employed in the market at 25 000, and it had a turnover in 1995 of some £1.3 billion (Brzezinski, 1995). Dziesieciolecia, it seems, has become the world's largest bazaar. Similarly, in Prague, Václavské námésti (Wenceslas Square), the site of some of the most important events of the November 1989 revolution, has, like many other capital cities in CEE, become a mix of corporate office space and informalized retailing and hawking – primarily, in this case, for the Western tourist market (*see* Cook, 1995, for example).

## MAFIAS, PROTECTION AND ILLEGAL SPECULATION

> Capitalism will always be wild here. . . . It is corrupt from the top to the bottom. (Henryk Niewidomski (a.k.a. Dziad), Polish mafia leader quoted in Borger, 1994: 11)

Illegal activity is at the core of both the economic and political crisis in Russia (Cohen, 1995): some 5000 gangs, 3000 criminals, 300 mob bosses and 150 legal organizations control an increasing web of pseudo-militaristic protection, racketeering and economic violence. Some 40 per cent of Russian privatization shares are owned by the Russian mafia, and enterprises spend an average of 20 per cent of annual turnover on protection money. Mafia and other associated gang activity has also been reported in Prague (Cook, 1995), Warsaw (Church, 1994; Brzezinski, 1995), Bulgaria (Bogetic and Hillman, 1995) and Lithuania (Girnius, 1995). The effects of mafia activities originating from CEE have been felt internationally, with reports of involvement in a variety of shady economic deals in the UK (Blundy, 1994a) and the USA (Blundy, 1994b; Cohen, 1995) and with the recycling of Russian and Ukrainian mafia money into the British public school system (Blundy, 1994c); and the illegal export of nuclear materials has been seen as a threat to global security (*Guardian*, 1994; Tomforde and Meek, 1994; Koudi, 1994).

While 'corruption' was a central part of the state socialist system – the widespread use of personal networks to achieve power (Simis, 1982), the embezzlement of goods from state enterprises, particularly to fuel the second economy (Varese, 1994), and a kind of tribute–reward economy in which favours (particularly the supply of scarce goods) were often repaid in kind through special access to recreational facilities (Grabher, 1994b) and local political power – it *was* part of the functioning of the state socialist system, and activity that took place was largely through the state or its organs. It therefore had a legitimate function in the 'shortage economy' (Kornai, 1992) as a means of oiling the wheels of economic development. What makes the mafia different in the current context is its resort to illegal and pseudo-legal mechanisms to extort money and capital. Indeed, we cannot understand the emergence of the CEE mafias without placing them within the context of the political economy of restructuring in the region. Varese (1994), for example, has emphasised the lack of clearly defined property rights as the key condition opening a space for the mafia to operate, alongside the effective privatization of military and paramilitary forces as armies and the secret police are 'downsized'. However, while this is part of the story, it is important to recognize that such activity thrives first and foremost in situations where the accumulation of capital and the privatized means of violence (the former domain of the state through the protection of private property) are legitimized by a state no longer able to wield sufficient power (a kind of 'delegitimated state'), a rapid process of relatively unregulated privatization, and the existence of unstable 'presents' and 'futures' in which populations resort to desperate tactics in order to survive. In other words, rather than being some product of the 'backward' and feudal nature of CEE economies (however important this broad context must be), the mafias are a product of the contemporary attempt to engineer a fast transition to capitalism which builds upon the systems of paternalism and social relations established under the command economy, and they also have good examples to draw upon of illegal economic activities under Western capitalism (Dear, 1990).

## Small-firm and Large-firm Economies: 'Flexspec' or Mass Production?

In the face of industrial collapse in CEE, there has been a search for an alternative paradigm for industrial organization. Conveniently, the crisis of mass production in the West was seen to have its parallels in the East, and so, it has been argued (Joffe, 1990; Murray, 1992; Bianchi, 1992; Said, 1995), the key method of constructing an alternative is to borrow the model that has been so successful in the West: flexible specialization and small-firm development.[19] For example, the European Union PHARE programme committed some 40 per cent of its 1994 budget to the development of small and medium enterprises (SMEs). Such claims arise from the spurious argument that what existed in the industrial economies of CEE represented 'Soviet Fordism' (Joffe, 1990; Murray, 1992). While mass production did exist, this claim overlooks the character of Western Fordism in intensifying accumulation, while in the East mass production and 'giganticism' were a result of an extensive accumulation model and were not tied to mass, privatized consumption as in the West (*see* A. Smith, 1994a, 1996b, for more detailed discussions).

Aside from these conceptual problems, there are also a number of empirical reasons for being sceptical about the rise of a new, flexible SME economy giving rise to dynamic industrial districts. In the context of the Slovak space-economy, Said (1995) has argued that the presence of SMEs specializing in R&D and business services suggests the emergence of a kind of industrial district model based upon a 'new industrial paradigm'. What he fails to recognize, however, is that many of these firms are rent-seeking mercantilist firms jostling for position and control over a large local heavy engineering firm (*see also* A. Smith, 1994a, 1994b). Similarly, data collected on SMEs in two regions in Slovakia (Smith, 1996b, 1997) have highlighted the competitive (rather than co-operative) nature of the sector, the low-tech nature of workshop production, the use of poorly paid labour with few employment rights, and the somewhat limited evidence for any form of locally embedded and integrated system of small-scale production. The evidence for the emergence of industrial districts therefore appears to be limited. 'Flexspec' is not providing the new model of development that some Western commentators would wish of it. Rather, small-firm economies are regional concentrations of mediating practices between past relations (the 'second' economy) and new accumulation possibilities.

Amid all the rhetoric regarding the potential for SME development, the large-enterprise economy established under state socialism remains dominant, although severely eroded. Two main processes have reduced the size of the large-enterprise sector. First, large state enterprises have been at the forefront of the deindustrialization process discussed earlier, experiencing closure and downsizing, particularly in peripheral regions. Second, decentralization from planning ministries to enterprises has been replicated in the internal structures of large enterprises. Vertical disintegration has occurred as sections of large firms, often the most profitable divisions, have been 'spun off' and given semi-independence (and sometimes independence) from the core firm. Stark (1993; 1996) has called this process 'decentralized reorganization', where state forms of ownership are often retained and recast in new forms – 'recombinant property' – suggesting that the notion of fast privatization and the transformation of property rights to a clearly delineated private capitalist sector is somewhat wanting.

The debate about enterprise restructuring, however, has not addressed the question of constructing alternative and more democratic forms of ownership away from the paternalism of the state socialist model or the likelihood of foreign capital buy-outs. The dualistic nature of the debate – state ownership or privatization, a return to the past or the creation of a capitalist future – precludes a discussion of alternatives that might democratize the workplace. Weisskopf (1993) has argued that democratic worker self-management represents the best way of ensuring a pathway between the centralized plan and the market in post-state socialist societies. Democratic self-management, it is claimed, is particularly appropriate for the post-state socialist economy because there is only a weak environment for the development of market capitalism, whereas workers were accorded a significant degree of control under state socialism over the day-to-day labour process (*see also* Clarke *et al.*, 1993), and there is an increasing desire to construct the new system on the basis of 'fairness, stability, and security' (Weisskopf, 1993: 139). Schweickart (1993) has also argued the case for workplace democratization by increasing worker motivation through tying wages to productivity, and through combining self-management with market forms of co-ordination. Thus,

> On the one hand, planning and producers' democracy (self-management) replace the market function in determining and coordinating economic decisions. But on the other hand, not all economic processes can be planned, and the market remains indispensable within a complex democratic system of regulation. (Altvater, 1993: 254)

Various forms of co-ordination are thus required, ranging from competitive forms of regulation in the market, through collaborative systems, to democratic planning of general development trajectories and key economic sectors.[20] As Sayer and Walker (1992: 262) have argued, in their discussion of the successes of co-operative co-ordination in the Spanish Mondragon system, 'the creation of networks of supporting institutions that provide education and training, management services, research and development, and special credit provisions' is essential. Democratic planning as a way of reconstructing 'the new' in CEE must ensure the co-ordination of trajectories of development but not at the expense of the autonomy of socially owned productive units.

# CONCLUSIONS: REGIONAL FRAGMENTATION, UNEVEN DEVELOPMENT, AND THE SEARCH FOR ALTERNATIVE REGULATORY FIXES

> There will be a growing number of cities with decreasing economic activities [deindustrialization] and populations on the one hand, and a growing number of cities with marked growth [regional and global cities]. (Musil, 1993: 904)

The preceding discussion has highlighted some of the new economies of CEE. What is clear is that there is only limited evidence of any sort of coherent system of regulation and accumulation emerging in post-socialism. Rather, 'bricolage', the mixing of old and new, and the evolution of sets of economic survival strategies at both household and enterprise levels seem to dominate the experience of transformation. The complexity of change is also apparent when we consider the dynamics of uneven development in the region. I have already discussed some of the divergent experiences between nations. Yet we are also witnessing the profound fragmentation and reconstitution of sub-national economies. This divergence can be seen as the result of the implementation of increasingly zero-sum games in which regions and their populations are forced into aggressive competitive struggles over access to increasingly scarce resources, ranging from foreign investment to Western-sponsored local development programmes (A. Smith, 1996b). The question of who wins and who loses in this competitive war remains open. But as Musil (1993) argues, polarization is occurring between increasingly globally integrated regional economies and cities, and increasingly marginalized peripheries. This has already taken a broad east–west pattern reflecting the re-emergence of the inequalities of pre-state socialist CEE (Good, 1994). Yet it is also the result of the way in which many of the relatively unsustainable regional economies created under state socialist 'forced industrialization' have been differentially able to restructure under pressures from the global capitalist economy (A. Smith, 1995, 1996a).

The bold project currently being engineered to enable CEE to make a transition to a newly formed capitalist system is thus highly problematic. I have pointed to some of the very profound material dislocations resulting from the attempted fast dismantling of the economies and political structures of state socialism. I have also pointed to the unevenness of change in the space-economies of the region to argue that landscapes of capitalist production and consumption cannot be produced overnight and are themselves constituted out of the past social relations of CEE societies. However, amid all this seeming messiness there is an element of coherence in the changing nature of power in the region. The regulation of economic life is increasingly dominated by neo-liberal discourses wielded by global institutions, particularly the World Bank, the IMF and the EU (Gowan, 1995). The 'local response' to such a globalization of institutional and economic control has been varied. In Russia, Slovakia and Poland we have seen reactions to these processes in the forms of nationalism and the re-election of various reformed (and unreformed) Communist parties. Elsewhere, such as in Hungary and the Czech Republic, we have witnessed a kind of 'global engagement' in which 'stronger' economies have been brought into the fold of a global division of labour.[21] The result, however, is the creation of 'insiders' and 'outsiders', new cores and peripheries through uneven economic and geo-political integration (Lipietz, 1993a).

This argument, then, suggests that we need to rethink the project of 'transition' and look for alternative ways to (re)construct the economies *and* regions of CEE. What is central, and often overlooked, is a project of constructing both economic and political democracy in the East. The project of political democratization has been pursued apace (although not without its contradictions). However, what has been overlooked is the question of economic democratization. Indeed, if the alternatives with which the populations of the region are presented revolve around a discourse of state versus market, a crucial role for the creation of spaces of empowerment and the construction of new, more democratic regulatory fixes and spaces is overlooked. *Modernization* is clearly a key element in the process of transformation, but *modernization theory*, with its focus upon uni-linear pathways to capitalism, does not aid in the construction of democratic ways out of the collapse of state socialism.

This chapter has also argued that social relations are central to the trajectory of systemic change. The social relations of state socialism have seen both dramatic transformation and important continuity after 1989. Yet what is clear is that, along with the polarization of socio-economic well-being and the complexity of economic change, one finds a clear emergence of class positions and relations. There has been a heated debate about the nature of class relations in state socialist societies (Cliff, 1974; Konrád and Szelenyi, 1979; Binns and Haynes, 1980; Feher *et al.*, 1983; S. Clarke, 1993; Bahro, 1977). While inequalities under state socialism were related to how the state both extracted surpluses from the working population and recycled them through privileged and uneven access, the new sets of class relations are clearly related to one's ability to accumulate and control *private* and *pseudo-private capital*. While the (re)emergent bourgeoisie have been given the task of stabilizing liberal democratic and capitalist institutions, it is important to recognise that the bourgeoisie is itself a very diverse group and barely formed – what Piirainen (1994b) has called the 'imaginary middle class' ranging from former *nomenklatura* elements to groups of former state-sector workers who have established new firms (a more traditional entrepreneurial group), to the illegal mobsters of the 'black' economy (*see* Piirainen, 1994a, 1994b; Kivinen, 1994). Piirainen (1994b) has identified an additional two 'proto-classes' which form counter-class positions to the emergent thin layer of the new middle class: (a) 'survivors', for whom the future is difficult but not impossible and who may become part of an enlarged bourgeoisie; and (b) the 'proletariat', for whom the future appears to mean increasing levels of poverty. However, the precise playing out of these class positions is uncertain at the present time and they are likely to remain unevenly developed as the space-economies of CEE undergo a continued polarization. Piirainen (1994b) identifies three possible trajectories: an American variant, where one finds the emergence of two-thirds of society with access to levels of living considered middle class; a polarized 50/50 class society of immiserization and inequality; and an 'Argentinian' pathway of mass proletarianization and economic and political disenfranchisement which is likely to breed either the politics of a revolutionary alternative or the politics of hatred akin to that of Vladimir Zhirinovsky. These alternatives (or some unforeseen hybrid of them) are, however, undetermined and will only be constructed through social action and conflict over pathways out of the collapse of the 'old' and the birth of the 'new'. If nothing else, the 'revolutions' of 1989 showed us that.

## ACKNOWLEDGEMENTS

This chapter aims to draw together a wide variety of work on restructuring and transformation in Central and Eastern Europe and the former USSR. As such it has benefited from discussions over the past four years or so with Mick Dunford, Mary Kaldor, Petr Pavlínek, John Pickles, James Sidaway and Adam Swain. I would like to thank Roger Lee, Petr Pavlínek, James Sidaway, Adam Swain and Jane Wills for their helpful comments on earlier versions. All errors of fact and interpretation are, of course, my own.

## NOTES

1. This soberness is reflected in the recent publication of a number of works arguing for the inevitability of complex and uneven social and economic change in the region (Good, 1994; Bryant and Mokrzycki, 1994; Amsden *et al.*, 1994; Gowan, 1995; Poznanski, 1995; Pickles and Smith, 1997).
2. *See also* Bryant and Mokrzycki (1994) for a treatment of this argument.
3. Sachs' 1990 article in *The Economist* was entitled 'What is to be done?', echoing Lenin's (1917) 'pamphlet' addressing the situation in Russia.
4. *See* Lloyd's (1996) critique of Gowan's argument and Gowan's (1996) rebuttal.
5. The European Bank's *Transition report* goes on to argue that many CEE economies 'share a number of the key features underpinning the outstanding growth of East Asia over the last few decades'.

Emerging export strength, macro-economic stability, an educated labour force and large potential neighbouring markets in the EU are pointed to as being key similar features. Amsden *et al.* (1994), however, have argued that what made the East Asian 'miracle' was careful state planning of development trajectories. Similarly, Cumings (1987) has argued that repression and high levels of exploitation underlined the dynamic growth of the Asian newly industrializing countries. Unsurprisingly, no mention of these crucial factors is made.

6. This is not least in South-East Europe and the former Yugoslavia, where war and its associated regional impacts have resulted in a fall of per capita GDP in Bosnia-Herzegovina from $2719 in 1992 to $250 in 1994 and Bulgaria is said to have lost 25 per cent of its 1994 GNP from the impact of trade sanctions against Serbia (Sussex European Institute, 1995).
7. Pavlínek (1995), for example, has found that employment loss has been regionally uneven and sectorally specific. Parts of the northern Bohemia coal region saw up to 25 per cent loss of employment, and estimates suggest that in the coal sector in northern Moravia, 51 per cent of jobs were lost between 1989 and 1993 and that this figure may increase to 80 per cent in 1996.
8. One of the most virulent forms of reaction to crisis has also been in the rise of nationalism and fascist movements in the countries of CEE. Žižek (1990), for example, argues that the rise of nationalism in the former Yugoslavia (and elsewhere) is the direct result of uneven economic development.
9. Médecins sans Frontières estimates put Moscow's homeless population alone at around 30 000 (V. Clark, 1993).
10. In Ozd, 51.3 per cent of the workforce were employed in the steel industry, and some 62 per cent of the local population were dependent on incomes from the sector (Research Institute of Industrial Economics, 1992).
11. Illegal activity is not the only form that such activity takes. Recent reports have argued that the siphoning-off of Western aid by Western consultants could amount to a wastage of some 90 per cent of aid commitments (Traynor, 1994). *See* Sidaway and Power (1995) for a similar argument that national *comprador* classes use dependency upon foreign aid to reposition themselves in the emerging social and power structures of post-state socialist Mozambique.
12. The astonishing expansion of the Warsaw exchange was in part due to the important role of foreign investment, accounting for a quarter of average daily turnover.
13. There have been several clear cases of speculative accumulation, booms and busts in the new financial markets. The most famous case is that of the MMM finance company, which created a pyramid selling operation to internally up-bid share prices without any shares reaching 'open' investment markets. Consequently, share prices eventually plummeted, leading to a suspension of 'trading' in which a large proportion of investors lost huge amounts of savings (Lloyd, 1994a, 1994b). Another example of speculative capital market accumulation is that of the Czech privatization investment fund (mutual fund) the Harvard Group (McIntyre *et al.*, 1994; Balaž, 1995b). While the entrepreneurial 'brains' behind the company, Viktor Kožený', has been lauded as a hero of the emerging capitalist economy, he was forced to flee to Switzerland after allegations that insider information had been provided to his company on privatization options from secret-police databases (Balaž, 1995b). Similar scandals have emerged over bribes taken by the Czech Republic's head of the Centre for Coupon Privatization, who was sentenced to seven years in prison, and in Slovakia, where scandals have emerged around the role of government control over mutual funds (Balaž, 1995a).
14. Staniszkis (1991: 132–6) identifies six forms of political capitalism which variously combine different forms of power relations and control and utilization of capital: (a) the dual status of fixed capital (use of redundant or under-utilized capital in a state firm by its workers to produce goods for other firms on their own account); (b) the spinning-off of state enterprise functions to new firms run by former managers in the firm (a form of vertical and horizontal disintegration); (c) the leasing-out of departments from state enterprises to private companies run by former managers; (d) new companies which were established in collusion or combination with key personnel in national treasury departments to 'protect' the interests of such firms; (e) the participation of managers and local state officials in 'protected' and subsidized foreign investments; and (f) 'organization ownership', in which parties, trade unions and social organizations develop companies to exploit for profit their existing assets.
15. In this article they argue a case for *merchant capitalism*, which probably overstates the strength of the role of mercantilist relations. In a related article, however, they distance themselves from this term, which implies a system of exchange *and* production, to favour the term used here, *merchant capital*, as a mode of exchange rather than a dominant and hegemonic structure of productive relations (Burawoy and Krotov, 1993a: n. 6, p. 54).
16. Pickles (1995) is here playing upon Bahro's (1977) notion of 'actually existing socialism'.
17. Noticeably the reference here is to women performing these semi-formalized retail tasks, pointing towards the gendered nature of household survival strategies as formal female employment declines.
18. Such practices are not dissimilar from new consumption practices in the UK such as the car boot sale, so popular in part because of its low overhead

costs (Gregson and Crewe, 1994; Jackson and Thrift, 1995).

19. However, *see* Dunford *et al.* (1993) and Amin and Robins (1990) for a critique of the flexibility school.
20. Nove (1991), for example, argues that central control of areas such as electricity production, oil and railways seems 'natural'.
21. The latest instance of this is the joining of the OECD by the Czech Republic and Hungary. Poland has also recently joined.

# Chapter Twenty-seven

# CALIFORNIA RAGES: REGIONAL CAPITALISM AND THE POLITICS OF RENEWAL

RICHARD WALKER

California has long played a leading role in the political economy of capitalism, but at the end of the millennium it is a microcosm of American – even global – malaise, with a misguided economy, a disintegrative social order and decadent politics. After a triumphal epoch of growth, it must rid itself of the accumulated deadweight of the past: not just fixed capital and redundant labour, but all manner of social practices put in question by the shifting tides of world capitalism. This requires a wrenching process of economic, political and social restructuring, yet the state's ruling class (overwhelmingly white and male) is unprepared to cope with the profound tasks of industrial retooling, closing the class and race divides, and reviving the democratic polity. California suffers from many plagues, but three general social contradictions stand out: the worst economic crisis since the Great Depression, thorough race and class recomposition of the people, and a political system unable to govern. California's dilemmas, like its prior successes, have deep political and social causes, which we can only begin to touch. But it owes much to the political triumph of reaction since the 1960s, and the resulting failure of class, race and political renewal which changing circumstances demand.

This portrait of California is wedged awkwardly between the global and the local, geography and political economy, politics and economy. Despite the current obsession with globalization and continental integration, economic geographers keep working their way back toward the local. The old bugbear of uneven development refuses to go away despite the blurring of borders and extension of transnational corporations. It keeps coming back in new forms, whether as declining regions in the UK, thriving industrial districts in Italy, state-led growth in East Asia, or the global shift towards the Pacific basin. We might go so far as to refer to it as 'multiple capitalisms', with development trajectories often working against the grain of the dominant trends in the world system, the American age, or high Fordism.[1] These are not simply variations on global themes, but often the origins of differences in wages, technologies or organizations that prove critical to the operation of world capitalism as a whole.[2] Conversely, global opportunities and challenges are necessary but insufficient to explain growth and decline; internal social relations and production conditions are the crucibles of growth. California stands principally as an exemplar of extraordinary success in one place, a locality that has altered the global scene in remarkable ways by exporting innovations from the atomic bomb to micro-electronics.[3]

Economic geographers have also been working their way back to political economy through the reassertion of the local and uneven development. It is not so long ago, after all, that the peculiarities of place were demoted in an effort to get at the larger processes of capitalism, corporations and industrial restructuring. Since then, an appreciation of the social order of places has grown through various strands of locality studies, regulation theory, industrial governance, regional assets and networks, and the cultural turn. Most such work comes under the heading of thick description (the empirical turn), the search for middle-range theories, and the avoidance of economic determinism. This has been all to the good in one sense, but in another it has meant a loss of direction and of the analytic insight provided by Marxist political economy.[4]

This does not mean simply a return to economic foundations – although capitalist dynamics do have logic of their own once set loose upon the world (despite their prior and repeated social construction). But if institutions and embedded practices are crucial to economic action, what of the embeddedness of institutional forms in the power relations of property and labour? If regional assets are social in nature, let us investigate their relation to social structures of class, race and gender. If working people help produce labour geographies, then ruling-class practices must be responsible in large part for the differences in regional capitalisms. Finally, if we are to talk of governance, how about speaking plainly about government and politics? This is most difficult, especially for one prone to structural and historical accounts as I am, for it means capturing the critical moments in the froth of political currents when important changes in the political economy of capitalism are made manifest.[5] Moreover, these currents may surge over the banks of the local, wreaking havoc on the terrain of global political economy, as in Reaganism's terrifying impact on the globe or the contemporary impact of California's assault on immigrants.

## THE HARDER THEY COME: ECONOMIC GROWTH AND CRISIS

California has grown bigger than all but six countries in income and output, with a gross domestic product of $700 billion in 1990 and $900 billion in 1995 (it runs neck and neck with Great Britain). It enjoyed a spectacular boom from 1975 to 1990, seemingly immune to the American and European disease of falling profits, foreign competition, plant closures and stagnant employment.[6] When the circus was over, the big top fell in with a spectacular crash that triggered widespread panic about the future of the state and left the poor and the dark exposed to the winds of economic destruction and political scapegoating for the débâcle.

During the long boom, 5.5 million new jobs were added, employment peaking in 1990 at 14 million. Average income per capita doubled from 1980 to 1990 (18 per cent in real income). Well positioned on the eastern flank of the Pacific Rim, California became the national leader in exports to the global market, going from 10 to 20 per cent of US foreign trade, and the biggest recipient of direct foreign investment. Southern California manufacturing employment peaked in 1988 at over 1.25 million jobs, making Los Angeles the biggest industrial centre in the USA (twice the size of Chicago), while the Bay Area doubled its employment and Silicon Valley became the densest manufacturing site in the country.[7]

California took over as the principal engine of US economic growth and its high-tech sectors were trumpeted as the model for a nation losing its knack in manufacturing. On one side was electronics, where Silicon Valley was hailed as the world centre of the new computer-information age and emblem of American innovation and entrepreneurship at its best. On the other side was mighty aerospace, the American trump card for beating back the Soviets and economic decline; as defence spending shot up to $300 billion, California's share of prime contracts peaked at 23 per cent. A new generation of 'smart war machines' was ushered in, and Orange County avionics became the biggest electronic cluster on earth, while the Bay Area received huge contracts for satellites, guidance systems and Star Wars lasers. Everyone rushed to study the new technopoles. (Markusen *et al.*, 1991; Saxenian, 1994; Scott, 1993a).

Then there was finance capital: California entered the 1980s with the world's largest bank (Bank of America) and credit card company (Visa), the country's biggest Savings and Loans (led by impresario Charles Keating), and the nerve centre of the junk bond market (presided over by Michael Milken). New branches of foreign banks sprang up like mushrooms and loans were easy to come by. As regulations fell, fast-buck operators shuffled a deck of dubious assets, backed by the wizardry of Wall Street. Inflated by fire-sale finance, construction ballooned to $40 billion in 1989 (five times the previous peak in 1973). Excess piled upon excess, and the California economy became white-hot. (Pizzo *et al.*, 1989; Hector, 1988; Robinson, 1990; construction figures from Walker and Lizárraga, 1998).

Economic growth through the 1970s and 1980s was fed by the influx of almost 5 million new people from around the world. Not only were California jobs plentiful, they paid well – better than in the rest of the USA and an order of magnitude higher than in Mexico or China. During the recent boom, both wages and incomes remained well above the national average. California's wage and income advantage has not disappeared despite the arrival of millions of new workers, however. California has enjoyed a virtuous circle of investment, employment and spending in a highly diversified economy, and its skilled labour and ample capital have sustained a high rate of innovation that keeps California products in demand far and wide. (Walker and Lizárraga, 1998; Walker, 1996).

The crisis of 1990–5 slammed the high-flying California economy harder than anything since the Great Depression. The state was forced into collective downsizing in the wake of a decade of over-accumulation of factories, workers, securities, real estate and executive fat. Wealth shrank, thanks chiefly to real estate values shrivelling by 25–30 per cent. After leading the country in new business formation, California's failure rate soared to 20 per cent of national bankruptcies in 1992. Construction came grinding to a halt almost everywhere in the state, with housing starts hitting the lowest point since the Second World War. Southern California was the worst hit. The post-Cold War military cutbacks cost the state some 250000 of 400000 jobs in defence. Greater LA accounted for over a quarter of all job losses in the country in the period 1990–3, losing one-quarter to one-third of its manufacturing workforce. (Scott, 1993b; Davis, 1993a, 1993b). The south hit the financial skids as paper empires sank without a trace, Milken, Keating and other con-men went to jail, bank lending stagnated, venture capital plummeted and Japanese investment dried up. Then LA watched helplessly as a revived Bank of America and Wells Fargo bought its two largest banks.

Workers felt the brunt of the catastrophe. Gross job loss amounted to almost 1.5 million (10 per cent) from 1990 to 1992, including 900000 in wholesale and retail, 200000 in manufacturing, 150000 in construction, 70000 in agriculture. Net job growth was negative from 1991 to 1993, with unemployment nudging 10 per cent by 1993 and remaining at 8 per cent in 1995 – two points higher than the national average. Not surprisingly, immigrants stopped coming. In-migration plunged after 1990, out-migration increased, and net migration hit zero in 1992–3 as the recession bottomed out.[8]

California sits on the cusp of an epochal change in the geography of capitalism in which its place is no longer secure. (Council on California Competitiveness, 1992; Levy and Arnold, 1992). It has seen such economic sea-changes before, and survived through a combination of new technologies, political initiatives and cultural change. This time, one cannot be sure. An uptick in the business cycle is restoring some of the bloom to the Golden State, with strengths in electronics, entertainment and exports to East Asia. But long-run industrial leadership may be passing irreversibly across the Pacific. No region (or nation) is ever immune from the inevitable downswings of accumulation and shifts in the fortunes of places. (Storper and Walker, 1989).[9]

Faced with a crisis of such dimensions, California's leaders have been content with neo-liberal delusions.[10] Predictably, California business backs the North American Free Trade Agreement and the World Trade Organisation. While the rest of the Pacific Rim has a strong hand of government at the helm, local capitalists are still beating the bible of the entrepreneurial spirit. They have been encouraged in this by the way Silicon Valley, Hollywood and the biotech industry have become global icons of free-wheeling initiative, as well as by how American military prowess is second to none, as if there were nothing more to it all than wily businessmen, weaponry, imagineering and venture capital. Modern economies are profoundly social economies, whose productivity is a function of widely available and deeply embedded scientific knowledge and technical skills, worker education and practical know-how, organizations and managerial competence, and governmental laws and regulations. These things are neither cheap, nor instantly procured in the market-place; they require long-term and expensive investments in physical infrastructure, education, institutions and people. Moreover, the social economies of capitalism are markedly localized, resident in such favoured places as California. Today's bourgeoisie is living off the fruits of such investment in social assets, all the while singing the praises of free markets and instant fortunes. They have little notion of the benefits of an industrial policy (that is, a set of investment strategies and mechanisms of sectoral governance, geared to upgrading technology, business organization, labour skills, collective infrastructure and markets) for keeping the state on the high road of development. (Storper, 1992; Scott and Storper, 1995).[11]

The state's long history of industrial policies is little known. In the Gold Rush era, water and land laws were invented on the spot. Subsequently, railroads were planned, aqueducts surveyed, urban land development systematized, ports carved out of the coastline, electric power systems put in place, and new universities founded – all with massive government intervention at every level, collective effort led by key capitalists, and carefully engineered class alliances to ensure wide popular backing. All this happened *before* anyone heard of the New Deal and the Welfare State. After the Second World War, California's developmental policies, backed generously by government muscle and finance, included the world's largest and most advanced highway network, higher education system, water projects, and university and public research apparatuses for agribusiness, medicine and electronics (among others), as well as important port modernization schemes, land use management reforms and the like (Nash, 1964).

The military industries have been a form of state planning and subsidy on a colossal scale. The Second World War tilted the national economy sharply towards the Pacific. Reagan's Second Cold War underwrote the boom of the 1980s. But military industrialization has been remarkably sterile in terms of industrial process and consumer product development, given the logic of arcane weaponry. Modest post-Cold War cuts have left the state frantically trying to induce some kind of 'reconversion' of military bases and weapons companies to peaceful uses. So far, projects such as the electric car, trolleys and base-recycling have shown little promise for economic revitalization. (Markusen and Yudken, 1992).

The hottest industrial debate of the recession was geographic, over the out-migration of factories and offices such as Lockheed's Burbank assembly plant or Bank of America's credit card operations. Howls have gone up that the costs of doing business in California are too high, driving away jobs and investment – a complaint based on the notion that local cost structures determine industrial location. Yet California has always been a high-cost production area, and is among the world's technical and creative leaders in a range of vital industries, such as biotech and pharmaceuticals, medical and scientific instruments, entertainment, clothing and petrochemicals. California exemplifies, better than perhaps anywhere on earth, the virtues of a high road of industrialization based on skilled labour, abundant capital, advanced science and engineering, technical innovation, and marketing to a prosperous citizenry. A region on the high road can afford higher wages, rents and standards of performance. (Storper and Walker, 1989; Saxenian, 1994; Scott, 1993b).

But fears of fleeing investment have gained the upper hand. The last two Governors established expensive programmes to advertise California to outside investors, as if no one knows it is there (and even though most investment in the state's development has been internal; Walker, 1996). Enterprise zones have been carved out to promote urban economic renewal, as if sweatshops were not already proliferating in the 1980s. Yet the business wolves kept baying after more tax cuts, less spending and fewer regulations. So the legislature called a State Economic Conference in 1993 as a forum for business belly-aching, then passed a package of bills to limit workers' compensation claims, blunt environmental regulations and lower business taxes by $400 million. The president of the California Manufacturers Association called it 'an unbelievably happy occasion for all of us' (*San Francisco Examiner*, 7 October 1993, p. B-1).

The last generation has seen a ballooning of cheap labour in manufacturing, construction, domestic service, agribusiness, hotels and restaurants. This trend may well send California on a disastrous slide off the high road of development: cost-cutting in the name of competition is the low road, and is always vulnerable to low-cost production from poor countries. (Davis, 1993a, 1993b; Scott, 1993b). It flies in the face of California's experience with high-road growth. But the rhythm of California's past is not a simple upbeat or downbeat; it has been, rather, a tendentious mixture of high-wage skilled work and low-wage degraded work. California's garment, electronics and agribusiness sectors have, for example, been highly productive and dynamic while depending heavily on cheap labour for lower-level tasks. This two-track labour market tradition goes back over a century. (e.g. Villarejo and Runsten, 1993; Siegal and Markoff, 1985; Hayes, 1989). But can the whole economy function as a high-tech sweatshop or tomato field? It is unlikely that a mean-spirited yet innovative economy is viable, and that its intense class and race contradictions can be contained. Such containment was achieved in post-war East Asia, to be sure, but not without dramatic land reforms, reduction of income disparities and managerial hierarchies, highly authoritarian state systems, and the Cold War sword hanging over everyone's heads (Wade, 1990; Tabb, 1995; Amsden, 1989). In California it would require an ugly counter-revolution; but that may be exactly what is under way.

## RACE AND CLASS SCHISMS: THE FALL OF THE WHITE REPUBLIC

California has been a state of immigrants from the Spanish conquest in the late eighteenth century. It has never known a decade when the number of newly arriving people did not exceed the number of those born within the state. (Gordon, 1954; Walker and Lizárraga, 1998). Since the extermination of the indigenous people, the vast majority of Californians have been of European origin. Nevertheless, Asians and Mexicans have been a constant presence, and African-Americans finally arrived in large number in the 1940s for wartime work. Race in California and the West has never been a black and white screen, but a colour wheel with many axes (Limerick, 1987; Almaguer, 1994).

California became whiter in the mid-twentieth century, thanks to the exclusionary quotas of 1924 and the inter-war break-up of the European world-system. The gradual return to economic globalism, capitalist penetration of the Third World, and loosening of immigration restrictions in 1965 have returned California to something nearer its appearance in the nineteenth century. But the geography of new arrivals has tilted sharply towards the Pacific Basin, in line with the shifting centre of gravity of world capitalism. In the past two decades, California displaced New York as the chief receiving area for immigrants (35 per cent versus 14 per cent of the US total in the 1980s). Some 400000 migrants per year poured in during the past decade (versus 300000 births), and the state's population surged past 30 million by 1990, up 12 million in the previous 20 years.

This influx has transfigured the face of California. It will become a majority-minority state in the next century. Latinos rose sharply in number during the 1980s (by 70 per cent), Asians more precipitously (by 127 per cent). By 1990, whites had fallen to roughly 57 per cent of the populace, while Latinos jumped to 26 per cent and Asians nearly 10 per cent (Africans holding at about 7 per cent and indigenous people at 1 per cent). The number of foreign-born residents went up by 80 per cent. An economic earthquake moved 2.5 million Mexicans northwards, where they joined half a million Filipinos, a quarter-million Salvadorans, Vietnamese, Koreans and Chinese, and over 100000 Guatemalans, Canadians, Britons and Iranians. The central cities have undergone the most dramatic recomposition. In 1970, Los Angeles was 75 per cent white, by 1990 only 38 per cent white. San Francisco went from 75 to 43 per cent white in the same period.[12]

Racial recomposition of California went hand in hand with class recomposition. The working class of the 1990s is overwhelmingly Latino (mostly Mexicans and Central Americans) and South-East Asian (mostly Filipinos, Vietnamese and southern Chinese). Overall, 79 per cent of Mexican-origin men were in blue-collar jobs in 1980 versus 55 per cent of Anglo men. Because of labour market segmentation, immigrants are largely confined to specific occupational 'niches' (Waldinger and Bozorgmehr, 1996). Mexicans dominate southern California manufacturing, Salvadorans stock the furniture industry and gardening crews, Guatemalan and Salvadoran women are domestics, Chinese and Thai women fill the garment sweatshops, Little Vietnam supplies the electronics belt of Orange County, Silicon Valley electronics feeds off men and women of many origins, Chinese and Filipinos labour in the restaurants and tourist hotels of San Francisco, and agribusiness in the interior valleys makes hay on the backs of mestizos, Mixtecans and Zapatecans.

Contrary to popular images of hordes of unskilled peasants jumping border fences, today's immigrants include many who are competitive for technical, professional and managerial jobs, as well as well-capitalized business owners and entrepreneurs. The numbers of skilled immigrants are particularly large among East Asians, South Asians and Middle Easterners. Fields such as medicine, engineering and computing have become immigrant niches for Iranians, Chinese, Filipinos and Indians (Waldinger and Bozorgmehr, 1996: Light and Bonacich, 1988). Less visible are the many Canadians and Europeans in electronics, banking and teaching. These favoured migrants usually arrive already trained, and with considerable acculturation to English and American commercialism, and they have permeated the petty bourgeois layers of California society. California has always received an extraordinary bounty from its skilled and well-capitalized migrants and, unlike the rest of the USA, has seldom felt the full impact of mass migrations of the rural poor (Issel and Cherny, 1986; Gordon, 1954).

California has always depended on long-distance migration to feed its growth. In fact, the percentage of foreign-born residents was considerably higher a century ago: 39 per cent in 1860 and 25 per cent in 1900 versus 22 per cent in 1990. The recent wave of migration was no larger, nor of longer duration, than the great post-Second World War influx, and fits closely to a pattern of 15–25 year 'long swings' of migration going back a century. (Gordon, 1954; Walker and Lizárraga, 1998). The logistics of coping with millions of new people are imposing, and require money, ideas and commitment to rebuilding the state. But it has been done time and again. Schools, houses and infrastructure were built in ample number for the baby-boomers of the post-war era of in-migration. So what has changed? The racial composition, to be sure, but also class welfare and politics.

As the economy was roaring, a yawning chasm between the classes was opening up that left the USA the most unequal of all wealthy countries. California led the pack. Those who owned capital did spectacularly well. California's jetstream of fast-track entrepreneurs and *rentier* families more than doubled in the 1980s to over 340000 millionaires, and its richest families – Hearsts, Packards, Waltons, Gettys, Haases, Bechtels – disproportionately fill the top ranks of America's *haute bourgeoisie*.[13] Mike Milken earned the highest personal salary in history, while Richard Riordan, now Mayor of Los Angeles, made $100 million through leveraged buy outs. In the Bay Area the number of million-dollar executive paycheques jumped by a factor of ten (*San Francisco Chronicle*, 23 May 1994, p. B1). The professional and managerial class prospered: average income for the top fifth of families rose by 15 per cent to $107000, and the Bay Area, centre of the Yuppie lifestyle, remained the richest and most expensive large metropolitan area in the country.

Meanwhile, the working class lost ground. The real income of the middle 20 per cent remained flat through the decade (and declined by 10 per cent in the costly Bay Area). For the lower 40 per cent the bottom fell out.[14] Wages stagnated in full-time jobs, while temping and part-time work increased. Working people kept up their income by sending more family members out to work, holding two or three jobs, and working more overtime. Chronically high unemployment averaged 7.5 per cent over the past 20 years. A staggering gap opened up between total state product and total wages (including salaried professionals), expanding from $155 billion to $330 billion over the decade. This crude measure of growing aggregate surplus value helps explain why the rich did so well.

While wages for all workers ebbed, non-whites fared the worst. Latino workers earn 70 per cent of the earnings of white workers and per capita Latino incomes are 45 per cent of those of whites because of larger families. Blacks and Asians do somewhat better, with per capita incomes 61 per cent and 72 per cent of whites, respectively. Unemployment rates have been continually higher for non-whites. Immigrants have been a new mother-lode of economic surplus pocketed by the upper classes. Considering Latinos only, the excess profits from hiring them instead of better-paid white workers was about $85 billion in 1990. All the same, the widening class schism shows up *within* every race or nationality – European, African, Asian or Latin – muddying the race–class alignment. Throughout the mean-spirited 1980s, new battalions were added to the armies of the poor. The poverty rate stood at 12.5 per cent in 1990, before the recession sent it skyward to 18.2 per cent. This put California into the top 10 poor states in this most impoverished of rich nations. Saddest of all was the astronomical 25 per cent rate of poverty among children, 33 per cent of those under 6.[15]

In short, working-class comfort and security have declined in tandem with a massive engorgement of the rich. Anxiety over unemployment, bad wages, poverty, job competition, housing and health care is rife. All this would have been true regardless of immigration, because the erosion of working-class incomes and welfare has been taking place throughout the country and, indeed, the world (Harrison and Bluestone, 1988; Phillips, 1990; Goron, 1996; Fischer *et al.*, 1996; Danziger and Goltschalk, 1996; Schor, 1991). But in the venal rhetoric of Governor Wilson and the right, the poor, criminals, welfare mothers and immigrants are to blame for California's descent into the maelstrom – never the white millionaires.

California led the way in the long wave of hysteria over crime in the USA. Beginning with the Nixon presidency, the War on Crime was unleashed in the face of mass social unrest, above all the Watts Rebellion. Federal support added battalions to police legions, new armaments of repression, more bite in criminal penalties, and hundreds of prisons. All this was intensified under Reagan's War on Drugs, when LA's 'gang wars' were engraved on public consciousness and northern California marijuana fields came under assault (Davis, 1990). The anti-crime wave has been newly topped up by the $30 billion Crime Bill passed by Congress in 1994 – with LA's 1992 riots and the murder of the Bay Area's Polly Klass (the Lindbergh baby of our time) in the forefront. Since 1980 California has made prison construction its main form of infrastructural investment, spending over $5 billion on 19 new prisons, and has become the carceral state *par excellence*, with 125 000 prisoners today (200 000 including local jails and Youth Authority camps) The recent Three Strikes law may require another 15 to 20 prisons. Crime has been radically racialized, so that people of colour are now incarcerated at six times the rate of whites, and California jails more young black men than South Africa (Henwood, 1994; T. Platt, personal communication). California also ranks with Mississippi among the worst states for prison brutality.

The counter-attack on the poor began with Nixon's dismantling of the War on Poverty. Reagan followed up with a virulent campaign to punish welfare mothers, subsidized renters, and free-lunching schoolchildren. Social assistance cutbacks threw millions into poverty and hundreds of thousands on to the streets without shelter (Bloch *et al.*, 1987; Burt, 1992; Gans, 1995). California's recession and stinginess continued Reagan's evil work. Statewide Aid to Families with Dependent Children cases rose by 40 per cent between 1988 and 1993, even though benefits had fallen 20 per cent in real terms since 1973. General Assistance has been cut to the nub in the name of budget balancing, and Riverside County runs a nationally touted programme to force welfare recipients into low-wage work. LA police sweeps to rid the downtown area of homeless people began in the late 1980s, while San Francisco carried out a draconian crackdown in the 1990s. Local anti-panhandling ordinances have been passed up and down the state (figures from the *San Francisco Chronicle*, 13 December 1993, p. A1; also Wolch and Dear, 1993, and Winocur, 1994).

Now it is immigrants' turn to feel the sting of the venom; witness Proposition 187 (to refuse all state aid to undocumented immigrants). The nativist refrain is always the same: former immigrants were good, hard-working assimilators, while the new ones are inferior, parasitic and implacably foreign. California has an ignoble history of this sort of distinction. By the 1920s, Lewis Terman of Stanford, co-developer of the IQ test, was calling

Mexicans 'uneducable' and practical eugenicists were sterilizing more 'defectives' in California mental hospitals and prisons than anywhere in the USA. In the 1960s, William Shockley of semiconductor fame and Arthur Jensen of the University of California at Berkeley were the foremost exponents of African genetic inferiority. Today we have the Federation for American Immigration Reform (FAIR), backed by biologists Paul Ehrlich of Stanford and Garrett Hardin of the University of California, Santa Barbara, writing Proposition 187.[16]

Nativism is underpinned by economics, recruiting working people, including many African-Americans, to the view that immigrants take away jobs. No doubt there is competition, but labour market segmentation channels immigrants heavily into jobs expressly meant for them, with surprisingly little spillover and with high unemployment among the new arrivals themselves. Migrants are drawn overwhelmingly by labour demand rather than pushed out of their home countries by poverty, as the close correspondence of business cycles and migration cycles shows. Immigrants were not to blame for 'glutting' labour markets in the recession; lay-offs were the real culprit and unemployment rates rose equally for all races. At the same time, blacks remained unemployed even at the height of the boom, having lost thousands of union jobs in heavy industries and being uninvited to the party in the new technopoles (Piore, 1979; Thomas, 1974, Gordon, 1954; Walker and Lizárraga, 1998).[17]

What pushed the anti-immigrant agitation to centre stage was not mass racism or job competition, but the budget crisis and political opportunism. Pete Wilson rode out ahead of the anti-immigrant posse because Proposition 187 offered fiscal and electoral salvation. When Democrats jumped aboard Wilson's personnel carrier, Proposition 187 became a model for national policy – closing the door further on immigration.

## GOVERNMENTAL RIGOR MORTIS

Faced with the gravest crisis in half a century, Californians find themselves without an effective political system to provide direction. The treasury is empty, the Governor refuses to lead, the legislature is paralyzed, money buys everyone, and the citizenry are disenfranchised. The state has gone from having the cleanest and best machinery of government in the post-war USA to having some of the worst, setting trends for the country as it went.

The fiscal crisis of the state overwhelmed everything in the first half of the 1990s. Gargantuan deficits threatened the state's ability to govern. The 1992–3 budget had an $11 billion shortfall out of a total of $50 billion, and when the Governor and legislature refused to compromise on a strategy to cover the shortfall, government workers were issued paper IOUs and California was effectively bankrupt. Virtually all of the key functions of state government have suffered expenditure cutbacks of one-fourth to one-third. For example, California has plummeted from being one of the highest-ranked states in the USA in per pupil public school spending to being dead last. Higher education has been so severely cut that it now takes up less of the budget than cops and prisons.[18]

The burden of cutbacks has been borne largely by local governments (cities, counties and special districts), which have never recovered from Proposition 13 of 1978 – the beacon of the US 'tax revolt'. While slashing spending, local governments switched to a regressive system of parking tickets and sales taxes, making up the difference with state and federal revenue transfers. But the latter have whittled billions off transfer funds while shunting enforcement of many programmes on to state and local officials. Orange County went spectacularly bankrupt while trying to cover revenue shortages by speculating in the financial markets. The deficit is treated as an act of God, but assuredly is not. Proposition 13 and other tax cuts allow a huge proportion of the state's wealth to go untaxed. As a consequence, California has fallen from fourth among the states in tax rates to twentieth. Simply restoring the income tax and the property tax to previous levels would eliminate the deficit (Goldberg, 1991).[19]

The failure of leadership begins in the chief executive office. Reagan launched his political career as Governor from 1966 to 1974 and helped eliminate the moderate Republican lineage of the Progressives and usher in the New Right. But the electorate had not yet given up on liberalism, handing Democrat Jerry Brown landslide victories in 1974 and 1978. Brown was more fiscally conservative than Reagan, however, and he allowed a budget surplus to build up in the face of the revolt for lower taxes; then his social liberalism faded under fire from business critics of environmental regulation. His successor, a latter-day Calvin Coolidge named George Deukmejian, set about dismantling regulatory structures, vetoing legislation, and letting government fall into disuse for the next eight years. Pete Wilson was then elected as a liberal

Republican, but did an abrupt right-turn under pressure from the budget deficit and the Christian Right.

California government suffers from legislative sclerosis as well. As in Congress, the Democrats held the majority for decades, but without any clear direction in the past twenty years. An unruly system of personal fiefdoms was held together by liege lord Speaker Willy Brown, by the careful dispensation of state Democratic Party funds. The legislature grew increasingly venal over the years, and several key legislators have been caught accepting bribes. A progressive Proposition 73 in 1990 put ceilings on campaign contributions, but was overturned in the courts – money having been granted rights of free speech by the Supreme Court in 1976 in *Buckley* v. *Valeo*. Then a conservative Proposition 163 installed term limits on legislators in 1994, further reducing the professionalism and power of the legislature.

Legislative and executive sclerosis has led to bypass surgery, in the form of the Ballot Initiative. Many of the most vital issues of the past twenty years have been decided by this means. Conservatives have had a field day with the initiative game, passing Proposition 13, a Victims' Bill of Rights, Three Strikes and funds for prison-building. Liberals have responded in the same coin, as with toxic substance control, insurance control and coastal preservation, but have lost more often. Initiatives are even more subject to the vagaries of money flows and media campaigns than ordinary legislation, and ballots have become incomprehensible to any but the most dedicated of voters.

Judicial law has filled in the gaps. Here again, the conservative agenda has had the upper hand and helped to freeze government in its tracks. The Right (led by LA's Richard Riordan) pulled off a stunning *coup* in 1986 by removing the three most liberal justices of the California Supreme Court, targeting for particularly vitriolic attack the first woman Chief Justice, Rose Bird. Unbelievably, the liberal justices and the Democratic Party sat on their hands throughout the campaign, hoping it would go away. California's high court, once a paragon of legal scholarship and activism, is now notorious for mediocrity of argument, refusal to hear hard cases, favouritism towards business and unparalleled eagerness to uphold death sentences (*San Francisco Chronicle*, 16 November 1993, p. A1).

Political parties have become shells, opening the way for media-driven campaigns that focus on free-range candidates rather than issues. California long ago gave the world 'yellow journalism' through William Randolph Hearst and the Hollywood image campaign. Meanwhile, 'non-partisanship' covered for business-class dominance by other means; candidates for many state and local offices are prohibited from wearing party labels. So candidates run on their bank accounts rather than party principles, and media careers are considered ample experience for office. In the 1994 Senatorial campaign, Michael Huffington spent over $30 million, while winner Diane Feinstein spent $24 million (versus an average nationally of $4 million).[20] The Democrats have ceased to represent a viable alternative – and did so only for twenty years after Pat Brown's election as Governor in 1958. When radical movements have arisen in California, they have created their own organizations, such as the Workingmen's Party, the Union Labor Party and the Peace and Freedom Party.

Voting has declined steadily, reaching all-time lows in the 1990s of 54 per cent of eligible adults in presidential elections, 40 per cent in gubernatorial elections and 25 per cent in primaries. Voter apathy reflects the euthanasia of government: a non-functional state is hardly worth bothering over; the big parties are indistinguishable; lists of propositions are daunting; anti-government rhetoric is rife; vitriolic campaigns teach voters not to trust politicians; and the open purchase of office cheapens voting. Another reason for the receding electorate is that the size of representative districts has ballooned. A state senator today represents a populace of 800000. Worse, immigrants, people of colour and the working class are hardly represented. The California electorate today is two-thirds white, two-thirds over 40, and two-thirds earning more than $40000. People of Asian, African and Latino origin represent 47 per cent of the total population, 43 per cent of adults, 30 per cent of citizens eligible to vote, 24 per cent of registered voters and only 17 per cent of voters in 1992. Minority office-holders are scant at every level of government (figures from the *San Francisco Chronicle*, 22 September 1994, p. A4).

The death of government and electoral politics has sources deep in California's moneyed culture, rootless people and middle-class libertarianism, but has more to do with the contemporary political imagination of the bourgeoisie, the rightward flow of politics, and the euthanasia of the working-class franchise. Yet the prevailing discourse about how to end 'gridlock' in Sacramento targets the mechanics of government instead (Schrag, 1994). A Constitutional review commission was appointed, with some commentators urging a unicameral legislature, others a parliamentary system, still others a new Constitutional Convention. The Commission wants to remove the two-thirds rule for appropriations and budget approval, and favours eliminating all state-wide elective offices other than Governor. The most radical proposal in the air is to split California into two or three states (Levenson, 1993). While extreme, the secessionist movement points to the utopian hopes buried beneath political discourse today. People want the commonwealth to be made whole again. They imagine a community with a common interests and bonds of citizenship; they want to be working and protected from the buffeting of international economic winds; and they want government to function and be responsive. The tragedy is that these impulses have too often been channelled into retrograde forms: seeking after the white republic, walling up the border, or imprisoning the werewolves of social disintegration. What is needed is a radical turnabout in political life, which a change in geographic scale might assist but cannot guarantee.

## COUNTER-REVOLUTION IN ONE STATE?

How did things come to this impasse? Certainly the problems facing the state would challenge any ruling elite or government. None the less, the chief response of California's leaders has been to make matters worse. Why such immobility, arrogance and fear in the face of California's tripartite crisis? Today's malaise must be laid at the feet of the political counter-revolution that staunched the radical critique of American society which peaked in the 1960s. Every capitalist order requires political and social renewal; that is, new people must rise to the top, new ideas be heard, and new institutions put in place.[21] Market adjustments alone are insufficient to retool the economy, heal social divisions, or keep the machinery of state functioning. This is where class and race relations, government and politics, and the mechanics of capital accumulation come together, the strands becoming so twisted upon each other that something must give.

The course of upheaval in the 1960s is too well known to bear rehearsing. Among the bourgeoisie a reaction quickly set in against the central achievements and ideological gains of the radical protesters and liberal reformers caught up in the surge of popular dissent. The Vietnam syndrome, Affirmative Action, feminism and abortion, drugs and sexual freedom were anathema to the reactionaries, who found a thousand points to fight. Three currents converged as the tide of counter-revolution rose through the 1970s: take back the commanding heights of the state through a political offensive, recapture the past hegemony of Anglo-American racial and cultural dominance, and restore the rule of money-making and business legitimacy.

The rollback hit with full force with Reagan's ascension to the Presidency. Reagan's geographic origins are crucial: his agenda was crafted in California. He cut his teeth as a spy for the FBI while head of the Screen Actors Guild during the purge of Hollywood's left in the 1940s. Southern California has spawned a viperous strain of reaction with many sources: war industries, Veterans of Foreign Wars, the LA police department, triumphant Anglo-Saxonism, a Mexican proletariat, evangelicalism, and more (*see* McWilliams, 1946; Davis, 1986). The southland's proliferation of right-wing groups, such as the John Birch Society, the Church of Scientology and Concerned Women for America, is legendary. Meanwhile, a similar social base produced a different political tradition in the San Francisco Bay Area. Militant labour reached a stand-off with the capitalist class, the children of the robber barons leaned toward Progressivism, and the middle class took to secular libertarianism and nature worship. During the social and cultural revolution of the 1950s to the 1970s, the Bay Area achieved the closest thing to popular opposition to capitalism in America, showing both the civilizing potential of rebellion and the limits of the left in the country of Babbitry. San Francisco came to represent everything the right hates: organized labour, militant blacks, race-mixing, beatniks, hippies, anti-war students, free love, drugs, homosexuals and bleeding-heart liberals (R. Walker, 1990; Ashbolt, 1989).

The counter-revolution marched north from its strongholds in southern California to wreak its vengeance, with Reagan leading the attack on radical students and Fair Housing laws. His first military intervention against a small, unarmed populace took place in Berkeley in 1969. While skirmishes continued for the next twenty years, the struggle for the political soul of California was won by the right. An organized movement of the business class won over key sectors of the petit bourgeoisie and the workers in the name of 'liberation' from government, freedom to strike it rich, defence of law and order, and maintenance of the white republic. This was a political war declared on the left, the poor, the dark-skinned and organized labour, with the Democratic Party as the Maginot Line of the benighted. After that, the Bay Area was no more than a thorn in the side of conservative America.

What happened was a power shift towards a new centre of accumulation, southern California. The counter-revolution tested on California was subsequently visited on the whole country, and the rest of the globe.[22] It has meant the dismantling of the social welfare state, the enrichment of the top 5 per cent, reduced social investment, greater militarization and arms sales to fuel internecine slaughters, increased trade in drugs, further repression of labour and liberation movements, and greater poverty and social disintegration throughout the world. Its success has been a global tragedy, and now another wave of right-wing initiatives on immigration, incarceration, welfare cuts and affirmative action in California is spreading dangerously across the land.

Counter-revolutions do more than punish the poor and decimate the opposition; they turn back the hands of the clock, even as the inner works of history keep ticking onward. In the end, no ruling class can freeze time and space and hold on to its golden moment of triumph for ever. The counter-revolutionaries erupting from the fumaroles of California's economic uplift have deluded themselves that everything is well with the capitalist order of America. Every historic epoch is a mortal thing that eventually sickens. It may give way to rebirth of the social order or to perpetual senility – or barbarity and calamity, as in the racialized gulag of California's prisons, brutalization of African-American ghetto youth, the mass uprising of black and brown poor in the LA riots, the return to capital punishment and castration, or the rising body count of Mexicans fleeing border police (Davis, 1990, 1995).

California's extraordinary record of expansion has left it with a massive set of strains on its economy, governance and social cohesion. It has enormous reservoirs of talent, capital and exploitable labour from which to draw. Yet the state seems unable to make a breakthrough into new configurations of political participation, social renewal and economic restructuring. Californians are unable to see their way to reconversion of the war industries, racial integration, reconstruction of the cities, corporate reconfiguration, salvage of public education, full employment or universal health care. The failure of imagination rests partly on bourgeois ideological reflexes, but behind that lies a political impasse born of the right-wing hold on government and the public agenda. The right's victory left the structures of governance in disarray but those of class and race power intact, and that power is the thing that matters. It will not be given up voluntarily, and if we are not to be mourning the passing of the late, great Golden State as economic miracle and field of hope for millions, there will have to be a change caused by boiling energy and anger below.

## ACKNOWLEDGEMENT

A longer version of this article is R. Walker, 1995: California rages against the dying of the light, *New Left Review* 209, 42–74.

## NOTES

1. On alternatives roads to capitalism, the classics are Lenin, (1964 [1906]), Trotsky (1969 [1931]), Gerschenkron (1962) and Moore (1966). More recent classics are Brenner (1977), Post (1982), Wade, (1990) and Evans (1995).
2. For a dynamic theory of economic geography, *see* Storper and Walker (1989). On reflection, our book's strength is its attention to the unexpected and unsettling impacts of peripheral development, but it underplays the social and political bases of economic growth and change.
3. I explore this further in Walker (1996). *See also* Brechin (1998), Scott and Soja (1986, 1996). I agree with Soja and Scott that southern California is a critical model for the world, but by no means the future in microcosm.
4. For critical reflections on these moves, *see* Harvey (1989a), Pred and Watts (1992), Massey (1994), Sayer (1995), Walker (1995a) and Storper (1995c).
5. That is, bridging the gap between capital logic as worked out by Harvey (1982) and the current political analysis done so brilliantly by Davis (1990).
6. On the falling rate of profit and uneven development in the world economy since 1970, *see* Brenner (1997) and Webber and Rigby (1996).
7. On Los Angeles, *see* Davis (1990) and Soja (1989); on San Francisco, *see* R. Walker (1990).
8. Figures from *San Francisco Examiner*, 4 November 1992, p. C1, and *New York Times*, 19 December 1995, p. C22. On discouraged immigrants, *see San Francisco Chronicle*, 2 September 1993, p. Al, and *San Francisco Examiner*, 9 January 1994, p. B3.
9. On the recent upswing, *see* Ayres (1995).
10. For example, statements by the Governor's Council of Economic Advisors consisting of Milton Friedman, Bruce Boskin and George Schultz.
11. Concerning East Asia, *see* Wade (1990), Tabb (1995) and Amsden (1989).
12. All figures from the United States Census Bureau.
13. With 11 per cent of US adults in 1990, California had 17 per cent of US millionaires and 20 per cent of the 400 richest Americans.
14. California ranked 13th of 51 states in inequality in the 1980s; *San Francisco Examiner*, 22 August 1994, p. A6.
15. Child poverty figures from a study by Victor Fuchs and Diane Reklis of Stanford University, reported

in the *San Francisco Chronicle*, 3 January 1992, p. A-1. Wage figures and calculations from Walker and Lizárraga (1998). On intra-racial classes, *see* studies in Waldinger and Bozorgmehr (1988).

16. On eugenics, *see* Brechin (1996); the state reintroduced 'chemical castration' for repeat sex offenders in 1996. On FAIR, *see* the *San Francisco Examiner*, 12 December 1993, p. A10, Davis (1995), and Kadetsky 1994).
17. On black unemployment, *see* Davis (1992b) and Scott (1988a).
18. From 1978 to 1994, prisons went from 3.9 per cent to 9.8 per cent of the state budget, while higher education fell from 14.4 per cent to 9.8 per cent; *San Francisco Bay Guardian*, 31 August 1994, p. 7.
19. The deficit would disappear if real estate were taxed at the average national rate; Advisory Commission on Intergovernmental Relations (1988).
20. National figure from Raskin (1994). On Feinstein, *see* Benske (1994).
21. Cf. the analysis of the post-war UK in Anderson (1987).
22. Neo-liberalism became the dominant ideology and policy of the global bourgeoisie with the triumphs of Reagan in the USA and Thatcher in the UK.

# Concluding reflections on *Geographies of Economies*

ROGER LEE AND JANE WILLS

This book has begun to explore the intersections of economic geography with social, cultural and political theory and, in so doing, it seeks to enrich both theoretical and empirical understandings of the economic. By widening and 'stretching' notions of economy to incorporate other axes of social life and relations, the authors of this volume have shown that conceptions of the 'economic' are themselves strengthened. And while attempts to understand the economy are enhanced by such a geographical approach, so too the geographical discipline is augmented by careful attention to the economic. For understanding the human landscape is impossible without unpacking the processes of economic activity in all their diversity.

The impact of social theory and cultural studies has been positive for geographical scholarship in this field, securing both analytical and epistemological advances. Indeed, economic geographers illustrate the potential fruits of such new ways of thinking at a time when the reverberations of all things 'post-modern' have yet to erode the disciplinary island of economics itself. As the search for objectivity, neutrality and unpositioned research has been banished with the currents of post-Enlightenment thinking, economic geographers have been liberated to explore the interrelations of space and economy in new ways. Moreover, it is not just that culture, society and politics have been added to the economic for use by economic geographers; these intersections themselves reassert the significance of the geographies in, and through which, social reproduction takes place. The new conception of the economic insists that we recognize the social context of economic life, which is, of course, geographical. These new geographies also require a complex range of methodologies to capture the subtleties of the economic – a suite of research techniques which also have their own geographies. Researchers point to the importance of position, standpoint and definition, and the need to place our knowledge in its spatial and temporal location. But more than this, a widened notion of the economic also allows us to ask a huge range of questions, all pertinent to the world in which we live. So, for example, geographers might ask how we can integrate questions of subjectivity into understanding economic change; they might question the role of the economic in the constitution of culture, and culture in the economic; or explore the ways in which social divisions are reproduced through economic exchange.

In applauding these new developments in economic geography, however, we do not want to suggest that these 'new' economic geographies are simply replacing the scholarship of those working in the tradition of classical political economy. In some cases political economic analysis has made fruitful engagement with the challenge of new ideas (as David Harvey (1989a) has done for post-modernity or Andrew Sayer and Dick Walker (1992) have done in their understanding of the social economy); and in others, the tools of political economic understanding are ideally suited to the research task in hand. Indeed, it is hard to imagine the discipline of economic geography without political economy at its core, and many of the chapters in this book illustrate how economic geographers have successfully reconsidered and reimagined the traditions of political economy to reflect their concerns with social reproduction in the late twentieth century. In so doing, economic geographers have rightly been able to keep questions of politics and power at the heart of their enquiries.

But there is still much to be done. Capitalist social relations continue to be extended. Regulatory institutions such as the World Bank, the IMF and the World Trade Organization are propagating, even enforcing, neo-liberal prescriptions for capitalist economic development. Resistance to the imposition of such global values is by no means easy and has been extraordinarily diverse. Opposition may be recognized on the one hand in the fluctuating politics of parts of Eastern Europe, and on the other by the struggle of the Chiapas people in Mexico. But as the old certainties of the Cold War and post-war economy have been shattered, new progressive ideas are just whispers in a world dominated by a neo-liberal agenda. The centre-left across the world agonizes about the way forward from the neo-liberal agenda but with little indication of a clear way ahead. Political strategy is still in the process of being reconfigured to meet the demands of a world in which it often seems impossible to 'do' anything which can change things. In such a setting, there is a desperate need for ideological and practical alternatives; for new ways of imagining, of understanding and of practising economic geographies.

Being outside the imposed pragmatism of mainstream politics, the authors included in this volume have more room and freedom to explore their ideas. And although it is academic in orientation, we hope that this book has been read as a collective response to the new ideological juncture, as part of an attempt to revitalize debate about the significance of economic processes in everyday life. Without offering prescriptions, these authors have reasserted the interconnections between economy, politics and society. Moreover, it is hoped that our ability to make some sense of these processes might allow further exploration of alternative forms of economic life and relationships. When we are armed with greater insight, debate can commence about the practical interventions that might reconfigure the world in more equitable directions in future. But practice is still not a priority for economic geographers. As Peet and Thrift (1989: xiii) suggested in their overview of political economic analysis in geography, the emancipatory potential of critical research is still to be realized:

> in a world where millions of people are dying in famines or war, where more millions live in acute poverty and fear, and where there is an ecological crisis of grave proportions, it is surely important to hold on to that emancipatory vision. Here, at the cutting edge of capitalism, much new thinking and ideological face work remains to be done.

By using a richer understanding of the economic to capture the changing nature and territory of the world in which we live, the economic geographer can offer real insight into the contemporary order of things and work towards an emancipatory agenda. But the key task for the future must be in mobilizing analysis into action, linking theory and practice, turning prognosis to policy. Indeed, we still need a vision in order to reimagine economic geographies of the world in which we live. As images of alternative futures are fading from view, the revitalization of ideas is a real step on the road to re-visioning the world as we know it. It is hoped that this book has played a small part in this process.

# Bibliography


**Abowd**, J.M. and Freeman, R.B. 1990: The internationalization of the NBER U.S. labor market. Working Paper 3321. Cambridge, MA: NBER.

**Advisory Commission on Intergovernmental Relations** 1988: *State fiscal capacity and effort*. Washington, DC: ACIR.

**Aglietta**, M. 1979: *A theory of capitalist regulation: the US experience*. London: New Life Books.

**Agnew**, J. 1993: The United States and American hegemony. In Taylor, P.J. (ed.), *Political geography of the twentieth century*. London: Belhaven, 207–38.

**Agnew**, J. and Corbridge, S. 1989: The new geopolitics: the dynamics of geopolitical disorder. In Johnston, R.J. and Taylor, P.J. (eds), *A world in Crisis?*, 2nd edition. Oxford: Blackwell, 266–88.

**Agnew**, J. and Corbridge, S. 1995: *Mastering space: hegemony, territory and international political economy*. London: Routledge.

**Albert**, M. 1993: *Capitalism against capitalism*, trans. Paul Haviland. London: Whurr.

**Albion**, R.G. 1932: Yankee domination of New York port, 1820–1865. *New England Quarterly* 5: 665–98.

**Albo**, G. 1994: 'Competitive austerity' and the impasse of capitalist employment policy. In Miliband, R. and Panitch, L. (eds), *Between globalism and nationalism*. London: Merlin Press, 144–70.

**Allen**, J. 1995: Crossing borders: footloose multinationals. In Allen, J. and Hamnett, C. (eds), *A shrinking world?* Oxford: Oxford University Press, 55–102.

**Allen**, J. and du Gay, P. 1994: Industry and the rest: the economic identity of services. *Work, Employment and Society* 8, 255–71.

**Allen**, J. and Hamnett, C. (eds) 1995: *A shrinking world?* Oxford: Oxford University Press.

**Allen**, J. and Massey, D. (eds) 1988: *The economy in question*. London: Sage.

**Allen**, R. 1994: *Financial crises and recessions in the global economy*. London: Edward Elgar.

**Allen**, Z. 1835: *Sketches of the state of the useful arts . . . or the practical tourist*. Hartford: Beach & Beckwith.

**Almaguer**, T. 1994: *Racial fault lines*. Berkeley: University of California Press.

**Altvater**, E. 1993: *The future of the market: an essay on the regulation of money and nature after the collapse of 'actually existing socialism'*. London: Verso.

**Altvater**, E. 1994: Ecological and economic modalities of time and space. In O'Connor, M. (ed.), *Is capitalism sustainable? Political economy and the politics of ecology*. New York: Guilford, 76–90.

**Alvarez**, C. 1992: *Science, development and violence*. Delhi: Oxford University Press.

**Alvesson**, M. 1993: *Cultural perspectives on organisations*. Cambridge: Cambridge University Press.

**Amin**, A. 1993: The globalization of the economy: an erosion of regional networks? In Grabher, G. (ed.), *The embedded firm: on the socioeconomics of industrial networks*. London: Routledge, 278–95.

**Amin**, A. (ed.) 1994: *Post-Fordism: a reader*. Oxford: Blackwell.

**Amin**, A. (1996) Beyond associative democracy. New *Political Economy* 1, 3; 309–333.

**Amin**, A. and Dietrich, M. 1991: From hierarchy to 'hiearchy': the dynamics of contemporary corporate restructuring in Europe. In Amin, A. and Dietrich, M. (eds), *Towards a new Europe? Structural change in the European economy*. Aldershot: Edward Elgar, 49–73.

**Amin**, A. and Malmberg, A. 1992: Competing structural and institutional influences on the geography of production in Europe. *Environment and Planning A* 24, 401–16.

**Amin**, A. and Malmberg, A. 1994: Competing structural and institutional influences on the geography of production in Europe. In Amin, A. (ed.), *Post-Fordism: a reader*. Oxford: Blackwell, 227–48.

**Amin**, A. and Robins, K. 1990: The re-emergence of regional economies? The mythical geography of flexible specialisation. *Environment and Planning D: Society and Space* 8, 7–34.

**Amin**, A. and Thrift, N. 1992: Neo-Marshallian nodes in global networks. *International Journal of Urban and Regional Research* 16, 571–87.

**Amin**, A. and Thrift, N. 1993: Living in the global. In Amin, A. and Thrift, N. (eds), *Globalization, institutions and regional development in Europe*. Oxford: Oxford University Press, 1–22.

**Amin**, A. and Thrift, N. (eds) 1994: *Globalization, institutions and regional development in Europe*. Oxford: Oxford University Press.

**Amin**, A. and Thrift, N. 1995: Institutional issues for

the European regions: from markets and plans to socioeconomics and powers of association. *Economy and Society* 24, 41–66.

**Amin**, A. and Tomaney, J. 1995a: The challenge of cohesion. In Amin, A. and Tomaney, J. (eds), *Behind the myth of European union: prospects for cohesion*. London: Routledge, 10–50.

**Amin**, A. and Tomaney, J. 1995b: The regional dilemma in a neo-liberal Europe. *European Urban and Regional Studies* 2, 171–88.

**Amin**, S. 1996: The challenge of globalisation. *Review of International Political Economy* 3, 216–59.

**Amsden**, A. 1989: *Asia's next Giant: South Korea and late industrialization*. Oxford: Oxford University Press.

**Amsden**, A., Kochanowicz, J. and Taylor, L. 1994: *The market meets its match: restructuring the economies of Eastern Europe*. Cambridge, MA: Harvard University Press.

**Anderson**, J. 1992: *The territorial imperative: pluralism, corporatism and economic crisis*. Cambridge: Cambridge University Press.

**Anderson**, J. and Goodman, J. 1995: Regions, states and European union: modernist reaction or postmodern adaptation? *Review of International Political Economy* 2, 600–32.

**Anderson**, K. and Blackhurst, R. (eds) 1993: *Regional integration and the global trading system*. London: Harvester Wheatsheaf.

**Anderson**, K. and Norheim, H. 1993: From imperial preferences to regional trade preferences: its effects on Europe's intra- and extraregional trade. *Weltwirtschaftliches Archiv* 129, 78–102.

**Anderson**, P. 1964: Origins of the present crisis. *New Left Review* 23.

**Anderson**, P. 1984: *In the tracks of historical materialism*. Chicago: University of Chicago Press.

**Anderson**, P. 1987: The figures of descent. *New Left Review* 161, 20–77.

**Anderson**, R.B. and Bone, R.M. 1995: First nations economic development: a contingency perspective. *Canadian Geographer* 39, 120–30.

**Angel**, D. 1994: *Restructuring for innovation: the remaking of the U.S. semiconductor industry*. New York: Guilford Press.

**Angel**, D. and Engstrom, J. 1995: Manufacturing systems and technological change: the U.S. personal computer industry. *Economic Geography* 71, 81–104.

**Angel**, D. and Savage, L. 1996: Global-localization? Japanese R&D laboratories in the United States. *Environment and Planning A* 28, 819–33.

**Antonelli**, C. 1995: *The economics of localized technological change and industrial dynamics*. Dordrecht: Kluwer.

**Appadurai**, A. 1986: Introduction: commodities and the politics of value. In Appadurai, A. (ed.), *The social life of things: commodities in cultural perspective*. Cambridge: Cambridge University Press, 1–63.

**Arce**, A. and Marsden, T.K. 1993: The social construction of international food. *Economic Geography* 69, 293–311.

**Arce**, A., Villarreal, M. and de Vries, P. 1994: The social construction of rural development. In Booth, D. (ed.), *Rethinking social development*. London: Longman, 152–71.

**Archibugi**, D. and Michie, J. 1993: The globalisation of technology: myths and realities. *Research Papers in Management Studies*. Judge Institute of Management Studies, University of Cambridge.

**Archibugi**, D. and Michie, J. 1995: The globalisation of technology: a new taxonomy. *Cambridge Journal of Economics* 19, 121–40.

**Arendt**, H. 1958: *The human condition*. Chicago: University of Chicago Press.

**Arendt**, H. 1970: *On violence*. San Diego: Harvest.

**Arestis**, P. 1992: *The post-Keynesian approach to economics: An alternative analysis of economic theory and policy*. Aldershot: Edward Elgar.

**Arestis**, P. 1996: Post-Keynesian economics: towards coherence. *Cambridge Journal of Economics* 26, 111–35.

**Arrighi**, G. 1993: The three hegemonies of historical capitalism. In Gill, S. (ed.), *Gramsci, historical materialism and international relations*. Cambridge: Cambridge University Press, 148–85.

**Arrighi**, G. 1994: *The long twentieth century: money, power and the origins of our times*. London: Verso.

**Arrow**, K.J. 1962: The economic implications of learning by doing. *Review of Economic Studies* 29, 155–73.

**Arthur**, W.B. 1988a: Competing technologies: an overview. In Dosi, G., Freeman, C., Nelson, R., Silverberg, G. and Soete, L. (eds), *Technical change and economic theory*. London and New York: Pinter, 590–607.

**Arthur**, W.B. 1988b: Self-reinforcing mechanisms in economics. In Anderson, P., Arrow, K.J. and Pines, D. (eds), *The economy as an evolving complex system*. Redwood City, CA: Addison-Wesley, 9–27.

**Arthur**, W.B. 1989: Competing technologies, increasing returns, and lock-in by historical events. *Economic Journal* 99, 116–31.

**Asanuma**, B. 1989: Manufacturer–supplier relationships in Japan and the concept of relationship-specific skill. *Journal of the Japanese and International Economies* 3, 1–30.

**Ashbolt**, A. 1989: Tear down the walls: sixties radicalism and the politics of space in the San Francisco Bay area. Unpublished doctoral dissertation, Australian National University, 1989.

**Asheim**, B. 1996: 'Learning regions' in a globalised world economy: towards a new competitive advantage of industrial districts? Paper to the First European Urban and Regional Studies Conference, 11–14 April, University of Exeter.

**Ashley**, R.K. 1987: The geopolitics of geopolitical space: toward a critical social theory of international politics. *Alternatives* 12, 403–34.

**Ashley**, R.K. 1988: Untying the sovereign state: a double reading of the anarchy problematique. *Journal of International Studies* 17, 227–62.

**Ashley**, R.K. and Walker, R.B.J. 1990: Reading dissidence/writing the discipline: crisis and the question of sovereignty in international studies. *International Studies Quarterly* 34, 367–416.

**Australia**: Department of Industry, Technology and Commerce 1991: *Australian Processed Food and Beverages Industry*. Canberra: Australian Government Publishing Service.

**Australia**: East Asia Analytical Unit (Department of Foreign Affairs and Trade) 1994: *Subsistence to supermarket: food and agricultural transformation in South-East Asia*. Canberra: Australian Government Publishing Service.

**Avadikyan**, A., Cohendet, P. and Llerena, P. 1993: Coherence, diversity of assets and network learning. Mimeo, BETA, Université Louis Pasteur, Strasbourg.

**Aydalot**, P. (ed.) 1986: *Milieux innovateurs en Europe – Innovative environments in Europe*. Groupement de Recherche sur les Milieux Innovateurs en Europe. Paris: Université de Paris I (Sorbonne).

**Ayres**, B.D. 1995: California's economy shows signs of regaining its glitter. *New York Times*, 19 December, A1, C22.

**Bacon**, L. 1838: Address. In Brainerd and Brainerd, E.W. (eds), *The New England Society orations*, vol. 1. New York: The Century Company, 1901, 169–209.

**Bagguley**, P., Mark-Lawson, J., Shapiro, D., Urry, J., Walby, S. and Warde, A. 1990: *Restructuring: place, class and gender*. London: Sage.

**Bahro**, R. 1977: *The alternative in Eastern Europe*. London: New Left Books.

**Bakshi**, P., Goodwin, M., Painter, J. and Southern, A. 1995: Gender, race and class in the local welfare state: moving beyond regulation theory in analysing the transition from Fordism. *Environment and Planning A* 27, 1539–54.

**Balaž**, V. 1995a: A look at the capital markets of the V4 countries. Unpublished manuscript, Institute of Forecasting, Slovak Academy of Sciences, Bratislava.

**Balaž**, V. 1995b: Book review of *Managing in emerging market economies*. *European Urban and Regional Studies* 2, 273–4.

**Banuri**, T. and Schor, J. 1992: *Financial openness and national autonomy*. Oxford: Clarendon Press.

**Barnes**, B. 1988: *The nature of power*. Cambridge: Polity.

**Barnes**, C.B. 1915: *The longshoremen*. New York: Russell Sage Foundation.

**Barnes**, T.J. 1992: Reading the texts of theoretical economic geography: the role of physical and biological metaphors. In Barnes, T.J. and Duncan, J.S. (eds), *Writing worlds: discourse, text and metaphor in the representation of landscape*. London: Routledge, 118–35.

**Barnes**, T. 1995a: Political economy 1: 'the culture stupid'. *Progress in Human Geography* 19, 423–31.

**Barnes**, T. 1995b: *Geographies of dislocation*. New York: Guilford.

**Barnes**, T. 1996: *Logics of dislocation: models, metaphors, and meanings of economic space*. New York: Guilford.

**Barnet**, R. and Cavanagh, J. 1995: *Global dreams: imperial corporations and the New World Order*. New York: Simon & Schuster.

**Barr**, N. 1993: *The economics of the welfare state*, 2nd edition. Oxford: Oxford University Press.

**Barratt Brown**, M. 1988: Away with all the great arches. *New Left Review* 167.

**Barratt Brown**, M. 1995a: *Models in political economy*, 2nd edition. London: Penguin.

**Barratt Brown**, M. 1995b: The new orthodoxy. In Barratt Brown, M. and Radice, H., *Democracy versus capitalism: a response to Will Hutton with some old questions for New Labour*. Socialist Renewal, Pamphlet 4. European Labour Forum, 26–41.

**Barratt Brown**, M. and Radice, H. 1995: *Democracy versus capitalism: a response to Will Hutton with some old questions for New Labour*. Socialist Renewal, Pamphlet 4. European Labour Forum.

**Barro**, R. and Sala-i-Martin, X. 1995: *Economic growth*. New York: McGraw-Hill.

**Barthelemy**, P., Miguelez Lobo, F., Mingione, E., Pahl, R. and Wenig, A. 1988: *Underground economy and irregular forms of employment (travail au noir)*. Programme for research and action on the development of the labour market. Brussels: Commission of the European Communities, DG V.

**Bayart**, J.-F. 1992. *The state in Africa: the politics of the belly*. London: Longman.

**Becattini**, G. and Rullani, E. 1993: Sistema locale e mercato globale. *Economia e Politica Industriale* 80, 25–40.

**Beck**, U. 1992: *Risk society: towards a new modernity*. London: Sage.

**Beck**, U., Giddens, A. and Lash, S. 1994: *Reflexive modernization*. Cambridge: Polity.

**Becker**, G. 1976: *The economic approach to human behavior*. Chicago: University of Chicago Press.

**Begg**, I. and Mayes, D. 1993: The case for decentralised industrial policy. Mimeo, Department of Applied Economics, University of Cambridge.

**Bélis-Bergouignan**, M.-C., Bordenave, G. and Lung, Y. 1993: Le mirage de la territorialisation de la grande enterprise: le redeploiement spatial de Ford. *Revue d'Economie Regionale et Urbaine* 3, 527–44.

**Bélis-Bergouignan**, M.-C., Bordenave, G. and Lung, Y. 1996: Global strategies in the automobile industry. Paper presented to the European Science Foundation EMOT (European Management and Organisations in Transition) conference 'Learning and Embeddedness: Evolving Transnational Firm Strategies in Europe', University of Durham.

**Bell**, D, 1973: *The coming of post-industrial society*. New York: Basic Books.

**Bell**, D. 1979: *The cultural contradictions of capitalism*, 2nd edition. London: Heinemann.

**Bellandi**, M. 1986: *The Marshallian industrial district*. Marshallian Studies 1. Florence: University of Florence.

**Bellandi**, M. 1989: Capacità innovativa diffusa e distretti industriali. Unpublished paper, University of Florence, Department of Economics.

**Bellandi**, M. 1995: *Economie di scale e organizzazione industriale*. Milan: Franco Angeli.

**Belzer**, V. and Dankbaar, B. 1993: *The future of industry in Europe*, vol. 11: Automotive industry. Commission of the European Communities FAST programme, FOP 369, Brussels.

**Benhabib**, S. 1992: *Situating the self.* Cambridge: Polity.

**Benigno**, F. 1989: Famiglia mediterranea e modelli anglo-sassoni. *Meridiana* 6, 76–94.

**Benko**, G.S. 1995: Theory of regulation and territory: an historical review. In Benko, G. S. and Strohmayer, U. (eds), *Geography, history and social sciences.* Dordrecht: Kluwer, 193–210.

**Bennett**, R. 1990: *Decentralization, local governments and markets: towards a post-welfare agenda.* Oxford: Clarendon.

**Benske**, L. 1994: The best senator money can buy. *East Bay Express*, 18 November, 6–7.

**Benton**, L. 1990: *Invisible factories: industrial development and the informal economy in Spain.* Albany, NY: SUNY Press.

**Benton**, T. 1989: Marxism and natural limits. *New Left Review* 178, 51–86.

**Benton**, T. 1992: Ecology, socialism and the mastery of nature: a reply to Reiner Grundmann. *New Left Review* 194, 55–72.

**Benton**, T. and Redclift, M. (eds) 1994: *Social theory and the global environment.* London: Routledge.

**Bercovitch**, S. 1975: *The puritan origins of the American self.* New Haven: Yale University Press.

**Beresford**, P. 1990: Britain's rich: the top 200. *Sunday Times Magazine*, 8 April.

**Berger**, S. and Dore, R. (eds) 1996: *National diversity and global capitalism.* Ithaca, NY: Cornell University Press.

**Berggren**, C. 1994: Japan as number two: competitive problems and the future of alliance capitalism after the burst of the bubble boom. *Work, Employment and Society* 9, 53–95.

**Berman**, M. 1982: *All that is solid melts into air: the experience of modernity.* New York: Penguin.

**Bernstein**, J.M. 1991: Introduction. In Adorno, T.W. *The culture industry: selected essays on mass culture.* London: Routledge, 1–25.

**Bernstein**, R.J. 1983: *Beyond objectivism and relativism: science, hermeneutics and praxis.* Philadelphia: University of Pennsylvania Press.

**Best**, M. 1990: *The new competition: institutions of industrial restructuring.* Cambridge, MA: Harvard University Press.

**Beynon**, H. 1973: *Working for Ford.* Harmondsworth: Penguin.

**Beynon**, H. 1995: The changing experience of work: Britain in the 1990s. Paper to the Conference on Education and Training for the Future Labour Markets of Europe, 21–24 September 1995, University of Durham.

**Beynon**, H., Hudson, R., Lewis, J., Sadler, D. and Townsend, A. 1990: 'It's all falling apart here': Coming to terms with the future in Teesside. In Cooke, P. (ed.), *Localities.* London: Hutchinson.

**Bhabha**, H. 1994: *The location of culture.* London: Routledge.

**Bhaskar**, R. 1975: *A realist theory of science.* Leeds: Leeds Books.

**Bhaskar**, R. 1979: *The possibility of naturalism.* Brighton: Harvester Press.

**Bhaskar**, R. 1993: *Dialectic: the pulse of freedom.* London and New York: Verso.

**Bhaskar**, R. 1994: *Plato etc.* London and New York: Verso.

**Bianchi**, G. 1992: Combining networks to promote integrated regional development. In Vasko, T. (ed.), *Problems of economic transition: regional development in Central and Eastern Europe.* Aldershot: Avebury, 89–105.

**Bianchi**, P. and Miller, L.M. 1994: Innovation, collective action and endogenous growth: an essay on institutions and structural change. Mimeo, Department of Economics, University of Bologna.

**Bienefeld**, M. 1996: Is a strong national economy a utopian goal at the end of the twentieth century? In Boyer, R. and Drache, D. (eds), *States against markets: the limits of globalization.* London: Routledge, 415–40.

**Biersteker**, T. 1993: Evolving perspectives on international political economy: twentieth century contexts and discontinuities. *International Political Science Review* 14, 7–33.

**Biewener**, C. 1995: Marxism and feminism: towards a decentered Marxian politics. Unpublished paper, Simmons College, Boston, MA.

**Binns**, P. and Haynes, M. 1980: New theories of European class societies. *International Socialism* 2, 1–8.

**Bischof**, R. 1994: Why German cycles give better ride: debate. *Guardian*, 27 December, p. 13.

**Blaikie**, P. 1985: *The political economy of soil erosion.* London: Methuen.

**Blaikie**, P. and Brookfield, H. 1987: *Land degradation and society.* London: Methuen.

**Blau**, P. 1964: *Exchange and power in social life.* New York: Wiley.

**Bleeke**, J. and Ernst, D. 1993: The death of the predator. In Bleeke, J. and Ernst, D. (eds), *Collaborating to compete.* New York: Wiley, 1–16.

**Blim**, M. 1996: Cultures and the problems of capitalisms. *Critique of Anthropology* 16, 79–93.

**Bloch**, F., Coward, R., Ehrenreich, B. and Piven, F. 1987: *The mean season.* New York: Pantheon.

**Bloch**, M. 1950: Critique historique et critique du temoignage. *Annales: Economies, Sociétés, Civilisations* 5, 1–8.

**Block**, F. 1990a: Capitalism versus socialism in world-systems theory. *Review* 13, 265–71.

**Block**, F. 1990b: *Postindustrial possibilities: a critique of economic discourse.* Berkeley: University of California Press.

**Block**, F. 1991: Some contradictions of capitalist success. *Dissent*, Winter, 55–62.

**Block**, F. 1992: Capitalism without class power. *Politics and Society* 20, 277–303.

**Block**, F. 1993: Remaking our economy. *Dissent*, Spring, 166–71.

**Block**, F. 1994: The roles of the state in the economy. In Smelser, N.J. and Swedberg, R. (eds), *The handbook of economic sociology.* Princeton, NJ: Princeton University Press, 691–710.

**Block**, F. and Somers, M.R. 1984: Beyond the economistic fallacy: the holistic social science of Karl

Polanyi. In Skocpol, T. (ed.), *Vision and method in historical sociology*. Cambridge: Cambridge University Press, 47–84.

**Blundy**, A. 1994a: Russian organized crime in the UK. *Guardian*, 27 May, 2.

**Blundy**, A. 1994b: Russian organized crime in the US. *Guardian*, 27 May, 3.

**Blundy**, A. 1994c: Cash and caviar. *Guardian*, 8 September, 6.

**Blyton**, P. and Turnbull, P. (eds) 1992: *Reassessing human resource management*. London: Sage.

**Board of Inquiry** 1953: *Report to the President on the Labor Dispute Involving the International Longshoremen's Association and Associated Occupations in the Maritime Industry on the Atlantic Coast*, dated 4 December 1953. Washington, DC: United States Government.

**Boden**, D. 1993: *The business of talk*. Cambridge: Polity.

**Boer**, L. 1990: (In)formalization: the forces beyond. *International Journal of Urban and Regional Research* 14, 404–22.

**Bogetic**, Z. and Hillman, A. 1995: Privatizing profits of Bulgaria's state enterprises. *Transition* 6, 4–6.

**Boltanski**, L. and Thevenot, L. 1989: *De la justification*. Paris: Editions Gallimard.

**Bonanno**, A., Busch, L., Friedland, W., Gouveia, L. and Mingione, E. (eds) 1994: *From Columbus to Conagra: the globalization of agriculture and food*. Lawrence: University of Kansas Press.

**Booth**, D. 1985: Marxism and development sociology: interpreting the impasse. *World Development* 13, 761–87.

**Booth**, D. 1991: Timing and sequencing in agriculture policy reform: Tanzania. *Development Policy Review* 9, 353–79.

**Booth**, D. (ed.) 1994: *Rethinking social development*. London: Methuen.

**Bordenave**, G. and Lung, Y. 1996: New spatial configurations in the European automobile industry. *European Urban and Regional Studies* 3, 305–21.

**Borger**, J. 1994: After the wall: the Godfather. *Guardian*, 3 November, 11.

**Borts**, G.H. (1960) The equalization of returns and regional economic growth. *American Economic Review* 50, 319–47.

**Bosworth**, B. and Rosenfeld, S. 1993: *Significant others: exploring the potential of manufacturing networks*. Chapel Hill, NC: Regional Technology Strategies.

**Bourdieu**, P. 1977: *Outline of a theory of practice*. Cambridge University Press: Cambridge.

**Bourdieu**, P. 1984: *Distinction: a social critique of the judgement of taste*. London: Routledge & Kegan Paul.

**Bourdieu**, P. 1990: *In other words*. London: Polity.

**Bourdieu**, P. 1993: *The field of cultural production: essays on art and literature*. Cambridge: Polity.

**Bowen**, D. 1992: Business class connections. *The Independent on Sunday*, 4 October, pp. 14–17.

**Bowonder**, B. and Miyake, T. 1992: A model of corporate innovation management: some recent high technology innovations in Japan. *R&D Management* 22, 319–35.

**Boyer**, R. 1990: *The regulation school: a critical introduction*. New York: Columbia University Press.

**Boyer**, R. 1992: *La théorie de la régulation*. Paris: Economica.

**Boyer**, R. and Drache, D. (eds) 1996: *States against markets: the limits of globalization*. London: Routledge.

**Boyer**, R. and Durand, J.P. 1993: *L'après-Fordisme*. Paris: Syros.

**Bradley**, H. 1989: *Men's work, women's work*. Cambridge: Polity.

**Brainerd** and Brainerd, E.W. (eds) 1901: *The New England Society orations*, 2 vols. New York: The Century Company.

**Brass**, T. 1995: Old conservatism in new clothes. *Journal of Peasant Studies* 22, 516–40.

**Braverman**, H. 1974: *Labor and monopoly capital*. New York: Monthly Review Press.

**Brechin**, G. 1996: Conserving the race. *Antipode* 28, 229–45.

**Brechin**, G. 1998 forthcoming: *Imperial San Francisco*. Berkeley: University of California Press.

**Brenner**, R. 1977: The origins of capitalist development. *New Left Review* 104, 25–92.

**Brenner**, R. 1997 forthcoming: Uneven development and the long downturn. *New Left Review*.

**Broad**, R. 1993: *Plundering paradise*. Berkeley: University of California Press.

**Brooks**, V.W. 1936: *The flowering of New England, 1815–1865*. New York: E.P. Dutton.

**Brown**, P. 1988: *The body in society*. New York: Columbia University Press.

**Bruegel**, I. and Kean, H. 1995: The moment of municipal feminism: gender and class in 1980s local government. *Critical Social Policy* 15, 147–169.

**Bryan**, R. 1987: The state and the internationalisation of capital: an approach to analysis. *Journal of Contemporary Asia* 17, 253–9.

**Bryant**, C. and Mokrzycki, E. (eds) 1994: *The new great transformation: change and continuity in East-Central Europe*. London: Routledge.

**Brzezinski**, M. 1995: In the market for making a fast rouble. *Guardian*, 8 July, p. 39.

**Buck**, P., Drennan, M. and Newton, K. 1992: Dynamics of the metropolitan economy. In Fainstein, S., Gordon, I. and Harloe, M. (eds), *Divided cities: New York and London in the contemporary world*. Oxford: Blackwell, 68–103.

**Buck-Morss**, S. 1989: *The dialectics of seeing: Walter Benjamin and the Aracades Project*. London: MIT Press.

**Buck-Morss**, S. 1995: Envisioning capital: political economy on display. *Critical Inquiry* 21, 434–67.

**Budd**, L. and Whimster, S. 1992: *Global finance and urban living*. Routledge.

**Buell**, L. 1986: *New England literary culture: from revolution through renaissance*. Cambridge: Cambridge University Press.

**Burawoy**, M. 1979: *Manufacturing consent*. Chicago: University of Chicago Press.

**Burawoy**, M. 1985: *The politics of production*. London: Verso Press.

**Burawoy**, M. 1992: The end of Sovietology and the

renaissance of modernisation theory. *Contemporary Sociology* 21, 774–85.

**Burawoy**, M. and Krotov, P. 1993a: The economic basis of Russia's political crisis. *New Left Review* 198, 49–69.

**Burawoy**, M. and Krotov, P. 1993b: The Soviet transition from socialism to capitalism: worker control and economic bargaining in the wood industry. In Clarke, S., Fairbrother, P., Burawoy, M. and Krotov, P. *What about the workers? Workers and the transition to capitalism in Russia*. London: Verso, 56–90.

**Burchell**, G., Gordon, C. and Miller, P. (eds) 1991: *The Foucault effect: studies in governmentality*. Hemel Hempstead: Harvester Wheatsheaf.

**Burke**, J.A. 1954: Daily report regarding the New York waterfront situation, from John A. Burke (Commissioner, Federal Mediation and Conciliation Service) to Frank H. Brown (Regional Director, FMCS); box 2650; Federal Mediation and Conciliation Service Case Files, Dispute Case Files 1953; Category 1, Cases Referred to Presidential Board of Inquiry; Records of the FCMS, Record Group 280; National Archives, Washington, DC.

**Burt**, M. 1992: *Over the edge*. New York: Russell Sage.

**Butler**, J. 1990: *Gender trouble: feminism and the subversion of identity*. New York: Routledge.

**Butler**, J. 1993: *Bodies that matter: on the discursive limits of ' sex'*. New York: Routledge.

**Butler**, J. 1995: Against proper objects. *Differences* 6: 1–26.

**Buttel**, F. 1992: Environmentalization. *Rural Sociology* 57, 1–27.

**Byrnes**, D. 1994: *Australia and the Asia game*. Sydney: Allen & Unwin.

**Cable**, V. and Henderson, D. (eds) 1994: *Trade blocs: the future of regional integration*. London: Royal Institute of International Affairs.

**Cain**, P.J. and Hopkins, A.G. 1986: Gentlemanly capitalism and British expansion overseas I: The old colonial system. *Economic History Review*, 2nd series 39, 501–25.

**Cain**, P.J. and Hopkins, A.G. 1987: Gentlemanly capitalism and British expansion overseas II: New imperialism, 1850–1945. *Economic History Review* 2nd series 40, 1–26.

**Cain**, P.J. and Hopkins, A.G. 1993a: *British imperialism: innovation and expansion, 1688–1914*. Harlow: Longman.

**Cain**, P.J. and Hopkins, A.G. 1993b: *British imperialism: crisis and deconstruction, 1914–1990*. Harlow: Longman.

**Calleo**, D.P. 1987: *Beyond American hegemony: the future of the western alliance*. New York: Basic Books.

**Callon**, M. 1986: Some elements of a sociology of translation: domestication of the scallops and the fishermen of St Brieux Bay. In Law, J. (ed.), *Power, action and belief: a new sociology of knowledge*. London: Routledge & Kegan Paul, 196–233.

**Callon**, M. 1991: Techno-economic networks and irreversibility. In Law, J. (ed.), *A sociology of monsters: essays on power, technology, and domination*. London: Routledge, 132–61.

**Callon**, M. 1992: Variété et irréversibilité dans les reseaux de conception et d'adoption des techniques. In Foray, D. and Freeman, C. (eds), *Technologie et richesse des nations*. Paris: Economica, 275–324.

**Callon**, M. and Latour, B. 1981: Unscrewing the big Leviathan: how actors macro-structure reality and how sociologists help them to do so. In Knorr-Cetina, K. and Cicourel, A.V. (eds), *Advances in social theory and methodology: towards an integration of micro- and macro-sociologies*. London: Routledge & Kegan Paul, 277–303.

**Camagni**, R. (ed.) 1991: *Innovation networks: spatial perspectives*. London: Belhaven.

**Camilleri**, J. and Falk, J. 1992: *The end of sovereignty? The politics of a shrinking and fragmenting world*. Aldershot: Edward Elgar.

**Campanella**, M.P. 1993: The effects of globalization and turbulence in policy making processes. *Government and Opposition* 28, 190–205.

**Campbell**, B. 1993: *Goliath. London: Methuen.*

**Canada Consulting Group** 1992. *Underfunding the future*. Toronto: Report prepared for the National Advisory Board on Science and Technology, Government of Canada.

**Caporaso**, J. and Levine, D. 1992: *Theories of political economy*. Cambridge: Cambridge University Press.

**Cardoso**, F. and Faletto, E. 1979: *Dependency and development in Latin America*. Berkeley: University of California Press.

**Casey**, C. 1995: *Work, self and society*. London: Routledge.

**Casson**, M. (ed.) 1991: *Global research strategy and international competitiveness*. Oxford: Blackwell.

**Casson**, M. and Singh, S. 1993: Corporate research and development strategies: the influence of firm, industry, and country factors on the decentralization of R&D. *R&D Management* 23, 91–107.

**Castells**, M. 1989: *The informational city*. Oxford: Blackwell.

**Castoriadis**, C. 1991: *Philosophy, politics, autonomy*. New York: Oxford University Press.

**Castree**, N. 1996: Birds, mice and geography: marxisms and dialectics. *Transactions of the Institute of British Geographers* NS 21, 342–62.

**Central Statistical Office** 1992: *Regional Trends* 27. London: HMSO.

**Cerny**, P.G. 1990: *The changing architecture of politics: structure, agency and the future of the state*. London: Sage.

**Cerny**, P.G. 1991: The limits of deregulation: transnational interpenetration and policy change. *European Journal of Political Research* 19, 173–96.

**Cerny**, P. 1995: Globalization and the changing logic of collective action. *International Organization* 49, 595–625.

**Chaney**, D. 1994: *The cultural turn: scene setting essays on contemporary cultural history*. London: Routledge.

**Chapman**, S. 1992: *Merchant enterprise in Britain: from the Industrial Revolution to World War I*. Cambridge: Cambridge University Press.

**Charles**, D. and Feng, L. 1994: *Lean production, supply chain management and new industrial dynamics: logistics in the automobile industry*. PICT WP 13, Centre for Urban and Regional Development Studies, University of Newcastle upon Tyne.

**Chase-Dunn**, C. 1989: *Global formation: structures of the world economy*. Cambridge, MA: Blackwell.

**Cheevers**, G.B. 1842: The elements of national greatness. In Brainerd and Brainerd, E.W. (eds), *The New England Society orations*, vol. 1 (1901). New York: The Century Company, 289–320.

**Chesnais**, F. 1993: Globalization, world oligopoly and some of their implications. In Humbert, M. (ed.), *The impact of globalization on Europe's firms and industries*. London: Pinter, 12–21.

**Christopherson**, C. 1993: Market rules and territorial outcomes: the case of the United States. *International Journal of Urban and Regional Research* 17, 274–88.

**Christopherson**, S. 1994: The fortress city: privatized spaces, consumer citizenship. In Amin, A. (ed.), *Post-Fordism: a reader*. Oxford: Blackwell, 409–27.

**Church**, M. 1994: Poles apart. *Observer*, 16 January, p. 4.

**City Lives Project** 1991: Unpublished interviews with key decision makers in the City. National Sound Archive, London SW7.

**Clark**, E. 1986: *Ascetic piety and women's faith*. Lewiston, NY: Edwin Mellen Press.

**Clark**, G. 1992: Real regulation: the administrative state. *Environment and Planning A* 24, 615–27.

**Clark**, G.L. 1988: A question of integrity: the National Labor Relations Board, collective bargaining and the relocation of work. *Political Geography Quarterly* 7, 209–27.

**Clark**, G.L. 1992: 'Real' regulation: the administrative state. *Environment and Planning A* 24, 615–27.

**Clark**, G.L. and Dear, M. 1984: *State apparatus: structures and language of legitimacy*. Boston: Allen & Unwin.

**Clark**, G.L., Gertler, M.S., and Whitman, J. 1986: *Regional dynamics: studies in adjustment theory*. Boston: Allen & Unwin.

**Clark**, G.L. and Wrigley, N. 1995: Sunk costs: a framework for economic geography. *Transactions of the Institute of British Geographers* NS 20, 204–23.

**Clark**, G.L.C., McKay, J., Missen, G. and Webber, M. 1992: Objections to economic restructuring and the strategies of coercion: an analytical evaluation of policies and practices in Australia and the United States. *Economic Geography* 68, 43–59.

**Clarke**, J. 1991: The American way: processes of class and cultural formatiom. In *New times and old enemies: essays on cultural studies and America*. London: Harper Collins, 42–72.

**Clarke**, S. 1993: The crisis of the soviet system. In Clarke, S., Fairbrother, P., Burawoy, M. and Krotov, P., *What about the workers? Workers and the transition to capitalism in Russia*. London: Verso, 30–55.

**Clarke**, S., Fairbrother, P., Burawoy, M. and Krotov, P. 1993: *What about the workers? Workers and the transition to capitalism in Russia*. London: Verso.

**Clarke**, V. 1993: Dying to get a home of one's own. *Observer*, 28 November, p. 28.

**Clegg**, S. 1990: *Modern organisations*. London: Sage.

**Cliff**, T. 1974: *State capitalism in Russia*. London: Pluto.

**Clower**, R.W. 1969a: Foundations of monetary theory. In Clower, R.W. (ed.), *Monetary theory*. Harmondsworth: Penguin, 202–11. (First published in 1967.)

**Clower**, R.W. 1969b: The Keynesian counter-revolution: a theoretical appraisal. In Clower, R.W. (ed.), *Monetary theory*. Harmondsworth: Penguin, 270–97. (First published in 1965.)

**Coase**, R. 1937: The nature of the firm. *Economica* 4, 386–405.

**Cobb**, C., Halstead, T. and Rowe, J. 1995: If the GDP is up, why is America down? *Atlantic Monthly*, October, 59–78.

**Cochrane**, A. and Clarke, J. 1993: *Comparing welfare states: Britain in international context*. London: Sage.

**Cochrane**, A., Peck, J. and Tickell, A. 1996: Manchester plays games: exploring the local politics of globalization. *Urban Studies* 33, 1319–36.

**Cockburn**, A. and Ridgeway, J. (eds) 1979: *Political ecology*. New York: Times Books.

**Cockburn**, C. 1977: *The local state*. London: Pluto.

**Cohen**, A. 1994: *Self-consciousness: an alternative anthropology of identity*. London: Routledge.

**Cohen**, A. 1995: Growing crime and corruption threaten not only Russia, but the entire world. *Transition* 6, 7–10.

**Cohen**, J. and Arato, A. 1992: *Civil society and political theory*. Cambridge, MA: MIT Press.

**Cohen**, S. 1987: A labor process to nowhere. *New Left Review* 165, 34–51.

**Cohen**, S. 1991: Us and them: business unionism in America and some implications for the UK. *Capital and Class* 45, 105–28.

**Cohen**, S. and Zysman, J. 1984: *Manufacturing Matters*. New York: Basic Books.

**Cohendet**, P. and Llerena, P. (eds) 1989: *Flexibilité, information et décision*. Paris: Economica.

**Commager**, H.S., 1967: *The search for a usable past*. New York: Knopf.

**Commission of the European Communities** 1977: *McDougall Report*. Report of the Study Group on the Role of Public Finance in European Integration, vols 1 and 2. Brussels: CEC.

**Commission of the European Communities** 1992: *Socio-economic situation and development of the regions in the neighbouring countries of the community in Central and Eastern Europe*. Regional Development Studies 2. Brussels and Luxembourg: CEC.

**Commission of the European Communities** 1994a: *Communication on the European automobile industry* COM (94) 49, Brussels.

**Commission of the European Communities** 1994b: Private capital flow to CEECs and NIS. *European Economy* Supplement A, no. 3, March.

**Committee of the Social and Cultural Geography Study Group** 1991: De-limiting human geography: new social and cultural perspectives. In Philo, C. (ed.), *New words, new worlds: reconceptualising* *social and cultural geography*. Department of Geography, St David's University College Lampeter, 14–27.

**Comor**, E. (ed.) 1994: *The global political economy of communication.* New York: St Martin's Press.

**Connell**, R. 1987: *Gender and power.* Oxford: Blackwell.

**Connolly**, P. 1956: Testimony of Patrick Connolly, Executive Vice President, ILA-IND, dated 8 September 1956, before NLRB Region 2, in the matter of 'New York Shipping Association et al., NLRB case no. 2–RC-8388;' box 416; National Labor Relations Board 1935–, Administrative Division Files and Dockets Section; Transcripts and Exhibits of Selected Taft–Hartley Cases, 1950–59; Records of the NLRB, Records Group 25; National Archives, Washington DC.

**Conti**, S. and Enrietti, A. 1995: The Italian automobile industry and the case of Fiat: one country, one company, one market. In Hudson, R. and Schamp, E.W. (eds), *Towards a new map of automobile manufacturing in Europe? New production concepts and spatial restructuring*. Berlin: Springer, 117–46.

**Cook**, I. and Crang, P. 1996: The world on a plate: culinary culture, displacement and geographical knowledges. *Journal of Material Culture* 1, 131–53.

**Cook**, J. 1995: Seamy side of enterprise rises with Prague's profits. *Guardian*, 29 August.

**Cooke**, G.W. 1910: *Unitarianism in America.* Boston: American Unitarian Association.

**Cooke**, P. (ed.) 1990: *Localities*. London: Hutchinson.

**Cooke**, P. 1983: *Theories of planning and spatial development*. London: Hutchinson.

**Cooke**, P. 1994: The co-operative advantage of regions. Mimeo, Centre for Advanced Studies, University of Cardiff.

**Cooke**, P. and Morgan, K. 1991: *The intelligent region: industrial and institutional innovation in Emilia-Romagna*. Regional Industrial Research Report 7. Cardiff: University of Wales.

**Cooke**, P., Moulaert, F., Swyngedouw, E. and Weinstein, O. 1992: *Towards global localization: the computing and communications industries in Britain and France*. London: University College Press.

**Cooke**, P., Price, A. and Morgan, K. 1995: Regulating regional economies: Wales and Baden-Württemberg in transition. In Rhodes, M. (ed.), *The regions and the new Europe: Patterns in core and periphery development*. Manchester: Manchester University Press, 105–35.

**Cooper**, F. 1996: *Development and colonialism in Africa*. Cambridge: Cambridge University Press.

**Cooper**, R. 1992: Formal organisations as representations: remote control, displacement and abbreviation. In Reed, M. and Hughes, H. (eds), *Rethinking organisation*. London: Sage, 254–72.

**Corbridge**, S. 1990: Post-Marxism and development studies: beyond the impasse. *World Development* 18, 623–39.

**Corbridge**, S. 1993: Colonialism, post-colonialism and the political geography of the Third World. In Taylor, P.J. (ed.), *The political geography of the twentieth century*. London: Belhaven Press.

**Corbridge**, S. and Agnew, J. 1991: The US trade and budget deficits in global perspective: an essay in geopolitical-economy. *Environmental and Planning D: Society and Space* 9, 71–90.

**Coriat**, B. 1991: Technical flexibility and mass production: flexible specialisation and dynamic flexibility. In Benko, G. and Dunford, M. (eds), *Industrial change and regional development: the transformation of new industrial spaces*. London: Belhaven, 134–58.

**Council on California Competitiveness** 1992: *California's future and jobs*. Sacramento: Council on California Competitiveness.

**Cowen**, M. and Shenton, R. 1995: The invention of development. In J. Crush (ed.), *The power of development*. London: Routledge.

**Cowen**, M. and Shenton, R. 1996: *Doctrines of development*. London: Routledge.

**Cox**, A. 1986: *The state, finance and industry: a comparative analysis of post-war trends in six advanced industrial economies*. Brighton: Harvester.

**Cox**, K. and Mair, A. 1988: Locality and community in the politics of local economic development. *Annals of the Association of American Geographers* 78, 307–25.

**Cox**, K. and Mair, A. 1989: Urban growth machines and the politics of local economic development. *International Journal of Urban and Regional Research* 13, 137–46.

**Cox**, K. and Mair, A. 1991: From localised social structures to localities as agents. *Environment and Planning A* 23, 197–213.

**Cox**, K.R. 1992: The politics of globalization: a sceptic's view. *Political Geography* 11, 427–9.

**Cox**, K.R. 1993: The local and the global in the new urban politics: a critical view. *Environment and Planning D: Society and Space* 11, 433–48.

**Cox**, R. 1993: Structural issues of global governance: implications for Europe. In Gill, S. (ed.), *Gramsci, historical materialism and international relations*. Cambridge: Cambridge University Press, 259–89.

**Cox**, R. W. 1987: *Production, power, and world order: social forces in the making of history*. New York: Columbia University Press.

**Crang**, P. 1994a: It's showtime: on the workplace geographies of display in a restaurant in southeast England. *Environment and Planning D: Society and Space* 12, 675–704.

**Crang**, P. 1994b: Teaching economic geography: some thoughts on curriculum content. *Journal of Geography in Higher Education* 18, 106–13.

**Crang**, P. 1996: Displacement, consumption and identity. *Environment and Planning A* 28, 47–67.

**Crang**, P. (in press): Performing the tourist product. In Rojek, C. and Urry, J. (eds), *Touring cultures*. London: Routledge.

**Cressey**, P. and Scott, P. 1992: Employment, technology and industrial relations in the UK clearing banks: is the honeymoon over? *New Technology, Work and Employment* 3, 83–96.

**Crewe**, L. 1996: Material culture: embedded firms, organizational networks and the local economic development of a fashion quarter. *Regional Studies* 30, 257–72.

**Crewe**, L. and Davenport, E. 1992: The puppet show: changing buyer–supplier relations within clothing retailing. *Transactions of the Institute of British Geographers* NS 17, 183–97.

**Crompton**, R. 1993: *Class and stratification*. Cambridge: Polity.

**Cronon**, W. 1992: *Nature's metropolis*. New Haven, CT: Yale University Press.

**Crouch**, C. and Marquand, D. 1993: *Ethics and markets*. Oxford: Blackwell.

**Crush**, J. (ed.) 1995: *The power of development*. London: Routledge.

**Cséfalvay**, Z. 1994: The regional differentiation of the Hungarian economy. *GeoJournal* 32, 351–61.

**Cumings**, B. 1987: The origins and development of the Northeast Asian political economy: industrial sectors, product cycles and political consequences. In Deyo, F. (ed.), *The political economy of new Asian industrialism*. Ithaca, NY: Cornell University Press, 44–83.

**Dalby**, S. 1988: Geopolitical discourse: the Soviet Union as other. *Alternatives* 13, 415–42.

**Dalby**, S. 1990: *Creating the Second Cold War: the discourse of politics* London: Pinter.

**Dalby**, S. 1991: Critical geopolitics: discourse, difference, and dissent. *Environment and Planning D: Society and Space* 9, 261–83.

**Dalton**, D. and Genther, P. 1991: *The role of corporate linkages in U.S.–Japan technology transfer*. US Department of Commerce: Japan Technology Program. NTIS PB 91 165571.

**Dalton**, D. and Serapio, M. 1993: *U.S. research facilities of foreign companies*. Washington, DC: US Department of Commerce.

**Dalzell**, R.F. 1987: *Enterprising elite: the Boston Associates and the world they made*. New York: Norton.

**Daniels**, P. 1987: Technology and metropolitan office location. *Services Industry Journal* 7, 272–91.

**Daniels**, P.W., Leyshon, A. and Thrift, N. 1988a: Large accountancy firms in the UK: operational adaptation and spatial development. *Service Industries Journal* 8, 317–46.

**Daniels**, P.W., Leyshon, A. and Thrift, N. 1988b: Trends in the growth and location of professional producer services: UK property consultants. *Tijdschrift voor Economische en Sociale Geographie* 79, 162–74.

**Daniels**, P.W., Thrift, N. and Leyshon, A. 1989: Internationalisation of professional producers services: accountancy conglomerates. In Enderwick, P. (ed.), *Multinational service firms*. London: Routledge, 79–106.

**Danziger**, S. and Gottschalk, P. 1996: *American unequal*. Cambridge, MA: Harvard University Press for Russell Sage Foundation.

**DasGupta**, P. 1994: *An inquiry into wellbeing and destitution*. Cambridge: Cambridge University Press.

**David**, P.A. 1985: Understanding the economics of QWERTY. *American Economic Review* (Papers and Proceedings) 75, 332–7.

**Davis**, M. 1986: *Prisoners of the American dream*. London: Verso.

**Davis**, M. 1990: *City of quartz*. London: Verso.

**Davis**, M. 1992a: *Beyond 'Blade Runner': urban control, the ecology of fear*. Open Magazine Pamphlet Series 23. Westfield, NJ.

**Davis**, M. 1992b: The LA inferno. *Socialist Review* 92, 57–80.

**Davis**, M. 1993a: Who killed LA? Part 1. *New Left Review* 197, 3–28.

**Davis**, M. 1993b: Who killed LA? Part 2. *New Left Review* 199, 29–54.

**Davis**, M. 1995: California über alles. *Red Pepper*, January, 30–1.

**de Banville**, E. and Chanaron, J.-J. 1991: *Vers un système automobile européen*. Paris: Economica.

**de Certeau**, M. 1984: *The practice of everyday life*. Berkeley: University of California Press.

**de Janvry**, A. and Garcia, R. 1988: Rural poverty and environmental degradation in Latin America. Paper presented to the IFAD Conference on Smallholders and Sustainable Development, Rome.

**de Melo**, J. and Panagariya, A. (eds) 1993: *New dimensions in regional integration*. Cambridge: Cambridge University Press.

**De Waal**, A. 1994: The genocidal state: Hutu extremism and the origins of the 'final solution' in Rwanda. *Times Literary Supplement*, July, 3–4.

**Dear**, M. 1990: Geographies of corruption. *Environment and Planning D: Society and Space* 8, 249–53.

**Debord**, G. 1990: *Comments on the society of the spectacle*. London: Verso.

**dei Ottati**, G. 1994: Trust, interlinking transactions and credit in the industrial district. *Cambridge Journal of Economics* 18, 529–46.

**Deleuze**, G. 1988: *Foucault*, trans. S. Head. London: Athlone Press.

**Demos**, 1995: Missionary government. *Demos Quarterly* 7.

**DeNero**, H. 1993: Creating the hyphenated corporation: Japanese multinational corporations in the United States. In Bleeke, J. and Ernst, D. (eds), *Collaborating to compete*. New York: Wiley, 165–86.

**Denis**, C. 1995: 'Government can do whatever it wants': moral regulation in Ralph Klein's Alberta. *Canadian Review of Sociology and Anthropology* 32, 365–83.

**Der Derian**, J. 1989: Spy versus spy: the international power of international intrigue. In *International Intertextual relations: postmodern readings of world politics*. Lexington: Lexington Books.

**Der Derian**, J. 1990: The (s)pace of international relations: simulation, surveillance, and speed. *International Studies Quarterly* 34, 295–310.

**Desai**, M. 1979: *Marxian economics*. Totowa, NJ: Littlefield & Adams.

**Deutsche**, R. 1991: Boys town. *Environment and Planning D: Society and Space* 9, 5–30.

**Dewey Commission** 1953: Record of the Public Hearings Held by Governor Thomas E. Dewey on the Recommendations of the New York State Crime Commission for Remedying Conditions on the Waterfront of the Port of New York. Albany, NY: Governor's Office.

**Dicken**, P. 1986: *Global shift: industrial change in a turbulent world*, London: Harper & Row.

**Dicken**, P. 1992: *Global shift: the internationalization of economic activity*, 2nd edition. London: Paul Chapman.

**Dicken**, P. 1994a: Global–local tensions: firms and states in the global space-economy. *Economic Geography* 70, 101–28.

**Dicken**, P. 1994b: The Roepke lecture in economic geography. Global–local tensions: firms and states in the global space economy. *Economic Geography* 70, 101–28.

**Dicken**, P. 1995: How the world works. *Review of International Political Economy* 2, 197–204.

**Dicken**, P. 1998: *Global shift*, 3rd edition. London: Paul Chapman.

**Dicken**, P., Forsgren, M., and Malmberg, A. 1994: The local embeddedness of transnational corporations. In Amin, A. and Thrift, N. (eds), *Globalization, institutions, and regional development in Europe*. Oxford: Oxford University Press, 23–45.

**Dicken**, P. and Oberg, S. 1996: The global context: Europe in a world of dynamic economic and population change. *European Urban and Regional Studies* 3, 101–20.

**Dicken**, P. and Thrift, N. 1992: The organization of production and the production of organization: why business enterprises matter in the study of geographical industrialization. *Transactions of the Institute of British Geographers* NS 17, 279–91.

**Diggins**, J.P. 1978: *The bard of savagery*. New York: Seabury Press.

**Djellal**, F. and Gallouj, C. 1995: Innovation et développement regional: le cas des firmes de conseil en technologie d'information. Paper presented to the Conference Industrial Dynamics, Spatial Dynamics, Toulouse, August.

**Dodds**, K.-J. 1993a: Geopolitics, experts and the making of foreign policy. *Area* 25, 70–4.

**Dodds**, K.-J. 1993b: War stories: British elite narrative of the 1982 Falklands/Malvinas War. *Environment and Planning D: Society and Space* 11, 619–40.

**Dodds**, K.-J. 1994: Geopolitics and foreign policy: recent developments in Anglo-American political geography and international relations. *Progress in Human Geography* 18, 186–208.

**Doraisami**, A. 1995: An insider's view of the Asian success story. Mimeo available from the author, Department of Economics, Monash University, Clayton, Victoria, Australia.

**Dore**, R.P. 1987: *Taking Japan seriously*. London: Athlone Press.

**Dosi**, G. 1988: Institutions and markets in a dynamic world. *Manchester School* 56, 119–46.

**Dosi**, G., Freeman, C., Nelson, R., Silverberg, G., and Soete, L. (eds) 1988: *Technical change and economic theory*. London: Pinter.

**Dosi**, G., Pavitt, K. and Soete, L. 1990: *The economics of technical change and international trade*. New York: New York University.

**Dosi**, G. and Salvatore, R. 1992: The structure of industrial production and the boundaries between firms and markets. In Storper, M. and Scott, A. J. (eds), *Pathways to industrialization and regional development*. London: Routledge, 171–93.

**Dow**, S.C. 1990: Beyond dualism. *Cambridge Journal of Economics* 14.

**Drache**, D. and Gertler, M. (eds) 1991: *The new era of global competition: state policy and market power*. Montreal: McGill-Queen's University Press.

**Drew**, P. and Heritage, J. (eds), 1992: *Talk at work: interaction in institutional settings*. Cambridge: Cambridge University Press.

**Dreyfus**, H.C. and Rabinow, P 1982: *Michel Foucault: beyond structuralism and hermeneutics*. Brighton: Harvester Press.

**Driver**, F. 1985: Power, space and the body: a critical assessment of Foucault's *Discipline and punish*. Environment and Planning D: Society and Space 3, 425–46.

**Driver**, F. 1993: *Power and pauperism: the workhouse system, 1834–1884*. Cambridge: Cambridge University Press.

**Driver**, F. 1994: Bodies in space: Foucault's account of disciplinary power. In Jones, C. and Porter, R. (eds), *Reassessing Foucault: power, medicine and the body*. London: Routledge.

**du Gay**, P. 1996: *Consumption and identity at work*. London: Sage.

**du Gay**, P. and Salaman, G. 1992: The cult[ure] of the consumer. *Journal of Management Studies* 25(9), 615–33.

**Duncan**, J.S. 1994: After the civil war: reconstructing cultural geography as 'heterotopia'. In Foote, K.E., Hugill, P.J., Mathewson, K. and Smith, J.M. (eds), *Re-reading cultural geography*. Austin: University of Texas Press, 401–8.

**Dunford**, M. 1990: Theories of regulation. *Environment and Planning D: Society and Space* 8, 297–322.

**Dunford**, M. 1994: Winners and losers: the new map of economic inequality in the European Union. *European Urban and Regional Studies* 1, 95–114.

**Dunford**, M. 1996: Disparities in employment, productivity and output in the EU: the roles of labour market governance and welfare regimes. *Regional Studies* 30, 339–57.

**Dunford**, M., Fernandes, A., Musyck, B., Sadowski, B., Cho, M.R. and Tsenkova, S. 1993: The organisation of production and territory – small firm systems – an intensive ERASMUS course held in Lombardy, 4–16 July 1992. *International Journal of Urban and Regional Research* 17, 132–6.

**Dunford**, M. and Fielding, A.J. 1997 forthcoming: Greater London, the South East region and the wider Britain: metropolitan polarisation, uneven development and interregional migration. In Blotevogel, H. and Fielding, A. J. (eds), *People, jobs and mobility in the new Europe*. Chichester: Wiley.

**Dunford**, M. and Perrons, D. 1994: Regional inequality, regimes of accumulation and economic development in contemporary Europe. *Transactions of the Institute of British Geographers* NS 20, 163–82.

**Dunn**, J. 1979: *Western political theory in the face of the future.* Cambridge: Cambridge University Press.

**Dunning**, J. 1993: Globalisation: the challenge for national economic regimes. Mimeo, Department of Economics, University of Reading.

**Dunning**, J. 1994: Globalisation, economic restructuring and development. Mimeo, Department of Economics, University of Reading.

**Dunning**, J.H. 1979: Explaining changing patterns of international production: in defense of the eclectic theory. *Oxford Bulletin of Economics and Statistics* 41, 269–95.

**Eccleston**, B. 1989: *State and society in post-war Japan.* Cambridge: Polity.

**The Economist** 1996a. Le défi américain, again, 13 July, pp. 21–3.

**The Economist** 1996b. Showing Europe's firms the way, 13 July, p. 15.

**The Economist** 1996c. Stakeholder capitalism: unhappy families, 10 February, pp. 23–5.

**Elger**, T. and Smith, D. 1994: *Japanisation.* London: Routledge.

**Ellman**, M. 1994: The increase in death and disease under 'katastroika'. *Cambridge Journal of Economics* 18, 329–56.

**Elster**, J. 1983: *Explaining technical change.* Cambridge: Cambridge University Press.

**Elster**, J. 1985: *Making sense of Marx.* Cambridge: Cambridge University Press.

**Emerson**, R.W. 1870: Oration and response 1870. In Brainerd and Brainerd, E.W. (eds), *The New England Society orations*, vol. 2. New York: The Century Company, 1901, 373–96.

**Emerton**, E. (ed.) 1960: *The correspondence of Pope Gregory VII.* New York: Columbia University Press.

**Enderwick**, P. 1995: The contribution of foreign direct investment to the New Zealand economy. Mimeo available from the author, Department of Economics, University of Waikato, Hamilton, New Zealand.

**Engels**, D., and Marks, S. (eds), 1994: *Contesting colonial hegemony: state and society in Africa and India.* London: British Academic Press.

**Engels**, F. 1968 [1845]: *The conditions of the working class in England.* Stanford, CA: Stanford University Press.

**Environment and Planning A** 1995, 27(9).

**Epstein**, S. 1987: Gay politics, ethnic identity: the limits of social constructionism. *Socialist Review* 17, 9–54.

**Ernst**, J. and Webber, M. 1996: Ideology and interests: privatisation in theory and practice. In Webber, M. and Crooks, M.L. (eds), *Putting the people last: government, services and rights in Victoria.* Melbourne: Hyland House, 113–40.

**Escobar**, A. 1992a: Imagining a post-development era? Critical thought, development and social movements. *Social Text* 31/32: 20–56.

**Escobar**, A. 1992b: Culture, economics, and politics in Latin American social movements theory and research. In Escobar, A. and Alvarez, S.E. (eds), *The making of social movements in Latin America.* Boulder, Co: Westview Press, 56–78.

**Escobar**, A. 1995: *Encountering development.* Princeton, NJ: Princeton University Press.

**Escobar**, A. 1996: Constituting nature: elements for a poststructuralist political ecology. In Peet, R. and Watts, M. (eds), *Liberation ecologies: environment, development, social movements.* London: Routledge 46–68.

**Esping-Andersen**, G. 1990: *The three worlds of welfare capitalism.* Cambridge: Polity.

**Esser**, J. 1989: Does industrial policy matter? Land governments in research and technology policy in federal Germany. In Crouch, C. and Marquand, D. (eds), *The new centralism: Britain out of step in europe.* Oxford: Blackwell, 94–108.

**Esteva**, G. 1992: Development. In Sachs, W. (ed.), *The development dictionary: a guide to knowledge as power.* London: Zed Books, 6–25.

**Esty**, D. 1994: *Greening the GATT.* Washington, DC: IIE.

**Etzioni**, A. 1988: *The moral dimension.* New York: Free Press.

**European Bank for Reconstruction and Development** 1995: *Transition report.* London: HMSO.

**Evans**, B., Rueschemeyer, D. and Skocpol, T. (eds) 1985: *Bringing the state back in.* Cambridge: Cambridge University Press.

**Evans**, P. 1995: *Embedded autonomy.* New York: Cambridge University Press.

**Fagan**, R.H. 1990: Elders IXL Ltd: finance capital and the geography of corporate restructuring. *Environment and Planning A* 22, 647–66.

**Fagan**, R.H. 1995: Economy, culture and environment: perspectives on the Australian food industry. *Australian Geographer* 26, 1–10.

**Fagan**, R.H. 1996: Exploring economic integration with Asia: the Australian food industry and global change. In Robison, R. (ed.), *Pathways to Asia: the politics of engagement.* Sydney: Allen & Unwin, 205–25.

**Fagan**, R.H. and Le Heron, R.B. 1994: Reinterpreting the geography of accumulation: the global shift and local restructuring. *Environment and Planning D: Society and Space* 12, 265–85.

**Fagan**, R.H. and Rich, D.C. 1990: Industrial restructuring in the Australian food industry: corporate strategy and the global economy. In Wilde, P.D. and Hayter, R. (eds), *Industrial transformation and challenge in Australia and Canada.* Ottawa: Carleton University Press, 175–94.

**Fagan**, R.H. and Webber, M. 1994: *Global restructuring: the Australian Experience.* Melbourne: Oxford University Press.

**Fagerberg**, J. 1995: User–producer interaction, learning and comparative advantage. *Cambridge Journal of Economics* 19, 243–56.

**Featherstone**, M. 1990: Global culture: an introduction. *Theory, Culture and Society* 7, 1–14.

**Featherstone**, M. 1994: City cultures and post-modern lifestyles. In Amin, A. (ed.), *Post Fordism: a reader.* Oxford: Blackwell, 387–408.

**Featherstone**, M. 1995a: The autonomization of the cultural sphere. In *Undoing culture: globalization, postmodernism and identity.* London: Sage, 15–33.

**Featherstone**, M. 1995b: The heroic life and everyday life. In *Undoing culture: globalization, postmodernism and identity*. London: Sage, 54–71.

**Featherstone**, M., Lash, S. and Roberston, R. (eds) 1995: *Global modernities*. London: Sage.

**Feher**, F., Heler, A. and Markus, G. 1983: *Dictatorship over needs*. Oxford: Blackwell.

**Ferguson**, J. 1990: *The anti-politics machine*. Minneapolis: University of Minnesota Press.

**Ferguson**, M. 1992: The mythology about globalization. *European Journal of Communication* 7, 69–93.

**Ferrão**, J. and Vale, M. 1995: Multi-purpose vehicles, a new opportunity for the periphery? Lessons from the Ford/VW project (Portugal). In Hudson, R. and Schamp, E. (eds), *Towards a new map of automobile manufacturing in Europe?* Berlin: Springer, 195–217.

**Financial Times** 1994: A manager's tale, 23 May, p. 12.

**Fiorenza**, E.S. 1983: *In memory of her*. New York: Crossroad Publishing Company.

**Fischer**, C., Hout, M., Sanchez Jankowski, M., Lucas, S., Swidler, A. and Voss, K. 1996: *Inequality by design*. Princeton: Princeton University Press.

**Florida**, R. 1995: The industrial transformation of the Great Lakes Region. In Cooke, P. (ed.), *The rise of the Rustbelt*. University of London Press, 162–76.

**Florida**, R. and Kenney, M. 1990: *The breakthrough illusion*. New York: Basic Books.

**Florida**, R. and Kenney, M. 1993: The globalization of innovation: the economic geography of Japanese R&D in the United States. Working Paper 93–55, H. John Heinz III School of Public Policy and Management, Carnegie Mellon University, Pittsburgh, PA.

**Folsom**, M.B. and Lubar, S.D. 1982: *The philosophy of manufactures: early debates over industrialization in the United States*. Cambridge, MA: MIT Press.

**Foroutan**, F., and Pritchett, L. 1993: Intra-subSaharan trade: is it too little? *Journal of African Economies* 2, 74–105.

**Foster**, J. and Woolfson, C. 1989: Corporate reconstruction and business unionism: the lessons of caterpillar and ford. *New Left Review* 174, 51–66.

**Foucault**, M. 1972: *The archaeology of knowledge*. London: Routledge.

**Foucault**, M. 1977: *Discipline and punish: the birth of the prison*, trans. A. Sheridan. London: Allen Lane.

**Foucault**, M. 1979: *Discipline and punish: the birth of the prison*. New York: Vintage Books.

**Foucault**, M. 1980: *Power/knowledge*, (ed. Gordon, C.). Brighton: Harvester.

**Foucault**, M. 1981: The order of discourse. In Young, R. (ed.), *Untying the text: a post-structuralist reader*. London: Routledge & Kegan Paul, 48–78.

**Foucault**, M. 1982: The subject and power. In Dreyfus, H.L. and Rabinow, P. (eds), *Michel Foucault: beyond structuralism and hermeneutics*. Brighton: Harvester Press, 108–226.

**Foucault**, M. 1986: *The history of sexuality*, vol. 3: *The care of the self*. London: Penguin.

**Foucault**, M. 1988: The ethic of care for the self as a practice of freedom. In Bernauer, J. and Rasmussen, D. (eds), *The final Foucault*. Boston, MA: MIT Press.

**Foucault**, M. 1991: Governmentality. In Burchell, G., Gordon, C. and Miller, P. (eds), *The Foucault effect: studies in governmentality*. Hemel Hempstead: Harvester Wheatsheaf.

**Fox**, J. 1995: The politics of the World Bank's sustainable development reforms. Paper prepared for the Multilateralism and the Environment Conference, University of California, Berkeley.

**Francis**, J. 1993: *The politics of regulation: a comparative perspective*. Cambridge, MA: Blackwell.

**Franklin**, B. 1806: *The complete works in philosophy, politics and morals* London: Johnson.

**Fransman**, M. 1995: Is national technology policy obsolete in a globalised world? The Japanese response. *Cambridge Journal of Economics* 19, 95–119.

**Fraser**, N. 1981: Foucault on modern power: empirical insights and normative confusions. *Praxis International* 1, 272–87.

**Fraser**, N. 1995: From redistribution to recognition? Dilemmas of a 'post-socialist' age. *New Left Review* 212, 68–93.

**Freeland**, C. 1995: Russian companies strike barter deals as cash dries up. *Financial Times*, 22 November p. 1.

**Friedman**, J. 1994: *Cultural identity and global process*. London: Sage.

**Friedmann**, H. and McMichael, P. 1989: Agriculture and the state system: the rise and decline of national agricultures, 1870 to the present. *Sociologia Ruralis* 29, 93–117.

**Friedmann**, J 1986: The world city hypothesis. *Development and Change* 17, 69–84.

**Friedmann**, J and Wolff, G. 1982: World city formation: an agenda for research and action. *International Journal of Urban and Regional Research* 6, 309–344.

**Fröbel**, F., Heinrichs, J. and Kreye, O. 1980: *The new international division of labour*. Cambridge: Cambridge University Press.

**Fukuyama**, F. 1989: The end of history? *The National Interest*, Summer, 3–18.

**Furtado**, J. 1970: *Obstacles to development in Latin America*. New York: Anchor.

**Fuss**, D. 1989: *Essentially speaking: feminism, nature and difference*. London: Routledge.

**Fyfe**, N. 1996: Contested visions of a modern city: planning and poetry in postwar Glasgow. *Environment and Planning A* 28, 387–403.

**Gans**, H. 1995: *The war against the poor*. New York: Basic Books.

**Garnaut**, R. 1989: *Australia and the Northeast Asian ascendancy*. Canberra: Australian Government Publishing Service.

**Garrahan**, P. and Stewart, P. 1992: *The Nissan enigma: flexibility at work in a local economy*. London: Mansell.

**Garrett**, G. and Lange, P. 1991: Political responses to interdependence: what's 'left' for the left? *International Organization* 45, 539–64.

**Gayle**, L.C. 1963: Statement by Lawrence C. Gayle, Vice President and Director of Labor Relations, New Orleans Steamship Association, before the US House of Representatives Committee on Merchant Marine and Fisheries hearings on HR 1897, HR 2004, HR

2331 bills to amend the 1936 Merchant Marine Act, 88th Cong., 1st Sess., 1963.

**Geddes**, M. 1994: Public services and local economic regeneration in a post-Fordist economy. In Burrows, R. and Loader, B. (eds), *Towards a post-Fordist welfare state?* London: Routledge.

**Gendron**, B. 1986: Theodor Adorno meets The Cadillacs. In Modleski, T. (ed.), *Studies in entertainment: critical approaches to mass culture*. Bloomington: Indiana University Press, 18–36.

**Genther**, P. and Dalton, D. 1992: Japanese-affiliated electronics *geographical anatomy of industrial capitalism*. London: Allen & Unwin.

**Georgescu-Roegen**, N. 1971: *The entropy law and the economic process*. Cambridge, MA: Harvard University Press.

**Gereffi**, G. 1989: Development and the global factory. *Annals of the American Academy of Political and Social Science* 525, 92–104.

**Gereffi**, G. and Korzeniewicz, M. (eds) 1994: *Commodity chains and global capitalism*. Westport: Praeger.

**Gerschenkron**, A. 1962: *Economic backwardness in historical perspective*. Cambridge, MA: Harvard University Press.

**Gertler**, M. 1992: Flexibility revisited: districts, nation-states, and the forces of production. *Transactions of the Institute of British Geographers* NS 17, 259–78.

**Gertler**, M.S. 1993: Implementing advanced manufacturing technologies in mature industrial regions: towards a social model of technology production. *Regional Studies* 27, 259–78.

**Gertler**, M.S. 1995a: 'Being there': proximity, organization, and culture in the development and adoption of advanced manufacturing technologies. *Economic Geography* 71, 1–26.

**Gertler**, M.S. 1995b: Manufacturing culture: the spatial construction of capital. Paper presented in the special sessions on Sociey, Place, Economy at the Annual Meetings of the Institute of British Geographers, Newcastle upon Tyne, 3–6 January.

**Gertler**, M.S. 1996: Worlds apart: the decline of export dominance in the German machinery industry? *Small Business Economics* 8, 1–20.

**Gertler**, M.S. 1997a: Between the global and the local: the spatial limits to productive capital. In Cox, K.R. (ed.), *Spaces of globalization: reasserting the power of the local*. New York: Guilford Press.

**Gertler**, M.S. 1997b: In search of the new social economy: collaborative relations between users and producers of advanced manufacturing technologies. *Environment and Planning A* 27.

**Gibson-Graham**, J.K. 1995: Identity and economic plurality: rethinking capitalism and 'capitalist hegemony'. *Environment and Planning D: Society and Space* 13, 275–82.

**Gibson-Graham**, J.K. 1996: *The end of capitalism (as we knew it): a feminist critique of political economy*. Oxford: Blackwell.

**Giddens**, A. 1971: *Capitalism and modern social theory: an analysis of the writings of Marx, Durkheim and Max Weber*. Cambridge: Cambridge University Press.

**Giddens**, A. 1979: *Central problems in social theory: action, structure and contradiction in social analysis*. Basingstoke: Macmillan.

**Giddens**, A. 1984: *The constitution of society: outline of a theory of structuration*. Cambridge: Polity.

**Giddens**, A. 1990: *The consequences of modernity*. Cambridge: Polity.

**Giddens**, A. 1991: *Modernity and self-identity*. Cambridge: Polity.

**Giddens**, A. 1994: *Between left and right*. Cambridge: Polity.

**Gill**, S. 1990: *American hegemony and the trilateral commission*. Cambridge: Cambridge University Press.

**Gill**, S. 1992: Economic globalisation and the internationalisation of authority: limits and contradiction. *Geoforum* 23, 269–83.

**Gill**, S. (ed.) 1993: *Gramsci, historical materialism and international relations*. Cambridge: Cambridge University Press.

**Gill**, S. 1995: The global panopticon: the neoliberal state, economic life and democratic surveillance. *Alternatives* 20, 1–49.

**Gill**, S. and Law, D. 1988: *The global political economy: perspectives, problems and policies*. Hemel Hempstead: Harvester Wheatsheaf.

**Gill**, S. and Law, D. 1989: Global hegemony and the structural power of capital. *International Studies Quarterly* 33, 479–99.

**Girnius**, S. 1995: World Bank survey on bribery in Lithuania. *OMRI Daily Digest* 192, part 2, 3 October.

**Glacken**, C. 1967: *Traces on the Rhodian shore*. Berkeley: University of California Press.

**Gleason**, T.W. 1955: Testimony by Thomas Gleason, General Organizer, ILA, before the US House Committee on Merchant Marine and Fisheries, hearings on HR 5734, a bill to amend Section 301 (a) of the 1936 Merchant Marine Act, 84th Cong., 1st Sess.

**Glennerster**, H. and Le Grand, J. 1994: *The development of quasi-markets in welfare provision*. London: Suntory Toyota Centre Discussion Paper, London School of Economics.

**Glyn**, A. 1995: Social democracy and full employment. *New Left Review* 211, 33–55.

**Glyn**, A. and Sutcliffe, B. 1992: Global but leaderless? The new capitalist order. In Miliband, R. and Panitch, L. (eds), *New world order: the socialist register 1992*. London: Merlin Press, 76–95.

**Godelier**, M. 1986: *The mental and the material: thought, economy and society*. London: Verso.

**Goldberg**, L. 1991: *Taxation with representation*. Sacramento: California Tax Reform Association.

**Golub**, S. 1995: Comparative and absolute advantage in the Asia Pacific region. San Francisco: Federal Reserve Bank of San Francisco Working Paper.

**Good**, D. 1994: The economic transformation of central and eastern Europe in historical perspective: main themes and issues. In Good, D. (ed.), *Economic transformations in East and Central Europe: legacies from the past and policies for the future*. London: Routledge, 3–24.

**Goodhart**, D. 1994: *The reshaping of the German social market*. London: Institute for Public Policy Research.

**Goodman**, D. and Watts, M. 1994: Reconfiguring the rural or fording the divide? Capitalist restructuring and the global agro-food system. *Journal of Peasant Studies* 22(1).

**Goodman**, J. and Honeyman, K. 1988: *Gainful pursuits: the making of industrial Europe 1600–1914*. London: Edward Arnold.

**Goodman**, P. 1966: Ethics and enterprise: the values of a Boston elite, 1800–1860. *American Quarterly* 18, 437–51.

**Goodwin**, M., Duncan, S. and Halford, S. 1993: Regulation theory, the local state, and the transition of urban politics. *Environment and Planning D: Society and Space* 11, 67–88.

**Gordon**, D.M. 1988: The global economy: new edifice or crumbling foundations? *New Left Review* 168, 24–64.

**Gordon**, M. 1954: *Employment expansion and population growth*. Berkeley: University of California Press.

**Gordon**, R. 1989: Beyond entrepreneurialism and hierarchy: the changing social and spatial organization of innovation. Paper presented at the Third International Workshop on Innovation, Technological Change and Spatial Impacts, Selwyn College, Cambridge, UK, 3–5 September.

**Goron**, D. 1996: *Fat and mean*. New York: Free Press.

**Goss**, D. 1987: Instant print: technology and capitalist control. *Capital and Class* 31, 79–92.

**Gough**, J. 1996: Neoliberalism and localism: a reply to Peck and Tickell. *Area* 28, 392–8.

**Gould**, P. and White, R. 1974: *Mental maps*. Harmondsworth: Penguin.

**Gould**, S.J. 1987: The panda's thumb of technology. *Natural History* 1, 14–23.

**Gould**, S.J. and Eldredge, N. 1977: Punctuated equilibria: the tempo and mode of evolution reconsidered. *Paleobiology* 3, 115–51.

**Gourevitch**, P. 1986: *Politics in hard times: comparative responses to international economic crises*. Ithaca, NY: Cornell University Press.

**Gowan**, P. 1995: Neo-liberal theory and practice for Eastern Europe. *New Left Review* 213, 3–60.

**Gowan**, P. 1996: Eastern Europe, western power and neo-liberalism. *New Left Review* 216, 129–40.

**Grabher**, G. 1992: Eastern 'conquista': the 'truncated industrialisation' of east European regions by large west European corporations. In Ernste, H. and Meier, V. (eds), *Regional development and contemporary industrial response: extending flexible specialisation*. London: Belhaven, 219–33.

**Grabher**, G. (ed.) 1993a: *The embedded firm: on the socioeconomics of industrial networks*. London: Routledge.

**Grabher**, G. 1993b: Instant capitalism: the regional impacts of western investment in eastern Germany. Paper presented at the Conflict and Cohesion in the Single Market conference, Newcastle upon Tyne, November.

**Grabher**, G. 1993c: The weakness of strong ties: the lock-in of regional development in the Ruhr area. In Grabher, G. (ed.), *The embedded firm: on the socioeconomics of industrial networks*. London: Routledge, 255–77.

**Grabher**, G. 1994a: The disembedded regional economy: the transformation of East German industrial complexes into Western enclaves. In Amin, A. and Thrift, N. (eds), *Globalization, institutions and regional development in Europe*. Oxford: Oxford University Press, 177–95.

**Grabher**, G. 1994b: The elegance of incoherence: institutional legacies, privatization and regional development in East Germany and Hungary. Wissenschaftszentrum Berlin Discussion Paper FS 1 94–103.

**Graham**, J. 1990: Theory and essentialism in Marxist geography. *Antipode* 22, 53–66.

**Graham**, S. and Marvin, S. 1996: *Telecommunications and the City: electronic spaces, urban places*. London: Routledge.

**Granovetter**, M. 1985: Economic action and social structure: the problem of embeddedness. *American Journal of Sociology* 91, 481–510.

**Granovetter**, M. 1990: The old and the new economic sociology: a history and an agenda. In Friedland, R. and Robertson, A. F. (eds), *Beyond the marketplace: rethinking economy and society*. New York: Aldine de Gruyter, 89–112.

**Granovetter**, M. and Swedberg, R. (eds) 1993: *The sociology of economic life*. Boulder, CO: Westview, 29–52.

**Grant**, R. 1994: The geography of international trade. *Progress in Human Geography* 18, 298–312.

**Grant**, R. and Agnew, J. 1996: Representing Africa: the geography of Africa in world trade, 1960–92. *Annals of the Association of American Geographers* 86, 729–44.

**Gregg**, P. and Machin, S. 1994: Is the UK rise in inequality different? In Barrell, R. (ed.), *The UK labour market: comparative aspects and institutional development*. Cambridge: Cambridge University Press, 93–125.

**Gregory**, D 1994: *Geographical Imaginations*. Oxford: Blackwell.

**Gregson**, N. and Crewe, L. 1994: Beyond the high street and the mall: car boot fairs and the new geographies of consumption in the 1990s. *Area* 26, 261–7.

**Grey**, C. 1994: Career as a project of the self and labour process discipline. *Sociology* 28(2), 479–98.

**Griliches**, Z. 1991: The search for R&D spillovers. NBER Working Papers 3768. Cambridge, MA: NBER.

**Grimshaw**, J. 1986: *Feminist philosophers: women's perspectives on philosophical traditions*. Brighton: Wheatsheaf.

**Gros**, D. and Thygesen, N. 1992: *European monetary integration: from the european monetary system towards monetary union*. London: Longman.

**Grossberg**, L. 1995: The space of culture, the power of space. In Chambers, I. and Curti, L. (eds), *The post-colonial question: common skies, divided horizons*. London: Routledge, 169–88.

**Grosz**, E. 1990: Philosophy. In Gunew, S. (ed.), *Feminist knowledge: critique and construct*. London: Routledge, 147–74.

**Group of Lisbon** 1994: *Grenzen aan de concurrentie*.

Group of Lisbon/R Petrella, Brussels: University of Brussels Press.

**Grove**, R. 1992: Origins of western environmentalism. *Scientific American* 267(1).

**Grundmann**, R. 1992: The ecological challenge to Marxism. *New Left Review* 187, 103–20.

**Guardian** 1994: Buy now die later, 17 August, p. 17.

**Gudeman**, S. 1986: *Economics as culture: models and metaphors of livelihood*. London: Routledge & Kegan Paul.

**Guha, R. 1990:** *Unquiet woods*. Berkeley: University of California Press.

**Habermas**, J. 1978: *Knowledge and human interests*, 2nd edition. London: Heinemann.

**Habermas**, J. 1984: *The theory of communicative action*. Boston: Beacon Press.

**Hacking**, I. 1986: Making up people. In Heller, T.C. *et al.* (eds), *Reconstructing individualism*. Stanford: Stanford University Press, 222–36.

**Hadduck**, C.B. 1841: The elements of national greatness. In Brainerd and Brainerd, E.W. (eds), *The New England Society orations*, vol. 1. New York: The Century Company, 1901, 261–85.

**Hadjimichalis**, C. and Vaiou, D. 1990a: Whose flexibility? The politics of informalisation in Southern Europe. *Capital and Class* 42, 79–106.

**Hadjimichalis**, C. and Vaiou, D. 1990b: Flexible labour markets and regional development in Northern Greece. *International Journal of Urban and Regional Research* 14, 1–24.

**Hagedoorn**, J. 1993: Understanding the rationale of strategic technology partners: interorganizational modes of cooperation and sectoral differences. *Strategic Management Journal* 14: 371–86.

**Hagedoorn**, J. and Schakenraad, J. 1990: Inter-firms partnerships and cooperative strategies in core techologies. In Freeman, C. and Soete, L. (eds), *New explorations in the economics of technical change*. London: Pinter, 3–28.

**Hakansson**, H. (ed.) 1987: *Industrial technological development: a network approach*. London: Croom Helm.

**Hakansson**, H. 1989: *Corporate technological behavior: cooperation and networks*. New York: Routledge.

**Halford**, S. 1992: Feminist change in a patriarchal organisation: the experience of women's initiatives in local government and implications for feminist perspectives on state institutions. In Savage, M. and Witz, A. (eds). *Gender and bureaucracy*. Oxford: Blackwell, 155–85.

**Halford**, S. and Savage, M. 1995: Restructuring organisations, changing people: gender and restructuring in banking and local government. *Work, Employment and Society* 9, 97–122.

**Halford**, S., Savage, M. and Witz, A. 1997: *Gender, careers and organisations*. Basingstoke: Macmillan.

**Hall**, J. and Ikenberry, G. 1989: *The state*. Milton Keynes: Open University Press.

**Hall**, P. (ed.) 1989: *The political power of economic ideas: Keynesianism across countries*. Princeton, NJ: Princeton University Press.

**Hall**, S. 1988: Brave new world. *Marxism Today*, 24–29 October.

**Hamilton**, W.R. 1963: Institution. In Seligman, E.R.S. and Johnson, A. (eds), *Encyclopedia of the Social Sciences* 7, 84–9.

**Hanappi**, H. 1994: *Evolutionary economics: the evolutionary revolution in the social sciences*. Aldershot: Avebury.

**Hanson**, S. and Pratt G. 1995: *Gender, work and space*. London: Routledge.

**Haraway**, D. 1991: *Reinventing nature: simians, cyborgs and women*. New York: Routledge.

**Harbeson**, J.W. *et al.* 1994: *Civil society and the state in Africa*. Boulder, CO: Rienner.

**Harcourt**, G.C. 1985: Post-Keynesianism: quite wrong and/or nothing new. In Arestis, P. and Skouras, T. (eds), *Post-Keynesian economic theory: a challenge to neoclassical economics*. Brighton: Wheatsheaf, 125–45.

**Harding**, S.G. (ed.) 1976: *Can theories be refuted?* Dordrecht: Reidel.

**Hareven**, T. and Langenbach, R. 1978: *Amoskeag: life and work in an American factory town*. New York: Pantheon.

**Harris**, N. 1986: *The end of the Third World: newly industrializing countries and the decline of an ideology*. London: I.B. Taurus.

**Harris**, N. 1990: *National liberation*. Reno: University of Nevada Press.

**Harrison**, B. 1992: Industrial districts: old wine in new bottles? *Regional Studies* 26, 469–83.

**Harrison**, B. and Bluestone, B. 1988: The great U-turn. New York: Basic Books.

**Harvey**, D. 1973: *Social justice and the city*. London: Arnold.

**Harvey**, D. 1978: The urban process under capitalism: a framework for analysis. *International Journal of Urban and Regional Research* 2, 101–31.

**Harvey**, D. 1982: *The limits to capital*. Oxford: Blackwell.

**Harvey**, D. 1984: On the history and present condition of geography: an historical materialist manifesto. *Professional Geographer* 36, 1–10.

**Harvey**, D. 1985a: *Consciousness and the urban experience*. Oxford: Blackwell.

**Harvey**, D. 1985b: The geography of capitalist accumulation. In *The urbanization of capital*. Oxford: Blackwell.

**Harvey**, D. 1985c: The place of urban politics in the geography of uneven capitalist development. In *The urbanization of capital*. Oxford: Blackwell, 125–64.

**Harvey**, D. 1985d: *The urbanization of capital*. Oxford: Blackwell.

**Harvey**, D. 1987: Three myths in search of a reality in urban studies. *Environment and Planning D: Society and Space* 5, 367–76.

**Harvey**, D. 1989a: *The condition of postmodernity: an enquiry into the origins of cultural change*. Oxford: Blackwell.

**Harvey**, D. 1989b: From managerialism to entrepreneurialism: the transformation of urban governance in late capitalism. *Geografisker Annaler* 71B, 3–17.

**Harvey**, D. 1993: The nature of environment. *Socialist Register*, 1–51.

**Harvey**, D. 1994: Militant particularism and global ambition: the conceptual politics of place, space and environment in the work of Raymond Williams. Mimeo Department of Geography and Environmental Engineering, Johns Hopkins University, Baltimore, MD.

**Harvey**, D. 1996a: Globalization in question. Mimeograph, available from author at the Department of Geography and Environmental Engineering, Johns Hopkins University, Baltimore, MD.

**Harvey**, D. 1996b: *Justice, nature and the geography of difference*. Oxford: Blackwell.

**Harvey**, D. and Scott, A.J. 1988: The practice of human geography: theory and empirical specificity in the transition from Fordism to flexible accumulation. In Macmillan, W.D. (ed.), *Remodelling geography*. Oxford: Blackwell, 217–29.

**Harvey**, D. and Swyngedouw, E. 1993: Economic restructuring and grass-roots resistance. In Hayter, T. and Harvey, D. (eds), *The city and the factory*. London: Mansell.

**Harvie**, C. (1994) *The rise of regional Europe*. London: Routledge.

**Hassard**, J. and Parker, M. (eds) 1993: *Postmodernism and Organisations*. London: Sage.

**Hatch**, H.M. 1957: *New England Society in the city of New York, 1805–1957*. New York: The New England Society in the City of New York.

**Hausner**, J. 1995: Imperative vs interactive strategy of systematic change in Central and Eastern Europe. *Review of International Political Economy* 2, 249–66.

**Hay**, C. 1995: Re-stating the problem of regulation and re-regulating the local state. *Economy and Society* 24, 387–407.

**Hay**, C. and Jessop, B. 1995: Introduction: local political economy: regulation and governance. *Economy and Society* 24, 303–6.

**Hayek**, F.A. 1948: *Individualism and the economic order*. Chicago: University of Chicago Press.

**Hayes**, D. 1989: *Beyond the silicon curtain*. Boston: South End Press.

**Hebdige**, D. 1988: *Hiding in the light: on images and things*. London: Routledge.

**Hector**, G. 1988: *Breaking the bank*. Boston: Little, Brown.

**Heelas**, P. and Morris, P. 1992: *The values of enterprise culture*. London: Routledge.

**Held**, D. 1989: *Political theory and the modern state: essays on state, power and democracy*. Stanford, CA: Stanford University Press.

**Held**, D. 1991: Democracy, the nation-state and the global system. *Economy and Society* 20, 138–72.

**Helleiner**, E. 1993: When finance was the servant: international capital movements in the Bretton Woods order. In Cerny, P. (ed.), *Finance and world politics: markets, regimes and states in the post-hegemonic era*. Aldershot: Edward Elgar, 20–48.

**Helleiner**, E. 1994: *States and the re-emergence of global finance*. Ithaca, NY: Cornell University Press.

**Helm**, D. 1994: British utility regulation: theory, practice and reform. *Oxford Review of Economic Policy* 10, 17–39.

**Henwood**, D. 1994: *Left Business Observer* 62 (7 March), 3.

**Herbert**, C. 1991: *Culture and anomie*. Chicago: University of Chicago Press.

**Herbert**, E. 1989: Japanese R&D in the United States. *Research Technology Management* 32, 11–20.

**Herod**, A. 1991a: Local political practice in response to a manufacturing plant closure: how geography complicates class analysis. *Antipode* 23, 385–402.

**Herod**, A. 1991b: The production of scale in United States labour relations. *Area* 23, 82–8.

**Herod**, A. 1992: Towards a labor geography: The production of space and the politics of scale in the East Coast longshore industry, 1953–1990. Unpublished PhD dissertation, Department of Geography, Rutgers University, New Brunswick, NJ.

**Herod**, A. 1994: On workers' theoretical (in)visibility in the writing of critical urban geography: a comradely critique. *Urban Geography* 15(7), 681–93.

**Herod**, A. 1995: International labor solidarity and the geography of the global economy. *Economic Geography* 71, 341–63.

**Herod**, A. 1997a forthcoming: Back to the future in labor relations: from the New Deal to Newt's Deal. In Staeheli, L., Kodras, J. and Flint, C. (eds), *Transforming American government: implications for a diverse society*. Thousand Oaks, CA: Sage.

**Herod**, A. 1997b: Labor as an agent of globalization and as a global agent. In Cox, K. (ed.), *Spaces of globalization: reasserting the power of the local*. New York: Guilford Press, 167–200.

**Herod**, A. 1997c: Labor's spatial praxis and the geography of contract bargaining in the U.S. East Coast longshore industry, 1953–1989. *Political Geography* 16, 145–69.

**Herod**, A. 1997d: From a geography of labor to a labor geography: labor's spatial fix and the geography of capitalism. *Antipode* 29, 1–31.

**Herod**, A. 1998 forthcoming: The geostrategies of labor in post-Cold War Eastern Europe: an examination of the activities of the International Mineworkers' Federation. In Herod, A. (ed.), *Organizing the landscape: labor unionism in geographical perspective*. Minneapolis: University of Minnesota Press.

**Herod**, A., Ó Tuathail, G. and Roberts, S. (eds) 1997: *An unruly world? Geography, globalization and governance*. Minneapolis: University of Minnesota Press.

**Herr**, H. 1994: Two views on how a deformed monetary economy can be developed. In Perczynski, M. (ed.) *After the market-shock: Central and East-European economies in transition*. Aldershot: Dartmouth, 161–74.

**Herrigel**, G. 1994: Industry as a form of order: a comparison of the historical development of the machine tool industries in the United States and Germany. In Hollingsworth, J.R., Schmitter, P.C. and Streeck, W. (eds), *Governing capitalist economies: performance and control of economic sectors*. New York: Oxford University Press, 97–128.

**Hill**, C. 1961: *The century of revolution*. New York: Norton.

**Hill**, C. 1972: *The world turned upside down*. London: Maurice Temple Smith.

**Hill**, D. 1897: Address of Dr David J. Hill. Seventeenth Annual Report, The New England Society in the City of Brooklyn, 10–35.

**Hindess**, B. 1996: *Discourses of power: from Hobbes to Foucault*. Oxford: Blackwell.

**Hirsch**, J. 1991: From the Fordist to the post-Fordist state. In Jessop, B., Kastendiek, H., Nielson, K. and Pedersen, O.K. (eds), *The politics of flexibility: restructuring state and industry in Britain, Germany and Scandinavia*. Aldershot: Edward Elgar, 67–81.

**Hirschman**, A.O. 1982: Rival interpretations of market society: civilizing, destructive or feeble? *Journal of Economic Literature* 20, 1463–84.

**Hirst**, P. 1994: *Associative democracy*. Cambridge: Polity Press.

**Hirst**, P. and Thompson, G. 1992: The problem of 'globalisation': international economic relations, national economic management and the formation of trading blocs. *Economy and Society* 21, 357–96.

**Hirst**, P. and Thompson, G. 1995: Globalization and the future of the nation state. *Economy and Society* 24, 408–42.

**Hirst**, P. and Thompson, G. 1996: *Globalization in question: the international economy and the possibilities of governance*. Cambridge: Polity.

**Ho**, M.-W. 1993: *The rainbow and the worm: the physics of organisms*. London: World Scientific.

**Hobsbawm**, E. 1995: *The Age of Extremes: the short twentieth century, 1914–1991*. London: Michael Joseph.

**Hochschild**, A. 1983: *The managed heart: commercialization of human feeling*. Berkeley: University of California Press.

**Hodgson**, G.M. 1993: *Economics and evolution: bringing life back into economics*. Cambridge: Polity Press.

**Hodgson**, G.M. 1994: Corporate culture and evolving competencies: an 'old' institutionalist perspective on the nature of the firm. Mimeo, Judge Institute of Management Studies, University of Cambridge.

**Hodgson**, G.M., Samuels, W.J. and Tool, M.R. (eds) 1993: *The Elgar companion to institutional and evolutionary economics*. Aldershot: Edward Elgar.

**Hoffman**, M.E. 1966: *A contemporary analysis of a labor union: International Longshoremen's Association*. Labor Monograph 7. Philadelphia: School of Business Administration, Temple University.

**Holliday**, R. 1995: *Investigating small firms: nice work?* London: Routledge.

**Holloway**, J. 1987: The red rose of Nissan. *Capital and Class* 32, 142–64.

**Holmes**, O.W. 1855: Oration. In Brainerd and Brainerd, E.W. (eds), *The New England Society orations*, vol. 2. New York: The Century Company, 1901, 271–302.

**Holtham**, G., and Kay, J. 1994: The assessment: institutions of economic policy. *Oxford Review of Economic Policy* 10, 1–16.

**Homer-Dixon**, T., Boutwell, J. and Rathjens, G. 1993: Environmental change and violent conflict. *Scientific American*, February, 38–45.

**Honneth**, A. 1986: The fragmented world of symbolic forms: reflections on Pierre Bourdieu's sociology of culture. *Theory, Culture and Society* 3, 55–66.

**Horne**, T. 1983: Bourgeois virtue: property and moral virtue in America. *History of political thought* 4, 317–40.

**Hotch**, J. 1994: Theories and practices of self-employment: prospects for the labor movement. MS thesis, Labor Relations and Research, University of Massachusetts-Amherst.

**Howe**, D.W. 1970: *The Unitarian conscience: Harvard moral philosophy, 1805–1861*. Cambridge: Harvard University Press.

**Howells**, J. 1990: The internationalization of R&D and the development of global research networks. *Regional Studies* 24, 495–512.

**Howith**, R. 1993: 'A world in a grain of sand': towards a reconceptualisation of geographical scale. *Australian Geographer* 24, 33–44.

**Howley**, J. 1990: Justice for Janitors: the challenge of organizing in contract industries. *Labor Research Review* 15, 61–72.

**Hu**, Y.-S. 1992: Global firms are national firms with international operations. *California Management Review* 34, 107–26.

**Hubbard**, R. 1996: Gender and genitals: constructs of sex and gender. *Social Text* 46–7, 157–65.

**Huczynski**, A.A. 1996: *Management gurus: what makes them and how to become one*. London: International Thomson Business Press.

**Hudson**, R. 1988: Uneven development in capitalist societies: changing spatial divisions of labour, forms of spatial organization of production and service provision, and their impact upon localities. *Transactions of the Institute of British Geographers* NS 13, 484–96.

**Hudson**, R. 1989a: Labour market changes and new forms of work in 'old' industrial regions: maybe flexibility for some but not flexible accumulation. *Environment and Planning D: Society and Space* 7, 5–30.

**Hudson**, R. 1989b: *Wrecking a region: state policies, party politics, and regional change in North East England*. London: Pion.

**Hudson**, R. 1994: New production concepts, new production geographies? Reflections on changes in the automobile industry. *Transactions of the Institute of British Geographers* NS 19, 331–45.

**Hudson**, R. and Sadler, D. 1992: 'Just-in-Time' production and the European automotive components industry. *International Journal of Physical Distribution and Logistics Management* 22, 40–5.

**Hudson**, R. and Schamp, E. (eds) 1995: *Towards a new map of automobile manufacturing in Europe? New production concepts and spatial restructuring*. Berlin: Springer.

**Hudson**, R. and Williams, A.M. 1995: *Divided Britain*, 2nd edition. Chichester: Wiley.

**Hughes**, A.L. 1996: Changing food retailer–manufacturer power relations within national economies: a UK–USA Comparison. PhD dissertation, Department of Geography, University of Southampton.

**Humbert**, M. 1994: The globalisation of technology as a challenge for national innovation systems. Mimeo, CERETIM, University of Rennes.

**Humphreys**, D. 1794: A poem on industry. In Humphreys, D. (ed.), *The miscellaneous works*. Gainesville: Scholars Facsimiles and Reprints, 1968.

**Hurd**, R. and Rouse, W. 1989: Progressive union organizing: the SEIU Justice for Janitors campaign. *Review of Radical Political Economics* 21, 70–75.

**Hutchinson**, J. 1970: *The imperfect union: a history of corruption in American trade unions*. New York: E.P. Dutton.

**Hutton**, B., Denton, N. and Bobinski, C. 1994: Foreign tidal wave hits eastern markets. *Financial Times*, 4 February, p. 3.

**Hutton**, W. 1995a: *The state we're in*. London: Jonathan Cape.

**Hutton**, W. 1995b: Myth that sets the world to right. *Guardian*, June 12.

**IBL** (International Brotherhood of Longshoremen) 1956: Brief filed on behalf of the International Brotherhood of Longshoremen, dated 13 September 1956, in the matter of 'New York Shipping Association et al., NLRB case no. 2–RC-8388'; box 691; National Labor Relations Board 1935–, Administrative Division Files and Dockets Section; Selected Taft–Hartley Cases, 1947–59; Records of the NLRB, Record Group 25; National Archives, Washington DC.

**Ignatieff**, M. 1984: *The needs of strangers*. London: Vintage.

**ILA-IND** (International Longshoremen's Association-Independent) 1956: Brief of Intervenor, International Longshoremen's Association, Independent, before NLRB Region 2, in the matter of 'New York Shipping Association et al., NLRB case no. 2-RC-8388;' box 691; National Labor Relations Board 1935–, Administrative Division Files and Dockets Section; Selected Taft–Hartley Cases, 1947–59; Records of the NLRB, Records Group 25; National Archives, Washington DC.

**Illner**, M. 1992: Municipalities and industrial paternalism in a 'real socialist' society. In Dostál, P., Illner, M., Kára, J. and Barlow, M. (eds), *Changing territorial administration in Czechoslovakia: international viewpoints*. Amsterdam: Instituut voor Sociale Geografie, 39–47.

**Ingham**, G. 1996: Review essay: the 'new economic sociology'. *Work, Employment and Society* 10, 549–64.

**Ingham**, G: 1984: *Capitalism divided? The City and industry in British social development*. London: Macmillan.

**International Monetary Fund** 1990: *World economic outlook May 1990*. Washington DC: IMF.

**Isaac**, J.C. 1987: *Power and Marxist theory: a realist view*. Ithaca, NY: Cornell University Press.

**Issel**, W. and Cherny, R. 1986: *San Francisco: 1865–1930*. Berkeley: University of California Press.

**Jackson**, P. and Thrift, N. 1995: Geographies of consumption. In Miller, D. (ed.), *Acknowledging consumption: a review of new studies*. London: Routledge, 204–37.

**Jaffe**, A. 1986: Technological opportunity and spillovers of R&D: evidence from firms' patents, profits, and market value. *American Economic Review* 76, 984–1001.

**Jaffe**, A. 1989: Real effects of academic research. *American Economic Review* 79, 957–70.

**Jaffe**, A., Trachtenberg, M., and Henderson, R. 1993: Geographic localization of knowledge spillovers as evidenced by patent citations. *Quarterly Journal of Economics*, 577–98.

**Jaher**, F.C. 1982: *The urban establishment: upper strata in Boston, New York, Charleston, Chicago, and Los Angeles*. Urbana: University of Illinois Press.

**Jameson**, F. 1990: *The cultural logic of late capitalism*. London: Verso.

**Janicke**, M. 1990: *State failure: the impotence of politics in industrial society*. Cambridge: Cambridge University Press.

**Japanese Science and Technology Agency** 1994: *White Paper on Science and Technology*. Tokyo: Office of the Prime Minister.

**Jazouli**, A. 1992: *Les années banlieuses*. Paris: Seuil.

**Jeelof**, G. 1989: Global strategies of Philips. *European Management Journal* 7, 84–91.

**Jefferson**, T. 1788: *Notes on the State of Virginia*. Philadelphia: H.C. Carey & I. Lea.

**Jensen**, V. 1974: *Strife on the waterfront: the port of New York since 1945*. Ithaca, NY: Cornell University Press.

**Jenson**, J. 1989: 'Different' but not 'exceptional'; Canada's permeable Fordism. *Canadian Review of Sociology and Anthropology* 26, 69–84.

**Jermier**, J.M., Knights, D. and Nord, W.R. (eds) 1994: *Resistance and Power in Organizations*. London: Routledge.

**Jessop**, B. 1990a: Regulation theories in retrospect and prospect. *Economy and Society* 19, 153–216.

**Jessop**, B. 1990b: *State theory: putting capitalist states in their place*. Cambridge: Cambridge University Press.

**Jessop**, B. 1991: The welfare state in transition from Fordism to post-Fordism. In Jessop, B., Kastendiek, H., Nielson, K. and Pedersen, O.K. (eds), *The politics of flexibility: restructuring state and industry in Britain, Germany and Scandinavia*. Aldershot: Edward Elgar, 82–105.

**Jessop**, B. 1993: Towards a Schumpeterian workfare state? Preliminary remarks on post-Fordist political economy. *Studies in Political Economy* 40, 7–39.

**Jessop**, B. 1994a: Post-Fordism and the state. In Amin, A. (ed.), *Post-Fordism: a reader*. Oxford: Blackwell, 251–79.

**Jessop**, B. 1994b: The transition to post-Fordism and the Schumpeterian workfare state. In Burrows, R. and Loader, B. (eds), *Towards a post-Fordist welfare state?* London: Routledge, 13–37.

**Jessop**, B. 1995a: The future of the national state in Europe: erosion or reorganization? Lancaster Regionalism Group Working Paper 51, Department of Sociology, University of Lancaster.

**Jessop**, B. 1995b: Regional economic blocs, cross-border cooperation, and local economic strategies in postsocialism. *American Behavioral Scientist* 38, 674–715.

**Jessop**, B. 1995c: Towards a Schumpeterian workfare regime in Britain? Reflections on regulation, governance, and welfare state. *Environment and Planning A* 27, 1613–26.

**Jessop**, B., Bonnett, K. and Bromley, S. 1990: Farewell to Thatcherism? Neo-liberalism and 'New Times'. *New Left Review*, January, 81–102.

**Joffe**, A. 1990: 'Fordism' and 'post'-fordism in Hungary. *South African Sociological Review* 2, 67–88.

**Johansen**, J. and Mattson, L.G. 1987: Interorganizational relations in industrial systems: a network approach compared with the transaction-cost approach. *International Studies of Management and Organization* 17, 34–48.

**Johnson**, A.F., McBride, S. and Smith, P.J. 1994: Introduction. In Johnson, A. F., McBride, S. and Smith, P.J. (eds), *Continuities and discontinuities: the political economy of social welfare and labour market policy in Canada*. Toronto: University of Toronto Press, 3–21.

**Johnson**, R. 1993: Editor's introduction: Pierre Bourdieu on art, literature and culture. In Bourdieu, P., *The field of cultural production: essays on art and literature*. Cambridge: Polity, 1–25.

**Jonas**, A. 1994: The scale politics of spatiality. *Environment and Planning D: Society and Space* 12, 257–64.

**Jonas**, A.E.G. 1996: Local labour control regimes: uneven development and the social regulation of production. *Regional Studies* 30, 323–38.

**Jones**, A. 1996: (Re)producing gender cultures: the case of merchant banking recruitment. Unpublished MSc dissertation, University of Bristol.

**Journal of Commerce** 1953a: Lewis, UMW eye control of pier union, 22 December, pp. 1, 22.

**Journal of Commerce** 1953b: New AFL union claims 9000 dock workers, 26 October, p. 17.

**Journal of Commerce** 1954: Weigher strike threat ended as accord set, 7 January, p. 17.

**Journal of Commerce** 1956a: ILA broadens bargaining area, 17 September, pp. 1, 44.

**Journal of Commerce** 1956b: Both sides in pier strike try to end impasse today, 19 November, pp. 1, 28

**Journal of Commerce** 1956c: ILA to push for all-port wage accord, 19 October, pp. 1, 24.

**Kadetsky**, E. 1994: Bashing illegals in California. *The Nation*, 17 October, 416–22.

**Kaldor**, N. 1956: Alternative theories of distribution. *Review of Economic Studies* 23, 83–100.

**Kaldor**, N. 1966: *Causes of the slow growth of the United Kingdom: an inaugural address*. Cambridge: Cambridge University Press.

**Kaldor**, N. 1972: The case for regional policies. *Scottish Journal of Political Economy* 17, 337–47.

**Kaldor**, N. 1985: *Economics without equilibrium*. Cardiff: Cardiff University Press.

**Kaldor**, N. and Mirlees, J.A. 1962: A new model of economic growth. *Review of Economic Studies* 29, 174–92.

**Kanter**, R.M. 1995: Thriving locally in the global economy. *Harvard Business Review*, September–October, 151–60.

**Kaplan**, D. 1994: The coming anarchy. *Atlantic Monthly*, February, 44–76.

**Katzenstein**, P. 1987: *Policy and politics in West Germany: the growth of a semi-sovereign state*. Philadelphia: Temple University Press.

**Kauffman**, C. 1995: Thoughts on anticapitalist activism. *Socialist Review* 25, 65–82.

**Kay**, C. 1991: Reflections on the Latin American contributions to development theory. *Development and Change* 22, 31–68.

**Keat**, R. and Abercrombie, N. 1991: *Enterprise Culture*. London: Routledge.

**Keat**, R., Whiteley, N. and Abercrombie, N. 1994: *The authority of the consumer*. London: Routledge.

**Keck**, M. 1995: Social equity and environmental politics in Brazil. *Comparative Politics* 27, 409–24.

**Keck**, O. 1993: The national system for technical innovation in Germany. In Nelson, R. R. (ed.), *National innovation systems: a comparative analysis*. New York: Oxford University Press, 115–57.

**Keeble**, D. 1979: Industrial geography. *Progress in Human Geography* 3, 425–33.

**Keen**, D. 1995: When war itself is privatized: the twisted logic that makes violence worthwhile in Sierra Leone. *Times Literary Supplement*, December, 13–14.

**Kennedy**, P. 1988: *The rise and fall of the great powers: economic change and military conflict from 1500 to 2000*. London: Unwin Hyman.

**Kenney**, M. and Florida, M. 1992a: *Beyond mass production*. New York:

**Kenney**, M. and Florida, R. 1992b: The Japanese transplants: production organisation and regional development. *Journal of the American Regional Planning Association* 58, 21–38.

**Kerfoot**, D. 1993: An infinite number of 'monkeys': gendered jobs in Supabank. Mimeo, School of Business and Economic Studies, Leeds University.

**Keynes**, J. 1933: National self-sufficiency. *Yale Review* 22(4).

**Keynes**, J.M. 1951: *Essays in biography*, ed. G. Keynes. New York: Horizon Press.

**Keynes**, J.M. 1973: *The General Theory and after, collected writings*, vol. 14. London: Macmillan.

**Kimeldorf**, H. 1988: *Reds or rackets? The making of radical and conservative unions on the waterfront*. Los Angeles: University of California Press.

**King**, D. 1987: *The New Right: politics, markets and citizenship*. London: Macmillan.

**King**, D. 1995: *Actively seeking work? The politics of unemployment and wefare policy in the United States and Great Britain*. London: University of Chicago Press.

**Kivinen**, M. 1994: Class relations in Russia. In Piirainen, T. (ed.), *Change and continuity in Eastern Europe*. Aldershot: Dartmouth, 114–47.

**Knapp**, S.L. 1829: Address. In Brainerd and Brainerd, E.W. (eds), *The New England Society orations*, vol. 1. New York: The Century Company, 1901, 141–64.

**Knights**, D. 1992: Changing spaces: the disruptive impact of a new epsitemological location for the study of management. *Academy of Management Review* 17, 514–36.

**Knox**, P. and Agnew, J. 1994: *The geography of the world economy*, 2nd edition. London: Arnold.

**Koc**, M. 1994: Globalization as a discourse. In Bonanno, A. *et al.* (eds), *From Columbus to Conagra: the globalization of agriculture and food.* Lawrence: University of Kansas Press.

**Kodima**, F. 1992: Technology fusion and the new R&D. *Harvard Business Review*, 70–8.

**Konrád**, G. and Szelenyi, I. 1979: *The intellectuals on the road to class power*. Brighton: Harvester.

**Kornai**, J. 1992: *The socialist system: the political economy of state socialism.* Oxford: Clarendon.

**Korton**, D.C. 1995: *When corporations rule the world.* New York: Kumarian Press.

**Kotz**, D.M., McDonough, T. and Reich, M. (eds) 1994: *Social structures of accumulation: the political economy of growth and crisis.* Cambridge: Cambridge University Press.

**Koudi**, W. 1994: Inside Europe: the stuff from hell. *Guardian*, 18 August, p. 11.

**Kravaritou**, Y. 1988: *New forms of employment and activity: their repercussions for labour law and social security in the member states of the European Community.* Dublin: Report to the European Foundation for the Improvement of Living and Working Conditions.

**Krueger**, A.O. 1992: Global trade prospects for the developing countries. *The World Economy* 15, 457–74.

**Krugman**, P. 1991: *Geography and trade.* Cambridge, MA: MIT Press.

**Krugman**, P. 1993: Lessons of Massachusetts for EMU. In Torres, F. and Giavazzi, F. (eds), *Adjustment and growth in the European Union.* Cambridge: Cambridge University Press, 241–60.

**Krugman**, P. 1994a: *Peddling prosperity: economic sense and nonsense in the age of diminished expectations.* New York: Norton.

**Krugman**, P. 1994b: Competitiveness: a dangerous obsession. *Foreign Affairs* 73, 28–44.

**Krugman**, P. 1995: *Development, geography and economic theory.* Cambridge, MA: MIT Press.

**Kuhn**, T. 1970: *The structure of scientific revolutions*, 2nd edition. Chicago: University of Chicago Press.

**Kulik**, G., Parks, R. and Penn, T.Z. (eds) 1982: *The New England mill village, 1790–1860.* Cambridge, MA: MIT Press.

**Kuznets**, S. 1966: *Modern economic growth: rate, structure and spread.* New Haven, CT: Yale University Press.

**Kymlicka**, W. 1989: *Liberalism, community and culture.* Oxford: Clarendon Press.

**Ladanyi**, J. 1993: Patterns of residential segregation and the gypsy minority in Budapest. *International Journal of Urban and Regional Research* 17, 30–41.

**Lagendijk**, A. 1993: *The internationalisation of the Spanish automobile industry and its regional impact: the emergence of a growth–periphery.* Tinbergen Institute Research Series 59.

**Lagendijk**, A. 1997: Towards an integrated automotive industry in Europe: a 'merging *filière*' approach. *European Urban and Regional Studies* 4, 5–18.

**Lamming**, R. 1990: Strategic options for automotive suppliers in the global market. *International Journal of Technology Management* 5, 649–84.

**Larrowe**, C.P. 1955: *Shape-up and hiring hall: A comparison of hiring methods and labor relations on the New York and Seattle waterfronts.* Berkeley: University of California Press. (Reprinted in 1976 by Greenwood Press, Westport, CT.)

**Larrowe**, C.P. 1959: *Maritime labor relations on the Great Lakes.* East Lansing: Labor and Industrial Relations Center, Michigan State University.

**Lasch**, C. 1991: *The true and only heaven.* New York: Norton.

**Lash**, S. 1990: *Postmodernist sociology.* London: Routledge.

**Lash**, S., Szersynski, B. and Wynne, B. (eds) 1996: *Risk, environment and modernity.* London: Sage.

**Lash**, S. and Urry, J. 1994: *Economies of signs and space: after organized capitalism.* London: Sage.

**Laszlo**, E. 1987: *Evolution: the grand synthesis.* Boston, MA: New Science Library–Shambhala.

**Latouche**, S. 1993: *In the wake of an affluent society.* London: Zed Press.

**Latour**, B. 1986: The powers of association. In Law, J. (ed.), *Power, action and belief.* London: Routledge & Kegan Paul, 264–80.

**Latour**, B. 1987: *Science in Action.* Cambridge, MA: Harvard University Press.

**Latour**, B. 1991: Technology is society made durable. In Law, J. (ed.), *A sociology of monsters: essays on power, technology and domination.* London: Routledge.

**Latour**, B. 1993: *We have never been modern.* Cambridge, MA: Harvard University Press.

**Law**, J. 1994: *Organising modernity.* Oxford: Blackwell.

**Lawson**, V. 1992: Industrial subcontracting and employment forms in Latin America: a framework for contextual analysis. *Progress in Human Geography* 16, 1–23.

**Lazonick**, W. 1991: *Business organization and the myth of the market economy.* Cambridge: Cambridge University Press.

**Lazonick**, W. 1993: Industry clusters versus global webs: organisational capabilities in the American economy. *Structural Change and Economic Dynamics* 4, 1–24.

**Le Heron**, R.B. 1994: *Globalised Agriculture: Political Choice.* Oxford: Pergamon.

**Leacock**, E. and Safa, H. 1986: *Women's work.* South Hadley, MA: Bergin & Garvey.

**Lee**, R. 1995: Look after the pounds and the people will look after themselves: social reproduction, regulation and social exclusion in Western Europe. *Environment and Planning A* 27, 1577–94.

**Lefebvre**, H. 1974: *La production de l'espace.* Paris: Anthropos.

**Lefebvre**, H. 1976: *De l'état: les contradictions de l'état moderne.* Paris: Union Général d'Editions.

**Lefebvre**, H. 1978: *De l'état: l'état dans le monde moderne.* Paris: Union Général d'Editions.

**Leff**, E. 1995: *Green production.* New York: Guilford.

**Leidner**, R. 1993: *Fast food, fast talk.* Berkeley: University of California Press.

**Lenin**, V.I. 1917: *What is to be done?* Moscow: Progress Publishers.

**Lenin**, V.I. 1964 [1906]: *The development of capitalism in Russia.* Moscow: Progress Publishers.

**Leontidou**, L. 1990: *The Mediterranean city in transition.* Cambridge: Cambridge University Press.

**Leontidou**, L. 1993: Informal strategies of employment relief in Greek cities: the relevance of family, locality and jousing. *European Planning Studies* 1, 43–68.

**Leontief**, W. 1953: *Studies in the structure of the American economy.* New York: Oxford University Press.

**Leopold**, M. 1985: The transnational food companies and their global strategies. *International Journal of Social Science* 37, 315–30.

**Levenson**, R. 1993: State splitting sage rolls on. *California Historian*, September, pp. 5–23.

**Levins**, R. and Lewontin, R. 1985: *The dialectical biologist.* Cambridge, MA: Harvard University Press.

**Levitt**, T. 1983: The globalization of markets. *Harvard Business Review* May–June, 92–102.

**Levy**, S. and Arnold, R. 1992: *The outlook for the California economy.* Palo Alto: Center for the Continuing Study of the California Economy.

**Lewis**, M. 1989: *Liar's poker: two cities, true greed.* London: Hodder & Stoughton.

**Lewis**, M. 1991: *The money culture.* London: Hodder & Stoughton.

**Lewontin**, R. 1992: *Biology as ideology.* New York: Harper.

**Leyshon**, A. 1992: The transformation of regulatory order: regulating the global economy and environment. *Geoforum* 23, 249–67.

**Leyshon**, A. 1994: Under pressure: finance, geo-economic competition and the rise and fall of Japan's postwar growth economy. In Corbridge S., Martin, R. and Thrift, N. (eds), *Money, power and space.* Oxford: Blackwell, 116–46.

**Leyshon**, A. 1995: Missing words: whatever happened to the geography of poverty? *Environment and Planning A* 27, 1021–5.

**Leyshon**, A. and Thrift, N. 1992: Liberalisation and consolidation: the single European market and the remaking of the European financial capital. *Environment and Planning A* 24, 49–81.

**Leyshon**, A. and Thrift, N. 1993: The restucturing of the UK financial services industry in the 1990s: a reversal of fortune? *Regional Studies* 9, 223–41.

**Leyshon**, A. and Thrift, N. 1995: European financial integration: the search for an 'island of monetary stability' in the sea of global financial turbulence. In Hardy, S., Hart, M., Albrechts, L. and Katos, A. (eds), *An enlarged Europe: regions in competition?* London: Jessica Kingsley, 109–44.

**Leyshon**, A. and Thrift, N. 1997: *Money/space: Geographies of monetary transformation.* London: Routledge.

**Leyshon**, A. and Tickell, A. 1994: Money order?: the discursive construction of Bretton Woods and the making and breaking of regulatory space. *Environment and Planning A* 26, 1861–90.

**Lianzu**, C. 1991: L'impossible modèle urbain. *Le Monde Diplomatique*, special issue 'La ville partout et partout en crise'.

**Light**, I. and Bonacich, E. 1988: *Immigrant enterpreneurs.* Berkeley: University of California Press.

**Limerick**, P. 1987: *Legacy of conquest.* New York: W.W. Norton.

**Lindberg**, L. and Campbell, J. 1991: The state and the organisation of economic activity. In Campbell, J., Hollingsworth, J. and Lindberg, L. (eds), *Governance of the American economy.* Cambridge: Cambridge University Press, 356–95.

**Linkenbach**, A. 1994: Ecological movements and the critique of development. *Thesis Eleven* 39, 63–85.

**Lipietz**, A. 1986: New tendencies in the international division of labour: regimes of accumulation and modes of regulation. In Scott, A. J. and Storper, M. (eds), *Production, work, territory.* London: Unwin Hyman, 16–39.

**Lipietz**, A. 1987: *Mirages and miracles: the crisis of global Fordism.* London: Verso.

**Lipietz**, A. 1992: *Towards a new economic order: post-Fordism, ecology and democracy.* Cambridge: Polity Press.

**Lipietz**, A. 1993a: Social Europe, legitimate Europe: the inner and outer boundaries of Europe. *Environment and Planning D: Society and Space* 11, 501–12.

**Lipietz**, A. 1993b: The local and the global: regional individuality or interregionalism. *Transactions of the Institute of British Geographers* NS 18, 8–18.

**Lipietz**, A. 1995a: Enclosing the global commons. Unpublished paper, Paris: CEPREMAP.

**Lipietz**, A. 1995b: *Green hopes: the future of political ecology.* Cambridge: Polity.

**Lipovetsky**, G. 1994: *The empire of fashion: dressing modern democracy*, trans. C. Porter. Princeton, NJ: Princeton University Press.

**Livingstone**, D.L. 1995: The spaces of knowledge: contributions towards a historical geography of science. *Society and Space* 13, 5–34.

**Lloyd**, J. 1994a: MMM shares 'suspended' by Mavrodi. *Financial Times*, 2 November, p. 2.

**Lloyd**, J. 1994b: MMM chief targets UK for pyramid sales schemes. *Financial Times*, 4 November, pp. 1, 16.

**Lloyd**, J. 1996: Eastern reformers and neo-marxist reveiwers. *New Left Review* 216, 119–28.

**Locke**, R. 1995: *Remaking the Italian economy.* Ithaca, NY: Cornell University Press.

**Lolos**, S. 1989: Informal economy: critical remarks and quantitative approaches. *Synchrona Themata* 36, 34–9 (in Greek).

**Lowe**, A. 1951: On the mechanistic approach to economics. *Social Research* 18, 403–34.

**Luke**, T. and Ó Tuathail, G. 1996: Global flowmations, local fundamentalisms, and fast geopolitics: 'America' in an accelerating world order. Paper presented at the Crisis of Global Regulation and Governance Conference, Athens, Ga, 6–8 April.

**Luke**, T.W. 1993: Localised spaces, globalised places: tracing the Pacific Rim. *Journal of Pacific Studies* 17, 38–56.

**Luke**, T.W. 1994: Placing power/siting space: the politics of global and local in the New World Order.

*Environment and Planning D: Society and Space* 12, 613–28.

**Luke**, T.W. 1995: New World Order or Neo-World Orders: power, politics and ideology in informationalizing glocalities. In Featherstone, M., Lash, S. and Robertson, R. (eds), *Global modernities*. London: Sage, 91–107.

**Lundvall**, B.-A. 1988: Innovation as an interactive process: from user–producer interaction to the national system of innovation. In Dosi, G., Freeman, C., Silverberg, G. and Soete, L. (eds), *Technical change and economic theory*. London: Pinter, 349–69.

**Lundvall**, B.-A. 1990: User–producer interactions and technological change. Paper presented to the OECDs (Organization for Economic Co-operation and Development)–TEP Conference, Paris/La Villette, 4–6 June.

**Lundvall**, B.-A. (ed.) 1992: *National systems of innovation*. London: Pinter.

**Lundvall**, B.-A. 1994: The learning economy: challenges to economic theory and policy. Paper presented at the meetings of the European Association for Evolutionary Political Economy, Copenhagen, 27–29 October.

**Lundvall**, B.-A. and Johnson, B. 1992: The learning economy. Paper presented to the European Association for Evolutionary Political Economy Conference, Paris, 4–6 November.

**Lung**, Y. 1992: Global competition and transregional strategy: spatial reorganisation of the European car industry. In Dunford, M. and Kafkalas, G. (eds), *Cities and regions in the new Europe*. London: Belhaven, 68–85.

**Lury**, C. 1993: *Cultural rights*. London: Routledge.

**Lury**, C. 1994: Planning a culture for the people? In Keat, R., Whiteley, N. and Abercrombie, N. (eds), *The authority of the consumer*. London: Routledge, 138–53.

**Luttwak**, E.N. 1990: From geopolitics to geoeconomics. *The National Interest* 20, 17–24.

**Luttwak**, E.N. 1993: *The endangered American dream*. New York: Simon & Schuster.

**Lyon**, D. 1994: *The electronic eye: the rise of surveillance society*. Cambridge: Polity.

**McCloskey**, D. 1988: The consequences of rhetoric. In Klamer, A., McCloskey, D. and Solow, R. (eds), *The consequences of economic rhetoric*. Cambridge: Cambridge University Press, 280–93.

**McDermott**, G. 1994: Renegotiating the ties that bind: the limits of privatisation in the Czech Republic. Wissenschaftszentrum Berlin Discussion Paper FS 1 94–101.

**MacDougall**, D. and Hutt, R. 1954: Imperial preference: a quantitative analysis. *Economic Journal* 64, 233–57.

**McDowell**, L. 1994: The transformation of cultural geography. In Gregory, D., Martin, R. and Smith, G. (eds), *Human geography: society, space, and social science*. Minneapolis: University of Minnesota Press, 126–73.

**McDowell**, L. 1995: Body work: heterosexual gender performances in City workplaces. In Bell, D. and Valentine, G. (eds), *Mapping desire: geographies of sexualities*. London: Routledge, 75–95.

**McDowell**, L. 1997: *Capital culture: gender at work in the city*. Oxford: Blackwell.

**McDowell**, L. and Court, G. 1994a: Performing work: bodily representations in merchant banks. *Economic Geography* 70 (3), 253–78.

**McDowell**, L. and Court, G. 1994b: Missing subjects. *Economic Geography* 70, 229–51.

**McDowell**, L. and Massey, D. 1984: A woman's place? In Massey, D. and Allen, J. (eds), *Geography matters! A reader*. Cambridge: Cambridge University Press, 128–47.

**McDowell**, L. 1997: *Capital culture: money, sex and power at work*. Oxford: Blackwell.

**McIntyre**, J., Porter, L. and Wendelová, P. 1994: Ex-émigré entrepreneur Viktor Kozeny and the Harvard Group. In Fogel, D. (ed.), *Managing in emerging market economies: cases from the Czech and Slovak Republics*. Boulder, CO: Westview, 149–65.

**MacKay**, R. 1995: Non-market forces, the nation state and the European Union. *Papers in Regional Science* 74, 209–31.

**MacKay**, R.R. 1994: Automatic stabilizers, European union and national unity. *Cambridge Journal of Economics* 18, 571–85.

**MacLean**, J. 1994: The state-firm-society triad: aspects of globalisation. Mimeo, University of Sussex.

**McLuahn**, M. 1960: *Understanding media*. London: Routledge & Kegan Paul.

**McMichael**, P. 1994a: Global restructuring: some lines of inquiry. In McMichael, P. (ed.), *The global restructuring of agro-food systems*. Ithaca, NY: Cornell University Press, 277–99.

**McMichael**, P. (ed.) 1994b: *The global restructuring of agro-food systems*. Ithaca, NY: Cornell University Press.

**McMichael**, P. and Myhre, D. 1991: Global regulation versus the nation state, *Capital and Class* 43, 83–106.

**McNay**, L. 1994: *Foucault: a critical introduction*. Cambridge: Polity.

**McWilliams**, C. 1946: *Southern California country*. New York: Duell, Sloan & Pierce.

**Mahler**, V.A. 1994: The Lomé Convention: assessing a North–South institutional relationship. *Review of International Political Economy* 1, 233–56.

**Maillat**, D., Crévoisier, O. and Lecoq, B. 1990: Innovation and territorial dynamism. Paper presented at the workshop, 'Flexible Specialization in Europe', Zurich, 25–26 October.

**Maillat**, D., Quévit, M. and Senn, L. (eds) 1993: *Milieux innovateurs et réseaux d'Innovation: un défi pour le développment regional*. Neuchâtel: EDES.

**Mair**, A. 1994: *Honda's global–local corporation*. London: Macmillan.

**Mair**, A., Florida, R. and Kenney, M. 1988: The new geography of automobile production: Japanese transplants in North America. *Economic Geography* 64, 352–73.

**Malecki**, E.J. 1984: Technology and regional development: a survey. *APA Journal* 50, 262–6.

**Mann**, M. 1986: *The sources of social power*, vol. 1: A history of power from the beginning to AD 1760. Cambridge: Cambridge University Press.

**Mann**, M. 1993: *The sources of social power*, vol. 2: The rise of classes and nation states, 1760–1914. Cambridge: Cambridge University Press.

**Mansfield**, E. 1972: The contribution of R&D to economic growth in the United States. *Science* 175, 477–86.

**Mansfield**, E. 1988: Industrial R&D in Japan and the United States: a comparative study. *American Economic Review* 78, 223–8.

**Mantouvalou**, M. and Mavridou, M. 1993: Illegal building: one way path to an impasse? *Bulletin of the Association of Greek Architects* 7, 58–71.

**Mantouvalou**, M., Mavridou, M. and Vaiou, D. 1993: Informal activities and micro-landownership vs planning in Greece: local specificities in a unifying Europe. *Proceedings*, Syros Seminar on 'Geographies of Integration, Geographies of Inequality in a Post-Maastricht Europe', Athens and Thessaloniki: University Press.

**Manzo**, K. 1991: Modernist discourses and the crisis of development theory. *Studies in International Comparative Development* 26, 3–36.

**Marcus**, G. 1989: *Lipstick traces: a secret history of the twentieth century*. Cambridge, MA: Harvard University Press.

**Marcuse**, P. 1995: Glossy globalisation: unpacking a loaded discourse. In Bounds, M. (ed.), *Globalisation of the West: the impacts of global restructuring on local and regional development in western Sydney*. University of Western Sydney–Macarthur: Urban Studies Research Group, 136–49.

**Markusen**, A. 1985: *Regions: the economics and politics of territory*. Totowa, NJ: Rowman & Littlefield.

**Markusen**, A., Hall, P., Campbell, I. and Dietrick, S. 1991: *The rise of the gunbelt: the military remapping of industrial America*. Oxford: Oxford University Press.

**Markusen**, A. and Yudken, J. 1992: *Dismantling the Cold War economy*. New York: Basic Books.

**Marquand**, D. 1988: *The unprincipled society*. London: Fontana.

**Marsden**, T. and Wrigley, N. 1995. Regulation, retailing and consumption. *Environment and Planning A* 27, 1899–912.

**Marsh**, G.P. 1844: 'Address, 1844'. In Brainerd and Brainerd, E.W. (eds), The *New England Society orations*, vol. 1. New York: The Century Company, 1901, 373–416.

**Marshall**, A. 1910: *Elements of economics of industry*. London: Macmillan.

**Marshall**, A. 1961: *Principles of economics*, 8th edition. Cambridge: Cambridge University Press. (First published in 1890.)

**Martin**, R.L. 1994a: Economic theory and human geography. In Gregory, D., Martin, R. and Smith, G. (eds), *Human geography: society, space, and social science*. Minneapolis: University of Minnesota Press, 21–53.

**Martin**, R. 1994b: Stateless monies, global financial integration and national economic autonomy: the end of geography? In Corbridge, S., Martin, R. and Thrift, N. (eds), *Money, power and space*. Oxford: Blackwell, 253–78.

**Martin**, R. and Rowthorn, B. 1986: *The geography of de-industrialisation*. London: Macmillan.

**Martin**, R., Sunley, P. and Wills, J. 1994: The decentralization of industrial relations? New institutional spaces and the role of local context in British engineering. *Transactions of the Institute of British Geographers* NS 19, 457–81.

**Martinez-Alier**, J. 1990: *Poverty as a cause of environmental degradation*. Report prepared for the World Bank, Washington DC.

**Marx**, K. 1967: *Capital* vol. 1. New York: International Publishers.

**Maskell**, P. and Malmberg, A. 1995: Localised learning and industrial competitiveness. Mimeographed paper presented at the Regional Studies Association Conference on Regional Futures, Gothenburg, 6–9 May 1995.

**Massey**, D. 1984: *Spatial divisions of labour: social structures and the geography of production*. London: Macmillan.

**Massey**, D. 1988: Uneven development: social change and spatial divisions of labour. In Massey, D. and Allen, J. (eds), *Uneven re-development: cities and regions in transition*. London: Hodder & Stoughton, 250–76.

**Massey**, D. 1991: Flexible sexism. *Environment and Planning D: Society and Space* 9, 31–57.

**Massey**, D. 1992: Politics and space/time. *New Left Review* 196, 65–84.

**Massey**, D. 1993a: Power-geometry and a progressive sense of place. In Bird, J., Curtis, B., Putnam, T., Robertson, G. and Tickner, L. (eds), *Mapping the Futures: local cultures, global change*. London: Routledge, 59–69.

**Massey**, D. 1993b: Questions of locality. *Geography* 78, 142–9.

**Massey**, D. 1994: *Space, place and gender*. Cambridge: Polity.

**Massey**, D. 1995a: *Spatial divisions of labour: social structures and the geography of production*, 2nd edition. London: Macmillan.

**Massey**, D. 1995b: Reflections on gender and geography. In Butler, T. and Savage, M. (eds), *Social change and the middle classes*. London: UCL Press.

**Massey**, D. 1995c: Masculinity, dualisms and high technology. *Transactions of the Institute of British Geographers* NS 20, 487–99.

**Massey**, D. 1996: Space/power, identity/difference tension in the city. In Merrifield, A. and Swyngedouw, E. (eds), *The urbanization of injustice*, London: Lawrence & Wishart, 100–16.

**Massey**, D. and Meegan, R. 1979: *The anatomy of job loss: the how, why and where of employment decline*. London: Methuen.

**Mathews**, L.K. 1909: *The expansion of New England*. Boston: Houghton Mifflin.

**Mathews**, T. and Ravenhill, J. 1996: The neoclassical ascendancy: the economic policy community and Northeast Asian economic growth. In Robison, R. (ed.), *Pathways to Asia: the politics of engagement*. Sydney: Allen & Unwin, 131–70.

**Matless**, D. 1992: An occasion for geography: landscape, representation and Foucault's corpus. *Environment and Planning D: Society and Space* 10, 41–56.

**Matzner**, E. and Streeck, W. 1991: *Beyond Keynesianism: the socio-economics of production and full employment.* Aldershot: Edward Elgar.

**Mayer**, M. 1989: Restructuring and opposition in West German cities. In Feagin, J. and Smith, M.P. (eds), *The capitalist city: global restructuring and community politics.* Oxford: Blackwell, 343–63.

**Mayer**, M. 1993: The career of urban social movements in West Germany. In Fisher R. *et al.* (eds), *Mobilizing the community: local politics in the era of the global city.* London: Sage, 112–37.

**Mayer**, M. 1994: Post-Fordist city politics. In Amin, A. (ed.), *Post-Fordism: a reader.* Oxford: Blackwell, 316–37.

**Mayr**, E. 1985: How biology differs from the physical sciences. In Depew, D. J. and Weber, B. H. (eds), *Evolution at the crossroads: the new biology and the new philosophy of science.* Cambridge, MA: MIT Press, 43–63.

**Meeks**, W. 1974: The image of the androgyne. *History of Religions* 13, February.

**Merchant**, C. 1980: *The death of nature.* San Francisco: Harper & Row.

**Metcalfe**, J.S. 1988: Evolution and economic change. In Silbertson, A. (ed.), *Technology and economic progress.* London: Macmillan, 54–85.

**Meulders**, D. and Plasman, R. 1992: *Women in atypical employment.* Brussels: Commission of the European Communities, V/1426/89–FR.

**Michie**, R. 1992: *The City of London: continuity and change, 1850–1990.* London: Macmillan.

**Milanovic**, B. 1994: A cost of transition: 50 million new poor and growing inequality. *Transition* 5, 1–4.

**Mill**, J.S. 1869: The subjection of women. In *Three Essays.* Oxford: Oxford University Press, 1975.

**Miller**, D. 1989: *Mass culture and mass consumption.* Oxford: Blackwell.

**Miller**, D. 1992: The young and the restless in Trinidad: a case of the local and the global in mass consumption. In Silverstone, R. and Hirsch, E. (eds), *Consuming technologies: media and information in domestic spaces.* London: Routledge, 163–82.

**Miller**, D. (ed.) 1995a: *Acknowledging consumption.* London: Routledge.

**Miller**, D. 1995b: Consumption as the vanguard of history: a polemic by way of an introduction. In Miller, D. (ed.), *Acknowledging consumption: a review of new studies.* London: Routledge, 1–57.

**Miller**, P. 1953: *The New England mind: from colony to province.* Cambridge, MA: Harvard University Press.

**Miller**, P. and Rose, N. 1990: Governing economic life. *Economy and Society* 19, 1–31.

**Mingione**, E. 1985: Social reproduction and the surplus labour force: the case of Italy. In Redclift, N. and Mingione, E. (eds), *Beyond employment.* Oxford: Blackwell, 14–54.

**Mingione**, E. 1988: Work and informal activities in urban southern Italy. In R. Pahl (ed.), *On work.* Oxford: Blackwell, 548–78.

**Mingione**, E. and Magatti, M. 1995: The informal sector: follow-up to the White paper. Report to the European Commission's Employment Task Force (DG V). *Social Europe*, supplement 3/95, 67–111.

**Mingione**, E. and Morlicchio, E. 1993: New forms of urban poverty in Italy: risk path models in the North and South. *International Journal of Urban and Regional Research* 17, 413–28.

**Mirowski**, P. 1984a: Physics and the 'marginalist revolution'. *Cambridge Journal of Economics* 8, 361–79.

**Mirowski**, P. 1984b: The role of conservation principles in twentieth-century economic theory. *Philosophy of the Social Sciences* 14, 461–73.

**Mirowski**, P. 1987: Shall I compare thee to a Minkowski–Ricardo–Leontief–Metzler matrix of the Mosak–Hicks type? Rhetoric, mathematics and the nature of neoclassical theory. *Economics and Philosophy* 3, 67–96.

**Mirowski**, P. 1989a: *More heat than light: economics as social physics, physics as nature's economics.* Cambridge: Cambridge University Press.

**Mirowski**, P. 1989b: How not to do things with metaphors: Paul Samuelson and the science of neoclassical economics. *Studies in the History and Philosophy of Science* 20, 175–91.

**Mirowski**, P. 1994: Doing what comes naturally: Four metanarratives on what metaphors are for. In Mirowski, P. (ed.), *Natural images in economic thought: 'Markets read in tooth and claw'.* Cambridge: Cambridge University Press, 3–19.

**Mishra**, R. 1990: *The welfare state in capitalist society.* Brighton: Harvester Wheatsheaf.

**Mitchell**, D. 1998 forthcoming: The scales of justice: localist ideology, large-scale production and agricultural labor's geography of resistance in 1930s California. In Herod, A. (ed.), *Organizing the landscape: labor unionism in geographical perspective.* Minneapolis: University of Minnesota Press.

**Mitchell**, K. 1995: Flexible circulation in the Pacific Rim: capitalisms in cultural context. *Economic Geography* 71(4), 364–82.

**Miyoshi**, M. 1993: A borderless world? From colonialism to transnationalism and the decline of the nation state. *Critical Inquiry* 19, 726–51.

**Mohan**, J. 1995: *A National Health Service? The restructuring of health care in Britain since 1979.* London: St Martin's Press.

**Moore**, B. 1966: *The social origins of dictatorship and democracy.* Boston: Beacon Press.

**Morgan**, G. 1986: *Images of organisation*: Beverly Hills, Sage.

**Morgan**, G. 1990: *Organisations in society.* Basingstoke: Macmillan.

**Morgan**, G. and Knights, D. 1991: Management control in sales: a case study from the labour process of life insurance. *Work, Employment and Society* 4, 369–89.

**Morgan**, K. 1996: Learning-by-interacting: inter-firm networks and enterprise support. In *Local systems of small firms and job creation.* Paris: OECD.

**Morgan**, K. and Roberts, E. 1993: The democratic deficit: a guide to Quangoland. *Papers in Planning*

*Research*, N 144, Cardiff: Department of City and Regional Planning, University of Wales College of Cardiff.

**Morgan**, K. and Sayer, A. 1988: *Microcircuits of capital*. Cambridge: Polity.

**Morita**, A. *et al*., 1988: *Made in Japan*. New York: Dutton.

**Morley**, D. and Robins, K. 1991: Techno-orientalism: futures, foreigners and phobias. *New Formations* 16, 136–56.

**Morley**, H., and Schmid, G. 1993: Public services and competitiveness. In Hughes, K. (ed.), *European competitiveness*. Cambridge: Cambridge University Press.

**Morris**, M, 1992, The man in the mirror: David Harvey's 'condition' of postmodernity. *Theory, Culture and Society* 9, 253–79.

**Mouffe**, C. 1995: Post-Marxism: democracy and identity. *Environment and Planning D: Society and Space* 13, 259–65.

**Moulaert**, F. and Djellal, F. 1990: Les firmes de conseil en technologie de l'information: des économies d'agglomération en réseaux. Paper presented at the Conference on 'Métropôles en Déséquilibre', Lyon, France, 22–23 November.

**Mouriki**, A. 1991: Flexible forms of employment: blessing or anathema? *Topos* 3, 97–118 (in Greek).

**Mowery** D. and Teece, D. 1993: Japan's growing capabilities in industrial technology. *California Management Review* 36, 9–33.

**Muellbauer**, J. 1990: The housing market and the UK economy: problems and opportunities. In Ermisch, J. (ed.), *Housing and the national economy*. Aldershot: Avebury.

**Murgatroyd**, L., Savage, M., Shapiro, D., Urry, J., Walby, S., Warde, A. with Mark-Lawson, J. 1985: *Localities, class and gender*. London: Pion.

**Murray**, F. 1983: The decentralization of production: the decline of the mass-collective worker. *Capital and Class* 19, 74–99.

**Murray**, M. 1995: Correction at Cabrini-green: a socio-spatial exercise of power. *Environment and Planning D: Society and Space* 13, 311–27.

**Murray**, R. 1992: Flexible specialisation and development strategy: the relevance for eastern Europe. In Ernste, H. and Meier, V. (eds), *Regional development and contemporary industrial response: extending flexible specialisation*. London: Belhaven, 198–217.

**Musil**, J. 1993: Changing urban systems in post-communist societies in central Europe: analysis and prediction. *Urban Studies* 30, 899–905.

**Myerscough**, J. 1988: *The economic importance of the arts in Britain*. London: Policy Studies Institute.

**Naanen**, B. 1995: Oil producing minorities and the restructuring of Nigerian federalism. *Journal of Comparative and Commonwealth Politics* 3, 46–78.

**Nash**, G. 1964: *State government and economic development*. Berkeley: University of California Press.

**National Research Council** 1992: *U.S.–Japan strategic alliances in the semiconductor industry*. Washington DC: National Academy Press.

**NEDO** 1991: *The experience of Nissan suppliers: lessons for the UK engineering industry*. London: National Economic Development Office.

**Nelson**, K. 1986: Labor demand, labor supply and the suburbanization of low-wage office work. In Scott, A. J. and Storper, M. (eds), *Production, work, territory*. Boston: Allen & Unwin.

**Nelson**, R. (ed.) 1993: *National innovation systems*. New York: Oxford University Press.

**Nelson**, R.R. and Winter, S.G. 1982: *An evolutionary theory of economic change*. Cambridge, MA: Harvard University Press.

**Nemes Nagy**, J. 1994: Regional disparities in Hungary during the period of transition to a market economy. *GeoJournal* 32, 363–8.

**NESCB** 1882–1925: New England Society in the City of Brooklyn, Annual Reports. Brooklyn: New England Society in the City of Brooklyn.

**NESNY** 1875–1925: New England Society in the City of New York, Anniversary celebration of the New England society in the City of New York. New York: New England Society of New York.

**Neumann**, R. 1995: Local challenges to global agendas. *Antipode* 27, 363–82.

**New York State** 1953: New York State Crime Commission (Port of New York Waterfront) to the Governor, the Attorney General and the Legislature of the State of New York. Albany, NY: New York State Crime Commission.

**New York Times** 1954: Pier vote hearing called by NLRB, 18 February, p. 55.

**Newlands**, D. 1995: The economic role of regional governments in the European Community. In Hardy, S., Hart, M., Albrechts, L. and Katos, A. (eds) *An enlarged Europe: regions in competition?* London: Jessica Kingsley, 70–80.

**Nilsson**, J.E. and Schamp, E. 1996: Restructuring of the European production system: processes and consequences. *European Urban and Regional Studies* 3, 121–32.

**Nishiguchi**, T. 1994: *Strategic industrial sourcing: the Japanese advantage*. Oxford: Oxford University Press.

**Noble**, D. 1992: *A world without women: the Christian clerical culture of western science*. Oxford: Oxford University Press.

**Nonaka**, I. and Takeuchi, H. 1995: *The knowledge-creating company: how Japanese companies create the dynamics of innovation*. Oxford: Oxford University Press.

**Notermans**, T. 1993: The abdication from national policy autonomy: why the macroeconomic policy regime has become so unfavorable to labor. *Politics and Society* 21, 133–68.

**Notermans**, T. 1997 forthcoming: Social democracy and external constraints. In Cox, K.R. (ed.), *Spaces of globalization: reasserting the power of the local*. New York: Guilford Press.

**Nove**, A. 1991: *The economics of feasible socialism revisited*, 2nd edition. London: HarperCollins.

**NYSA** (New York Shipping Association) 1953: Brief of New York Shipping Association: Preliminary Statement before NLRB Region 2, in the matter of 'New York Shipping Association et al., NLRB case nos.

2–RC-6282 et al.' Located in box 366, National Labor Relations Board 1935–, Administrative Division Files and Dockets Section; Selected Taft–Hartley Cases, 1947–59; Records of the NLRB, Record Group 25, National Archives, Washington DC.

**O'Brien**, R. 1992: *Global financial integration: the end of geography*. London: Pinter.

**Obstfeld**, M. 1995: International capital mobility in the 1990s. In Kenen, P. (ed.), *Understanding interdependence: the macroeconomics of the open economy*. Princeton, NJ: Princeton University Press, 201–61.

**O'Connor**, J. 1973: *The fiscal crisis of the state*. New York: St Martin's Press.

**O'Connor**, J. 1988: Capitalism, nature, socialism: a theoretical introduction. *Capitalism, Nature, Socialism* 1, 11–38.

**O'Connor**, J. 1994: Is sustainable capitalism possible? In O'Connor, M. (ed.), *Is capitalism sustainable?* New York: Guilford, 152–75

**O'Connor**, M. 1993: On the misadventures of capitalist nature. *Capitalism, Nature, Socialism* 4, 7–40.

**Odagiri**, H. and Goto, A. 1993: The Japanese innovation system: past, present and future. In Nelson, R. (ed.), *National innovation systems: a comparative study*. New York: Oxford University Press.

**O'Donnell**, R. 1994: Decision-making in the 21st century: implications for national policy-making and political institutions. Mimeo, National Economic and Social Council, Dublin.

**OECD** 1991: *Historical statistics 1960–89*. Paris: OECD.

**OECD** 1992: *Globalisation of industrial activities: four case studies – auto parts, chemicals, construction and semiconductors*. Paris: OECD.

**Offe**, C. 1975: The theory of the capitalist state and the problem of policy formation. In Lindberg, L.N., Alford, R. Crouch, C. and Offe, C. (eds), *Stress and contradiction in modern capitalism*. Lexington, MA: Lexington Books, 125–44.

**Offe**, C. 1976: Political authority and class structures. In Connerton, P. (ed.), *Critical sociology*. London: Penguin, 388–421.

**Offe**, C. 1984: *Contradictions of the welfare state*, ed. J. Keane. London: Hutchinson.

**Offe**, C. 1985: *Disorganised capitalism: contemporary transformations of work and politics*, ed. J. Keane. Cambridge: Polity.

**Offe**, C. and Ronge, V. 1975: Theses on the theory of the state. *New German Critique*, 6; reprinted in Offe, C. 1984: *Contradictions of the welfare State*, ed. J. Keane. London: Hutchinson.

**Ogley**, R.C. 1994: International relations. In Outhwaite, W., Bottomore, T., Gellner, E., Nisbet, R. and Touraine, N. (eds), *The Blackwell dictionary of twentieth-century social thought*. Oxford: Blackwell, 295–7.

**O'Grady**, J. 1994: Province of Ontario, Canada: removing the obstacles to negotiated adjustments. In Sengenberger, W. and Campbell, D. (eds), *Creating economic opportunities: the role of labour standards in industrial restructuring*. Geneva: International Institute for Labour Studies, 255–78.

**Ohmae**, K. 1990: *The borderless world: power and strategy in the interlinked economy*. London: Fontana.

**Ohmae**, K. 1993: The rise of the region state. *Foreign Affairs* 72, 78–87.

**Ohmae**, K. 1995a: *The end of the nation state: the rise of regional economies*. London: HarperCollins.

**Ohmae**, K. 1995b: Putting global logic first. In Ohmae, K. (ed.), *The evolving global economy*. New York: Harvard Business Review, 129–40.

**O Huallachain**, B. 1992: Industrial geography. *Progress in Human Geography* 16, 545–52.

**Okamura**, C. and Kawahito, H. 1990: *Karoshi*. Tokyo: Mado-Sha.

**Olds**, K. 1995a: Globalization and the production of new urban spaces: Pacific Rim megaprojects in the late twentieth century. *Environment and Planning A* 27, 1713–44.

**Olds**, K. 1995b: Pacific Rim mega-projects and the global cultural economy: tales from Vancouver and Shanghai. Unpublished PhD dissertation, University of Bristol.

**Ollman**, B. 1993: *Dialectical investigations*. London: Routledge.

**O'Neill**, J. 1994: Humanism and nature. *Radical Philosophy* 66, 21–9.

**O'Neill**, P.M. 1994: Capital, regulation and region: restructuring and internationalisation in the Hunter Valley, NSW. Unpublished PhD thesis, Macquarie University.

**O'Neill**, P.M. 1996: In what sense a region's problem? The place of redistribution in Australia's internationalisation strategy. *Regional Studies* 30, 401–12.

**O'Neill**, P.M. 1997: Internationalisation and the nation state: Australia's changing accumulation strategy. *Environment and Planning A* (forthcoming).

**O'Neill**, P.M. and Fagun, R.H. 1995: The new regional policy: what chance of success? *Australian Quarterly* 67, 55–68.

**Ong**, W.J. 1959: Latin language as a renaissance puberty rite. *Studies in Philology* 56, April.

**Orans**, M. 1968: Maximising in Jajmaniland. *American Anthropologist* 70, 875–97.

**O'Reilly**, J. 1992a: Where do you draw the line? Functional flexibility, training and skill in Britain and France. *Work, Employment and Society* 6, 369–96.

**O'Reilly**, J. 1992b: Subcontracting in banking: some evidence from Britain and France. *New Technology, Work and Employment* 3, 107–15.

**Ó Tuathail**, G. 1992a: Putting Mackinder in his place: material transformations and myth. *Political Geography* 11, 100–18.

**Ó Tuathail**, G. 1992b: 'Pearl Harbor without bombs': a critical geopolitics of the US–Japan 'FSX' debate. *Environment and Planning A* 24, 975–94.

**Ó Tuathail**, G. 1993a: The effacement of place? US foreign policy and the spatiality of the Gulf Crisis. *Antipode* 25, 4–31.

**Ó Tuathail**, G. 1993b: Japan as threat: geo-economic discourses on the USA-Japan relationship in US civil society, 1987–91. In Williams, C. H. (ed.), *The political geography of the New World Order*. London: Belhaven, 181–209.

**Ó Tuathail**, G. 1996: *Critical geopolitics: the politics of writing global space*. Minneapolis: University of Minnesota Press.

**Ó Tuathail**, G. and Agnew, J. 1992: Geopolitics and discourse: practical geopolitical reasoning in American foreign policy. *Political Geography* 11, 190–204.

**Ó Tuathail**, G. and Luke, T. 1994: Present at the disintegration: deterritorialization and reterritorialization in the New Wor(l)d Order. *Annals of the Association of American Geographers* 84, 381–98.

**Ould-Mey** M. 1994: Global adjustment: implications for peripheral states. *Third World Quarterly* 15, 319–36.

**Paci**, M. 1992: *Il mutamento della struttura sociale in Italia*. Bologna: Il Mulino.

**Paci**, M. 1995: Welfare and fiscal policies and informal activities: the Italian model and its crisis. *Proceedings*, European Seminar 'Forms of Informal Employment in the European Union', Athens: Praxis.

**Page**, S. 1994: *How developing countries trade*. New York: Routledge.

**Pahl**, R. 1984: *Divisions of Labour*. Oxford: Blackwell.

**Pahl**, R. and Wallace, C. 1985: Household work strategies in economic recession. In Redclift, N. and Mingione, E. (eds), *Beyond Employment: household, gender and subsistence*. Oxford: Blackwell, 189–227.

**Painter**, J. 1991: The geography of trade union responses to local government privatization. *Transactions of the Institute of British Geographers* NS 16, 214–26.

**Painter**, J. 1997: Regulation, regime and practice in urban politics. In Lauria, M. (ed.), *Reconstructing urban regime theory: regulating urban politics in a global economy*. Thousand Oaks, CA: Sage, 123–44.

**Painter**, J. and Goodwin, M. 1995: Local governance and concrete research: investigating the uneven development of regulation. *Economy and Society* 24, 334–56.

**Palloix**, C. 1993: *Les marchands et l'industrie: un essai sur les rapports entre la société et l'économie*. Amiens: ERSI.

**Palmer**, B. 1990: *Descent into discourse: the reification of language and the writing of social history*. Philadelphia: Temple University Press.

**Papataxiarchis**, E. 1992: Introduction. From the perspective of gender: anthropological approaches of modern Greece. In E. Papataxiarchis and Th. Paradellis (eds), *Identities and gender in modern Greece*. Athens: Kastaniotis, 11–98 (in Greek).

**Parajuli**, P. 1991: Power and knowledge in development discourse. *International Social Science Journal* 127, 173–90.

**Pareto**, V. 1971: *Manual of political economy*, translated from the original 1927 French edition by A.S. Schwier. New York: Augustus Kelley.

**Parker**, D. 1993: Privatisation: Ten years on. In Healey, N. (ed.), *Britain's Economic miracle: myth or reality?* London: Routledge, 174–94.

**Parr**, J. 1990: *The gender of breadwinners: women, men, and change in two industrial towns 1880–1950*. Toronto: University of Toronto Press.

**Parsons**, P. 1960: *Structure and process in modern societies*. Glencoe: Free Press.

**Parsons**, T. 1963: On the concept of political power. *Proceedings of the American Philosophical Society* 107, 232–62.

**Pascale**, R. 1990: *Managing on the edge*. New York: Simon & Schuster.

**Pasinetti**, L.L. 1962: Rate of profit and income distribution in relation to the rate of economic growth. *Review of Economic Studies* 29, 267–79.

**Patel**, P. 1995: Localised production of technology for global markets. *Cambridge Journal of Economics* 19, 141–53.

**Patel**, P. and Pavitt, K. 1994: National innovation systems: why they are important and how they might be measured and compared. *Economics of Innovation and New Technology* 3, 77–95.

**Paukert**, L. 1995: Privatization and employment: labour transfer policies and practices in the Czech Republic. International Labour Office Labour Market Papers 4.

**Pavit**, K. and Patel, P. 1991: Large firms in the production of the world's technology: an important case of non-globalization. *Journal of International Business Studies* First Quarter, 1–21.

**Pavlínek**, P. 1995: Transition and the environment in the Czech Republic: democratization, economic restructuring and environmental management in the Most district after the collapse of state socialism. Unpublished PhD dissertation, Department of Geography, University of Kentucky.

**Pavlopoulos**, P. 1987: *The informal economy in Greece*. Athens: IOBE (in Greek).

**Peck**, J. 1994: Regulating labour: the social regulation and reproduction of local labour markets. In Amin, A. and Thrift, N. (eds), *Globalization, institutions and regional development in Europe*. Oxford: Oxford University Press, 147–76.

**Peck**, J. 1995: Moving and shaking: business elites, state localism and urban privatism. *Progress in Human Geography* 19, 16–46.

**Peck**, J. 1996: *Work-place: the social regulation of labor markets*. New York: Guilford.

**Peck**, J. and Jones, M. 1994: Training and enterprise councils: Schumpeterian workfare state or what? *Environment and Planning A* 27, 1361–96.

**Peck**, J.A., and Tickell, A. 1992: Local modes of social regulation? Regulation theory, Thatcherism and uneven development. *Geoforum* 23, 347–83.

**Peck**, J.A. and Tickell, A. 1994a: Jungle law breaks out: neoliberalism and global–local disorder. *Area* 26, 317–26.

**Peck**, J.A. and Tickell, A. 1994b: Searching for a new institutional fix: the after Fordist crisis and global–local disorder. In Amin, A. (ed.), *Post-Fordism: a reader*. Oxford: Blackwell, 280–316.

**Peck**, J. and Tickell, A. 1995a: Business goes local: dissecting the business agenda in Manchester. *International Journal of Urban and Regional Research* 19, 55–78.

**Peck**, J. and Tickell, A. 1995b: The social regulation of uneven development: 'regulatory deficit', England's south-east and the collapse of Thatcherism. *Environment and Planning A* 27, 15–40.

**Peet**, R. (ed.) 1987: *International capitalism and industrial restructuring*. London: Allen & Unwin.

**Peet**, R. 1996: A sign taken for history: Daniel Shays' memorial in Petersham, Massachusetts. *Annals Association of American Geographers* 85, 21–39.

**Peet**, R. and Thrift, N. (eds) 1989: *New Models in geography: the political-economy perspective*. London: Unwin Hyman.

**Peet**, R. and Watts, M. (eds.) 1996a: *Liberation ecologies: environment, development, social movements*. London: Routledge.

**Peet**, D. and Watts, M. 1996b: Liberation ecology: development, sustainability, and environment in an age of market triumphalism. In Peet, R. and Watts, M. (eds), *Liberation ecologies: environment, development, social movements*. London: Routledge, 1–45.

**Pendleton**, A. 1991: Barriers to flexibility: flexible rostering on the railways. *Work, Employment and Society* 5, 241–57.

**Perroux**, F. 1950a: Les Éspaces Économiques. *Economie Appliqué* 1, 25–44.

**Perroux**, F. 1950b: Economic space: theory and applications. *Quarterly Journal of Economics* 64, 89–104.

**Peters**, L. 1991: Technology strategies of Japanese subsidiaries and joint ventures in the United States. Working paper, Center for Science and Technology Policy, Rensselaer Polytechnic Institute.

**Peters**, T. and Waterman, R.H. 1982: *In search of excellence*. Harper Collins.

**Pfaller**, A., Gough, I., and Therborn, G. 1991: *Can the welfare state compete? A comparative study of five advanced capitalist countries*. London: Macmillan.

**Phillips**, K. 1990: *The politics of rich and poor*. New York: Random House.

**Philo**, C. 1989: Enough to drive one mad: the organization of space in nineteenth century lunatic asylums. In Wolch, J. and Dear, M. (eds.), *The power of geography: how territory shapes social life*. London: Unwin Hyman, 258–90.

**Philo**, C. (ed.), 1991: *New words, new worlds: reconceptualising social and cultural geography*. Department of Geography, St David's University College Lampeter.

**Philo**, C. 1992: Foucault's geography. *Environment and Planning D: Society and Space* 10, 137–61.

**Philo**, C. and Kearns, G. (eds) 1994: *Selling places*. Oxford: Pergamon Press.

**La piccola impresa** 1990: collective volume. Società e Lavoro, Simposi Interdisciplinari 4. Bologna: Jovene Editore.

**Picciotto**, S. 1991: The internationalisation of the statel. *Capital and Class* 43, 43–63.

**Pickles**, J. 1995: Restructuring state enterprises: industrial geography and eastern European transition. *Geographische Zeitschrift* 82, 114–31.

**Pickles**, J. and Smith, A. (eds) 1997 forthcoming: *Theorizing transition: the political economy of change in Eastern Europe*. London: Routledge.

**Pierson**, C. 1994: *Dismantling the welfare state? Reagan, Thatcher and the politics of retrenchment*. Cambridge: Cambridge University Press.

**Piirainen**, T. 1994a: Survival strategies in a transition economy: everyday life, subsistence and new inequalities in Russia. In Piirainen, T. (ed.), *Change and continuity in Eastern Europe*. Aldershot: Dartmouth, 89–113.

**Piirainen**, T. 1994b: Three scenarios for the future of eastern Europe. In Piirainen, T. (ed.), *Change and continuity in Eastern Europe*. Aldershot: Dartmouth, 214–42.

**Pine**, B. J. 1993: *Mass customization: the new frontier in business competition*. Cambridge, MA: Harvard University Press.

**Piore**, M. 1979: *Birds of passage*. New York: Cambridge University Press.

**Piore**, M. and Sabel, C. 1984: *The second industrial divide*. New York: Basic Books.

**Piven**, F.F. 1995: Is it global economics or neo-laissez-faire? *New Left Review* 213, 107–14.

**Pizanias**, P. 1993: *The urban poor: the know-how of survival in interwar Greece*. Athens: Themelio (in Greek).

**Pizzo**, S., Fricker, M. and Muolo, P. 1989: *Inside job*. San Francisco: McGraw-Hill.

**Polanyi**, K. 1957a: The economy as instituted process. Repr. in Granovetter, M. and Swedberg, R. (eds) 1993: *The sociology of economic life*. Boulder, CO: Westview, 29–52.

**Polanyi**, K. 1957b [1944]: *The great transformation: the political and economic origins of our time*. Boston: Beacon Press.

**Pollard**, J. 1995: The contradictions of flexibility: labour control and resistance in the Los Angeles banking industry. *Geoforum* 26, 121–3.

**Pollert**, A. 1988: Dismantling flexibility. *Capital and Class* 34, 42–75.

**Pooley**, S. 1991: The State rules, OK? The continuing political economy of nation-states. *Capital and Class* 43, 65–82.

**Popke**, E. J. 1994: Recasting geopolitics: the discursive scripting of the International Monetary Fund. *Political Geography* 13, 255–69.

**Porter**, M. 1990: *The competitive advantage of Nations*. London: Macmillan.

**Porter**, M.E. (ed.) 1986: *Competition in global industries*. Cambridge, MA: Harvard University Press.

**Post**, C. 1982: The American road to capitalism. *New Left Review* 133, 30–51.

**Poznanski**, K. 1995: Institutional perspectives on post-communist recession in Eastern Europe. In Poznanski, K. (ed.), *The evolutionary transition to capitalism*. Boulder, CO: Westview, 3–30.

**Pratt**, D.J. 1995: Re-placing money. Unpublished PhD dissertation, University of Hull.

**Prattis**, J.I. 1987: Alternative views of economy in economic anthropology. In Clammer, J. (ed.), *Beyond the new economic anthropology*. London: Macmillan, 8–44.

**Pred**, A. and Watts, M.J. 1992: *Reworking modernity: capitalisms and symbolic discontent*. New Brunswick, NJ: Rutgers University Press.

**Prestowitz**, C. 1988: *Trading places*. New York: Basic Books.

**Prestwich**, M. (ed.) 1985: *International Calvinism, 1541–1715*. Oxford: Clarendon.

**Pringle**, R. 1989: *Secretaries talk*. London: Verso.

**Pringle**, R. 1995: Destabilising patriarchy. In Caine, B. and Pringle, R. (eds), *Transitions: new Australian feminisms*. Sydney: Allen & Unwin, 198–211.

**Pritchard**, W. 1995: Uneven globalisation: the restructuring of Australian dairy and vegetable processing sectors. Unpublished PhD thesis, University of Sydney.

**Prokhvatilov**, V. 1994: Russians run from taxman. *Guardian*, 20 October, p. 15.

**Pryke**, M. 1991: An international city going 'Global': spatial change in the City of London. *Environment and Planning D: Society and Space* 9, 97–222.

**Pryke**, M. and Lee, R. 1995: Place your bets: towards an understanding of globalization, socio-financial engeineering and competition within a financial center. *Urban Studies* 32, 329–44.

**Putnam**, R. 1993: *Making democracy work*. Princeton, NJ: Princeton University Press.

**Pyke**, F. and Sengenberger, W. (eds) 1992: *Industrial districts and local economic regeneration*. Geneva: International Institute for Labour Studies.

**Radice**, H. 1984: The national economy: a Keynesian myth. *Capital and Class* 22, 111–40.

**Radice**, H. 1995: 'Globalisation' and the UK economy. In Barratt Brown, M. and Radice, H., *Democracy versus capitalism: a response to Will Hutton with some old questions for New Labour*. Socialist Renewal, Pamphlet 4, European Labour Forum, 14–25.

**Rainnie**, A. 1993: The reorganization of large firm subcontracting. *Capital and Class* 49, 53–76.

**Ramesh**, M. 1995: Economic globalization and policy choices: Singapore. *Governance* 8, 243–60.

**Rangan**, P. 1995: Contesting boundaries. *Antipode* 27 (4), 343–62.

**Rappaport**, R. 1967: *Pigs for the ancestors*. New Haven, CT: Yale University Press.

**Raskin**, J. 1994: Challenging the wealth primary. *The Nation*, 21 November, 610.

**Recio**, A., Miguelez Lobo, F. and Alos, R. 1988: *El trabjo precario en Catalunya: la industria textil lanera des valles occidental*. Barcelona: Comisio Obrera Nacional de Catalunya.

**Redwood**, J. 1988: *Popular capitalism*. London: Routledge.

**Rees-Mogg**, W. 1995: Division of the future. *The Times*, 20 July, p. 18.

**Regini**, M. 1995: *Uncertain boundaries: the social and political construction of European economies*. Cambridge: Cambridge University Press.

**Reich**, R. 1991: *The work of nations: preparing ourselves for the 21st century capitalism*. London: Simon & Schuster.

**Reich**, R. 1992: *The work of nations*. New York: Vintage.

**Reich**, R. 1995a: Who is us? In Ohmae, K. (ed.), *The evolving global economy*. New York: Harvard Business Review, 141–60.

**Reich**, R. 1995b: Who is them? In Ohmae, K. (ed.), *The evolving global economy*. New York: Harvard Business Review, 161–82.

**Reich**, R. and Mankin, E. 1986: Joint ventures with Japan give away our future. *Harvard Business Review* 86, 78–85.

**Reid**, N. 1990: Spatial patterns of Japanese investment in the United States automobile industry. *Industrial Relations Journal* 21, 49–59.

**Research Institute of Industrial Economics** 1992: *Regional patterns of structural adjustments in Hungary: the case of the iron and steel industry in an international comparison*. Budapest: Hungarian Academy of Sciences.

**Resnick**, S. and Wolff, R. 1987: *Knowledge and class: a Marxian critique of political economy*. Chicago: University of Chicago Press.

**Rich**, M. 1996: Britain's textile manufacturers cotton on to cheap labour. *Financial Times*, 16 April, p. 23.

**Richards**, P. 1985: *Indigenous agricultural revolution*. London: Hutchinson.

**Richardson**, H. 1973: *Regional growth theory*. London: Macmillan.

**Richardson**, J.D. 1995: Income inequality and trade: how to think, what to conclude. *Journal of Economic Perspectives* 9, 33–55.

**Richardson**, J.L. 1994: Problematic paradigm: liberalism and the global order. In Camilleri, J.A., Jarvis, P. and Paolini, A. J. (eds), *The state in transition: reimagining political space*. Boulder, CO: Rienner.

**Rip**, A. 1991: A cognitive approach to technology policy. Paper presented to the symposium 'New Frontiers in Science and Engineering', Paris, 27–29 May.

**Rist**, G. 1991: 'Development' as part of the modern myth. *European Journal of Development Research* 2, 10–21.

**Ristelhueber**, R. 1994: Setting sun: the slide of the Japanese. *Electronics Business Buyer* 20, 52–60.

**Ritzer**, G. 1993: *The McDonaldization of society*. Thousand Oaks, CA: Pine Forge Press.

**Robbins**, D. 1991: *The work of Pierre Bourdieu*. Boulder, CO: Westview Press.

**Roberts**, B. (guest ed.) 1994: Informal Economy and Family Strategies. special issue, *International Journal of Urban and Regional Research*, 18(1).

**Roberts**, E. 1993: Strategic benchmarking of technology. Paper presented to the Consortium on Competitiveness and Cooperation. Cambridge, MA: Harvard Business School.

**Roberts**, S. 1995: The world is whose oyster? The geopolitics of representing globalization. Mimeo.

**Robertson**, R. 1992: *Globalization: social theory and global culture*. London: Sage.

**Robertson**, R. 1995: Glocalization: time–space and homogeneity–heterogeneity. In Featherstone, M., Lash, S. and Robertson R. (eds), *Global modernities*. London: Sage, 91–107.

**Robinson**, A. 1981: *George Meany and his times*. New York: Simon & Schuster.

**Robinson**, D. 1985: *The unitarians and the universalists*. Westport, CT: Greenwood Press.

**Robinson**, J.V. 1964: *Economic philosophy*. Harmondsworth: Penguin.

**Robinson**, J.V. 1973a: A lecture delivered at Oxford by

a Cambridge economist. In *Collected economic papers* vol. 4. Oxford: Blackwell, 254–63.

**Robinson**, J.V. 1973b: An open letter from a Keynesian to a Marxist. In *Collected economic papers* vol. 4. Oxford: Blackwell, 264–68.

**Robinson**, J.V. 1979: History versus equilibrium. In *Collected economic papers* vol. 5. Oxford: Blackwell, 48–58.

**Robinson**, M. 1990: *Overdrawn*. New York: Dutton.

**Robinson**, R. (ed.) 1996: *Pathways to Asia: the politics of engagement.* Sydney: Allen & Unwin.

**Rocheleau**, D. *et al*., 1995: Environment, development, crisis and crusade, *World Development* 23, 1037–51.

**Roe**, E. 1995: *Narrative policy analysis*. Durham, NC: Duke University Press.

**Roldan**, M. 1987: Yet another meeting on the informal sector? Or the politics of designation and economic restructuring in a gendered world. *Proceedings* of the conference 'The Informal Sector as an Integral Part of the National Economy'. Roskilde: Roskilde University.

**Rollo**, J. and Stern, J. 1992: Growth and trade prospects for Central and Eastern Europe. *The World Economy* 15, 645–68.

**Romer**, P. 1986: Increasing returns and long-run growth. *Journal of Political Economy* 94, 1002–37.

**Rorty**, R. 1979: *Philosophy and the mirror of nature*. Princeton, NJ: Princeton University Press.

**Rorty**, R. 1985: Texts and lumps. *New Literary History* 17, 1–16.

**Rose**, N. 1990: *Governing the soul: the shaping of the private self.* London: Routledge.

**Rose**, N. and Miller, P. 1992: Political power beyond the State: problematics of government. *British Journal of Sociology* 43, 173–205.

**Rosenbaum**, E. 1954: The expulsion of the International Longshoremen's Association from the American Federation of Labor. Unpublished PhD dissertation, University of Wisconsin.

**Rosenberg**, N. 1982: *Inside the black box: technology and economics*. New York: Cambridge University Press.

**Rosenfeld**, S., Shapira, P., and Williams, J.T. 1992: *Smart firms in small towns.* Washington DC: Aspen Institute.

**Ross**, R.J.S. 1995: The theory of global capitalism: state theory and variants of capitalism on a world scale. In Smith, D.A. and Böröcz, J. (eds), *A New World Order? Global transformations in the late twentieth century.* Westport, CT: Praeger, 19–36.

**Ross**, R.J.S. and Trachte, K.C. 1990: *Global capitalism: the new Leviathan*. Albany, NY: SUNY Press.

**Routledge**, P. 1994: *Resisting and shaping the modern*. London: Routledge.

**Routledge**, P. 1995: Resisting and reshaping the modern: social movements and the development process. In Johnston, R.J., Taylor, P.J. and Watts, M.J. (eds), *Geographies of global change: remapping the world in the late twentieth century.* Oxford: Blackwell.

**Rueschemeyer**, D. and Evans, P.B. 1985: The state and economic transformation. In Evans, P.B., Rueschemeyer, D. and Skocpol, T. (eds), *Bringing the state back in*. Cambridge: Cambridge University Press, 44–77.

**Ruggie**, J. 1993: Territoriality and beyond: problematizing modernity in international regions. *International Organisation* 47, 39–174.

**Ruigrok**, W. and Van Tulder, R. 1995: *The logic of international restructuring*. London: Routledge.

**Russell**, M. 1966: *Men along the shore*. New York: Brussel & Brussel.

**Russian Economic Trends** 1994(3).

**Russo**, M. 1986: Technical change and the industrial district: the role of inter-firm relations in the growth and transformation of ceramic tile production in Italy. *Research Policy* 14, 329–43.

**Sabel**, C. 1989: Flexible specialization and the resurgence of regional economies. In Hirst, P. and Zeitlin, J. (eds), *Reversing industrial decline?* Oxford: Berg, 17–70.

**Sabel**, C. 1992: Studied trust: building new forms of cooperation in a volatile economy. In Pyke, F. and Sengenberger, W. (eds), *Industrial districts and local economic regeneration*. Geneva: International Institute for Labour Studies.

**Sabel**, C.F. 1994: Learning by monitoring: the institutions of economic development. In Smelser, N. and Swedberg, R. (eds), *Handbook of economic sociology*. Princeton, NJ: Princeton University Press.

**Sachs**, J. 1990: Eastern Europe's economies: what is to be done? *The Economist*, 13 January, pp. 21–6.

**Sachs**, W. (ed.) 1992: *The development dictionary*. London: Zed Press.

**Sachs**, W. 1993: Global ecology and the shadow of development. In Sachs, W. (ed.), *Global ecology: a new arena of political conflict*. London: Zed Books, 3–21.

**Sadler**, D. 1992: Beyond '1992': the evolution of European Community policies for the automobile industry. *Environment and Planning C: Government and Policy* 10, 229–48.

**Sadler**, D. 1994: The geographies of 'Just-in-Time': Japanese investment and the automotive components industry in Western Europe. *Economic Geography* 70, 41–59.

**Sadler**, D. 1996: Changing inter-firm relations in the automotive industry: implications of new buyer–supplier arrangements for the European production system. Unpublished manuscript, Department of Geography, University of Durham.

**Sadler**, D. and Amin, A. 1995: 'Europeanisation' in the automotive components sector and its implications for state and locality. In Hudson, R. and Schamp, E. (eds), *Towards a new map of automobile manufacturing in Europe?* Berlin: Springer, 39–61.

**Sadler**, D. and Swain, A. 1994: State and market in Eastern Europe: regional development and workplace implications of direct foreign investment in the automobile industry in Hungary. *Transactions of the Institute of British Geographers* NS 19, 387–403.

**Sadler**, D., Swain, A. and Hudson, R. 1993: The automobile industry and eastern Europe: new production strategies or old solutions? *Area* 25, 339–49.

**Said**, E. 1985a: Orientalism reconsidered. In Barker, F.

*et al.* (eds), *Europe and its others,* vol. 1. Colchester: University of Essex.

**Said**, E. 1985b: *Orientalism.* London: Penguin.

**Said**, Y. 1995: Defense conversion: regional and international aspects. Unpublished manuscript, HCA, Prague.

**Ste. Croix**, G.E.M. de 1981: *The class struggle in the ancient Greek world.* Ithaca, NY: Cornell University Press.

**Sakolosky**, R. 1992: Disciplinary power and the labour process. In Sturdy, A., Knights, D. and Wilmott, H. (eds), *Skill and Consent: Contemporary Studies in the Labour Process.* London: Routledge, 235–54.

**Sala-i-Martin**, X. and Sachs, J. 1991: *Fiscal federalism and optimum currency areas: evidence for Europe from the United States.* Working Paper 3855. Cambridge, MA: National Bureau of Economic Research.

**Samuelson**, P.A. 1972: Maximum principles in analytical economics. *American Economic Review* 62, 249–62.

**Sand**, P. 1995: Trusts for the earth. In Lang, W. (ed.), *Sustainable development and international law.* London: Graham & Trotman.

**Sassen**, S. 1991: *The global city: New York, London, Tokyo.* Princeton, NJ: Princeton University Press.

**Sassen**, S. 1994: *Cities in a world economy.* London: Pine Forge Press.

**Saunders**, P. 1981: *Social theory and the urban question.* London: Hutchinson.

**Saunders**, P. 1985: The forgotten dimension of central–local relations: theorising the 'regional state'. *Environment and Planning C: Government and Policy* 22, 133–45.

**Savage**, M., Barlow, J., Dickens, P. and Fielding. T. 1992: *Property, bureaucracy and culture: middle class formation in contemporary Britain.* London: Routledge.

**Savage**, M. and Witz, A. (eds) 1992: *Gender and bureaucracy.* Oxford: Blackwell.

**Sawyer**, M.C. 1989: *The challenge of radical political economy: An introduction to the alternatives to neoclassical economics.* Savage, MD: Barnes & Noble.

**Saxenian**, A. 1994: *Regional advantage: culture and competition in Silicon Valley and Route 128.* Cambridge, MA: Harvard University Press.

**Sayer**, A. 1982: Explanation in economic geography. *Progress in Human Geography* 6, 68–88.

**Sayer**, A. 1986: New developments in manufacturing: the just-in-time system. *Capital and Class* 30, 43–72.

**Sayer**, A. 1989: The 'new' regional geography and problems of narrative. *Environment and Planning D: Society and Space* 7, 253–76.

**Sayer**, A. 1992: *Method in social science: a realist approach,* 2nd edn. London: Routledge.

**Sayer**, A. 1994: Cultural studies and 'the economy, stupid'. *Environment and Planning D: Society and Space* 12, 635–7.

**Sayer**, A. 1995: *Radical political economy: a critique.* Oxford: Blackwell.

**Sayer**, A. 1996: Contractualisation, work and the anxious classes, paper presented to the International Symposium 'Work Quo Vadis', Department of Working Life Science, University of Karlstad, Sweden, 15–17 June.

**Sayer**, A. and Walker, R. 1992: *The new social economy: reworking the division of labour.* Oxford: Blackwell.

**Schamp**, E. 1991: Towards a spatial reorganisation of the German car industry? The implications of new production concepts. In Benko, G. and Dunford, M. (eds), *Industrial change and regional development.* London: Belhaven, 158–70.

**Schamp**, E.W. 1995: The German automobile industry going European. In Hudson, R. and Schamp, E.W. (eds), *Towards a new map of automobile manufacturing in Europe? New production concepts and spatial restructuring.* Berlin: Springer, 93–116.

**Schluchter**, W. 1989: *Rationalism, religion, and domination: a Weberian perspective.* Berkeley: University of California Press.

**Schmidt**, V. 1996: Industrial policy and policies of industry in advanced industrialized nations. *Comparative Politics* 28, 225–48.

**Schoenberger**, E. 1988: From Fordism to flexible accumulation: technology, competitive strategies, and international location. *Environment and Planning D: Society and Space* 6, 245–62.

**Schoenberger**, E. 1994: Corporate strategy and corporate strategists: power, identity and knowledge within the firm. *Environment and Planning A* 26, 435–51.

**Schoenberger**, E. 1996: *The cultural crisis of the firm.* Oxford: Blackwell.

**Schor**, J. 1991: *The overworked American.* New York: Basic Books.

**Schott**, K. 1984: *Policy, power and order: the persistence of economic problems in capitalist states.* New Haven, CT: Yale University Press.

**Schrag**, C.O. 1975: Praxis and structure: conflicting models in the science of man. *Journal of the British Society for Phenomenology* 6, 23–31.

**Schrag**, P. 1994: California's elected anarchy. *Harper's*, November, 50–7.

**Schurmann**, F. (ed.) 1993: *Beyond the impasse.* London: Zed Press.

**Schweickart**, D. 1993: *Against capitalism.* Cambridge: Cambridge University Press.

**Scott**, A. 1993a: *Technopolis.* Los Angeles: University of California Press.

**Scott**, A. 1993b: The new southern California economy: pathways to industrial resurgence. *Economic Development Quarterly* 7, 296–309.

**Scott**, A. 1996: Regional motors of the global economy. *Futures* 28, 391–411.

**Scott**, A. and Soja, E. 1986: Los Angeles: the capital of the late twentieth century. *Society and Space* 4, 249–54.

**Scott**, A. and Soja, E. (eds) 1996: *The city: Los Angeles and urban theory at the end of the twentieth century.* Los Angeles: University of California Press.

**Scott**, A. and Storper, M. (eds) 1986: *Production, work, territory: the geographical anatomy of contemporary capitalism.* Boston: Allen and Unwin.

**Scott**, A. and Storper, M. 1995: The wealth of regions. *Futures* 27, 505–26.

**Scott**, A.J. 1985: Location processes, urbanization, and territorial development: an exploratory essay. *Environment and Planning A* 17, 479–501.

**Scott**, A.J. 1988a: *Metropolis: from the division of labor to urban form*. Berkeley and Los Angeles: University of California Press.

**Scott**, A. J. 1988b: *New industrial spaces: flexible production organization and regional development in North American and Western Europe*. London: Pion.

**Scott**, A. J. 1988c: Flexible production systems in regional development: the rise of new industrial spaces in North America and Western Europe. *International Journal of Urban and Regional Research* 12, 171–86.

**Scott**, J. 1979: *Corporations, classes and capitalism*. London: Hutchinson.

**Scott**, J. 1991: *Who rules Britain?* Cambridge: Polity.

**Scott**, J. 1992: Experience. In Butler, J. and Scott, J. (eds), *Feminists theorize the political*. New York: Routledge, 22–40.

**Scott**, J.C. 1985: *Weapons of the weak: everyday forms of peasant resistance*. New Haven, CT: Yale University press.

**Scott-Kemis**, D. 1990: *Innovation and competitiveness in the Australian processed food industry*. Canberra: Australian Government Publishing Service.

**Seabrook**, J. 1993: *Victims of development*. London: Verso.

**Sebba**, G. 1953: The development of the concepts of mechanism and model in physical science and economic thought. *American Economic Review* 43, 259–68.

**Seidler**, V.J. 1994: *Unreasonable men: masculinity and social theory*. London: Routledge.

**Serapio**, M. 1994: Japan–U.S. direct R&D investments in the electronics industries, Japan Technology Program. Springfield, VA: US Department of Commerce, NTIS PB94–127974. NTIS.

**Sheppard**, E.S., and Barnes, T.J. 1990: *The capitalist space economy: geographical analysis after Ricardo, Marx and Sraffa*. London: Unwin Hyman.

**Sheridan**, K. 1993: *Governing the Japanese economy*. Cambridge: Polity.

**Shiva**, V. 1989: *Staying alive*. London: Zed Press.

**Shiva**, V. 1993: The greening of the global reach. In Sachs, W. (ed.), *Global ecology: a new arena of political conflict*. London: Zed Press.

**Shurmer-Smith**, P. and Hannam, K. 1994: *Worlds of desire, realms of power: a cultural geography*. London: Arnold.

**Sibley**, D. 1995: Families and domestic routines: constructing the boundaries of childhood. In Pile, S. and Thrift, N. (eds), *Mapping the subject: geographies of cultural transformation*. London: Routledge.

**Sidaway**, J. and Power, M. 1995: Sociospatial transformations in the 'postsocialist' periphery: the case of Maputo, Mozambique. *Environment and Planning A* 27, 1463–91.

**Siegal**, L. and Markoff, J. 1985: *The high cost of high tech*. New York: Harper & Row.

**Simis**, K. 1982: *USSR: the corrupt society*. New York: Simon & Schuster.

**Simon**, D. 1995: Introduction. In Simon, D.F. (ed.), *Corporate strategies in the Pacific Rim: global versus regional trends*. London: Routledge, xv–xxviii.

**Sinclair**, T.J. 1994: Passing judgement: credit rating processes as regulatory mechanisms of governance in the emerging world order. *Review of International Political Economy* 1, 133–59.

**Skocpol**, T. 1985: Bringing the state back in: strategies of analysis in current research. In Evans, P.B., Rueschemeyer, D. and Skocpol, T. (eds), *Bringing the state back in*. Cambridge: Cambridge University Press, 1–43.

**Smelser**, N. and Swedberg, R. 1994: *Handbook of economic sociology*. Princeton, NJ: Princeton University Press.

**Smith**, A. 1759: *The theory of moral sentiments*. Indianapolis: Liberty Fund.

**Smith**, A. 1976 [1776]: *The wealth of nations*, (ed.) E. Cannan. Chicago: University of Chicago Press.

**Smith**, A. 1994a: Regional development and the restructuring of state socialism: a regulationist approach to 'the transition' in Slovakia. University of Sussex Research Papers in Geography 15.

**Smith**, A. 1994b: Uneven development and the restructuring of the armaments industry in Slovakia. *Transactions of the Institute of British Geographers* NS 19, 404–24.

**Smith**, A. 1995: Regulation theory, strategies of enterprise integration and the political economy of regional economic restructuring in central and eastern Europe: the case of Slovakia. *Regional Studies* 29, 761–72.

**Smith**, A. 1996a: From convergence to fragmentation: uneven regional development, industrial restructuring and the 'transition to capitalism' in Slovakia. *Environment and Planning A* 28, 135–56.

**Smith**, A. 1996b: Industrial restructuring and uneven regional development in Slovakia: a regulationist approach to 'the transition' in Central and Eastern Europe. Unpublished DPhil thesis, Sussex European Institute, University of Sussex.

**Smith**, A. 1997: Constructing capitalism? Small and medium enterprises, industrial districts and regional policy in Slovakia. *European Urban and Regional Studies* 4 (forthcoming).

**Smith**, A. and Swain, A. 1997: The geographies of transformation: critical political-economy approaches to regional economic restructuring in east and central Europe. In Pickles, J. and Smith, A. (eds), *Theorizing transition: the political economy of change in Eastern Europe*. London: Routledge, forthcoming.

**Smith**, G. 1994: Towards an ethnography of idiosyncratic forms of livelyhood. *International Journal of Urban and Regional Research* 18, 71–87.

**Smith**, N. 1984: *Uneven development: nature, capital and the production of space*. Oxford: Blackwell.

**Smith**, N. 1988a: Regional adjustment or restructuring. *Urban Geography* 9, 318–24.

**Smith**, N. 1988b: The region is dead! Long live the region. *Political Geography Quarterly* 7, 141–52.

**Smith**, N. 1992: Geography, difference and the politics of scale. In Doherty, J. Graham, E. and Malek, M. (eds), *Postmodernism and the social sciences*. London: Macmillan, 57–79.

**Smith**, N. 1993: Homeless/global: scaling places. In Bird, J., Curtis B., Putnam T., Robertson, G. and Tickner, L. (eds), *Mapping the futures: local cultures, global change*. London: Routledge, 87–119.

**Smith**, N. and Dennis, W. 1987: The restructuring of geographical scale: coalescence and fragmentation of the northern core region. *Economic Geography* 63, 160–82.

**Smith**, S.K. 1995: Internal cooperation and competitive success: the case of the US steel minimill sector. *Cambridge Journal of Economics* 19, 277–304.

**Smith**, V. 1996: Employee involvement, involved employees: participative work arrangements in a white-collar service occupation. *Social Problems*, 43, 166–79.

**Soja**, E. 1989: *Post-modern geographies*. London: Verso.

**Solow**, R. 1987: A review of Cohen, S.S. and Zysman, J., *Manufacturing matters: the myth of the post-industrial economy*. New York Times, 12 July, p. 36.

**Soskice**, D. 1990: Reinterpreting corporatism and explaining unemployment: coordinated and uncoordinated market economies. In Brunetta, R. and Dell Aringa, C. (eds) *Labour relations and economic performance*. London: Macmillan.

**Staniszkis**, J. 1991: 'Political capitalism' in Poland. *East European Politics and Societies* 5, 127–41.

**Stanworth**, P. and Giddens, A. 1974a: An economic elite: a demographic profile of company chairmen. In Stanworth, P. and Giddens, A. (eds), *Elites and power*. Cambridge: Cambridge University Press, 83–101.

**Stanworth**, P. and Giddens, A. 1974b: The modern corporate economy: interlocking directorships in Britain, 1906–1970. *Sociological Review* 23, 5–28.

**Stark**, D. 1992: Path dependence and privatization strategies in East Central Europe. *East European Politics and Societies* 6, 17–54.

**Stark**, D. 1993: Recombinant property in East European capitalism. Paper presented at the Annual Meetings of the American Sociological Association, Miami, August.

**Stark**, D. 1996: Recombinant property in East European capitalism. *American Journal of Sociology* 101, 993–1027.

**Steen**, H., ed. 1992: *The origins of the national forests: a centennial symposium*. Durham, NC: Forest History Society.

**Stein**, A. 1984: The hegemon's dilemma: Great Britain, the United States and the international economic order. *International Organization* 38, 355–86.

**Steinmo**, S., Thelen, K. and Longstreth, F. 1992: *Structuring politics: historical institutionalism in comparative analysis*. Cambridge: Cambridge University Press.

**Stepan**, A. 1978: *The state and society: Peru in comparative perspective*. Princeton, NJ: Princeton University Press.

**Stern**, D. 1992: Do regions exist? Implications of synergetics for regional geography. *Environment and Planning A* 24, 1431–48.

**Stewart**, J. 1994: *The lie of the level playing field: industry policy and Australia's future*. Melbourne: Text Publishing.

**Stewart**, J.Q., and Warntz, W. 1958: Macrogeography and social science. *Geographical Review* 48, 167–84.

**Stewart**, M. 1983: *The age of interdependence*. Cambridge, MA: MIT Press.

**Stigler**, G. 1951: The division of labor is limited by the extent of the market. *Journal of Political Economy* 69, 213–25.

**Stigler**, G. 1971: The theory of economic regulation. *Bell Journal of Economics and Management* 2, 3–21.

**Stopford**, J. and Strange, S. 1991: *Rival states, rival firms: competition for world market shares*. Cambridge: Cambridge University Press.

**Storper**, M. 1992: The limits to globalization: technology districts and international trade. *Economic Geography* 68, 60–93.

**Storper**, M. 1994: Institutions of a learning economy. Mimeo, School of Public Policy and Social Research, UCLA, Los Angeles.

**Storper**, M. 1995a: Territorial economies in a global economy: what possibilities for middle-income countries and their regions? *Review of International Political Economy* 2, 394–424.

**Storper**, M. 1995b: The resurgence of regional economies, ten years later: the region as a nexus of untraded interdependencies. *European Urban and Regional Studies* 2, 191–221.

**Storper**, M. 1995c: Territories, flows and hierarchies in the global economy. *Aussenwirtschaft, 50. Jahrgang*, Heft II. Zurich: Ruegger, 265–93.

**Storper**, M. 1996: Innovation as collective action: convention, products, and technologies. *Industrial and Corporate Change* 5, 1–30.

**Storper**, M. 1997: *The regional world: territorial development in a global economy*. New York: Guilford Press.

**Storper**, M. and Salais, R. 1997: *Worlds of production: the action frameworks of the economy*. Cambridge, MA: Harvard University Press.

**Storper**, M. and Scott, A. 1988: The geographical foundations and social regulation of flexible production complexes. In Wolch, J. and Dear, M. (eds), *The power of geography: how territory shapes social life*. London: Unwin Hyman, 21–40.

**Storper**, M. and Walker, R. 1984: The spatial division of labor: labour and the local of industries. In Sawers, L. and Tabb, W. (eds), *Sunbelt/Snowbelt*. New York: Oxford University Press.

**Storper**, M. and Walker, R. 1989: *The capitalist imperative: territory, technology and industrial growth*. Oxford: Blackwell.

**Strang**, D. 1991: Anomaly and commonplace in European political expansion: realist and institutionalist accounts. *International Organization* 45, 143–62.

**Strange**, S. 1988: *States and markets*. London: Pinter.

**Strange**, S. 1991: An eclectic approach. In Murphy, C.N. and Tooze R., (eds), *The new international political economy*. Boulder, CO: Reinner, 33–49.

**Streeck**, W. 1985: Industrial relations and technical change in the British, Italian, and German automobile industry: three case studies. Discussion Paper IIM-LMP 83–85. Berlin: Wissenschaftszentrum.

**Sun**, M. 1989: Investors yen for U.S. technology. *Science*, 1238–41.

**Sunkel**, O. 1969: National development policy and external dependence in Latin America. *Journal of Development Studies* 6, 23–48.

**Sunley**, P. 1992: An uncertain future: a critique of post-Keynesian geographies. *Progress in Human Geography* 16, 58–70.

**Sunley**, P. 1993: Selective perceptions: evolutionary metaphors and economic geography. Unpublished manuscript available from the author: Department of Geography, University of Edinburgh.

**Sussex European Institute** 1995: *Draft of the final report for the European Commission project on post-war reconstruction in the Balkans*. University of Sussex: Sussex European Institute.

**Svedberg**, P. 1991: The export performance of sub-Saharan Africa. *Economic Development and Cultural Change* 39, 549–66.

**Svedberg**, P. 1993: Trade compression and economic decline in sub-Saharan Africa. In Blomstrom, M. and Lundahl, M. (eds), *Economic crisis in Africa: perspectives on policy responses*. London: Routledge.

**Swain**, A. 1996: A geography of transformation: the automotive industry in Hungary and eastern Germany, 1989–94. Unpublished PhD thesis, Department of Geography, University of Durham.

**Swedberg**, R. and Granovetter, M. 1992: Introduction. In Granovetter, M. and Swedberg, R. (eds), *The sociology of economic life*. Boulder, CO: Westview Press, 1–26.

**Sweeney**, G. 1994: *Learning efficiency, technical change and economic progress*. Dublin: SICA Innovation Consultants.

**Swyngedouw**, E. 1991: Homing in and spacing out: externalization of innovation, competitive practices and technology transfer. Mimeo, School of Geography, University of Oxford.

**Swyngedouw**, E. 1992a: The Mammon quest. 'Glocalization', interspatial competition and the monetary order: the construction of new scales. In Dunford, M. and Kafkalas, G. (eds), *Cities and regions in the new Europe*. London: Belhaven Press, 39–67.

**Swyngedouw**, E. 1992b: Territorial organization and the space/technology nexus. *Transactions of the Institute of British Geographers* NS 17, 417–33.

**Swyngedouw**, E. 1993: Communication, mobility and the struggle for power over space. In Giannopoulos, G. and Gillespie, A. (eds), *Transport and communication innovation in Europe*. London: Belhaven Press, 305–25.

**Swyngedouw**, E. 1996a: Reconstructing citizenship, the re-scaling of the state and the new authoritarianism: closing the Belgian mines. *Urban Studies* 33.

**Swyngedouw**, E. 1996b: Producing futures: international finance as a geographical project. In Daniels, P. and Lever, W. (eds), *The global economy in transition*. Harlow: Longman, 135–63.

**Swyngedouw**, E. 1997: Neither global nor local: 'glocalisation' and the politics of scale. In Cox, K. (ed.), *Spaces of globalization: reasserting the power of the local*. New York: Guilford Press, 137–66.

**Szasz**, F.M. 1993: Religious America in modern American culture: an introduction. In Gidley, M.M. (ed.), *Modern American culture: an introduction*. London: Longman, 23–44.

**Tabb**, W. 1995: *The postwar Japanese system*. New York: Oxford University Press.

**Tanzi**, V. 1995: *Taxation in an integrating world*. Washington DC: Brookings Institution.

**Tanzi**, V. and Schuknecht, L. 1996: The growth of government and the reform of the state in industrial countries. IMF Working Paper (December).

**Taylor**, C. 1989: *Sources of the self: the making of modern identity. Cambridge*, MA: Harvard University Press.

**Taylor**, C. and Gutman, A. 1992: *Multiculturalism and the politics of recognition*. Priceton, NJ: Princeton University Press.

**Taylor**, M. 1986: Industrial geography. *Progress in Human Geography* 10, 407–15.

**Taylor**, M. and Thrift, N. (eds) 1982a: *The geography of multinationals*. London: Croom Helm.

**Taylor**, M.J. and Thrift, N. 1982b: Industrial linkage and the segmented economy: 1. Some theoretical proposals. *Environment and Planning A*, 14, 1601–13.

**Taylor**, M.J. and Thrift, N. 1982c: Industrial linkage and the segmented economy: 2. An empirical reinterpretation. *Environment and Planning A* 14, 1615–32.

**Taylor**, M. and Thrift, N. (eds) 1986: *Multinationals and the restructuring of the world economy*. London: Croom Helm.

**Taylor**, P. 1985: *Political geography: world-economy, nation-state and locality*. London: Longman.

**Taylor**, P. 1996: *The way the modern world works: world hegemony to world impasse*. London: Wiley.

**Taylor**, P. and Buttel, F. 1992: How do we know we have environmental problems? *Geoforum* 23(3).

**Taylor Gooby**, P. and Lawson, R. 1993: *Markets and managers: new issues in the delivery of welfare*. Buckingham: Open University Press.

**Teague**, P. 1995: Europe of the regions and the future of national systems of industrial relations. In Amin, A. and Tomaney, J. (eds), *Behind the myth of the European Union*. London: Routledge, 149–73.

**Terlouw**, C. 1992: *The regional geography of the world system: external arena, periphery, semiperiphery, core*. Utrecht: Netherlands Geographical Studies.

**Thanopoulou**, M. 1992: *Women's employment or work in Greece: main tendencies and arguments in post-war literature*. Athens: National Centre for Social Research (in Greek).

**Théret**, B. 1994: To have or to be: on the problem of the interaction between State and economy in its 'solidarist' mode of regulation. *Economy and Society* 23, 1–46.

**Thoben**, H. 1982: Mechanistic and organistic analogies in economics reconsidered. *Kyklos* 35, 292–306.

**Thomas**, B. 1974: *Migration and economic growth*. Cambridge: Cambridge University Press.

**Thomas**, N. 1991: *Entangled objects: exchange, material culture and colonialism in the Pacific*. London: Harvard University Press.

**Thomas**, R. 1994: *What machines can't do*. Berkeley: University of California Press.

**Thompson**, G. 1990: *The political economy of the New Right.* London: Pinter.

**Thompson**, P. and Ackroyd, S. 1995: All quiet on the workplace front? A critique of recent trends in British industrial sociology. *Sociology* 29, 615–33.

**Thrift**, N. 1986: The geography of international economic disorder. In Johnston, R.J. and Taylor, P.J. (eds), *A world in crisis? Geographical perspectives.* Oxford: Blackwell, 12–67.

**Thrift**, N. 1987: The geography of late twentieth-century class formation. In Thrift, N. and Williams, P. (eds), *Class and space: the making of urban society.* London: Pion, 207–53.

**Thrift**, N. 1990a: Doing regional geography in a global system: the new international financial system, the City of London and the South East of England, 1984–87. In Johnston, R.J., Hauer, J. and Hoekveld, G. A. (eds), *Regional geography: current developments and future prospects.* London: Routledge, 180–207.

**Thrift**, N.J. 1990b: The perils of the international financial system. *Environment and Planning A* 22, 1135–7.

**Thrift**, N. 1991: Over wordy worlds? Thoughts and worries. In Philo, C. (ed.), *New words, new worlds: reconceptualising social and cultural geography.* Department of Geography, St David's University College Lampeter, 144–8.

**Thrift**, N.J. 1992: Muddling through: world orders and globalization. *Professional Geographer* 44, 3–7.

**Thrift**, N. 1994a: Globalisation, regionalisation, urbanisation: the case of the Netherlands. *Urban Studies* 31, 365–80.

**Thrift**, N. 1994b: On the social and cultural determinants of international financial centres: the case of the City of London. In Corbridge, S., Thrift, R. and Martin, R. (eds), *Money, Power and Space.* Oxford: Blackwell, 327–55.

**Thrift**, N. 1996a: New urban eras and old technological fears. Reconfiguring the goodwill of electronic things. *Urban Studies* 33, 1463–93.

**Thrift**, N. 1996b: Shut up and dance: or, is the world economy knowable? In Daniels, P. and Lever, W. (eds), *The global economy in transition.* Harlow: Longman, 11–23.

**Thrift**, N. 1996c: *Spatial formations.* London: Sage.

**Thrift**, N. 1996d: The rise of soft capitalism. Paper presented at Crises of Global Regulation and Governance Conference, Athens, GA, 6–8 April, 1996 (copy available from author, Department of Geography, University of Bristol, Bristol BS8 1SS, UK).

**Thrift**, N. and Leyshon, A. 1994: A phantom state? The detraditionalisation of money, the international financial system and international financial centres. *Political Geography* 13, 299–327.

**Thrift**, N. and Olds, K. 1996: Refiguring the economic in economic geography. *Progress in Human Geography* 20, 311–37.

**Thurow**, L. 1992: *Head to head: the coming economic battle among Japan, Europe and America.* New York: William Morrow.

**Tickell**, A. and Peck, J. 1992: Accumulation, regulation and the geographies of post-Fordism: missing links in regulationist research. *Progress in Human Geography* 16, 190–218.

**Tickell**, A. and Peck, J.A. 1995: Social regulation *after* Fordism: regulation theory, neo-liberalism and the global–local nexus. *Economy and Society* 24, 357–86.

**Tomforde**, A. and Meek, J. 1994: New seizure of smuggled plutonium. *Guardian*, 17 August, p. 18.

**Touraine**, A. 1988: *The return of the actor.* Minneapolis: University of Minnesota Press.

**Traynor**, I. 1994: When the east's dreams evaporate. *Guardian*, 19 November, p. 16.

**Treichler**, P. 1992: Beyond *Cosmo. Camera Obscura* 29, 21–76.

**Trotsky**, L. 1969 [1931]: *Permanent revolution.* New York: Pathfinder.

**Tsilenis**, S. and Hadjimichalis, C. (guest eds) 1991: Production Subcontracting and Informal Labour Relations in the City. Special issue, *Synchrona Themata* 45 (in Greek).

**UNCTAD** 1966–90: Commodity trade statistics. New York: United Nations. UNDP 1996: *Human development report.* New York: Oxford University Press.

**UNCTAD** 1993: *World investment report 1993: transnational corporations and integrated international production.* New York: United Nations.

**UNDP** 1992: *The human development report.* Geneva: United Nations Development Programme.

**UNDP** 1996: *Human development report.* New York: Oxford University Press.

**United Nations** 1992: *World investment report 1992: transnational corporations engines of growth.* New York: UN, Transnational Corporations and Management Division.

**United Nations Economic Commission for Europe** 1993: *Economic survey of Europe in 1992–1993.* New York: United Nations.

**United Nations Economic Commission for Europe** 1994: *Economic survey of Europe in 1993–1994.* New York: United Nations.

**United Nations Economic Commission for Europe** 1995: *Economic survey of Europe in 1994–1995.* New York: United Nations.

**United Nations Economic Commission for Europe and International Robotics Federation** 1994: *World industrial robots 1994: statistics 1983–93 and forecasts to 1997.* Geneva: United Nations.

**United States Department of Commerce, International Trade Administration** 1990: *The competitive status of the U.S. electronics sector.* Washington DC: Government Printing Office.

**Unseem**, M. 1990: Business and politics in the United States and Britain. In Zukin, S. and DiMaggio, P. (eds), *Structures of capital: the social organization of the Economy.* Cambridge: Cambridge University Press, 263–96.

**Upham**, C.W. 1846: The spirit of the day and its lesson. In Brainerd and Brainerd, E.W. (eds), *The New England Society orations*, vol. 1. New York: The Century Company, 419–66.

**Urry**, J. 1987: Some social and spatial aspects of services. *Society and Space* 5, 5–26.

**Urry**, J. 1990: Work, production and social relations. *Work, Employment and Society* 4, 271–80.

**Vaiou**, D., Golemis, H., Lambrianidis, L., Hadjimichalis, C. and Chronaki, Z. 1996: *Informal forms of production and work and urban space in Greater Athens*. Athens: Research Report for the General Secretariat for Research and Technology.

**Vaiou**, D. and Hadjimichalis, C. 1997: *With the sewing machine in the kitchen and the poles in the fields*. Athens: Themelio (in Greek).

**Vaiou**, D., Lambrianidis, L., Hadjimichalis, C. and Chronaki, Z. 1991: *Diffused industrialisation in the Greater Thessaloniki area*. Thessaloniki: Research Report for the Ministry of Housing, Physical Planning and Public Works (4 vols, in Greek).

**Vaiou**, D. and Stratigaki, M. (guest eds) 1989: Women's Work: Between Two Worlds. Special issue, *Synchrona Themata* 40 (in Greek).

**van Tulder**, R. and Ruigrok, W. 1993: Regionalisation, globalisation or glocalisation: the case of the world car industry. In Humbert, M. (ed.), *The impact of globalisation on Europe's firms and regions*. London: Pinter, 22–33.

**Varese**, F. 1994: Is Sicily the future of Russia? Private protection and the rise of the Russian Mafia. *Archives Européennes de Sociologie* 35, 224–58.

**Varikas, E. 1987:** *The revolt of the ladies: the birth of feminist conscience in Greece, 1833–1907*. Athens: Foundation for Research and Education of the Commercial Bank of Greece (in Greek).

**Vartanian**, P. 1971: The puritan as a symbol in American thought: A study of the New England Societies. PhD dissertation, University of Michigan.

**Vavouras**, J. (guest ed.) 1990: Paraeconomia. Special issue I, *Issues of Political Economy,* (in Greek).

**Veblen**, T. 1919: *The place of science in modern civilization and other essays*. New York: B.W. Huebsch.

**Veblen**, T. 1953 [1899]: *The theory of the leisure class: an economic study of institutions*. New York: New American Library.

**Večerník**, J. 1995: Changing earnings distribution in the Czech Republic: survey evidence from 1988–1994. *Economics of Transition* 3, 355–71.

**Veltz**, P. 1991: New models of production organisation and trends in spatial development. In Benko, G. and Dunford, M. (eds), *Industrial change and regional development: the transformation of new industrial spaces*. London: Belhaven, 193–204.

**Ventola**, E. 1987: *The structure of social interaction: a systematic approach to the semiotics of service encounters*. London: Pinter.

**Vergopoulos**, K. 1986: *De-development today? Essay of dynamic stasis in Southern Europe*. Athens: Exandas (in Greek).

**Villarejo**, D. and Runsten, D. 1993: *California's agricultural dilemma*. Davis: California Institute for Rural Studies.

**Vishvanathan**, S. 1995: The republic of Brundtland. Paper prepared for Multilateralism and the Environment Conference, University of California, Berkeley.

**Von Hippel**, E. 1987: Cooperation between rivals: informal know-how trading. *Research Policy* 16, 291–302.

**Von Hippel**, E. 1988: *The sources of innovation*. New York: Oxford University Press.

**Wade**, R. 1990: *Governing the market: economic theory and the role of government in East Asian industrialization*. Princeton, NJ: Princeton University Press.

**Wakasugi**, R. 1992: Why are Japanese firms so innovative in engineering technology? *Research Policy* 21, 1–12.

**Walby**, S. and Bagguley, P. 1991: Gender restructuring: a comparative analysis of five local labour markets. Lancaster Regionalism Group Working Paper 28.

**Waldinger**, R. and Bozorgmehr, M. (eds) 1996: *Ethnic Los Angeles*. Newbury Park: Sage Publications.

**Waldinger**, R. and Lapp, M. 1993: Back to the sweatshop or ahead to the informal sector? *International Journal of Urban and Regional Research* 17, 6–29.

**Walker**, R. 1989: A requiem for corporate geography: new directions in industrial organisation, the production of place and the uneven development. *Geografiska Annaler* 71B, 43–68.

**Walker**, R. 1990: The playground of US capitalism? In Davis, M., Hiatt, S., Kennedy, M., Ruddick, S. and Sprinker, M. (eds), *Fire in the hearth*. London: Verso, 3–82.

**Walker**, R. 1995a: Regulation and flexible specialization as theories of capitalist development: challengers to Marx and Schumpeter? In Liggett, H. and Perry, D. (eds), *Spatial practices: markets, politics, and community life*. Thousand Oaks, CA: Sage, 167–208.

**Walker**, R. 1995b: California rages against the dying of the light. *New Left Review* 209: 42–74.

**Walker**, R. 1996: Another round of globalization in San Francisco. *Urban Geography* 17, 60–94.

**Walker**, R. and Lizárraga, J. 1998 forthcoming: California in Flux. In Yanez, A. and Maldonado, L. (eds), *Immigration: the panic and the promise*. Albany, NY: SUNY Press.

**Walker**, R.A. 1978: Two sources of uneven development under advanced capitalism: spatial differentiation and capital mobility. *Review of Radical Political Economics* 10, 28–38.

**Walker**, R.B.J. 1990: Security, sovereignty and the challenge of world politics. *Alternatives* 15, 3–27.

**Walker**, R.B.J. 1991: State sovereignty and the articulation of political space/time. *Millennium: Journal of International Studies* 20, 445–61.

**Walker**, R.B.J. 1993: *Inside/outside: international relations as political theory*. Cambridge: Cambridge University Press.

**Warren**, M. 1994: Exploitation or co-operation: the political basis of regional variation in the Italian informal economy. *Politics and Society* 22, 89–122.

**Waterman**, P. 1996: Beyond globalism and developmentalism. *Development and Change* 27, 165–80.

**Waters**, M. 1995: *Globalization*. London: Routledge.

**Watts**, M. 1983: *Silent violence: food, famine and peasantry in northern Nigeria*. Berkeley: University of California Press.

**Watts**, M. 1991: Visions of excess: African development in an age of market idolatry. *Transition* 51, 124–41.

**Watts**, M. 1992: Living under contract: work, production politics, and the manufacture of discontent in a peasant society. Ch. 3 in Pred, A. and Watts, M.J. *Reworking modernity capitalisms and symbolic discontent*. New Brunswick NJ: Rutgers University Press.

**Watts**, M. 1995: A new deal of the emotions. In Crush, J. (ed.), *Power of development*. London: Routledge.

**Watts**, M. 1997: Black gold, white heat. In Pile S. and Keith, M. (eds), *Geographies of resistance,* London: Routledge.

**Weaver**, P. and Hudson, R. 1995: Economic restructuring and public expenditure for sustainable development: an eco-Keynesian model. In McClaren, D. (ed.), *Working Futures*. London: Friends of the Earth, 60–74.

**Webb**, M.C. 1991: International economic structures, government interests, and international coordination of macroeconomic adjustment policies. *International Organization* 45, 309–42.

**Webber**, M. 1994: Enter the dragon: lessons for Australia from Northeast Asia. *Environment and Planning A* 26, 71–94.

**Webber**, M., Clark, G.L., McKay, J. and Missen, G. 1991: Industrial restructuring: definition. Working Paper Monash–Melbourne Joint Project on Comparative Australian–Asian Development, 91–3. Monash University Development Studies Centre, Melbourne.

**Webber**, M. and Rigby, D. 1996: *The golden age of illusion*. New York: Guilford Press.

**Webber**, M., Sheppard, E. and Rigby, D. 1992: Forms of technical change. *Environment and Planning A* 24, 1679–709.

**Weber**, M. (ed.) 1958 [1904–5]: *The Protestant ethic and the spirit of capitalism.* New York: Charles Scribner's Sons.

**Weber**, M. 1968: *Economy and society*, (ed. G. Roth, and C. Wittich). New York: Bedminster Press.

**Weiss**, L. 1989: Regional economic policy in Italy. In Crouch, C. and Marquand, D. (eds), *The new centralism.* Oxford: Blackwell, 109–24.

**Weiss**, L. and Hobson, J. 1995: *States and economic development: a comparative historical analysis.* Cambridge: Polity.

**Weisskopf**, T. 1993: Democratic self-management: an alternative approach to economic transformation in the former Soviet Union. In Silverman, B., Vogt, R. and Yanowitch, M. (eds), *Double shift: transforming work in postsocialist and postindustrial societies – a U.S. – post-Soviet dialogue*. New York: Sharpe, 127–43.

**Welch**, C. 1995: The Ogoni and self determination. *Journal of Modern African Studies* 33, 635–50.

**Wells**, P. and Rawlinson, M. 1992: New procurement regimes and the spatial distribution of suppliers: the case of Ford in Europe. *Area* 24, 380–90.

**Wells**, P. and Rawlinson, M. 1994: *The new European automobile industry.* London: Macmillan.

**Westney**, D.E. 1993: Cross-Pacific internationalization of R&D by US and Japanese firms. *R&D Management* 23, 171–81.

**Westney**, D.E. 1994: Country patterns in R&D organization. In Kogut, B. (ed.), *The organization of work.* Oxford: Oxford University Press.

**Wever**, K.S. 1995: *Negotiating competitiveness: employment relations and organizational innovation in Germany and the United States.* Boston: Harvard Business School Press.

**Whatmore**, S. 1995: From farming to agribusiness: the global agro-food system. In Johnston, R., Taylor, P. and Watts, M. (eds), *Geographies of Global Change.* Oxford: Blackwell, 36–49.

**Whitehead**, A.N. 1938: *Modes of thought.* Cambridge: Cambridge University Press.

**WHO 1996:** *The World Health Report.* Geneva: United Nations.

**Wickens**, P. 1986: *The road to Nissan.* London: Macmillan.

**Wilks**, S. and Wright, M. 1991: *The promotion and regulation of industry in Japan.* Basingstoke: Macmillan.

**Williams**, C. and Windebank, J. 1995: Black market work in the European Community: Peripheral work for peripheral localities? *International Journal of Urban and Regional Research* 19, 23–39.

**Williams**, F. 1994: ILO warning on Russian jobless. *Financial Times*, 1 November, p. 3.

**Williams**, K. *et al.* 1994: *Cars: history, analysis.* London: Berghahn Books.

**Williams**, R. 1958: *Culture and society.* Harmondsworth: Penguin.

**Williams**, R. 1976: *Keywords: a dictionary of culture and society.* Oxford: Oxford University Press.

**Williams**, R. 1980: *Problems in materialism and culture.* London: New Left Books.

**Williams**, R. 1981: *Culture.* Glasgow: Fontana.

**Williams**, R. 1983: *Keywords: a vocabulary of culture and society*, revised edition. Glasgow: Fontana.

**Williamson**, O. 1985: *The economic institutions of capitalism.* New York: Basic Books.

**Willis**, S. 1991: Work(ing) out. In *A primer for daily life.* London: Routledge, 62–85.

**Wills**, J. 1995: Geographies of trade union tradition. Unpublished PhD dissertation, Department of Geography, Open University, Milton Keynes.

**Winocur**, S. 1994: Frank Jordan's war on the homeless. *San Francisco Examiner* Magazine, 6 November, 14–30.

**Winthrop**, R.C. 1838: Address. In Brainerd and Brainerd, E.W. (eds), *The New England Society orations*, vol. 1. New York: The Century Comapny, 213–60.

**Wolch**, J. and Dear, M. 1993: *Malign neglect.* San Francisco: Jossey-Bass.

**Wolf**, E. 1972: Ownership and political ecology. *Anthropological Quarterly* 45, 201–5.

**Womack**, J., Jones, D. and Roos, D. 1990: *The machine that changed the world.* New York: Rawson.

**Wood**, A. 1994: *North–South trade, employment and inequality: changing fortunes in a skill-driven world.* Oxford: Clarendon Press.

**Woolgar**, S. 1988: *Science: the very idea.* London: Tavistock.

**World Bank**. 1992: *The development report.* Washington DC: World Bank.

**World Commission on Environment and Development** 1987: *Our common future.* New York: Oxford University Press.

**Worster**, D. 1977: *Nature's economy*. Cambridge: Cambridge University Press.
**Wright**, E.O. 1985: *Classes*. London: Verso.
**Wright**, S. 1995: *The anthropology of organisations*. London: Routledge
**Wrigley**, N. and Lowe, M. 1996: *Retailing, consumption and capital: towards the new retail geography*. Harlow: Longman.
**Wrigley**, N. J. 1992: Antitrust regulation and the restructuring of grocery retailing in Britain and the USA. *Environment and Planning A* 24, 727–49.
**Yannitsis**, T. 1988: *Greek industry: development and crisis*. Athens: Gutenberg (in Greek).
**Yates**, R.L. 1959: *Forty years of foreign trade*. London: Allen & Unwin.
**Young**, I.M. 1990a: The ideal of community and the politics of difference. In Nicholson, L. (ed.), *Feminism/postmodernism*. London: Routledge.
**Young**, I.M. 1990b: *Justice and the politics of difference*. Princeton, NJ: Princeton University Press.
**Young**, M. 1991: Structural adjustment of mature industries in Japan: legal institutions, industry associations. In Wilks, S. and Wright, P. (eds), *The promotion and regulation of industry in Japan*. Basingstoke: Macmillan.
**Young**, R. 1995: Culture and the history of difference. In *Colonial desire: hybridity in theory, culture and race*. London: Routledge, 29–54.
**Zevin**, R.B. 1992: Are financial markets more open? If so, why and with what effects? In Banuri, T. and Schor, J. B. (eds) *Financial openness and national autonomy*. Oxford: Oxford University Press.
**Zimmerman**, H. 1990: Fiscal federalism. In Bennett, R. (ed.), *Decentralization, local governments and markets: towards a post-welfare agenda*. Oxford: Clarendon, 245–64.
**Žižek**, S. 1990: The republics of Gilead. *New Left Review* 183, 50–62.
**Zukin**, S. 1996: Cultural strategies of economic development and the hegemony of vision. In Merrifield, A. and Swyngedouw, E. (eds), *The urbanization of injustice*. London: Lawrence & Wishart, 223–43.
**Zukin**, S. and DiMaggio, P. (eds) 1990a: *Structures of capital: the social organization of the economy*. Cambridge: Cambridge University Press.
**Zukin**, S. and DiMaggio, P. 1990b: Introduction. In Zukin, S. and DiMaggio, P. (eds), *Structures of capital: the social organization of the economy*. Cambridge: Cambridge University Press, 1–36.
**Zysman**, J. 1994: National roots of a 'global' economy? Mimeo, BRIE, University of California, Berkeley.

# Index


academic asceticism 33
accumulation process 100
accumulation theories 231–47
active citizen 24
actor-network theory 152
advanced manufacturing technologies and processes 48
aesthetic reflexivity 13
aestheticization 18
AFL-ILA 188–95
Africa
  exports 225, 227
  'failure' 221
  'falling-out' theories 220–1
  GDP 226
  imperial trading system 223–4
  national development 224–5
  neo-liberalism 226
  position in world trade 219–28
  structural adjustment regime 226–8
  syncretic solutions to theoretical impasse 221–2
  theories of dominant practices 222–3
  trading partners 224
ahistoricism 233
Albania 333
Albert, M. 139
Allen, Z. 42
alternative development community 77
Amazonian rubber-tappers 79
American Federation of Labor (AFL) 188
Amin, A. 63, 67, 68
anti-development community 77
anti-development thinking 73
anti-essentialism 98–107
*archaeology of knowledge*, The 98–9
Arendt, H. 60, 62
Arthur, W.B. 242
Asia Pacific Economic Cooperation (APEC) 198
assembly sourcing 313–14
Association of Russian Banks 338
Athens, young women in 328
Australia, food processing industry 197–208
authority 66
automobile production system 311–20
autopoiesis and local politics 101–3

Baden-Württemberg 55, 56
Bagguley, P. 110
Balaž, V. 337
Bank of International Settlements (BIS) 297
banking 112–17
  *see also* City of London
Barings 122, 127, 128
Barnes, B. 66
Barnes, T. 12, 101
Barratt Brown, M. 140
Bell, D. 13
Benedict of Nursia 34
Benedictines 32
biogenetic hazards 71
biological metaphors 235–6
biology and theories of accumulation and regulation
    235–45
biotechnology 81
black economy 342
Blair, Tony 56, 159
Block, F. 50–1, 294
Bourdieu, P. 8, 9, 10, 19, 25, 85
Brazilian rubber-tappers 79
Broad, R. 77
Bruce Plan 104
Bruegel, I. 105
Brundtland Report 75
Budd, L. 122
Bulgaria 333, 336
Burawoy, M. 66, 338
business culture 47, 55
Butler, J. 28, 32

California 345–55
  class recomposition 348–51
  counter-revolution 353–4
  economic growth and crisis 346–8
  governmental 'rigor mortis' 351
  race and class schisms 348–51
  racial recomposition 348–51
Callon, M. and Latour, B. 67
Calvinism 40, 41, 44, 45
capital markets 337
capital-labour relations 20, 102, 174, 304–5
capitalism 18, 20, 80–4, 83

forms of 278
historical geography 170
morality of 41
*Capitalism against capitalism* 139
capitalist colonization of culture 10
capitalist economy 302
capitalization of nature 80–4
capitalocentrism 90, 91, 95
Castell, M. 61
Castoriadis, C. 46
causal powers of organizations 110
celibacy 31
Central and Eastern Europe (CEE) 331–44
alternative regulatory fixes 341–2
capital markets 337
deindustrialization 335–6
economic and geographical change 333–5
economic collapse and uneven growth 334
geo-economies 333–41
kiosk economy 338–9
merchant capital 337–8
merchant economies 336
money markets 337
percentage employment change 335
political capitalism 337–8
regional fragmentation 341–2
speculative economies 336
stock markets 337
centralization 186
Cerny, P.G. 296, 298
Chaney, D. 3
change implementation 115–16
China, special economic zones 308
Chipko 79
Christianity 29–31
Christopherson, S. 24
church, women's role 29, 31
citizenship 175
City Lives Project 123
City of London 118–29, 140, 141, 173
'new' City 123–6
old and new times 122
'old' City 123
civilizing effects 21
Clark, E. 29
Clarke, S. 340
class
and locality 92–5
and non-capitalist practices 92–5
as dimension of economic difference 91
definition 90–1, 95
redefined 91–2
class connections 126–7
class landscape 88–90
class politics 87–97
class pretensions 124
class recomposition, California 348–51
class structure 123
class struggles 83, 95, 101
class transformation 95
classical Fordism 239
Clegg, S. 110
clients/patients/students 18
Clinton administration 56
Club of Paris 172
Clyde Valley Regional Plan (CVRP) 105
Cockburn, C. 101
cognitive labour 13
cognitive structures 252
Cohen, A. 117
Cold War 136, 137–41
collaboration 50
collaborative systems 341
collective activity 286
Colonial Development and Welfare Act (1940) 76
commercial gain 271–3
Committee of the Social and Cultural Geography Study Group of the Institute of British Geographers 4
commodification of culture 22–3
communal conception 25
*Communist manifesto* 20, 23
competition 83, 150
competition states 137
competitive austerity 181
competitiveness 47, 250, 285
component sector restructuring 313
component sourcing policies in Western Europe 315–19
*condition of postmodernity, The* 7
Confucian morality 51
conservation movement 82
consumption and culture 22–3
contextuality 46
control, power as 60
conventional or relational (C-R) transactions 253
convergence 261–4, 281
co-ordination 252–3, 341
core–periphery competition 336
Cornwall
class absence and class presence 93
non-capitalist economic activity 95
corporate culture 20
corporate governance 150
corporate paternalism 51
corporate restructuring 310
corporate strategies 204–6
corruption 339
cost minimization 179
Court, G. 111
Cowen, M. 76
Crang, P. 65
Croatia 333
cultural commonality 49
cultural durables 7
cultural embeddedness 11, 120
cultural hegemony 39
cultural influences 48
cultural materialism 22
cultural materialization 13–14
cultural norms and values 17
cultural order of production 12, 45–6
cultural practice, economic rationality of 9
cultural production 8
cultural regulation theory 39–40
cultural relationships 161

cultural sociology 8
cultural specialist ideologies 8
cultural traits 47
cultural turns 3–15, 16
  definition 5
  embedding the economic in 10–12
  exporting the economic in 7–10
  opposing the economic to 5–7
  representing the economic through 12–13
  resistance to 5
  response to 4–5
  within political economy 21–2
culturalization of economy 16, 22, 23
culture
  and consumption 22–3
  and economy 17–19, 23, 34–5, 37–46
  capitalist colonization of 10
  commodification of 22–3
  dialectic of 16–26
  economic colonization of 7
  economic determination of 7
  economic operation to 7
  economization of 21–3
  of organizations 119–21
*Culture and society* 24
culture change 113–14
cumulative strengths 147
customer care 108
Czech Republic 333

decentralization 186, 187, 313
decentred autonomous spaces 78
de Certeau, M. 69
deconstruction 6
de-differentiation of economy and culture 13
degradation 72
deindustrialization, CEE 335–6
democratic planning 341
dependency theory 220, 222
Der Derian, J. 68
designer work cultures 21
deskilling 179–80
destandardization 251
determinacy 233
determinism 8
development 76
  alternatives 77–80
  and capitalization of nature 80–4
  and environment 73–7, 84
  policy 309
  theory and practice 73, 76
Dicken, P. 63, 110, 311
DiMaggio, P. 120
discursive formation 12, 38–9
discursive reflexivity 13
discursive regulation 12, 39
disembedding 19–20
distance matters 65–8
distanciation 281
distribution 299–300
divergence 259–64
Divine Providence 45
division of labour 179
domination 299
downsizing 333
Drama, subcontractors 328
du Gay, P. 111, 116
Dunn, J. 24

East Germany 336
Eastern Europe 138, 311
  *see also* Central and Eastern Europe (CEE)
ecological deterioration 72
economic activities and processes 17
economic behaviour 48
economic colonization 9
economic competition 152
economic decline 139
economic determination 9
economic development 140, 147
economic difference, class as dimension of 91
economic differentiation 88–90
economic embeddedness 11
economic geographies
  and social relations xiii–xiv
  asserted xiv
  constituted xiii
  globalization 133
  interpreted xii–xiii
  neo-classical 12
  'new' xv–xvi, 87–97
  observed xi
  'old' xv–xvi
  (re)constitution 3–15
economic identities 46
economic imaginary, hermeneutics of 44–5
economic rationality 38, 50
  of cultural practice 9
economic relationships 161
economic restructuring 77, 311
economic sociology 119–21
economic violence 339
economistic fallacy 50–1
economization of culture 16, 21–3
economy
  and culture 17–19, 23, 34–5, 37–46
  and local politics 106–7
  and the state 291–3
  culturalization of 22, 23
  dialectic of 16–26
  globalization 148–9
  rescaling 170–2
electronics industry, R&D 209–18
embedded organizations 122
embeddedness 19–20, 164
  and embodiment 121
  forms of 120
  levels of 120
  political 120
  structural 11, 120
  transformation 331–2
embodied work 119–21
embodied workers 127–8
employment 265–6, 333

Great Britain 266–9
practices 118
*Encountering development* 73, 75, 76
endogenous strengths 147
entrepreneurship 150
environment, and development 73–7, 84
environmentalism 77–80
Escobar, A. 73, 74–5, 78, 79, 81, 82, 83, 84
essentialism, breaking with 101–3
Esteva, G. 73
Etzioni, A. 19, 24, 152
European Commission (EC) 321
European Monetary Union 171
European Union (EU) 21, 174, 186, 200, 204, 259, 297, 309, 311, 321–30
Europeanization of production systems 311–20
evolutionary economics 240–3
exclusion 259–60, 265–6, 269–75
expert systems 148
exploitative accumulation 81

family role, Greece 327
Featherstone, M. 25
feelgood factor 23
feminine qualities 114
femininity 127
feminism 35
municipal 105
feminist economists 121
feminization 129
financial gain 271–3
financial markets 122
financial products 179
financial sector 173–4
financial services 112
Financial Services Act 1986 118, 122
financial structure 148
financial transactions 182
firms, institutional characteristics of 22
First World 178, 181–3, 308
fish processing, Kavalla 328
Flexspec 340–1
food consumption 206
food processing industry, Australia 197–208
Fordism 38, 182, 239, 284, 302, 345
foreign exchange markets 337
Foreign Investment Review Board (FIRB) 207
forest reserves 82
Foucault, M. 46, 63, 64, 66, 98–9, 136
free-market triumphalism 73
Friedman, J. 5
full employment 273–5, 305, 308
Fyfe, N. 104

G–7 172
Gambia 94–5
GDP
Africa 226
Great Britain 261–4, 266
OECD 182
gender connections 114, 126–7
gender relations 34, 174
gender structure 123
Gendron, B. 9
General Agreement on Tariffs and Trade (GATT) 172, 176, 297
General Motors 109
generation of variety 251
genesis discourse 144
geo-economic discourses 137
Geo-economic intellectuals 137–41
geopolitical economy 135–7
Giddens, A. 46, 62, 68
Glasgow, post-war planning 104–5
global change 83
global competition 144, 162, 197
global demographic growth 71
global ecological crisis 71
global economic change 142
global economic inequalities 71, 73
global environmental crisis 71
global environmental discourses 75
Global Environmental Facility (GEF) 74, 80
global exploitation of technology 151
global–local integration 202–3
global–local production organization 313
global orientation of state strategies 149
global restructing 198–202
*Global shift* 133
global warming 71
globalism *see* globalization
globalization 25, 133–46, 178, 181, 182, 281, 292, 313, 318
and internationalization 161–2
and nation-states 149–50
and state strategy 206–7
approaching 158–60
as caricature 158
complex of interrelated processes 166
concept 133, 147, 158, 197
constrained 217–18
diagnostic features 160–1
dialectical 159
discourses 141–5
dynamics 165
economic geographies 133
economy 148–9, 161
extent, novelty and implications 160
impact of 162
interpretation 147
myth 160
new phase 160
politics of 183
principles of 165
processes 159, 161, 164, 219
repacking 161–5
scripting 141–5
terminology 159
unpacking 160
versus the state 281–3
'glocalization' 170–2
Glyn, A. 182, 183
GNP and educational base of workforce 148
Godelier, M. 6, 8

Gold Standard 182
Gould, S.J. 236
Gowan, P. 333
Grabher, G. 336
Granovetter, M. 49
grassroots initiatives 78
Great Britain
  employment 266–9
  GDP 261–4, 266
  non-employment 265
  regional dynamics 259–77
  regional economies 260–9
  unemployment 264
  urban economies 260–9
Greece
  family role 327
  firms and employment by size of firm 326
  informal activities 324–6
  informal cities 329
  productive structure of cities 324
  small firms 326–7
  urban development 327
  women's work 327–8
greening capitalism 80–4
Gregorian reforms 32
Gregory, D. 61
Grey, C. 116
Group of 77, 172
*Growth, competitiveness and employment* (White Paper) 321

Habermas, J. 46
Hacking, I. 64
Hall, S. 19
Haraway, D. 74
Harrison, B. 49
Harvey, D. 7, 8, 61, 68, 100–2, 237–8
Harvey, W. 237
Hay, C. 297
hermeneutics of economic imaginary 44–5
high-technology workplace 27–8, 34, 35
  *see also* technology
high-volume production (HVP) 303–10
Hill, C. 30
Hirst, P. 297, 298, 299
Ho, M.-W. 35
Hodgson, G.M. 236, 243
homeworking, Thessaloniki 328
Huczynski, A.A. 144
human agency 110
human–environment relations 84
human resource management (HRM) 305–8
Humphreys, D. 42
Hungary 333
Hutton, W. 139–41, 143, 293

Ibn-Rushd 33
ideal realities 6
idealization 299
ideological exceptionalism xvi-xvii
ILA-Independent (ILA-IND) 188–95
Illegal activity 339
Illner, M. 336
indigenous growth 282
individual character 44
industrial cultures 47, 51, 55
industrial networks 48
industrial practices 55–7
industrial relations 151
industrial restructuring 90
industrialization and morality 42
inequality 259–61
informal activities 321–30
  definitions and dichotomies 322
  specifity of place 323
informally constituted assets 155
information as binding agent of networks 152
information storage 13
informational capitalism 197
Inkatha Freedom Party 194
innovation 151, 179, 248
  national systems 150
  regional systems 48–50
innovation–development link 153
in-person services 142
instability 259–60, 269–75
institutional characteristics of firms 22
institutional economics 152, 240–3
institutions 55–7
instrumentalization of work culture 20–1
integration 281
intelligible genders 32
inter-firm collaboration 55
inter-firm relations 55
internal market 311
International Brotherhood of Longshoremen (IBL) 188
International Conference of Free Trades Unions 309
international financial transactions 182
International Longshoremen's and Warehousemen's Union (ILWU) 193
International Longshoremen's Association (ILA) 188
International Monetary Fund (IMF) 331, 332
international organization 309
international political economy 149
international relations (IR) 135
International Standards Organization (ISO) 297
internationalization 292, 297, 300, 311
  and globalization 161–2
  Japan's R&D capability 210–12
  of technology 148
interpersonal conversational product elements 14
interpersonal production processes 14
interpersonal service 13
interpretative communities 141
interrelated tendencies 159, 161
intimidation 299
Investment Fund for Economic and Social Development (1946) 76
Isaac, J.C. 60–1

Japan
  electronics industry R&D 209–18
    in USA 213–15
  technology partnerships 215–17

Japanese culture 22
Japanese industry 22
Jessop, B. 103, 162–3, 292–3, 300
joint ventures 67, 151
just-in-time (JIT) 312

Kaldor, N. 244
Kanter, R.M. 50
Kavalla, fish processing 328
Kean, H. 105
Keynes, J.M. 243, 273
Keynesian welfare state (KWS) 284
  space economy in 279–81
*Keywords* 76
Kiosk economy 338–9
knowledge and local politics 104–6
knowledge structure 148
Korton, D.C. 160
Krotov, P. 338
Krugman, P. 285
Kuhn, T. 232

labour demand 112
labour force 129
labour government 140
Labour–Management Relations (Taft–Hartley) Act 187
labour market 100, 101, 105, 151, 309
labour migration 102
Labour Party 140, 141, 159
labour politics 186–96
labour power 102
labour relations 186
labour unions 188–93
land ownership 44
large-firm economy 340–1
Lasch, C. 76
Lash, S. 13, 23, 165
lateral modes 68
Latin 32
Latour, B. 67
Law, J. 12
learning 155, 250, 252
learning-based activities 250
learning-based competitiveness 155
learning-based economy 148
learning capability 155
learning economy 250
*lebenstraum* 235
Leeson, N. 122, 128
Lefebvre, H. 104
liberalism
  and political culture 23–5
  *see also* neo-liberalism
life histories 111–12
local community 78
local economic governance 282
local economy 106
local government 103, 105
  municipal feminism in 105–6
local politics 98–107
  and autopoiesis 101–3
  and knowledge 104–6
  and the economy 106–7
  arena constitution 103–4
  making space for 99–101
locality and class 92–5
Luhmann, N. 103
Lundvall, B.-A. 49, 52
Luttwak, E.N. 138, 139, 141

McCloskey, D. 12
McDermott, G. 332
McDonaldization of mass culture 9
McDowell, L. 51, 111
machinery
  and equipment maintenance 52–3
  complexity and ease of operation 53–4
  culture 51–4
  implementation 55
MacKay, R. 285
McNay, L. 64
macro-economic policy 181–3
Mafia 339
maintenance 52–3
management style 114
managerial practice 144
Mandinka society 94
manipulation of workers 67
Mann, M. 62
marginality 72
market adjustment 273–4
market relations 249–50
markets 255, 294
Marquand, D. 24
Marshall, A. 154, 231–3
Marxism 3, 8, 10, 12, 20, 46, 73–4, 81, 82, 88, 89, 100, 102, 236–7, 240, 274
masculinism 127
mass collective worker 307–9
mass culture 22
  McDonaldization of 9
mass production 302–10, 340–1
Massey, D. 61, 66, 69, 92, 108
material exceptionalism xvi–xvii
material spatialization 34
matters of distance 65–8
Mayr, E. 235
mechanistic metaphors 234–5
Meegan, R. 108
Meeks, W. 29
merchant banking 120–1, 127
merchant capital 337–8
merchant economies 336
mergers and acquisitions 126
metaphors 232–5
micro-economic policies 181
micro-regulation 286–7
migration 175
Miller, D. 10
Mind–Body dualism 34
Mirowski, P. 234
mixed economies, socio-economic order of 260
mobile powers 68–70
modernization theory 342

monasticism 31
money markets 337
morality
  and industrialization 42
  of capitalism 41
Morgan, G. 110
Morgan Grenfell 128
MOSOP 79
multiculturalism 25
multidimensionality 78
mulitple class processes 92
municipal feminism in local government 105–6
mutual orientation 152

nation-state
  and globalization 149–50
  changing structure and influence of 162
  in global economy 162
  qualitative reorganization 162
  restructuring 164
  *see also* state
national economic governance 150–1
national economic policy priorities 150
National Labor Relations Board (NLRB) 188, 190–1
national restructuring 198–202
National Statistical Service (NSS) 324
national systems of innovation 150
nature
  as artefact 71–86
  as artifice 71–86
  capitalization of 80–4
Nelson, R.R. 241–2
neo-classical economic geographies 12
neo-liberalism 24, 272, 275, 291–2, 309, 331
  Africa 226
  regulatory regimes 310
neo-Marshallian nodes 63
network state 286–7
networks 62, 152–3, 164
New England 38–42
New England Societies 42–5
new growth theory 242–3
New Industrial Division of Labour 181
New International Division of Labour 178
New Right 140, 281, 283
new social movements (NSM) 77–80
New York Steamship Association (NYSA) 188–90
newly industrializing countries (NICs) 178, 181, 182, 198, 199, 203
NGO activities 77, 78
Nietzschean post-structural theory 46
Nilsson, J.E. 311
Nissan
  European Technology Centre (NETC)317
  in North-East England 315–19
  partnership philosophy 319
  Supplier Development Team (SDT) 317–18
Noble, D. 28–30, 32, 33
*nomenklatura* 337, 342
non-capitalist practices and class 92–5
non-employment 264–5
non-essentialist approach 102
normative prescription 299
North Amercian Free Trade Agreement (NAFTA) 172, 297
Northern Europe 322
Notermans, T. 183
nuclear war 136

Ó Tuathail, G. 137, 138, 139
Oberg, S. 311
obligation 152
OECD 182
Offe, C. 293, 296
Ogoni 79
old times 112–13
OPEC 172
oppositional groups 176
organization theory 152
organizational culture 119–21
organizational restructuring 108–17
organizations 248, 251–3, 255–6
  causal powers of 110
*Our common future* 75
overengineering 51–4
ozone depletion 71

Pacific Rim 198, 199
Palloix, C. 272
Palmer, B. 5
Paracelsus 30
*paraeconomia* 324
partnership philosophy 319
Pascale 109
path-dependent evolutionary change 153
Peck, J.A. 292
Pendleton, A. 110
performative meritocracy 114–15
permeable Fordism 239
PHARE programme 340
Pickles, J. 336, 338
Piirainen, T. 342
Pilgrims, 43, 44
Piven, F.F. 184
place-specific assets 155
Poland 333
Polanyi, K. 10–11, 18, 50, 294
policy networks 287
political capitalism 337–8
political communities 144
political culture and liberalism 23–5
political ecology 72, 84–5
political economy
  cultural turn within 21–2
  state-centred approaches to 278
political embeddedness 120
political relationships 161
politics
  of distribution 16
  of globalization 183
  of recognition 16
  of renewal 345–55
  of rescaling 172–5
  *see also* local politics

Pooley, S. 297
post-Cold War economic discourse 137–41
post-Fordism 20, 83, 231, 239, 250, 275, 284, 302–3
post-Keynesianism 243–5, 278–89
post-modern politics 85
post-structural political ecology 84
post-structuralism 82
post-structuralist political theory 136
post-war planning, Glasgow 104–5
poverty 72, 73
power
  as capacity 60
  as control 45
  as medium 62–3
  as technology 63–5
  conceptions of 60
  economies of 59–70
  overlapping modes of 66
  realist notion of 60
  relationships 90, 94, 162
  spatial vocabulary of 59
  spatiality of 65
*Power and Marxist theory: a realist view* 60
powers of mobility 68–70
Pred, A. 90
preventive maintenance 52
private sector 106, 143
privatization 181
problem-solving 52
product based technological learning (PBTL) 49, 179
product development 179
product markets 255
production
  cultural order of 45–6
  regional systems 48–50
  social relations of 72
production geography 178–81
production systems 311
productive order restructuring 271–3
protection 339
Protestant ethic 40–1
  and morality of capitalism 41
pseudo-productive services 271–3
public choice theory 291
public goods state 294
Puritanism 40, 43, 44

qualitative state 294–7
quasi-global strategies 163
quasi-markets 285
Quesnay, F. 237
QWERTY system 236, 242

racial recomposition, California 348–51
Radice, H. 279
Reason 34, 35
reciprocal recognition 20
redistribution 293
reductionist methodological strategy 233
reflections 27
reflexivity 249–50, 252
region-states 282
regional capitalism 345–55
regional convergence 281
regional culture 47–58
regional dynamics in Great Britain 259–77
regional economics 248–58
regional economies, in Great Britain 260–9
regional systems of innovation and production 48–50
regionalism 149
regions as worlds 255–6
regulation 341
  theories 231–47, 292–3
regulationists 238–40
regulatory regimes 310
Reich, R. 141, 142, 143
relational assets 248
renewal politics 345–55
research and development (R&D) 27–8, 34, 35, 61, 151
  electronics industry 209–18
resources, pressure of production on 72
restructuring paradigm 108–11
*Reworking modernity* 93
Richards, P. 78
right to work laws 187
Ritzer, G. 9
Roberts, S. 133–4, 143
Roe, E. 76
Rorty, R. 232
Route 128 50
routine production services 142
Royal Society 33
Russia 333, 339

Sachs, W. 72, 84
Said, Y. 340
Salaman, G. 111, 116
Saxenian, A. 50, 51
Sayer, A. 4, 51, 60, 66, 341
scalar metamorphoses 169–70
scale and scaled politics 167–76
  mobilizing 175–6
scales
  and lines of power 168–70
  of inclusion/exclusion 172–5
Schamp, E. 311
Schumpeterian workfare state (SWS) 284, 285
Schweickart, D. 340
science 32–3, 35
Scott, A. 231
Seidler, V.J. 34
self-regenerating growth poles 147
service-sector restructuring 111
sex and gender 28
sexual dissent 29–30
Shenton, R. 76
Shiva, V. 74, 78
shortage economy 39
Shurmer-Smith, P. and Hannam, K. 5
sign value of commodities 18
Silicon Valley 50
skilled labour 56
Slovakia 333
small and medium enterprises (SMEs) 340

small firms
  economy 340–1
  Greece 326–7
Smith, A. 20, 23, 154, 180, 338
Smith, N. 237–8
social capital 56
social dissent 29–30
social foundations of economic processes 50–1
social issues 174
social learning 179
social movements 77–80
  as alternative approach 78
social relations xiii–xiv, 183, 342
  of production 72
societal paradigm 39
socio-economic development 281
socio-economic order of mixed economies 260
socio-economic spatial integration 280
socio-economic spatial structure 93
socio-economics 155
  and territoriality 151–5
  perspectives from 152–3
socio-spatial relations 168
South Africa 194
Southern Europe 322, 327
space economy 59–70, 278–89, 336
  under Keynesian welfare state 279–81
space–time compression 197
Spanish Mondragon system 341
spatial centralization 279
*Spatial divisions of labour* 92
special economic zones of China 308
speculative economies 336
stakeholder capitalism 56
Staniszkis, J. 337, 338
Stark, D. 340
state
  and the economy 291–3
  form and functions 293
  'hollowed-out' 292–3
  in economic geography 278–9, 290–301
  macro-economic stabilizing form 294
  qualitative 290, 294–7
  rescaling 170–2
  response to supra-national scale issues 297–8
  structures and mechanisms 296
  versus globalization 281–3
  *see also* nation-state
state-centred approaches to political economy 278
state intervention 286
state-organized networks 287
state strategy
  and globalization 206–7
  global orientation of 149
*State we're in, The* 139
stock markets 337
Storper, M. 49, 154
strategic alliances 67
structural embeddedness 11, 120
structured coherence 100–1
subcontractors, Drama 328
subordination 299
supply chain management 311–20
sustainability, deconstructing 73–7
sustainable development 80
sustained capital accumulation 80–4
symbolic analytic services 142

tax-evading activities 324
technology 81, 255–6
  and technological change 248, 250–1
  global exploitation of 151
  internationalization of 148
  partnerships 215–17
  *see also* high-technology workplace
teleology 234
territorial economies 255
territoriality and socio-economics 151–5
territories 248, 253–6
Thessaloniki, homeworking 328
Third World 76, 78, 89, 183, 308, 321
Thomas, N. 9
Thompson, G. 297, 298, 299
Thrift, N. 19, 63, 67, 68, 69, 110, 144
Tickell, A. 292
time–space compression 281
Touraine, A. 46
trade unions 309, 324
traded interdependencies 252
transformation, embeddedness 331–2
transnational corporations (TNCs) 151, 163, 164, 197, 200, 201, 204, 205
transnational economic diplomacy 149
transnational oligopolies 148
transportation costs 180
trouble-shooting 52

*Underground economy and irregular forms of employment (travail au noir)* 321
unemployment 259–60, 264–5, 333
uneven development theory 237–8
Union Representation 191–2
Unitarianism 40
United Mine Workers of America (UMWA) 191–2
university education system 123
untraded interdependencies 252
urban development, Greece 327
urban economies in Great Britain 260–9
urban studies 323
Urry, J. 13, 23, 111, 165
USA
  Japanese electronics R&D in 213–15
  Longshore industry 186–96
USSR 138

value-driven company 20
Varese, F. 339
Veblen, T. 234, 240–1
vertical modes 68

Walker, R. 51, 66, 109, 341
Walras' law 274
Watts, M. 90, 92, 93, 94
Weber, M. 40, 41

Weisskopf, T. 340
welfare state 283–6
Wever, K.S. 55
Whimster, S. 122
Williams, K. 21
Williams, R. 5, 8, 24, 76
Winter, S.G. 241–2
witchcraft 33
women
  and the church 29, 31
  in small firms 326
  young, in Athens 328
women's work 323
  Greece 327–8
work
  formal and informal forms 323
  forms of 322
work culture, instrumentalization 20–1
workers' resistance 177–8
workers' struggle 177–85
workfare geographies 283–6
working-class solidarity 92
World Bank 74, 75, 198, 332
World Trade Organization (WTO) 297
*World without women, A* 28

*Young and the Restless, The* 10

Zukin, S. 120
Zysman, J. 153

7529